C0-AMX-755

Tissue Culture

METHODS AND APPLICATIONS

TISSUE CULTURE

Methods and Applications

Edited by

PAUL F. KRUSE, JR., and M. K. PATTERSON, JR.

Biomedical Division
The Samuel Roberts Noble Foundation, Inc.
Ardmore, Oklahoma

ACADEMIC PRESS New York San Francisco London 1973

A Subsidiary of Harcourt Brace Jovanovich, Publishers

ACADEMIC PRESS, INC.
111 Fifth Avenue, New York, New York 10003

United Kingdom Edition published by
ACADEMIC PRESS, INC. (LONDON) LTD.
24/28 Oval Road, London NW1

Library of Congress Cataloging in Publication Data
Main entry under title.

Tissue culture: methods and applications.

Includes bibliographical references.
1. Tissue culture. I. Kruse, Paul F
DATE ed. II. Patterson, Manford Kenneth,
DATE ed.
[DNLM: 1. Tissue culture. QS 525 K94t 1973]
QP88.T57 574'.0724 72-9978
ISBN 0-12-427150-2

PRINTED IN THE UNITED STATES OF AMERICA

Contents

Section III. Cell Harvesting

Section IV. Replicate Culture Methods

Section V. Single Cell Isolations and Cloning

Section VI. Perfusion and Mass Culture Techniques

Section VII. Cell Propagation on Miscellaneous Culture Supports

SECTION VIII. EVALUATION OF CULTURE DYNAMICS

SECTION IX. RECENT TECHNIQUES FACILITATING MICROSCOPIC OBSERVATION OF CELLS

Section X. Cell Hybridization

Section XI. Virus Propagation and Assay

Section XII. Production of Hormones and Intercellular Substances

Section XIII. Diagnosis and Understanding of Disease

Section XIV. Quality Control Measures

List of Contributors

Numbers in parentheses indicate the pages on which the authors' contributions begin.

NABIL ABAZA (452), *Department of Pathology, The Medical College of Pennsylvania, Philadelphia, Pennsylvania*

MARLENE ABSHER (395), *Department of Medical Microbiology, College of Medicine, University of Vermont, Burlington, Vermont*

BYRON TH. AFTONOMOS (185), *Department of Internal Medicine, University of Nebraska, College of Medicine, Omaha, Nebraska*

CORINE R. ANDERSEN (589), *Department of Dairy Science, University of Illinois, Urbana, Illinois*

FRANCES E. ARRIGHI (773), *Department of Biology, The University of Texas M. D. Anderson Hospital and Tumor Institute at Houston, Houston, Texas*

S. AVRAMEAS (379), *Institut de Recherches Scientifiques Sur Le Cancer (CNRS), Villejuif, France*

RALPH BAKER (735), *Botany and Plant Pathology Department, Colorado State University, Fort Collins, Colorado*

MICHAEL F. BARILE (729), *Division of Bacterial Products, Bureau of Biologics, Food and Drug Administration, Department of Health, Education and Welfare, Rockville, Maryland*

G. BARSKI (469, 659), *Tissue Culture and Virus Laboratory, Institut Gustave Roussy, Villejuif, France*

ANNIE R. BEASLEY (12, 133), *Department of Microbiology, University of Miami School of Medicine, Miami, Florida, and Lerner Marine Laboratory, Bimini, Bahamas*

J. BELEHRADEK, JR. (659), *Institut Gustave Roussy, Villejuif, France*

CHARLES W. BOONE (677), *Cell Biology Section, Viral Biology Branch, National Cancer Institute, National Institutes of Health, Bethesda, Maryland*

J. BORNSTEIN (331), *Department of Biochemistry, Monash University, Clayton, Victoria, Australia*

MURRAY B. BORNSTEIN (86), *The Saul R. Korey Department of Neurology, and the Rose F. Kennedy Center for Research in Mental Retardation and Human Development, Albert Einstein College of Medicine of Yeshiva University, Bronx, New York*

CHAIM BRAUTBAR (758), *Department of Immunology, The Hebrew University-Hadassah Medical School, Jerusalem, Israel*

W. E. BRIDSON (570), *Endocrinology Service, Reproduction Research Branch, National Institutes of Health, Bethesda, Maryland*

B. R. BRINKLEY (438), *Division of Cell Biology, Department of Human Biological Chemistry and Genetics, The University of Texas Medical Branch, Galveston, Texas*

LAROY N. CASTOR (298), *The Institute for Cancer Research, Fox Chase, Philadelphia, Pennsylvania*

JEFFREY P. CHANG (438), *Division of Cell Biology, Department of Human Biological Chemistry and Genetics, The University of Texas Medical Branch, Galveston, Texas*

JESSE CHARNEY (750, 753), *Department of Biochemistry, Institute for Medical Research, Camden, New Jersey*

WILLIAM R. CHERRY (97), *Lilly Research Laboratories, Indianapolis, Indiana*

H FRED CLARK (147), *Wistar Institute of Anatomy and Biology, Philadelphia, Pennsylvania*

D. O. CLIVER (224), *Food Research Institute and Department of Bacteriology, University of Wisconsin, Madison, Wisconsin*

E. CONWAY (665), *Biomedical Division, The Samuel Roberts Noble Foundation, Inc., Ardmore, Oklahoma*

JOHN E. K. COOPER (266), *Department of Genetics, The Children's Hospital of Winnipeg, Winnipeg, Manitoba, Canada*

LEWIS L. CORIELL (62, 671, 718), *Institute for Medical Research, Camden, New Jersey*

RODY P. COX (639), *Department of Medicine and Pharmacology, New York University Medical Center, New York, New York*

V. J. CRISTOFALO (204), *The Wistar Institute, Philadelphia, Pennsylvania*

C. H. CUNNINGHAM (527), *Department of Microbiology and Public Health, Michigan State University, East Lansing, Michigan*

H. DALEN (443), *Department of Pathology, Karolinska Institutet, Stockholm, Sweden*

W. F. DALY (398), *Lederle Laboratories, Pearl River, New York*

JOSEPH DANCIS (639), *Department of Pediatrics, New York University Medical Center, New York, New York*

B. SHANNON DANES (627), *Division of Human Genetics, Department of Medicine, Cornell University Medical College, New York, New York*

P. S. DANNIES (561), *Laboratory of Pharmacology, Harvard School of Dental Medicine, and Department of Pharmacology, Harvard Medical School, Boston, Massachusetts*

A. D. DEITCH (463), *Department of Pathology, Columbia University, New York, New York*

R. T. DELL'ORCO (231), *Biomedical Division, The Samuel Roberts Noble Foundation, Inc., Ardmore, Oklahoma*

D. K. DOUGALL (261), *W. Alton Jones Cell Science Center, Lake Placid, New York*

GERALD M. EDELMAN (29), *The Rockefeller University, New York, New York*

MAGDALENA EISINGER (65), *Sloan-Kettering Institute for Cancer Research, New York, New York*

WILLIAM J. EISLER, JR. (351), *Division of Biological and Medical Research, Argonne National Laboratory, Argonne, Illinois*

VIRGINIA J. EVANS (211), *Tissue Culture Section, National Cancer Institute, National Institutes of Health, Bethesda, Maryland*

L. FALK, JR.* (532), *Rush-Presbyterian St. Luke's and University of Illinois Medical Centers, Chicago, Illinois*

* Present address: St. Francis Hospital, Evanston, Illinois.

Rosann A. Farber (797), *Department of Genetics, University of Washington, Seattle, Washington*

S. Fedoroff (782), *Department of Anatomy, University of Saskatchewan, Saskatoon, Canada*

H. Firket (378), *Laboratoire d'Anatomie Pathologique, Université de Liege, Liege, Belgium*

John F. Foley (185), *Department of Internal Medicine, University of Nebraska, College of Medicine, Omaha, Nebraska*

Jerome J. Freed (14, 123), *The Institute for Cancer Research, Fox Chase, Philadelphia, Pennsylvania*

James G. Gallagher (102), *Department of Microbiology, University of Vermont College of Medicine, Burlington, Vermont*

O. L. Gamborg (500), *Prairie Regional Laboratory, National Research Council, Saskatoon, Saskatchewan, Canada*

Stanley M. Gartler (797), *Departments of Medicine and Genetics, University of Washington, Seattle, Washington*

K. M. Gautvik (578), *Laboratory of Pharmacology, Harvard School of Dental Medicine and Department of Pharmacology, Harvard Medical School, Boston, Massachusetts*

Nazareth Gengozian (93), *Medical Division, Oak Ridge Associated Universities, Oak Ridge, Tennessee*

Richard E. Giles (475), *Department of Biology, Yale University, New Haven, Connecticut*

G. C. Godman (463), *Department of Pathology, Columbia University, New York, New York*

B. Goldberg (586), *Department of Pathology, New York University School of Medicine, New York, New York*

Arthur E. Greene (69, 750, 753), *Department of Cell Biology, Institute for Medical Research, Camden, New Jersey*

P. M. Gullino (321), *Laboratory of Biochemistry, National Cancer Institute, National Institutes of Health, Bethesda, Maryland*

Ralph B. L. Gwatkin (3), *Merck Institute for Therapeutic Research, Rahway, New Jersey*

A. J. Hackett* (539), *The School of Public Health, University of California, Berkeley, California*

Richard G. Ham (254), *Department of Molecular, Cellular and Developmental Biology, University of Colorado, Boulder, Colorado*

M. Harris (400), *Department of Zoology, University of California, Berkeley, California*

Leonard Hayflick (43, 220, 722, 822), *Department of Medical Microbiology, Stanford University School of Medicine, Stanford, California*

Charles Heidelberger (644), *Department of Oncology, McArdle Laboratory for Cancer Research, University of Wisconsin, Madison, Wisconsin*

René Heller (387), *Department of Physiologie, Végétale, Université de Paris, Paris, France*

A. C. Hildebrandt (215, 244, 549), *Department of Plant Pathology, University of Wisconsin, Madison, Wisconsin*

S. Robert Hilfer (16), *Department of Biology, Temple University, Philadelphia, Pennsylvania*

H. T. Holden (408), *Department of Microbiology, Laboratory of Virology, University of Miami School of Medicine, Miami, Florida*

* Present address: Naval Biomedical Research Laboratory, Naval Supply Center, Oakland, California.

WILLIAM HOUSE (338, 739), *Imperial Cancer Research Foundation Laboratories, Lincoln's Inn Fields, London, England*

T. C. HSU (764), *Department of Biology, The University of Texas M. D. Anderson Hospital and Tumor Institute at Houston, Houston, Texas*

ROBERT N. HULL (97), *Lilly Research Laboratories, Indianapolis, Indiana*

JOEL HUTZLER (639), *Department of Pediatrics, New York University Medical Center, New York, New York*

JOAN KABAKJIAN (204), *The Wistar Institute, Philadelphia, Pennsylvania*

R. H. KAHN (114), *Department of Anatomy, The University of Michigan, Ann Arbor, Michigan*

M. EDWARD KAIGHN (54), *W. Alton Jones Cell Science Center, Lake Placid, New York*

S. S. KALTER (524), *Division of Microbiology and Infectious Diseases, Southwest Foundation for Research and Education, San Antonio, Texas*

K. N. KAO (500), *Prairie Regional Laboratory, National Research Council, Saskatoon, Saskatchewan, Canada*

FREDERICK H. KASTEN (72, 291, 458), *Department of Anatomy, Louisiana State University Medical Center, New Orleans, Louisiana*

L. N. KEEN (448), *Physiology Department, University of Colorado Medical School, Denver, Colorado*

HAROLD KINNAMAN (735), *Botany and Plant Pathology Department, Colorado State University, Fort Collins, Colorado*

R. A. KNAZEK (321), *Laboratory of Biochemistry, National Cancer Institute, National Institutes of Health, Bethesda, Maryland*

P. O. KOHLER (570), *Endocrinology Service, National Institutes of Health, Bethesda, Maryland*

G. M. KOLODNY (226), *Harvard Medical School, Massachusetts General Hospital, Radiology Research Laboratory, Boston, Massachusetts*

DAVID F. KRAHN (644), *Department of Oncology, McArdle Laboratory for Cancer Research, University of Wisconsin, Madison, Wisconsin*

B. KRAMARSKY (535), *Institute for Medical Research, Camden, New Jersey*

ROBERT S. KROOTH (633), *Department of Human Genetics and Development, College of Physicians and Surgeons, Columbia University, New York, New York*

P. F. KRUSE, JR.* (231, 327, 448), *Biomedical Division, . The Samuel Roberts Noble Foundation, Inc., Ardmore, Oklahoma*

W. G. W. KURZ (359), *Prairie Regional Laboratory, National Research Council, Saskatoon, Saskatchewan, Canada*

B. L. LARSON (589), *Department of Dairy Science, University of Illinois, Urbana, Illinois*

E. Y. LASFARGUES (45), *Department of Cytological Biophysics, Institute for Medical Research, Camden, New Jersey*

JOHN C. LEE (135), *Department of Microbiology, University of Miami School of Medicine, Miami, Florida and Lerner Marine Laboratory, Bimini, Bahamas*

JOSEPH LEIGHTON (367, 452), *Department of Pathology, Cancer Bioassay Laboratory, The Medical College of Pennsylvania, Philadelphia, Pennsylvania*

W. LICHTER (408), *Laboratory of Virology, University of Miami School of Medicine, Miami, Florida*

* Deceased.

C. C. LIN (778), *Departments of Pediatrics and Pathology, McMaster University, Hamilton, Ontario, Canada*

JACK LITWIN (188, 383), *Department of Applied Microbiology, Karolinska Institutet, Stockholm, Sweden*

LAURA S. MCKENZIE (143), *School of Anatomy, University of New South Wales, New South Wales, Australia*

E. CHURCHILL MCKINNEY (135), *Department of Microbiology, University of Miami School of Medicine, Miami, Florida and Lerner Marine Laboratory, Bimini, Bahamas*

IAN MACPHERSON (241, 276), *Department of Tumor Virology, Imperial Cancer Research Foundation Laboratories, London, England*

A. MACIEIRA-COELHO (379, 412), *Institut de Cancerologie et Immunogénétique (INSERM), Villejuif, France*

MARVIN L. MACY (712, 804), *American Type Culture Collection, Rockville, Maryland*

SUNDER MANSUKHANI (452), *Department of Pathology, The Medical College of Pennsylvania, Philadelphia Pennsylvania*

E. P. MARKS (153), *Metabolism and Radiation Research Laboratory, U. S. Department of Agriculture, Agricultural Research Service, Fargo, North Dakota*

G. M. MARTIN (39, 264), *Department of Pathology, University of Washington School of Medicine, Seattle, Washington*

M. C. MATHES (161), *Department of Biology, College of William and Mary, Williamsburg, Virginia*

YUJI MATSUOKA* (599), *Institute for Cancer Research, Osaka University Medical School, Osaka, Japan*

M. D. MAXWELL (682), *Biomedical Division, The Samuel Roberts Noble Foundation, Inc., Ardmore, Oklahoma*

W. J. MELLMAN (623), *Department of Human Genetics, School of Medicine, University of Pennsylvania, Philadelphia, Pennsylvania*

P. J. MELNICK (808), *Veterans Administration Hospital, Martinez, California*

J. G. MERTENS (231), *Biomedical Division, The Samuel Roberts Noble Foundation, Inc., Ardmore, Oklahoma*

LISELOTTE MEZGER-FREED (123), *The Institute for Cancer Research, Fox Chase, Philadelphia, Pennsylvania*

R. A. MILLER (500), *Prairie Regional Laboratory, National Research Council, Saskatoon, Saskatchewan, Canada*

A. F. MIRANDA (463), *Department of Pathology, Columbia University, New York, New York*

J. M. MOEHRING (593), *Department of Medical Microbiology, University of Vermont School of Medicine, Burlington, Vermont*

MARGARET A. J. MOFFAT (611), *Department of Bacteriology, Virus Diagnostic Laboratory Medical School, Foresterhill, Aberdeen, Scotland*

HÉCTOR MONTES DE OCA (8), *Department of Surgery, University of Miami School of Medicine, Miami, Florida*

PAUL S. MOORHEAD (58, 768), *Department of Medical Genetics, University of Pennsylvania School of Medicine, Philadelphia, Pennsylvania*

JOSEPH F. MORGAN (686), *Department of Biochemistry, University of Saskatchewan, Saskatoon, Saskatchewan, Canada*

HELEN J. MORTON (686), *Division of Biological Sciences, National Research Council of Canada, Ottawa, Ontario, Canada*

* Present address: Department of Biochemistry, Faculty of Medicine, Fukuoka University, Fukuoka, Japan.

T. MURASHIGE (170, 698), *Department of Plant Sciences, University of California, Riverside, California*

CARL R. PARTANEN (791), *Department of Biochemistry, University of Pittsburgh, Pittsburgh, Pennsylvania*

M. K. PATTERSON, JR. (192, 665, 682), *Biomedical Division, The Samuel Roberts Noble Foundation, Inc., Ardmore, Oklahoma*

ROLAND A. PATTILLO (575), *Cancer Research and Reproductive Endocrinology Laboratories, Medical College of Wisconsin, Milwaukee, Wisconsin*

CARL PERAINO (351), *Division of Biological and Medical Research, Argonne National Laboratory, Argonne, Illinois*

HUGH J. PHILLIPS (406), *Department of Physiology, Creighton University School of Medicine, Omaha, Nebraska*

JAN PONTÉN (50), *Department of Pathology, University of Uppsala, Uppsala, Sweden*

RAYMOND P. PORTER* (93), *Medical Division, Oak Ridge Associated Universities, Oak Ridge, Tennessee*

F. J. A. PROP (21), *Division for Experimental Cytopathology, Pathological Anatomy Laboratory, University of Amsterdam, Amsterdam, Holland*

R. W. PUMPER (674), *Department of Microbiology, University of Illinois at the Medical Center, Chicago, Illinois*

LEONARD J. QUADRACCI (82), *Department of Medicine, Division of Nephrology, University of Washington, Seattle, Washington*

Y. RABINOWITZ† (25), *Research Service, Veterans Administration Hospital, Department of Medicine, Loyola University School of Medicine, Maywood, Illinois*

F. RAPP (653), *Department of Microbiology, College of Medicine, The Milton S. Hershey Medical Center of The Pennsylvania State University, Hershey, Pennsylvania*

R. REYNOLDS (448), *Department of Pathology, The University of Texas Southwestern Medical School, Dallas, Texas*

CATHRINE A. REZNIKOFF (644), *Department of Oncology, McArdle Laboratory for Cancer Research, University of Wisconsin, Madison, Wisconsin*

ALAN RICHTER (274), *Laboratory Cell Suppliers, Frederick, Maryland*

JAMES A. ROBB (270, 517), *Department of Pathology, University of California, San Diego, La Jolla, California*

E. ROBBINS (436), *Department of Cell Biology, Albert Einstein College of Medicine, Yeshiva University, Bronx, New York*

GEORGE G. ROSE (283), *Department of Medicine, The University of Texas Dental Branch, Houston, Texas*

LEONARD J. ROSENTHAL* (509), *Huntington Laboratories of Harvard University, Massachusetts General Hospital, Boston, Massachusetts*

D. E. ROUNDS (129), *Pasadena Foundation for Medical Research, Pasadena, California*

H. RUBIN (119), *Department of Molecular Biology, University of California, Berkeley, California*

FRANK H. RUDDLE (475), *Department of Biology, Yale University, New Haven, Connecticut*

K. K. SANFORD (237), *Tissue Culture Section, National Cancer Institute,*

* Present address: Life Sciences Center, Nova University, Fort Lauderdale, Florida.

† Deceased, 7 December 1972.

* Present address: Département de Biologie Moléculaire, Université de Geneve, Geneve, Suisse.

National Institutes of Health, Bethesda, Maryland

M. A. SAVAGEAU (316), *Department of Microbiology, The University of Michigan, Ann Arbor, Michigan*

JOSEPH B. SCHLEICHER (333), *Abbott Laboratories, North Chicago, Illinois*

IMOGENE SCHNEIDER (150, 788), *Walter Reed Army Institute of Research, Washington, D. C.*

BERNARD SCHWARTZ (309), *Department of Ophthalmology, Tufts-New England Medical Center, Boston, Massachusetts*

P. L. SCHWARTZ (331), *Department of Clinical Biochemistry, University of Otago, Dunedin, New Zealand*

ELIZABETH K. SELL (633), *Department of Human Genetics, University of Michigan Medical School, Ann Arbor, Michigan*

SYDNEY SHALL (195, 198), *Biochemistry Laboratory, University of Sussex, Sussex, England*

JOHN E. SHANNON (712, 804), *American Type Culture Collection, Rockville, Maryland*

ISAAC L. SHECHMEISTER (509), *Department of Microbiology, Southern Illinois University, Carbondale, Illinois*

CHARLES SHIPMAN, JR. (5, 709), *Department of Oral Biology (School of Dentistry), Department of Microbiology (School of Medicine), The University of Michigan, Ann Arbor, Michigan*

G. SHRAMEK* (532), *Rush-Presbyterian St. Luke's Medical Center, Chicago, Illinois*

M. MICHAEL SIGEL (12, 133, 135, 408), *Department of Microbiology, University of Miami School of Medicine, Miami, Florida, and Lerner Marine Laboratory, Bimini, Bahamas*

W. F. SIMPSON (744), *The Child Research Center of Michigan, Children's Hospital of Michigan, Detroit, Michigan*

ROBERT F. SMITH (431), *Department of Microscopy and Photomicrography, Brookhaven National Laboratory, Upton, Long Island, New York*

T. A. STEEVES (179), *Department of Biology, University of Saskatchewan, Saskatoon, Saskatchewan, Canada*

N. G. STEPHENSON (143), *School of Biological Sciences, University of Sydney, New South Wales, Australia*

W. R. STINEBRING (593), *Department of Medical Microbiology, University of Vermont School of Medicine, Burlington, Vermont*

H. E. STREET (173), *Botanical Laboratories, University of Leicester, Leicester, England*

GARY E. STRIKER (82), *Department of Medical Pathology, University of Washington, Seattle, Washington*

C. S. STULBERG (744), *The Child Research Center of Michigan, Children's Hospital of Michigan, Detroit, Michigan*

J. A. SYKES (303), *Research Department, Southern California Cancer Center, California Hospital Medical Center, Los Angeles, California*

A. H. TASHJIAN, JR. (561, 578), *Laboratory of Pharmacology, Harvard School of Dental Medicine, and Department of Pharmacology, Harvard Medical School, Boston, Massachusetts*

WILLIAM G. TAYLOR (211), *Tissue Culture Section, National Cancer Institute, National Institutes of Health, Bethesda, Maryland*

T. A. TEDESCO (623), *Department of*

* Present address: St. Francis Hospital, Evanston, Illinois.

Human Genetics, School of Medicine, University of Pennsylvania, Philadelphia, Pennsylvania

PHILIP S. THAYER (345), *Life Sciences Division, Arthur D. Little, Inc., Cambridge, Massachusetts*

JOHN G. TORREY (166), *Cabot Foundation, Harvard University, Petersham, Massachusetts*

IRENE A. UCHIDA (778), *Department of Pediatrics, McMaster University, Hamilton, Ontario, Canada*

CARLO VALENTI (617), *Department of Obstetrics and Gynecology, State University of New York, Downstate Medical Center, Brooklyn, New York*

J. VAN'T HOF (423), *Biochemistry Department, Brookhaven National Laboratory, Upton, New York*

A. L. VAN WEZEL (372), *Rijks Instituut voor de Volksgezondheid, Bilthoven, The Netherlands*

I. K. VASIL (157), *Department of Botany, University of Florida, Gainesville, Florida*

JAMES L. VAUGHN (692), *Insect Pathology Laboratory, ARS, Beltsville, Maryland*

H. C. WANG (782), *Department of Anatomy, University of Saskatchewan, Saskatoon, Saskatchewan, Canada*

CHARITY WAYMOUTH (703), *The Jackson Laboratory, Bar Harbor, Maine*

GEORGE WEAVER (97), *Lilly Research Laboratories, Indianapolis, Indiana*

R. E. H. WETTENHALL (331), *Clinical Chemistry Department, Queen Elizabeth Medical Centre, Edgbaston, Birmingham, England*

W. L. WHITTLE (327, 448), *Biomedical Division, The Samuel Roberts Noble Foundation, Inc., Ardmore, Oklahoma*

G. J. WIEPJES (21), *Division for Experimental Cytopathology, Pathological Anatomy Laboratory, University of Amsterdam, Amsterdam, Holland*

L. L. WINTON (161), *Division of Natural Materials and Systems and Department of Biology, The Institute of Paper Chemistry, Appleton, Wisconsin*

KEN WOLF (138), *Eastern Fish Disease Laboratory, Bureau of Sport Fisheries and Wildlife, Kearneysville, West Virginia*

DAVID YAFFE (106), *Department of Cell Biology, Weizmann Institute of Science, Rehovot, Israel*

Preface

Tissue culture research has been expanding explosively during the past decade, and the initial concept of this book was simply to collate some of the newer tissue culture methods, particularly those involving some degree of automation. This plan was altered in favor of a treatise that not only brings tissue culture methodology up-to-date but also includes representative protocols for the application of these techniques. Accordingly, over a hundred authors have pooled their efforts to produce this volume. In so doing, the mutual hope is that it will serve as a reference portfolio to help both the novice and veteran researcher "get on with the job" in an inspired, efficient, and productive manner.

Since there are several excellent books available that describe many of the basic techniques of tissue culture, some of the classic procedures, e.g., hanging-drop cultures, are not included. Methodology in tissue culture has progressed dramatically, however, and much of this gain has not been summarized in book form. In fact, over eighty percent of the procedures included are derived from research conducted in the last eight years. Moreover, tissue culture application has mushroomed so in recent years that it has become a part of diverse biological inquiries. Thus, tissue culture methods per se and applications are scattered profusely throughout the contemporary scientific literature. This book attempts to present an overview of up-to-date procedures for working with cells in culture and for using them in a wide variety of scientific disciplines.

The reader will find a series of protocols useful for dissociating tissues (Section I) followed (Section II) by exemplary procedures for establishing primary cultures of cells from vertebrate, invertebrate, and plant sources. Once cultures are established, it is important to harvest them (Section III), reproduce them (Section IV), or clone them (Section V) efficiently and properly. In some instances it is most desirable to automate systems, regulate culture environments, or work with massive quantities of cells (Section VI), and for certain purposes not be limited necessarily to culture supports only of glass or plastic (Section VII).

Standardized procedures for evaluating the progress and behavior of cells in culture are presented in Section VIII, including enumerations of cells and nuclei and analyses of cell cycle events. Some of the more recent "tricks of the trade" in microscopy applied to tissue culture are described in Section IX, including aspects of light, electron, and fluorescence microscopy.

The next four sections (X–XIII) are intended to indicate the tremendous latitude that tissue culture has in science. More importantly, they may provide inspiration for further work. Although a dozen or more subjects might have been chosen for illustration, these sections have been confined to descriptions of procedures for cell hybridization, virus propagation and assay, production of hormones and intercellular substances, and the use of tissue culture in the diagnosis and understanding of disease.

The last portion of this book (Section XIV) deals with quality control measures in the tissue culture laboratory, a first consideration for anyone employing cells in culture. This aspect of tissue culture has been surveyed often in the literature and in the relatively few existing books on the subject. However, just as in all other phases of tissue culture methodology and application, the techniques of sterility testing, cell species (and intraspecies) identifications, standardizations of media preparation, etc., have much improved during the past decade. Thus in some cases, and especially for those relatively unfamiliar with working with cell cultures, this last section may be read first.

Most tissue culture work to date has dealt with mammalian, avian, and plant material. Especially in recent years, though, there has been considerable pioneering work with amphibian and fish tissues and with cells from invertebrate species. The latter are of interest, to cite just one reason, because insects are vectors for a number of human diseases. Also in recent years some very important observations have been made on characteristics of cells in culture. These include cells which continue to show specialized functions or express genetic abnormalities *in vitro*. All of these and other recent and exciting developments in methodology and application are represented by one or more well-documented procedures.

There are thousands of people in the world of science working with cells in culture who have only a peripheral interest in methodology per se. This is because their primary research expertise lies in another discipline; for them, the use of cells in culture is but one possible exploitable technology. These investigators may find this book in the same vein as the young high school girl in biology class who was assigned a book about penguins for outside reading. When asked about the book by her instructor, she replied in effect that it was interesting but that it told her more about penguins than she really wanted to know. On the other hand, there are thousands of students of cells in culture for whom it is impossible, of course, to include in one volume everything they always wanted to know about tissue culture. Our hope is that something in this book will be of value to all using tissue culture.

We express our deep appreciation for the enthusiastic cooperation of the contributors, many of whom felt as we did that a compilation on tissue culture of this nature would be valuable. Thanks are due also to the staff of Academic Press for their encouragement and cooperation. In addition, we are much indebted to other people who have been directly connected with this endeavor. Assistance in editing, retyping manuscripts, and indexing was supervised by Gwen Taft, who had very capable help from Jane Lawrence, Marcia Remondino, and Linda Platt. Necessary printed informational material for the large number of contributors and assistance in photography were provided by Glenn Elmore and Ivan Lawson. Finally, we express our appreciation to the Board of Trustees of The Samuel Roberts Noble Foundation, Inc., who authorized a subvention which helped reduce the cost of the book in the hope that by so doing the information generously provided by the many contributors might be more readily available to students and life science researchers alike throughout the world.

If the reader finds this volume a useful guide and reference work, or, more importantly, derives stimulation to conduct *in vitro* studies of vertebrate, invertebrate, or plant cells more meaningful than those described to date, then the contributions of all will have been worthwhile.

PAUL F. KRUSE, JR.
M. K. PATTERSON, JR.

Dr. Paul F. Kruse, Jr.

(1921–1973)

On January 31, 1973, Dr. Paul F. Kruse, Jr., died after a brief illness. Because of his dedication and contributions to tissue culture and to this book, the coeditor and contributors dedicate this volume to his memory.

Dr. Kruse was Director of the Biomedical Division and Vice President of The Samuel Roberts Noble Foundation, Inc., Ardmore, Oklahoma. He was Treasurer of the Tissue Culture Association (1966–1971) during the period of construction of The W. Alton Jones Cell Science Center. He served on the Executive Committee, was Chairman of the Publicity and Endowment Committee, and was a visiting staff lecturer at the Center. He was a member of the Association's committees on NASA programs and International Science Fairs, he was on the Advisory and Peer Review Committee of the American Type Culture Collection, and was a member of its Board of Trustees. He was a member of numerous site review teams and a consultant on governmental, university, and industrial programs.

In spite of the hours demanded of him for administrative duties, Dr. Kruse found time for research. The perfusion culture system he developed was the prototype of the systems commercially available today. He applied the method to the study of metabolism and growth of cells under "more physiological conditions."

Dr. Kruse insisted that this book be edited in such a manner that it would be useful to investigators at all levels. His interest in tissue culture was exemplified by his vigorous pursuit of a subvention so the cost of this book would not be prohibitive for anyone. All editorial royalties are toward this end.

In all things, Paul Kruse truly expressed his beliefs in both words and actions.

M. K. PATTERSON, JR.

SECTION I

Primary Tissue Dissociation

Editors' Comments

The dissociation of organized tissues into single cell suspensions requires a dissolution or weakening of the extracellular matrix (stroma). The choice of method thus depends on the composition of the stromal material. Other parameters, however, must be considered, namely, the quality and quantity of cells desired. There is little doubt that the use of hydrolytic enzymes to effect dispersion of cells alters the cellular exterior. Such enzymes are used in most cases not only because there appears to be no substitute for them, but also because with their use the cells appear to repair minor membrane damage. To obtain quality cells, it is apparent that a minimum of mechanical trauma and exposure time to the hydrolytic enzymes must be achieved. Larger quantities of cells with a lower viable population can be obtained, of course, by longer exposure times.

The methods described in this section refer to specific tissues, and the investigator should not infer that they can successfully be used with every tissue. For example, it is illustrated that the saline vehicle which is satisfactory for preparations of cells from marine fish tissues is different in composition from that used with tissues of other animals, including fresh-water fish.

This section also includes two methods for nonenzymatic dissociation or, perhaps more appropriate, fractionation of cell types. Both make use of the fact that certain cell types differ in the degree to which they attach to various surfaces; these novel techniques might serve as the basis for developing cell separation procedures for other tissues composed of heterogeneous cell populations.

The reader is referred also to other descriptions of cell preparations elsewhere in the book, particularly Section II in which dissociation of specific tissues in preparing primary cultures is discussed, e.g., for separating cardiac myocardial and endothelial cells.

CHAPTER 1

Pronase

Ralph B. L. Gwatkin

Pronase, an enzyme preparation from *Streptomyces griseus*, consists of a mixture of neutral and alkaline proteases, aminopeptidases, and carboxypeptidases.[1] It has been used to remove the zona pellucida of mammalian eggs,[2-4] to liberate chondrocytes from cartilage prior to freezing,[5] and in tissue culture.

Applications of Pronase in tissue culture include the dissociation of chicken and mouse embryo cells prior to cultivation and subcultivation.[6,7] Pronase acts much more rapidly than trypsin and, unlike trypsin, yields cell suspensions without large clumps of cells[7] (see Fig. 1). It is also superior to trypsin in digesting dead cells and cellular debris.[8] When Pronase is used to subculture cell monolayers obtained from mouse embryo cells,[7] primary cultures of human diploid fibroblasts,[9-11] and epithelial cells derived from monkey kidney tissue[11] the cells are brought rapidly into suspension. However, certain continuous cell lines, e.g., HeLa and KB are removed in flakes, indicating incomplete dispersion.[11,12] The reason for the relative ineffectiveness of Pronase with these cell lines is not known. Prolonged subcultivation of diploid human fibroblasts with Pronase was found to have no detectable effect on karyotype or on their growth potential.[9,10] However, since potent inhibitors of Pronase comparable to antitrypsins are not present in serum, the cells should be washed free of residual enzyme before initiating new cultures.[11]

Primary Cell Cultures from 14-Day Mouse Embryos

Pronase Solution (0.25%). Add 1.25 g standard Pronase (Cat. No. 53702, Calbiochem Corp., 10933 N. Torrey Pines Road, La Jolla, California 92037) to

[1] Y. Narahashi, K. Shibuya, and M. Yanagita, *J. Biochem.* (*Tokyo*) **64**, 427 (1968).
[2] B. Mintz, *Science* **138**, 594 (1962).
[3] R. B. L. Gwatkin, *J. Reprod. Fert.* **6**, 325 (1963).
[4] R. B. L. Gwatkin, *J. Reprod. Fert.* **7**, 99 (1964).
[5] A. U. Smith, *Nature* (*London*) **205**, 782 (1965).
[6] B. W. Wilson and T. L. Lau, *Proc. Soc. Exp. Biol. Med.* **114**, 649 (1963).
[7] R. B. L. Gwatkin and J. L. Thomson, *Nature* (*London*) **201**, 1242 (1964).
[8] C. C. Stewart and M. Ingram, *Blood* **29**, 628 (1967).
[9] J. C. Sullivan and I. A. Schafer, *Exp. Cell Res.* **43**, 676 (1966).

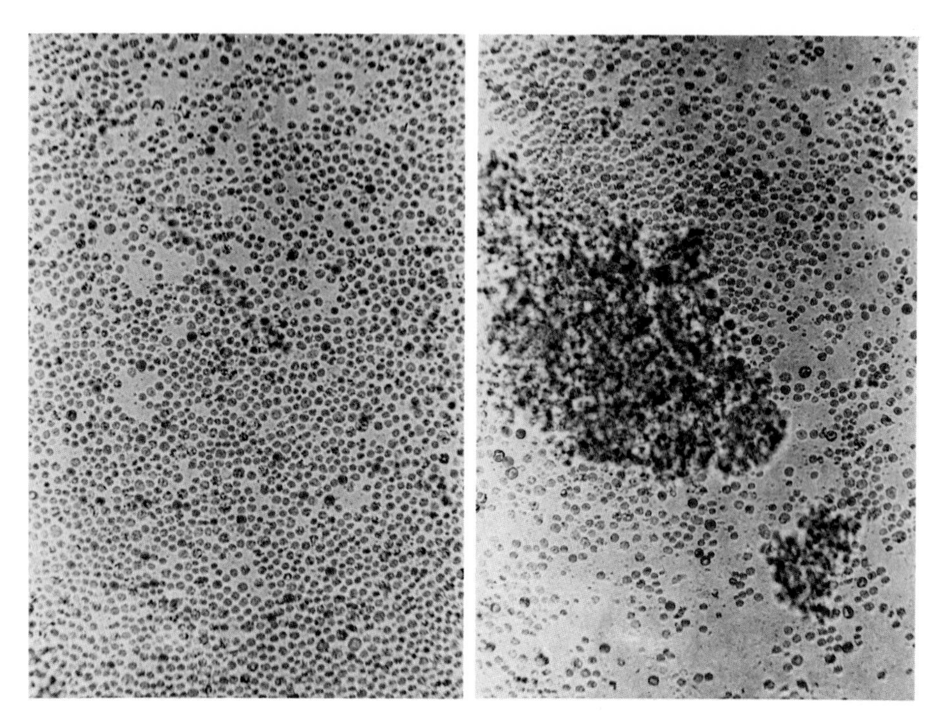

Fig. 1. Cells released after digestion of 13-day mouse embryo tissue with 0.25% Pronase (left) and 0.25% trypsin (right). Note that Pronase yields a monodisperse cell suspension, while trypsin leaves large clumps of cells intact.

500 ml phosphate-buffered saline (PBS).[13] Mix and allow to stand for 30 minutes at room temperature. Centrifuge (100 *g* for 10 minutes) to remove undissolved material,[14] then sterilize by filtration (Millipore GS, 0.22 pore size). Store 50-ml aliquots frozen.

Dissociation of the Tissue. Kill mice on thirteenth to fourteenth day of pregnancy (12–13 days after a copulation plug is found in the vagina) by quickly stretching their necks. Swab abdomens with 70% ethanol, open abdomen, and remove uterine horns to a Petri dish containing PBS. Slit open uterine horns to obtain embryos. Wash them thoroughly in PBS and place 10 embryos in a 10-ml hypodermic syringe (without a needle). Force embryos through the syringe into a 125-ml Erlenmeyer flask containing 50 ml 0.25% Pronase in PBS (see above). This single passage through the orifice of the syringe reduces the embryos to a pulp with minimal damage to their constituent cells. Drop a sterile magnetic bar into the flask, place the flask on a magnetic stirrer, and stir slowly

---

[10] D. Weinstein, *Exp. Cell Res.* **43**, 234 (1966).

[11] J. F. Foley and B. Aftonomos, *J. Cell. Physiol.* **75**, 159 (1970).

[12] J. Kahn, M. J. Ashwood-Smith, and D. M. Robinson, *Exp. Cell Res.* **40**, 445 (1965).

[13] R. Dulbecco and M. Vogt, *J. Exp. Med.* **99**, 167 (1954).

[14] A completely soluble Pronase (Cat. No. 537011) and a nuclease-free Pronase (Cat. No. 537088) are now available from Calbiochem Corp. Further study is needed to determine whether these preparations have any advantages over standard Pronase.

at room temperature for 2 hours. Add 5 ml sterile calf serum; stir again for 5 minutes and pipette up and down several times. Cells now appear as seen on left of Fig. 1. Decant into a centrifuge tube (50 ml) and centrifuge cells at 100 *g* for 12 minutes. Pour off the supernatant and resuspend the cells in 20 ml growth medium (Eagle's Minimal Essential Medium[15] supplemented with 10% calf serum). Centrifuge and resuspend the cells twice to remove Pronase.

Primary Cultures. Add 5-ml aliquots of the cell suspension (approximately 2×10^8 cells) to each of four plastic culture flasks (75 cm^2). These are prepared beforehand by adding 10 ml growth medium to each and gassing them with 5% CO_2 in air. Incubate the cultures at 37°C. After 24 hours change the medium to remove cellular debris. Monolayers of cells are usually complete and ready for subcultivation after 5 days.

Subcultivation. Pour off medium and wash monolayers once with PBS. Add 10 ml Pronase solution (diluted 1:5 with PBS to give a final Pronase concentration of 0.05%). Cells become detached as single cells in 3–5 minutes. Add 1 ml calf serum, pipette up and down several times, and transfer to a 20-ml centrifuge tube. Centrifuge and resuspend in growth medium twice to remove Pronase as already described. Finally, resuspend in growth medium and use this suspension to prepare new cultures, e.g., monolayers for virus titrations.[16]

[15] H. Eagle, *Science* **130**, 432 (1959).
[16] R. B. L. Gwatkin, *Fert. Steril.* **17**, 411 (1966).

CHAPTER 2

Trypsin

A. Mammalian Tissues

Charles Shipman, Jr.

Trypsin is a pancreatic proteolytic enzyme which preferentially catalyzes the hydrolysis of peptide bonds between the carboxy group of arginine or lysine and the amino group of another amino acid. For an extensive review of this enzyme see Desnuelle.[1]

[1] P. Desnuelle, *In* "The Enzymes" (P. D. Boyer, H. Lardy, and K. Myrback, eds.), Vol. 4, p. 119. Academic Press, New York, 1960.

The enzymatic release of monodisperse cells from tissue fragments historically has its origin with the early work of Rous and Jones.[2] These monodisperse living cells can be obtained from animal tissues by trypsinization because the enzyme differentially degrades the protein matrix which binds the cells in the tissue and releases these cells in suspension before they are seriously damaged in the process.

Discontinuous batch methods were developed later[3-6] and are still in widespread use. Because of the inherent problems in batch methods of trypsinization, continuous operating systems of varying degrees of sophistication have been developed.[7-9] The continuous methods of trypsinization obviate to some degree overdigestion due to prolonged mean residence time (MRT) of monodisperse cells in the enzyme solution. Those devices which rely upon a sintered-glass filter for cell sizing are easily clogged[7,9,10] whereas other devices do not minimize MRT[8] or are so mechanically complex that their construction is not feasible in most laboratories.[11]

Recently a simple glass device has been developed wherein enzymatically released monodisperse cells can be separated and isolated from tissue fragments by means of a discontinuous fluid velocity gradient.[12] The operation of this unit will be described in this subsection. Discontinuous batch methods of trypsinization are described in other sections.

Preparation of 0.25% Trypsin in HEPES-Buffered Saline (HBS)

Since the pH of a balanced salt solution buffered by HEPES (see Section XIV) is unaffected by the composition of the overlaying atmosphere there seems little reason to use a $NaHCO_3$-buffered salt solution to dissolve trypsin.

The formula of a HEPES-buffered salt solution[13] suitable for the compounding of a trypsin solution is given in Table I.

To prepare 1 liter of 0.25% trypsin, 2.5 g of commercially available trypsin (1:250)[14] are added to 1 liter of HBS and stirred overnight at 4°C. The following day the solution is warmed to room temperature, the pH readjusted to 7.4 (corresponding to approximately 7.2 at 37°C), and sterilized by means of mem-

[2] P. Rous and F. S. Jones, *J. Exp. Med.* **23**, 549 (1916).

[3] D. Bodian, *Virology* **2**, 575 (1956).

[4] R. Dulbecco and M. Vogt, *J. Exp. Med.* **99**, 167 (1954).

[5] J. L. Melnick, C. Rappaport, D. D. Banker, and P. N. Bhatt, *Proc. Soc. Exp. Biol. Med.* **88**, 676 (1955).

[6] J. S. Youngner, *Proc. Soc. Exp. Biol. Med.* **85**, 202 (1954).

[7] G. Barski, *Ann. Inst. Pasteur (Paris)* **91**, 103 (1956).

[8] L. W. J. Bishop, M. K. Smith, and A. J. Beale, *Virology* **10**, 280 (1960).

[9] C. Rappaport, *Bull. WHO* **14**, 147 (1956).

[10] L. Nicol, O. Girard, R. Corvazier, M. Cheyroux, P. Reculard, and P. Sizaret, *Ann. Inst. Pasteur (Paris)* **98**, 149 (1960).

[11] G. B. Gori, *Appl. Microbiol.* **12**, 115 (1964).

[12] C. Shipman, Jr. and D. F. Smith, *Appl. Microbiol.* **23**, 188 (1972).

[13] C. Shipman, Jr., *Proc. Soc. Exp. Biol. Med.* **130**, 305 (1969).

[14] 1:250 = One part of trypsin will convert 250 parts of casein to protease, peptones, and amino acids under the conditions of the N.F. assay for pancreatin.

TABLE I
Formulation of 0.01 M HEPES-Buffered Saline (HBS)[a]

NaCl	8.00 g
KCl	0.40 g
Na_2HPO_4	0.10 g
Dextrose	1.00 g
4-(2-Hydroxyethyl)-1-piperazineethanesulfonic acid (HEPES)	2.38 g
Distilled water	1000 ml
Adjust pH to 7.40 at 22°C with NaOH	

[a] The tonicity of this balanced salt solution is approximately 290 mOsm/kg.

brane filtration. If larger quantities are prepared it is often desirable to prefilter the solution through a 0.8 μm (mean pore diameter) filter so as not to clog the 0.2 μm sterilizing filter.

Use of the Continuous Flow Velocity Gradient Trypsinizing Cylinder[12,15]

The device consists of a single Pyrex reactor which can be constructed with or without a water jacket (Fig. 1). The inner cylinder is 38 mm (OD), whereas the water jacket is 160 mm (OD). Reactor volume is approximately 75 ml.

Operationally, three velocity patterns of fluid movement are involved. In the region below the flutes, the minced tissue fragments are suspended in a vortex

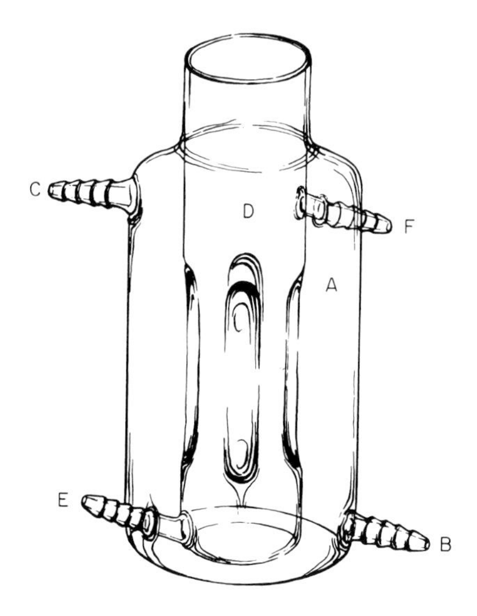

Fig. 1. Drawing of the automatic trypsinizing device. A, Water jacket; B, inlet, water jacket; C, outlet, water jacket; D, trypsinizing chamber with four flutes; E, inlet, trypsinizing chamber (connected via regulating valve to trypsin reservoir); F, spillover, trypsinizing chamber. From Shipman and Smith.[12]

[15] Available from Bellco Glass Inc., 340 Edrudo Road, Vineland, New Jersey 08360.

by the action of the magnetic stir bar. In the fluted region, aggregates of cells are released by the combined action of the turbulence and the enzymatic digestion. Above the flutes, there is neither a vortex nor turbulence but only an upward component of velocity. Monodisperse cells occupy this upper region of the vessel prior to their egress from the chamber. It is the small volume (approximately 10 ml) and the upward velocity component which allows one to achieve a minimum MRT.

The unit is normally sterilized with inlet and outlet hoses attached and a magnetic stir bar in the inner cylinder. After sterilization the chamber is aseptically connected to a reservoir of trypsin–HBS and to a delivery vessel for the collection of monodisperse cells. To stop the action of the trypsin, the receiving vessel contains a small amount of serum and is surrounded by crushed ice.

Tissue fragments prepared using a mincing tube and stainless steel barber's shears are transferred in HBS to the trypsinizing chamber by means of a large orifice volumetric pipette.[15] The magnetic stirring device is turned on and water at 37°C is circulated in the jacket. The flow of trypsin is now started from the reservoir. A Teflon stopcock at the reservoir is used to control flow rate. Alternatively, flow rate can be controlled by using a peristaltic pump. The speed of the stir bar and the flow rate of the trypsin must be empirically determined and are to a large extent dependent upon the type of tissue being trypsinized.

The automatic device has been used with monkey kidney, rabbit kidney, puppy salivary gland, and chick embryo tissues. It has also been utilized to free secretory cells from human thyroid tissue. Cell viability, as measured by trypan blue dye exclusion, and total yield of cells were always as high and often higher than using discontinuous batch methods of trypsinization.

CHAPTER 2

Trypsin

B. High Yield Method for Kidney Tissue

Héctor Montes de Oca

The method to be described here was reported in part elsewhere[1,2] and is a combined method employing trypsin and disodium ethylenediaminetetraacetic

[1] H. Montes de Oca, P. Probst, and B. Grubbs, *In Vitro* **2**, 127 (1966) (Abstr.).

[2] H. Montes de Oca, P. Probst, and B. Grubbs, *Appl. Microbiol.* **21**, 90 (1971).

acid (disodium EDTA) as the dispersing agents. The advantages of this method, as compared with other standard dispersion methods, are marked increase in the cell yield, reduction in the time required to accomplish complete dissociation of the tissue, and production of a culture with a uniform and "clean" monolayer that facilitates detection of viral cytopathogenic effect.

Source of Tissue

Description of the present technique is based on the work with simian, rabbit, and human kidneys. The animal kidneys are obtained from animals anesthetized with Pentothal sodium and killed by exsanguination. Human kidneys are obtained from infant cadavers under sterile conditions. The kidneys are aseptically removed and placed in Hanks' Balanced Salt Solution with 0.5% (w/v) lactalbumin hydrolysate, 2% (v/v) calf serum, 200 units of penicillin, and 200 μg of streptomycin per milliliter or, preferably, in CMRL 1415 ATM[3] without serum and kept at 4°C until processing.

Media

Five Percent (w/v) Trypsin Stock Solution. Trypsin (Difco 1:250) is added to warm (37°C) Dulbecco's phosphate buffer saline[4] without calcium or magnesium (pH 7.7, 330 mOsm) and stirred vigorously at 37°C for 3–4 hours, until the suspension is almost clear. This suspension is then sterilized through a membrane filter (Millipore Corporation, Bedford, Massachusetts) assembly with a prefilter, a 0.8, a 0.45, and a 0.22 μm pore size membranes and sterilized together as a unit. Tryton X-100 free membranes are used and, in addition, the filters are washed with hot (90°C) triple-distilled water immediately before use, to eliminate any other soluble contaminant which may be present.[5] The sterile trypsin solution can be kept frozen for 3 to 6 months at −18°C without appreciable loss of activity. The small precipitate which may appear when thawed will usually disappear when the solution is heated to 37°C.

Ten Percent (w/v) Stock Disodium EDTA Solution. Disodium ethylenediaminetetraacetic acid (ACSS0311 Fisher) is prepared with triple-distilled water. This solution is sterilized by vacuum filtration with an MT-VFA 7-500 Selas Flotronic filter unit with 0.27 μm disposable candles.

Dispersing Solution. The solution contains 0.25% (w/v) trypsin and 0.02% (w/v) EDTA in Dulbecco's BSS without calcium or magnesium ions (pH 7.2, 290 mOsm). This solution is prepared immediately before use from the stock solutions.

[3] G. M. Healy and R. C. Parker, *J. Cell Biol.* **30,** 531 (1966).
[4] R. Dulbecco and M. Vogt, *J. Exp. Med.* **99,** 167 (1954).
[5] R. D. Cahn, *Science* **155,** 195 (1967).

Dispersing Procedure

1. Remove and discard the renal capsule and pelvis, and weigh the remaining tissue.
2. Mince the tissue very thoroughly with scissors, using a beaker of appropriate size, until all pieces are approximately 2–3 mm or less.
3. Wash the minced tissue with warm (37°C) Dulbecco's BSS or CMRL 1415 ATM to eliminate the red blood cells and tissue debris until the supernatant is clear.
4. Wash once with the dispersing solution and place the minced tissue in a trypsinization flask (Bellco Glass, Inc., Vineland, New Jersey) with a Teflon-coated magnetic bar.
5. Add approximately 125 ml of warm (37°C) dispersing solution for each 10 g of tissue. Place the flask on a magnetic stirrer at room temperature and stir as vigorously as possible without producing foam.
6. Stir for intervals (runs) of 15–20 minutes. At the end of each run, stop the stirring for 1 to 2 minutes to allow for the nondispersed fragments to settle, and then decant the supernatant. Repeat these runs until all of the tissue is dispersed. To keep constant the volume relation between the remaining tissue and the added dispersing agent, reduce its amount by approximately 10% for each successive run.
7. After each run, decant and filter the cell suspension through several layers of gauze into a 200-ml round bottom centrifuge bottle. Add an equal volume of growth media with 2 to 10% calf or fetal calf serum to the bottle.
8. Centrifuge the cell suspension immediately for 20 minutes at 90 *g* at room temperature or at 4°C. Immediately after centrifugation remove the supernatant by suction, and resuspend the cell pellet in growth media by vigorous pipetting.
9. Pool all the cells obtained, as described above, in approximately 100 ml of growth media per each 5–8 g of dispersed tissue, and stir gently with a magnetic stirrer to keep the cells in suspension.
10. Take samples for cell count and cell viability determination. Obtain the percentage of viable cells by using the trypan blue exclusion method.[6]

Results

Simian Kidney. Results obtained with the method described above, using different dispersing agents, are tabulated in Table I: (1) trypsin 0.25% (w/v); (2) trypsin 0.25% (w/v) disodium EDTA 0.02% (w/v) and, (3) by using the Youngner technique[7] with trypsin 0.25% (w/v) as dispersing agent.

The higher percentage of dead cells in the Youngner's method is due to the centrifugation performed in a conical tube, with all of the dispersed cells, to

[6] J. M. Hoskins, G. G. Meynell, and F. K. Sanders, *Exp. Cell. Res.* **11**, 297 (1956).

[7] J. S. Youngner, *Proc. Soc. Exp. Biol. Med.* **85**, 202 (1954).

TABLE I
Cell Yield with Rhesus Monkey Kidneys[a]

Method of tissue dispersion	Cells per gram of tissue (± standard error)[b]	Amount of dead cells (%) (± standard error)
(1) Trypsin 0.25% in PBS	96 ± 15	15 ± 9
(2) Trypsin 0.25%; EDTA 0.02% in PBS	146 ± 20	8 ± 6
(3) Trypsin 0.25% in PBS (Youngner)	76 ± 15	30 ± 10

[a] Courtesy of American Society for Microbiology.
[b] Values to be multiplied by 10^6.

determine their total volume. When this step is omitted and, instead, cell counts are performed, the number of dead cells is reduced by approximately one-half.

In this and all similar experiments in which different techniques were compared, the minced tissue was distributed in various samples of equal weight and each sample was treated with one of the dispersing agents.

Table II summarizes the results reported in the literature,[8-13] and also presents

TABLE II
Monkey Kidney Cell Yields Reported by Different Authors[a]

Author	Average cell number per pair of kidneys[b]	No. of cells per gram of tissue[c]	Seeding cell/ml[c]	Amount of cell suspension (ml per pair of kidneys)
Present study	1310	132	0.3	4,400
			0.15	8,800
			0.07[d]	18,700
Cancevici *et al.*[8]	565.6	56.5	0.09	7,070
Dobrova[9]	900	100	0.3	3,000
Wallis *et al.*[10]	705	83	0.3	2,380
Mironova *et al.*[11]	1017	113.7	0.3	3,400
Rappaport[12]	900	100	0.3	2,400
				3,000
Kammer[13]	1470[e]	90	0.21	7,000

[a] Courtesy of American Society for Microbiology.
[b] Nine grams of tissue per pair of kidneys. Values to be multiplied by 10^6.
[c] Values to be multiplied by 10^6.
[d] The flasks were fully covered between 5 and 7 days.
[e] Fifteen grams of tissue per pair of kidneys.

[8] G. Cancevici, M. Dima, I. Stoian, and R. Crainic, *Arch. Roum. Pathol. Exp. Microbiol.* **23**, 239 (1964).
[9] I. N. Dobrova, *Vop. Virusol.* **4**, 118 (1959) (Transl.).
[10] C. Wallis, R. T. Lewis, and J. L. Melnick, *Tex. Rep. Biol. Med.* **19**, 194 (1961).
[11] L. L. Mironova, N. E. Goldrin, and L. L. Mamonenko, *Acta Virol.* **7**, 189 (1963).
[12] C. Rappaport, *Bull. WHO* **14**, 147 (1956).
[13] H. Kammer, *Appl. Microbiol.* **17**, 524 (1969).

the results obtained by this author during routine dispersion of 3800 g of simian kidneys. No significant difference was observed between Vervet and Rhesus monkey kidneys. The average cell yield for the first was 133×10^6 cells/g and for the second was 131×10^6 cells/g.

Rabbit Kidney. An increase in cell yield was obtained with rabbit kidney using trypsin–disodium EDTA, as compared with trypsin alone.

With young rabbits (3 g per pair of kidneys) 72×10^6 cells/g of tissue were obtained by using trypsin alone as a dispersing agent, and 120×10^6 cells/g were obtained with trypsin–disodium EDTA as the dispersing agent following the technique described above. With older rabbits (6 g per pair of kidneys) cell yields of 37×10^6 cells/g and 76×10^6 cells/g were obtained respectively.

Human Kidney. Kidneys from refrigerated cadavers of newborn babies were procured aseptically no later than 12 hours after death.

Using the above described technique, a cell yield of approximately 10×10^6 cells/g of tissue was obtained.

In experiments presently being conducted, the addition of 0.1% (w/v) of collagenase (Sigma Chemical Co., St. Louis, Missouri) to the trypsin–EDTA dispersing solution, increases the cell yield to 20×10^6 cells/g of tissue. This effect of the collagenase is apparently due to the fact that it dissolves a viscous material which appears during the dissociation of these human kidneys. This viscous material traps inside many cells and thus, when dissociated, the cell yield increases.

CHAPTER 2

Trypsin

C. Marine Teleost Fish Tissues

M. Michael Sigel and Annie R. Beasley

The following procedure is employed for obtaining dispersed cells from fins of grunts (*Haemulon* sp.) and snappers (*Lutjanus* sp.). It is essentially an adaptation of the technique described by Clem *et al.*[1]

[1] L. W. Clem, L. Moewus, and M. M. Sigel, *Proc. Soc. Exp. Biol. Med.* **108**, 762 (1961).

TABLE I
Comparative NaCl Concentrations in Standard Reagents and Solutions for Marine Fish Cells

	NaCl in standard solution		NaCl in solution modified for marine fish cells	
	g/liter	Molarity	g/liter	Molarity
Calcium- and magnesium-free phosphate-buffered saline	8	0.137	11.51	0.197
Basal medium Eagle, Hanks' base, with 10% calf serum and 10% human serum	6.4	0.109	9.91	0.169

It should be emphasized that saline and growth medium formulated for other types of tissue culture must be modified for use with *marine* teleost fish cells. This modification consists of increasing the final content of NaCl by a constant of 0.06 *M*, as shown in Table I. (This is simply effected by the addition of 1.35 ml of 26% NaCl/100 ml of conventional solution, such as Hanks' BSS.) The requirement for extra salt, a reflection of the high osmolarity and NaCl content of marine fish sera, is only for marine teleost fish cells and not for cells of freshwater fishes.

PROCEDURE

Fish are killed by severing the spinal column immediately posterior to the head, and all of the fins are excised just above the muscled areas. (If this is done in the field, the specimens are placed in chilled marine calcium- and magnesium-free phosphate-buffered saline (M-CMF-PBS) containing 10% calf serum plus 400 units penicillin, 200 μg streptomycin, and 2 μg amphotericin B/ml, and kept on ice for transport to the laboratory.) The tissue is soaked in Dakin's solution[2] for 3 minutes then vigorously agitated in M-CMF-PBS. It is further sequentially washed for 10-minute periods in 2 additional aliquots of the saline with antibiotics, and all fins from a single fish are transferred to a flask containing 35–50 ml of 0.25% trypsin (1:250 activity) solution in M-CMF-PBS. After incubation at room temperature for 15 minutes, with continuous stirring, the trypsin is discarded and replaced with another aliquot. Twenty minutes later the fluid is harvested and more trypsin is added to the tissue, a procedure which is repeated three times, all at room temperature. As each harvest is made, calf serum is added to it to give a concentration of 10% and the cell suspension is placed in an ice bath. After the final harvest, the tissue should be reduced to bones and a slimy residue of connective tissue.

When trypsinization is completed, the harvests are pooled, filtered through four layers of sterile gauze, and centrifuged in the cold at 200 *g* for 10 minutes.

[2] Dakin's solution: 36 ml concentrated solution (approximately 12%) NaOCl, 1 ml 6 *N* HCl, 8 g $NaHCO_3$/liter in H_2O.

The supernate is decanted and the cells are resuspended in Marine Basal Medium Eagle, Hanks' base, supplemented with 10% calf serum plus 10% human serum, and containing 200 units penicillin, 100 μg streptomycin and 1 μg amphotericin B per milliliter. An aliquot of the suspension is used for a cell count and additional growth medium is added to the remainder to give the desired cell concentration.

By this procedure, 90–120 ml of suspension containing 10^6 cells/ml can be obtained from the fins of a single adult fish.

CHAPTER 2

Trypsin

D. Amphibian Tissues[1]

Jerome J. Freed

Primary monolayer cell cultures have been prepared by dissociating adult amphibian tissue with trypsin in a Ca^{2+}- and Mg^{2+}-free balanced salt solution of about 200 mOsmoles osmotic pressure. The following procedure is one of several published for kidney cultures from the grass frog, *Rana pipiens*.[2–6]

Animals should be healthy; better success is obtained with freshly captured animals than with those stored in artificial hibernation. Frogs are pithed, washed in tap water, and pinned to a dissecting board. The skin is swabbed with 70% alcohol, the abdomen opened and pinned out, and the viscera moved aside. The elongated, purplish mesonephric kidneys lie behind the dorsal peritoneum on either side of the midline. Whitish nodules or masses may be the virus-associated renal adenocarcinoma (Lucke tumor) common in this species.[7] The kidneys

[1] The techniques described were developed with support from Grant AT(11-1)3110 from the Atomic Energy Commission (U.S.A.E.C. Report No. C00-3110-5), U.S.P.H.S. Grants CA-05959, CA-06927, and RR-05539 from the National Institutes of Health, and by an appropriation from the Commonwealth of Pennsylvania.

[2] W. Auclair, *Nature* (*London*) **192**, 467 (1961).

[3] V. C. Shah, *Experientia* **18**, 239 (1962).

[4] K. Wolff and M. C. Quimby, *Science* **144**, 1578 (1964).

[5] J. J. Freed and S. J. Rosenfeld, *Ann. N. Y. Acad. Sci.* **126**, 99 (1965).

[6] D. Malamud, *Exp. Cell Res.* **45**, 277 (1967).

[7] See papers *In* "Biology of Amphibian Tumors" (M. Mizell, ed.). Springer-Verlag, 1970 (Special Supplement, Recent Results in Cancer Research).

are dissected free and transferred to a Petri dish containing wash medium with antibiotics.[8]

Kidneys from 6 frogs are collected, transferred to a dry dish, minced into 1-mm fragments with scissors, and suspended in wash medium. With a pipette, the mince is transferred to a screw-capped 125-ml Erlenmeyer flask. The fragments are allowed to settle, the fluid removed with a pipette, and fresh wash medium added. When the tissue has been rinsed three times with 20-ml portions of fluid, it is resuspended in a 0.5% solution of trypsin (Difco 1:250) in wash medium, using 3 ml per pair of kidneys. A belted spin-bar is added, and the flask is stirred magnetically for 30 minutes at from 60 to 75 rpm. Two 50-ml centrifuge tubes are placed in crushed ice, and in the neck of each is placed a 3-inch square of sterile gauze, pushed in with a pipette to form a funnel. The tissue in the flask is allowed to settle, the fluid is drawn off with a pipette, and added through the gauze to the centrifuge tube. Fresh trypsin solution is added to the flask (5 ml per pair of kidneys) and stirring continued for 15 minutes. This fluid is added to the cells in the ice bath, and the procedure repeated twice more.

The gauze is removed, the tubes capped, and the cells sedimented by centrifugation for 10 minutes at 500 *g*. The pellets are each suspended in about 1 ml of wash medium, the suspensions are combined, and the cells counted with a hemocytometer. Red blood cells (ovoid and nucleated) should not be scored. Plastic tissue culture flasks (25 cm^2) or 2-ounce prescription bottles are charged with 4–5 ml of growth medium[8] and inoculated with 2×10^6 tissue cells. After 3 days incubation at 25°C, the medium and unattached cells are removed, and fresh growth medium is added. From 10 to 30% of the inoculated cells will be attached, and they will begin to multiply after about the third day of culture. The cultures will gradually become confluent and may be maintained at high cell density by twice-weekly changes of medium. An agar overlay for detection of virus plaques may be applied.[9]

We have not been successful in subculturing adult anuran kidney cells, although Balls and Ruben[10] have reported success in subculturing adult *Xenopus* kidney cells, and Gravell[11] has isolated a cell line (3 AKRP) from adult *Rana pipiens* kidney.

[8] See Section II.

[9] M. Gravell, A. Granoff, and R. W. Darlington, *Virology* 36, 467 (1968).

[10] M. Balls and L. N. Ruben, *Exp. Cell Res.* 43, 694 (1966).

[11] M. Gravell, *Virology* 43, 730 (1971).

CHAPTER 3

Collagenase Treatment of Chick Heart and Thyroid

S. Robert Hilfer

Enzymatic dissociation with collagenase and collagenase-containing mixtures has become the method of choice for the preparation of clonal cell cultures (i.e., Coon,[1] Spooner,[2] and Konigsberg[3]) because it seems to cause the least damage to cells and allows the greatest plating efficiencies. The procedures, furthermore, have been used successfully to separate the cells of many different embryonic and adult tissues by minor adjustments in the basic technique. The method which will be described is based upon the procedure of Cahn *et al.*[4] as described in Spooner,[2] Hilfer and Brown,[5] and Shain.[6] It produces viable single cell suspensions from both heart muscle and thyroid at a wide range of embryonic ages.

Source and Preparation

Collagenase is available from a number of chemical supply houses as a crude preparation and in purified form from at least one source (Worthington, Freehold, New Jersey, Cat. No. CLSPA). For most purposes the crude preparations seem to be satisfactory but the added expense of the purified preparation may be justified for particularly delicate cell types. Even the use of the purified preparation does not assure that enzyme activity will be confined to collagen. Although relatively free of proteolytic contamination (Worthington Enzyme Manual, 1972), proteolytic,[5,7] polysaccharidase, and esterase[8-10] activities are difficult to remove from the collagenase activity. Partial purification of the crude preparation can be accomplished by chromatography,[3] proteolytic activity can be removed with *N*-ethyl maleimide[10] but so far it has been impossible to re-

[1] H. G. Coon, *Proc. Nat. Acad. Sci. U. S.* **55**, 66 (1966).
[2] B. S. Spooner, *J. Cell. Physiol.* **75**, 33 (1970).
[3] I. R. Konigsberg, *Develop. Biol.* **26**, 133 (1971).
[4] R. D. Cahn, H. G. Coon, and M. B. Cahn, *In* "Methods in Developmental Biology" (F. H. Wilt and N. K. Wessells, eds.), pp. 493. Thomas Y. Crowell, New York, 1967.
[5] S. R. Hilfer and J. M. Brown, *Exp. Cell Res.* **65**, 246 (1971).
[6] W. G. Shain, Doctoral dissertation, Temple University, Philadelphia, Pennsylvania, 1971.
[7] W. Mitchell and W. Harrington, *J. Biol. Chem.* **243**, 4683 (1968).
[8] M. R. Bernfield and N. K. Wessells, *Develop. Biol. Suppl.* **4**, 195 (1970).
[9] M. R. Bernfield, S. D. Banerjee, and R. H. Cohn, *J. Cell Biol.* **52**, 647 (1972).
[10] B. Peterkofsky and R. Diegelmann, *Biochemistry* **10**, 988 (1971).

move the esterase and polysaccharidase activities,[11] and the polysaccharidase activity can be minimized only by dilution.[9]

The enzyme solution should be prepared in a calcium- and magnesium-free saline (CMF) such as Hanks' saline and sterilized by pressure filtration through a sterile microcellulose filter (i.e., a GS Millipore filter in a Swinny holder, Millipore Corp., New Bedford, Massachusetts). A sterile concentrate can be stored frozen and diluted as needed. The ability to dissociate seems to be more consistent, however, if a small portion is prepared for immediate use. The final concentration can range from 0.1 to 2.0% in CMF. For embryonic thyroid and heart, 0.25% collagenase in CMF Hanks' or a mixture of 0.25% collagenase, 0.1% trypsin (such as Bactotrypsin, Difco, Detroit, Michigan) and 10.0% chicken serum (CTC, Coon[1]) in CMF Hanks' saline have both proved to be satisfactory.

Dissociation

The following procedure has been used successfully for 3- to 20-day thyroids and 8- to 16-day heart ventricles from chick embryos. It is recommended that the original method[4] be read for detailed information on each step of the procedure. The tissue is removed to a 60-mm glass Petri dish containing Hanks' saline and adherent tissue is dissected away. The ventricles or thyroids are transferred to separate dishes of fresh saline and the larger glands or hearts are minced. Injury can be minimized by using fine, sharp knives and by making the fragments no smaller than 1 mm^3. The fragments are transferred to a 25-ml Erlenmeyer flask containing approximately 5 ml CMF Hanks'. Care must be taken not to place too much tissue in each flask or the cell yield will be reduced. The flasks are loosely stoppered (i.e., with an aluminum cap) and incubated 10 minutes in a 5% CO_2/95% air atmosphere at 37°C. The CMF Hanks' is withdrawn and replaced with an equal volume of the dissociation medium. The flasks are stoppered tightly with silicone rubber stoppers and incubated in a shaking water bath at 37°C (80 strokes/minute) for 10 minutes. Adherent mesenchyme can then be stripped from the tissue by either repeated pipetting with a Pasteur pipette or by agitation with a vortex mixer for no more than 10 seconds. The dissociation medium is changed and the flasks incubated under the same conditions. After 15 minutes the flasks are agitated again and the progress of dissociation inspected with a dissecting microscope. The cells which have been released can be withdrawn and held in nutrient medium while the remaining clumps are subjected to additional treatment. It is more practical to leave the suspended cells (which are actually mixed with many partially dissociated cell clumps) in the flask and to incubate for 1 or 2 additional 10-minute periods. Microscopic examination after agitation should be made at the end of each period and, when sufficient separation has occurred, the suspension is transferred to a centrifuge tube. The enzyme action is stopped by filling the tube with cold

[11] M. R. Bernfield, personal communication.

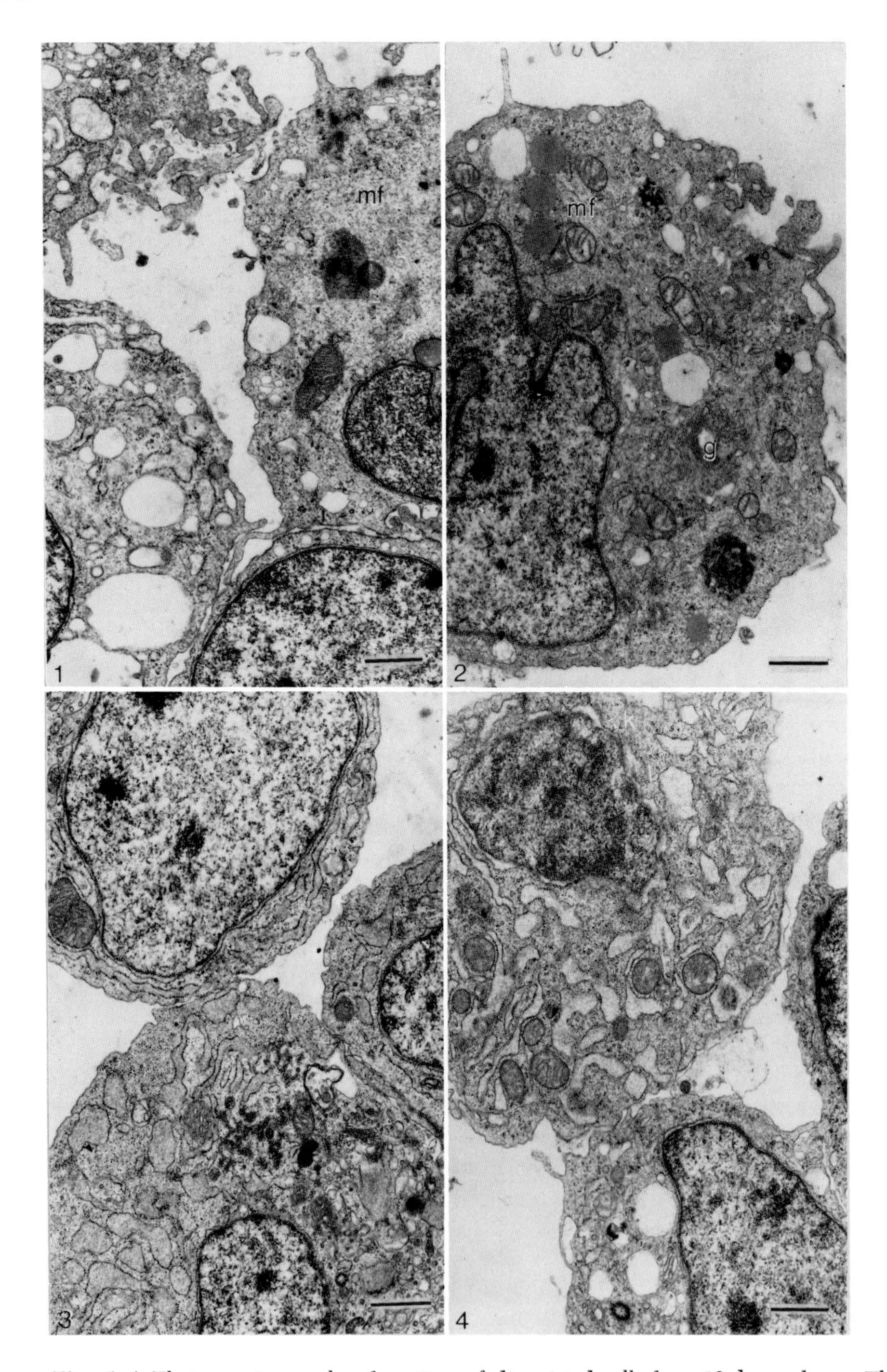

Figs. 1–4. Electron micrographs of portions of dissociated cells from 12-day embryos. The cells were centrifuged to form a pellet, fixed in glutaraldehyde and postfixed in osmium tetroxide. Sections were stained in aqueous uranyl acetate and lead citrate. Bar represents 1.0 μm. ×8000.

(4°C) nutrient medium containing serum. The cells are packed by spinning for 3 to 5 minutes at 1000 rpm in a clinical centrifuge. The supernatant is decanted and the cells suspended by pipetting in a small volume of fresh, cold nutrient medium. The larger cell clumps are removed from the suspension by centrifugation at 1000 rpm for 20 to 30 seconds. The suspension is transferred carefully to a fresh tube and brought to a convenient volume for counting. Greater than 98% of the cells should be present as singles. The few remaining doubles and small clusters can be removed by filtration through sterilized nylon cloth (20-μm mesh Nitex; Tobler, Ernst, & Traber, Inc., New York).

Viability

The only real test of the quality of a dissociation procedure is the viability of the cells it produces. Viability can be estimated by a dye exclusion test (i.e., Cahn *et al.*[4]) but the methods are not necessarily reliable.[12] Inspection by phase microscopy for the absence of irregular outlines and the presence of smooth cell surfaces is a useful procedure. Thyroid cells are usually smoother after collagenase alone and show blebbing and greater size differentials after CTC. Heart cells, on the other hand, tend to be round after CTC and irregular to elongate after collagenase alone. A better estimate of cell damage can be obtained by electron microscopy. Both heart (Fig. 1) and thyroid (Fig. 2) dissociated with CTC show less evidence of internal disruption and surface blebbing than the same tissues separated with collagenase alone (Figs. 3 and 4). Both methods disrupt myofilaments, but the rough endoplasmic reticulum tends to remain intact with CTC.

The critical test of viability, however, is the ability of the cells to attach and grow in culture. This parameter was assessed by making mass and clonal plates of the cell suspensions in modified Ham's F12 Medium supplemented with 10% fetal calf serum.[2] Cells were plated in 5 ml of medium in 60 × 15 Falcon tissue culture dishes (Falcon Plastics, Oxnard, California). Mass cultures were made with 10^5 cells per dish and clonal cultures with 10^2 and 10^3 cells per dish.

[12] C. Levinson and J. W. Green, *Exp. Cell Res.* **39**, 309 (1965).

Fig. 1. Heart dissociated with CTC. Surfaces contain fine projections but are relatively smooth. Cytoplasm is vesiculated although a few channels of endoplasmic reticulum are intact. Myofilaments form a disoriented mass (mf). The mitochondrial structure is normal and the nuclei are relatively rounded in outline.

Fig. 2. Heart dissociated with collagenase. The cell surface is uneven and the cytoplasmic membranes are vesiculated, including the Golgi region (g). Myofilaments (mf) are disoriented; mitochondria contain few cristae; and the nuclei tend to be distorted.

Fig. 3. Thyroid dissociated with CTC. The cells have smooth surfaces and the extensive rough endoplasmic reticulum shows little sign of disruption. The interconnected channels contain fine particulate material. The mitochondria and nuclei show no signs of alteration.

Fig. 4. Thyroid dissociated with collagenase. The cell surfaces are blebbed; the rough endoplasmic reticulum shows some vesiculation; and its contents appear to be lost in comparison to Fig. 3 and the gland *in vivo*. The mitochondria and nuclei appear normal.

TABLE I
Viability of Cells Dissociated with Collagenase and CTC

	Heart		Thyroid	
	Collagenase	CTC	Collagenase	CTC
Mass culture (16-day embryonic cells, 10^5 per dish)				
No. of cells released per organ	1,320,000	2,340,000	110,000	135,000
Cells/dish at 2 days	120,000	150,000	180,000	140,000
Clonal culture (12-day embryonic cells, 10^2 and 10^3 per dish)				
No. of cells released per organ	800,000	3,600,000	140,000	110,000
Plating efficiency	0.2	3.2	2.6	1.3[a]

[a] Plating efficiencies as high as 4% for this embryonic age can be obtained by selection of the serum used in the medium (W. G. Shain[6]).

The results of typical experiments are summarized in Table I. A higher percentage of the thyroid cells seem to attach after CTC, but the yield (number of cells released per gland) and the rapid increase in cell numbers indicate that collagenase alone is better for thyroid cells. The plating efficiencies of the clonal cultures support this conclusion. It should be noted that the difference in cell structure (cf. Figs. 2 and 4) does not seem to be significant for survival although it might explain the difference in initial attachment. It is interesting that trypsin and EDTA treatment produces a greater cell yield but results in greater cytoplasmic damage and lower percentage attachment.[13] Heart cells appear to respond better to CTC than to collagenase alone. The cell yield is greater, attachment seems to be better, and in most experiments cell growth is more prolific. Heart cells separated with collagenase alone showed little ability to clone even at 10^4 cells per dish. It seems clear that both procedures should be tested on any new tissue to be dissociated.

Mode of Action

Since collagenase alone will separate at least some cell types that are not surrounded by large amounts of collagen, it seems unlikely that the basis of cell separation is entirely collagenolytic. The contaminating protease, esterase, and polysaccharidase activities may account for the effect on such cells. The demonstration that it is these contaminants to collagenase that cause the collapse of the branched organization of the salivary[8,9] is in agreement with this suggestion. However, the absence of fibrils near the surfaces of epithelial cells does not exclude the possibility that collagen is present in some other form.

[13] S. R. Hilfer and E. K. Hilfer, *J. Morphol.* **119**, 217 (1966).

CHAPTER 4

Sequential Enzyme Treatment of Mouse Mammary Gland

F. J. A. Prop and G. J. Wiepjes

For obtaining suspensions of viable cells from normal mouse mammary glands, existing methods using trypsin or Pronase or collagenase, etc., were unsatisfactory as to cell yield and/or cell viability. Therefore a method was devised[1] that first weakens—without completely digesting—the structural constituents of the stroma by collagenase and hyaluronidase, and then, in a second step, a general proteolytic enzyme like Pronase or trypsin brings forth the complete dissolution of the stroma and also of cell adhesion, resulting in a monocellular suspension.

An important feature of the method, furthermore, is that no means of mechanical disruption is applied during the enzyme treatments. Vigorous repeated pipetting used for the final disruption in many enzymatic methods is the main reason for low viability. The only mechanical step in the method is the initial dissection of the tissue.

The cell yield by this method approximates the maximum attainable.

Preparation of Cell Suspension

A 9- to 10-week-old virgin CBA mouse is killed by cervical luxation. It is then immersed in 70% ethanol. The skin and the mammary glands are dissected free under the usual aseptic precautions. The mammary glands are taken off and the lymph nodes removed from them. They are placed in a Petri dish which contains a calcium- and magnesium-free balanced salt solution[2] [CMF: (in g/liter) 8.00 NaCl, 0.30 KCl, 0.05 $NaH_2PO_4 \cdot H_2O$, 0.025 KH_2PO_4, 1.00 $NaHCO_3$, 2.00 glucose]. The tissue is chopped in CMF using two crossed scalpels fitted with surgical blades (No. 11) using a cutting movement, avoiding as much as possible compression of the tissue. Subsequently, the tissue fragments are incubated for 15 minutes in 15 ml of CMF in a 25-ml Erlenmeyer flask at 36°C on a gyratory shaker (80 rpm, ¾ inch rotation diameter) in order to remove most of the free Ca and Mg ions and to get rid of debris. All through the following steps the gyratory shaker is operated at the same specifications; this means a very gentle

[1] G. J. Wiepjes and F. J. A. Prop, *Exp. Cell Res.* **61**, 451–454 (1970).
[2] A. Moscona, *Exp. Cell Res.* **22**, 455–475 (1961).

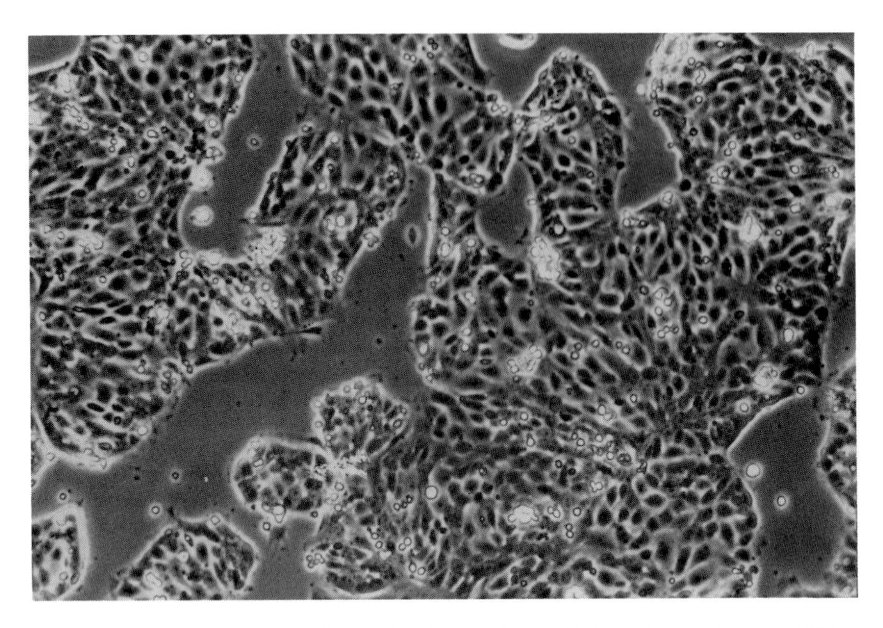

Fig. 1. + Pronase 2.5%; chiefly epithelial monolayer; 5 days living culture. ×100.

movement. The tissue fragments floating at the surface due to the fatty stroma are then transferred by means of a wide-mouthed pipette into 5 ml of a solution containing 0.125% collagenase (Worthington code CLS, activity 200 U/mg) and 0.1% hyaluronidase (Sigma, Type I) dissolved in CMF with 4% demineralized bovine serum albumin (Poviet Producten N. V.). This solution may be kept

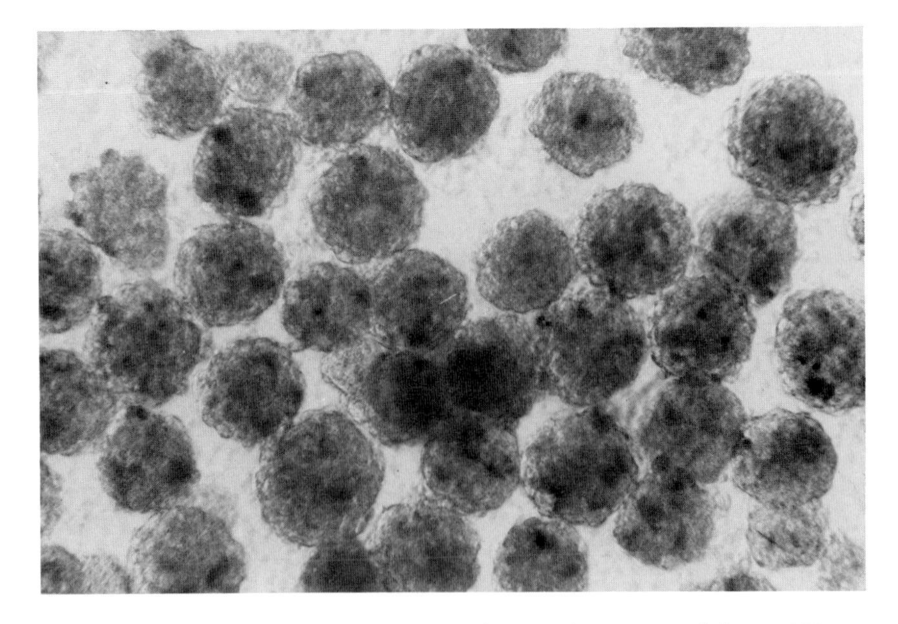

Fig. 2. + Pronase 1.25%; reaggregates, 3 days on the gyratory shaker. ×100.

frozen or prepared fresh and filtered sterile through a 0.45-μm Millipore filter. After incubation on the gyratory shaker for 45 minutes, the much loosened but still adherent tissue is transferred to 5 ml of Medium 199 containing 1.25 or 2.50% Pronase (CalBiochem, B grade). Pronase contains insoluble impurities; sterile filtration is greatly facilitated when these are first removed by spinning at 3000 rpm for 10 minutes. It is extremely important at this stage of the enzyme treatment to avoid any mechanical damage to the cells. One hour incubation on the gyratory shaker effects a complete dissolution of the tissue structure, the result being a suspension of cell clumps and single cells. The procedure is then carried out at room temperature. Twice the volume of serum is added to stop enzyme action and the cells are spun at 1000 rpm for 5 minutes. The supernatant and the floating fat cells are discarded. The pellet is resuspended gently in 20 ml of culture medium (Hanks' balanced saline, 0.5% lactalbumin hydrolysate, 20% human serum). Should a mucouslike substance prevent resuspension, three drops of a 0.04% DNase solution (Worthington, D) can be added. Resuspension is followed by filtration through a No. 3 sintered-glass filter (pore width 15–40 μm); suction is applied by a water jet pump; filtration should be accomplished before any appreciable vacuum has been built up. The filtrate is a single-cell suspension. The cells are washed twice in culture medium by centrifugation and resuspension. Cell number and viability are checked in a hemocytometer with 0.25% trypan blue in Hanks' saline. The viability is generally better than 90%. Cell yield from 300 mg wet weight of mammary gland is approximately 5.5×10^6.

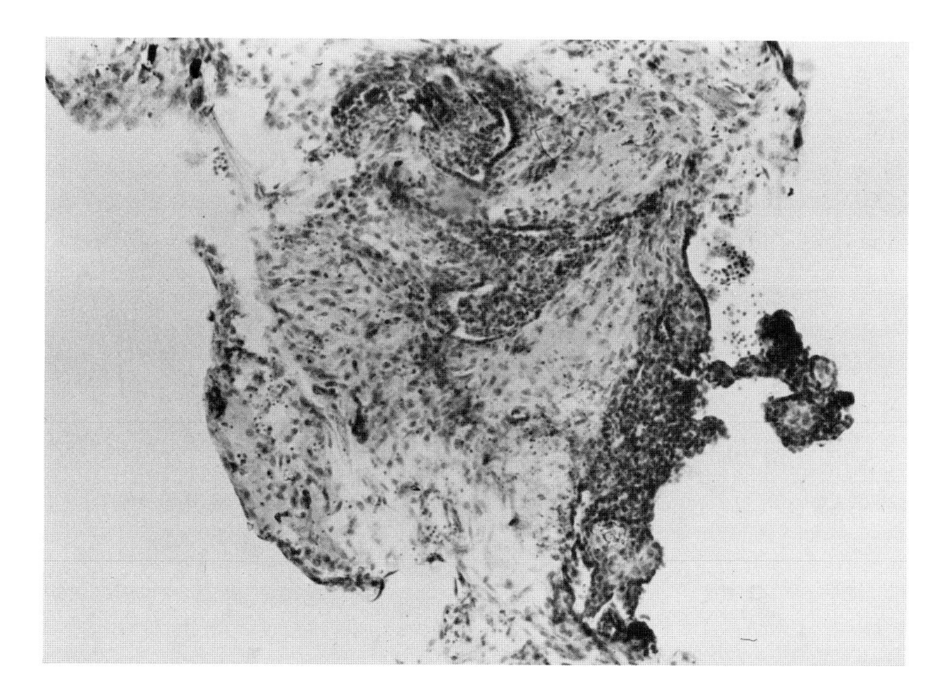

Fig. 3. Structural development in aggregate; 3 days on the gyratory shaker. H. E. ×100.

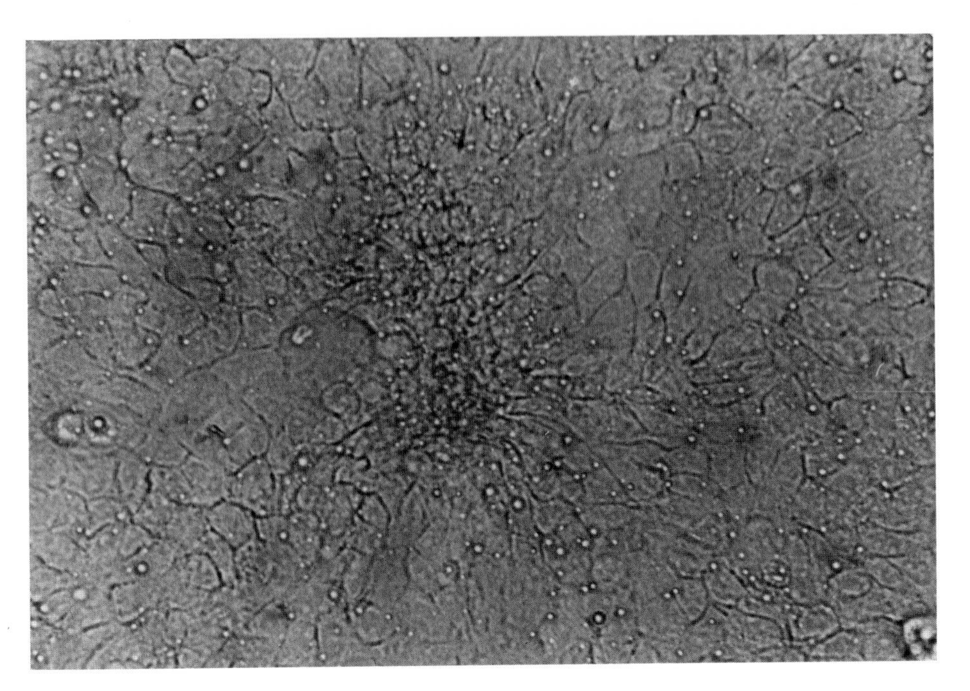

Fig. 4. Monolayer of primary culture from C3H mammary carcinoma. ×100.

APPLICATIONS

If cultured on glass, the 2.5% Pronase-isolated cells show epithelium-like growth (Fig. 1); in cultures of 1.25% Pronase-isolated cells there is admixture of fibroblastlike cells. Starting from this latter material, Visser *et al.*[3] found hormone effects in primary cell cultures of mouse mammary glands.

Using Moscona's technique for making aggregates on the gyratory shaker[2] the cells obtained by the method described form globular aggregates with an average diameter of 0.2 mm (Fig. 2); in sections these may show histological structures (Fig. 3).

The procedure has also been used for the preparation of cell suspensions from mammary tumors. Good yields were obtained from C3H mouse mammary adenocarcinomas (15×10^6 to over 200×10^6 from tumors weighing 500–2000 mg). Figure 4 shows the growth pattern in culture of these cells.

Cultures could also be prepared from human breast cancers. For these, enzyme concentrations and incubation times have to be adapted in relation to the structure of the tumor as revealed in a frozen section made immediately after receiving the material from the operating theater. A procedure for preparing cultures from human breast tumors is given by Lasfargues (Section II, Chapter 3).

[3] A. S. Visser, W. R. E. de Haas, C. Kox, and F. J. A. Prop, *Exp. Cell Res.* 73, 516 (1972).

CHAPTER 5

Nonenzymatic Dissociations

A. Leukocyte Cell Separation on Glass

Y. Rabinowitz*

Viable normal[1] or leukemic[2] leukocytes can be partially or completely separated in glass bead columns on the basis of differences in adherence of the various cell types to glass in the presence or absence of fresh plasma or serum, Mg^{2+}, and Ca^{2+}.

Blood Collection

Heparin without preservative[3] (phenol) is used as the anticoagulant. Panheprin (Abbott No. 6945), 4–5 U.S.P. units per ml of blood, is used for small samples, while 500 ml samples are collected in plastic bags containing 2115 units of heparin (Abbott No. 4686). Addition of 500 extra units of Panheprin to the plastic bags has proved to be desirable to prevent clotting in the columns.

Preparation of Leukocyte-Rich Plasma

Sedimentation of the red blood cells (RBC) is hastened by the addition of 6% dextran in saline (clinical grade H, average mol. wt. 184,000, Pharmachem Corp., Bethlehem, Pennsylvania)—1 ml of dextran per 5 ml of blood. After RBC sedimentation is completed, the supernatant leukocyte-rich plasma is collected by aspiration. The volume of the cell suspension is adjusted, if necessary, to fit onto a suitable stock column (5–75 ml). The white blood cell count is kept under 20,000 per mm^3 to avoid overloading the columns.

Since exposure to cold or heat affects adherence of the cells, all centrifugations are carried out in a refrigerated centrifuge set to maintain temperatures of 17° to 19°C. Centrifugal forces over 150 *g* are avoided since they also are injurious to the cells and may affect adherence.

* Deceased December 7, 1972.

[1] Y. Rabinowitz, *Blood* **23**, 811 (1964).

[2] Y. Rabinowitz, *Blood* **26**, 100 (1965).

[3] J. E. Garvin, *J. Exp. Med.* **114**, 51 (1961).

The Glass Bead Columns

The columns (Fig. 1) are made from Pyrex glass tubing. Columns 16 cm in length by 1 cm in diameter permit the retention of up to 5 ml of cell suspension, while columns 40 cm in length and 2.5 cm in diameter permit the retention of up to 75 ml. The bottom of the column is sealed with a glass shelf except for a short piece of glass outlet tubing (2 cm long, 3 mm ID). A rubber stopper, with a hypodermic needle inserted through it, is fitted to the top of the column to serve as the inlet. Cell suspensions or wash solutions are added through the needle from syringes or by means of infusion bottles and tubing for large volumes.

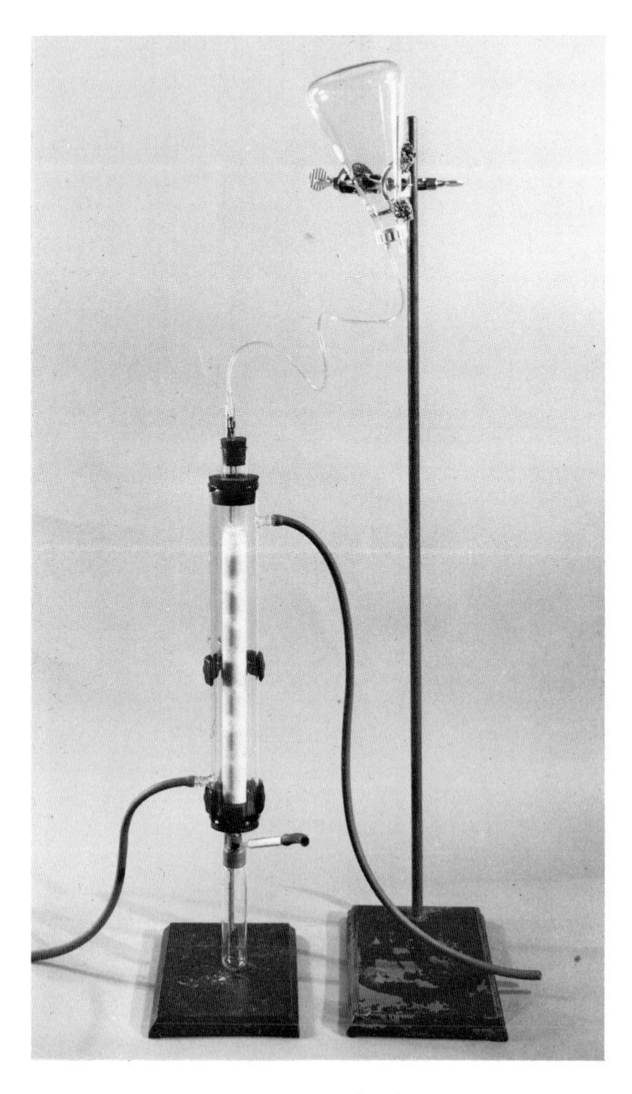

Fig. 1. Glass bead column.

A sleeve-type rubber stopper (Arthur A. Thomas 8826) is used at the outlet. A test tube is inserted into the sleeve to provide a closed system which helps to retain the cell suspension on the column during incubation. The inlet needle is conveniently stoppered with a small disposable sterile syringe. An air vent is inserted during collection of effluents. This consists of the glass portion of a Wintrobe pipette (Scientific Products B4467-1) filled with cotton and inserted through the lower stopper into the collecting tube with a hypodermic needle.

A thin layer of glass wool is placed at the bottom of the column to prevent loss of the beads. The column is filled 80–85% with No. 100 (0.2 mm diameter) "Superbrite" glass beads (3M Co., St. Paul, Minnesota).

The glass beads are washed with nitric acid and exhaustively rinsed. All glass used is siliconized with Siliclad (Clay-Adams, Inc., New York).

The Column Procedure

The entire procedure is perhaps best carried out in a 37°C room, but good results are obtainable by simply placing the column into an incubator. Another alternative is a column with a water jacket (Fig. 1). The leukocyte-rich plasma suspension is added to the column by syringe through the top inlet needle and retained on the glass beads during a 30-minute incubation at 37°C. The column is then washed with 20% autologous plasma (or serum) in Hanks' BSS (pre-warmed to 37°C) and the effluents collected in column-volume samples. Most nonadherent cells (lymphocytes, RBC, etc.) appear in the first two column volumes (Fig. 2A). Weakly adhering cells (platelets) and residual nonadherent cells (Fig. 2B) are then washed completely from the columns with 8 to 10 additional column volumes of wash. Completion of the wash is simply determined

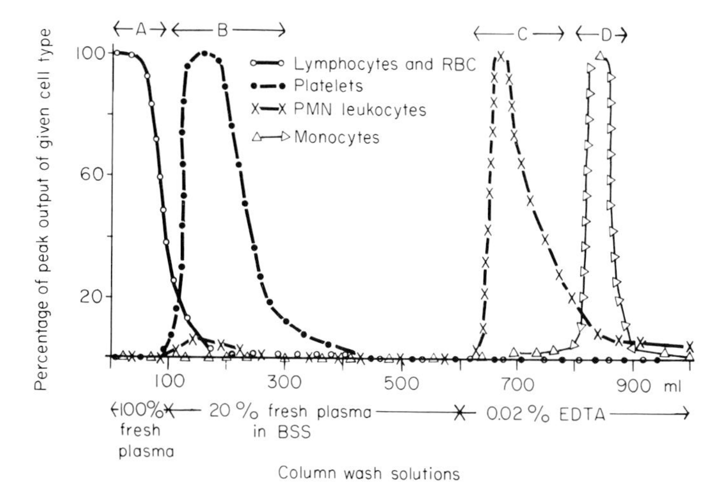

Fig. 2. Peaks of cell elution from a 40-cm glass bead column. Leukocyte-rich cell suspension concentrated to 75 ml added to column after dextran sedimentation of 500 ml of blood. Curves represent composite of eighteen runs. Uniformity of results permits prediction of location of peaks. Behavior of smaller columns is proportional.

by the absence of a pellet when an aliquot of effluent is centrifuged for 1 minute in an Adams Serofuge. In the presence of fresh plasma (or serum) few adherent cells are lost from the column. Addition of 0.02% EDTA in Ca^{2+}- and Mg^{2+}-free BSS to the column results in the release of the strongly adherent cells (granulocytes, monocytes, etc.) from the glass (Fig. 2C and 2D).

The chief cause of failure of the separation is the use of aged serum which lacks the required fresh serum adherence factor. When there is a shortage of autologous plasma, as may occur in cases of chronic lymphatic leukemia with high cell counts, good separations may be obtained with the use of only 5% plasma.

Cellular Adherence Patterns

Normal or leukemic cell types are for the most part divisible into adherent or nonadherent classes. There is overlap in the case of granulocytes which become more adherent as they mature.

The cells which do not adhere to glass in the presence of fresh plasma or serum include: normal or leukemic lymphocytes, lymphosarcoma cells, blasts, promyelocytes, most myelocytes, some metamyelocytes, and both nucleated and nonnucleated erythrocytes.

The cells which adhere to glass in the presence of fresh plasma or serum, and which require EDTA for release include: neutrophilic, eosinophilic, and basophilic granulocytes, bands, most metamyelocytes, an occasional myelocyte, and monocytes. Monocytes are usually concentrated in a small peak in the second EDTA column volume.

The procedure permits the collection of normal or chronic lymphatic leukemic lymphocytes contaminated by 0 to 2% granulocytes, and variable numbers of RBC's depending on the efficiency of the initial dextran sedimentation. Normal granulocytes, virtually free of all other cellular contaminants, are collected in the first EDTA column volume. The small monocyte peak in the second EDTA column volume may be found by monitoring the effluent with phase microscopy. A concentration of 25% monocytes may thus be obtained.

The complex cell mixtures of chronic granulocytic leukemia or bone marrow can be partially separated on the columns. Further separation then would depend upon the use of other procedures such as centrifugation on density gradients. When all the column volumes are collected, the procedure provides a series of tubes with varying cell differential counts from the same specimen. These may be useful in a variety of applications.[4]

Hemolysis of Red Blood Cells

After column separation, the NH_4Cl procedure of Roos and Loos[5] gives a complete hemolysis of contaminating RBC prior to cell culture. Good lymphocyte

[4] Y. Rabinowitz, *Blood* **27**, 470 (1966).
[5] D. Roos and J. A. Loos, *Biochim. Biophys. Acta* **222**, 565 (1970).

response to PHA is obtained, while avoiding the difficulties of hemolysis of RBC in cells clumped by PHA.

Lymphocyte and Monocyte Cultures

In cultures of glass-separated cells with PHA, blast cells develop only from lymphocytes, while macrophages are obtained only in cell preparations containing monocytes. Pure cultures of macrophages are obtainable with cultures of column-separated monocytes contaminated by granulocytes. After a week of culture, monocytes develop into macrophages which adhere to the glass of the container while most of the granulocytes die and are washed away with a change of medium.

CHAPTER 5

Nonenzymatic Dissociations

B. Specific Cell Fractionation on Chemically Derivatized Surfaces

Gerald M. Edelman

Although a number of physical methods for the fractionation of eukaryotic tissues and cells are in current use, there are few methods utilizing the chemical properties and specificities of the cell as a means for both cell fractionation and cell manipulation. The main requirements of a chemical approach to the manipulation of cell populations are specificity, general applicability, high yield, and maintenance of cell viability. The requirement for specificity suggests the use of a method employing solid supports coupled to proteins capable of binding to cell surface components. With this principle in mind, specific methods for the isolation of antibody-forming cells on columns of derivatized beads have been developed with varying degrees of success.[1-3] Subsequent specific dissociation of cells from such supports is limited, however, because the structures of the surface components are usually not known nor are they generally available in

[1] H. Wigzell and B. Andersson, *J. Exp. Med.* **129**, 23 (1969).
[2] P. Truffa-Bachi and L. Wofsy, *Proc. Nat. Acad. Sci. U. S.* **66**, 685 (1970).
[3] J. M. Davie and W. E. Paul, *Cell Immunol.* **1**, 404 (1970).

soluble form for use as competitive inhibitors of cell binding. Moreover, these methods do not readily lend themselves to direct visualization and quantitation of the cells and their surface reactions.

In order to circumvent some of these difficulties, a method of fractionation has been devised which is based on the interaction of cell surfaces with chemically derivatized fibers.[4,5] A variety of molecules such as lectins, antigens, or antibodies can be covalently coupled to the fibers to provide the requisite binding specificity for the cells. Adsorbed cells are removed by various means to complete the process of fractionation. The fixation technique may also be used with flat surfaces, and cells may be made sessile in tissue culture in arrangements which are under control of the investigator.

Before describing the details of this method, a clarification of the criteria of purity is in order. In chemical fractionation, purity is defined operationally in terms of a number of criteria based on independent properties of a species or component. Ultimately, a stoichiometric analysis in terms of elemental composition can be used. There is no possibility, however, of defining an ultimate criterion of purity of a cell population in terms of its molecular structure alone. Although one approach would be to define a cell population as pure if it is homogeneous with respect to its genotype and phenotype, more practically an operational criterion must be adopted based on a limited number of markers unique to a phenotype. For example, a set of lymphocytes might be fractionated according to the specificity of their receptors for a particular antigen. Nevertheless, these cells may still be heterogeneous with respect to their differentiation as thymus-derived or bone marrow-derived cells, their stage of maturation, the class of their Ig receptors and the binding affinity of their receptors. Similar strictures may apply to cells of any type and therefore a fractionated population will be "pure" only with respect to the specificity of the molecules used to attach to their surface receptors.

The basic principles of fiber fractionation are illustrated in Fig. 1. After deciding upon the molecule or macromolecule to be used for the fractionation, it is coupled in suitable chemical form to nylon fibers arranged in a convenient geometry. Dissociated cells are then shaken in the appropriate medium with the fibers after which unbound cells are washed away. The bound cells may then be transported on the fiber to other media or be removed by three methods: competitive binding of free ligand, mechanical release by plucking the fiber, or chemical cleavage of the linkage between the cell and the fiber.

A water-soluble carbodiimide [1-cyclohexyl-3-(2-morpholinoethyl)carbodiimide metho-*p*-toluenesulfonate; Aldrich Chemical, Milwaukee, Wisconsin] can be used to couple ligands to the fiber and the extent of substitution can be controlled by varying the initial concentration of the reactants.[4] A certain amount of noncovalent binding of the reagents can occur and it is important to determine the actual extent of covalent linkage. For example, a fiber derivatized with a

[4] G. M. Edelman, U. Rutishauser, and C. F. Millette, *Proc. Nat. Acad. Sci. U. S.* **68**, 2153 (1971).

[5] U. Rutishauser, C. F. Millette, and G. M. Edelman, *Proc. Nat. Acad. Sci. U. S.* **69**, 1596 (1972).

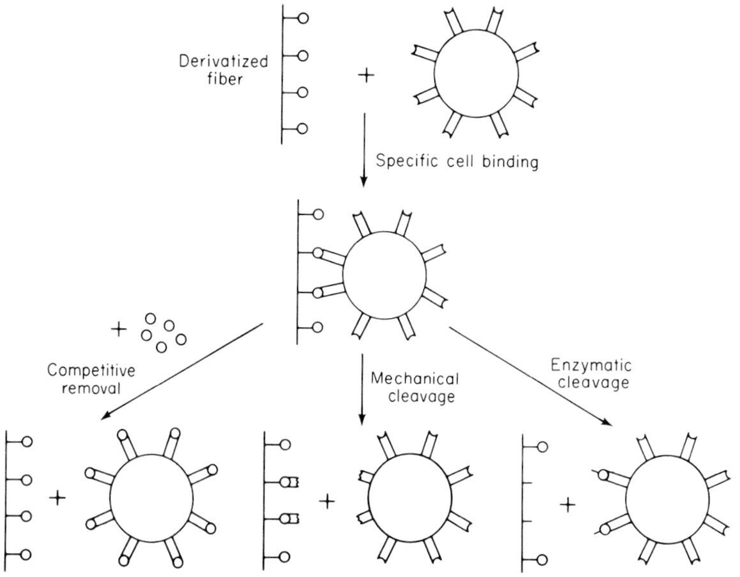

Fig. 1. General scheme for the fiber fractionation of cells.

radioactive protein may be incubated with 5 *M* guanidine hydrochloride and the amount of free and bound label quantitated. In most cases, about 70% of the bound protein appears to be covalently linked.[6] This is usually sufficient, particularly since the noncovalently bound material is not released under physiological conditions.

Preparation of Derivatized Fibers, Meshes, Surfaces, and Beads

The procedures for fibers and meshes are essentially identical except for the frame used to hold the meshes.[5] A detailed description is given for fibers and protein ligands only; modifications may be easily made for other ligands.[4]

Transparent nylon monofilament (size 50 sewing nylon; Dyno Merchandise Corp., Elmhurst, New York) is strung into polyethylene collars (cut from hollow S-6 stoppers, Mallinckrodt, New York). These fit snugly into 35 × 10 mm Petri dishes (NUNC, Vanguard International Inc., Red Bank, New Jersey) and hold the fibers under tension (see Fig. 2a). Surface contaminants are removed by 10-minute extractions of the strung fibers first with petroleum ether and then with carbon tetrachloride. In some cases, it is necessary to increase the reactivity of the nylon fibers by partial hydrolysis with 3 *N* HCl for 30 minutes at room temperature.[4] After soaking in a large volume of H_2O for 1 hour, the fibers are placed in a freshly prepared solution of the protein and carbodiimide in 0.15 *M* NaCl, pH 6.0. Useful protein concentrations range from 0.1 to 10 mg/ml and carbodiimide is added at carbodiimide:protein ratio of 5:1 (w/w). The reaction mixture is shaken at room temperature for 30 minutes. The polyethylene collars

[6] G. M. Edelman, J. L. Wang, P. G. Spear, and I. Yahara, unpublished observations.

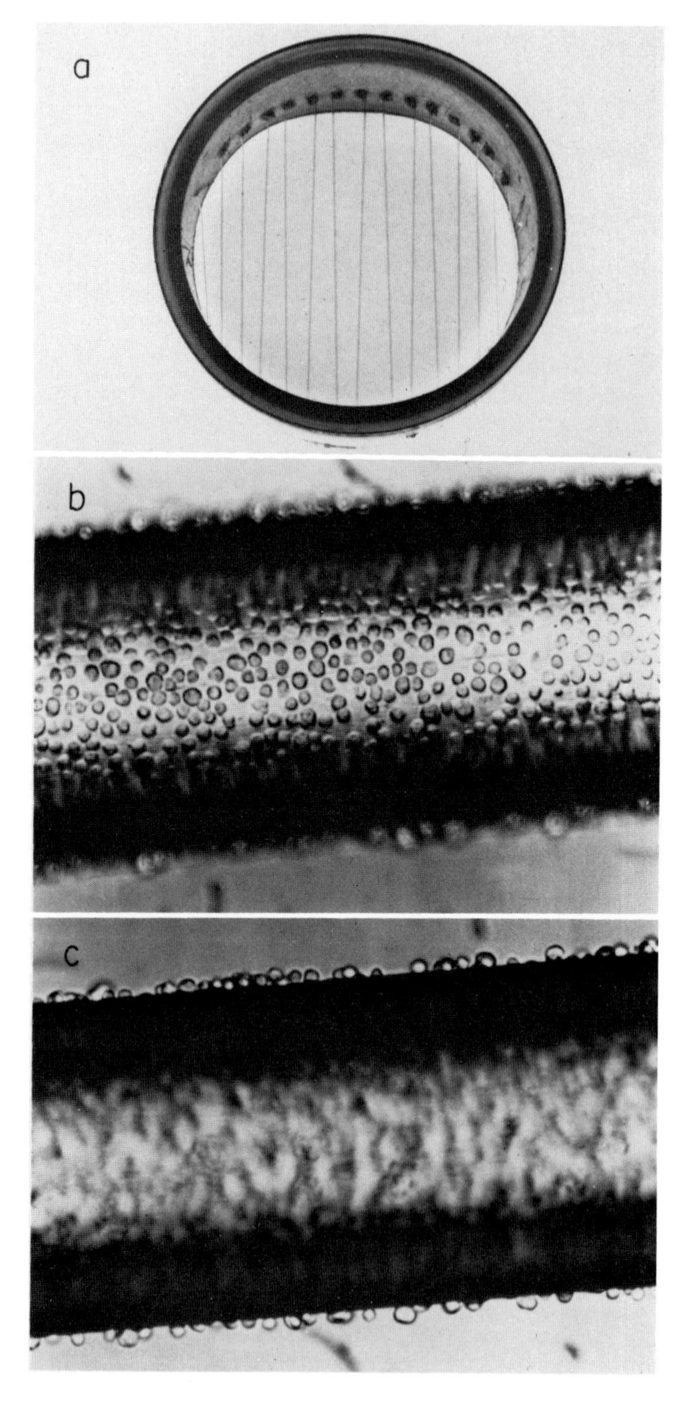

Fig. 2. (a) Petri dish containing polyethylene collar strung with nylon monofilament. (b) Mouse thymocytes bound to a Con A-derivatized nylon fiber. The field was focused on the face of the fiber at ×480 magnification by bright-field microscopy. (c) Mouse thymocytes bound to a Con A-derivatized nylon fiber. The field was focused on the edge of the fiber at ×360 magnification by bright-field microscopy.

are then washed in phosphate-buffered saline, pH 7.4 (PBS; 8.00 g NaCl, 0.20 g KCl, 0.20 g KH_2PO_4, 0.15 g Na_2HPO_4 per liter) and transferred to fresh Petri dishes for use.

Cell Binding, Removal, and Quantitation

Cell binding to derivatized nylon filaments is carried out by adding 10^7 to 10^8 cells in 4 ml total volume to the 35 × 10 mm dish containing the nylon fibers. To minimize cell aggregation, DNase (10–50 μg/ml) may be added to the cell suspension. The dishes are placed on a reciprocal horizontal shaker for 15–180 minutes at 4° to 25°C with the fibers arranged perpendicular to the direction of shaking. Unbound cells are removed by immersion of the dish in a large volume of medium. During this and all subsequent procedures care should be taken not to remove the fibers from the liquid, because interfacial forces result in the release and loss of the cells. If meshes are used, this stricture does not apply, for sufficient liquid remains in the interstices to prevent the loss of cells. Bound cells are most easily released into medium containing 10% serum by gently plucking the fibers with a needle near the frame. As an alternate procedure, coupling of ligands via linker molecules permits subsequent chemical removal of cells by enzymatic hydrolysis of the linker molecule.[6] In some cases, cells may be slowly released by a competitive inhibitor, although in general, secondary interactions between the cells and the fibers interfere with this type of specific dissociation.[4]

Bound cells may be quantitated by using radiolabeled cells, by counting cells after their removal from the fiber, as well as by direct observation of the fiber under the microscope (Fig. 2b). This last method takes advantage of the fact that the edge of the fiber may be put in sharp focus (Fig. 2c) and the bound cells in this focal plane can be then counted over a sufficient length of the fiber. As shown in Fig. 3, the factor to convert edge counts to total counts can be determined for any given set of experimental conditions. Large numbers of cells may be obtained using fiber meshes and this approach may be also used to deplete a population of cells of a specific subpopulation.[5] Cells are incubated in a 35 × 8 mm air-free chamber separated into two compartments by a derivatized nylon mesh (308 gauge, square weave, Nitex; Tobler, Ernst & Traber, New York). Eight milliliters of cell suspension (1.25–2.5 × 10^7 cells/ml) are added, and the chamber is placed on a horizontal rotary shaker at 200 rpm for 1 hour at 4°C. The chamber is inverted every 15 minutes so that the cells filter through the mesh under unit gravity. The mesh is then washed by immersion in medium. After their removal by pipetting 4 ml of medium repeatedly through the mesh, bound cells can be counted in a hemocytometer. The extent of depletion is tested by assaying the supernatant cells for fiber-binding cells using strung fibers.

Specific Fractionation of Immune Cells

The most extensive use of the fiber method has been in the separation of lymphoid populations. Three examples of the use of the method may serve to

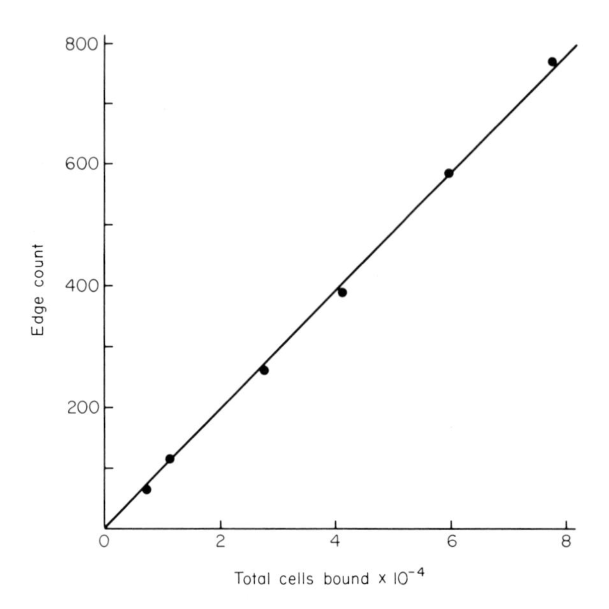

Fig. 3. Calibration curve for determining the total number of fiber-binding cells in an incubation dish from a count of cells bound along both edges of a 2.5-cm fiber segment. Total cell numbers were determined by γ radiation spectrometry of ^{51}Cr-labeled cells or by direct counting after plucking and centrifugation onto the bottom of a Petri dish.

indicate how it may be applied in other circumstances: (1) isolation of cells according to the *specificity* of their receptors, which are immunoglobulins directed against a particular antigen, (2) the isolation of cells according to the *affinity* of their receptors, and (3) the determination of *receptor distribution* on the surface of fiber bound cells.

Table I shows the isolation of specific lymphocytes using three different antigens. The specificity of the interaction (Table II) is indicated by the inhibition of cell binding, which occurs when free antigen is present in the medium during incubation of the cells with the fibers derivatized with the same antigen. By addition of increasing amounts of soluble antigen to separate dishes, cells may be fractionated according to the affinity of their receptors.[5]

Antibodies to immunoglobulins also inhibit the binding of lymphoid cells

TABLE I
Number of Mouse Spleen Cells Bound to Antigen Derivatized Fibers

	Number of cells bound[b]	
Fiber antigen[a]	Immunized	Unimmunized
Dnp-BSA	1004	301
Tosyl-BSA	353	143
BSA	173	65
Stroma	160	75

[a] Abbreviations: Dnp, the 2,4-dinitrophenyl group; Tosyl, the *p*-toluenesulfonyl group; BSA, bovine serum albumin.

[b] Expressed as number of cells bound to both edges of a 2.5 cm fiber segment.

TABLE II
Percentage Inhibition of Mouse Spleen Cell Binding to Derivatized Fibers[a]

Antigen on fiber	Inhibition (%)		
	Dnp	Tosyl	Anti-Ig
Dnp-BSA	91	1	93
Tosyl-BSA	3	87	90

[a] 300 μg/ml Dnp-BSA, Tosyl-BSA, and anti-Ig were used as inhibitors. Abbreviations: Dnp, the 2,4-dinitrophenyl group; Tosyl, the *p*-toluenesulfonyl group; BSA, bovine serum albumin; Anti-Ig, the gamma globulin fraction of a rabbit antiserum to mouse immunoglobulin.

(Table II) suggesting that it occurs via immunoglobulin receptors. The relative number of immunoglobulin receptors on fiber-fractionated cells can be estimated by incubating fiber-bound cells with increasing amounts of ^{125}I-labeled rabbit anti-mouse immunoglobulin for 30 minutes at room temperature. After washing away the unbound antibody, the cells are removed from the fibers by plucking[4] into medium containing 10% fetal calf serum. Ten million carrier spleen cells are added to facilitate further washing by centrifugation, and the radioactivity of the cells is determined.

The distribution of receptors on fiber-bound cells can be tested by the formation of specific rosettes by binding of hapten-coupled erythrocytes to lymphoid cells previously bound to a fiber derivatized with the same hapten.[5] Fluorescein-labeled antibodies may also be used for the same purpose.[6] In this case, care must be taken that the nylon fibers have not been treated with fluorescent additives.

Applications of the Method to Surfaces: Topoculture and Cytochemistry

In addition to fixation on fibers, cells may be attached to surfaces and to beads,[4] and each geometry has special advantages and disadvantages. Nylon surfaces which are sufficiently transparent to be viewed in the microscope have been prepared from Chieftain nylon film (American Hospital Supply, Evanston, Illinois).[6] This material is stretched in a frame and then by hand to assure sufficient transparency. Such a cold drawing process still leaves anisotropic background areas, but they are sufficiently small so that they do not interfere with the detailed microscopic observation of cells on the surface. The nylon may be derivatized as described for the fibers on a circular stretcher frame forming a chamber which may be sealed with a coverslip. The entire assembly is filled with cells and placed vertically on a rotating wheel. This prevents nonspecific attachment of cells by gravity and surface adherence. The fractionated cells may then be moved *in situ* to tissue culture.

Such an arrangement allows fixation of cells in various patterns according to their specificity and examination of their behavior in tissue culture. The geometrical effects in culture (topoculture) open new possibilities for studies of cell–cell interactions. In such studies, which are still in an elementary stage, the perturbing effect of the specific ligand and of adherence to the carrier (nylon in this case) must both be considered. In some cases, it may be useful to employ "fading" or "dying" ligands which break down or lose activity after a certain time.

In certain applications such as topoculture, it is often unnecessary and indeed damaging to remove the cell from the surface. Fibers, meshes, beads, and surfaces may be moved intact with their specifically attached cells from medium to medium for biochemical and cytochemical studies, and cells on two separate fibers may be moved into close proximity. By this means, cells in culture may be rapidly moved from reagent to reagent or environment to environment. Fluorescent markers, radiolabels, and other indicators may be used to advantage *in situ*, and fixation and sectioning of the surface-bound cells allows electron microscopy to be carried out. Addition of agar to the medium should permit detection of cell products by immunodiffusion techniques.

In several kinds of experiments,[7-9] the surface can be used to ensure that a covalently bound ligand (hormone, antibody or other effector molecules) *cannot* penetrate the cell but can instead exert their action at localized areas of the cell surface. Finally, the ability to fix living cells in a predetermined position permits an opportunity to carry out cytochemical experiments on single cells.

Advantages and Disadvantages of Fiber Fractionation Techniques

The advantages of fiber fractionation include the ability to visualize and quantitate binding, to utilize mechanical, competitive and enzymatic cell removal depending on the experimental requirements, to isolate cells ranging in number from several hundred to as many as 10^9, to apply cytochemical and single cell techniques of biochemistry, and to manipulate the geometry of specific cells in culture.

Clearly, one of the major disadvantages is the possible perturbing effect of the ligand on cell metabolism or function. The effect of cell release must also be considered. Furthermore, even limited experience with the method indicates that a number of variables must be controlled and studied in each case before success can be obtained. These variables include the choice of ligand, the temperature, the medium, the degree of derivatization, the number of cells added, the speed of oscillation or rotation during incubation, the viscosity and composition of the separation medium, and the control of cell viability and metabolism. At present, the data suggest that fiber fractionation is best used in conjunction with physical methods as well as other specific methods, such as columns of derivatized beads. Even with the limited experience accrued, however, it is clear that in addition to the methods detailed above, viruses, bacteria, and eukaryotic cells may be affixed to surfaces and used as "reagents" for detection of interactions with other cells as well as for the purposes of cell fractionation.

[7] J. Andersson, G. M. Edelman, G. Möller, and O. Sjöberg, *Eur. J. Immunol.* **2**, 233 (1972).
[8] P. Cuatrecasas, *Proc. Nat. Acad. Sci. U. S.* **63**, 450 (1969).
[9] M. F. Greaves and S. Bauminger, *Nature (London) New Biol.* **235**, 67 (1972).

SECTION II

Preparation of Primary Cultures

Editors' Comments

This section includes established methods for the preparation and continuation of primary cultures. In some methods emphasis has been placed on the biopsy or surgical techniques required to obtain the tissues for culture; in others, specialized media or specific environmental conditions are stressed.

It is quite clear that a standardized procedure cannot be established for the preparation of primary cultures. Each tissue, indeed each stage of development, e.g., adult versus embryonic, presents specialized problems and requires variations in methodology.

Most of the methods include procedures for subculturing the primary cultures as well as storage properties. For the former, the reader is also referred to Section III; for the latter, general procedures for freezing, storage, and recovery of cell stocks are given in Section XIV. Certain other aspects of primary culture preparations will be found elsewhere, so the reader should consult the index for tissues not specifically referred to in this section.

CHAPTER 1

Human Skin Fibroblasts

G. M. Martin

The great advantage of diploid human skin fibroblastoid cell cultures is that they permit controlled studies of individual strain variation to an extent not possible with the better characterized human embryo lung lines described in this volume. Except for a variable but significant degree of tetraploidy[1] and occasional nondisjunctional progeny,[2] the genotype of the cells established in culture is that of the donor. To my knowledge, "spontaneous transformation" to heteroploid cell lines has never been proved; if it occurs at all, it must be an exceedingly rare event. Without exception, human skin fibroblasts have a limited replicative life span.[3,4] While this phenomenon is of interest to gerontologists, such clonal senescence places limits on the types of experiments which can be performed.

Biopsy Techniques

Many methods of skin biopsy provide material satisfactory for the establishment of predominantly diploid fibroblastoid cultures.[5-9] Vigorous cultures can often be established from a single dermal–epidermal pinch biopsy less than a millimeter in greatest diameter. One is more certain of success, however, with somwhat more generous specimens, such as are provided by 4 mm punch biopsies or 10 mm surgical ellipse biopsies; these provide enough tissue for replicate cultures. Whenever possible, it is advisable to obtain sufficient material to permit the cryobiological preservation of several small intact explants; these can be em-

[1] G. M. Martin and C. A. Sprague, *Science* **166**, 761 (1969).
[2] G. M. Martin, C. A. Sprague, and J. S. Bryant, *Nature* (*London*) **214**, 612 (1967).
[3] L. Hayflick, *Exp. Cell Res.* **37**, 614 (1965).
[4] G. M. Martin, C. A. Sprague, and C. J. Epstein, *Lab. Invest.* **23**, 86 (1970).
[5] J. H. Edwards, *Lancet* **i**, 496 (1960).
[6] J. Masterson, *Amer. J. Clin. Pathol.* **29**, 593 (1963).
[7] R. G. Davidson, S. W. Brusilow, and H. M. Nitowsky, *Nature* (*London*) **199**, 296 (1963).
[8] R. DeMars and W. E. Nance, *Wistar Inst. Monogr.* **1**, 35 (1964).
[9] R. M. Loder and G. M. Martin, *J. Invest. Dermatol.* **47**, 603 (1966).

ployed in the establishment of a second series of young cultures in the same or in another laboratory months or years later. While quantitative studies concerned with the optimal cryobiological parameters for preservation of such explants have yet to be carried out, we have had good results using conventional techniques for the preservation of established cell cultures, such as those described by Shannon and Macy in this volume (Section XIV, Chapter 8).

The biopsy need include only superficial dermis. Although good growth can be obtained with explants which include subcutaneous tissue, as well as from explants solely of subcutaneous tissue, it is possible that the inclusion of variable amounts of such fatty areolar connective tissue in predominantly dermal explants may influence the extent of cellular heterogeneity of the established cultures, especially during early passages.[10,11] The minimum depth of biopsy necessary to obtain superficial dermis obviously varies with the thickness of the epidermis; the degree of epidermal keratinization is dependent upon the site of biopsy and varies somewhat from subject to subject. At our standard site of biopsy, the mesial aspect of the mid-upper arm, a depth of 1 mm will usually suffice.[9] This site was chosen because of cosmetic considerations, the relative ease of access, the minimal keratinization, and the sparseness of hair follicles. As far as I am aware, however, successful cultures can be established with dermal explants from virtually any region of skin. For well-controlled experiments, however, consistency in the site of the biopsy is advised. For example, Pinsky and his colleagues[12] have shown that fibroblasts cultured from genital skin metabolize testosterone much faster than cultures derived from nongenital skin.

Our laboratory employs a sterile skin prep of pHisoHex, sterile water rinse, and 70% ethanol. Local anesthetics are ordinarily not necessary for pinch, wedge, or punch biopsies; when used for surgical ellipse biopsies, the tissue which is to be used for culture should not be infiltrated and the procaine should probably not contain epinephrine.

Autopsies provide an excellent source of culture material. In our laboratory, lines have been established from well over 100 refrigerated subjects, usually at postmortem intervals between 12 to 24 hours, but often at intervals between 24 to 72 hours. Except for rare microbial contamination, the success rate is virtually 100%.

Transport and Storage of Tissue

The biopsy tissues are best stored and transported in a standard sterile tissue culture medium with 10 to 20% newborn or fetal calf serum. Skin biopsies have been transported by air express at ambient temperatures from various parts of the world and have been successfully established in culture after intervals of several days. Pious *et al.* have established cultures of foreskin fibroblasts after

[10] G. M. Martin, *Exp. Cell Res.* **44**, 341 (1966).

[11] T. G. Papayannopoulou and G. M. Martin, *Exp. Cell Res.* **45**, 72 (1966).

[12] L. Pinsky, R. Finkelberg, C. Straisfeld, B. Zilahi, M. Kaufman, and G. Hall, *Biochem. Biophys. Res. Commun.* **46**, 364 (1972).

storage at 4°C for several weeks.[13] The main precaution in shipment is the avoidance of freezing temperatures. Although there is not much experience as yet concerning long-term storage of tissues in the newer nonvolatile buffers,[14] they probably should be employed as the major buffer system for improved pH control.

Initiating Explant Cultures

While enzymatic digestion has been successfully employed for both mass cultures and primary cloning,[13] explant cultures are much more convenient and are probably more reliable in the case of small specimens of skin. For the past 10 years, our laboratory has successfully employed two basic methods, depending upon the amount of available tissue; both are variations of previously described techniques.[8,15–18]

Small Specimens (Pinch or Wedge Biopsies). Given a limited amount of material, we prefer a "sandwich" explant technique.[17] Cultures might be established more rapidly with other methods, but in our experience it is the surest method of obtaining growth since it minimizes the possibility of loss of explants because of detachment, desiccation, pH fluctuation, or contamination.

The specimen is washed in a 50-ml centrifuge tube by shaking in three changes of 15 to 20 ml of Hanks' balanced salt solution minus bicarbonate and buffered with 0.02 *M* Tris-HCl, pH 7.4, at room temperature. The tissue is then transferred to a 100-mm sterile glass or plastic Petri dish previously loaded with a sterile standard (1 × 3 inch) glass microscopic slide. (The slides are prewashed with warm 50% nitric acid, exhaustively rinsed with cold tap water and with several final rinses of single distilled water, blotted dry with Whatman No. 1 filter paper, and dry-heat sterilized at 165°C for 2 hours.) Approximately 2 ml of culture medium is placed over the glass slide and the tissue is diced into approximately 1 × 1 × 1 mm fragments using a new sterile scalpel (No. 10 blade with a No. 3 Bard-Parker handle). The tissue should by cut sharply rather than crushed; avoid scissors and dull knives. With a sterile watchmaker's forceps, transfer pairs of moist explants to each of a series of glass slides precut with a glass cutter from 2 × 3 inch slides to dimensions of approximately 10 × 50 mm. Fit the slides, tissue side down, into short style Leighton tubes (Bellco Co., Vineland, New Jersey, Cat. No. 1962, tube size 19 × 105 mm). The tissue fragments are, therefore, sandwiched between the glass slides and the bottom surfaces of the Leighton tubes. We prefer glass slides to coverslips because, in our experience, the coverslips made to fit these tubes are not heavy enough to ensure that the explants will not float loose. Add 1 ml of medium to each tube and place

[13] D. A. Pious, R. N. Hamburger, and S. E. Mills, *Exp. Cell Res.* **33**, 495 (1964).
[14] H. Eagle, *Science* **174**, 500 (1971).
[15] T. C. Hsu and D. S. Kellogg, Jr., *J. Nat. Cancer Inst.* **25**, 221 (1960).
[16] D. G. Harnden, *Brit. J. Exp. Pathol.* **41**, 31 (1960).
[17] A. J. Therkelsen, *Acta Pathol. Microbiol. Scand.* **61**, 317 (1964).
[18] V. C. Miggiano, M. Nabholz, and W. F. Bodmer, *In* "Histocompatibility Testing" (P. I. Terasaki, ed.), p. 623. Munksgaard, Copenhagen, 1970.

in the 37°C incubator. With medium containing conventional bicarbonate buffers, the tubes should be equilibrated with the appropriate concentration of carbon dioxide before the screw caps are tightened. Although a great variety of commercially available tissue culture media provide adequate growth, there seem to be definite differences in the rapidity with which a culture can be established as function of the medium. While we have made no systematic studies to evaluate such parameters as strain variation, it is our experience, for example, that the modification of the Dulbecco-Vogt medium prepared in our laboratory[19] is superior to Waymouth's medium[20] when both are supplemented with 10 to 20% fetal calf serum. Currently, we employ this Dulbecco-Vogt medium in an atmosphere of air with 40 m*M* HEPES (pH 7.4),[14] an initial concentration of 9 m*M* bicarbonate, and 16% heat-inactivated (56°C for 30 minutes) fetal calf serum. This formulation provides excellent growth and pH control for both the establishment and maintenance of cultures. Our routine is to employ penicillin (50 units/ml) and streptomycin (50 μg/ml) during the period of time when the culture is becoming established in the Leighton tubes; only penicillin is employed in media for routine passaging. Many laboratories avoid antibiotics altogether as a means of minimizing the possibility of latent microbial contamination. However, with rare and precious clinical material, one is probably justified in the use of prophylactic antibiotics. In any case, cultures should be monitored for microbial contaminants, preferably by laboratories with special experience in the mycoplasma field. The problem of exogenous mycoplasmic contamination via animal sera is probably largely eliminated by heat inactivation; contamination via human operators should be minimized by the avoidance of all mouth pipetting and by the use of laminar flow hoods.

In the early stages of cell migration from the explants one often observes occasional compact "caps" of epithelioid cells, probably derived from the epidermis. However, these are invariably and rapidly outgrown by the more actively migrating spindle-shaped fibroblastoid cells which eventually cover the surfaces of the Leighton tubes and slides. The time for confluency varies greatly depending upon the strain and the medium, but is generally of the order of several weeks in the case of Waymouth's medium when the medium is changed (1 ml) twice weekly after an initial 7-day interval in the original explant medium. Our routine is to transfer the cultures to 4- or 6-ounce prescription bottles with trypsin-Versene[21] when the Leighton tubes reach approximately 40% confluency, at which time there is typical yield of about 100,000 cells.

Large Specimens (*Newborn Foreskins*). If there is sufficient material to provide dozens of small explants, cultures can be established if only a proportion of explants attach to the surfaces of the culture vessels. Plasma clots or "sandwich" techniques are therefore unnecessary. One need only distribute an excess of explants with the minimum amount of medium necessary to keep them moist for a period of 1 to 3 days, during which time most of them will become attached to

[19] H. Ginsburg and D. Lagunoff, *J. Cell Biol.* **35**, 685 (1967).
[20] C. Waymouth, *J. Nat. Cancer Inst.* **22**, 1003 (1959).
[21] G. M. Martin, *Proc. Soc. Exp. Biol. Med.* **116**, 167 (1964).

the glass or plastic culture surfaces. If too much medium is added, the explants will "float"; with too little medium, explants may become desiccated.

The tissue is washed and diced essentially as described above, except that proportionately larger volumes of Hanks' balanced salt solution are used for washing. One should prepare several dozen segments of approximately 1 × 1 × 1 mm, again taking care to avoid crushing trauma. About one to two dozen fragments are distributed in each of several 6-ounce prescription bottles or plastic tissue culture flasks. It is convenient to employ a wide-bore pipette and a minimum amount of tissue culture medium (about 1 to 1.5 ml for 6-ounce bottles), which is enough medium to keep the fragments moist and to aid in their random distribution. If the explants float, the excess medium should be withdrawn. With conventional bicarbonate buffers, the caps should not be tightly closed until there is equilibration with the appropriate concentration of CO_2. The cultures are left undisturbed for approximately 48 hours to permit attachment of the explants. Small increments of medium (0.5–1 ml) may then be gently added along the shoulders of the bottles at intervals during the next 1 to 2 weeks, depending upon the rate of growth of the cultures. A few explants will be lost, but there is virtually always sufficient growth to permit complete medium changes (10 ml with 6-ounce bottles) by 2 to 3 weeks. The time to confluency varies with the strain, the number of viable explants, and the culture medium. When bicarbonate-buffered media are used, our experience has been that open Petri dishes (in a humidified atmosphere with the appropriate concentration of CO_2) are superior to closed systems; this is probably because of better pH control. There is a greater probability of contamination with such wet, open systems, however, and one always runs the risk of a catastrophe if there is an undetected depletion of the CO_2 supply.

CHAPTER 2

Fetal Human Diploid Cells[1]

Leonard Hayflick

The cultivation of human diploid cell strains from primary tissue does not differ from the traditional techniques used to initiate any monolayer culture

[1] Supported, in part, by Research Grant HD 04004 from the National Institute of Child Health and Human Development, National Institutes of Health, Bethesda, Maryland.

from intact tissue. Contrary to popular misconception, there are no "tricks" necessary to maintain normal human cells in the diploid state—providing, of course, that the tissue has been obtained from a normal donor. Indeed, the trick, if there is any, is how to transform cells *in vitro* to cells having, among other properties, a heteroploid karyology. These concepts have been discussed extensively elsewhere.[2–6]

Reagents

Growth Medium

Eagle's Basal Medium
10% calf serum
50 μg aureomycin/ml

The medium should be prewarmed to 37°C before use. The final pH should be brought to 7.2 before serum addition. The pH of the medium after equilibration of the culture at 37°C must be less than 7.4.

Aureomycin (Lederle Product No. 4691-96, intravenous) is bottled in 500-mg amounts. Reconstitute in 50 ml of warm (37°C) sterile, distilled water. Agitate to ensure a clear amber solution. Prepare 5-ml aliquots and store at −20°C. Use one 5-ml aliquot per liter of medium. Final concentration is, therefore, 50 μg/ml. Nine thousand cell cultures grown in aureomycin in lieu of penicillin and streptomycin have been found, upon testing, to be free of mycoplasmas over a 10-year period.

Trypsin. 0.25% in phosphate-buffered saline (PBS) or any BSS plus aureomycin as above. Final pH of trypsin solution *must be at least* 7.4.

Limitation of Cultures

The human fetal tissue chosen is minced with paired scissors or scalpels under aseptic conditions in a Petri dish containing a minimum amount of growth medium. The resulting macerated tissue mass is then transferred to an Erlenmeyer flask containing sufficient trypsin so that the tissue mass is freely suspended. The suspension is placed at 37°C and constantly stirred using a magnetic mixer. At approximately hourly intervals, the mixing is stopped, large fragments allowed to settle, and the supernatant decanted. This procedure is repeated until the entire tissue mass is "digested." Immediately after collection

[2] L. Hayflick and P. S. Moorhead, *Exp. Cell Res.* **25**, 585 (1961).
[3] L. Hayflick, *Exp. Cell Res.* **37**, 614 (1965).
[4] L. Hayflick, *In* "Perspectives in Experimental Gerontology" (N. W. Shock, ed.), Chapt. 14, p. 195. Thomas, Springfield, Illinois, 1966.
[5] L. Hayflick, *Sci. Amer.* **218**, 32 (1968).
[6] L. Hayflick, "Aging and Development." In press. Academy of Science and Literature, Mainz, Germany, F. K. Schattauer Verlag, Stuttgart, 1973.

supernatant trypsin fluids are centrifuged at low speeds (1000–5000 g), the trypsin discarded, and the cell pellet resuspended in appropriate culture vessels. Inoculation densities of the order of 10^5 viable cells/ml are optimum.

The cultures are incubated at 37°C, and fed at 3- to 4-day intervals until a confluent cell monolayer is formed at which time proceed as in this volume, Section IV, Chapter 3.

CHAPTER 3

Human Mammary Tumors[1]

E. Y. Lasfargues

The first successful explantation and cultivation of human breast carcinoma cells was obtained in our laboratory in 1958 from a multicentric, invading, medullary carcinoma.[2] The procedure to be described is fundamentally the same as was then used but with technical improvements to eliminate fibroblasts and to maintain the epithelial cells in suspension. Primary cultures of human mammary tumors are a desirable source of viable cells on which to test the effect of hormones or to be used in a variety of immunological, biochemical, or viral studies.

Processing of Clinical Specimens

Collection of the tumor is done directly in the operating room. Prior to mastectomy, the surgeon is supplied with sterile 130-ml specimen jars containing 30–40 ml of Eagle's MEM or RPMI-1640 Medium supplemented with 10% fetal calf serum and 0.050 mg gentamicin/ml.[3] Immediately after resection and before the specimen is sent to pathology, a pertinent part of the tumor is aseptically

[1] The techniques reported in this article were developed with the support of U. S. Public Health Service Grant CA-08515, Contract PH 43-68-1000 from the National Cancer Institute, General Research Support Grant FR-5582 from N. I. H., and Grant-in-Aid Contract M-43 from the State of New Jersey.

[2] E. Y. Lasfargues and L. Ozzello, *J. Nat. Cancer Inst.* **21,** 1131 (1958).

[3] Gentamicin Reagent Solution, Schering Corporation, 40 Markley Street, Port Reading, New Jersey 07064.

cut and placed in one of the jars. The tumor is either immediately processed or stored in a refrigerator at 4°C up to 24 hours.

The first step of processing involves cleaning; the tumor is transferred from the jar to a large 100-mm diameter Petri dish containing 10 ml of Hanks' saline for a rapid examination under a dissecting microscope. The original cut indicates whether the neoplasm is actively invading or confined by fibrous envelopes, whether the tissues are soft or hard, and whether made of masses of tumor cells or necrotic material. The excess adipose and connective tissues are then dissected out until only the tumor mass remains. The specimen, rapidly rinsed in two or three changes of fresh Hanks' saline, is finally transferred to a Petri dish containing 5 to 10 ml of the same culture medium as used in the collecting jars.

The tumor, firmly anchored between the prongs of a dissecting forceps, is then cut in a series of thin slices with a new Bard-Parker blade (B-P 10 or 22). Care should be taken at this point not to tear or scrape the tissues but to make sharp clean cuts. According to the type of tumor, slicing releases a variable number of cells mixed with particulate matter. A drop of this suspension placed between slide and coverslip and observed by direct light microscopy at 100× magnification shows the morphology of the cells and their relative ratio to the debris.

The viable healthy cells either single or in groups appear globular and highly refringent (Fig. 1). The necrotic cells and debris are grayish, shell-like, without

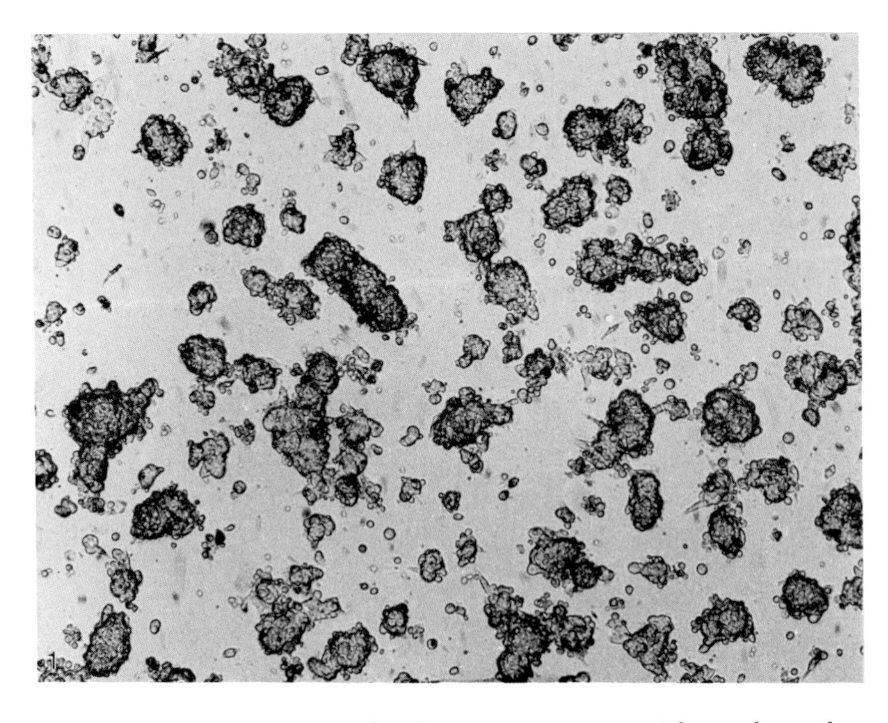

Fig. 1. Spilled cells from an intraductal mammary carcinoma 18 hours after explantation. Single epithelial cells and aggregates can be observed; very few fibroblasts can be seen already spreading on the floor of the culture flask. ×400.

substance. Small pieces of tissue and bundles of elastic or collagen fibers might also be present. The procedure to follow depends on whether: (a) more than 50% of the particulate suspension is formed by viable cells or (b) whether very few viable cells are released but remain essentially free of necrotic elements or debris.

In the first case, the success in explantation is obviously determined by the amount of debris and extent of necrosis. In the second case, the neoplastic cells entrapped in a tight fibrocytic network have to be released by enzymatic action.

Culture of Spilled Cells

When a large amount of spilled material is obtained, the debris is partially removed by passing the suspension through a small glass funnel containing three to four layers of sterile gauze. The filtrate, free of the largest pieces of tissue, is collected in a 15-ml conical centrifuge tube plugged with gauze and left standing vertically for 2 or 3 minutes in a test tube holder.

The viable cells sediment rapidly while most of the cell debris and small particulate matter remain in suspension. The turbid supernatant, carefully removed with a long-tip Pasteur pipette, is replaced by fresh culture medium. This procedure is repeated until the supernatant is clear; the cells resuspended in 10 ml of complete culture medium are then placed in a 50-ml Erlenmeyer flask sealed with a rubber snap-cap.

Complete Culture Medium. The complete culture medium is basically composed of Eagle's MEM Medium (Hanks' base) or RPMI-1640 complemented with 20% inactivated fetal calf serum. A 0.2% solution of crystallized bovine insulin (Schwarz-Mann, Orangeburg, New York) made in 0.1 *N* hydrochloric acid is incorporated into the medium for a final concentration of 0.010 to 0.020 mg of hormone per milliliter. Gentamicin is also added in amounts of 0.020 mg/ml because of its antibiotic activity as well as its fungacidal and antimycoplasmal properties.[4] The pH is adjusted at about 7.3 and the culture flasks incubated.

Renewal of the Medium. Whenever the pH of the culture drops below 7 as shown by the phenol red indicator incorporated in the basic solution, fresh complete medium is added in 5-ml aliquots up to a maximum volume of 20 ml. Beyond this point, part of the culture medium should be replaced. The flask is then placed at a 45° slant on a rack to allow the cells to sediment in the angle formed by the floor and the wall of the container. The supernatant is aspirated with a pipette and replaced by an equal volume of fresh complete medium.

[4] Because gentamicin is not effective against the whole spectrum of mycoplasmas it is suggested that tetracycline hydrochloride be added in the concentration of 0.010 mg/ml of culture medium. Achromycin is the trade name for the tetracycline manufactured by Lederle Laboratories, Division of the American Cyanamid Company, Pearl River, New York.

Subcultures. When the cells and aggregates in suspension have increased to approximately twice their original amount, the flask is vigorously shaken to obtain as homogeneous a suspension as possible and the total volume divided between two new flasks. The volume is completed to 20 ml in each flask by addition of fresh culture medium.

The high concentration of serum often induces a good proportion of cells to settle on the floor of the culture flask where they grow as monolayers. These cells can be resuspended following enzymatic dissociation as described in Section IV and seeded into new flasks.

Cultivation of Hard Tumors

When, upon slicing the tumor, very few cells spill in the surrounding medium, direct cell explantation is not possible. The most successful course in that case is an enzymatic dissociation with collagenase.

Collagenase isolated from *Clostridium histolyticum* is a noncytotoxic hydrolytic enzyme which maintains maximum efficiency in isotonic media with a pH range from 6.5 to 7.8. Unlike trypsin, collagenase retains full activity in the presence of calcium and magnesium ions. It is not inactivated by serum, which at concentration of 5 to 10% does not impede removal of the native collagen from the surface of the epithelial cells.[5] Because collagenase has not shown any damaging effect to the external cell membrane regardless of the time of exposure, it can be incorporated directly to the culture medium in which thin slices of breast tumors are dissociated without loss of cell viability.

The collagenase manufactured by Worthington Biochemical Corporation (Freehold, New Jersey) is listed as "CLS tissue culture grade" and has an average titer of 150 units/mg.[6] The crystals are soluble in Hanks' saline and this solution can be sterilized by filtration through a 0.45 μm Millipore filter. The collagenase stock solution is then incorporated into RPMI-1640 or other media enriched with 5% fetal calf serum at a concentration of 1 mg/ml.

Upon receipt, the tumor is cleaned as described earlier, then sliced. As many slices as possible are made and placed in a small 50-ml Erlenmeyer flask containing 10 ml of collagenase medium with 0.05 mg of gentamicin per milliliter. The flask, sealed with a screw cap, is incubated at 37°C for 48 hours. At this point, gentle pipetting breaks down most of the large pieces of tumor tissues. From this point the resulting suspension is handled in the same manner as the "spilled cells."

Maintenance. Large amounts of viable cells can be obtained from hard tumors but, as a result of the dissociation of the tumor stroma, the ratio of mesenchymal cells to the epithelium is high. The fibroblasts replicate much more

[5] E. Y. Lasfargues and D. H. Moore, *In Vitro* 7, 21 (1971).

[6] One unit is the amount of enzyme which liberates 1 m*M* leucine equivalents from purified collagen in 18 hours at 37°C.

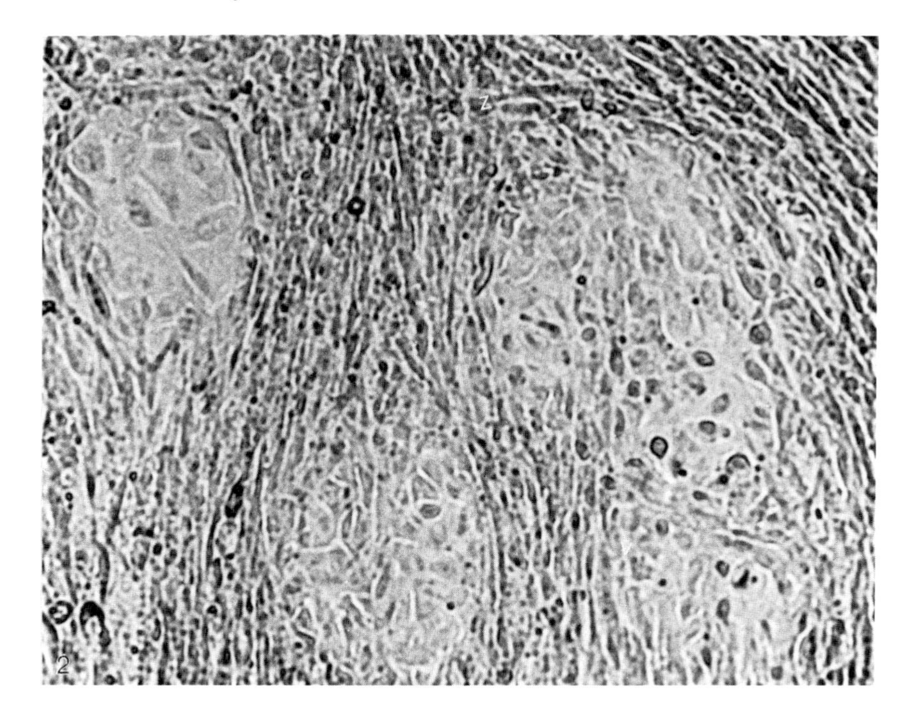

Fig. 2. Epithelial cells invested by fibroblasts. 7-Day-old culture obtained from the dissociation of a fibrous tumor with collagenase. ×780.

actively than the epithelial cells which they invest and coat with collagen (Fig. 2).

Trypsinization of the culture dissociates the cellular components successfully but does not remove the collagen polymerized on the surface of the epithelial cells. The small epithelial colonies smothered by accumulated collagen overlays are soon overtaken and eliminated.

To avoid the loss of the epithelium, the cultures are fed once every 3 weeks with a complete medium containing 0.5 mg collagenase/ml. Three days later, the epithelium, freed from the polymerized collagen by withdrawal of the enzyme-supplemented medium, can again expand between the fibroblasts. Given the opportunity by periodic collagenase treatments to grow to an appreciable ratio, the epithelium can eventually be separated from the fibroblasts. Upon reexplantation of the trypsinized cultures the fibroblasts settle down first while the epithelial cells remain globular and float. This difference in the speed of attachment of the two cell types permits one to pipette off the epithelial suspension and place it in a separate culture flask.

General Remarks

The methods described above have been applied to the processing of 102 human mammary tumors obtained from local hospitals; 48 have been successfully explanted. Of these, 12 originated from hard fibrous tumors. The causes for failure were, in the majority of cases, due to: (a) the absence of tumor material in the tissue specimen; (b) dead cells resulting from preventive hormonal or chemotherapeutic treatments; (c) extensive necrosis, and (d) calcified tumors.

CHAPTER 4

Human Glia Cells

Jan Pontén

Adult human brain tissue gives rise to serially propagable lines of glia cells. The cultures have mainly been used in contact inhibition studies. The glia cells are susceptible to a variety of cytopathic viruses and have been transformed by Rous sarcoma virus, feline sarcoma virus, and simian virus 40.[1-3]

Glia lines behave as other stable human lines, i.e., they remain diploid during their finite life span and show no spontaneous transformation into established heteroploid lines.

The Primary Explant

Pieces of gray or white matter are transferred from the operation theater in sterile tissue culture medium. After careful dissection of membranes and grossly visible vessels, the soft tissue is made up into a suspension by vigorous pipetting in a suitable volume of medium (about 10 ml per 10–20 mm^3 of tissue). The suspension containing small clumps and isolated cells is fed to plastic Petri dishes which are incubated in CO_2 and air at 37°C.

Medium. We have used Eagle's MEM with 10% baby calf serum as a standard, but Ham's F-10 and human serum work equally well.

Medium changes are individually adjusted to the conditions of the culture.

[1] J. Pontén and E. Macintyre, *Acta Pathol. Microbiol. Scand.* **74**, 465 (1968).

[2] J. Pontén, B. Westermark, and R. Hugosson, *Exp. Cell Res.* **58**, 393 (1969).

[3] E. Macintyre, J. Pontén, and A. E. Vatter, *Acta Pathol. Microbiol. Scand.* **A80**, 267 (1972).

During the first 24 hours a rapid drop of pH often necessitates one change. Thereafter we renew medium on the average of two to three times a week.

THE PRIMARY OUTGROWTH

A majority of the cells will never attach but remain trapped in the floating lipid-rich acellular portions of the fragmented brain tissue. Adjustment to cell culture is slow and very often no migration or cell division is noted within 2–3 weeks among the few successfully attached cells. This does not imply failure; we have obtained viable lines from 50/50 biopsies in spite of apparent initial poor growth. Once proliferation and migration have started they proceed rather rapidly and the culture becomes confluent within 1–2 weeks after the beginning of vigorous cell multiplication.

The predominant cell is star-shaped, nonpolar, with moderately long branched extensions. It is very thin and may be difficult to observe unstained in the inverted microscope.

Sometimes fibroblasts characterized by their bipolar refractile cell bodies are also seen. These are easily distinguished from the glia cells by their arrangement in bundles and whorls compared to the random pattern of the glia cells.

SERIAL CULTIVATION

Confluent cultures are treated with a thin film of trypsin (0.25% in phosphate-buffered saline) after removal of old medium. Dishes are incubated at 37°C until gross detachment from the solid support has commenced (ordinarily 5–10 minutes). Cells are then made into a monodisperse suspension by vigorous pipetting with a few milliliters of added fresh medium. It is essential to break up virtually all clumps without disrupting the individual cells and a microscopic check is recommended. The suspension from one plate is dispensed to two new dishes to which supplementary medium is finally added.

Healthy cells should attach by almost 100% efficiency within an hour and begin to divide within 20 hours.

GENERAL CHARACTERISTICS OF CULTIVATED GLIA CELLS

After two to three passages all macrophages, fibroblast- and endotheliumlike cells will have disappeared and the cultures are then, for practical purposes, uniformly composed of astrocytelike glia cells. A typical growth curve is given in Fig. 1. The exponential phase has a growth fraction close to 1.0. The cell cycle periods in hours are: $G_1 = 8.5$, $S = 8$, $G_2 = 4.5$, and $M = 1$. Under routine maintenance with two medium changes per week a terminal cell density of 80,000 cells/cm^2 is attained. This level is increased to 130,000 cells/cm^2 under "steady state" conditions achieved either by perfusion or extensively frequent medium renewal. This value cannot be exceeded by any known manipulations

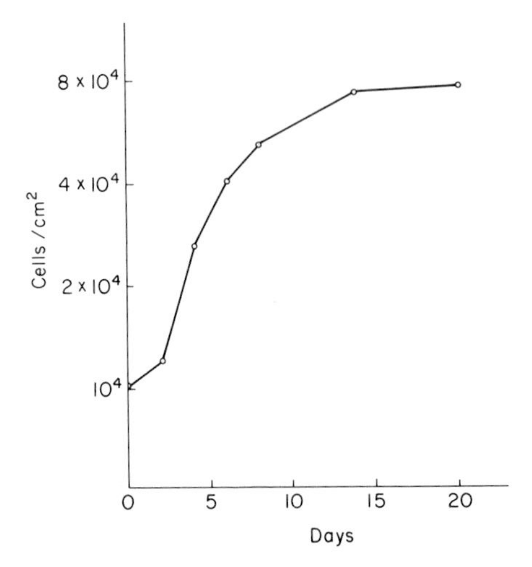

Fig. 1. Typical growth curve of early passage human glia cells under a standard regimen. Eagle's minimal essential medium changed three times a week. Note terminal density at 80,000 cell/cm^2.

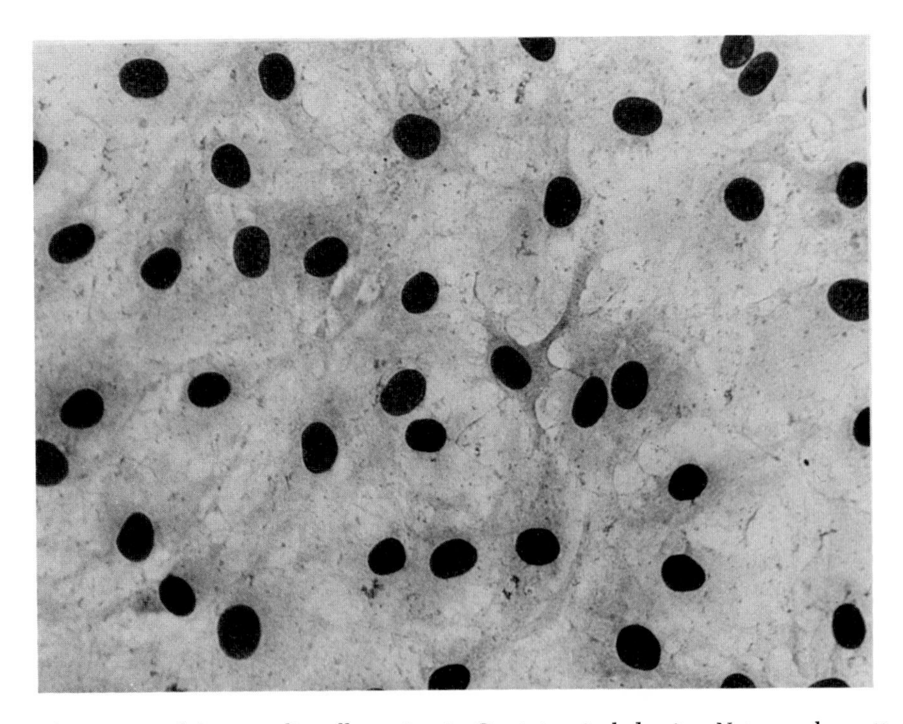

Fig. 2. Normal human glia cells resting in G_1 at terminal density. Note regular pattern with well-spread cells showing no nuclear overlapping. Time-lapse cinematography of such a culture would show no net locomotion and no ruffling of the lateral cell borders. May Grünwald Giemsa. Approx. mag. ×250.

of the medium. It is an expression of the strong degree of topoinhibition among these cells which then only show a labeling index of 0.5–1% (^{3}H-TdR, 24 hours) compared to 30% during the exponential phase.[4]

Glia cells at terminal density are moderately tightly packed in a monolayer with some cytoplasmic overlapping but very few superimposed nuclei (Fig. 2). The cells remain in fixed positions and show no significant membrane movements in time-lapse cinematography.[2]

The glia cells adjust to their environment by intracellular specialization. The lower surface develops desmosomelike attachment devices which fasten the cells at specified points to the solid substrate. Between these devices endocytosis is active. The upper surface is smooth with a dense continuous sheath of microfilaments just under the bilaminar outer membrane. Intercellular contacts are achieved by desmosomes.[5]

Topoinhibited cells rest in G_1 and need about 18 hours to reach S phase after stimulation. RNA synthesis (uridine incorporation) is reduced to about 50% of the exponential growth phase value, whereas protein synthesis (leucine incorporation) is not significantly diminished.

Glia cells are identified as neuroectodermal elements by the following criteria: (a) simplified astrocytic morphology, (b) presence of brain specific S-100 protein, (c) absence of collagen fiber formation, and (d) no synthesis of hyaluronic acid. Fibroblasts, which are the main source of confusion, display criteria (c) and (d) in culture and do not produce S-100.

If the glia cells are subcultivated at a 1:2 ratio, growth rate, terminal density, and growth fraction begin to decline around passage 10–15. Thereafter only a few additional passages are possible before irreversible degeneration terminates the finite life span of the cells. From one surgical biopsy one can obtain 2000–100,000 60-mm dishes which are useful for experiments, depending on the passage number at which spontaneous "senescence" sets in.

Glia cells can be frozen using any standard method. We use Eagle's MEM with 10% DMSO (dimethyl sulfoxide) and keep the cells in liquid nitrogen. Thawing is done in a 37°C water-bath. Medium is changed as soon as the cells have settled to avoid DMSO toxicity. Recovery should be close to 100%.

[4] B. Westermark, *Exp. Cell Res.* **69**, 259 (1971).

[5] U. Brunk, J. Ericsson, J. Pontén, and B. Westermark, *Acta Pathol. Microbiol. Scand.* **A79**, 309 (1971).

CHAPTER 5

Human Liver Cells

M. Edward Kaighn

The following procedures for culturing human liver cells are based on those developed by Coon[1,2] for the clonal culture of differentiated avian cells. The technique was later applied to primary cultures of rat liver.[3] There is a definite advantage in cloning directly from freshly dissociated liver since this approach permits isolation of specific cell types from a heterogeneous mixture. If clonal cultures are prepared at an early passage level, the plating efficiency will be higher than with primary cultures. However, the possibility of overgrowth by nonspecialized cells may be increased. Successful, repeated passage of clones that retain a normal karyotype and at least one demonstrable, distinctive liver function are accepted as criteria for success.

Source of Tissue

Specimens of infant, fetal, and adult liver obtained at autopsy, surgical biopsy, or hysterotomy are placed in chilled nutrient medium and transported to the culture laboratory in an ice bath. Although it is advisable to prepare the cultures without delay, viable cells have been obtained even after 8 hours under these conditions.

Dissociation of Tissue

The liver fragment is weighed, washed in several changes of CSS (1% chicken serum in Hanks' balanced saline without calcium and magnesium), drained on sterile gauze, then minced into fragments (1–2 mm^3). Curved scissors are used for young specimens, whereas with adult liver, especially if cirrhotic, crossed, disposable scalpel blades are more effective. The mince (1–2 g) is transferred to a 50-ml Ehrlenmeyer flask with a spatula and washed several times with CSS. With each wash, the fragments of tissue are allowed to settle and the solution removed with a 14-gauge cannula affixed to a syringe. Ten volumes of dissociation medium (CTC, 0.1% trypsin, 1:300, Nutritional Bio-

[1] H. G. Coon, *Proc. Nat. Acad. Sci. U. S.* **55**, 66 (1966).

[2] R. D. Cahn, H. G. Coon, and M. B. Cahn, *In* "Methods in Developmental Biology," (F. H. Wilt and N. K. Wessells, eds.), p. 493. Crowell, New York, 1967.

[3] H. G. Coon, *J. Cell Biol.* **39**, 29A (1968).

chemicals Inc., and 0.1% collagenase, Worthington, dissolved in CSS) are added and the flask is agitated gently in a reciprocal shaker bath for 20 minutes at 36.5°C. Agitation should be sufficient to keep the fragments in gentle motion. After the fragments have settled briefly, the supernatant solution which contains suspended cells is removed and transferred to a chilled centrifuge tube. An equal volume of chilled nutrient medium is then added and the tube is returned to an ice bath. Sometimes the viscosity of the solution impedes rapid settling of the unsuspended tissue fragments and makes removal of the dissociated cell suspension difficult. Since this effect is due to the release and uncoiling of DNA from ruptured cells, it can be mitigated readily by incubation for 1 to 2 minutes with crystalline deoxyribonuclease (0.1 ml at 4 mg/ml is usually sufficient for 20 ml of CTC). The cells are sedimented (600 *g*, 6 minutes), then suspended in chilled nutrient medium (10 ml/g liver). Viable cells are counted in a hemocytometer as those excluding trypan blue. This should be done rapidly since the fraction of stained cells increases with time. This extraction procedure is repeated several times. It is advisable to examine each extract of dissociated cells since the yield of viable cells is quite variable especially if older specimens are used. In most preparations about $1\text{–}2 \times 10^6$ viable cells are obtained per gram of tissue.

Cloning

Petri dishes (100 mm, Falcon Plastics) containing 6 ml of nutrient medium are allowed to reach equilibrium in a humidified incubator at 36.5°C, continually flushed with sufficient CO_2 to maintain the medium at pH 7.4. Since the primary plating efficiency is of the order of 0.01 to 0.1%, the inoculum per plate should vary between 10^4 and 10^6 viable cells. The medium is replaced three times a week.

Passage of Clones

After 2 or 3 weeks when the colonies are about 1 to 2 mm in diameter, single colonies are encircled with porcelain cylinders (Fisher Scientific) coated with silicone (Dow-Corning). The colony is washed with CSS and 0.1 ml CTC is added. It is convenient to use a 1-ml plastic syringe with a 20-gauge needle for these operations. The plate is then returned to the incubator. At intervals, the cells are examined at low power. When they have rounded up (15–30 minutes), the excess CTC is removed and the cylinder contents are transferred in a drop of nutrient medium to a 60-mm Petri dish containing 2 ml of nutrient medium. These dishes are incubated until they have become nearly confluent. At this time they are passed to 100-mm dishes at 10^5 cells per 10 ml of medium. When confluent, these cells, now at third passage, are ready for freezing. They are removed from the plates with CTC and suspended in medium at 5×10^6–1×10^7 per milliliter. An equal volume of nutrient medium containing 20% dimethyl sulfoxide is added dropwise. Ampoules containing 1 ml of this suspension are frozen at a controlled rate and stored in liquid nitrogen. Alternatively, the ampoules can be placed in a freezer at −73°C for 1 hour and then transferred directly to liquid nitrogen.

Notes on Media and Solutions

The successful propagation of liver cells is dependent on the selection of nontoxic reagents. Individual lots of different reagents such as collagenase and fetal calf serum differ in their potential toxicity and efficacy to promote cell growth.

A 1% solution of collagenase (Worthington, CLS) is dialyzed against calcium- and magnesium-free Hanks' saline for 2 hours at room temperature and stored frozen. Each lot is then tested for toxicity by the trypan blue exclusion method.

The nutrient medium consists of modified Ham's F12 (F12K) and optional antibiotics (penicillin G, K^+, 100 units/ml; kanamycin, 100 μg/ml; amphotericin B, 6.25 μg/ml), supplemented with fetal calf serum. It is most important that each batch of serum be tested by cloning a frozen stock of liver cells at 5 and 15% serum concentrations. The batch that gives the highest plating efficiency is selected and stored frozen at −73°C until used. With most lots, 15–17% fetal calf serum gives satisfactory results with human liver cells both in primary culture and in subsequent passage. All lots of serum are filtered (0.22 μm Millipore) and stored at −73°C.

Modified F12 (F12K) is prepared from a fourfold concentrate[4] containing eightfold concentrations of amino acids, sodium pyruvate, and putrescine·2HCl; fourfold concentrations of choline choride and all vitamins except biotin which is 40×; fourfold concentrations of $CuSO_4 \cdot 5H_2O$ and $FeSO_4 \cdot 7H_2O$. The concentration of $ZnSO_4 \cdot 7H_2O$ is 0.68× that specified by Ham. All other inorganic salts and glucose are as specified by Ham (see Table I). Linoleic acid is omitted when a serum supplement is used.

F12K is prepared from fourfold concentrate by dilution of 1 part concentrate with 3 parts K-Hanks'.[5]

Sufficient $NaHCO_3$ is added to bring the final concentration of F12K to 2.5 g/liter. The solution is then equilibrated with 5% CO_2 in air and sterilized by passage through a 0.22 μm Millipore filter.

Minor differences in the concentration of inorganic salts in F12K result from the use of K-Hanks' as diluent (see Table I). Ca^{2+} is increased threefold; $ZnSO_4 \cdot 7H_2O$ is 5×10^{-7} *M* as in Ham's F12M; glucose is slightly reduced; and $NaHCO_3$ is increased 2.2-fold. The pH should be 7.4 when equilibrated with 5% CO_2 in air. Ascorbic acid may be included as an option at 15 mg/liter.

Growth and Specific Properties

The fraction of viable cells that yields colonies depends upon the condition and age of the specimen. Usually it is about 0.01 to 0.10%. Once clonal strains

[4] H. G. Coon and M. C. Weiss, *Proc. Nat. Acad. Sci. U. S.* **62**, 852 (1969).

[5] I. R. Konigsberg, *Science* **140**, 1273 (1963).

TABLE I

Composition of Ham's[a] F12 and Modified Version (F12K)

Compound	Original F12		F12K	
	Moles/liter	Wt./liter	Moles/liter	Wt./liter
L-Arginine·HCl	1.0×10^{-3}	210.7 mg	2.0×10^{-3}	422.0 mg
Choline chloride	1.0×10^{-4}	13.8 mg	1.0×10^{-4}	13.8 mg
L-Histidine·HCl	1.0×10^{-4}	20.9 mg	2.0×10^{-4}	41.9 mg
L-Isoleucine	3.0×10^{-5}	3.94 mg	6.0×10^{-5}	7.88 mg
L-Leucine	1.0×10^{-4}	13.1 mg	2.0×10^{-4}	26.2 mg
L-Lysine·HCl	2.0×10^{-4}	36.5 mg	4.0×10^{-4}	73.0 mg
L-Methionine	3.0×10^{-5}	4.48 mg	6.0×10^{-5}	8.96 mg
L-Phenylalanine	3.0×10^{-5}	4.96 mg	6.0×10^{-5}	9.92 mg
L-Serine	1.0×10^{-4}	10.5 mg	2.0×10^{-4}	21.0 mg
L-Threonine	1.0×10^{-4}	11.9 mg	2.0×10^{-4}	23.8 mg
L-Tryptophan	1.0×10^{-5}	2.04 mg	2.0×10^{-4}	4.08 mg
L-Tyrosine	3.0×10^{-5}	5.44 mg	6.0×10^{-5}	10.88 mg
L-Valine	1.0×10^{-4}	11.7 mg	2.0×10^{-4}	23.4 mg
L-Cysteine·HCl·H_2O	2.0×10^{-4}	35.02 mg	4.0×10^{-4}	70.04 mg
L-Asparagine·H_2O	1.0×10^{-4}	15.01 mg	2.0×10^{-4}	30.02 mg
L-Proline	3.0×10^{-4}	34.53 mg	6.0×10^{-4}	69.06 mg
L-Glutamine	1.0×10^{-3}	146.2 mg	2.0×10^{-3}	292.0 mg
L-Alanine	1.0×10^{-4}	8.91 mg	2.0×10^{-4}	17.8 mg
L-Aspartic acid	1.0×10^{-4}	13.31 mg	2.0×10^{-4}	26.6 mg
L-Glutamic acid	1.0×10^{-4}	14.71 mg	2.0×10^{-4}	29.4 mg
Glycine	1.0×10^{-4}	7.51 mg	2.0×10^{-4}	15.0 mg
Sodium pyruvate	1.0×10^{-3}	110.1 mg	2.0×10^{-3}	220.0 mg
Putrescine·2HCl	1.0×10^{-6}	0.161 mg	2.0×10^{-6}	0.322 mg
Biotin	3.0×10^{-8}	0.007 mg	3.0×10^{-7}	0.07 mg
Calcium pantothenate	1.0×10^{-6}	0.477 mg	1.0×10^{-6}	0.477 mg
Niacinamide	3.0×10^{-7}	0.037 mg	3.0×10^{-7}	0.037 mg
Pyridoxine·HCl	3.0×10^{-7}	0.062 mg	3.0×10^{-7}	0.062 mg
Thiamine·HCl	1.0×10^{-6}	0.337 mg	1.0×10^{-6}	0.337 mg
Folic acid	3.0×10^{-6}	1.32 mg	3.0×10^{-6}	1.32 mg
Riboflavin	1.0×10^{-7}	0.0376 mg	1.0×10^{-7}	0.04 mg
Vitamin B_{12}	1.0×10^{-6}	1.36 mg	1.0×10^{-6}	1.36 mg
Hypoxanthine	3.0×10^{-5}	4.08 mg	3.0×10^{-5}	4.0 mg
Myo-Inositol	1.0×10^{-4}	18.02 mg	1.0×10^{-4}	18.0 mg
Lipoic acid (thioctic acid)	1.0×10^{-6}	0.206 mg	1.0×10^{-6}	0.206 mg
Thymidine	3.0×10^{-6}	0.727 mg	3.0×10^{-6}	0.700 mg
Linoleic acid	3.0×10^{-7}	0.084 mg	—	—
NaCl	1.3×10^{-1}	7.60 g	1.3×10^{-1}	7.53 g
KCl	3.0×10^{-3}	0.224 g	3.8×10^{-3}	0.285 g
$Na_2HPO_4{\cdot}7H_2O$	1.0×10^{-3}	0.268 g	8.1×10^{-4}	0.218 g
KH_2PO_4	—	—	4.3×10^{-4}	0.059 g
$MgSO_4{\cdot}7H_2O$	—	—	1.6×10^{-3}	0.393 g
$MgCl_4{\cdot}6H_2O$	6.0×10^{-4}	0.122 g	5.2×10^{-4}	0.106 g
$CaCl_2{\cdot}2H_2O$	3.0×10^{-4}	0.044 g	9.2×10^{-4}	0.135 g
$CuSO_4{\cdot}5H_2O$	1.0×10^{-8}	0.002 mg	1.0×10^{-8}	0.002 mg
$ZnSO_4{\cdot}7H_2O$	3.0×10^{-6}	0.863 mg	5.0×10^{-7}	0.144 mg
$FeSO_4{\cdot}7H_2O$	3.0×10^{-6}	0.834 mg	3.0×10^{-6}	0.800 mg
Glucose	1.0×10^{-2}	1.80 g	7.0×10^{-3}	1.26 g
$NaHCO_3$	1.4×10^{-2}	1.18 g	3.0×10^{-2}	2.50 g
Phenol red	3.3×10^{-6}	1.2 mg	8.3×10^{-6}	3.0 mg

[a] R. G. Ham, *Proc. Nat. Acad. Sci. U. S.* **53**, 288–293 (1965).

have been isolated, growth curves show a 2- to 3-day lag period followed by exponential growth, with a generation time of about 24 hours. Under favorable conditions, these cells clone with a plating efficiency of up to 50%. Several factors have been tested for their ability to increase growth rate and plating efficiency. Chicken embryo extract, sera of species other than bovine, and tryptose phosphate broth have been found to be deleterious. In general, the growth rate is proportional to the concentration of fetal calf serum. Cell strains initiated from individual clones differ in the spectra of liver properties they exhibit.[6]

[6] M. E. Kaighn and A. M. Prince, *Proc. Nat. Acad. Sci. U. S.* **68**, 2396 (1971).

CHAPTER 6

Human Blood Leukocytes

Paul S. Moorhead

Cultivation of leukocytes from the blood of man and other animals has been of value principally in karyological determinations. However, in recent years such short-term cultures of leukocytes have become important in the study of the initiation of cell division, especially as related to immunological phenomena. In addition to research studies requiring cell replication, leukocyte "cultures" of very short duration may be useful for observations on cell behavior and cell motility. Long-term cultivation of blood leukocytes is a subject of great interest to both virologists and karyologists since these permanent lines may possibly prove to be dependent upon the presence of a virus for their continued proliferative capacity.[1,2]

Blood Sampling

Materials

Syringes, 10 ml, sterile (as B-D Glaspak, disposable)
Needles, 20 G, sterile

[1] W. Henle and G. Henle, *In* "Perspectives in Virology" (M. Pollard, ed.), Vol. VI. Academic Press, New York, 1968.
[2] G. E. Moore, R. F. Gerner, and H. A. Franklin, *J. Amer. Med. Ass.* **199**, 519 (1967).

Heparin, sodium, aqueous, 1000 units/ml
Ethyl alcohol, 70%
Gauze pads, sterile
Rubber tubing for tourniquet, 2 feet

Procedure. Clean the inner area of the forearm by rubbing vigorously with a gauze pad soaked in 70% ethyl alcohol. Place a rubber tubing tourniquet around the upper arm using a slipknot for quick release. Wipe the serum cap of the vial of heparin with 70% ethyl alcohol and using the 10-ml syringe withdraw at least 0.2 ml of the heparin solution. Wet the barrel of the syringe and discharge most of the heparin from the syringe. Replace the needle exposed to heparin with another (20 G) or wipe it clean with a sterile gauze pad so as to avoid introduction of any heparin at the site of needle entry. Allow alcohol on skin to dry thoroughly. Choosing a less superficial vein, and with beveled opening up, insert the needle tip at an angle of about 15° with the surface of the arm and parallel to the vein. The more obvious veins tend to roll or move under the skin when puncture is attempted. Withdraw plunger gently to determine that the needle tip is in the lumen of the vein and draw about 5 ml into the syringe. Five milliliters will provide enough leukocytes for about five cultures. If less than 2 ml of blood is obtainable, use the microculture procedure given later. Before removing the needle place a dry sterile gauze pad over the site and release the tourniquet. Holding the syringe low and parallel to the arm, remove needle while maintaining pressure on the site of needle entry. Donor should press firmly on site of puncture for at least 1 minute.

Setting Leukocyte Cultures

Materials

Culture blanks, with complete medium
Incubator, 37°C
Fetal calf serum (or human AB serum)
Test tube, or other flask, sterile (for collection of plasma containing leukocytes)
Pasteur pipettes, with rubber bulbs
Forceps, large (stout enough to bend needle)

Preparation of Culture Blanks. Each culture blank consists of a small vessel with 5.0 ml of complete medium. Medium is obtained commercially or is prepared in 1- or 2-liter lots and contains the mitogen, antibiotics, serum, basic salts, amino acids, and vitamins (plus heparin, optional). Medium can be obtained with or without the mitogen (as Gibco's McCoy 5A). Medium can be prepared as below; 5-ml aliquots are dispensed into culture vessels to be frozen until use.

Leukocyte culture medium	Amount (ml)
2× conc. Basal Medium Eagle's	800
Fetal calf serum	200
Phytohemagglutinin, Difco M form	8.0
or Burroughs-Wellcome's PHA	10.0
Heparin, sodium (1000 units/ml)	20.0
Antibiotics solution (100,000 units/ml penicillin plus 100,000 μg/ml streptomycin)	1.0

Before the addition of serum, the pH should be approximately 7.1–7.3 as judged from the color (phenol red indicator in BME, Eagle's). Adjustments can be made with dilute sodium bicarbonate or with dilute acid (0.1 *N* HCl).

Mitogens available include Pokeweed mitogen (Gibco), M and P forms of PHA (Difco), and PHA (Burroughs-Wellcome). The Difco P form requires filtering for sterilization, the others are sterile as received. Pokeweed mitogen (Gibco) is added as 1.0 ml/liter, as is the P form (Difco).

For the culture blanks any small vessel similar to the following would be suitable. These four have approximately the same proportions so that area of floor of vessel (upright) and general dimensions are nearly the same: 2-ounce Brockway prescription bottle; Falcon flask, 15 ml, with screw cap; screw-cap vials, 60 × 28 mm, A. H. Thomas No. 9803-D; 25 × 65 mm disposable glass vial, Kimble "Opticlear."

Procedure. Use aseptic precautions and procedures in the handling of needles, blood sample, and the culture medium. However, with the use of antibiotics contaminations seldom are manifest by the time of harvesting (at 48 or 72 hours). In any case, the complication of infection, overt or not, is to be avoided.

Aspirate a small volume of air into the barrel of the syringe to clear blood from the needle. Then draw into the syringe a volume of fetal calf serum (or human AB serum) equal to the blood volume. Invert the syringe a few times to mix serum with blood. Using heavy forceps of which the tips have been dipped in alcohol and flamed, bend the needle at two points, each to about 40° angle. A single bend may break off the needle. This is done to permit extruding the supernatant plasma containing leukocytes without disturbing the interface between red blood cell-containing plasma and the cleared plasma. One may prefer not to bend the needle by collecting plasma in a serum bottle with a rubber serum cap secured in a vertical position.

Secure the syringe with tape in an upright position, needle up, to allow the supernatant fluid to clear of red cells. This may be done at room temperature, or more rapidly at 37°C. This gravity sedimentation usually requires from 20 to 60 minutes and frequent observation is recommended because of the variability. When the supernatant fluid constitutes about 40% of the total volume in the syringe, push the plunger gently to express the plasma, now largely free of red blood cells, into a sterile vial or test tube. If the amount cleared is less than 40% of total volume after 1½ hours, remove and proceed. Since after sedimentation the majority of white cells will lie close to the interface (not quite a buffy coat), be sure of obtaining all the cleared volume by pushing plunger until a very tiny amount of red blood cells appear.

With Pasteur pipette thoroughly mix the extruded supernatant fluid containing the leukocytes and dispense about 0.8 ml into each culture blank. Replace lid or top of culture blank firmly so that it is gas tight and swirl the contents (e.g., draw 5 ml blood; add 5 ml of fetal calf serum; about 4 ml volume clears after sedimentation by gravity; 4 ml provides enough for five cultures, 5×0.8 ml). If culture is to be transported or there will be any other delay before incubation, it may be held at 5°C for up to 48 hours and sometimes longer. It is best to delay cultivation at this stage rather than as whole blood. Incubate at 37°C for 3 days (68–72 hours). Alternatively one may wish to obtain for chromosome studies only the metaphase cells which are representative of the initial round of division, in which case cultures should be incubated for only 2 days (48 hours). In our experience the amount of division at 48 hours is much less.

Microculture of Leukocytes from Whole Blood

If necessary small amounts of blood can be obtained from finger or heel pricking which permit setting one to two cultures.[3] It is sometimes of value to use blood obtained by venipuncture to set a large number of microcultures so that one may determine which period of cultivation is best suited to that subject, or the effect of other variables in the culture process.

Clean the skin surface with soap and water and with 70% ethyl alcohol. Let alcohol dry and make puncture in non-index finger using sterile lancet. Hold a capillary collection tube (Drummond Microcap tubes)[4] horizontally to collect blood. Extrude collected blood into the culture blank directly or collect in sterile test tube or Kahn tube which contains about six drops of heparin. Optimal growth is obtained on setting 200 to 400 lambda (0.2–0.4 ml) whole blood per 5 ml culture. Subsequent procedures for harvesting the cultures (see also, Section XIV, this volume) are identical to those for ordinary cultures as described earlier, with the exception that these cultures may best be harvested on day 4 or day 5 rather than day 3. If numerous cultures are set from a larger sample of whole blood a few can be sacrificed to determine the peak of cell division. Where only two or three cultures are obtainable it may be advisable to pool the contents at harvest to provide a sufficient size of cell pellet for the harvesting procedures. If blood is collected for transport before being set, add some of the culture medium to the sample immediately after taking blood and transport at 5°C approximately. Rapid retrieval of blood from the puncture site can also be made efficiently by using a sterile Pasteur pipette.

[3] D. T. Arakaki and R. S. Sparkes, *Cytogenetics* **2,** 57 (1963).

[4] Drummond Microcap tubes, 100/vial, 100 lambda vol., heparinized. Drummond Scientific Co., 500 Parkway South, Broomall, Pennsylvania.

CHAPTER 7

Amniocentesis

Lewis L. Coriell

Amniocentesis for diagnostic purposes was described in 1930[1] and was used quite extensively in the 1960's in diagnosis and treatment of Rh disease.[2-6] More recently it has been revived for the detection of congenital fetal abnormalities early in pregnancy[7] in high risk families with the expectation that the pregnancy could be interrupted if the defect was present, and conversely, could continue to term with confidence if the tests were normal. Removal of 10 ml of amniotic fluid by transabdominal aspiration with a sterile syringe and needle early in the second trimester was not associated with any maternal or fetal complications in 155 patients studied by Nadler.[7] Sufficient experience has now been obtained in a number of clinics to confirm that cells from amniotic fluid can be successfully grown in tissue culture using a number of variations of culture media, culture vessel, and procedures.[8] Over thirty familial metabolic disorders and a number of chromosome abnormalities, sex, and blood group of the fetus can be determined on cultivated amniotic fluid cells.[9,10] Chromosome analysis cannot be made directly on cells suspended in amniotic fluid because they are not undergoing mitosis. Some metabolic defects can be detected by direct analysis of the amniotic fluid supernate but the results are unreliable and it is, therefore, recommended that all analyses be made on cultured cells. One to two weeks is the minimum time for culture of amniotic cells and so it is important to do the amniocentesis early in the second trimester and to use a culture method that will ensure rapid growth of cells.

Procedure

The volume of amniotic fluid specimen is measured, ½ ml retained for sterility test (see Section XIV) and for estimating the number of viable cells by trypan blue dye exclusion, and the remainder placed in a centrifuge tube.

[1] T. O. Menees, J. Miller, and L. E. Holly, *Amer. J. Roentgenol.* **24**, 363 (1930).
[2] D. C. A. Benis, *Lancet* **i**, 395 (1952).
[3] A. W. Liley, *Pediatrics* **35**, 836 (1965).
[4] V. J. Freda, *Bull. N. Y. Acad. Med.* **42**, 474 (1966).
[5] J. T. Queenan, *Clin. Obstet. Gynecol.* **9**, 491 (1966).
[6] R. G. Burnett and W. R. Anderson, *J. Iowa Med. Soc.* **58**, 130 (1968).
[7] H. L. Nadler and A. B. Gerbie, *N. Engl. J. Med.* **282**, 596 (1970).
[8] G. S. Marchant, *Amer. J. Med. Technol.* **37** (10), 391 (1971).
[9] H. L. Nadler, *Pediatrics* **49** (3), 329 (1972).
[10] M. W. Steele and W. R. Breg, *Lancet,* **i** 383 (1966).

Cells are sedimented in an International PR2 refrigerated centrifuge at +5°C for 5 minutes at 160 *g*. The supernatant fluid is removed, the cells resuspended in 5 ml of growth media, and transferred to a 25-cm^2 (30 ml) Falcon plastic tissue culture flask for incubation at 37°C in an atmosphere of 5% CO_2 in air. In 66 specimens collected between 14 to 20 weeks gestation, 15–30% of cells were viable (total viable cells ranging from 10 to 100,000 per milliliter).

Growth Medium. Eagle's-Earle's with 2x vitamins and 2x amino acids,[11] penicillin 100 μg/ml, streptomycin 100 μg/ml, plus 30% uninactivated fetal calf serum, pH 7.3.

Inspect for cell growth on the 6th or 7th day, remove the growth media, and replace with fresh media. Inspect daily thereafter and refeed each week. In our experience the surface of the flask is completely covered with cells in 8 to 21 days, with an average of 14 days. Cells can be analyzed for chromosome abnormalities or metabolic defects during this primary growth period, but it is more practical to conduct these examinations on cells in first tissue culture passage. In primary culture both epithelial and fibroblastlike cells are seen and the latter tend to adhere firmly to the culture flask. Removal by scraping or with 0.25% trypsin causes death of many cells. We recommend the use of Puck's saline–EDTA–trypsin[12] for gentle removal (Table I).

Subculture

Remove the culture media. Add 5 ml of Puck's saline containing 0.02% of EDTA (Solution B, Table I) and observe under the inverted microscope. At

TABLE I
Preparation of Saline, EDTA, and Trypsin Solutions[a]

Saline (Solution A)	
Distilled water	1000 ml
NaCl	8.0 g
KCl	0.4 g
Glucose	1.0 g
$NaHCO_3$	0.35 g
Sterilize in the autoclave	
EDTA (Solution B) 0.02%	
Solution A	1000 ml
EDTA	0.2 g
Sterilize in the autoclave	
Trypsin (Solution C) 0.04%	
Solution B	984 ml
2.5% Trypsin (filtered)	16.6 ml

[a] R. P. Cox, M. R. Krauss, M. E. Balis, and J. Dancis.[12]

[11] H. Eagle, *Science* **130**, 432 (1959).

[12] R. P. Cox, M. R. Krauss, M. E. Balis, and J. Dancis, *Proc. Nat. Acad. Sci. U. S.* **67**, 1573 (1970).

the end of 10 minutes or sooner if cells start to float free, invert the flask, remove the EDTA solution by aspiration, and replace with 3 ml of Puck's saline containing 0.02% EDTA and 0.04% trypsin (Solution C, Table I). Observe until the cells come loose from the substrate (2–10 minutes). As soon as the cells are loose, tap the flask gently to dislodge all cells and immediately add 3 ml of growth medium. Transfer with a pipette to three culture flasks of the same size, each containing 3 ml of culture medium, or to one larger Falcon tissue culture flask (75 cm^2–250 ml) containing 13 ml of growth medium.

Cells in subculture are fed twice a week, proliferate rapidly, and usually can be subcultured again in 5–7 days. Cells in first and second subculture provide good material for chromosome examination and a quantity of cells for metabolic assay.

Serial passage of amnion cells may be continued with 1:3 splits each week and vigorous growth continues for at least ten passages. After the 3rd or 4th passage the morphology of cultured cells resemble fibroblasts. Cells may be stored in liquid nitrogen at the 2nd or 3rd subculture and recovered without detectable change in chromosomes, or in isozyme patterns for glucose-6-phosphate dehydrogenase, lactic dehydrogenase, malic dehydrogenase, 6-phosphoglyconic dehydrogenase, esterase, alkaline phosphatase, and acid phosphatase.

Procedures for Liquid Nitrogen Storage

When the cell sheet is confluent after 5 to 7 days incubation, refeed with fresh growth medium, 24 hours later remove the cells with Puck's Versene-trypsin as for subculture and suspend in culture medium without serum in a wet ice bath. Mix by gentle aspiration and remove ½ ml for viability test by trypan blue dye exclusion. Centrifuge at 160 g for 5 minutes at +5°C, discard the supernate, and resuspend the cells in sufficient growth medium plus 5% glycerol to yield 500,000 to 1,000,000 viable cells per milliliter. Dispense 1.0 ml in 1.2 ml borosilicate glass ampoules, and make a pulled seal in an oxygen flame. The ampoules are tested for leaks by immersion in 70% alcohol-methylene blue at 4°C and then frozen in liquid nitrogen vapor[13] at the rate of 5° to 24°C per minute, and stored in liquid nitrogen liquid or vapor phase. After recovery from liquid nitrogen storage cells grow vigorously for at least ten passages. Additional passages have not been attempted in this laboratory.

[13] A. E. Greene, R. K. Silver, M. Krug, and L. L. Coriell, *Proc. Soc. Exp. Biol. Med.* **116**, 462 (1964).

CHAPTER 8

Human Lymph Nodes

Magdalena Eisinger

Cells cultured from normal human lymph nodes, in common with cells from other tissues, usually cease to divide after a variable period of active growth and thus fail to give rise to continuous cell lines. Occasionally, however, lymphoblastoid cell lines capable of indefinite propagation *in vitro* have been established from "normal" human lymphoid tissue.[1-4] More success has been achieved with material from human neoplastic sources, e.g., African Burkitt's lymphoma,[5] lymphocytic lymphoma,[6] leukemias,[7,8] Hodgkin's disease.[9-11] However, electron microscopy has shown that almost all human lymphoblastoid cell lines, whether originating from "normal" or neoplastic tissue, contain "herpeslike" viral particles, suggesting that the successful establishment of continuous cell lines in culture may depend on the presence of the hypothetical human "leukovirus."[5] Even apparently virus-free cell lines may carry them in a cryptic form, since the latter may be "activated" to produce a virus or viral-specific antigens by appropriate chemical treatment.[12]

Types of Methods

Three major types of methods resulting in successful long-term tissue cultures from human lymph nodes can be distinguished.

[1] F. A. Levy, M. Virolainen, and V. Defendi, *Cancer* **22**, 517 (1968).
[2] J. Pontén, *Bibl. Haematol.* (*Basel*) **31**, 319 (1968).
[3] K. Nilsson, J. Pontén, and L. Philipson, *Intern. J. Cancer* 3, 183 (1968).
[4] K. Nilsson, *Intern. J. Cancer* **8**, 432 (1971).
[5] J. Pontén, "Spontaneous and Virus Induced Transformation in Cell Culture." Springer-Verlag, New York and Wien, 1971.
[6] J. M. Trujillo, J. J. Butler, M. J. Ahearn, C. C. Shullenberger, B. List-Young, C. Gott, H. B. Anstall, and J. A. Shively, *Cancer* **20**, 215 (1967).
[7] J. G. Sinkovics, J. A. Sykes, C. C. Shullenberger, and C. D. Howe, *Texas Rep. Biol. Med.* **20**, 446 (1967).
[8] J. G. Sinkovics, *Med. Rec. Ann.* **61**, 50 (1968).
[9] J. Pontén, *Intern. J. Cancer* **2**, 311 (1967).
[10] Y. Ito, O. Shiratori, S. Kurita, T. Takahashi, Y. Kurita, and K. Ota, *J. Nat. Cancer Inst.* **41**, 1368 (1968).
[11] M. Eisinger, S. M. Fox, E. DeHarven, J. L. Biedler, and F. K. Sanders, *Nature* (*London*) **233**, 104 (1971).
[12] P. Gerber, *Proc. Nat. Acad. Sci. U. S.* **69**, 83 (1972).

Type 1. Mixed Cultures. Mixed cultures[6] containing both "reticular" cells attached to the substrate and lymphocytelike elements have been obtained by cultivating small pieces of tissue that were allowed to attach to the surface of tissue-culture flasks, and subcultures made by gently scraping the cells off the surface, at all times avoiding the use of enzymes. In our experience, trypsinization of such cultures results, after a few transfers, in cultures entirely of fibroblasts.

Type 2. Lymphoblastoid Cultures. Lymphoblastoid cultures, consisting predominantly of free-floating cells with a tendency to grow in clumps are grown using a modified Trowell type of organ culture.[4-9] Pieces of lymph node are explanted on the top of lens paper[9] or gelatin foam[4] overlying the metal grid of an organ culture dish. Cultures were originated by serial transfer, without trypsinization of the cell populations shed from the pieces through the grid onto the surface of the dish.

Type 3. Cultures of "Reticular" Cells, or Histiocytes, with Subsequent Lymphoblastoid Transformation.[11] These were developed as a means of studying lymph nodes from patients with pathologically diagnosed Hodgkin's disease, since it has been suggested that such cells constitute the neoplastic elements of this disease.[13] The essentials of the method used are illustrated in Figs. 1 and 2. Fresh biopsy specimens were washed two or three times in tissue culture medium (Eagle's MEM + 10% fetal calf serum) after which they were carefully decapsulated and cut into 1-mm^3 pieces while still immersed. After further washing (three times), two or three pieces were placed directly on the grids of organ culture dishes (Falcon Plastics, Division of B-D Laboratories, Inc., Los Angeles, California), which were then filled to the level of the grid with a medium consisting of 75% MEM + 10% fetal calf serum, and 15% filtered human serum (individual sera from young male donors, pretested for tissue culture toxicity). Dishes were incubated at 37°C in a water-saturated atmosphere containing 5% CO_2 in air in a device to be described elsewhere.[14] Cultures were examined every

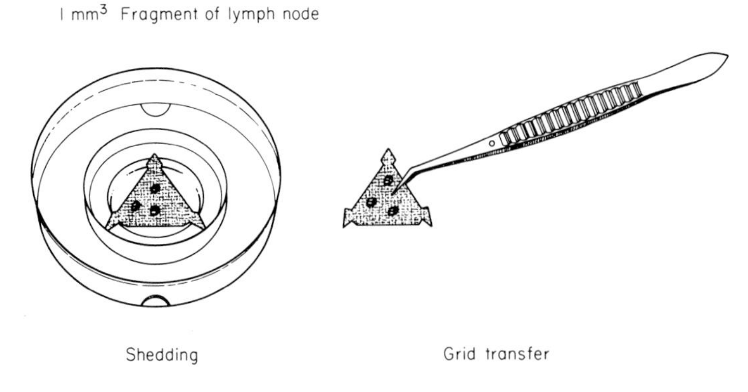

Fig. 1. Shedding and grid transfer.

[13] R. F. Lukes and J. J. Butler, *Cancer Res.* **26**, 1063 (1965).
[14] F. K. Sanders, B. O. Burford, and M. Monaghan, In preparation.

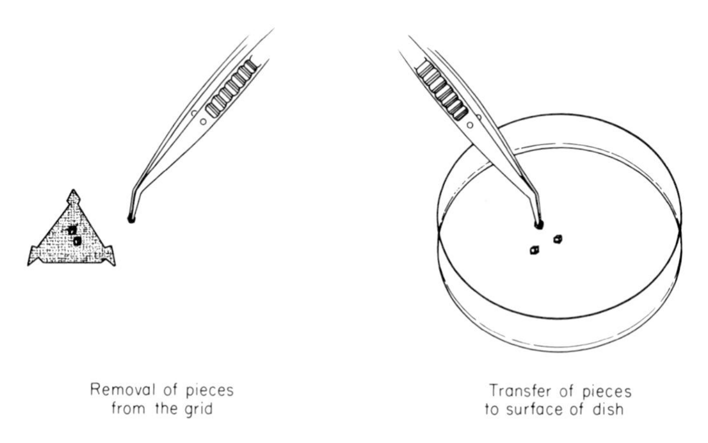

Fig. 2. Tissue transfer from grid to Petri dish.

2–3 days with an inverted microscope, the density of shed cells noted, and the medium changed twice a week. When the bottom of the dish became covered with a confluent layer of shed cells, the grid with its pieces of tissue was transferred to a new dish and the process repeated until cells were no longer being shed. The number of such transfers, and the interval between them, depended on the total quantity of cells shed. In general, three to four transfers at approximately weekly intervals were required.

The cells shed from the pieces on the grid at first consisted predominantly of a sparse population of macrophages, many with adherent lymphocytes, as well as a few fibroblasts (Fig. 3A). The duration of this phase was related to the histopathological nature of the lymph node used. It did not generally last more than 2 weeks and included two or three "sheddings" and subsequent grid transfers. It was succeeded by a second phase (one or two further "sheddings") in which the characteristics of the population of shed cells changed markedly. The increased cellularity of the grid "filtrate" then consisted of a mixture of multi- and mononucleated giant cells, "lymphoblastoid" cells, and a few lymphocytes and fibroblasts; macrophages appeared to be absent (Fig. 3B). This second population of cells closely resembled what is seen in pathological imprints from the original node, and it is from this second phase of "shedding" that most authors have succeeded in initiating "lymphoblastoid" cell lines.[9,10] Cell populations from the second shedding also resembled the "free-floating" cell cultures observed later following "lymphoblastoid transformation" of reticular cell cultures (see below).

When shedding was complete, the pieces were removed from the grid and placed on the surface of a dry Petri dish (or 250-ml T-flask) and left for 10–15 minutes to attach, before the addition of fresh medium (Petri dish, 3–5 ml; T-flask, 10–20 ml) (Fig. 2). Approximately 1 week later, when there was extensive outgrowth of cells, the latter could be subcultured by trypsinization and thereafter transferred by conventional methods.

When the above schedule was followed using lymph node fragments from patients exhibiting any one of the four histopathological types of Hodgkin's

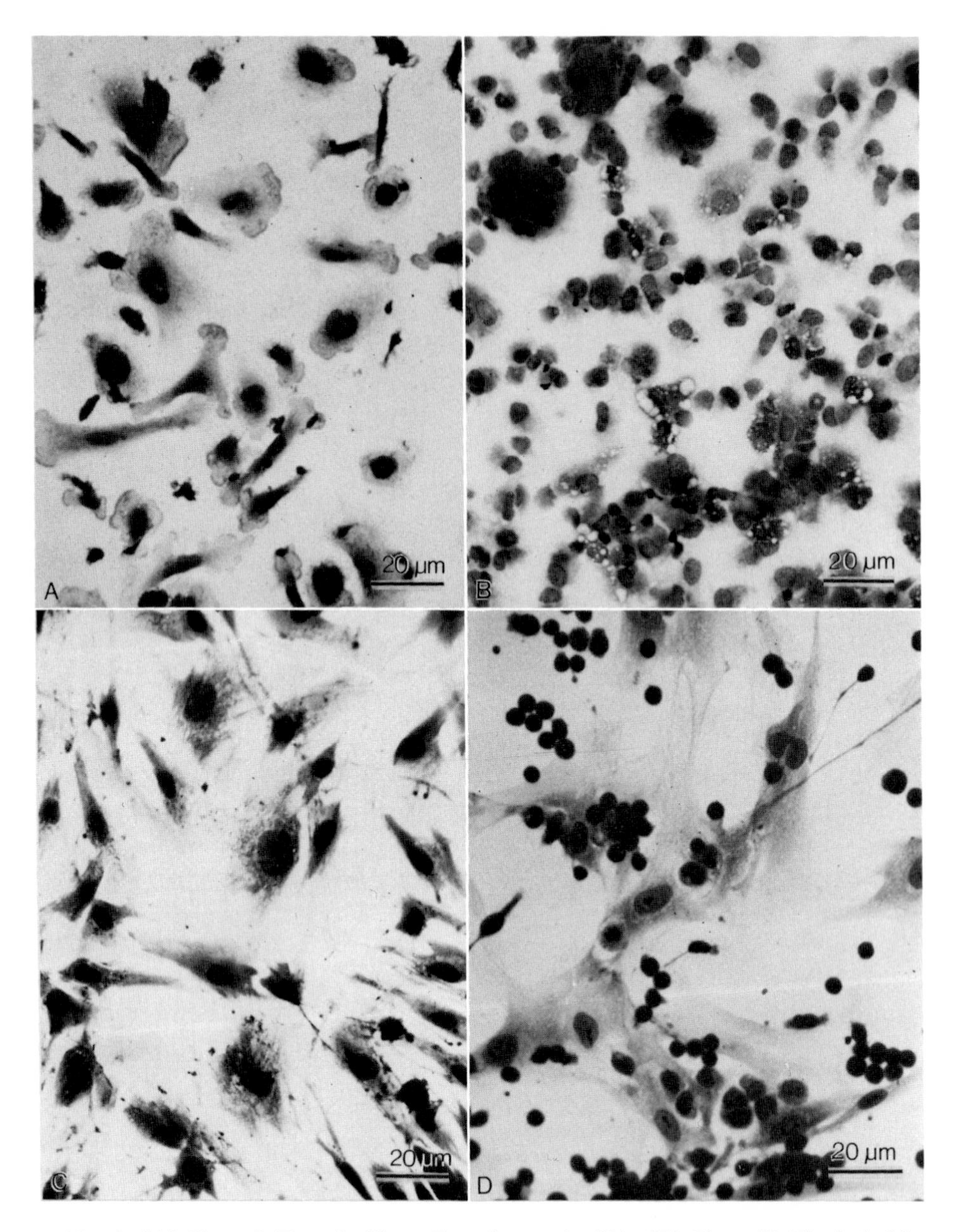

Fig. 3. (A) Phase I. First shedding. Tetrachrome (×50). (B) Phase II. Third shedding. Tetrachrome (×50). (C) Phase III. Confluent monolayer. Tetrachrome (×50). (D) Lymphoblastoid transformation in progress. May-Grünwald-Giemsa (×50).

disease (classified according to Lukes and Butler[13]) what is basically a single predominant cell type emerged. The cells forming the monolayer cultures were very large and pleomorphic, growing individually without parallel orientation. They were mostly mononucleated, though sometimes bi- or multinucleated. They had prominent nucleoli, a markedly reticulated cytoplasm, many lysosomes, and sometimes eosinophilic inclusions (Fig. 3C). Electron microscopic and histochemical evidence suggested their identity with the reticular cells of the parent node. The distinctive morphology of these cells was retained for at least seven to eight successive trypsinization transfers, following which some of them underwent changes which resulted in their "transformation" into "floating cultures" resembling cultures initiated by Type 2 methods (see above). After a decline in growth rate lasting 2–3 weeks, though not accompanied by any of the obvious cytopathology usually seen when fibroblasts from normal human individuals undergo the phase of "crisis," foci of highly basophilic, rounded "blastoid" cells appeared within the reticular cell cultures when observed by time-lapse photomicrography or stained preparations. Many of these rounded cells displayed emperipolesis, moving, and even undergoing mitosis, within the cytoplasm of the large attached cells (Fig. 3D). Within 4 to 5 weeks, the cultures became entirely composed of free-floating cells, lymphoblasts, and lymphocytes. At the same time there was both an increase in overall growth rate and in saturation density. In stationary cultures many of the free-floating cells became reattached to the surface of the flasks but could be easily detached therefrom. As in the case of other free-floating cultures originating from human lymphoid material, there was also biochemical, immunological, and electron microscopic evidence that a herpeslike virus was present.[11]

In conclusion, different versions of the grid culture technique seem to offer the best opportunity for initiating continuous cell lines from lymph nodes. Type 3 methods have the advantage of providing a means by which an investigator can attempt to cultivate selectively one particular cell type of personal interest.

CHAPTER 9

Human Biopsy Material from Genetic Abnormalities[1]

Arthur E. Greene

Among the most useful tissue cultures available to the geneticist are those derived from human skin. Skin biopsies are easily obtained, the cultures derived from them are easy to grow, they can be frozen for years in liquid nitrogen, and

[1] The techniques reported in this article were developed with the support of U. S. Public Health Service Grant CA-04953-13, Contract NIH-NIGMS-72-2070, General Research Support Grant FR-5582 from N.I.H., and Grant-in-Aid Contract M-43 from the State of New Jersey.

recovered in a highly viable state. Skin fibroblast cultures are well suited for studies in the field of somatic cell genetics and have contributed to the definition of congenital malformation syndromes and the diagnosis of a number of chromosomal aberrations, both numerical and structural, caused by translocations, deletions or chromosomal breakage.[2] Human diploid fibroblast cell lines developed from skin biopsies are increasingly utilized today for the identification of the biochemical basis of these hereditary diseases.[3]

Collecting the Biopsy

The specimen may be collected by simple punch biopsy.[4] Stainless steel punches together with forceps and scissors are cleaned, wrapped, and autoclaved. The skin is swabbed with alcohol, rinsed with sterile saline, and dried with sterile gauze. Almost any area of the body can be used including the buttocks, the back medial to the upper border of the scapula, the flexor surface of the forearm, and the deltoid region. The procedure is simple, rapid, and an anesthetic is usually not required. If it is considered essential to use one, an intradermal injection of several milliliters of 2% procaine solution will anesthetize the skin. Skin biopsies with a diameter of 2 to 8 mm may be taken by placing the punch on the skin area and quickly rotating the punch to loosen the piece of tissue to a depth of about 2 to 3 mm or to the subcutaneous fat. The core of skin is then gently lifted and snipped off quickly with the scissors. Bleeding is either absent or slight and easily stopped by brief application of pressure. The site is covered with a band-aid and scarring rarely occurs, especially with small-diameter biopsies. The removal of larger samples of tissue may require suturing, although a number of investigators report that bleeding can be arrested by a simple pressure dressing. The punch biopsy has been adapted to an electric rotary drill[5] and this modification has been most beneficial because the depth of the skin tissue excised can be controlled more accurately than by hand punch. The electric punch drill is a simple, rapid method which is suitable for the collection of skin biopsies in large-scale family or population studies.

Establishing the Primary Culture

The skin biopsy is placed in a sterile Petri dish and washed with three changes of McCoy's 5a Modified Medium[6] with 100 units/ml of penicillin and 100 μg/ml of streptomycin. The tissue is minced with two sterile scalpel blades into a number of small pieces. Approximately eight to ten skin fragments are

[2] E. Passarge, *Birth Defects Orig. Art. Ser.* **IV**, 26 (1968).

[3] W. Mellman, *In* "Advances in Human Genetics" (H. Harris and K. Hirschhorn, eds.), Vol. II, p. 259. Plenum, New York, 1971.

[4] D. H. Pillsbury, W. B. Shelley, and A. M. Kligman, "A Manual of Cutaneous Medicine," p. 71. Saunders, Philadelphia, Pennsylvania, 1961.

[5] R. G. Davidson, S. W. Brusilow, and H. M. Nitowsky, *Nature (London)* **199**, 296 (1963).

[6] S. Iwakata and J. T. Grace, Jr., *N. Y. J. Med.* **64**, 2279 (1964).

transferred with a sterile Pasteur pipette to a 25-cm^2 Falcon plastic tissue culture flask and positioned at different areas in the flask. The excess culture fluid surrounding the area of the explant is removed with the pipette and the flask is inverted with the tissue pieces adhering to the top of the vessel. Growth medium which consists of 5 ml of McCoy's Medium, 20% uninactivated fetal bovine serum, 100 units/ml penicillin, and 100 μg/ml streptomycin is added to the flask which remains inverted for at least 1 hour at room temperature to allow the tissues to adhere firmly to the surface. The flask is then carefully turned so that the culture medium gently flows over and now bathes the tissue explants and the flask is incubated at 37°C in an atmosphere of 5% CO_2 in air.

Fibroblast cells readily grow in abundance from the edges of the small bits of human skin and a confluent sheet of cells is obtained in approximately a month. Cultures can be analyzed for chromosome abnormalities and some metabolic defects during this primary growth period. However, we find it more practical to conduct these examinations on secondary cultures because of the increased yield of cells.

Subculturing the Primary Culture

Primary fibroblast cells adhere firmly to the culture flask. We find that treatment of the cells with Puck's saline A–EDTA–trypsin[7] is the most gentle and efficient way to dislodge them from the surface. Remove the old culture medium and pipette 5 ml of 0.02% EDTA in Puck's saline A onto the cells and incubate the flask for 10 minutes at room temperature. The culture is periodically observed in the inverted microscope and when the cells start to float free, invert the flask and remove the EDTA. Add 3 ml of Puck's saline A containing 0.02% EDTA and 0.04% trypsin and incubate at 37°C until the cells are released from the surface (approximately 2–10 minutes). Gently tap this flask until all the cells are in the fluid and then immediately add 3 ml of growth medium. Remove 0.5 ml of the total volume of cells and fluid to determine cell viability by trypan blue dye exclusion. Distribute the remaining 5.5 ml into three culture flasks of the same size (25 cm^2), each containing 3 ml of culture medium or into one larger Falcon tissue culture flask (75 cm^2) containing 13 ml of growth medium. Further passages are prepared, following the previously outlined procedure, with the exception that 6 ml of growth medium is added to the 3 ml of EDTA–trypsin solution. Transfer 3 ml of cells and medium to three new tissue culture flasks (75 cm^2) containing 17 ml of growth medium and incubate the cultures at 37°C in an atmosphere of 5% CO_2 in air.

Cells in subculture are fed twice a week; they proliferate rapidly and can usually be subcultured again in 5–7 days. Most cultures can be passaged at least thirty times.[3] Cells in the first and second subculture provide good material for chromosome examination and a quantity of cells for a variety of isozyme and biochemical tests.

[7] R. P. Cox, M. R. Krauss, M. E. Balis, and J. Dancis, *Proc. Nat. Acad. Sci. U. S.* **67**, 1573 (1970); cf. Section II, Chapter 7, this volume for formulation.

Storage in Liquid Nitrogen

Skin fibroblast cell cultures from patients with genetic abnormalities should be stored at low passage levels in liquid nitrogen to prevent loss from microbial contamination and eventual senescence through *in vitro* aging.[8] It has also been reported that cloning efficiency is most effective on cells in early passage.[9] For freezing, the cells are grown to confluence, and refed with fresh growth medium. Twenty-four hours later they are removed with Puck's EDTA-trypsin solution as for subculture and suspended in the growth medium. A 0.5 ml aliquot is removed for a viability count by the trypan blue dye exclusion test. Cells are centrifuged at 900 rpm for 5 minutes, the supernate discarded, and the cells resuspended in growth medium plus 10% glycerol to yield approximately 5×10^5 to 1×10^6 viable cells per ml; 1.0 ml is dispensed in thick-walled 1.2 ml borosilicate glass ampoules and sealed in an oxygen flame. Ampoules are safety tested for leaks by immersion in 70% alcohol–methylene blue at 4°C and then frozen in liquid nitrogen vapor[10] (freezing rate approximately 5°–24°C per minute) and stored in or over liquid nitrogen.

[8] L. Hayflick, *Exp. Cell Res.* **37**, 614 (1965).

[9] M. Hayakawa, *Tohoku J. Exp. Med.* **98**, 171 (1969).

[10] A. E. Greene, R. K. Silver, M. Krug, and L. L. Coriell, *Proc. Soc. Exp. Biol. Med.* **116**, 462 (1964).

CHAPTER 10

Mammalian Myocardial Cells[1]

Frederick H. Kasten

Great progress has been made in recent years in studies of cultured myocardial cells. This progress is due largely to the development of simplified and reproducible techniques for isolating and growing these differentiated cells.

[1] This research was supported in part by USPHS Research Grants NS-09524 from the National Institute of Neurologic Diseases and Stroke and CA-12067 from the National Cancer Institute. It has also benefited from USPHS Training Grant 5-T01-DE-00241-04 from the National Institute of Dental Research. I am pleased to acknowledge the technical assistance of Cindy Arey, Sylvia S. El Kadi, Ann Matthews, James Rodriguez, and Fredy Strasser.

Much of the early work on cultured myocardial cells was done using chick embryo heart tissue. Credit should be given to Burrows[2] for initiating work in this area. He explanted fragments of chick heart in a plasma clot. This technique was used by many other workers for the next 20 years to study myocardial outgrowths. The technique suffers from the disadvantages that there is a minimal migration of differentiated myocardial cells from the explant as well as difficulties in achieving good optical resolution. Apparently, the first film records of living myocardial cells *in vitro* were made by Goss,[3] again using primary explants. His film must be regarded as a milestone for the period in which he worked, when phase-contrast optics were not yet available and the method of cultivation was not optimal.

Since then, whole salamander hearts have been maintained in a contracting state in culture for up to 6 months.[4] Likewise, whole fetal mouse hearts have been cultivated for 3 weeks under conditions where they beat rhythmically and spontaneously throughout this period.[5] A significant technical advance was reported in 1955 by Cavanaugh,[6] who used trypsin to dissociate chick embryo heart fragments to single cells. This approach to the cultivation of myocardial cells proved to be the one of choice by later investigators. For example, Rumery *et al.*[7] used a modified version of Cavanaugh's technique to permit observations of living myocardial cells by phase-contrast and polarization microscopy. Trypsinized chick heart cells have been employed for many tissue culture studies. As has been pointed out,[8] trypsinized chick heart cells in culture actually are a mélange of cell types, not all of which are easily distinguished from each other. In my opinion, the only cell type which one can be sure of identifying at the light microscope level is the differentiated myocardial cell, with its characteristic myofibrils and/or its contractile activity. The rat heart cell system *in vitro* is an easier one to study from this standpoint, since there is only one nonmyocardial cell type present and is easy to distinguish.[8-10] The next significant advance was reported in 1960 by Harary and Farley,[11] who trypsinized whole hearts from newborn rats and cultured the cells in plastic Petri dishes. The possibilities of maintaining spontaneous beating mammalian heart cells under controlled conditions whetted the appetite of numerous workers. At our laboratory, the procedure was modified so as to permit ventricular myocardial cells of the newborn rat to be cultivated in Rose chambers.[9,10] The use of such chambers for perfusion studies of contracting myocardial cells is described elsewhere in this volume.[12]

[2] M. T. Burrows, *J. Amer. Med. Ass.* **55**, 2057 (1910).
[3] C. M. Goss, *Proc. Soc. Exp. Biol. Med.* **29**, 292 (1931).
[4] E. W. Millhouse, Jr., J. J. Chiakulas, and L. E. Scheving, *J. Cell Biol.* **48**, 1 (1971).
[5] K. Wildenthal, *J. Mol. Cell. Cardiol.* **1**, 101 (1970).
[6] M. W. Cavanaugh, *J. Exp. Zool.* **128**, 573 (1955).
[7] R. E. Rumery, R. J. Blandau, and P. W. Hagey, *Anat. Rec.* **141**, 253 (1961).
[8] F. H. Kasten, *Acta Histochem. Suppl.* **9**, 775 (1971).
[9] F. H. Kasten, R. Bovis, and G. Mark, *J. Cell Biol.* **27**, 122A (1965).
[10] G. E. Mark and F. F. Strasser, *Exp. Cell Res.* **44**, 217 (1966).
[11] I. Harary and B. Farley, *Science* **131**, 1674 (1960).
[12] See Section VI, Chapter 2.

We have reported the results of other cellular and substructural investigations based on cultures growing in Rose chambers.

With respect to the methodology involved in culturing mammalian myocardial cells, I do not propose to summarize all the variations one can employ. The method to be described has been in use in our laboratory for about 8 years and has proved its worth in many studies. The basic technique was enunciated some years ago[10] and was since modified, mainly so as to permit one to get a selective isolation of 95% mammalian myocardial cells from a mixed population of myocardial and endothelial cells. The technique, which is based on the principle of differential cell adhesion to the supporting surface, can be used when it is desired to isolate a 100% population of endothelial cells. The author presented the differential heart cell isolation technique at the June, 1969, meeting of the Tissue Culture Association[13] and described it briefly in a recent publication.[14] A similar procedure using chick embryo heart cells was reported.[15] Before describing the isolation and culture procedures in detail, attention is directed to a number of reports related to methodology.

A method was reported[16] which is said to restrict the tendency for myocardial cells to aggregate or agglutinate during isolation. This may be useful, although cellular reaggregation has not proved to be a problem with the procedure we use. Because of the difficulty in culturing adult myocardial cells, attention is directed to recent reports of the successful isolation of contracting adult rat and mouse cells.[17,18] However, in both cases cells remained viable and in a contracting state for only 2 to 4 hours. This is long enough to do certain short-term studies but would not be considered as a cell culture system. Apparently, the rapid addition of fibrous tissue to the growing heart makes it difficult to isolate and culture large numbers of contracting myocardial cells from adult animals. I have been able to culture beating rat heart cells from animals up to 10 days of age, but the recovery rate is low. The optimal age and highest yield from neonatal rats is at 3 to 4 days. Ventricular cells taken at this age are fully differentiated, from morphological and functional points of view and offer no limitations to most experimental studies. Other reports describe unusual growth patterns of myocardial cells cultivated on polystyrene sheets[19] or surfaces coated with collagen-agar or palladium.[20]

Isolation and Culture Technique for Mammalian Cardiac Cells

Animals. Three- to four-day rats are employed routinely. While younger rats and those in late stages of embryonic development are suitable, the amount of

[13] F. H. Kasten, F. Strasser, and A. Matthews, *In Vitro* **4**, 87 (1969).
[14] F. H. Kasten, *In Vitro,* in press.
[15] I. S. Polinger, *Exp. Cell Res.* **63**, 78 (1970).
[16] W. O. Gross, E. Schöpf-Ebner, and O. M. Bucher, *Exp. Cell Res.* **53**, 1 (1968).
[17] G. V. Vahouny, R. Wei, R. Starkweather, and C. Davis, *Science* **167**, 1616 (1970).
[18] S. Bloom, *Science* **167**, 1727 (1970).
[19] S. P. Halbert, R. Bruderer, and T. M. Lin, *J. Exp. Med.* **133**, 677 (1971).
[20] M. Lieberman, A. E. Roggeveen, J. E. Purdy, and E. A. Johnson, *Science* **175**, 909 (1972).

ventricular cardiac tissue available is small and more animals are required for a given run. For the same reason, rats are preferred over mice. As mentioned earlier, rats older than 4 days are not desirable because of a failure of many of the isolated myocardial cells to develop rhythmic spontaneous contractions in culture. The yield of viable myocardial cells falls rapidly, beginning with 5-day-old animals. Cardiac cells can be isolated from other newborn mammals, such as hamsters,[21] rabbits, kittens, and dogs.[22] I have as yet been unable to obtain as high a yield of contracting cells *in vitro* from these species as from the rat. Undoubtedly, the organization of myocardial tissue and amount of connective tissue varies from species to species at birth. For the present, the use of neonatal rat ventricular tissue remains the one of choice for studies of isolated, mammalian cardiac cells in culture.

Solutions. Since myocardial cells are more sensitive than most other cell types to noxious impurities, the choice of trypsin employed in the disaggregation procedure becomes especially important. I have observed no toxic effects using trypsin (1:250) from Difco Company and routinely employ this preparation in all our myocardial cell isolations. It should be noted that in any case, a certain amount of intracellular damage to myofibrils occurs during trypsinization as a consequence of enzyme penetration of the cell membrane.[8,23] This type of damage is reversible, so that treated cells rapidly regenerate new myofibrils and begin contracting in less than 15 hours. With certain other brands of trypsin, the damage is so severe that few cells regain the ability to contract. Trypsin is prepared in 100-ml aliquots in an 0.125% concentration in Ca- and Mg-free Hanks' balanced salt solution (referred to as incomplete BSS), sterilized by filtration through a 0.22 μm Millipore filter, and stored at $-20°$C. Washing of the heart tissue and trypsinized cells is done with incomplete BSS. The nutrient heart medium is a modified Eagle's MEM, essentially the same as described previously,[10] except that gentamycin is used instead of penicillin and streptomycin. The medium consists of 85% Gey's BSS supplemented with 1% MEM 100× vitamin stock, 2% MEM 50× amino acids, 1% 100× nonessential amino acids, 1% 200 m*M* L-glutamine, 50 mg/liter L-glycine, 12.5 mg/liter hypoxanthine, 10% fetal calf serum, 50 mg/liter gentamycin, and 0.01 g/liter phenol red. It may be necessary to test different batches of fetal calf serum to find one which permits spontaneous contractility *in vitro.* The phenol red used should have a high dye content and not contain any insoluble product which may be toxic. The 0.5% concentrated phenol red solution supplied by Microbiological Associates or Flow Laboratories is routinely used in this laboratory.

Preparation of Cells. Six to eight 3- to 4-day rats are decapitated and handled with aseptic techniques. In this laboratory, all handling procedures from this point on are done in a laminar flow hood. While this method of helping to maintain a clean working area is desirable, it is by no means essential. Noncontaminated heart cultures are obtained as well under less sophisticated con-

[21] W. DeW. Andrus and F. F. Strasser, *Exp. Cell Res.* **47**, 613 (1967).

[22] F. H. Kasten, unpublished studies.

[23] F. H. Kasten, *J. Cell Biol.* **31**, 131A (1966).

ditions with the use of ordinary aseptic procedures. Whole hearts are removed and placed in a small Petri dish containing incomplete BSS. Atria, attached tissue from the lung, and other excess tissue are carefully removed with forceps from the ventricular surfaces. Ventricles are transferred successively to three other plastic dishes containing fresh, incomplete BSS. In the second dish, ventricles are squeezed with forceps to remove any large blood clots. After about a minute in this dish, ventricles are transferred and washed briefly in each of the other two dishes. After the final bath, the ventricles are put in a depression slide containing incomplete BSS. With fine scissors, ventricles are cut into 1-mm pieces and transferred into a 75-ml trypsinizing flask with a DeLong style neck, containing a magnetic stirring bar and 15 ml of freshly thawed trypsin solution. Tissues are digested for 10 minutes at 37°C with the aid of a magnetic stirrer at a very low speed. Most stirrers cannot be adjusted to a sufficiently low speed without the addition of a powerstat. The supernatant is discarded since it contains erythrocytes, nonviable cells, debris, and nonbeating elements. This step is probably the most critical one up to this point because of the tendency to pour off pieces of undigested heart with the supernatant. Ten milliliters of fresh trypsin is added to the flask and stirred at 37°C for 15 minutes. The supernatant is poured into a sterile 15-ml centrifuge tube and labeled supernatant I (S-1). Two milliliters of heart medium is added to S-1 to inactivate the trypsin. Ten milliliters of fresh trypsin is added to the trypsinizing flask and stirred again at 37°C for 15 minutes. The supernatant is decanted into a second 15-ml centrifuge tube and labeled S-2. Two milliliters of heart medium is added to S-2. Tubes S-1 and S-2 are centrifuged for 5 minutes, position 6, in an International clinical centrifuge (approximately 1300 rpm). Centrifuging causes a certain amount of damage to myocardial cells.[8,23] Set the centrifuge speed so as to give the minimum force necessary to produce a light pellet of cells at the bottom. Remove and discard the supernatant carefully from S-1 and S-2 with the aid of a Pasteur pipette so as not to disturb the pellets. Break up the pellets by forcefully squirting them repeatedly with a few milliliters of medium. Pool the cells into a single 15-ml centrifuge tube containing a total of 10 ml. Label this tube S-A; cool in an ice bath.

While tubes S-1 and S-2 are being centrifuged, 10 ml of fresh trypsin are added to the residual heart tissue in the trypsinizing flask and the trypsinizing procedure repeated so as to yield two other centrifuge tubes of cells, labeled S-3 and S-4. These are combined into a single centrifuge tube containing 10 ml of medium, labeled S-B. Tubes S-A and S-B are centrifuged lightly and pooled into a single tube with 10 ml of medium. This final suspension is washed twice with fresh medium with alternate centrifugation so as to remove and inactivate the last traces of trypsin as well as to reduce cellular debris and clumping. It is possible that some undigested heart tissue may still be present after the five successive trypsinizations; further digestions can be carried out. For routine studies, most of the potential contractile cells are already removed and it is not worth the effort to obtain the others. However, it should be noted that when one examines each trypsinized fraction successively, it is evident that the yield of myocardial cells relative to endothelial cells increases markedly in the later fractions.[22] As is well known, excessively long digestion with trypsin is toxic, hence, the need to use successive, short exposures to the enzyme. Even using the multiple trypsinization technique leads to preferential enzyme digestion of Z bands in myofibrils and dissolution of the contracting units into myofilaments.[8,23]

When using 6 newborn rats, 3 to 4 days of age, the final cell suspension usually contains 6 to 7 $\times 10^5$ cells per milliliter, or a total of 6 to 7 $\times 10^6$ cells. The population includes two major cell types, myocardial and endothelial, in the ratio of approximately 3:2. Such a mixed population can be assayed for viability by the trypan blue dye exclusion test, counted, and diluted to give a total viable cell count of 1.5×10^5 cells per milliliter for cultivation. However, to obtain a significantly higher yield of myocardial cells relative to endothelial cells, the differential attachment technique is interjected at this point.

Differential Attachment Technique for Obtaining Relatively Pure Cultures of Myocardial Cells[13,14]

The technique is based on an observation I made several years ago that when a suspension of myocardial and endothelial cells settles on a substrate such as glass, the endothelial cells begin attaching to the surface within several minutes. Myocardial cells, on the other hand, remain rounded for approximately 15 hours before they exhibit signs of membrane extension and subsequent attachment to the surface. These observations are based on direct observations combined with cinematography. Numerous experiments have been done which will be reported in detail elsewhere. Since many colleagues of mine and other workers have had access to the details of this technique, it seems desirable to describe it at this time for the benefit of others. The essential steps are listed below.

Method A (See Fig. 1)

1. Erlenmeyer flask has had cells added to it from tubes S-A and S-B (see *Preparation of Cells*) and the suspension is stirred.
2. The cells are centrifuged and washed twice with fresh medium.
3. Final cell suspension in 10 ml of medium.
4. Mixed cell population is poured into a sterile 50-ml Erlenmeyer flask (or a deep Petri dish) and permitted to settle at 37°C. In the illustration, myocardial cells are shown as dots; endothelial cells appear as lines. Alternatively, the mixed cell suspension can be added to several layers of small glass beads in a beaker or Petri dish, which provide a greater surface area than a flat surface, but the results are no better than with the flask procedure alone.
5. After 90 minutes, the suspension of myocardial-rich cells (95% purity) is withdrawn, leaving endothelial cells attached to the floor of the flask. If desired, the endothelial cells can be isolated immediately by trypsinization or cultured as a pure endothelial population. The myocardial-rich suspension is counted, checked with trypan blue for percentage viability (usually 90%), and diluted to give 1.5×10^5 viable cells per milliliter. If higher cell concentrations

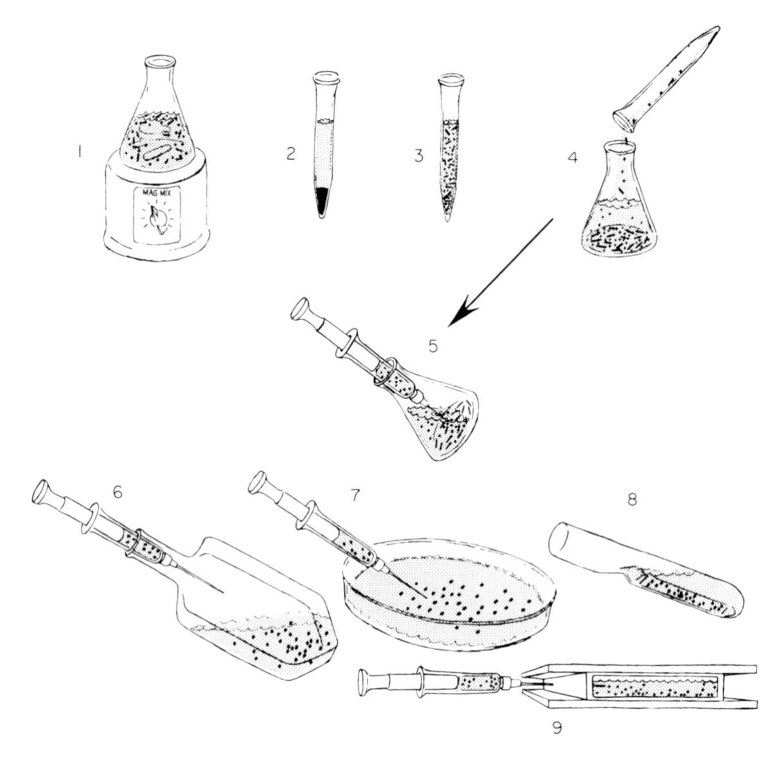

Fig. 1. Illustration of differential attachment technique (Method A) used to separate myocardial from endothelial cells. Principle depends on the rapid and preferential adhesion of endothelial cells to the substrate. Various methods are shown of culturing the isolated myocardial cells. See text for description.

are used, such as 3 to 4 × 10^5 cells per milliliter, there is less chance of residual endothelial cells overgrowing the myocardial cells in long-term cultures.

6–9. The myocardial cells can be injected into a T-flask, a Petri dish, a Leighton tube, or a Rose chamber.

Method B (See Fig. 2)

1–3. Same procedures as described for Method A in Fig. 1.

4. Mixed cell population is counted, checked with trypan blue for percentage viability and diluted to give 3.0 × 10^5 cells per milliliter (see 5 under Method A regarding effect of higher cell densities). Two milliliters are injected into each Rose chamber.

5. The mixed cell population is allowed to settle at 37°C.

6. At the end of 90 minutes, the chamber is inverted so as to permit the rounded myocardial cells to fall to the opposite coverslip. I call this the "flip-flop" technique. There is a way at this point to improve the yield of myocardial cells. Just prior to flipping the chamber over I hold it tightly in the palm of the hand and hit the edge of the hand hard against a firm surface, such as the lab table. Such a "karate-chop" dislodges most of the remaining myocardial cells.

7. The chamber now contains a practically pure endothelial population hang-

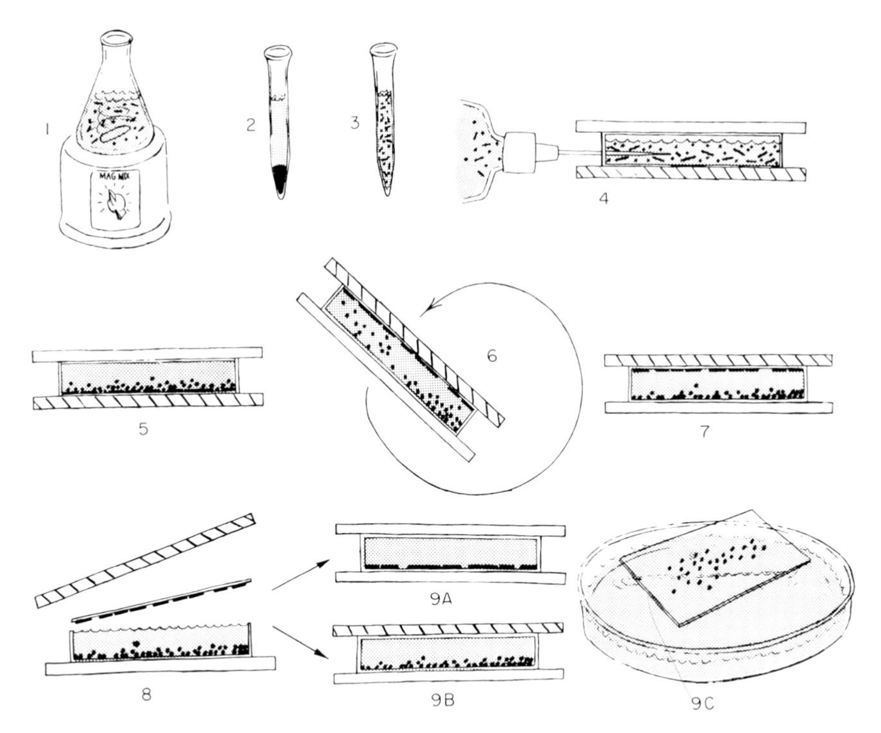

Fig. 2. Illustration of differential attachment and "flip-flop" techniques (Method B) used to separate myocardial from endothelial cells directly in the Rose chamber. Either myocardial or endothelial-rich chambers can be obtained. See text for description.

ing "upside down" on the upper coverslip and a myocardial-rich population on the bottom coverslip, with both groups exposed to the same milieu. The chamber is incubated at 37°C.

8. Twenty-four hours later, the chamber is partly disassembled under aseptic conditions. The upper coverslip containing the endothelial cell population is removed. Medium is changed at this time to remove debris and toxic cellular by-products.

9A. The endothelial-rich coverslip is used in a new Rose chamber if it is desired to study or utilize these cells.

9B. The original Rose chamber containing the myocardial-rich coverslip is reassembled using a fresh, sterile coverslip to replace the one which had been removed. At this time a strip of perforated cellophane (Microbiological Associates) may be placed over the coverslip containing the cells. This is useful in cutting down the amount of proliferation by residual endothelial cells and allowing long-term maintenance of beating heart cells. I add the cellophane only to chambers which will be kept more than 10 days.

9C. If desired, the coverslip containing the myocardial-rich population can be removed and placed in a Petri dish for cultivation.

The differential attachment technique permits us to obtain initially a 95% myocardial and/or a 100% endothelial cell population, which is a considerable

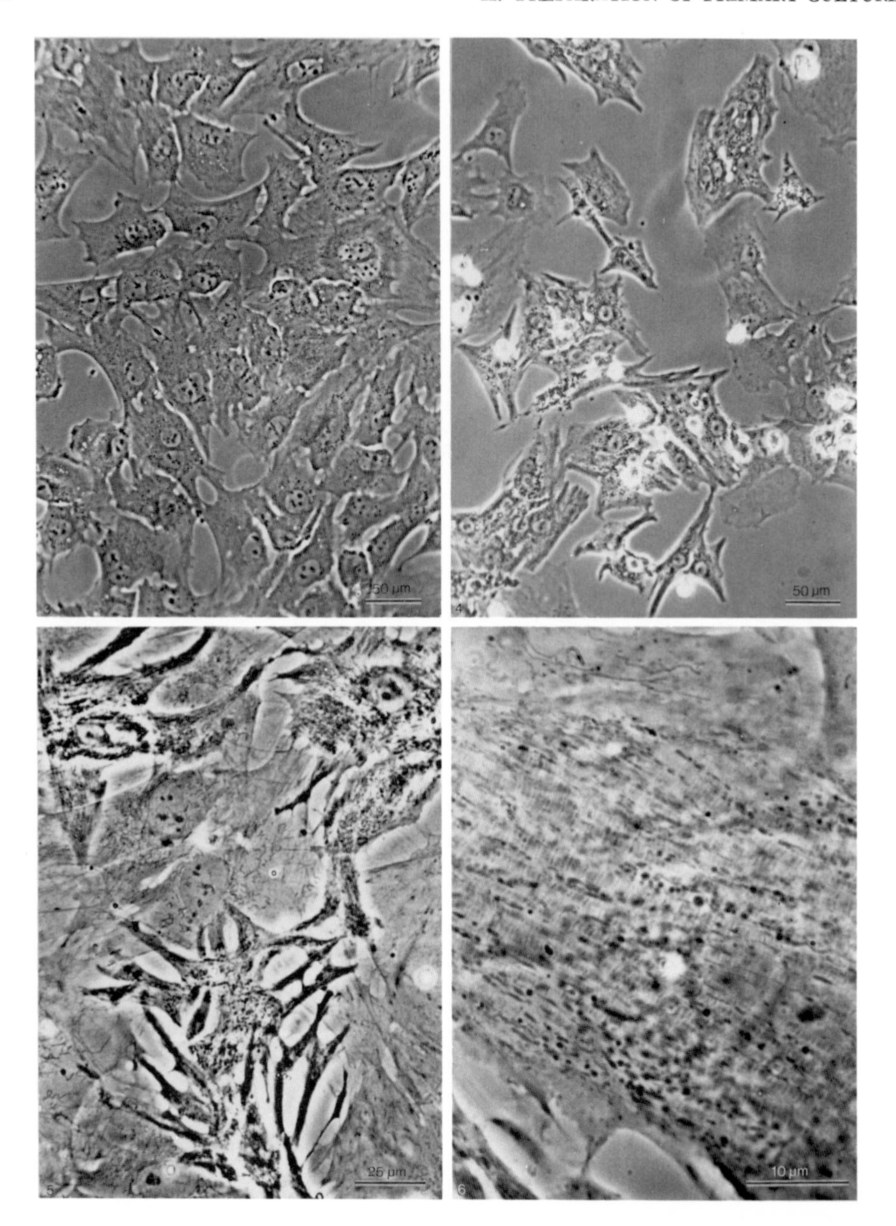

Fig. 3. Pure culture of endothelial cells obtained by Method B. Three-day culture from newborn rat heart. Phase-contrast optics (×150).

Fig. 4. Myocardial-rich culture obtained by Method B. Typical field from opposite coverslip to one shown in Fig. 3. Note large numbers of dense myocardial cells compared with endothelial cells. Same chamber and culture day as Fig. 3. Phase-contrast optics (×150).

Fig. 5. Typical field of well-established, spontaneously contracting myocardial cell networks in Rose chamber. Dense myopodial processes are attached to endothelial cell layer underneath. Six-day culture of newborn rat heart cells. Phase-contrast optics (×385).

Fig. 6. Large numbers of myofibrils with typical sarcomere structures are seen in this myocardial cell from an 8-day culture of newborn rat heart. Phase-contrast optics (×1155, ref. 8).

improvement over a mixed population of 60:40 (myocardial:endothelial). Figures 3 and 4 are typical examples of results to be expected after 3 days in culture. The technique permits biochemical analyses to be made of myocardial cells during the first few days of cultivation without risking misinterpretation of results because of the endothelial population. However, the initial 5% of contaminating endothelial cells proliferates rapidly so that by 4–5 days *in vitro*, the ratio of myocardial:endothelial cells is changed from 10:1 to 3:2. This is largely due to an eightfold increase in endothelial cells.

Tissue Culture

Myocardial-rich suspensions are injected into Rose chambers[12] with glass coverslips, which permit cell growth, mitosis, formation of synchronized networks, and spontaneous contractility of the primary cultures for as long as 90 days. Cells form synchronized networks within a few days and most experiments are carried out during the first week. As indicated earlier, the medium is changed about 24 hours after seeding to remove dead cells and other debris as well as to permit perforated cellophane to be added[24] for long-term cultures and the endothelial-rich coverslip to be removed (if Method B is used). Oxygen deficiency accounts for the death of many myocardial cells in older cultures; the presence of an air bubble in the chamber helps here. Also, the use of gas-permeable polystyrene sheets instead of glass on one surface permit a better gaseous environment for the cells.[24] Medium is routinely changed twice a week. When endothelial cells dominate the culture, the pH falls rapidly from one day to the next. Otherwise, the pH remains at 7.3 for up to a week, even without a medium change. Cultures are maintained at 37°C in an incubator or on the microscope stage with the aid of an air curtain and thermistor probe. The use of such cultures for drug perfusion studies is given elsewhere in this volume.[12] The methods described above are generally applicable to all types of culture vessels, as indicated in Method A (Fig. 1). The morphology of myocardial cells at medium and high magnifications is shown in Figs. 5 and 6. It seems to me that the two most important criteria for successful myocardial cell cultivation and usage are:

1. The development of healthy looking cells which contract spontaneously and rhythmically.
2. An understanding of the nature of the heart cell population in regard to its possible cellular heterogeneity, which could otherwise lead to a misinterpretation of results.

[24] G. E. Mark, J. D. Hackney, and F. F. Strasser, *In* "Factors Influencing Myocardial Contractility" (R. D. Tanz, F. Kavaler, and J. Roberts, eds.), p. 301. Academic Press, New York, 1967.

CHAPTER 11

Mammalian Glomerular Cells

Leonard J. Quadracci and Gary E. Striker

Various kidney cell lines derived from renal tubular epithelium have long been used on a routine basis in many tissue culture laboratories. Isolation of viable glomeruli and their maintenance *in vitro* was recently reported by Bernik.[1] All three glomerular cell types survived in culture for prolonged periods, but cellular proliferation was minimal and the cells were not subcultured.

The method of Krakower and Greenspon[2] has been used in this laboratory for preparing large numbers of glomeruli from various species of animals and humans. The procedure to be described is a further modification of this technique. Large numbers of viable glomeruli which can be maintained in cell culture can be obtained with this method.[3]

Species Studied

Viable glomeruli suitable for culture have been obtained from rats, dogs, monkeys, adult humans, and human fetuses. Attempts to use this modification in rabbits have not been successful. Rat glomerular cells are easy to prepare but are difficult to maintain under the conditions tested. Human fetal glomeruli are easily prepared and have luxuriant cellular outgrowths, appear early, and are easily maintained and transferred. Glomeruli obtained from adult human nephrectomy specimens have been used successfully. It also has been possible to maintain glomeruli obtained from autopsy kidneys approximately 3 hours after death.

Preparation of Glomeruli

The kidneys are carefully removed using sterile technique. The capsule is stripped from the cortex. With large kidneys the cortex is shaved from the medulla and diced into small pieces. With very small kidneys, the pelvic structures are wedge resected in an attempt to remove the majority of the medullary tissue. The removed cortex is then minced and 2–3 ml of Eagle's Minimal Essential Media (MEM) is added to the tissue. The fragments are buttered

[1] M. B. Bernik, *Nephron* **6**, 1 (1969).

[2] C. A. Krakower and S. A. Greenspon, *Arch. Pathol.* **58**, 401 (1954).

[3] L. J. Quadracci and G. E. Striker, *Proc. Soc. Exp. Biol. Med.* **135**, 947 (1970).

through an 80 mesh stainless steel screen which is rinsed with MEM containing penicillin and streptomycin. The material is collected in a beaker and centrifuged at 30 *g* for 2 minutes. The supernatant is aspirated and discarded. If the tissue is heavily contaminated with red blood cells, the sedimented red cells will form a ring on top of the sedimented glomeruli, which may be aspirated with a Pasteur pipette. The sediment is resuspended in the MEM and recentrifuged for 2 minutes at 30 *g* twice. The final sediment is then resuspended and poured over a 150 mesh stainless steel screen and washed with liberal amounts of MEM. The remaining glomeruli are carefully washed from the screen with MEM. They may have to be washed vigorously off the screen using a Pasteur pipette. The suspension is centrifuged for 2 minutes and the supernatant discarded. The sediment contains a large number of glomeruli which are generally free of capsular and tubular debris. Debris present usually stains with vital dyes.

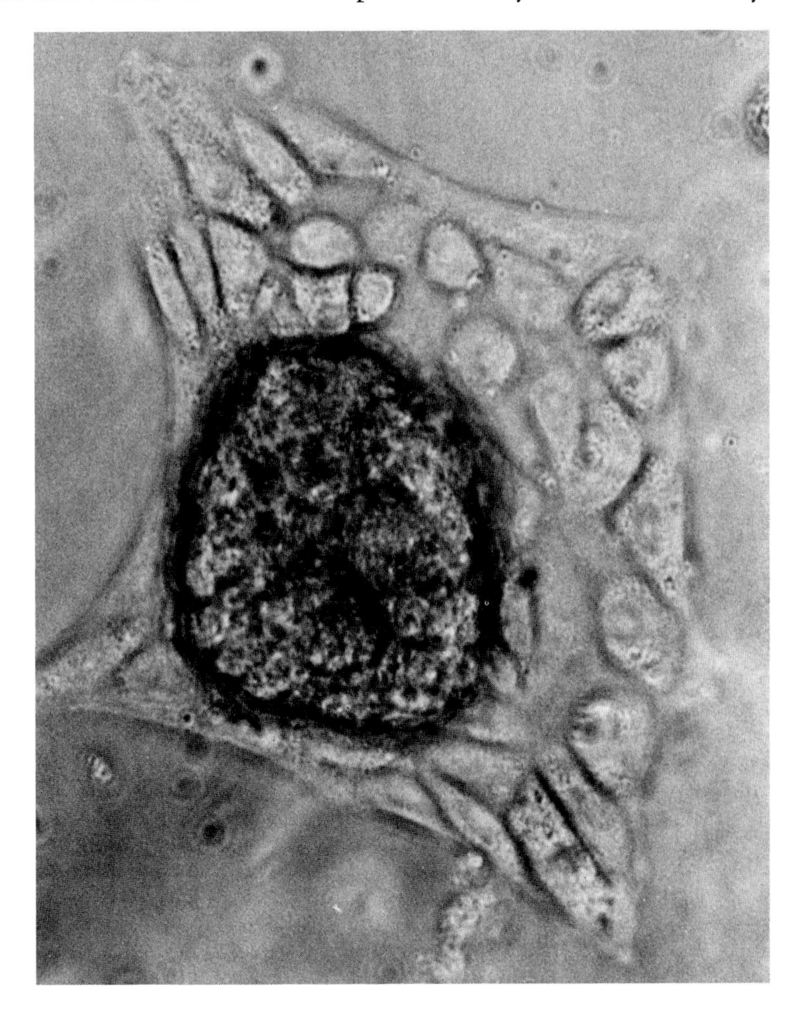

Fig. 1. Phase-contrast micrograph of a human glomerulus maintained in culture for 14 days (×100).

Culture Techniques

The glomerular suspensions are placed in Waymouth's Media 752/1 supplemented with nonessential amino acids, sodium pyruvate, penicillin, streptomycin, and 30% heat-inactivated newborn calf serum. The suspension is pipetted into sterile 30-mm plastic tissue culture flasks or glass bottles. Enough glomeruli should be added per flask so that individual glomeruli can easily be seen in the suspension. Enough media is added to just cover the flat surface to facilitate attachment of the glomeruli to the flask surface. The flasks are incubated in a 5% CO_2 gassed tissue culture incubator at 37°C. Care is taken not to disturb the culture plates so as to dislodge the glomeruli from the culture surfaces for the first week. Media (0.5 ml) is added daily to prevent drying of the cultures.

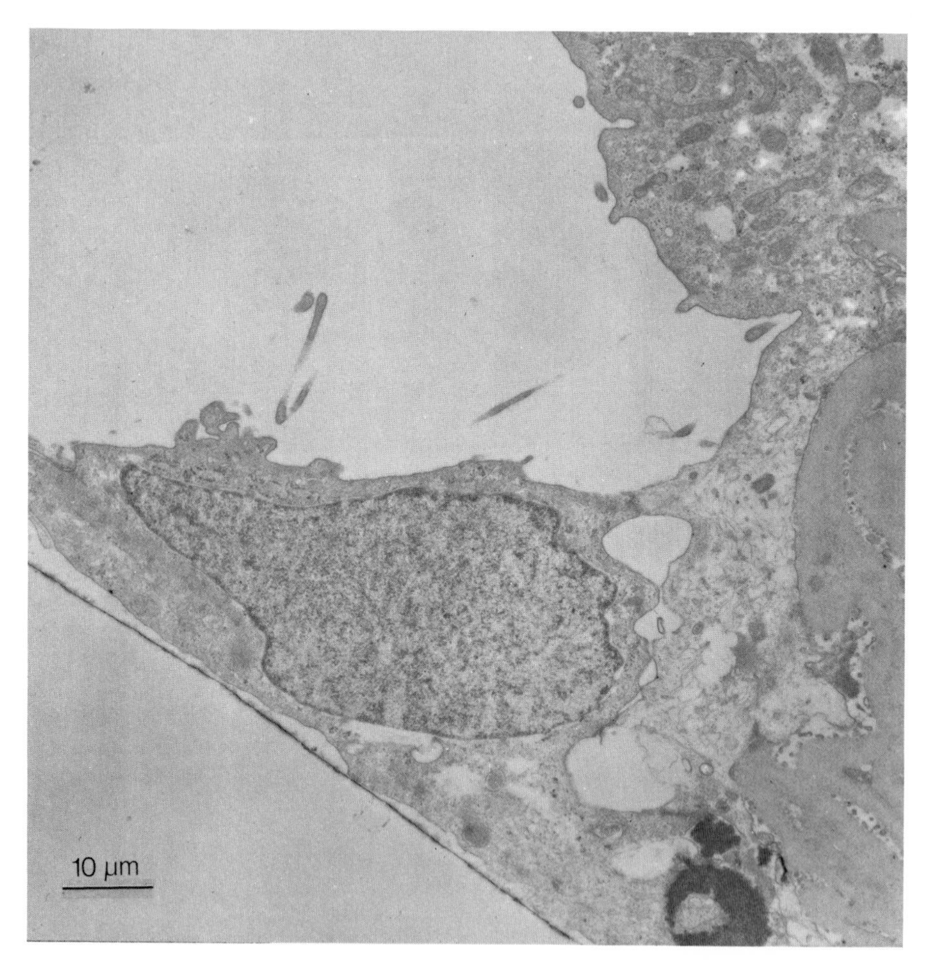

Fig. 2. Interdigitation and similarity between cells resident on the glomerulus and the monolayer (×1200).

The tissue culture media is changed only when the glomeruli have become firmly attached to the culture surfaces after which it may be routinely changed.

Attachment and cell proliferation is most rapid in human fetal glomeruli. They attach to tissue culture flasks in 3–5 days and growth of cells is apparent in 5–7 days. Adult human, monkey, and dog glomeruli attach in approximately 7 days showing definite cellular outgrowth in 10–14 days. Proliferating cells cover the culture flask by about 25 days, the exact time depending on the number of glomeruli explanted and the type of glomeruli used. Rat glomerular cells proliferate much more slowly, requiring 3–5 weeks for attachment and outgrowths to appear. Monolayers have not been obtained using rat glomeruli under the culture conditions as outlined above.

The ease of preparation of glomeruli with intact cells depends on the kidney tissue used. Normal tissue yields large quantities of satisfactory glomeruli. Tissue obtained from patients with chronic renal disease is much more difficult to process, since the cortical medullary junction is obscure and the tissue is fibrous and difficult to pass through the screen. Despite these problems, glomeruli with viable cells and cell outgrowths can be obtained from all but the most shrunken and fibrotic kidneys.

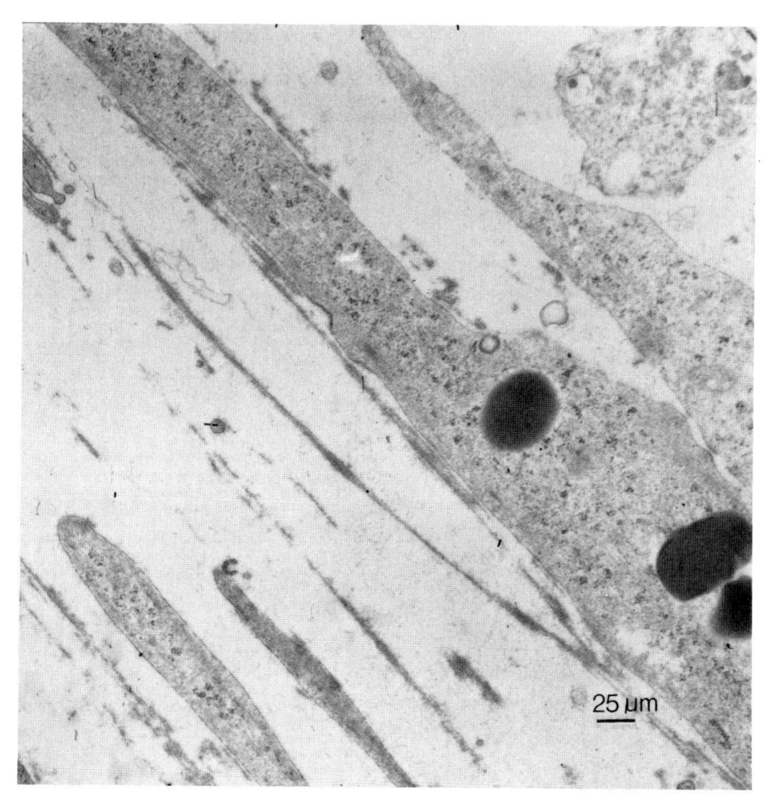

Fig. 3. Multilayered cellular outgrowth of cells which resemble smooth muscle cells (×5000).

Cell Growths

The cells on the glomerular basement membranes are large and irregularly shaped (Fig. 1). Near the glomerulus they often appear to overlap and are often several layers deep. The appearance of cells resident on the exterior of the glomerular tufts by electron microscopy is quite similar to that of normal glomerular epithelial cells. Complex interdigitations with adjacent cells are frequent and the cells maintain a close relationship to the glomerular basement membrane. Free ribosomes are prominent cytoplasmic features as are profiles of Golgi apparatus. The cells of the monolayer share many features with those on the external aspects of the glomeruli (Fig. 2). A second cell type is commonly seen in the outgrowths. The cells are elongate, contain a substantial amount of endoplasmic reticulum, form multiple layers, and often have a regular external lamina (Fig. 3). Peripheral dense bodies and microfilaments are present but not prominent. These features suggest that the cells are related to smooth muscle cells as may be derived from mesangial cells.

CHAPTER 12

Organotypic Mammalian Central and Peripheral Nerve Tissue[1]

Murray B. Bornstein

The maintenance of mammalian nerve tissues in the Maximow slide assembly offers the investigator a method which favors differentiation and maturation, exposes the cultured tissue to continuous observation and photographical recording of active and reactive events at relatively high-power microscopy, and permits facile manipulation of the tissues' environment. Our current method of maintaining central and peripheral nerve tissues is an adaptation of the one developed and originally used for dorsal root ganglia by Peterson and Murray[2]

[1] The research in the author's laboratory is supported by grants from the National Institute of Neurological Diseases and Stroke, The National Multiple Sclerosis Society, and The Sloan Foundation.

[2] E. R. Peterson and M. R. Murray, *Amer. J. Anat.* **96**, 319 (1955).

in the Laboratory of Cell Physiology at Columbia's College of Physicians and Surgeons.

Preparation and Sterilization of Glassware

Only Goldseal coverslips (Clay Adams, Inc., New York, Cat. No. A-1420) are used for the intimate glassware, i.e., 22-mm round coverslips which are No. 1 thickness and 40-mm square, No. 3 thickness. The Maximow depression slide is plate glass, 75 × 45 mm and 7–8 mm thick (Cat. No. 1477, Clay Adams, New York, Cat. No. 7048C, Arthur H. Thomas Co., Philadelphia, Pennsylvania). All other glassware is borosilicate (Pyrex, Corning code 7740; Kimax, Kimble brand KG-33). When new, coverslips are cleaned by soaking in reagent grade nitric acid at 69 to 71% concentration for 24 hours. The Chen staining rack (A. H. Thomas, Cat. No. 9202-B3) in a covered glass dish (A. H. Thomas, Cat. No. 9208) is convenient for this procedure. Round coverslips are used only once. Square coverslips are reused until they break. To recycle the latter, wax and vaseline are first removed by boiling repeatedly (twice is usually adequate) for 10 minutes in Ivory soap solution and rinsing in running hot tap water for 10 minutes. They are then soaked for 24 hours in nitric acid, rinsed in cold running deionized water for ½ hour, and soaked in glass-distilled water overnight. Deionized water is supplied by a regenerative type mixed bed demineralizer (Barnstead Still and Sterilizing Co., Boston, Massachusetts, Model MM-3) and a Model BROF-24 prefilter. Glass-distilled water is obtained by passing the deionized water through a Corning AG-10 still. Only Tygon (U. S. Stoneware Co., Akron, Ohio) tubing and stainless steel pipes are used to deliver deionized or distilled water.

All new glassware, including the Maximow slides, are first soaked in 10% reagent grade HCl overnight. When reused, Maximow slides are boiled in Ivory soap solution and rinsed with hot tap water to remove vaseline and wax before entering the usual wash procedure. Except for coverslips, which are cleaned in nitric acid as stated above, all glassware is cleaned by boiling in sodium metasilicate (Fisher Scientific Cat. No. S-408). A 20% stock solution of metasilicate is prepared in deionized water and 40 ml of this is added to each 2 liters of deionized water for the boiling solution. The glassware is then rinsed once in cold deionized water and soaked overnight in 1% HCl in deionized water. It is then repeatedly rinsed with hot deionized water in a Turbomatic (Better-Built Machinery Corp., Saddle Brook, New Jersey, Model 3000) washing machine. After an overnight soak in glass distilled water, it is dried by exposure to 250°C for 2 hours. Drying at this temperature also serves to burn any organic material which may have come into contact with and remained on the glass during the cleaning procedures. The material is then packaged for sterilizing. Small items are put into culture dish holders (A. H. Thomas, Cat. No. 4352). Larger pieces are wrapped in aluminum foil (1235-0-dry, fully annealed fat-free, dead soft finish, 0.001 inch thick, Industrial Distributors, Inc., 8234 S. Miles Parkway, Cleveland, Ohio 44128) and a heat-resistant paper ("Steriroll," A. J. Buck and

Sons, Inc., 10534 York Road, Cockeysville, Maryland 21030). For dry sterilization, the material is heated at 165°C for 2 hours. For wet sterilization of stoppers, solutions, etc., we use an American Sterilizer Co. autoclave (type QDS 1624 CS) which employs the steam supplied by the institution to heat the chamber and to convert our glass-distilled water into steam which is delivered to the interior of the chamber. The steam supplied by the hospital may carry some of the chemicals used to pretreat the water as a deterrent to scale formation in the boilers. The special autoclave prevents the potentially toxic chemicals from contacting our solutions, etc. Material is autoclaved at 120°C at 16 pounds pressure for times varying from 15 to 40 minutes, depending on the volume of solutions, etc. Both dry and wet methods are controlled by Thermotubes sensitive to 166° and 121°C, respectively (Paper Thermometer Co., 10 Stagg Drive, Natick, Massachusetts). Sterile supplies are stored in cabinets and drawers until used.

Preparation of Solutions and Media

Simms' Balanced Salt Solution (BSS). This solution is prepared, 2 liters at a time, by dissolving 2 bottles of TC Simms' X7 solution dried powder (Difco 5799-37) in 1975 ml of fresh glass-distilled water. It is stirred until dissolved while warmed to 37°C. Sodium bicarbonate solution, 25.0 ml of 7.5% (Grand Island Biological Co., Cat. No. 508) is added and the solution gassed with CO_2 for about 4 minutes, using a large-mouth 10 ml pipette, until the solution turns a distinct yellow. The solution is then passed under vacuum through a Millipore all glass filter (Millipore Corporation, Bedford, Massachusetts, Cat. No. XX15 047 00) fitted with a GS, 0.22 mean pore size filter, into a specially modified Bellco 1000-ml funnel which is designed to accept the Millipore filter at its top and to deliver the sterilized solution through a Teflon stopcock at the bottom. A sterility check is performed by adding 3–4 drops of filtered BSS to a tube of thioglycolate medium (Baltimore Biological Laboratories, Cat. No. 11716). After sterilization, 100-ml aliquots of the BSS are dispensed into 125-ml Erlenmeyer flasks, closed with a No. 5½ floating crepe rubber stopper (Standard Scientific Supply Co., New York, Cat. No. 867300) and refrigerated until use.

Serum. Unfiltered, sterile fetal calf serum is purchased from various suppliers. Human placental cord serum is prepared by collecting blood from the umbilical vein at the base of the clamped umbilical cord of a placenta brought from the delivery room in a sterile bowl. The blood is slowly drawn into a 50-ml sterile, disposable syringe through a large needle (No. 18). When no more blood can be withdrawn easily (one may obtain from 10 to 60 ml from a placenta) the needle is removed and the blood is gently expressed down the wall of a 125-ml Erlenmeyer flask. Rapidly withdrawing and squirting the blood into the flask produces hemolysis. Large amounts of hemoglobin appear to be toxic to nerve tissues. The stoppered flask is refrigerated overnight. The red blood cells are removed by pipetting the serum into 12-ml centrifuge tubes and spinning at

2000 rpm in a refrigerated centrifuge (Sorvall R. C. 2B; G2 head) for ½ hour. A few drops are removed for a sterility check. The serum is then withdrawn by pipette, placed in glass tubes (Corning No. 9820) stoppered (A. H. Thomas, Cat. No. 8826, plug diameter 13 mm, cap diameter 18 mm), and refrigerated until use.

Eagle's Minimum Essential Medium (MEM) with glutamine is purchased from commercial suppliers.

Nutrient Solutions. The nutrient solutions consist essentially of serum–saline mixtures with various additives. For the most part, they have been empirically determined and vary from tissue to tissue and, sometimes, from investigator to investigator. Those currently employed in our laboratory are as follows:

1. For spinal cord, dorsal root ganglion, brain stem, muscle, cord-ganglion and cord-muscle combinations: 33% human placental cord serum heat-inactivated at 56°C for ½ hour; 10% 9-day chick embryo extract (EE-50); 50% Eagle's MEM plus glutamine; 7% BSS; 600 mg% glucose; approximately 1.3 μg/ml achromycin.
2. For cerebellum: 40% fetal calf serum heat-inactivated at 56°C for ½ hour; 25% Eagle's MEM with glutamine; 35% BSS; 600 mg% glucose; 0.1 units/ml of low zinc glucagon-free insulin, supplied through the courtesy of The Squibb Institute for Medical Research, New Brunswick, New Jersey; approximately 1.3 μg/ml achromycin.
3. For cerebrum: The nutrient medium is the same as that for cerebellum except that human placental cord serum is used in place of fetal calf serum. Recently, HEPES [4-(2-hydroxyethyl)-1-piperazineethanesulfonic acid, Mann Research Laboratories, Cat. No. 6599] has been added to all nutrient solutions at a concentration of 10^{-2} *M* for its buffering ability.

Collagen-Coated Coverslips. The collagen solution is prepared from the tendons obtained from the tails of 400 to 500 g rats. The tail is wrapped in gauze and soaked with 80% ethyl alcohol for 15 minutes. By successive fracture from the tip to the root, the tendons are pulled out, cut loose, and collected in a small amount of sterile distilled water in a Petri dish. They are then transferred to a 150-ml Corex (Corning, Cat. No. 1265) bottle containing 1:1000 acetic acid solution. The volume of acetic acid solution is adjusted to 75 to 100 ml depending on the total mass of tendon. The bottle is stoppered and stored at 4°C for 24 hours. It is then centrifuged at 6000–7000 rpm at 4°C for an hour. If the tendons have not separated from the supernate, more acetic acid solution, approximately 25 ml, is added and the contents shaken. The mixture is spun again and the supernate removed. The supernate is stored at refrigerator temperature (4°C) until dialysis. A dialysis set-up is prepared by tying about 6 inches of ⅝-inch dialyzer tubing (Fisher Scientific, Cat. No. 8-667-D) onto a glass tube which is then held in the center hole of a stopper of a 1-liter flask containing 700 ml of fresh glass-distilled water. The top is wrapped in Steriroll paper and the unit autoclaved for 30 minutes. After it has cooled, about 7 ml of collagen–acetic acid solution is placed into the uncovered tube and allowed to run down into the bag of dialyzer tubing. The solution is dialyzed for 24 hours at 4°C and then removed by tearing the top of the bag and pipetting the thickened solution into tubes. If it has not reached an adequate viscosity, it is dialyzed again. The dialyzed collagen can again be stored at 4°C. To prepare collagen-coated coverslips, 1–2 drops of the dialyzed collagen solution are placed on a round coverslip and spread with a glass rod. It is then exposed to ammonia fumes for 2 minutes which gels the solution. The coverslips are repeatedly washed in sterile distilled water untii free of ammonia, as indicated by phenol red in the water. They are then stored 7 coverslips to a Columbia staining jar (A. H. Thomas, Cat. No. 9201) containing about 7 ml BSS with glucose added to a final concentration of 600 mg% and three drops of fetal calf serum, which promotes the wettability of the collagen gel surface. They are stored at room temperature until use.

Sterile Room

All dissections of tissue as well as maintenance operations are performed in sterile rooms in which personnel wear cap, mask, and gown. The room is supplied with temperature controlled air which has passed through an "absolute filter" calculated to be 96% effective in removing particulate matter, followed by a Precipitron electrostatic filter. The air is delivered into the rooms under positive pressure in relationship to the surrounding laboratory and at a rate sufficient to effect six exchanges an hour. The sterile room walls are coated with epoxy paint. Each morning the rooms are stocked with sterile supplies and all surfaces are wiped down with 80% ethyl alcohol. When not in use, the room is exposed to UV irradiation (Model No. ST 2830, Hanovia Lamp Division of Engelhard Industries, Inc., Newark, New Jersey).

Preparation of Cultures

All levels of the mammalian neuraxis from the cerebral neocortex to the neuromuscular junction are being cultured. There is an optimal time, empirically determined, for explantation at each level.

The following list indicates the ages of tissue as usually employed in this laboratory:

Mouse cerebral cortex: 18-day embryo to 5-day postnatal.
Mouse and rat cerebellum: newborn
Mouse and rat brain stem: 13- to 15-day embryo
Mouse and rat spinal cord: 13- to 15-day embryo
Mouse and rat dorsal root ganglion: 15- to 18-day embryo
Mouse and rat cord-ganglion combinations: 13- to 14-day embryo
Mouse and rat cord-muscle: 13- to 14-day embryo

Dissections are performed under sterile conditions. Newborn and older animals are etherized and soaked in 80% alcohol for 10 to 15 minutes. The desired tissue, e.g., cortex cerebellum, is then removed en bloc and transferred to a Petri dish containing Simms' BSS. To obtain embryos, the pregnant animal is anesthetized and soaked in 80% ethyl alcohol for 15 minutes. Sterile instruments are used to reflect the skin, open the peritoneum, and remove the uterus and its contained embryos. This is placed into a sterile Petri dish. All further dissections are performed under microscopic control. The usual fine forceps, scissors, etc., are used for the gross dissections. For the finer dissections, jeweler's forceps (Dumont No. 5 stainless steel), iridectomy scissors, and new No. 11 scalpel blades are used. Care is taken to avoid as much trauma to the fragments as is possible. In preparing fragments, it is important to remember that nutrients and oxygen are supplied to the culture by diffusion. Therefore, one dimension of the fragment should not exceed 0.5 mm. The prepared fragments are usually kept in a small volume, 1–2 ml, of their nutrient solution in a Stender dish (Greiner Scientific Corp., Cat. No. 29-580, 370D, height 25 mm) until the dissection has been completed. To set up the Maximow slide assembly, two 40-mm square coverslips are picked up with coverglass forceps (Clay-Adams Cat. No. A-1906, or stainless steel, if available) put down on a background piece of black filter paper (A. H. Thomas, Cat. No. 5226, 11 cm diameter) in a Petri dish (150 × 20 mm). The collagen-coated round coverslip is removed from its storage fluid by means of the same forceps, drained of excess fluid, and centered face-up on the square coverslip. With a Pasteur pipette (Bellco Glass, Inc., Vineland, New Jersey, Cat. No. 12-1401) a fragment or two of the prepared tissue is placed on the coverslip with a single drop of nutrient medium. The Maximow slide is touched with sterile (autoclaved) vaseline at the four corners surrounding the depression and pressed on to a square coverslip which adheres and covers the depression. The assembly is lifted out and sealed with a paraffin–vaseline mixture. The sealed slides are placed in the lying-drop position into a rack which holds eight slides. The cultures are maintained in the lying-drop position and incubated at 34° to 35°C. (The usual 37°C incubator temperature is not well tolerated by nerve tissue.)

Maintenance of Cultures

As mentioned above, the cultures are incubated at 34° to 35°C in the lying-drop position. Since they are sealed into the Maximow slide assembly, there is no need to gas the incubator with CO_2 or to humidify it. The cultures may be removed from the incubator daily for observation in the light microscope. These periods at room temperature may be extended to at least an hour with no apparent harmful effect on the development of the tissue. We have found one lens of particular value in these observations or for photography. Bausch and Lomb, Rochester, New York, markets a 40× fluorite oil microscope objective which they will, on request, modify to have a working distance up to 1.5 mm. With this modification, one can penetrate both coverslips, the collagen gel, and the depth of the tissue without having to change any of the routine procedures.

Twice a week, the cultures are fed a fresh drop of nutrient medium. This performance involves removing the wax seal, which can be done outside the sterile room. Under sterile precautions, the round coverslip is lifted off the old square coverslip by means of a flame-sterilized needle and a coverglass forceps (Clay-Adams, Cat. No. A-1906). (Since these are also flame-sterilized during the feeding, it would be better to have stainless steel rather than chrome plated forceps, but these are difficult to find in the United States. Occasionally, one can find suppliers in Europe.) The cultures are either drained of their drop of old nutrient or washed in BSS by placing them for a time, 5–15 minutes, in a Columbia staining jar (A. H. Thomas, Cat. No. 9201), placed on a clean, sterile square coverslip, fed a drop of fresh medium and, finally, incorporated and sealed into the Maxinow slide assembly.

The cultured fragments have been maintained in this manner for times ranging up to a year or more. At any time, the cultural environment can be manipulated by either adding or withdrawing substances from the nutrient medium. In addition, they can be selected, by direct observation, for study and examination by bioelectrical, biochemical, histological, or ultrastructural techniques. Further work concerning these applications can be found elsewhere.[3–13]

[3] M. B. Bornstein and M. R. Murray, *J. Biophys. Biochem. Cytol.* **4**, 499 (1958).

[4] M. B. Bornstein and S. H. Appel, *J. Neuropathol. Exp. Neurol.* **20**, 141 (1961).

[5] S. M. Crain, *Intern. Rev. Neurobiol.* **9**, 1 (1966).

[6] S. M. Crain, *In* "Bioelectric Recording Techniques: Cellular Processes and Brain Potentials" (R. F. Thompson and M. Patterson, eds.), Vol. 1A. Academic Press, New York, in press.

[7] L. A. Feldman, R. D. Sheppard, and M. B. Bornstein, *J. Virol.* **2**, 621 (1968).

[8] J. E. Leestma, M. B. Bornstein, R. D. Sheppard, and L. A. Feldman, *Lab. Invest.* **20**, 70 (1969).

[9] G. M. Lehrer, M. B. Bornstein, C. Weiss, and D. J. Silides, *Exp. Neurol.* **26**, 595 (1970).

[10] G. M. Lehrer, M. B. Bornstein, C. Weiss, M. Furman, and C. Lichtman, *Exp. Neurol.* **27**, 410 (1970).

[11] E. R. Peterson and S. M. Crain, *Z. Zellforsch.* **106**, 1 (1970).

[12] C. S. Raine and M. B. Bornstein, *J. Neuropathol. Exp. Neurol.* **29**, 177 (1970).

[13] C. S. Raine, L. A. Feldman, R. D. Sheppard, and M. B. Bornstein, *J. Virol.* **8**, 318 (1971).

CHAPTER 13

Marmoset Bone Marrow[1]

Raymond P. Porter and Nazareth Gengozian

Hematopoietic tissue from most animal species proliferates poorly *in vitro* (see review by Woodliff).[2] Some success has been achieved using stimulating factors[3] or gradients[4] or when leukemic cells are cultured.[5] Our findings show that except for the erythrocyte precursors, blood cells from the marmoset, genus *Saguinus,* readily proliferate and differentiate *in vitro.* Marmoset hematopoietic cells grow in liquid medium designed to propagate nondifferentiating human lymphoid cells.[6] Cultures have been initiated from the marrow, peripheral blood,[7] fetal liver, and placental tissue (unpublished data). As with other primary cultures, attempts to propagate these cells *in vitro* have not uniformly succeeded. The successful cultures exhibit qualitative and quantitative differences among themselves.

The marmosets, small New World primates (adult weight 300 to 500 g), are also interesting to the experimental hematologist because (1) they are natural mixed-blood cell chimeras,[8] and (2) the placenta serves as an extra-embryonic site for hematopoiesis.[9]

Cell Inoculum Preparation

To prepare the marrow for culture, the loosely associated tissue expelled from the long bones of a freshly killed animal is suspended in culture medium. This suspension is repeatedly drawn through progressively smaller (18 through 25 gauge) needles until cells are nearly monodispersed. Alternately, marrow is obtained from a lightly anesthetized (Sernylan 2 mg/kg body weight) marmoset

[1] Supported by United States Public Health Service Grant ROI AM09289-06, 07, 08 AIB, from the National Institutes of Health.

[2] H. J. Woodliff, "Blood and Bone Marrow Cell Culture." Lippincott, Philadelphia, Pennsylvania, 1964.

[3] D. Metcalf, *In* "Bone-Marrow Conservation, Culture and Transplantation," Proceedings of Panel, Moscow, 22–26 July, 1968, pp. 3–12. International Atomic Energy Agency, Vienna, 1969.

[4] E. E. Osgood and J. H. Brooke, *Blood* **10**, 1010 (1955).

[5] A. Todo, A. Strife, J. Fried, and B. D. Clarkson, *Cancer Res.* **31**, 1330 (1971).

[6] G. E. Moore and J. Minowada, *In Vitro* **4**, 100 (1969).

[7] R. P. Porter and N. Gengozian, *J. Cell. Physiol.* **79**, 27 (1972).

[8] N. Gengozian, J. S. Batson, and P. Eide, *Cytogenetics* **3**, 384 (1964).

[9] G. B. Wislocki, *Amer. J. Anat.* **64**, 445 (1939).

by introducing a 21-gauge needle into the medulla of a femur through the distal epiphysis and collecting 1–2 ml of marrow-rich blood in a glass syringe which contains 500 units of preservative-free heparin. Before they are suspended in 1–2 ml of medium, the cells are washed in Hanks' balanced salt solution (HBSS) that is free of Ca^{2+} and Mg^{2+}; this is to prevent further clotting. Since the red blood cells (RBC) are undesirable, they should be separated from bloody marrow aspirates by fixed-gradient centrifugation over a mixture of Isopaque and Methocel.[7] The nucleated cells are harvested from the medium-gradient interface and thoroughly washed in HBSS. The cell number is estimated by standard methods and the cells are dispensed at the appropriate concentration in fluid medium. White blood cells (WBC) from heparinized venous blood of some marmosets spontaneously transform and proliferate into marrow cultures *in vitro*. Nucleated blood cells, separated from the RBC on a fixed gradient, are washed, counted, and placed in culture medium. Precleaned placental or fetal tissue is disassociated by first finely mincing it with blades and then forcing the pieces suspended in 5–10 ml of medium through needles (18–22 gauge). The suspension is allowed to stand in a tube for 5 minutes to allow the larger pieces of tissue to settle out. Small pieces of the erythropoietic islands and monodispersed cells are retrieved from the supernatant, washed in HBSS, and placed in culture medium. All cell manipulations are carried out at room temperature under sterile conditions.

Initiation of Cultures

Cultures are initiated by placing $1–3 \times 10^6$ nucleated marrow or blood cells in each milliliter of Roswell Park Memorial Institute (RPMI) Medium-1640 that contains 20% fetal bovine serum (various lots of serum should be tried); 100 units penicillin/ml; 100 μg streptomycin/ml; and 2.5 μg fungizone/ml. Cells inoculated into stationary flasks, dishes, or tubes are incubated at 37°C in a humidified gas-flow incubator with 5% CO_2 and 95% air. Primary liver and placental tissue suspensions are initially cultured in small dishes; the nonadherent cells are then transferred to flasks when they become abundant.

General Growth Characteristics

Cultures derived from marrow diminish in cellularity during the initial week or so *in vitro*, but an increase in cell number usually begins by the second or third week. An occasional marrow culture does not show an initial lag in growth while others do not increase in proliferation until the fifth or sixth week. Cultures initiated from other than the peripheral blood cells are seeded with nonhematopoietic adherent cells. However, this does not pose a problem. Reduction in the unwanted cell population is achieved each time the nonadherent hematopoietic stem cells suspended in medium are transferred to a new culture vessel. During the first week the nonadherent cell population should be transferred two or three times. When cell metabolism drops the pH of the medium to below 7 the nonadherent cells are sedimented in a tube at 1000 g for 10 minutes, resuspended in fresh medium, and passed to a new vessel. The passage schedule varies depending on the rate of proliferation. About 50% of successful cultures go through a stage of extensive proliferation during the third and fourth week *in vitro*. During this period cultures require passing from one flask to two every 3–4 days.

Perpetual lines have not been produced from marmoset hemic cells, perhaps because of differentiation of precursor cells; the average lifetime of cells *in vitro* of over 150 attempted cultures was about 2 months. Monocyte–macrophage elements are the longest-lived cell type. These cells attach most avidly after being passed into fresh medium. Daily transfer of the nonadherent cells to a new vessel has helped prolong the *in vitro* life of two cultures up to 4 months; a steady state of proliferation appears to develop, nonadherent precursors replenishing differentiated adherent cells that are discarded daily. These monocyte–macrophage series cells also help keep the cultures clean. Polymorphonuclear granulocytes phagocytize bacteria and help to protect the cultures against contamination.

Notes on Cell Morphology and Use of Cultures

The floating hematopoietic cells can be viewed directly through the bottom of the culture vessel by means of an inverted microscope (Fig. 1A). To study morphology and staining characteristics of the mixed cell types, a very thick slurry of nonadherent cells, prepared by sedimentation and draining off of the supernatant, is smeared on a slide as is done for blood films. The adherent population, i.e., the attached polymorphs, macrophages, fibroblasts, and other cells, are viewed *in situ* after the nonadherent cells are removed. A coverslip introduced into the culture vessel at the time of passage can be retrieved and the adherent cell types fixed and stained on the coverslip. Cells cluster onto the cytoplasm of a large adherent type of cell (Fig. 1B). In some cultures these clusters are seen in relatively large numbers (up to 20/cm^2 surface) and are observed only when megakaryocytes are present. The cluster of cells is believed to represent the adherence of other hemic cells onto an attached and spread out megakaryocyte.

Metaphase spreads can be obtained for cytogenetic analyses by removing a few milliliters of culture medium containing nonadherent cells and processing them according to the method of Moorhead *et al.*[10] without adding a mitogen. Both diploid and polyploid metaphases can be scored male or female by a sex chromosome marker. The majority of proliferating nonadherent diploid cells have morphological and staining characteristics of myeloblasts or promyelocytes (Fig. 1C). These differentiate into polymorphonuclear elements (Fig. 1D). The polyploid megakaryocytes (Fig. 1E) progressively enlarge while they undergo the stages of maturation that approach but do not include thrombocytopoiesis

[10] P. S. Moorhead, P. C. Nowell, W. J. Mellman, D. M. Battips, and D. A. Hungerford, *Exp. Cell Res.* **20**, 613 (1960); cf. Section XIV, Chapter 15.B, this volume.

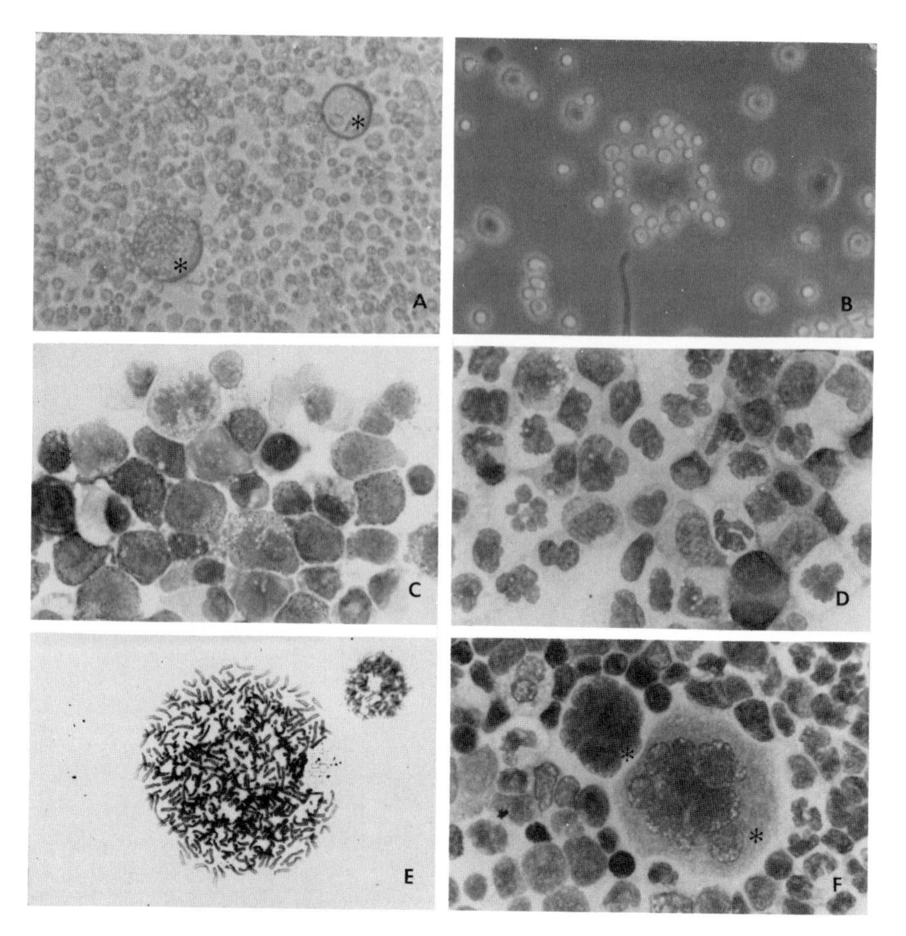

Fig. 1. Morphology of marmoset bone marrow cells cultured *in vitro*. (A) Microscopic view of a 24-day-old culture; nonadherent cells including megakaryocytes (*) grow suspended in the medium. (B) Round cells are seen clustered on a large adherent cell. Note nuclei of adherent cell in center of cluster (see text) (phase optics). (C) Granulocyte precursors are commonly seen during the most proliferative phase of growth. Wright-stained smear, 22-day culture. (D) Polymorphonuclear cells become abundant after proliferative phase diminishes. Wright-stained smear, 46-day culture. (E) Megakaryocyte metaphase (about $16n$), 17-day culture. (F) Megakaryocytes (*) identified by their size and staining characteristics proliferate and mature in culture for over 2 months. Wright-stained smear, 46-day culture.

(Fig. 1F). These large cells have been further identified as megakaryocytes by their specific incorporation of sulfur-35 (unpublished data).

Marmoset hemic cells form spherical colonies in semisolid medium (Bacto-agar, 0.4%, in our previously described liquid medium). Granulocytic and monocyte-macrophagic colonies or colonies of mixed cell types develop and persist for a period that corresponds to that observed for cultures in liquid medium (unpublished data). Megakaryocytes have not formed colonies in soft agar; however, if fluid medium is added to the dish they will proliferate in the fluid medium. Fibroblastoid cells grow under the agar on the dish surface. White blood cells ($10–15 \times 10^3/0.4$ ml liquid medium) in tissue culture tray wells (6 mm diameter) as well as hemic cells from other tissues, have produced cultures ideally suited to sequential microscopic observation. The small dimensions of the culture wells allow the tray to be transported without disturbing the spatial relationship of cells which proliferate as isolated nests on the flat bottom of the well.

CHAPTER 14

Monkey Kidney Cells

Robert N. Hull, George Weaver, and William R. Cherry

Monolayer cultures of trypsin-dispersed monkey kidney cells were used extensively during the 1950's for virus propagation and made possible the production of poliomyelitis and other virus vaccines. These cells from *Macaca,* and later from *Cercopithecus* monkeys had a broad sensitivity to many viruses of interest at that time, and still rival most cell types in respect to the breadth of their viral spectra. Primary monkey kidney cells (pMKC) are still used in some laboratories, but alternatives, now available for growth of most viruses, circumvent the hazards and other undesirable characteristics of pMKC. This chapter describes techniques for the preparation of pMKC. Other methods have been reported in the literature.[1–4]

Animal or Tissue Procurement

Kidney cultures may be prepared from any species of subhuman primates by the procedures outlined below, but the choice should be based on the requirements of the research and knowledge of the viral sensitivity and of the problems associated with the cells of various species. Kidney cells from rhesus, or African

[1] J. S. Younger, *Proc. Soc. Exp. Biol. Med.* **85**, 202 (1954).
[2] C. Rappaport, *Bull. WHO* **14**, 147 (1956).
[3] G. C. Taylor, H. R. Wetmore, R. J. Cotton, and J. F. Winn, *J. Lab. Clin. Med.* **65**, 518 (1965).
[4] D. Bodian, *Virology* **2**, 575 (1956).

green monkeys (AGM) are the most widely susceptible to viruses that infect man, although other Asiatic and African species, have been used. Cells from South American monkeys are more limited in their range of susceptibility to these viruses. In addition to problems associated with economics and husbandry, the chief disadvantage to the use of pMKC is that latent viruses may be encountered in their tissues. These agents, in some instances, present health hazards to laboratory personnel,[5] but more often they serve as a nuisance factor by interfering with, or complicating, observations and data collected from studies in which latently infected cultures were used. These viruses, known as simian viruses, have been reviewed by Hull,[6] and by Hsiung.[7]

SV40 virus occurs in 90%, or more, of the rhesus monkeys and is almost certain to be present in pMKC prepared from this species. Cynomolgus monkeys also have a high incidence of SV40 infection. The hemadsorbing virus, SV5, and the foamy viruses, while not as common as SV40, occur with some frequency in pMKC of both rhesus and AGM. Cytomegaloviruses may be encountered with some frequency, if cultures of either species are held for long periods of time (4–8 or more weeks). A human pathogen, B virus, occurs under natural conditions in Asiatic macaques. The AGM presents the least problem in respect to latent virus infection, and the viral sensitivity of kidney cells from this species equals, or exceeds, that of the rhesus monkey. Thus, the AGM is the animal of choice in most instances.

Most species of monkeys may be purchased from United States dealers, but South American species can only be obtained from these sources, due to yellow fever quarantine regulations. Asiatic and African monkeys, however, may be obtained from foreign dealers in the country of origin, and generally, specific arrangements can be made with these people for the handling and shipment of animals. On occasion, surplus animals are offered by research laboratories or primate centers. Sexually immature animals are most desirable as tissue donors, and contamination problems associated with kidney removal may be less with males than with females.

Monkeys purchased for tissue donors should be placed in a quarantined isolation area for at least 8 weeks prior to use. It is advisable to tuberculin test these animals at the time of arrival and again 2 weeks prior to use. Positive reactors should be destroyed. This quarantine period will tend to reduce, but will not eliminate, the hazards to personnel and the problem with latent viruses. Mixing of species should be avoided at all times to eliminate the possibility of cross-infections.

There are alternatives to purchasing and housing monkeys as tissue donors. At least one United States source offers monkey kidneys for sale, while others supply frozen suspensions of trypsin-dispersed monkey kidney cells. Numerous biological supply houses furnish pMKC cultures ready for immediate use, and arrangements can be made for a steady supply of such cultures.

[5] R. N. Hull, *Ann. N. Y. Acad. Sci.* **162**, 472 (1969).
[6] R. N. Hull, *Monogr. Virol.* **2**, 1 (1968).
[7] G. D. Hsiung, *Bacterial. Rev.* **32**, 185 (1968).

Kidney Removal

The kidneys are removed aseptically while the animal is still alive and under anaesthesia. The kidney should be drained as free of blood as possible for best cultivation results. The following procedure is recommended: Place a clamp (hemostat) on the renal artery, and either a clamp or ligature on the renal vein about an inch distal to the kidney. Sever the vein just proximal to the ligature and allow the blood to drain from the kidney. When the bleeding has stopped, remove the kidney and deposit it in a chilled, sterile container with sufficient medium, or balanced salt solution to provide a moist atmosphere. Do not submerge the kidney. The flask containing the kidney should be held in crushed ice until delivered to the culture preparation room. Ideally, this should be as soon as possible; however, good cultures have been obtained from kidneys chilled for several hours in crushed ice.

Preparation of Tissue for Trypsinization

The kidneys (from one animal) should be placed in a container, such as a wide-mouth 250-ml centrifuge bottle, and washed three times with 50 ml of chilled culture medium containing antibiotics. The medium for this, and other procedures, may be any one of several commercially available preparations (Medium 199, Eagle's MEM, Waymouth's 752/1, Ham's F12 or others), but Medium 199 will be used throughout this sample procedure. The serum supplement may be horse, fetal bovine, or other species that have been found acceptable. Antibiotics generally consist of 100 units of sodium penicillin G, and 100 μg of streptomycin sulfate per milliliter, although others may be used, either alone or in combinations with these.

The washed kidneys are placed in a sterile Petri dish and the connective tissue capsules are removed. Each kidney is bisected longitudinally with a sharp scalpel, and the renal pelvis, major and minor calyces, and medulla are trimmed away with curved iris scissors leaving essentially the dark red cortices. The cortex tissue is again washed three times with 50-ml volumes of chilled medium in a centrifuge bottle to remove as much residual blood as possible. The tissue is then placed in a 30×90 mm thick-walled tube and minced to a particle size of 2–3 mm^3 with scissors. Barber shears work well for this procedure in this size container. The minced tissue is washed into a 500-ml fluted trypsinizing flask fitted with a side arm. The mince is washed two additional times with 150 ml of chilled medium and the supernatant medium is decanted through the side arm.

Trypsinization

Difco 1:250 trypsin made up to 0.25% in Tris buffer, with the pH adjusted to 7.8 and sterile filtered, is used. The trypsin solution should be stored frozen, but must be thawed and prewarmed to 37°C before use. The minced tissue is washed for 3 minutes with 80 ml of trypsin with frequent agitations (shaking). After the tissue has settled, this trypsin is decanted and discarded. Two hundred milliliters of trypsin solution, plus a sterile magnetic stirring bar, are added to the mince, and the flask is placed on a Magnestirrer base of the nonheating type. The speed of the stirring bar should be adjusted to just below the level that produces foaming. The trypsinization is continued for 25 minutes at room temperature. At the end of this period, the remaining pieces of tissue should be allowed to settle and the supernatants withdrawn (decanted) and passed through a sterile gauze filter into a chilled 250-ml centrifuge bottle. The filter is made by placing three layers of gauze over a short-stemmed, 75-mm glass funnel. The stem of the funnel should be inserted through a rubber stopper (fitted with a cotton plugged breather) in the mouth of the collecting centrifuge bottle. The lid of a Petri dish may be placed over the gauze filter as an aid to asepsis. Additional 200 ml volumes of trypsin solution are added to the fluted flask and the process is repeated two or more times. Generally, after three trypsinizations, the cells of the cortical tissue will have been dispersed, and only strands of connective tissue will remain in the trypsinizing flask; if not, additional processing may be required.

Preparation of Cell Inoculum for Planting Cultures

The chilled supernatants collected from the trypsinization, and containing the kidney epithelial cells, are centrifuged at 1500 rpm for 10 minutes at 4°C. (This speed is based on the use of a PR2 International Centrifuge with the No. 259 head.) The cell pellets from all tubes are pooled and resuspended in 200 ml of Medium 199 containing 5% horse serum (or other species) plus antibiotics. The serum will neutralize residual trypsin. The cells are again sedimented by spinning at 1000 rpm for 10 minutes, and the supernatant is discarded. Repeat the process two additional times, then resuspend the cells in 200 ml of medium.

The cell suspension should be uniformly dispersed by repeated up and down pipetting. A 1.0-ml sample is withdrawn and diluted 1:10 in 0.15% trypan blue stain made up in a physiological saline. This stain is selectively absorbed by nonviable cells. A hemocytometer is charged with the diluted cell suspension and the viable kidney cells in the four large corner squares are counted. These counts are totaled and the sum is multiplied by 25,000 which gives the number of cells per milliliter in the original 200 ml of cell suspension. [Count in 0.4 $mm^3 \times 2500$ = count/$cm^3 \times 10$(dilution) = count/ml in original suspension.] Based upon the cell count, a volume of complete medium is added to the cell suspension that yields a final density of 10^5 cells/ml. The kidneys from a 2.5-kg monkey should provide about 5 liters of cell suspension. The pH of the planting medium should be 7.0 to 7.2, as this enhances the attachment of cells to glass. This can be obtained in Medium 199 by the addition of 1.12 g of $NaHCO_3$/liter. (The pH of Medium 199 without $NaHCO_3$ is about 3.5.) A 2% horse serum supplement is adequate for planting medium, but if fetal bovine serum is used, 5–10% is preferred. The cell suspension is inoculated into culture tubes, or flasks, in volumes suitable to the vessels used. An automatic syringe rigged to a flask in which the cell suspension is stirred by a magnetic stirrer facilitates the planting operation, especially when large numbers of tubes are to be planted. Conventional roller tubes, or Leighton tubes, are inoculated with 1.5 ml per tube. Sixteen-ounce prescription bottles with a floor area of 98 cm^2 require 30 ml of inoculum. Glass T-30 and T-60 flasks are inoculated with 5 and 10 ml, respectively, while plastic F-25 and F-75 flasks receive 4 or 15 ml. Appropriate inocula for other types of vessels may be calculated from these examples based on the available growth area. Approximately 20,000 cells/cm^2 of growth area provides a satisfactory inoculum.

Roller tubes should be incubated in a stationary position until the cell sheets have formed. Thereafter, rotation appears to be beneficial to both cell survival and virus propagation. Conventional roller tube apparatuses rotate the cultures at a rate of about 11 to 12 rph. These, and other cultures, should be incubated at 35° to 37°C. If Petri dishes, or other nonsealable vessels are used, it will be necessary to incubate these in a 5% CO_2 in air atmosphere to control pH. Cultures in any of these vessels should be confluent, and ready for virus studies within 6–7 days from time of planting. Medium changes prior to this time are generally not necessary, but the medium should be changed at time of use. If the cultures are not used within 6–8 days, or if the cell sheets have not developed fully in this time a refeed is necessary. The pH of this refeed medium should be in the 7.4–7.6 range and can be obtained in Medium 199 with 1.68 g/liter of $NaHCO_3$.

Multiple pairs of kidneys may be processed by these procedures, providing the volumes of medium and trypsin solution are adjusted proportionally. Multiple processing is not recommended, however, since it enhances the probability of latent viral, or other microbial contamination. If more cultures are required than can be obtained from the kidneys of a single monkey, it is best to process each pair of kidneys separately, and to plant the cultures in sections that are related to the donor animal. Thus, one infected animal will not contaminate the entire cell suspension.

Precautionary Measures

Monkeys, their tissues, their cells, and everything coming into contact with them should be handled as though known to be infectious. After the kidneys have been removed, the carcass should be placed in a plastic bag and taken immediately to an incinerator. All instruments, gloves, gowns, etc., used in the procedure should be autoclaved, and the area should be cleaned with a good disinfectant. Similarly, all solutions and equipment used in the preparation of the kidneys for trypsinization, or during the trypsinization, must be decontaminated before discarding. This includes all wash solutions, tissue debris, etc. Any fluids withdrawn from cultures of pMKC at any time should be collected and autoclaved, as should the cultures themselves before discarding, regardless of whether or not they have been inoculated with known viruses. Human fatalities have occurred from infections contracted from both rhesus and AGM, or from their tissues. Also, other less harmful viruses may be present, which, at the least, may be a source of cross-infection to other cultures maintained in the laboratory.

CHAPTER 15

Rabbit Kidney and Skin

James G. Gallagher

The preparation of primary cell cultures from various animal tissues is a relatively simple and inexpensive means of obtaining cells for research or diagnostic purposes. Such cells can be used for chromosome studies, enzyme analysis, virus production, and a host of other procedures. This procedure is designed to illustrate the processing of tissue for preparation of primary cultures of both epithelial and fibroblast cells.

Primary Culture of Kidney Epithelial Cells

Embryonic or neonatal kidney tissue can be used to obtain large quantities of epithelial cells. While these cells do not have a particularly long life *in vitro,* they are especially useful for virus isolation since the various types of epithelial cells present in the *primary* culture ensure a wide spectrum of sensitivity to viruses, whereas long-term *established* cell lines usually have a narrow spectrum of viruses to which they are susceptible. Human embryonic kidney, rabbit kidney, and monkey kidney are perhaps the most frequently used primary epithelial cells. Rabbit kidney will be used here but the same procedure has been used successfully with other species.

Excision of the Kidneys. Sacrifice a young rabbit (4–6 weeks of age or less) by injecting 5 ml of air into an ear vein. Secure the animal on a dissecting board, shave the abdomen, and disinfect the skin with merthiolate or tincture of iodine. Using a scalpel (scissors, if preferred) and forceps, a midline incision in the abdomen is made. Lateral incisions are made on both sides above and below the level of the kidneys. This produces two flaps in the abdominal wall allowing access to the kidneys. Care should be taken to avoid lacerating the intestinal or other organs while cutting through the abdominal wall. Gloves and mask should be worn to avoid contaminating the kidney tissue. Expose one kidney at a time by pushing the intestines to one side. Using sterile scissors and forceps remove the kidney by cutting through blood vessels, connective tissue, and perirenal fat at the point of attachment opposite the renal pelvis. The excised kidney is placed in a sterile 100-mm glass Petri dish containing 10 ml of Hanks' Balanced Salt Solution (HBSS). Fat, excess mesentery, and other debris is removed until the kidney looks "clean." Take care not to lacerate the tough white fibrous capsule that invests the kidney. Next, remove the kidney to a separate 100-mm

Petri dish containing HBSS and peel off the capsule. This is easily done if the capsule is slit along the entire length of the kidney. Start the incision at the renal pelvis and cut to each pole.

Preparation of Kidney Tissue for Cell Culture. Remove the kidneys to another Petri dish containing HBSS and quarter them. Wash the pieces gently in the HBSS to remove blood from the tissue. Each quarter is removed in turn to a dry Petri dish to be minced to pieces approximately 1–2 mm^3 in size. (Curved-blade scissors or two opposed scalpels can be used for mincing the tissue.) After each mince the fragments are suspended in 10 ml HBSS and poured into a 250-ml flask containing 150 ml of HBSS. When all of the kidney tissue has been minced and transferred to the 250-ml flask the fragments are washed repeatedly in HBSS until the wash fluid remains clear. This often requires approximately ten washes of about 100 ml each. After each wash fragments must be allowed to settle in the flask before wash fluid is decanted.

Washed kidney tissue fragments in 150 ml of CMF (calcium- and magnesium-free) trypsin[1,2] (0.25%) at 37°C are next placed in a sterile 250-ml trypsinizing flask (No. 2259, Bellco Glass Inc., Vineland, New Jersey) containing a 1-inch molded Teflon magnetic stirring bar. The fluted trypsinizing flask is placed on a magnetic stirrer. Stirring should be at as fast a rate as can be obtained without excessive splashing or foaming. After 20 to 30 minutes of trypsinization, the stirrer is shut off, the undigested fragments allowed to settle, and the supernatant from this trypsinization discarded. This eliminates a toxic factor released during initial processing of the tissue. Another 150-ml aliquot of trypsin is added and stirred for 30 minutes. The supernatant from this and all subsequent trypsinizations is poured through a gauze-covered funnel into a 1000-ml flask set in an ice bath. Trypsinization is repeated with fresh aliquots of trypsin every 30 minutes until all cells have been released and only clumps of connective tissue remain. The total cell suspension obtained by trypsin treatment is placed into 250-ml centrifuge bottles and centrifuged at 600 *g* for 10 minutes (higher *g* forces kill many cells). Supernatant trypsin is poured off and the sedimented cells are resuspended in 100 ml of growth medium (Eagle's Minimum Essential Medium with Hanks' salts supplemented with 10% fetal calf serum). One milliliter of the resuspended cell suspension is pipetted into 9.0 ml of HBSS (1:10 dilution) and a total cell count and a viable cell count are performed.[1] If care is taken during the processing of tissue, 85% of the cells will be found to be viable.

By addition of growth medium the cell suspension is adjusted to 500,000 viable cells/ml. Various cell culture vessels containing appropriate volumes of growth medium are seeded with cells from the adjusted suspension. Table I lists the cell inocula and media requirements for a number of frequently used cell culture vessels. Confluent monolayers of epithelial cells will be obtained after

[1] D. J. Merchant, R. H. Kahn, and W. H. Murphy, "Handbook of Cell and Organ Culture." Burgess, Minneapolis, Minnesota, 1964.

[2] The concomitant use of trypsin and disodium ethylenediaminetetraacetate (EDTA) as described by H. Montes De Oca in Section I will produce excellent cell yields.

TABLE I
Suggested Inocula and Media Volumes for Primary Rabbit Kidney Cell Cultures

	Cell inoculum		
	Volume (ml)	Number	Total volume of media required (including cell inoculum)(ml)
Test tube (16 × 125 mm)	1	500,000	1
T-flask 30 ml, 25 cm²	2	1,000,000	5
T-flask 250 ml, 75 cm²	8	4,000,000	20
8-ounce Prescription or milk dilution bottle	5	2,500,000	15
32-ounce Prescription bottle	10	5,000,000	40

4 to 7 days of incubation at 37°C and preferably under 5% CO_2 tension. If a CO_2 incubator is not available cultures will have to be tightly stoppered to maintain CO_2 tension and proper buffering of the pH of the medium. A change of medium may be required after 4 days of incubation if the medium becomes too acidic. This technique can be modified to provide for production of satisfactory primary cell cultures from other organs and tissues.

Culture of Primary Skin Fibroblasts

Culture of skin fibroblasts obtained by punch biopsy provides an excellent means of obtaining cells for special studies. While the technique is not applicable in experiments requiring large numbers of cells in a short time, skin biopsy is a quick, painless method of obtaining tissue from numerous individuals. The technique allows rapid accumulation of a battery of genotypes for testing cellular variability as a function of the genotype and/or phenotype of the donors. This procedure is also useful in experiments where specimens from several different skin areas of an animal are desired. The same animals may be used over a period of time if serial samples are needed since the punch wound heals rapidly. A local anesthetic such as Xylocaine given subcutaneously may be used during the biopsy procedure without depressing the growth of the cells from the explanted tissue.

In humans this technique has been used to great advantage in the study of cytogenetic and enzymologic disorders. Skin fibroblast cultures obtained from afflicted individuals can be maintained *in vitro* for in depth studies on the nature of genetic defects at the cellular level. (Cf. Sections II and XIII, this volume.)[3-6]

[3] D. Yi-Yung Hsia, *Clin. Genet.* **1**, 5 (1970).

[4] D. Yi-Yung Hsia, *Metabolism* **19**, 309 (1970).

[5] A. Milunsky, W. Littlefield, J. N. Kanfer, E. H. Kolodny, V. E. Shih, and L. Atkins, *N. Engl. J. Med.* **283**, 1370–1381, 1441–1447, 1498–1504 (1970).

[6] The assistance of Naomi Hayes and David Newcombe, M.D. in preparation of this section is gratefully acknowledged. A videotape illustrating all the cell culture techniques described is available from the author.

Preparation of the Animal. Secure the animal on a dissecting board, positioned so as to expose the lateral surface of the thigh. Shave the area cleanly first using clippers and then a safety razor. Wash the prepared area thoroughly with 95% ethanol. (Merthiolate and other stronger agents may inhibit cell growth.)

The Punch Biopsy Technique. Skin samples can easily be removed using sterile skin punches. Multiple biopsies may be taken from the cleansed area. To remove a sample, grasp the thigh with one hand and press the area to be removed firmly with the sharp end of the punch. Push in, twisting the punch with a swift wrist motion. (DO NOT GRIND.) The punch should penetrate about 2 mm on the first try and leave a circular indentation. Using sterile instruments, lift the circular plug of tissue with forceps and snip off.

Preparation of Biopsy Tissue for Culture. Place the tissue plugs in a glass Petri dish containing HBSS supplemented with penicillin (50 units/ml), streptomycin (100 μg/ml), and amphotericin B (5 μg/ml). Wash the tissue with HBSS three to five times to remove any hair which may adhere to it. Red blood cells associated with the tissue should be removed also. When the wash is clear, cut each tissue plug into two to three uniformly sized pieces using sterile scalpel and forceps. The instruments should be dipped into 70% alcohol and flamed as needed to preserve aseptic conditions. Transfer the biopsy pieces to 35-mm plastic Petri dishes containing 1 ml of Eagle's Basal Medium (diploid modification) supplemented with 10% fetal calf serum and antibiotics, placing no more than three pieces in each dish.

Maintenance of the Culture. The cultures should be maintained at 37°C in a 5% CO_2 in air atmosphere. The explants should be left undisturbed for an initial period of 2 to 3 days to allow firm anchoring to the dishes. After this initial attachment period, medium should be replenished every 2 days. Cells may be observed to migrate from the tissue pieces in 3 to 5 days. Initially both epithelioid and fibroblastlike cells can be seen. Epithelioid cells usually predominate in the culture for at least 1 week. Following the early epithelioid migration, fibroblasts become the main cell type in the culture. Epithelioid cells persisting until the time of the first subculture (usually in 2 to 5 weeks) do not survive transfer and are not seen thereafter.

CHAPTER 16

Rat Skeletal Muscle Cells[1]

David Yaffe

The use of muscle cell cultures for the study of various aspects of cell differentiation has been considerably extended during the last several years. The transition from proliferating mononucleated cells to nondividing multinucleated fibers is associated with very distinct changes in the synthetic activities. These changes in cell biosyntheses include the cessation of DNA synthesis, large changes in the activity or synthesis of many enzymes, and intense synthesis of the contractile proteins. These can readily be followed on the morphological as well as biochemical levels. The two most common sources of skeletal muscle cells are the chick embryo and the newborn rat. The advantages of using the chick embryo are the large amount of tissue obtainable and the ease of preparation of the cultures. Rat primary cultures are dependent on a breeding colony of rats. This increases the expense and limits the amount of available material. However, several features make rat muscle cells more favorable for certain studies, i.e., differentiation in these cultures commences in a rather reproducible pattern, and myoblasts are morphologically distinct from the nonmyogenic cells (Fig. 1). The process of fiber formation is more easily visible than with chick cells; there is a clear separation between an initial period of cell proliferation and very little fusion, and a subsequent phase of intense cell fusion. The present article summarizes procedures for the cultivation of rat skeletal muscle cells practiced in our laboratory during recent years. Various uncontrolled parameters and variations between laboratories, as well as differences in the goals of experiments, necessitate continuous exploration and modification of methods. Therefore, the procedures described here should be taken merely as suggestions upon which the investigator should build his own protocol.

Most of the basic procedures can be applied to chick cultures as well. However, differences in the requirements and properties between these two kinds of cells should be considered. Examples of current application of chick muscle cell cultures can be found elsewhere.[2-7]

[1] This work was supported by Grant No. DRG 10007 from the Damon Ruyon Memorial Fund for Cancer Research, a grant-in-aid from the Muscular Dystrophy Association of America, Inc., and a long-term fellowship from EMBO. The valuable comments and suggestions of Sara Neuman, Erika Krull, H. Dym, G. Lavie, and S. Kaufman during the preparation of this manuscript are gratefully acknowledged.

[2] R. Bischoff and H. Holtzer, *J. Cell Biol.* **44**, 134 (1970).

[3] J. R. Coleman and A. W. Coleman, *J. Cell. Physiol.* **72** (Suppl. 1), 19 (1968).

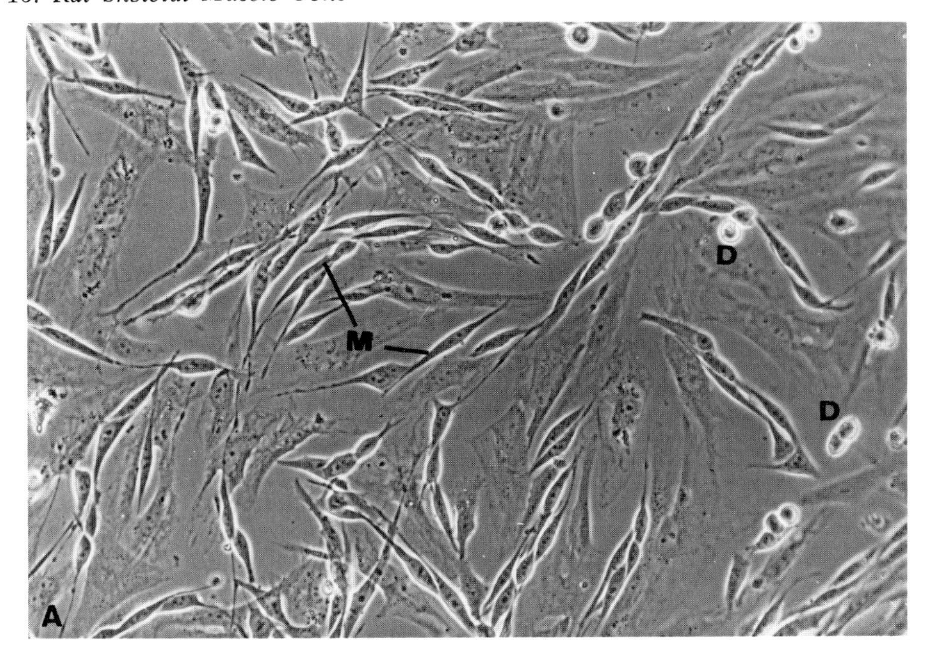

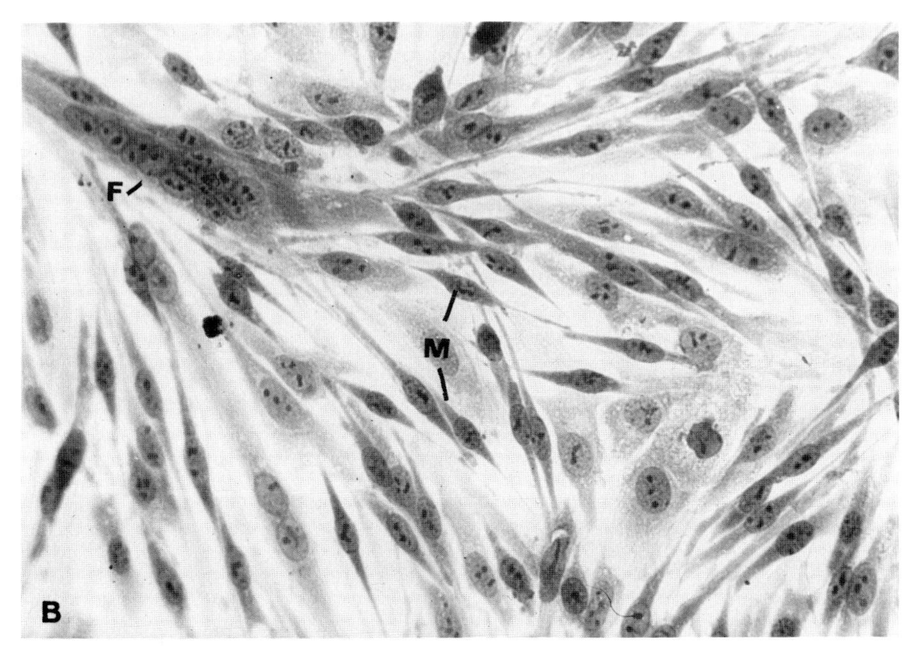

Fig. 1. Primary skeletal muscle cell culture at the onset of fusion. A. Phase contrast micrograph of living cells (magnification ×225). B. Giemsa stained (magnification ×350). M, myoblasts; F, nascent fiber; D, dividing cells.

Primary Cultures

Dissociation of Tissue. Cultures may be prepared from thigh muscles of rat embryos or neonates. One- or two-day-old rats are optimal. Preparation of cultures from embryos necessitates killing of the mother whereas, when newborns are used, the mothers can immediately be put back to breed. As the newborn becomes older, the proportion of mononucleated myoblasts diminishes rapidly due to progressive fusion into muscle fibers. Cultures prepared from 10-day and older rats are usually fibroblastic and exhibit very little fiber formation. We also observed that some strains (e.g., Wistar rats) yield better cultures than others (e.g., Lewis rats).

The animals are first washed in 70% alcohol, killed by cervical dislocation, and fixed to a dissection board. The hind legs are skinned, the thigh muscles are separated with the aid of forceps, and washed twice with Tyrode's solution or phosphate-buffered saline containing antibiotics (100 units/ml penicillin + 100 μg/ml streptomycin). They are then put into a trypsinization flask containing a magnet and 0.25% trypsin in Ca^{2+}- and Mg^{2+}-free Hanks' or Earle's solution. The dissociation of cells can be done at room temperature but 37°C is preferable. A similar trypsinization procedure for the preparation of embryonic fibroblasts is followed. At 20-minute intervals stirring is stopped, the undigested fragments are allowed to settle, the suspended cells are collected by decantation, and the action of trypsin is stopped by the addition of horse serum to a final concentration of 10%. The cells are centrifuged (1000 *g* for 3 to 5 minutes), resuspended in a small amount of the nutritional medium, counted, and diluted for plating. For quantitative experiments it is necessary that the cell suspension be filtered before counting to remove aggregates and tissue fragments. This is done by passing the cell suspension through a double layer of sterile lens cleaning tissue (Brinkmann, New York) fixed in a Swinex Millipore filter holder mounted on a 20-ml syringe.

Standard Growth Conditions and Differentiation. The following are culture conditions which usually promote reproducible growth and differentiation. Modifications should be made according to the requirements of the specific experiments. Cells can be grown either in Petri plates in a 10% CO_2 and 100% humidity atmosphere or in stoppered bottles. Disposable tissue culture plates offer great convenience in handling and microscopic observations of the cultures. Coating of the plates with collagen[8] or gelatin enhances differentiation considerably. For most practical purposes these two methods give comparable results.

---

[4] I. R. Konigsberg, *Develop. Biol.* **26**, 133 (1971).

[5] M. O'Neill and F. E. Stockdale, *J. Cell Biol.* **52**, 52 (1972).

[6] M. O'Neill and R. C. Strohman, *J. Cell. Physiol.* 73, 61 (1969).

[7] S. D. Hauschka and N. K. White, *In* "Research in Muscle Development and the Muscle Spindle" (Banker *et al.*, eds.), p. 53. Excerpta Medica, Amsterdam, 1972.

[8] S. D. Hauschka and I. R. Konigsberg, *Proc. Nat. Acad. Sci. U. S.* **55**, 119 (1966).

Since the preparation of gelatin-coated plates is much simpler, they were used routinely.[9] Gelatin (Baltimore Biological Lab.) 0.01% dissolved in hot glass-distilled water, is autoclaved. Three milliliters of the solution is put into each 60-mm diameter plate and the plates are left in the cold for about 2 hours. The gelatin solution is then aspirated and the medium containing the cells is introduced.

As standard growth medium we have used a mixture of M199[10] and Dulbecco's Modified Eagle Medium[11] in a proportion of 1:4, supplemented with 10% horse serum and 1% embryo extract. Embryo extract is prepared from 10-day-old chick embryos. The embryos are washed twice with Hanks' salt solution and homogenized for 60 seconds with a 2-liter Braun Blender at maximum speed in an equal volume of Hanks' salt solution. The homogenate is left at 4°C for 1 hour, centrifuged in the cold for 20 minutes at 35,000 *g*, and stored frozen overnight at −20°C. After thawing, the extract is centrifuged again, as above, and the supernatant is subdivided and frozen in small aliquots until used.

The recommended cell density at plating for most purposes is 3×10^6 per 60-mm plate (1.5×10^5 cells/cm^2); however, cell density should be adjusted according to the plan of the experiment. For example, when autoradiography is planned and a flat, not too dense network of fibers is required, plating cells at a lower density is recommended (2×10^6/plate) whereas, for cultures to be used in biochemical assays, plating at 3.5×10^6 cells per plate may be the choice.

Several hours after plating most of the cells are attached to the plate. The plating efficiency is about 50%. After settling, the cultures consist of mononucleated cells. The spindle-shaped, refractile myoblasts can be distinguished from the nonmyogenic cells (Fig. 1). During the first 2 days in culture the cells multiply, but no significant cell fusion takes place. At about 50 to 52 hours after plating the cultures enter a period of rapid cell fusion which results in the formation of a network of rapidly growing multinucleated fibers. After a few days rapid cell fusion ceases and between 40–60% of the nuclei are found within fibers. Spontaneous contraction of the fibers can be observed about 1 day after the onset of fusion, however, cross-striation becomes prominent usually 1 or 2 days later. Contraction and cross-striation increase during the following days. Often the contractions cause the detachment of the cell layer from the plates. Sometimes the contractions of a detaching cell layer can be observed with the naked eye. The changes in the synthetic activities associated with the fusion can be followed by autoradiography,[12,13] cytochemically,[14–16] and with biochemical assays of extracts of cultures.[3,6,16]

[9] C. Richler and D. Yaffe, *Develop. Biol.* **23**, 1 (1970).
[10] J. F. Morgan, H. J. Morton, and R. C. Parker, *Proc. Soc. Exp. Biol. Med.* **73**, 1 (1950).
[11] J. D. Smith, G. Freeman, M. Vogt, and R. Dulbecco, *Virology* **12**, 185 (1960).
[12] D. Yaffe and M. Feldman, *Develop. Biol.* **11**, 300 (1965).
[13] D. Yaffe and S. Fuchs, *Develop. Biol.* **15**, 33 (1967).
[14] G. De La Haba, G. W. Cooper, and V. Elting, *J. Cell. Physiol.* **72**, 21 (1968).
[15] E. W. Emmart, D. R. Komintz, and J. Miguel, *J. Histochem. Cytochem.* **11**, 207 (1963).
[16] A. Shainberg, G. Yagil, and D. Yaffe, *Develop. Biol.* **25**, 1 (1971).

Selective Transfer of Myoblasts

The attachment of myoblasts to the tissue culture plate is slower than that of the accompanying nonmyogenic cells. This can be used to obtain cultures very much enriched with myogenic cells.[17] Young primary cultures, prior to the onset of fusion (e.g., 30–40 hours after establishment of the primary culture), are washed and gently trypsinized with 0.025% trypsin. As soon as the cells come off, the activity of trypsin is stopped by the addition of horse serum. The cells are collected, washed, and suspended in complete nutritional medium and immediately plated onto "intermediate plates" for 40 minutes. During this time most of the nonmyogenic cells become attached to the plate while most of the myoblasts are still floating or are settled but unattached. The medium containing the unattached cells is then collected by aspiration; the cells are counted and plated at the desired concentration. This procedure may produce cultures consisting of more than 90% myoblasts. The same procedure can also be applied to cells obtained directly from trypsinized muscle tissue, however, a better separation is obtained with cells obtained by trypsinization of cultures. This also works to enrich cultures of chick skeletal myoblasts,[5] beating heart cells,[18] and liver cells.[19]

Controlled Fusion

As mentioned earlier, during the first period of development of primary cultures, cells proliferate, but very little cell fusion occurs. When the cultures are grown in the standard condition, i.e., in medium supplemented with 10% horse serum and 1% embryo extract, the phase of rapid cell fusion starts between 50 and 52 hours. The duration of the initial period prior to the onset of fusion is rather constant and is affected very little by a variety of manipulations such as changing the cell density or resuspending and replating the cells.[20] This apparently reflects the requirement for some intrinsic changes to take place in the cells before they are able to fuse. Although it was not possible to shorten this prefusion period, the timing of initiation of fusion could be delayed and controlled by two apparently different methods. The first one involves changing the source of serum and the concentration of embryo extract; in the other method the concentration of Ca^{2+} in the medium is lowered. While in the first method cell fusion is prevented apparently by interfering with the developmental process preceding fusion, the second method seems to act directly on the process of fusion while it takes place.

FE Medium. When horse serum in the nutritional medium is replaced with fetal calf serum (FCS) the onset of fusion is delayed and becomes dependent

[17] D. Yaffe, *Proc. Nat. Acad. Sci. U. S.* **61**, 477 (1968).
[18] I. S. Polinger, *Exp. Cell Res.* **63**, 78 (1970).
[19] G. M. Williams, E. K. Weisburger, and J. H. Weisburger, *Exp. Cell Res.* **69**, 106 (1971).
[20] D. Yaffe, *Exp. Cell Res.* **66**, 33 (1971).

to a greater extent on cell density. A somewhat similar effect is obtained by increasing the concentration of horse serum to 20%. Increasing the concentration of embryo extract to 8%, or more, has a much greater effect on delaying the time of initiation of fusion. Cells grown in nutritional medium supplemented with 20% FCS and 10% embryo extract (FE medium) proliferate but do not fuse until they become very crowded. Fusion can be delayed for several more days if the overcrowding is prevented by serial subculturing.[20]

In cultures in which fusion was prevented by FE medium, a change to the standard medium (S) induces a period of very intense cell fusion which starts about 18 hours after the change of medium. This enables one to determine at will, within an accuracy of 1 to 2 hours, the time of initiation of a phase of intensive cell fusion (Fig. 2).

Myoblasts grown in FE medium continue to proliferate for a longer period. There is also at least one more cell division following the change to FE medium.[20,21] Thus, by the time fusion starts the cell density is higher than in cultures grown in standard medium and, therefore, the cell density at plating should be reduced accordingly ($1–2 \times 10^6$ cells/plate).

We found it convenient to plate the cultures first in S medium (which is also more economical) and change to FE medium after about 20 hours. Changing back to S medium is done the next day, 18 hours prior to the desired time

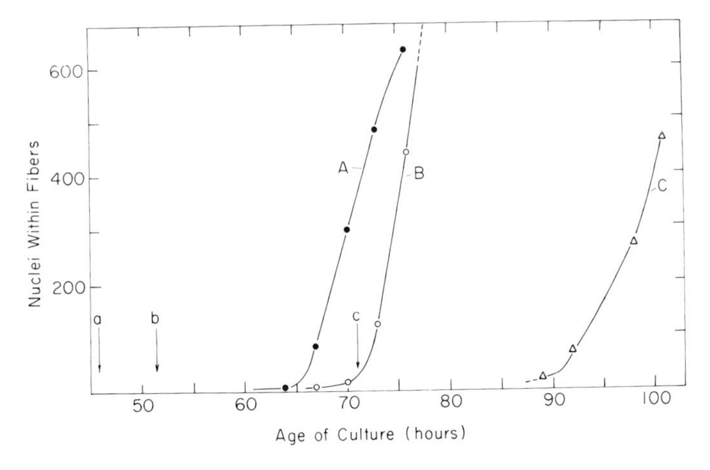

Fig. 2. The onset of cell fusion following a change from FE to S medium. Primary cultures were grown in FE medium and were changed to S medium at: A, 46 hours, B, 52 hours and, C, 71 hours after plating. Subsequently, aliquots of the cultures were fixed at the indicated time intervals, stained, and the number of nuclei within the fibers in randomly selected fields were counted. Each point represents the number of nuclei in five fields (average of two plates). Note the temporal relation between the change to S medium and cell fusion. In all groups fusion started between 17 and 18 hours following change to the permissive medium. a, b, c, Time of change of medium in A, B, C, respectively. [From D. Yaffe, A. Shainberg, and H. Dym, *In* "Research in Muscle Development and the Muscle Spindle" (Banker *et al.*, eds.), pp. 110–121. Excerpta Medica, Amsterdam, 1972.]

[21] G. Lavie and D. Yaffe, to be published.

for the initiation of fusion. It is recommended that the medium of cultures grown in FE medium should be changed daily.

Ca^{2+}-Controlled Fusion. Cell fusion requires a higher level of Ca^{2+} than that needed to maintain cell replication.[22] Cells grown in Ca^{2+}-deficient medium unlike those in FE medium will proceed through the prefusion developmental process but will be unable to fuse.[20] Addition of Ca^{2+} to such cultures results in the initiation of cell fusion within a very short time (fiber formation is detectable within 1 to 2 hours). Conversely, lowering the Ca^{2+} concentration of cultures after they start to form fibers will immediately inhibit further cell fusion. Thus, while the change from FE medium to the standard medium enables control and analyses of the events preceding cell fusion, changing the Ca^{2+} concentration enables a fine control of the duration of cell fusion.

Cultures may be plated in standard medium. Twenty-four hours after plating, or anytime prior to the onset of fusion, they are washed with Ca^{2+}-free phosphate-buffered saline and fed with medium containing 35 μM Ca^{2+} supplemented with serum that was dialyzed twice against 50 volumes of Ca^{2+}-free phosphate-buffered saline. At the desired time, a concentrated $CaCl_2$ solution is added to the plates to give a final concentration of 1400 μM. Conversely, to stop fusion at a definite time, the standard medium is replaced with the Ca^{2+}-deficient medium. It should be noted that when myoblasts are very crowded, some cell fusion may occur even in Ca^{2+}-deficient medium.

Cloning. When myoblasts are plated in very low densities, single isolated cells multiply and produce colonies of muscle cells.[8,9] By cloning populations of cells enriched by the selective transfer technique described above, it is possible to obtain cultures in which almost all colonies form muscle fibers.

Cells obtained from primary cultures, or after incubation in an intermediate plate, are counted and diluted to give a final concentration of 25 to 100 cells per plate. These are plated on gelatin-coated plates, containing about 5×10^4 to 1×10^5 feeder-layer cells. The latter are obtained from 5- to 6-day-old primary or secondary cultures exposed to 6000 R of X-irradiation. It is possible to obtain colonies in the absence of feeder-layer cells, but such cultures are more affected by minor uncontrolled variations and the reproducibility of cloning is thus reduced.

Under favorable conditions, the cells multiply with a generation time of 11 to 13 hours and start to fuse after about 3 to 6 days. Clones grown in medium containing horse serum start to fuse earlier than those grown in fetal calf serum although the generation time is similar. FE medium will further delay the onset of fusion.

Single clones can be isolated for further passages. During microscopic observations clones are marked with a wax pencil, the medium is then aspirated, and a glass ring is sealed around the clone with silicon grease. A drop of 0.025% trypsin solution is introduced into the well. After several minutes the cell suspen-

[22] A. Shainberg, G. Yagil, and D. Yaffe, *Exp. Cell Res.* **58**, 163 (1969).

sion is aspirated and transferred onto a new plate. Clones of cells obtained from primary cultures usually degenerate after one to two passages. Cells obtained from established myogenic lines can easily be serially passaged and recloned many times (see below).

Myogenic Cell Lines

Cell lines offer several advantages over primary cultures, e.g., independence of the availability of animals, a considerable saving of labor, and the possibility of working with standard, uniform cells. However, serial passage of cells obtained from various tissues very often results in cessation of growth and degeneration. The establishment of most cell lines is associated with changes in the properties of the cells, and most available lines do not maintain specific differentiation properties. Serial *in vitro* passage of myoblasts of rat or mouse origin most often resulted in the cessation of proliferation after several passages; however, sometimes it resulted in the establishment of a cell line. Although these cells differ from myoblasts of primary cultures in their ability to proliferate for extended periods in culture, they can retain the capacity to differentiate.[17] Thus, while the line is maintained by the continuous proliferation of mononucleated cells, these cells at the same time inherit the capacity to enter a pathway of terminal differentiation which leads to the formation of multinucleated muscle fibers. Unfortunately, there is no reproducible method to establish such cell lines. The lines available at present were obtained by serial selective passages of myoblasts, plated at low densities (10^4–10^5 cells/plate) with or without feeder-layers. Various media and sera were used with equivalent results. In several experiments a pellet of a carcinogen was included in the first several passages, but cultures not exposed to the carcinogen yielded cell lines as frequently.[9]

Maintenance and Differentiation. The lines are maintained by serial passage. Cultures approaching confluence are suspended with trypsin (0.05%) and replated at low density (10^4–10^5 cells/plate) on gelatin-coated plates. The addition of a small amount of irradiated feeder-layer cells (5×10^4/plate) improves growth but is not essential. The medium used for primary cultures, or Waymouth's Medium,[23] supplemented with 10% horse serum, 0.5–1.0% embryo extract, 100 μg/ml streptomycin, 100 units/ml penicillin, and 10 μg/ml kanamycin can be used. Fusion usually starts when the cultures approach confluence. The rate and extent of fusion vary considerably among the different lines. They are also influenced by many variables such as the type of media and source of serum. It is therefore rather essential to test each lot of serum for its effect on the differentiation of the lines. Under good culture conditions, most of the myoblasts participate in fiber formation, but a certain proportion always remain unfused.[9,24] Spontaneous contractions start several days after the onset of fusion but are less common than in primary cultures and there is a difference in this respect among

[23] C. Waymouth and R. B. Jackson, *J. Nat. Cancer Inst.* **22**, 1003 (1959).

[24] D. Yaffe, *Curr. Top. Develop. Biol.* **4**, 37 (1969).

the lines. As with primary cultures, fiber formation is associated with distinct changes in metabolic activities, i.e., cessation of DNA synthesis, and large increases in the activity of enzymes such as creatine kinase, myokinase, and phosphorylase.[16,17,24]

Once in several passages it is recommended that the lines be recloned to avoid possible accumulation of undesirable variants in the population. Several cultures are plated at a density of 25–50 cells per plate, and when fusion starts, a few well differentiated clones are individually isolated and transferred to two to three plates. The differentiation of these cells is followed, and a clone which exhibits good differentiation is selected for further transfer.

Storage of Cells. The cells can be preserved frozen. The myoblasts are suspended at a density of about 10^6/ml in medium containing 20% horse serum + 12% dimethyl sulfoxide and are sealed in ampoules (e.g., 1–2 ml cell suspension in 5-ml ampoules). The ampoules are placed at 4°C for several hours. Slow cooling during freezing is achieved by placing the ampoules in a styrofoam container precooled to 4°C (thickness of insulating walls should be 1 cm or more). The container is then tightly closed and put in a −80°C freezer. After 1 to 2 days the ampoules can be transferred to a smaller container for storage in liquid air or nitrogen. Defrosting is done by melting an ampoule in water prewarmed to 37°C; the cell suspension is immediately aspirated with a pipette, diluted in fresh medium, and plated on two to three plates. After the cells become attached, the plates are washed once and fresh medium is added.

CHAPTER 17

Rat Pituitary Explants

R. H. Kahn

The histotypical maintenance of endocrine tissue *in vitro* provides a useful model for elucidating mechanisms of hormone biosynthesis and for analyzing the physiological interactions that regulate their manufacture, storage, and release.[1] As an example of this organ culture procedure, the technique for removing pituitary tissue aseptically and the maintenance of the intact pars distalis is included. In recent years, this culture method has been useful for demon-

[1] E. M. Rivera and R. H. Kahn, *In Vitro* **5**, 28 (1970).

strating a variety of physiological actions and has been adapted to several target organ systems.[2-8]

Removal of Pituitary

The aseptic removal of the rat or mouse pituitary is achieved by a series of successive steps, each requiring flaming of instruments or a new set of sterile instruments. The animal is killed by cervical dislocation or decapitation and is thoroughly wetted down with 70% alcohol to prevent hair and bacterial contamination of the area to be exposed. The skin over the skull is removed by making a small incision at the base of the neck just behind the occipital protuberance. Two incisions of the skin on either side extending anteriorly along the lateral crest of the skull (over the temporal bone) frees the overlying tissues from the skull and creates a flap which can be reflected anteriorly (Fig. 1). Using sterile bone scissors, the roof of the skull is removed by making two incisions

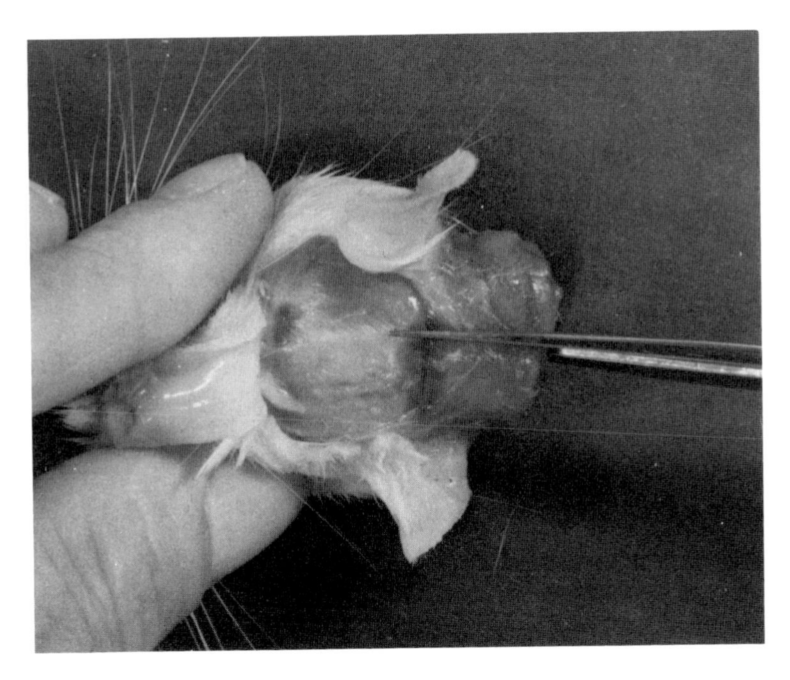

Fig. 1

[2] D. J. Merchant, R. H. Kahn, and W. H. Murphy, Jr., "Handbook Of Cell And Organ Culture," 2nd ed. Burgess, Minneapolis, Minnesota, 1964.

[3] L. R. Murrell, *Exp. Cell Res.* **41**, 350 (1966).

[4] T. Fainstat, *Fert. Steril.* **19**, 317 (1968).

[5] D. C. Klein and J. Rowe, *Mol. Pharmacol.* **6**, 164 (1970).

[6] J. D. Feinblat and L. G. Raisz, *Endocrinology* **88**, 797 (1971).

[7] D. L. Odor and R. J. Blandau, *Amer. J. Anat.* **131**, 387 (1971).

[8] E. M. Rivera, *In* "Methods In Mammalian Embryology" (J. C. Daniel, Jr., ed.), p. 442. Freeman & Co., San Francisco, California, 1971.

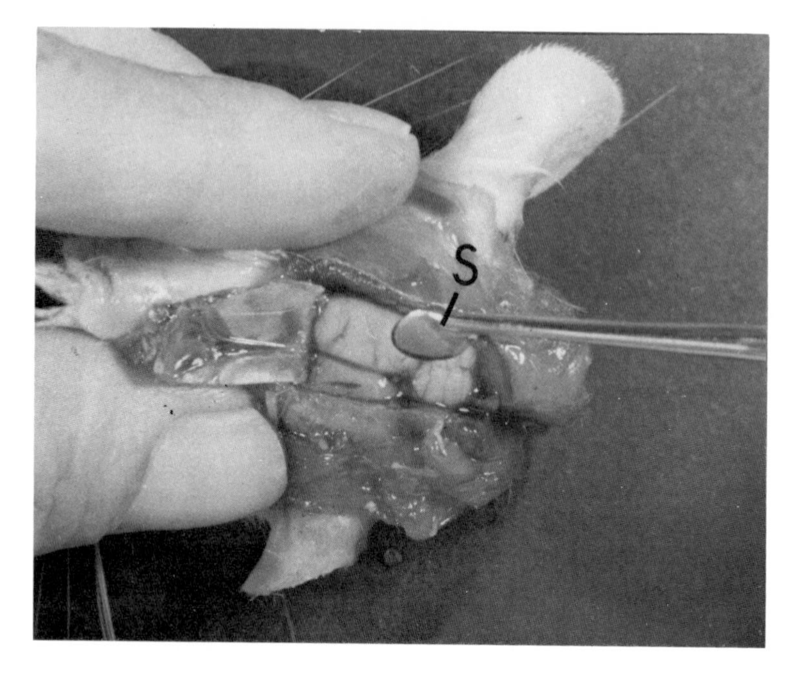

Fig. 2

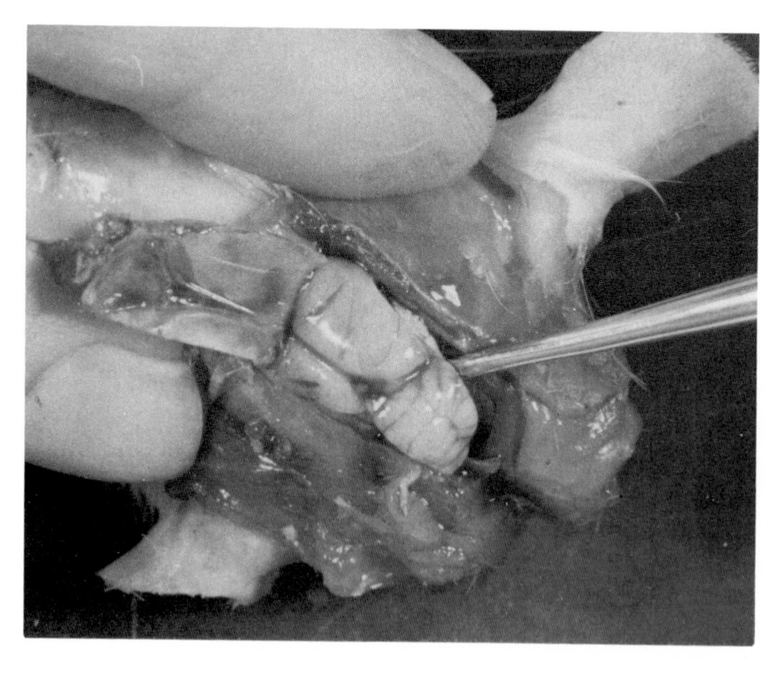

Fig. 3

along the parietal-temporal suture, starting at the foramen magnum (Fig. 1) and extending forward just anterior to the coronal suture. With sterile forceps, the roof of the skull is cracked forward and removed (Fig. 2). The brain itself is lifted from the cavity by slipping a spatulalike instrument (S, Fig. 2) underneath the cerebellum and cerebrum (Fig. 3) and gently lifting and removing the entire brain, exposing the pituitary (arrow) lying in the hypophyseal fossa (Fig. 4). In removing the brain, the hypophyseal stalk will have been broken just above the pituitary and normally remains attached to the brain.

Since the pituitary is covered by a thin connective tissue membrane, it will be necessary to take a sterile sharp instrument (e.g., needle) and scribe around the periphery of the gland at its junction with the skull wall to remove this layer. Once exposed, the pituitary (arrow) should be lifted from the fossa by a small spatulalike instrument (Fig. 5). Do not attempt removal with a forceps which would unquestionably squash the tissue. The pituitary should be transferred to sterile balanced salt solution or complete media, previously warmed to 36°C.

If desired, the whitish, centrally located pars nervosa may be removed from the darker, more vascular pars distalis by inserting a fine-tipped forceps between the two lobes and simply allowing the forceps to expand. The posterior lobe will normally separate from the pars distalis without much trouble. The intermediate lobe which may be identified as the gray thin line between these two lobes will separate randomly between them.

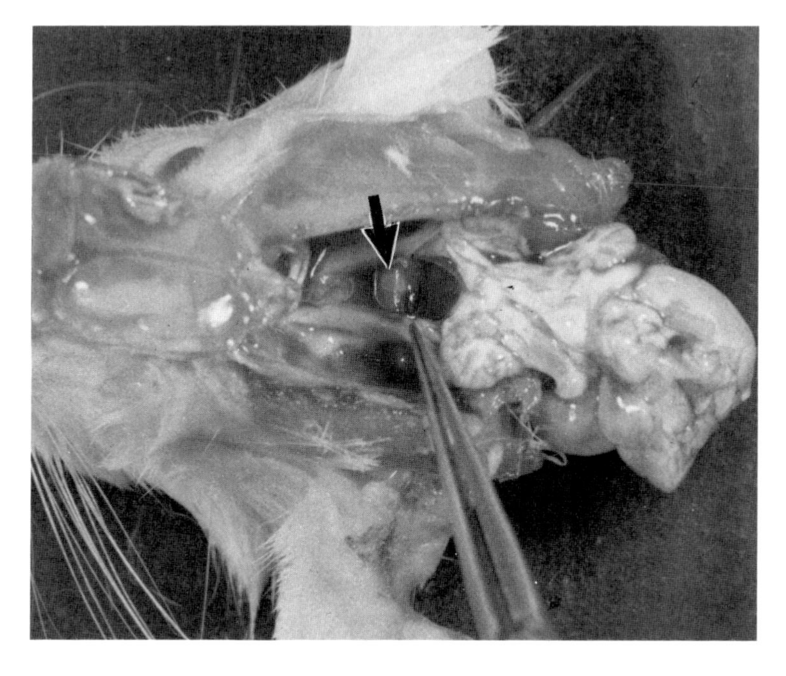

Fig. 4

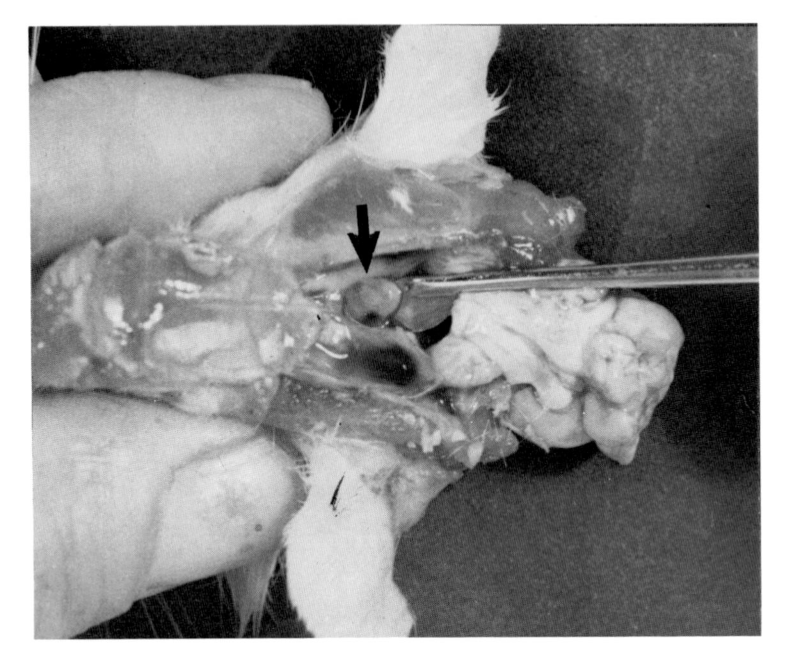

Fig. 5

Culture Preparation

The most convenient vessel for maintaining explants of the hypophyseal pars distalis is the presterilized plastic organ culture dish,[9] used in conjunction with a triangular stainless steel grid.[10] These plastic dishes are provided with an absorbent disk which should be aseptically moistened with 5 ml of sterile saline.

Explants are prepared by cutting the pars distalis into small pieces (less than 2 mm^3) with a pair of sterile scalpels (iris or Bard-Parker No. 10), using opposed scalpel blades in a scissorlike fashion. It is important to size the explants fairly accurately since too large of an explant will develop central necrosis while too small of an explant will lose its histological identity. Moreover, recognize that sharp clean cuts are essential to prevent tearing of the tissue.

Explants are transferred from the media by means of a pipette and gently placed on the surface of a grid, housed in a sterile Petri dish. After a number of explants are in position, the grid is placed in the center well of a Falcon organ culture dish.[9] Medium is added to the center well in sufficient amount to just touch the lower surface of the stainless steel grid. While several media may be used, we have had the most experience with Trowell's T-8[11] which is available commercially. Other media that have been used are Waymouth's MB752/1[12] and

[9] Falcon Organ Culture Dish, 60 × 15 mm. BioQuest (Cat. No. 3010).

[10] Falcon Organ Culture Grid. BioQuest (Cat. No. 3014).

[11] O. A. Trowell, *Exp. Cell Res.* **16**, 118 (1959).

[12] C. Waymouth, *J. Nat. Cancer Inst.* **22**, 1003 (1959).

NCTC 109.[13] In this organ culture method, the explants are located at the interface between the gaseous and liquid phases when incubated in 95% oxygen and 5% CO_2. This gas mixture may be obtained commercially and should be constantly perfused in the incubator unless a tightly sealed container is available. In our laboratory we use a sealable Plexiglas box with stainless steel connectors arranged so that we can pass the gas mixture through a membrane filter, subsequently through a gas washing bottle containing sterile saline to humidify the gas before it is forced into the box containing the organ culture dishes. It should be noted that for some organs it may be advantageous to submerge the explant under the media to allow for a lesser oxygen content.[14] In our laboratory, the cultures are maintained at 36°C and every second or third day the media is removed and replaced by a fresh solution, the old media being saved for assay.

Explants of the pituitary may be fixed in a variety of fixatives and stained either by classical procedures and/or the peroxidase-labeled antibody staining technique.[15,16]

[13] V. J. Evans, J. C. Bryant, W. T. McQuilkin, M. C. Fioramonti, K. K. Sanford, B. B. Westfall, and W. R. Earle, *Cancer Res.* **16**, 87 (1956).

[14] T. Fainstat, *In Vitro* 7, 300 (1972).

[15] A. E. Swope, R. H. Kahn, and J. L. Conklin, *J. Histochem. Cytochem.* **18**, 450 (1970).

[16] B. L. Baker, *In* "Handbook Of Physiology. Section 7, Endocrinology: The Pituitary Gland and Its Control Adenohypophysis" (E. Knobil, ed.). American Physiological Society, Washington, D. C., in press.

CHAPTER 18

Chick Embryo Cells

H. Rubin

The chick embryo has been a favorite source material for tissue culture since the inception of this technique. Almost all of the work done with chick embryo culture in the first half of the century employed explants of pieces of tissue into plasma clots, and is described in Fischer's classic treatise on the subject.[1] Dulbecco popularized the use of dispersed chick embryo cell cultures in plaque formation for virus research in 1952,[2] and most of the work done since then has

[1] A. Fischer, "The Biology of Tissue Cells." Glydenalske Boghandel Nordisk Forlag, Copenhagen, 1946.

[2] R. Dulbecco, *Proc. Nat. Acad. Sci. U. S.* 38, 747 (1952).

made use of such dispersed cultures. The technique described here for preparing dispersed cultures of chick embryo cells has been used in the author's laboratory for more than 10 years.[3]

Source of Embryos

For ordinary tissue culture work fertile eggs can be obtained from any local hatchery. For specialized work with avian tumor viruses, embryos of a particular genotype are often required, and these must be shipped from specialized sources such as Kimber Farms, Niles, California, and the Regional Poultry Laboratory, East Lansing, Michigan. The fertilized embryos can be stored at 4°–10°C for up to 2 weeks without major lethality. The eggs are incubated at 37°C in a high humidity incubator with a device for turning the eggs at regular intervals. For most general purposes, embryos are used after 10 days of incubation. At earlier times, the embryos are too small to yield large numbers of cells; beyond 11 days, bone, muscle, and feathers begin interfering with the procedures.

Composition of Growth Media

Medium 199 (Morgan *et al.*[4]) has given the most satisfactory results under the widest variety of circumstances over a 15-year period in this laboratory. For the growth of primary cultures it is combined with tryptose phosphate broth (Difco) 2% and calf and chicken serum (each 1%). Enough $NaHCO_3$ is added to the medium to maintain the pH at 7.2 to 7.4 in an atmosphere of 5% CO_2 in air at 38°C. Temperatures as low as 35°C may be used if the cultures are to be kept for up to 7 or 8 days before use. Chick embryo cells will grow well at temperatures up to 42°C, but will die if kept at temperatures greater than 43°C for more than a few minutes. The recommended concentrations of sera are lower than those used in earlier work because it has been found that micronutrients are depleted less rapidly in lower serum concentrations. If experiments with avian tumor viruses are contemplated, the chicken serum should be heated at 60°C for 30 minutes to inactivate endogenous avian leukosis virus.

Handling of Eggs

Before incubation, the eggs are fumigated to destroy surface parasites and organisms. They are placed in a 3.3 ft^3 cardboard box. A large Petri dish containing 2 g of $KMnO_4$ is placed in the box and 4 ml of HCHO (formaldehyde) is added to the dish. This is done in a chemical fume hood, and a hair drier type blower is used to circulate the air in the box for 20 minutes. The eggs are then placed in the incubator. This procedure has been found to reduce the risk of contaminating the embryo with shell surface organisms when the embryo is removed.

[3] A. Rein and H. Rubin, *Exp. Cell Res.* 49, 666 (1968).
[4] F. Morgan, H. J. Morton, and R. C. Parker, *Proc. Soc. Exp. Biol. Med.* 73, 1 (1950).

Preparation of Primary Cultures

After a 10-day incubation period, the eggs are removed from the incubator and candled to determine viable embryos. Only well-developed, active embryos with good blood supply are used. The eggs are exposed to a strong ultraviolet light for 10 minutes, with the blunt, airsac end uppermost. The entire egg is washed with 5% Zephiran chloride in water. The egg is broken at the airsac end with an egg punch (TriR Instrument Co., Jamaica, New Jersey). In the following steps, a separate set of instruments is used for each egg to avoid cross-contamination with avian leukosis viruses, or the instruments are carefully sterilized by flaming. The punched out shell is removed with forceps and the shell membrane and chorioallantoic membrane on the floor of the airsac are peeled away. A surgical hook is used to fish out the embryo by the neck, and the embryo is placed in a 100-mm glass Petri dish. The head and internal organs are removed with forceps and the remaining carcass placed in a 50-ml Erlenmeyer flask containing 20 ml of Tris–saline buffer. The embryo is swirled in the buffer to wash off adherent blood, and the buffer discarded. A metal spatula with spoon-shaped end is used to mince the embryo as finely as possible, and 10 ml of 0.25% trypsin (1:250 Difco) in Tris–saline at 37°C is added. A small Teflon-covered magnetic stirring bar is inserted into the flask which is then placed on a magnetic stirrer at low speed for 10 minutes. The flask is set on a slanted rack for 2 to 3 minutes to allow the large clumps to settle. The turbid supernatant containing dispersed cells is poured into a chilled 40-ml centrifuge tube containing 10 ml of Medium 199 plus 4 ml of calf serum. Repeat the trypsinization with 8 ml of the trypsin solution, for 10 minutes, and a third time with 6 ml for 8 minutes. At the last treatment the entire content of the flask is poured into the centrifuge tube. The cells are centrifuged at 1000 rpm for 8 minutes in a floor model clinical centrifuge, and the supernatant discarded. The cells are gently resuspended in 20 ml of Medium 199. The large clumps are allowed to settle, and the cell suspension is removed to a large test tube. A sample is diluted 1:10 in Tris–saline, and counted at a 10× or 40× magnification in a hemocytometer. Red blood cells and disrupted cells are not included in the count. Approximately 10^8 cells will be obtained from each embryo.

About 8×10^6 cells are seeded in 100-mm diameter plastic tissue culture dishes (Falcon Plastics, Los Angeles, California). The dishes should contain 12 ml of the growth medium (see above) previously equilibrated to 37°C and to the CO_2 tension of the incubator. The cultures are then placed in the incubator. Only about half the seeded embryo cells attach and grow. If the cell concentration is lowered, the fraction which attach decreases, particularly at seedlings below about 1×10^6 per 100-mm dish.

Secondary Cultures

The cells in the heavily seeded cultures grow slowly at first. If the medium is changed on the third day, the growth rate increases. Primary cultures are used for transfer to make secondary cultures only between the third and seventh day after explantation from the embryo. In the transfer procedure, the medium is removed by suction, the culture washed once with Tris–saline, and 5 ml of 0.05% trypsin (1:250) in Tris–saline are added for 5 to 10 minutes. The cells are dispersed by alternately drawing up and discharging from a 5-ml hand pipette with rubber bulb. An aliquot of cells is counted in a hemocytometer or in a Coulter electronic counter, and the appropriate number seeded in dishes containing the growth medium. For most of our work, the cells are seeded in 60-mm plastic tissue culture dishes containing 5 ml of medium, although for special purposes, 35- or 100-mm dishes are used. The medium may be the same as that used for the primary cultures, although calf serum is frequently omitted in the secondary cultures.

To attain optimal growth, 10^5 cells or more should be seeded per 60-mm dish. At concentrations of less than 10^4 per 60-mm dish, the growth rate is usually markedly reduced.[5] If only 10^2–10^3 cells are seeded per dish, a large fraction will fail to grow into colonies. In mass cultures cell growth occurs at a rapid rate until confluency is reached (a concentration of about 1.8×10^6 cells per 60-mm dish) at which point the growth rate decreases. With higher serum concentrations, rapid growth can be sustained beyond confluency, but the medium is rapidly depleted.[6]

Further cell transfers are carried out in the same manner as are the transfers from the primary cultures. Chick embryo cultures are difficult to sustain for more than four or five transfers without undergoing a marked reduction in growth rate. No one has yet obtained a permanent tissue culture line of chick embryo cells.

Special techniques are required for the assay of the Rous sarcoma virus and the avian leukosis viruses.[7] Also, transformed Rous sarcoma cells require very frequent medium changes as they deplete the medium of glucose rapidly, and quickly acidify the medium.

[5] H. Rubin, *Exp. Cell Res.* **41,** 138 (1966).

[6] H. Rubin, *Ciba Found. Symp. Growth Control Cell Cultures,* pp. 127–149 (1971).

[7] P. Vogt, *In* "Fundamental Techniques in Virology" (K. Habel and N. P. Salzman, eds.), pp. 198–211. Academic Press, New York, 1969.

CHAPTER 19

Frog Embryos (Haploid Lines)[1]

Jerome J. Freed and Liselotte Mezger-Freed

The methods to be described here can be used to initiate lines of either haploid or diploid amphibian cells. The frog, *Rana pipiens*, is a good source of haploid cell lines because a number of techniques are available for the production of haploid embryos. Although many of these lines become diploid in 10 to 20 subcultures, one line, ICR 2A, is still haploid after 160 subcultures (700 doublings).[2]

Culture Conditions for Anuran Cells

From experience in this and other laboratories, the following conditions are appropriate for embryo cells of *Rana*, *Bufo*, *Hyla*, and *Xenopus*. Solutions used in contact with living cells should be adjusted to about 200 mOsmoles osmotic pressure; this can generally be done by diluting about 60 parts of solutions designed for mammalian cells with 40 parts glass-distilled water. The maximum rate of cell multiplication occurs between 25° and 28°C, with lethal effects at higher temperatures; we keep cultures in low temperature (biological oxygen demand) incubators. The most favorable pH is 7.5. Multiplication is more rapid at pH 7.0, but the cells are more prone to lysis.

Production and Rearing of Embryos

Large frogs suitable for breeding are obtained from commercial suppliers and maintained in the laboratory in tap water at 5°C.[3] Gravid females captured in the fall are the best source of healthy embryos. Ovulation is hormonally induced by injecting a single frog pituitary gland intraperitoneally through a No. 23 hypodermic needle; at the same time, during November to January, 0.2

[1] The techniques were developed with support from Grant AT(11-1)3110 from the Atomic Energy Commission (U. S. A. E. C. Report No. COO-3110-4), U. S. P. H. S. Grants CA-05959, CA-06927, and RR-05539 from the National Institutes of Health, and by an appropriation from the Commonwealth of Pennsylvania.

[2] J. J. Freed and L. Mezger-Freed, *Proc. Nat. Acad. Sci. U. S.* **65**, 337 (1970).

[3] E. L. Gibbs, G. W. Nace, and M. B. Emmons, *Bioscience* **21**, 1027 (1971); M. DiBerardino, *In* "Methods in Developmental Biology" (F. H. Wilt and N. K. Wessels, eds.). p. 53. Crowell, New York, 1966.

ml to 0.05 ml of mammalian progesterone (Lilly, 25 mg/ml) is injected.[4] After January, the addition of mammalian progesterone is not required and may even be fatal. The injected females should be placed in tap water (about 5 cm deep) at about 15°C for 48 to 72 hours. To determine if ovulation has occurred the frog's sides are gently pressed to express eggs from the cloaca.

Androgenetic haploids are obtained by the technique of Porter.[5] An active sperm suspension is prepared by mincing both testes of a male frog in 5–10 ml 10% Ringer's solution (see *Solutions*) and allowing this preparation to stand at room temperature for 10 minutes. Eggs are expressed from the female in a spiral pattern on the bottom of a clean, dry Syracuse watchglass, and the sperm suspension is added dropwise to wet the eggs. Ten minutes later, the eggs are covered with 10% Ringer's solution, added with care so as not to detach the eggs from the dish. About 20 minutes after fertilization, the appearance of a black dot or hole corresponding to the migration of pigment from the area over the meiotic spindle marks the location of the maternal nucleus in the dark animal hemisphere. A glass needle is inserted into the egg under the dot, and the nucleus is flipped out. A stereo microscope (30X) with two focusing overhead lamps is required for this operation; the surface of the animal hemisphere should be under bright illumination directed downward at an angle of about 45°. A detailed description of this enucleation procedure has been given by King.[6]

Gynogenetic haploids may be obtained in large numbers by photodynamic inactivation of sperm as described by Briggs.[7] Two testes are minced in 5 ml Ringer's solution and mashed with a glass rod; coarse tissue fragments are then removed. Liberated sperm in suspension are transferred to a centrifuge tube and sedimented at 500 *g* for 5 minutes. The supernatant is discarded and the sperm suspended in a buffered solution of toluidine blue O (see later *Solutions*). The tube is placed in a beaker of water adjusted to 20°C, and exposed for 25 minutes to the fluorescent lamp illumination of the room (about 200 ft-c) with occasional stirring. The sperm are then sedimented by centrifugation, re-suspended in 10% Ringer's solution, and used to fertilize eggs in the usual way. Time, temperature, and intensity of illumination may be adjusted to provide a yield of haploid embryos that is nearly 100% of these eggs that cleave.

The quality of the eggs appears to be a factor in the success of the culture procedures and can be determined in advance by the frequency of normal cleavage of diploid embryos (obtained by fertilization without subsequent enucleation). Good eggs should give better than 90% cleavage.

Twenty haploid or diploid embryos are reared in each aquarium (10-cm finger bowl) containing about 100 ml 0.25% sodium sulfadiazine in spring water or 10% Ringer's solution. They are maintained at about 18°C during development.

[4] P. A. Wright and A. R. Flathers, *Proc. Soc. Exp. Biol. Med.* **106**, 346 (1961).

[5] K. R. Porter, *Biol. Bull.* **77**, 233 (1939).

[6] T. J. King, *In* "Methods in Cell Physiology" (D. M. Prescott, ed.), Vol. 2, p. 1. Academic Press, New York, 1966.

[7] R. Briggs, *J. Gen. Physiol.* **35**, 761 (1951-52).

When the embryos have reached stage 15 to 18 (tailbud),[8] the jelly and membranes are removed with watchmaker's forceps. Those haploid embryos that are most like a normal tailbud stage are selected for setting up cultures. In addition, the haploid condition of each embryo should be verified by a comparison of the size of the dorsal epidermal cells in haploid and diploid embryos; a haploid cell is half the size of a normal diploid cell at the same stage.

Setting Up Cultures

The selected embryos are surface-sterilized by exposing a number of them for 15 minutes to sterile spring water to which is added 100 μg/ml Merthiolate (Lilly); subsequent steps are then carried out aseptically. With a cotton-plugged 8 × 200 mm Pasteur pipette (tip orifice about 6 mm), the embryos are transferred to a 500-ml Erlenmeyer flask filled with sterile spring water. They are released just under the surface and allowed to fall to the bottom. This wash is carried out twice more, and the embryos are finally transferred to a Petri dish containing 20 ml dissociating solution (see *Solutions*).

The embryos are left in dissociating solution for up to 30 minutes, to remove epidermis and ventral yolky cells. This process is completed by pipetting each embryo several times with a Pasteur pipette with a 1.0 mm tip orifice. Removal of most of the darkly pigmented epidermal cells is required to prevent their "healing" around small tissue masses and preventing attachment to the growth surface. The cleaned "carcasses" are transferred to a Petri dish containing 20 ml balanced salt solution (see *Solutions*).

To initiate a culture in a 35-mm Petri dish five embryos are pipetted to a dry dish and minced with needles to give a suspension of free cells and small tissue masses. The best results are obtained from cultures that contain 1–2 mm pieces of tissue.

The minced embryo tissue is dispersed in 2.5 ml complete growth medium plus antibiotics (see *Solutions*); observations indicate that the best outgrowth occurs when the pieces are near but not touching each other. The dishes are incubated at 25°C in a moist chamber that has been sterilized to prevent mold growth. Growth medium should be aspirated, and fresh medium added, at the end of the first 4 days and once or twice each week thereafter.

Behavior of Primary Cultures

Within 24 hours, in successful cultures, migration of cells from tissue clumps should be observed. Many of these cells will contain the ovoid, refractile yolk platelets characteristic of frog embryo cells. On further incubation, yolk-free cells having the typical appearance of animal cells in monolayer cultures are

[8] W. Shumway, *Anat. Rec.* **78**, 139 (1940); see also R. Rugh, "Experimental Embryology," 3rd Ed., p. 56. Burgess, Minneapolis, Minnesota, 1962.

found. By about the eighth day, areas of such cells should be of sufficient size to permit subculture (see below).

Although cells apparently expressing differentiated functions are observed in these primary outgrowths, they have not been recognized after subculture.

Subculture Procedure

When primary cultures are passaged, many cells fail to reattach to the growth surface and are lost. In order to assure a reasonable cell population density after subculture, we usually combine three primary cultures in one plastic culture flask (25 cm^2) containing 5 ml of growth medium. In successful subcultures, attached cells are only a few cell diameters from each other. After prolonged passage, lower population densities may be used.

To subculture, the medium is removed and the monolayer rinsed for 1 minute with 2 ml of divalent cation-free wash medium (see *Solutions*). Approximately 2.5 ml of trypsin (0.5% in wash medium) is added, allowed to remain in contact with the cells for 1 minute, and then removed. Residual enzyme will loosen the cells during 20-minute incubation at room temperature (20°–23°C); 2 ml of complete growth medium is then added and the cells dispersed by gentle pipetting. The cell suspension is diluted as necessary with additional growth medium and transferred to new culture vessels.

Propagation of Cell Lines; Monitoring Haploidy

The subculture technique above is used to subdivide the cultures when they reach saturating densities; the reduction in growth rate is conveniently detected by microscope counts of cell density on the growth surface.[9] In cell lines of overlapping fibroblastlike morphology (e.g., the haploid line ICR 2A), multiplication may be retarded before a physically confluent cell layer is attained.

The population doubling time for *Rana pipiens* cell lines in exponential growth at 25°C is between 40 and 50 hours. For routine propagation, we try to adjust the subdivision ratio so that the daughter flasks become saturated in 1 week; they are then subcultured in turn.

Most haploid cell lines propagated in this way show an increasing frequency of diploid variants that progressively overgrow the culture. The extent of diploid overgrowth can be assessed by scoring chromosome preparations for the frequency of cells with thirteen chromosomes (haploid). A quicker, less accurate method for assessing the ploidy composition of a population is counting the frequency of interphase cells with two nucleoli: since the tenth chromosome bears the single nucleolar organizer, a haploid cell has one nucleolus. Diploids show a maximum nucleolar number of two, but this may be reduced to one in many cells by fusion of the nucleoli. Determination of nucleolar number can

[9] J. J. Freed and L. Mezger-Freed, *In* "Methods in Cell Physiology" (D. M. Prescott, ed.), Vol. 4, p. 19. Academic Press, New York, 1970.

be made with an inverted phase contrast microscope in lines of well-spread cells where the nucleus is visible. Other cell lines, in which the cell body is rounded, are preferably fixed and stained with Giemsa or Unna's methyl green–pyronin for this kind of observation.

For some purposes, it may be convenient to distinguish haploid from diploid cultures on the basis of cell volume, using the Coulter counter. The cells are trypsinized to form a single-cell suspension, diluted with saline, and the cell volume distribution plotted with a calibrated counter. The modal cell volume in exponentially multiplying *Rana pipiens* haploid cultures lies in the range 1000–1300 μm^3; diploid cell strains yield a value of about 2500 μm^3.

Applications

Through use of cells with a single chromosome set, it should be possible to detect and therefore isolate the maximum range of genetic mutations that lead to loss of a functional gene product (recessives). However, it should be pointed out that studies of drug resistance using the haploid frog cells have raised the question of whether many of the stable variants arising in culture are the result of epigenetic events rather than gene mutation.[10]

A unique advantage of frog cultured cells is the option of using the Briggs and King nuclear transplant method[11] to substitute the nucleus of a variant cell for the zygote nucleus in order to test whether the variant is a result of gene mutation. The possibility also exists of using the same method to study the effects on embryonic differentiation of a mutation induced in culture.

Solutions

Ringer's Solution. NaCl, 0.66 g, KCl, 0.015 g, and $CaCl_2$, 0.015 g (or $CaCl_2 \cdot 2\,H_2O$, 0.0174 g) are dissolved in glass-distilled water and made up to 100 ml. Using a glass electrode, adjust pH to 7.8 by dropwise addition of 5% $NaHCO_3$ solution.

Toluidine Blue Solution. Prepare Tris buffer, pH 7.6, by adding 50 ml of a 0.2 *M* solution of 2-amino-2-hydroxymethyl-1,3-propanediol to 38.4 ml of 0.2 *M* HCl. Prepare a 10^{-4} *M* solution of toluidine blue O, certified (Allied Chemical Corp.). These solutions may be stored in the refrigerator. For use, add 0.1 ml of the dye solution to 10 ml of buffer. Prepare fresh daily.

Dissociating Solution. Solution A: dissolve in 500 ml glass-distilled water: NaCl, 2.943 g and KCl, 0.050 g. Solution B: dissolve in 250 ml glass-distilled water: Na_2HPO_4, 1.300 g and KH_2PO_4, 0.116 g. Solution C: dissolve in 250 ml glass-distilled water: $NaHCO_3$, 0.200 g.

[10] L. Mezger-Freed, *J. Cell Biol.* **51**, 742 (1971); L. Mezger-Freed, *Nature (London) New Biol.* **235**, 245 (1972).

[11] R. Briggs and T. J. King, *Proc. Nat. Acad. Sci. U. S.* **38**, 455 (1952).

Mix the three solutions above, and add 0.700 g disodium ethylenedinitrilotetraacetate (EDTA). Sterilize by membrane filtration and store frozen.

Balanced Salt Solution. Dilute 65 ml Hanks' balanced salt solution, 10 X (Grand Island Biological Co.) to 1 liter with glass-distilled water. Adjust pH to 7.5 by addition of 7.5% $NaHCO_3$ solution. Sterilize by membrane filtration and store at 5°C.

Complete Growth Medium. To 4 liters of glass-distilled water, add 4 liter-size packets of Leibovitz L-15 powder medium (Grand Island Biological Co.). Dissolve on magnetic stirrer.

Thaw 800 ml fetal bovine serum (pretested for ability to support clonal growth of frog cell lines) and clarify by centrifugation if a fine precipitate forms. Add to medium.

Add 3.2 liters glass-distilled water. The pH should require no adjustment, but should be near 7.5.

Sterilize by membrane filtration and store at 5°C.

A medium defined with respect to constituents of low molecular weight may be prepared by substituting for the whole serum the macromolecular fraction from 800 ml fetal bovine serum and adding 1.36 mg/liter hypoxanthine (10^{-5} *M* final concentration).

Antibiotics (50×). To 100 ml kanamycin solution (Grand Island Biological Co.), 10,000 μg/ml, add 0.5 g dihydrostreptomycin sulfate·HCl (Squibb), 0.93 g Na penicillin G (Merck), and 1.0 ml 0.2% phenol red solution. Adjust pH to about 7.0 with 0.2 *N* NaOH, and sterilize by membrane filtration.

Add 2 ml aseptically to each 100 ml growth medium before using in cultures.

Wash Medium. To about 800 ml glass-distilled water, add in order and dissolve: NaCl, 2.750 g, KCl, 0.267 g, Na_2HPO_4, 0.769 g, KH_2PO_4, 0.072 g, and glucose, 2.89 g.

Add 10 ml phenol red solution 0.2%.

Dissolve 1.5 g bovine serum albumin (Fraction V) in 10 ml water and add to above.

Adjust pH to 7.5 with 0.2 *N* NaOH and make up to 2 liters with glass-distilled water.

Sterilize by membrane filtration and store frozen.

Trypsin Solution (0.5%). To 50 ml of wash medium, add 2.5 g trypsin (1:250, Difco Laboratories); stir for 20 minutes on a magnetic stirrer; centrifuge for 20 minutes at 2000 *g*.

Decant the supernatant into 450 ml of wash medium and mix on magnetic stirrer, being careful to avoid foaming.

Adjust pH to 7.5 with $NaHCO_3$ solution (7.5%).

Sterilize by membrane filtration and store frozen in 50-ml portions.

CHAPTER 20

Adult Amphibian Tissues and Leukocytes

D. E. Rounds

Amphibian tissue is composed of cells which are generally larger than those from avian or mammalian sources. In addition, they frequently contain fewer and larger chromosomes. Moreover, since the growth of these poikilothermic tissues is optimal at room temperature, successful maintenance of amphibian cultures can be achieved without an incubator.

In our laboratory, a variety of soft-organ tissues have been used in an attempt to compare their growth rates *in vitro*.[1] In work with adult frogs, successful cultures were obtained from kidney, heart, and lung tissues. Newt heart and lung tissue was also found to be an excellent material; whereas, liver, spleen, and intestine of both animals gave poor results. Of these, frog kidney and salamander lung have been most frequently employed as culture material. Besides these more commonly used tissues, frog tongue, lenses, and newt coracoid cartilages have been occasionally used in culture, and have also proved to be very satisfactory materials.[2]

Procedure

Soft Tissues. Prior to the surgical operation, in the culture room, the animal is rinsed in dilute potassium permanganate solution (approximately 5–10 mg/liter) for 30 minutes, and the entire body surface is wiped with 70% alcohol in order to eliminate the prevalent bacteria, fungi, and parasites. After the animal is narcotized with 1 part of tricaine methanesulfonate (MS 222-Sandoz) in 2000 parts of spring water, or is immobilized by severing the spinal column, the selected organ is removed by careful aseptic techniques.

The tissue is placed in Eagle's MEM, made with Hanks' balanced salt solution and supplemented with 10% fetal calf serum. The initial maintenance medium also contains 100 mg/liter aureomycin, 100,000 units/liter penicillin G, and 62.5 mg/liter streptomycin sulfate. The tissue is minced into 1–2 mm^3 pieces with knife blades or scissors, and the tissue fragments are then suspended in fresh medium in order to remove blood clots or adherent mucus. Petri dishes containing the medium should not be used for more than one rinsing procedure.

[1] T. Seto, *Jap. J. Genet.* 39, 268 (1964).

[2] T. Seto and D. E. Rounds, *In* "Methods in Cell Physiology" (D. Prescott, ed.), Vol. 3, pp. 75–94. Academic Press, New York, 1968.

The tissue fragments can be set up in plasma clot in either depression slides, Maximow chambers, roller tubes, Leighton tubes, or T-flasks. However, the system which is most frequently used in our laboratory is to maintain the explants under "cellophane" strips in the Rose multipurpose culture chamber.[3]

The cellophane strips are derived from slitting Visking dialysis tubing along their length to provide single thickness semipermeable membranes of about 1.5 cm in width. These are sterilized by soaking for 20 to 30 minutes in 70% ethanol. The alcohol is removed by two rinses for 5 to 30 minutes each in sterile balanced salt solution. The strips are then transferred to Eagle's medium until ready for use.

Generally three explants are arranged in a row in the center of a sterile 43 × 50 mm coverglass which has been positioned over the center hole of a metal plate. The cellophane strip is laid over the explants with both ends extending over the chamber margins. They are allowed to rest on the table surface. The silicone gasket is next positioned over the coverglass, and the chamber is closed with a second coverglass placed on the gasket. A second metal plate is positioned on this "sandwich" of materials so that the holes in the four corners come into alignment with the threaded corner holes in the bottom plate. Screws, introduced in these holes, are gently tightened until the assembly is held firmly together. Then the extended ends of the cellophane strips are carefully pulled to hold the explants securely against the coverglass wall and to remove any existing wrinkles in the membrane. The chamber is filled with approximately 2 ml of Eagle's medium through an inlet needle which penetrates the gasket, while an opposite outlet needle serves as an air vent. The assembly is completed by tightening the four screws, removing the needles, and trimming off the extended ends of the cellophane strips.

The subsequent exchange of fluids is accomplished at approximately weekly intervals by inserting two opposing hypodermic needles. The expended medium is removed with negative pressure produced by one syringe, then fresh medium is injected into the chamber from a second syringe. After the first week, the aureomycin-containing medium is replaced with nutrient fluid containing only penicillin and streptomycin. If it is desired to fix and stain the cultures, the cells on the bottom coverslip are washed with balanced salt solution and fixed before the retaining plates have been removed.

Our trials with the frog kidney cell strain and newt lung cultures showed the growth in a limited series of cultures to be definitely retarded below 22°C and inhibited at 37°C. If frog cell cultures are gradually adjusted from 26° to 37°C, with an elevation of 1° for each passage, they can adapt to the environment and survive, but the cultures demonstrate little cell proliferation at temperatures higher than 26°C. Recent studies repeatedly showed the upper limit of the optimal temperature range from amphibian cultures to be 26°C, with few exceptions.

A common type of cell originating in cultures of newt lung is the epithelial-like cell. Under the dialysis membrane in the Rose chamber, the cells in the

[3] G. G. Rose, *Tex. Rep. Biol. Med.* **12**, 1074 (1954); cf. Section VI, this volume.

outgrowth become flattened to form a monolayer of pavement epithelium on the coverglass. Within 6 or 7 days of cultivation, mitotic figures can be found with a frequency of less than 2%. The newt cell is recognized as a particularly suitable material for study of cell division, since the beginning of mitotic prophase is more distinguishable among interphase cells than is found in mammalian mitotic cells.

Nearly every mitotic figure can be observed with good resolution and structural detail with the phase-contrast microscope because of the cellophane technique employed. In cultures of lung tissue, most cells show two nucleoli, and normally binucleate or trinucleate cells appear in low numbers. Ciliated cells can be observed immediately after the cells migrate out from an explant if an appropriate region of the lung is utilized. Ciliary movements continue as long as the cell is alive under minimal culture conditions. This motion can serve as a relative indicator of physiological activity of the cell. Pinocytosis can be observed very clearly in peripheral cells of the epithelial pavement, and usually the activity increases in the older cultures. For pulmonary epithelium the growth rate *in vitro* is rather slow, and usually primary cultures can be maintained for at least 3 weeks. If cells are left without any changes in culture conditions, they begin to gradually degenerate after a month of incubation.

Leukocytes. Another cell type, which has been used successfully for chromosome analyses, is the leukocyte. These procedures of amphibian leukocyte culture are based fundamentally on the technique of the human leukocyte culture. However, the conventional method occasionally has the disadvantage that more blood is required than can be obtained from small animals such as salamanders and frogs.

When relatively large-sized animals are used, such as *Ambystoma, Necturus,* and toads, blood can be drawn from blood vessels. Large blood vessels, e.g., those of the arterial trunk and femoral artery, are exposed by careful dissection. A 27-gauge needle, having a sharp tip, is then injected into the vessel. Blood is slowly drawn into a 5-ml syringe containing about 0.05 ml of heparin sodium (1000 U.S.P. units/ml). The heparinized blood is placed in a Kahn tube (7 mm diameter, 100 mm long), and allowed to stand for about 2 hours in a refrigerator (2°–4°C). The use of a small tube contributes to the ease of separation of white blood cells from small amounts of blood. Erythrocytes become agglutinated and settle to the lower part of the tube. About 1 ml of the supernatant suspension of leukocytes can be separated from the 3 ml of heparinized blood, which is aspirated into a syringe and transferred into a culture bottle. To 0.5–0.7 ml of the plasma, containing the leukocytes, 5.0 ml of culture medium and 0.2 ml of phytohemagglutinin M (PHA-M, Difco) are added.

In case the animal is not large enough to draw blood from a vessel, the amount of acquired blood is inadequate for the above method. Therefore, a technique which requires no separation of leukocytes from whole blood must be employed to avoid the loss of the mitotically competent cells. After animals are washed in the potassium permanganate solution and their body surface cleansed with 70% alcohol, the tip of a hind limb or tail is rapidly amputated with sterilized scissors. Blood is drained from the section into a culture bottle containing heparin (0.025 ml), phytohemagglutinin M (PHA-M) (0.2 ml), and the nutrient fluid (5 ml). An alternative source of blood is from the heart. A cardiac puncture is made by passing the needle through the upper abdominal wall and diaphragm into the heart using a 27-gauge needle attached to a 1-ml syringe wetted with heparin. However, this technique requires more experience and skill in order to ensure success.

To 5 ml of the medium, about 0.1 to 0.2 ml of blood is directly inoculated into a T-15 flask. The blood and culture medium are mixed and placed in an incubator at 26°C. This method is less successful than the one described above, but has the advantages that only very small amounts of blood are needed for each study, and the procedure is simple, requiring no special handling or treatment of the blood sample prior to initiating the culture of the leukocytes.

The concentration of phytohemagglutinin M in the culture medium can significantly influence the mitotic index.[1] Satisfactory results for the production of leukocytic mitosis were only obtained at concentrations of 3.0 to 6.0%. An increased concentration of colchicine was also required to produce an effect on the amphibian cell mitoses. Inhibition of mitosis at metaphase required three or four times the concentration of colchicine usually employed for mammalian cells *in vitro*.

The length of the culture period must also be considered in obtaining an adequate number of mitotic cells. Although 3 days of cultivation prior to the 24-hour treatment with colchicine is sufficient to secure some mitotic figures, 6–8 days of cultivation, including the time of colchicine treatment, will produce better results. Experience has shown that the mitosis of amphibian leukocytes begins within 4 days after setting up the culture at a temperature of 26°C, with the number of mitoses increasing in the following 3 or 4 days.

In order to prepare cells for chromosome analysis, colchicine is added to 5-day leukocyte cultures at a final concentration of 2×10^{-5} g/ml for a period of 8 to 24 hours. The cells are then harvested from the bottom of the flask by scraping with a rubber policeman, transferred to a 10-ml conical tube, and centrifuged at 600 rpm for 5 minutes. The culture medium is discarded, the cells are resuspended in a hypotonic solution (1.12% sodium citrate solution), and allowed to stand at room temperature (20°–24°C) for 30 minutes. The KCl solution (0.075 *M* in distilled water), described by Hungerford,[4] is also recommended as a suitable hypotonic solution for obtaining satisfactory spreads of chromosomes. A small amount of Carnoy's fixative (3 parts of absolute methanol and 1 part of glacial acetic acid) is added carefully to the hypotonic cell suspension so that cells are lightly fixed before centrifugation The hypotonic solution is then replaced with a freshly made fixative. The pellet of cells must be dispersed into the fixative at once by gentle agitation with a pipette. After 30 minutes or more, the cells are centrifuged again and resuspended in fresh fixative. This is repeated twice to fix the cells completely.

[4] D. A. Hungerford, *Stain Technol.* **40**, 333 (1965).

Chromosome preparations are made by the air-drying method.[5] Chromosomes are stained for 1 hour with acetic orcein (1% orcein in 50% acetic acid) or Giemsa's solution (Merck, diluted into 5% of the stock solution), dehydrated in 95% ethanol, then in two changes of absolute ethanol, and mounted in a synthetic resin (Hartman-Leddon Co.). Phase-contrast microscopy can be occasionally used for such preparations, which are lightly stained. Carbol fuchsin stain has also been favorably employed in our laboratory for the autoradiographical study of chromosomes.

[5] K. H. Rothfels and L. Siminovitch, *Stain Technol.* 33, 73 (1958); cf. Section XIV, this volume.

CHAPTER 21

Marine Teleost Fish Tissues

M. Michael Sigel and Annie R. Beasley

Tissue cultures of marine fish origin have been used in this laboratory since 1960 as host cells in the search for viruses of marine fishes. Tissues from grunts (*Haemulon* sp.) have been most frequently employed due to their abundance in south Florida waters. At various times, however, tissues have been cultured from the sand perch (*Diplectrum formosum*), black angelfish (*Pomacanthus* sp.), pork fish (*Anisotremus virginicus*), porgy (*Calamus* sp.), and snappers (*Lutjanus* sp.).

Explant Cultures

Fish are killed by severing the spinal cord just posterior to the head. Scales are removed and the skin is wiped with 70% ethanol. The tissue to be cultured is removed aseptically and consecutively washed for 10-minute periods in three aliquots of modified Hanks' balanced salt solution (NaCl content increased to 11.51 g/liter) containing 2X antibiotics. The tissue is minced with sterile scalpels and three to four explants are placed in each culture tube. The fragments may be cultured directly on glass, but superior results are obtained when they are placed on a coagulum of equivalent amounts of 50% chick embryo extract and chick plasma. (In view of our experience with cultures of dissociated cells, cited below, it may be advantageous to use plastic culture vessels. This has not been extensively tested in this laboratory, however.) To each culture is added 1 ml of marine basal medium Eagle, Hanks' base (sodium chloride content increased by 0.06 *M*)[1] supplemented with 10% calf serum plus 10% human serum and containing 200 units penicillin, 100 μg streptomycin, and 1 μg amphotericin B/ml. Cultures are incubated at 20°–25°C.

As has been reported,[2] such cultures of grunt and sand perch fin show a sequence of outgrowths. A sheet of epithelial-like cells appears around the explants within 24 hours, but this cell type degenerates after 5 to 8 days. A fibroblastlike cell type appears 5 to 8 days and, after 3 to 5 weeks, forms an extensive monolayer which can be subcultured. Explants of grunt heart and ovary also yield monolayers of fibroblastlike cells which can be subcultured. As stated below, however, cultures derived from enzymatically dissociated grunt fin cells are epithelial-like in character. It should therefore be kept in mind that epithelial-like cells rather than fibroblasts may become established following explantations.

Cultures of Dissociated Cells

Our experience with dissociated cells has been limited to cultures derived from fins of grunts and snappers. For these, the culture medium is the same as that employed for explant cultures.

Trypsin-dispersed cells[1] at a concentration of 10^6 cells/ml are seeded in plastic tissue culture flasks (15 ml/250-ml flask; 4 ml/30-ml flask), soft glass prescription bottles (5 ml/3-ounce bottle; 10 ml/8-ounce bottle), or 16 × 125 borosilicate or plastic tubes (1 ml/tube). Cultures are incubated at 20°–25°C, and the medium is renewed twice weekly. The optimal pH is 7.2; cultures tolerate a lower pH, but excess alkalinity is deleterious.

The type of substrate used appears to be critical. If prescription bottles are employed, they should be rinsed and then boiled in distilled water prior to sterilization. Borosilicate tubes from various sources show a wide divergence in their suitability in that little or no growth occurs in those from some manufacturers. Those from Bellco Glass Inc. have been consistently satisfactory. Plastic vessels are optimal for use, as cells occasionally slough from glass substrates when confluent growth is reached.

In the initial stages of growth, both fibroblastlike and epithelial-like cells are seen in cultures derived from grunt fin. Cultures usually form confluent monolayers within 2 to 3 weeks, at which time essentially all cells are epithelial-like. In the snapper fin cultures, the cells remain fibroblast-like, even after serial subculture.

Unlike cultures derived from normal tissues of other animals, those initiated from marine fish fin can be passaged with ease and, to date, show no evidence of a finite life. One cell line (designated GF; American Type Culture Collection cell repository No. CCL 58) has been maintained for over 10 years and has undergone over 300 subcultures. During this period, there has been no alteration in

[1] See Section I, this volume.

[2] L. W. Clem, L. Moewus, and M. M. Sigel, *Proc. Soc. Exp. Biol. Med.* **108**, 762 (1961).

morphology or in growth rate. When last tested, after 8 years in culture, the karyotypical status of this cell line was essentially like that of primary grunt fin cultures.[3] This is in noteworthy contrast to the frequently noted major aberrations in established mammalian cell lines.

We have routinely cultured the marine fish cells within their optimal temperature range of 20°–25°C. Experimentally, the GF cells have been grown at 30°C. For cultivation at a supraoptimal temperature, however, it may be necessary to adapt the cells by stepwise acclimation.

[3] J. D. Regan, M. M. Sigel, W. H. Lee, K. A. Llamas, and A. R. Beasley, *Can. J. Genet. Cytol.* **10**, 448 (1961).

CHAPTER 22

Fish Lymphocytes and Blastogenesis

M. Michael Sigel, E. Churchill McKinney, and John C. Lee

There is ample documentation that fish are capable of synthesizing humoral antibodies in response to antigenic stimulation.[1-3] It is not known, however, whether the immune response of fishes requires the cooperation of specialized cells as does that in mammalian and avian species where thymus-conditioned T cells and bone marrow-derived B cells interact to certain antigenic stimuli.[4-6] In fact, it is not even known whether fishes possess analogs of T and B cells. Moreover, fishes lack a true bone marrow and their thymus glands have a more primitive structural organization than that of the mammalian thymus.

Studies in this laboratory have revealed that in 2 teleost fishes—the grey snapper (*Lutjanus griseus*) and the grouper (*Mycteroperca bonaci*)—antibody synthesis occurs in cells of the thymus, spleen, and pronephros (head kidney).[7] More recent studies have demonstrated that cellular immunity, as manifested by

[1] M. M. Sigel and L. W. Clem, *Nature (London)* **197**, 315 (1963).

[2] R. A. Good and B. W. Papermaster, *Advan. Immunol.* **4**, 1 (1964).

[3] G. Ridgway, H. O. Hodgins, and G. W. Klontz, *In* "Phylogeny of Immunity" (R. T. Smith, P. A. Milsher, and R. A. Good, eds.), p. 199. Univ. Florida Press, Gainesville, Florida, 1966.

[4] R. D. Corley and N. R. Joseph, *Proc. Soc. Exp. Biol. Med.* **122**, 1167 (1966).

[5] G. F. Mitchell and J. F. A. P. Miller, *J. Exp. Med.* **128**, 821 (1968).

[6] W. P. McArthur, J. Chapman, and G. J. Thorbecke, *J. Exp. Med.* **134**, 1036 (1971).

[7] G. Ortiz-Muniz and M. M. Sigel, *J. Reticuloendothelial Soc.* **9**, 42 (1971).

in vitro blastogenic transformation, is present in representatives from two classes of fishes, Chondrichthyes and Osteichthyes. Thus far we have shown that peripheral blood lymphocytes of the nurse shark (*Ginglymostoma cirratum*) immunized with bovine gamma globulin or with poliovirus respond specifically to the respective antigens with increased synthesis of DNA. Lymphocytes of gars (*Lepisosteus platyrhincus*), whose sera contain natural antibodies to *Salmonella typhosa* H antigen,[8] also respond *in vitro* to this antigen. Similarly, lymphocytes of the grey snapper immunized with rubella virus respond specifically to *in vitro* stimulation.

The procedures described below in detail have been successfully used with lymphocytes from sharks, snappers, and gars. Whenever modifications may be employed, they are included as alternatives.

Media

Medium 199 containing 10 IU/ml heparin and 0.01 *M* HEPES buffer[9] is the major constituent of the media for the fish lymphocytes. For shark and snapper cells, extra sodium chloride is added to give a content of 0.2 *M*. In addition, the basal medium for shark tissue also contains 350 m*M* urea. These basic solutions, to be referred to as ShM, SnM, and GaM, are used for washing during cell preparations. For final cell resuspension and growth, the solutions are supplemented with 10 to 20% homologous sera. (For gar cells, 10% agamma newborn calf serum is an equally satisfactory supplement.) All growth media used contained 100 units penicillin, 100 μg streptomycin, and 50 μg amphotericin B/ml, while twice these amounts of antibiotics are added to the basal media. All media are adjusted with NaOH to pH 7.2 prior to use.

Preparation of Cells

Peripheral Blood Leukocytes. All fish are anesthetized with tricaine methanesulfonate[10] diluted 1:4000 in sea water for the sharks and the snappers; use fresh water for the gars. Depending upon the size of the fish, 0.5–50 ml or more blood is drawn from the caudal blood vessel into a syringe containing enough heparin to give a final concentration of 100 IU/ml.

The blood of either the shark or the gar is transferred to a tube and allowed to sediment at 25°C for 1 to 2 hours. Additional sedimentation, to 15 hours, will increase the cellular yield from shark blood. The leukocyte-rich plasma is removed with a Pasteur pipette, and transferred to another tube. The cells are washed twice with ShM or GaM containing 2X antibiotics (see above), centrifuged at 200 to 300 *g* for 10 minutes, and the supernatant discarded.

When only small volumes of blood are available, the packed cells are resuspended in 0.05 ml of growth medium. The suspension is drawn into a piece

[8] C. M. Bradshaw, L. W. Clem, and M. M. Sigel, *J. Immunol.* **103,** 496 (1969).

[9] A grade, *N*-2-hydroxyethylpiperazine-*N*′-2-ethanesulfonic acid. Calbiochem No. 391338.

[10] Ethyl-*m*-aminobenzoate methanesulfonic acid. Sigma Chemical Co. No. E-1626.

of capillary polyethylene tubing (presterilized with ethylene oxide) by a vacuum created with a Cornwall syringe to which is attached a 23-gauge needle which fits into one end of the tubing. The vacuum can be precisely controlled by slowly turning the volume adjusting screw of the Cornwall. The syringe is removed and the tubing is sealed by heating and pinching shut one end. It is put into a Wintrobe capillary needle (Scientific Products) and centrifuged at 300 *g* for 10 minutes.[11] The tubing is removed from the needle and cut with a razor blade just above the packed red blood cells. The Cornwall apparatus is reattached and the white cells are expelled into growth medium by pressure from the syringe.

Separation of blood cells is more difficult with the snapper. The heparinized blood is transferred to a tube and centrifuged at 200 to 300 *g* for 10 minutes. With a Pasteur pipette the extremely thin buffy coat layer is removed along with large numbers of red cells. These cells are washed once with basal medium. The procedure for small blood volumes is then followed.

Alternatively, it is possible to achieve good cellular separation of snapper blood by mixing the blood with an equal volume of 6% dextran (mol. wt. 200,000–300,000) in 0.2 *M* NaCl and allowing it to stand for 8 to 16 hours at 25°C, or centrifuging it at 70 *g* for 10 minutes. The plasma–dextran mixture is aspirated and the cells which it contains are washed with SnM. At this point if further purification is desired one can proceed with the polyethylene tubing step or with separation of cells by means of the Isopaque Ficoll technique.[12] The washed cells are resuspended in growth medium, and a viable cell count is made.[13] All cells should be adjusted to a final concentration of 1×10^7 cells/ml. As a note of caution: cell viability can be diminished by excessive centrifugation; however, washings must be performed to remove the high molecular weight polymers.

Tissue Lymphocytes. The following procedure has thus far been tried extensively only with snappers. Thymus, spleen, and pronephros cells have been used individually or in combination when a single source of cells was inadequate.

Organs are dissected aseptically and placed in separate Petri dishes containing SnM. They are minced with scissors and transferred to a tube in which the fragments are agitated with a magnetic mixer in order to disperse the cells. This proceeds at 5°C for 1 to 2 hours. After the large fragments have settled, the supernatant is collected and centrifuged at 300 *g* for 10 minutes. The cells are then further purified using either polyethylene tubing or centrifugation in a 1-ml plastic syringe. The cells are mixed with SnM and drawn into the syringe. The extended plunger is cut with a hot scalpel. The syringe, needle pointing up, is centrifuged at 200 to 300 *g* for 10 minutes. The syringe is again cut with a hot scalpel below the needle so that the buffy coat may be removed with a Pasteur pipette. Cell counts are monitored at start and finish using the dye exclusion technique.[13]

[11] H. T. Holden, G. Ortiz-Muniz, and M. M. Sigel, unpublished, 1971.

[12] A. Boyum, *Scand. J. Clin. Lab. Invest.* **21** (Suppl. 97), 31 (1968).

[13] D. J. Merchant, R. H. Kahn, and W. H. Murphy, Jr., *In* "Handbook of Cell and Organ Culture," p. 157. Burgess, Minneapolis, Minnesota, 1964.

Assay for Blastogenesis

To each culture tube, 0.9 ml of growth medium and 0.05–0.1 ml of antigen or phytohemagglutinin in appropriate concentrations, or of growth medium, are added, followed by the addition of 0.1 ml of the suspension containing 1×10^7 cells/ml. For dealing with viral and other complex antigens or phytohemagglutinin, the optimal dose must be quantified by each investigator, because individual preparations differ in potency. With an antigen such as bovine gamma globulin, optimal stimulations have been obtained with concentrations of 10 to 100 μg.

The cultures are incubated at 25°C for 2½ days, at which time 2 μCi ^{3}H-thymidine are added to each. Sixteen hours later the experiment is terminated. From each culture, 0.9 ml is transferred to a glass tube; a viable cell count can be made on the residue. The cells are washed two to three times with 0.85% saline then subjected to two freeze-thaw cycles. To each tube, 0.025 ml of calf serum is added, and the acid-insoluble material is precipitated by the addition of 2 ml 5% trichloracetic acid (TCA). The precipitates are sedimented by centrifugation at 500 *g* for 10 minutes, the supernatants are decanted, and precipitates washed once with 5% TCA. The pellets are washed with abolute ethanol and the tubes inverted until dry. The resulting pellets are dissolved by the addition to each tube of 0.5 ml 10X hydroxide of hyamine.[14] The tubes are sealed with parafilm and incubated overnight at 37°C. Each fluid is combined with 8 ml toluene-based PPO and POPOP scintillation fluid, and the radioactivity determined by counting in a scintillation counter.

[14] *p*-(Diisobutyl-cresoxyethoxyethyl) dimethylbenzyl ammonium hydroxide. Packard Instrument Co. No. 6003005.

CHAPTER 23

Freshwater Fishes

Ken Wolf

Since 1955, this laboratory has helped develop methods of fish tissue culture and eventually initiated the first established fish and amphibian cell lines. The methods have been taught here since 1960. The ultimate goal has always been virological application in fish disease research. Although cell lines provide the mainstay of virological needs, there is still some need for primary cultures of organ or structural tissues and leukocyte cultures find application in cytology and genetics.

It is from such a background that this chapter has been prepared. For a comprehensive review and discussion of fish cell and tissue culture, the reader is referred to Wolf and Quimby.[1]

Physiological Salines and Media

The inorganic constituents of sera from mammals, teleosts, and even from cyclostome fishes in freshwater environments are qualitatively and quantitatively similar. There is then a physiological basis for the many reports which show that mammalian type salt solutions and media are appropriate for fish cell culture.

On the basis of its widespread use, Hanks' BSS is recommended as the most popular physiological saline for fishes. Earle's and the Cortland BSS[1] are equally appropriate, but pH control with the former is more difficult, and the latter is not available commercially.

Eagle's MEM (with Earle's BSS) is similarly recommended, but BME, NCTC 109, L-15, and Medium 199 are also appropriate and have been used successfully.

Without reservation, fetal bovine serum is the nutrient supplement of choice, and a 10% level has generally been found to be a compromise between effectiveness and economy.

Antibiotics routinely used in mammalian tissue culture are suitable at comparable levels for fish cells. Penicillin at 100 IU, streptomycin at 100 μg, and nystatin at 25 IU/ml are recommended for routine work, but tenfold higher levels of the first two may be employed with tissues such as gills and fins.

Temperature, pH, and Gases

On the basis of physiological requirements and environmental preferences, fishes are categorized as "coldwater" or "warmwater" species, and the temperature factor applies to *in vitro* culture conditions. Growth of salmonid and other coldwater fish cells extends from about 5° to 25°C with optima being near 20°C. Cells from warmwater fishes generally do well at 25° to 30°C but may grow at 15° through 35°C and higher.

The initial pH of culture media should be about 7.3 to 7.4 for optimal results. Many fish cell cultures will grow, but with somewhat reduced activity, at pH 7.0 to 7.2 and as high as 7.8–8.0.

Fish cells have a requirement for CO_2; this is usually met by atmospheric gas

[1] K. Wolf and M. C. Quimby, *In* "Fish Physiology" (W. S. Hoar and D. J. Randall, eds.), Vol. 3, p. 253. Academic Press, New York, 1969.

and/or bicarbonate. Poor cultures result if neither is present. Salmonid leukocyte cultures are mitotically most active in the presence of 100% oxygen.

Precautions and Preparation of Fish

All fish tissues are living, but some more than others harbor microbial flora, protozoan, and metazoan parasites. At times, apparently healthy fish may have a low level bacteremia while others are virus carriers. All should be taken into consideration when primary cultures are to be established on a routine basis, for one or more can seriously interfere.

Consistent success is usually obtained when internal tissues are removed aseptically from healthy and specific pathogen-free donors. To reduce risk of fecal contamination, food should be withheld for several days prior to use. External decontamination is achieved by holding a freshly killed donor in 500 ppm chlorine for several minutes, draining it, then sponging the side to be opened with 70% ethanol (Fig. 1).

Embryos may be obtained after similar treatment of gravid live-bearing fishes or suitably developed eggs.

Fin, corneal, and gill tissues may be decontaminated with bactericidal antibiotics active against gram-negative organisms. Hour-long treatment with 500 IU polymyxin B, 500 μg neomycin, and 40 IU bacitracin per milliliter is a worthwhile precaution, but gills are very difficult to decontaminate.

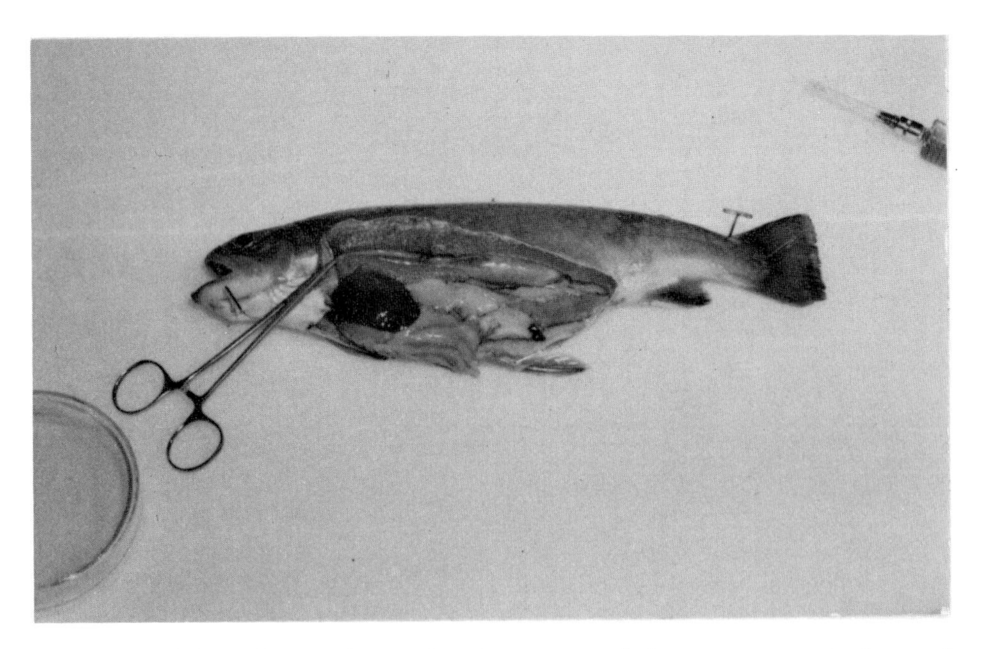

Fig. 1. Adult brook trout showing lateral opening used for aseptic removal of internal organs. Hemostat is secured to anterior of left ovary to facilitate removal of that organ without contamination.

Choice of Tissues and Their Preparation

Ovarian and embryonal tissues consistently provide good culture material. Immature testes, kidneys, spleen, swim bladder, heart, and liver may also be used, but young animals are preferred. Caudal peduncle, the trunk posterior to the anus, will usually yield successful cultures providing that young animals are used and that they are adequately decontaminated.

Three to ten grams of tissue provide a convenient starting mass. If necessary, pool several tissues or organs from several specimens. Do not wash. Transfer to a 50-ml beaker and add only enough BSS to allow fragments to flow during mincing. Reduce fragments to about 1 mm^3. Seven to ten minutes mincing with sharp scissors will usually be required.

Method 1. Direct Planting. Wet a bent tip Pasteur pipette with BSS and transfer minced tissue to glass or plastic culture vessels at the rate of 4 to 6 drops per 25 cm^2 area. Spread tissue fragments evenly over growth area using *back* of pipette tip to reduce scratching of plastic. Secure closure and let stand on edge or bottom to allow tissue adherence and liquid drainage. After 1 to 1.5 hours at suitable temperature, remove drained fluid, position vessels with tissues up, add culture medium, close, and move to place of incubation. There, the vessels should be carefully righted and the medium allowed to move smoothly over the attached tissues. This is the easiest and simplest way of establishing primary cultures and is well suited for flasks. Compared to tryptic digests, the procedure requires longer time to achieve confluency.

Method 2. Trypsin Dispersion. Wash minced tissue with BSS to remove blood cells and tissue debris. Transfer tissue to trypsinizing flask containing a Teflon-covered magnetic stirrer and add about five volumes of digestion mixture[2] at 0°–5°C. Place on magnetic stirrer at about 5°C and stir at low speed for 30 to 90 minutes. Remove flask and allow fragments to settle for several minutes. Decant supernatant fluid and discard. (Although cultivable cells will have been released, the yield is usually low and the cultures establish slowly, if at all.)

Add fresh digestion mixture to the original volume and stir at 5°C for at least several hours. It is very convenient to start final digestion at the end of the workday and to terminate it early the next day. As an alternative, released cells and small tissue fragments may be harvested at 3- to 6-hour intervals depending on the cell mass needed and upon the work schedule.

Ovarian, renal, hepatic, and swim bladder tissues contain ciliated cells which provide an immediate index of viability at any time during the procedures.

Some tissues trypsinize more readily than others. Embryonic and young animal tissue is dispersed rather easily. Therefore, one should depend less upon a time–temperature schedule than upon the progress of digestion. When the digestion

[2] Digestion mixture: 87.5% phosphate-buffered saline at pH 7.2–7.4; 10.0% trypsin solution (2.5% concentration; 1:250 activity); 0.5% serum. Optionally add 100 IU penicillin plus 100 μg streptomycin per milliliter.

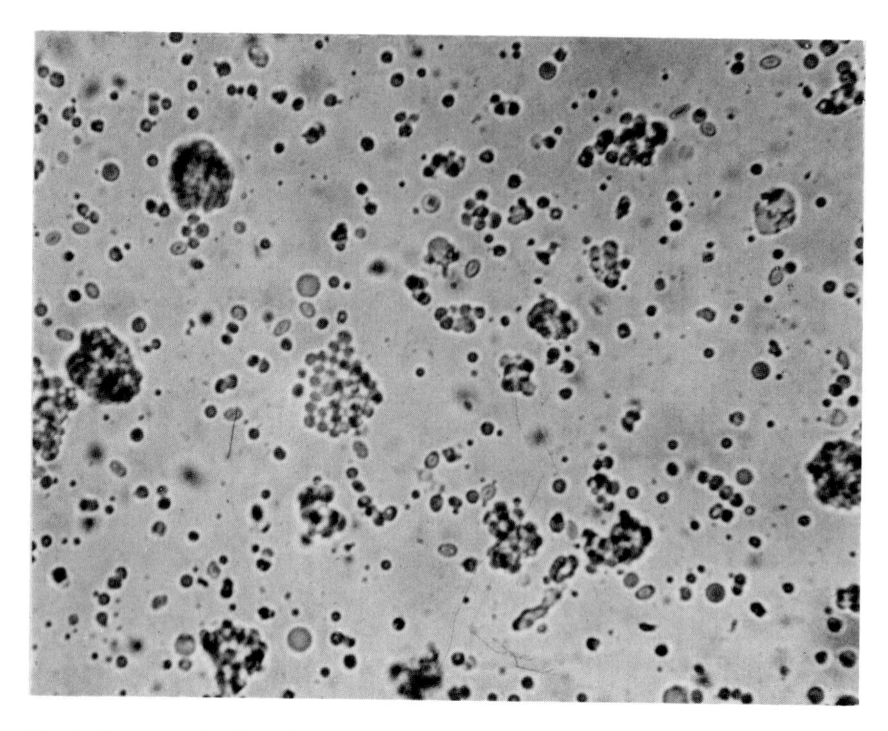

Fig. 2. Example of cell and tissue fragment suspension obtained by trypsinization of immature trout ovaries. This degree of dispersion is ideal for successful planting of primary monolayer cultures. Ciliated cells in this preparation were beating actively and trypan blue dye exclusion indicated that viability exceeded 95%.

reaches the point where there is a mixture of individual cells among a preponderance of very small tissue remnants consisting of anywhere from 10–1000 cells, a satisfactory disaggregation has been achieved. An example of such a preparation is shown in Fig. 2.

There are about 23,000 types of fishes but the cold trypsinization is appropriate for many and possibly for all but the thermophilic species. Normal ovarian tissue has universally given successful cultures; moreover, it may be used at any stage of maturation. Immature ovaries are preferred, but after removing near-term eggs, the mature ovary can be depended upon to provide cultivable cells.

At harvest, cells and small tissue fragments are sedimented in the cold at 200 *g* for 10 minutes. Supernatant fluid is decanted or aspirated. Washing is unnecessary. The pelleted material is carefully but thoroughly resuspended in 400 to 600 volumes of growth medium and planted. Greater density will speed attainment of confluency, but some workers have satisfactorily used dilutions of 1:1000 and greater. Because the best preparations contain small tissue fragments, counting is neither accurate nor worthwhile.

Salmonid ovarian tissue digests should provide active monolayers at near confluency after 3 to 5 days at 20°C. Depending upon the tissue, seeding density, and temperature of incubation other materials may achieve the same result in greater or lesser time.

There is no need to change culture medium before monolayers become confluent.

Primary Monolayer Cultures of Freshwater Fish Leukocytes

Healthy fish are anesthetized with 1:5000 to 1:25,000 tricaine methanesulfonate.[3] A small area surrounding the lateral line is topically disinfected with 70% ethanol or isopropanol and aseptically bled with a plastic syringe wetted with a 100 mg% solution of heparin in Hanks' BSS. Blood is gently mixed in the syringe, the needle removed, and replaced with a cap. The syringe is fitted with a lock (the cylindrical portion of a plastic syringe of the same size but from which about one-fourth the circumference has been removed; this locks the syringe plunger in place) and centrifuged at 200 *g* for 10 minutes in the inverted position. Cells of the buffy coat are harvested for planting by replacing the cap with an 18- to 20-gauge needle bent 90°. Plasma is discarded and the visibly turbid leukocyte layer dropped into growth medium. Salmonid leukocytes prosper under an atmosphere of 100% oxygen, but that is generally not necessary for cells from other fishes. Phytohemagglutinin may be added to stimulate mitosis. Seeding densities of 0.1 to 1.0 × 10^6 cells/ml are suggested.

[3] F. J. Bové, MS-222 Sandoz, p. 20. Sandoz Pharmaceuticals, Hanover, New Jersey 07936.

CHAPTER 24

Goldfish Tissues[1]

Laura S. McKenzie and N. G. Stephenson

Our attempt to culture fish cells under conditions identical to those used for culture of mammalian cells was prompted by the results of two groups of workers.

First, Wolf and his associates showed that Medium 199, which had been developed for growth of mammalian cells, could be used in modified form for culturing tissues of rainbow trout at temperatures up to 26°C and goldfish below 22°C.[2] Rainbow trout gonadal cells grown in this way were used to establish the

[1] Procedure at 37°C.

[2] K. Wolf and C. F. Dunbar, *Proc. Soc. Exp. Biol. Med.* **95**, 455 (1957).

first fish cell line, which could be continuously cultivated in a number of standard mammalian-type media, with a mammalian serum supplement.[3]

Second, it was demonstrated that tissues from some exothermic animals can be cultured at temperatures up to 40°C,[4,5] and that the temperature tolerance of cultured explants is correlated with the latitudinal range (i.e., environmental temperature range) of the tissue donors.[6]

It, therefore, seemed likely to us that selection of fish with anticipated high temperature tolerance as tissue donors would allow fish cells to be cultured at the mammalian temperature of 37°C, as well as in mammalian-type media. Goldfish can survive temperatures up to 40°C in aquariums.

Procedure

Tissues to be cultured were removed under aseptic conditions. Anesthetized fish were swabbed with 70% alcohol before and after removal of scales and skin in order to destroy surface microorganisms. All salines and media used contained the following antibiotics: sodium penicillin G, 40 IU/ml; streptomycin sulfate, 50 μg/ml; Fungizone (amphotericin B) 2.5 μg/ml.

Adult male and female goldfish were selected and organs of each type pooled; ovaries and testes were kept separate. Organs to be cultured were dissected out and placed in sterile Hanks' saline in covered Petri dishes. Pooled organs were rinsed twice in saline to wash off blood, and minced with fine scissors into approximately 1-mm cubes.

The minced tissues were placed in centrifuge tubes containing 0.25% trypsin in Hanks' calcium- and magnesium-free saline, and agitated with a wide-mouthed pipette at 5-minute intervals for 15 minutes. The trypsin solution was removed by centrifuging the tissue suspension at 500 to 1000 rpm for 3 minutes, and discarding the supernatant. The minced tissue was washed by resuspending in culture medium, centrifuging, and discarding the supernatant. This enzymatic digestion weakens intercellular connections in the explant and greatly enhances cell outgrowth.

The culture medium used was Eagle's Minimum Essential Medium supplemented with 20% calf serum (from 3 to 6 month calves) and 10% chick embryo extract (from 9-day chick embryos). The usefulness of the embryo extract was tested and found to greatly improve cell growth. Medium 199 with the same supplements was also used successfully for cultures grown at 31.5°C.

After trypsinization and washing, the minced tissues were set up as primary cultures in small covered Petri dishes. Explants were placed 1–2 mm apart in dishes which had dried collagen films over the bottom, to increase the number of explants that attached and to stimulate growth. Sufficient medium was added to cover the bottom of each dish, but not enough to cause the explants to float

[3] K. Wolf and M. C. Quimby, *Science* **135**, 1065 (1962).
[4] N. G. Stephenson and J. K. N. Tomkins, *J. Embryol. Exp. Morphol.* **12**, 825 (1964).
[5] E. M. Stephenson, *J. Embryol. Exp. Morphol.* **17**, 147 (1967).
[6] E. M. Stephenson, *Aust. J. Biol. Sci.* **21**, 741 (1968).

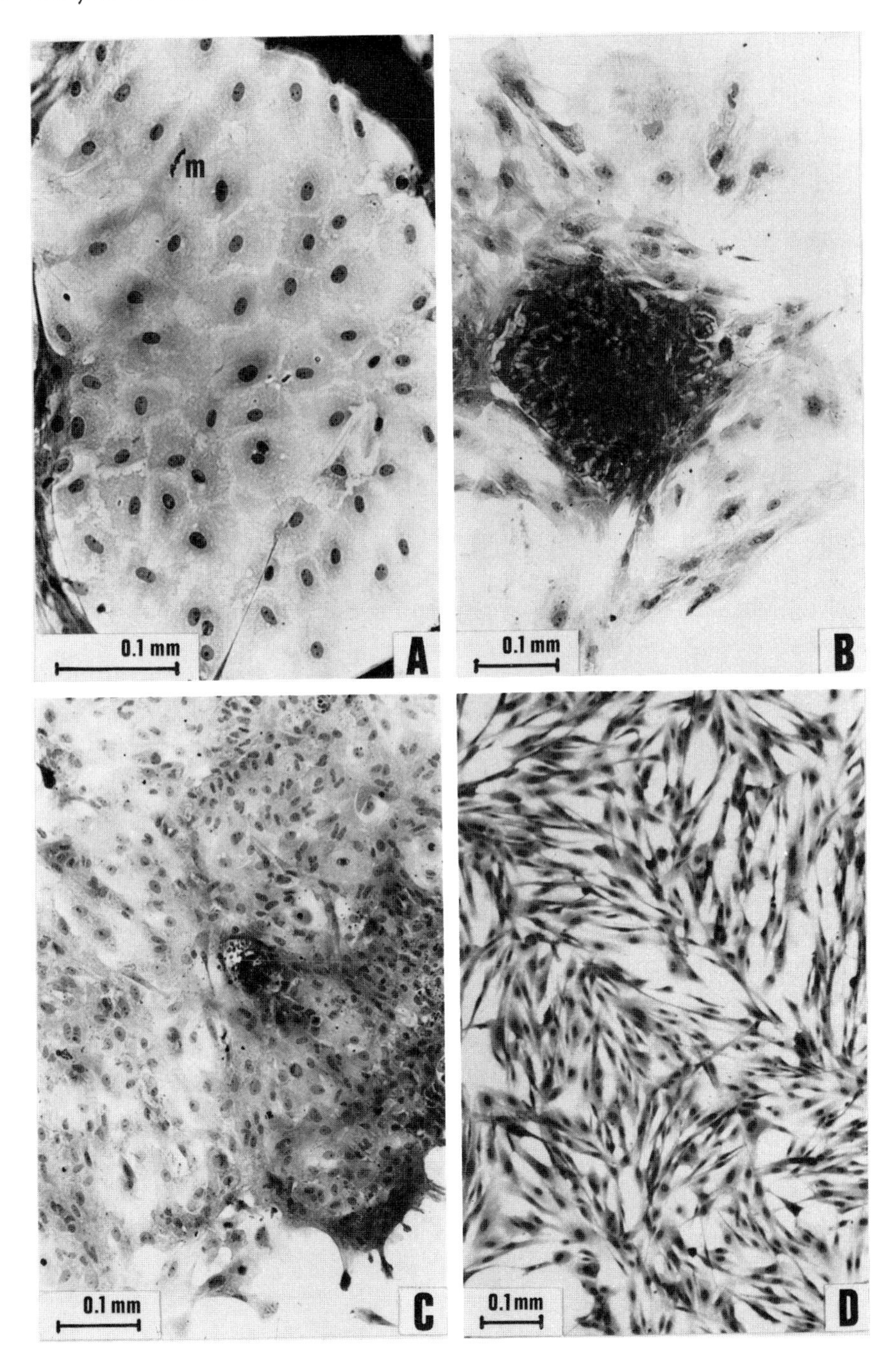

Fig. 1. (A)–(C) show primary cultures of goldfish tissues after 5 days growth at 37°C. Fixation, 24 hours, 10% formol saline. Staining, 20 minutes, Mallory's aqueous hematoxylin. (A) Epithelial outgrowth from heart explant (removed). Colcemid added 2 hours before fixation; m, cell arrested in metaphase. (B) Liver explant, with outgrowth. (C) Kidney explant with outgrowth. (D) Cells of SZGT-37 line, 94th passage at 37°C. Fixation and staining as before.

rather than become attached. The cultures were incubated at 37°C in a gas mixture (5% CO_2, 20% O_2, balance N_2). After 2 to 3 hours 4 ml medium were added to each culture and the Petri dishes returned to the incubator and regassed.

Cultures were observed daily under an inverted microscope. Medium in each dish was changed after the first day to remove blood cells and unattached explants, and changed each 1–2 days thereafter.

Application

Successful cultures were obtained from heart, kidney, swim bladder, liver, ovary, and testis. Cell outgrowth from explants could be seen after 24 hours, and dividing cells were evident in fixed and stained cultures.

Cultures for histological examination and photography were grown on coverslips coated with collagen and placed in the bottoms of Petri dishes. These cultures could be fixed and stained in the Petri dishes, and the coverslips then mounted on glass slides for examination (Fig. 1).

Concurrently with this work, a fibroblastic fish cell line designated SZGT-31[7] was established at 31°C. At the 70th passage some cells were transferred to 37°C; they have been continuously cultured at this temperature for 96 passages so far and are now designated as SZGT-37. Both the 31° and 37°C lines have been maintained in culture in Eagle's Minimum Essential Medium with 15% calf serum. Both show a doubling time of about 18 hours, which is comparable to mammalian cells, and have hyperdiploid chromosomal complements.

It seems likely that the tissues of other exothermic chordates could also be cultured at relatively high temperatures providing these do not exceed those experienced by the animal under natural conditions.[8] Primary cultures and cell lines which may be cultivated over a wide temperature range should prove particularly useful in research.

[7] SZGT was chosen to indicate the laboratory (Sydney Zoology) and tissue (goldfish testis) of origin.

[8] M. Gravell and R. G. Malsberger, *Ann. N. Y. Acad. Sci.* **126**, 555 (1965).

CHAPTER 25

Reptiles

H Fred Clark

Methods for the cultivation of reptilian cells have been investigated in our laboratory since 1961[1] and methods for the primary culture of Grecian tortoise (*Testudo graeca*) cells were reported as early as 1962.[2,3] There is, however, no single technique by which success may be assured with any species of reptile. This can be attributed not only to the wide diversity of reptilian species exhibiting a diversity of cellular responses to various culture conditions, but also to the paucity of reports of investigations in this area. The attempt here is to describe cell culture techniques which have been successfully applied to a limited number of reptilian species and to suggest principles that may aid the investigator in developing new methods appropriate for species not previously studied. The preparation of cultures derived from explants and from trypsin-dispersed cells will be considered separately.

Reptilian species vary greatly in the ease with which their cells may be cultured by techniques presently available. We have obtained the most vigorous growth with tissues of the turtle *Terrapene carolina,* the lizards *Iguana iguana, Gekko gecko,* and *Eublepharis macularis,* the snake *Vipera russelli,* and the crocodile *Alligator mississippiensis.* Others have reported consistent success with tissues of the European turtle *Testudo graeca*[2,3] and with the lizard *Anolis carolinensis*[4] (regenerating tissues only). Although adult tissues are readily cultured, whole embryos may also be useful, and with some snakes and lizards[5] may yield superior cell growth.

With each species, the cell incubation temperature is critical, despite the fortunate fact that poikilothermic vertebrate cells can replicate over a wider range of temperatures than mammalian or avian cells.[6] In general, we have experienced maximum success with cell growth of tropical species at 30°C and temperate climate species at 23°–25°C. Primary cell cultures grow most rapidly at warmer temperatures up to 37°C,[1,2] but prolonged maintenance of cell cultures (with or without subcultivation) is usually most successful at temperatures slightly below those supporting the maximum rate of cell multiplication.

[1] H F. Clark, M. M. Cohen, and D. T. Karzon, *Proc. Soc. Exp. Biol. Med.* **133**, 1039 (1970).

[2] L. Shindarov, *Dokl. Bolg. Akad. Nauk.* **15**, 539 (1962).

[3] B. Fauconnier and M. Pachopos, *Ann. Inst. Pasteur (Paris)* **102**, 661 (1962).

[4] S. B. Simpson and P. G. Cox, *Science* **157**, 1330 (1967).

[5] J. Somogyiova, *Biol. Bratislava* **19**, 257 (1964).

[6] H F. Clark and L. Diamond, *J. Cell. Physiol.* **77**, 385 (1971).

ESTABLISHMENT OF CULTURES FROM TISSUE EXPLANTS

Animals are most conveniently sacrificed by decapitation, followed by exsanguination through the severed cervical blood vessels. Prior to sacrifice, anesthesia may be administered by inhalation of halothane (Fluothane, Ayerst Labs Inc., New York) or more simply, the animal may be rendered stuporous by chilling overnight in a refrigerator (4°C). It is recommended that food and water be withheld from turtles and lizards for several days prior to sacrifice; otherwise considerable difficulty may be encountered in removing the kidneys without piercing the distended overlying cloacal sac.

The animal is pinned on its back to a dissecting board, the ventral surface is disinfected by liberal washing with 70% ethanol, and the visceral organs are exposed by means of a ventral incision performed with sterile instruments (the plastron of turtles must be removed with a hacksaw). The organs selected for cell culture are removed and placed in sterile 50-mm Petri dishes, each containing ca. 5 ml of sterile amphibian Ringer's solution. Visceral organs suitable for explant cell culture, listed in the descending order of frequency of success, are the heart, lungs, spleen, kidneys, liver, and gonads. Success has also been obtained occasionally with subcutaneous connective tissue and muscle tissues taken from lizard tails.[7] The organs may be processed immediately or stored in Ringer's solution at 4°C, for several days with little loss in cell viability.

The tissues are coarsely minced with fine scissors to yield fragments 2–3 mm in diameter and then thoroughly washed with Ringer's solution to remove suspended erythrocytes. The tissue fragments are then very finely minced in a minimum amount of fluid until no pieces exceed 1 mm in diameter. These small tissue fragments are resuspended in growth medium at a concentration of 10 to 20 fragments per milliliter and plated in sealed glass or plastic tissue culture vessels placed at appropriate incubation temperatures (see above). When initiating cultures from previously unstudied reptilian species, it is recommended that replicate flasks be incubated at several different temperatures. The volumes of tissue suspension explanted are 5.0 ml for a 25-cm^2 plastic flask or 4-ounce glass prescription bottle or 20 ml for a 75-cm^2 plastic flask or 225-ml "milk dilution bottle."

The best growth medium is Eagle's Basal Medium containing 0.15% HCO_3^- or $NaHCO_3$, and supplemented with 10% fetal bovine serum. The addition of antibiotics to the medium is essential, as primary reptilian cell cultures are frequently contaminated with microorganisms apparently harbored in grossly normal visceral organs. The recommended concentrations of antibiotics are 200 units/ml of penicillin with 200 μg/ml of streptomycin. While the use of medium diluted from mammalian electrolyte concentration to the approximate tonicity of reptilian body fluids, and the inclusion of homologous serum supplements, have both been

[7] F. Michalski, The Wistar Institute, Philadelphia, Pennsylvania, unpublished observations.

recommended occasionally for reptilian cell cultures (recommendations reviewed by Clark[8]), we find no advantage in these procedures.

Explant cultures should be left undisturbed for several days to allow for complete attachment of explants to the substrate. Cell outgrowth from explants, commonly of mixed morphological types, can be expected within 7 to 10 days of incubation. The medium should be replaced whenever acidic conditions are observed, usually about once a week. Cultures may be maintained in this manner until the cell outgrowth becomes confluent, a condition commonly requiring at least several weeks. Confluent cultures may be obtained more rapidly by dispersing the cells of the explant outgrowths when the cell cover has extended over 20–50% of the culture vessel surface. The culture medium is decanted, and the cells are bathed in a small volume (approximately one-fifth the volume of growth medium employed) of 0.25% trypsin (Difco) and 0.02% Versene (ethylenediaminetetraacetic acid) (EDTA) in calcium- and magnesium-free phosphate-buffered saline solution, at the incubation temperature used for cell growth. When the cells have separated from the substrate, they are further dispersed by vigorous mixing with a pipette, replated in the original culture vessel, and refed with fresh growth medium. Confluent cultures of such "second passage" cells are often obtained within 1 week. Further subcultures employing the same techniques and "cell splits" at a 1:2 ratio are often successful.

Establishment of Cultures from Dispersed Cells

The culture of enzymatically dispersed cells offers the potential advantage of providing (at one time) a large number of uniform cell cultures from a single source. However, the conditions of dispersion and cell culture must be more rigidly controlled than in explant cell culture, since individually isolated cells are more rapidly affected by deleterious culture conditions than are tissue explants. We have had success primarily with dispersed cell cultures prepared from the kidneys of turtles and lizards, the American alligator and, in one instance, a snake of the genus *Python*.

Shindarov's method[2] for the culture of kidney cells of *T. graeca* has been most satisfactory in our studies. As much of the capsular and pelvic connective tissue as possible is removed before the remaining parenchymatous tissue is cut into pieces approximately 2 mm in diameter. The tissue pieces are thoroughly washed in amphibian Ringer's solution and then resuspended in approximately 25 volumes of a solution of 0.05% trypsin and 0.075% Versene in calcium- and magnesium-free phosphate-buffered saline solution in an Erlenmeyer flask. The suspension is stirred with a magnetic mixer at room temperature. At 30- and 60-minute intervals, depending on the time of the appearance of cloudiness in the dispersing fluid indicative of large numbers of free cells, the supernatant fluid is removed

[8] H F. Clark, *In* "Growth, Nutrition and Metabolism of Cells in Culture" (G. Rothblat and V. Cristofalo, eds.), Vol. II, 287. Academic Press, New York, 1972.

and placed at 4°C. After each cell harvest, new dispersing solution containing serial twofold decreases in the concentration of Versene is added.

Cell dispersion is normally complete following three or four such tissue treatments. The pooled cells accumulated at 4°C are sedimented by centrifugation for 10 minutes at 100 to 200 *g*. The supernatant fluid is discarded and the packed cells are resuspended at a concentration of 2 to 5×10^5 cells/ml in growth medium (see above—explant cultures), and plated in glass or plastic vessels at the appropriate incubation temperature. Subsequent care of the cultures is similar to that of explant cultures, except that because of the smaller total acid-producing tissue mass in dispersed cell cultures more care must be taken to prevent the cultures from becoming too alkaline. Under optimal conditions a uniform epithelial-like cell monolayer is formed 7 to 14 days after cell plating.

CHAPTER 26

Dipteran Embryos and Larvae (Diploid Lines)

Imogene Schneider

Dipteran cell cultures have been used primarily for the cultivation of arboviruses (in mosquito cells)[1] and for studies in developmental genetics and cytogenetics (*Drosophila* cells).[2] The method given below was introduced in 1967 by K. R. P. Singh[3] and leads to the establishment of cell lines which are predominantly diploid. Some aneuploidy and polyploidy is evident, but in no instance do the numbers of chromosomes approach those sometimes found in hemipteran lines and invariably found in lepidopteran lines. An additional advantage of this method is that there is usually little or no lag period between the initiation of the primary culture and the subsequent transition to subculturing.

Surface Sterilization of Embryos

Mosquito eggs are collected within 12 hours of deposition and allowed to develop to the point of red eyespot formation. The eggs are transferred to a small diameter plastic cylinder with a screen of fine wire (200–400 mesh) or nylon

[1] C. E. Yunker, *Curr. Top. Microbiol. Immunol.* **55**, 113 (1971).
[2] C. Halfer, L. Tiepolo, C. Barigozzi, and M. Fraccaro, *Chromosoma* **27**, 395 (1969).
[3] K. R. P. Singh, *Curr. Sci.* **36**, 506 (1967).

monofilament cloth affixed to one end. The cylinder is placed in a solution of 2.5% NaOCl (a 1:1 solution of Clorox and distilled water suffices) and the eggs continually agitated for 2 minutes. The cylinder is then inverted and the eggs washed into a small Petri dish with 70% ethanol. Any debris, including the exochorions removed by the Clorox treatment, is discarded with a Pasteur pipette. After 10 minutes in 70% ethanol, the eggs are immersed in 0.05% $HgCl_2$ in 70% ethanol for an equal length of time. Following sterilization, the eggs are thoroughly rinsed in sterile saline or culture medium and transferred to a Petri dish containing a layer of sterile filter paper wetted with culture medium. The dish is placed overnight in a 22°–25°C incubator and the embryos allowed to hatch.

The procedure for *Drosophila* eggs is virtually identical except that it is preferable to use embryos which are a minimum of 6 hours old. The Clorox treatment should last no longer than 1–1½ minutes; otherwise the eggs become very sticky and extremely difficult to handle. After the rinsing step it is advisable to place the eggs on black filter paper (Fisher Scientific). Against the dark background the stages of embryonic development are readily seen and nonviable embryos easily detected. Embryos between the ages of 16 and 22 hours at 25°C appear to give the best results in primary cultures.

Inoculum Preparation and Trypsin Treatment

Neonate mosquito larvae or *Drosophila* embryos are transferred a few at a time to a depression slide containing a small amount of medium. They are each cut into two or three pieces with a tungsten needle[4] and then pipetted into a 10-ml beaker containing a few milliliters of medium. The beaker should be placed in ice water. Once a sufficient number of fragments have been collected (a minimum of 100 *Drosophila* embryos and from 20 to 50 mosquito larvae for each T-9 flask, Bellco or Kontes) they are transferred to a sterile cylinder (see above) and placed in 0.2% trypsin (1:250, Difco) in Rinaldini's salt solution[5] for 10 to 15 minutes at 37°C or from 45 to 60 minutes at room temperature. After inactivation of the trypsin with fetal bovine serum the fragments are rinsed two to three times with medium and then placed in a T-9 flask containing 1.25 ml complete medium. The cultures are incubated at 22°–25°C with a gaseous phase of ambient air.

Culture Media

Most media have been patterned after the composition of the appropriate hemolymph, if known, or have been derived in an empirical fashion. In neither instance are the media as optimal as they might be in meeting the physical and nutritional needs of the cells (see J. L. Vaughn's article, Section XIV). At present, relatively adequate media are available for both primary cultures and established

[4] H. Ursprung, *In* "Methods in Developmental Biology" (F. H. Wilt and N. K. Wessells, eds.). Crowell, New York, 1967.

[5] L. M. Rinaldini, *Nature* (*London*) **173**, 1134 (1954).

cell lines initiated from various species of *Aedes*,[6] *Anopheles*,[7,8] *Culex*,[9,10] and *Drosophila*.[11-13] The media are supplemented with 10 to 20% fetal bovine serum.

Development of Primary Cultures and the Establishment of Diploid Lines

The trypsin serves only to loosen the matrix between the cells as visually there is no difference in the appearance of the fragments before and after treatment. The explants behave in one of two ways: either (1) individual cells or groups of cells drop to the floor of the flask and form colonies or (2) growth takes place in the form of hollow, cellular spheres or vesicles issuing from the cut ends of the embryonic or larval fragments. If the former condition prevails, the colonies are allowed to grow until they reach a few millimeters in diameter. The medium is then partially renewed and the colonies pipetted off the bottom of the flask. If the individual cells reattach and resume multiplying it is usually possible to subculture within 7–10 days. Initially a density of about 5×10^5 cells per milliliter should be used for seeding a new flask. After a few subcultures this figure can often be reduced tenfold.

If growth is in the form of cellular vesicles, they may dislodge from the fragments by themselves or must be excised with tungsten needles. In the latter case, both fragments and vesicles are returned to the original flask to increase the cell density. Healthy fragments will continue to produce additional spheres which equal or surpass in number and size those of the initial growth. It is not known how long this can continue but some fragments have produced spheres at an undiminished pace more than 2 years after the primary cultures were initiated. By a budding process the spheres themselves may give rise to new spheres. Thus one T-9 flask may contain hundreds of these vesicles within the space of a week or two, some having perhaps 50 cells, others, thousands of cells. (This is especially true for explants from *Aedes* species. Vesicle formation and growth is considerably slower in *Anopheles* and *Drosophila* explants and colony formation is the rule with explants from *Culex* species.)

Usually sufficient numbers of spheres attach to the bottom of the flask, flatten out, and form colonies thereby assuring adequate cell densities for subculturing. Otherwise, they must be manually ruptured or treated with trypsin to release the cells. Subculturing is often possible within 4–7 days using the same cell density given above. The doubling time for most dipteran cell lines initiated by this method varies between 16 and 30 hours.

[6] J. Mitsuhashi and K. Maramorosch, *Contrib. Boyce Thompson Inst.* **22**, 435 (1964).
[7] I. Schneider, *J. Cell Biol.* **42**, 603 (1969).
[8] M. Pudney and M. G. R. Varma, *Exp. Parasitol.* **29**, 7 (1971).
[9] S. H. Hsu, W. H. Mao, and J. H. Cross, *J. Med. Entomol.* **7**, 703 (1970).
[10] S. Kitamura, *Kobe J. Med. Sci.* **16**, 41 (1970).
[11] V. T. Kakpakov, V. A. Gvosdev, T. P. Platova, and L. C. Polukarova, *Genetika* **5**, 67 (1969).
[12] G. Echalier and A. Ohanessian, *In Vitro* **6**, 162 (1970).
[13] I. Schneider, *J. Embryol. Exp. Morphol.* **27**, 353 (1972).

CHAPTER 27

Cockroach and Grasshopper Embryo Tissue

E. P. Marks

Among the more primitive groups of insects are several that lay their eggs in protective structures known as oothecae. These include cockroaches, mantises, grasshoppers, and some primitive termites. At least two of these (cockroaches and grasshoppers) are of considerable interest from an economic standpoint, and numerous attempts to culture their tissues have been made. Landureau[1] has established several strains of cockroach cells, but to date no strain of grasshopper cells has been reported. The preparation of cell cultures from the embryonated eggs of these insects is made easier because each ootheca may contain as many as 50 eggs, all approximately the same age and in the same stage of development. This permits obtaining a fairly large amount of partially differentiated tissue. Furthermore, these eggs are so packaged that they can be readily surface-sterilized without damaging the enclosed embryos.

Obtaining and Preparing the Embryos

Cockroaches. The Madeira cockroach, *Leucophaea maderae* (Fabr.), with which we have had the most success carries its ootheca internally in a brood pouch. The eggs are enclosed by a thin oothecal membrane and can be removed from the brood pouch by gently squeezing the abdomen of the gravid female. The stage of development of the embryos can be seen through the transparent oothecal membrane. We have found that the best stage for culture is before the eyespots form and just prior to dorsal closure. At this stage, the embryos can easily be lifted off the yolk mass and broken up to produce a suspension of cells. Older embryos can also be used, but each embryo must be opened and the enclosed yolk removed.

To prepare embryos for culture, three or four oothecae are selected in about the same stage of development and care taken not to prematurely rupture the oothecal membrane. The unopened oothecae are washed gently for about 3 minutes in a dish of sterilizing solution (0.2% solution of Hyamine[2] 1622 in 70% ethyl alcohol) rinsed carefully in three changes of sterile distilled water, and placed in a dish of culture medium. The membranes of the oothecae are then

[1] J. C. Landureau, *Exp. Cell Res.* **41**, 545 (1966).

[2] Mention of a proprietary product does not constitute an endorsement by the U. S. Department of Agriculture.

torn open and the embryos removed with a pair of fine forceps and placed in a dish of fresh medium.

Grasshoppers. The grasshoppers with which we have had the most success are *Melanoplus differentialis* (Thomas) and *M. bivitattus* (Say). The egg pods are laid in the soil during late summer and early fall and can either be collected in the field or the gravid females can be brought into the laboratory, fed on lettuce leaves and provided with moist sand in which to deposit their eggs. The pods should be held at room temperature for about 10 days after deposition and then packed in moist sand and stored in a refrigerator (5°C) until needed. Under these conditions, the partially developed embryos will remain in diapause for up to 3 months.

To prepare the embryos for culture, the individual eggs must be removed from the egg pods and cleaned before sterilizing. The eggs are sterilized by placing them in a small, flat-bottomed sieve which is then dropped into a dish of sterilizing solution and agitated for about 3 minutes to remove any adhering material that might prevent complete wetting of the eggs. The eggs are then rinsed three times in sterile distilled water, and any cracked or otherwise imperfect eggs are discarded. The eggs are then placed in a small dish of culture medium, cracked open, and the embryos removed to a separate dish.

Culture Medium. The culture medium of Landureau[1] has been used to maintain several cell strains from cockroaches (*P. americana*). In our work with *L. maderae* and *M. bivitattus*, we used the holidic Medium M14 (Marks *et al.*[3]), and cultures can be maintained in good condition for 60 to 90 days. More recently, we have used a modification of this medium (designated M20, Table I) with 7.5% fetal calf serum supplement. The M20 Medium permits the maintenance of *L. maderae* and *M. bivitattus* cells for up to 1 year.

Preparing the Tissue for Explantation

From this point, the same procedure will apply to both cockroach and grasshopper embryos.

Before culturing the embryo tissue, it is desirable to remove as much of the yolk material as possible. Embryos in which dorsal closure has not yet occurred are gently lifted off the surface of the yolk mass and placed in a dish of culture medium. Such embryos are broken up by forcing them through a Pasteur pipette, and the resulting cell suspension is counted in a hemocytometer, diluted to 1×10^6 cell/ml, and seeded in 20-ml plastic flasks (3 ml/flask). Problems are encountered with older embryos. First, the embryo must be torn open and as much of the yolk-filled digestive tract as possible removed. The embryos can then be transferred to another dish and minced with a sharp scissors and the tissues further separated by vigorous pipetting. The cell suspension is then pipetted off, leaving the fragments of body wall behind, and centrifuged in a clinical centrifuge at 1000 rpm for 2 minutes. The supernatant containing any yolk ma-

[3] E. P. Marks, J. P. Reinecke, and J. N. Caldwell, *In Vitro* 3, 85 (1967).

TABLE I
Culture Medium M20

Solution[a]	mg/100 ml	Solution	mg/100 ml
A. Amino acids		C. Organic acids[b]	
DL-Cystine·HCl	10	Malic	50
L-Leucine	20	α-Ketoglutaric	30
L-Phenylalanine	20	Succinic	20
L-Tryptophan	5	Fumaric	10
Taurine	5	Citric	10
L-Tyrosine·HCl	10	D. Anions	
L-Isoleucine	10		
L-Methionine	30	Na_2SO_4 anhydrous	10
L-Valine	30	$NaH_2PO_4 \cdot H_2O$	40
β-Alanine	20	$NaHCO_3$	20
L-Lysine·HCl	20	$NaH_2PO_2 \cdot H_2O$	20
L-Histidine·HCl	60	NaH_2PO_3	20
L-Arginine·HCl	80	E. Sugars	
L-Aspartic acid	40		
L-Threonine	10	Glucose	1500
L-Asparagine	50	Sucrose	1000
L-Serine	20	Trehalose	500
L-Proline	50	F. Vitamins	
Glycine	40		
L-Alanine	10	Thiamine·HCl	0.002
L-Glutamic acid	100	Riboflavin	0.002
L-Glutamine	100	Ca·pantothenate	0.002
B. Cations		Pyridoxine·HCl	0.002
		p-Aminobenzoic acid	0.002
$CaCl_2$	45	Folic acid	0.002
KCl	40	Biotin	0.001
$MgCl_2 \cdot 6H_2O$	20	Choline·HCl	0.020
NaCl	150	Isoinositol	0.002
		Niacin (nicotinic acid)	0.002
		Carnitine	0.001
		G. Antibiotics	
		Streptomycin (SO_4^{2-})	2.5
		Penicillin 6(K^+)	1.5

[a] Set pH at 7.5 with dilute NaOH. Add distilled water to volume; add 7.5% fetal calf serum and sterilize by filtration.
[b] Add NaOH (dilute) until acids go into solution.

terial or fat is discarded, and the cells are resuspended in fresh culture medium. The cells are then counted in a hemocytometer, diluted to a concentration of 1×10^6 cells/ml, and 3 ml of suspension seeded in each 20-ml plastic flask.

Examination of the Explanted Cells

The flasks are incubated at 27°C. At this temperature, initial attachment of the cells should be apparent within 1 hour after explantation, and maximum attachment should be obtained by 72 hours. Refeeding should not be attempted for at least 1 week. By this time, individual cells and cells emerging from small clumps should be migrating to produce islands of cells. These, in turn, will gradually spread until by 21 days after explantation, a partial monolayer has formed. Many types of cells will be found in the monolayer. Nerve cells can be recognized by their small, round perikarya and "jointed" axons; myoblasts will produce multinucleated myotubules that contract irregularly; polygonal epithelioid cells will form localized cell sheets; and phagocytes will migrate about, picking up any remaining yolk material and cellular debris.

Subculturing

In the cultures derived from *M. bivitattus,* we have consistently found a type of small cell in the original inoculum. These cells divide rapidly to form clumps of small rounded cells either on the surface of the flask or on the surface of a monolayer made up of phagocytes and epithelioid cells. The resulting cell clumps can be repeatedly subcultured by washing them loose with a pipette and transferring them to a fresh flask.

In the cultures derived from *L. maderae,* the cells that divide most consistently are epithelioid in appearance. In primary cultures, these cells form delicate vesicles that bud off and can be subcultured without disturbing the monolayer. However, attempts to carry these beyond the first subculture have been unsuccessful.

Possible Applications

Several uses for cultures of grasshopper and cockroach tissue have been proposed. Landureau and Jolles[4] used cultures of *P. americana* cells for studying the growth of intracellular symbionts and found that cultured cells produced enzymes capable of lysing bacteria. Marks *et al.*[3] used primary cultures of *L. maderae* cells to study the action of the molting hormone at the cellular level. At the present time, considerable interest has been shown in the possible use of various pathogens for the control of grasshopper populations. Cell cultures such as those described would provide an ideal environment for rearing and studying such pathogenic organisms in the laboratory.

[4] J. C. Landureau and P. Jolles, *Nature (London)* **225**, 968 (1970).

CHAPTER 28

Plants: Haploid Tissue Cultures

I. K. Vasil

A variety of haploid gametophytic tissues of gymnosperms and angiosperms have been cultured *in vitro* with variable success depending on the species and the tissue used. Tissue cultures from the female gametophytes[1-4] and pollen grains[5-7] of some gymnosperms have been used for physiological and morphogenetic studies. Anthers,[8-10] microspore mother cells,[11] and mature pollen grains[12,13] of a number of angiosperms have been extensively used for biochemical, cytological, genetical, morphogenetic, and physiological studies. Methods and applications of microspore mother cell and anther culture of angiosperms are described below.[14]

Anther Culture for Studies of Meiosis

Microspore mother cells within various anthers of a flower are generally synchronous during early stages of meiosis due to the presence of massive (0.5–1.5 μm) intercellular connections in the form of plasma channels.[10] It is advisable, therefore, to remove one anther from each bud and make a quick squash preparation to determine the stage of meiosis, rather than depend on the length

[1] C. D. LaRue, *Brookhaven Symp. Biol.* **6**, 187 (1954).
[2] W. Tulecke, *Nature* (*London*) **203**, 94 (1964).
[3] W. Tulecke, *In* "Reproduction: Molecular, Subcellular and Cellular" (M. Locke, ed.). Academic Press, New York (1965).
[4] N. Sankhla, D. Sankhla, and U. N. Chatterji, *Naturwissenschaften* **54**, 203 (1967).
[5] W. Tulecke, *Amer. J. Bot.* **44**, 602 (1957).
[6] W. Tulecke and N. Sehgal, *Contrib. Boyce Thompson Inst.* **22**, 153 (1963).
[7] R. N. Konar, *Phytomorphology* **13**, 170 (1963).
[8] I. K. Vasil, *Phytomorphology* **7**, 138 (1957).
[9] I. K. Vasil, *J. Exp. Bot.* **10**, 399 (1959).
[10] I. K. Vasil, *Biol. Rev.* (*Cambridge*) **42**, 327 (1967).
[11] M. Ito and H. Stern, *Develop. Biol.* **16**, 36 (1967).
[12] B. M. Johri and I. K. Vasil, *Bot. Rev.* **27**, 325 (1961).
[13] H. Linskens and M. Kroh, *Curr. Top. Develop. Biol.* **5**, 89 (1970).
[14] The triploid endosperm tissue of angiosperms has also been cultured: J. Straus, *Plant Physiol.* **35**, 645 (1960); K. Norstog, W. E. Wall, and G. P. Howland, *Bot. Gaz.* **130**, 83 (1969); D. J. Brown, D. T. Canvin, and B. F. Zilkey, *Can. J. Bot.* **48**, 2323 (1970). In some cases triploid endosperm tissues have been induced *in vitro* to regenerate triploid plants: B. M. Johri and S. S. Bhojwani, *Nature* (*London*) **208**, 1345 (1965); B. M. Johri and K. K. Nag, *Curr. Sci.* **37**, 606 (1968)

of the bud[15,16] which can be misleading.[10] Whole flower buds are then surface sterilized in 2.5–5.0% calcium hypochlorite, with a trace of some wetting agent, for about 5 minutes and rinsed twice in sterile distilled water. The interior of such buds are not contaminated by any microorganisms. Anthers are then carefully dissected out under a dissection microscope, transferred to agar nutrient media,[17] and incubated in the dark at 25°C (*Trillium* cultures are incubated at 4° to 6°C since meiosis in Nature also takes place at about the same temperature[18]). Handling of anthers during and after dissection is perhaps the most critical step in the entire procedure as even the slightest pressure or injury at this time can result in the death of the entire anther. One anther at a time can be removed aseptically from each culture vessel at frequent intervals to study the progress of cultures. Most such experiments are of short duration and can be concluded within 2 weeks.

In some cases flower buds, with or without the perianth, have been cultured in rather simple nutrient media consisting primarily of Hoagland's inorganic salt mixture.[15,16] During such procedures, 2–5 mm of the bud stalk is retained and the bud is placed in 0.2 to 0.5 ml of the nutrient medium in a narrow culture tube. Such experiments are complicated by the presence of massive amounts of somatic tissues which have salutory effects on the survival and development of the gametic tissues, but can not be of much value in understanding the nutritional and physiological requirements of anthers.

The tapetum, which is a highly glandular and often polyploid somatic tissue surrounding the gametic tissue, is known to influence the development of gametic cells in various ways.[10] Some attempts have been made to culture isolated strands of microspore mother cells without any surrounding tapetal or diploid anther wall tissues.[11] Whole buds are sterilized by immersion in 70% ethanol for 1 minute and then dried with sterile filter paper. Individual anthers are then removed aseptically and cut open at one end with a sharp sterile knife. A gentle squeezing of the anther end distal to the cut results in the extrusion of the microspore mother cells either in a coherent filament or as a viscous suspension of free cells depending on the stage of development of the anther. The isolated microspore mother cells are not contaminated by any tapetal or anther wall cells

[15] Y. Hotta and H. Stern, *J. Cell Biol.* **16**, 259 (1963).

[16] Y. Hotta and H. Stern, *Proc. Nat. Acad. Sci. U. S.* **49**, 648 (1963).

[17] Modified White's nutrient medium as used by Vasil[9]: (i) *Major element solution* (g/liter)—$MgSO_4 \cdot 7\ H_2O$, 3.6; $Ca(NO_3)_2 \cdot 4\ H_2O$, 2.6; Na_2SO_4, 2.0; KNO_3, 0.8; KCl, 0.65; $NaH_2PO_4 \cdot H_2O$, 1.65. (ii) *Trace element solution* (mg/liter)—$MnSO_4 \cdot 4\ H_2O$, 3000; $ZnSO_4 \cdot 7\ H_2O$, 500; H_3BO_3, 500; $CuSO_4 \cdot 5\ H_2O$, 25; $Na_2MoO_4 \cdot 2\ H_2O$, 25; cobalt chloride, 25; and 0.5 ml conc. sulfuric acid (sp. gr. 1.83). (iii) *Ferric citrate solution,* 2.5 g/liter. (iv) *Vitamin and amino acid solution* (mg/100 ml)—glycine, 150; niacin, 25; thiamine hydrochloride, 5; pyridoxine hydrochloride, 5; calcium pantothenate, 5. (v) *Indoleacetic acid,* 100 mg/100 ml ethanol. The final medium consisted of: solution (i) 100 ml, solution (ii) 1 ml, solution (iii) 4 ml, solution (iv) 5 ml, solution (v) 2 ml, bacto-agar 8 g, and sucrose 10 g, made up to 1 liter, at pH 5.5.

[18] A. H. Sparrow, V. Pond, and S. Kojan, *Amer. J. Bot.* **42**, 384 (1955).

and can be cultured in agar or liquid nutrient media.[19] As expected, the absence of any somatic tissues around the microspore mother cells necessitates providing a more complex nutrient medium and experimental environment than required by whole buds. Certain meiotic abnormalities, particularly in chromosome segregation, arise in such cultures. This may be because the tapetal and other tissues surrounding the microspore mother cells have been eliminated from these cultures but their usefulness, or the "permissive environment" provided by these somatic tissues, has not been fully replaced by the nutrient medium used.

Anther Culture for Producing Haploid Callus and/or Plants

Although the first successful production of haploid callus and haploid plants was reported with *Datura*,[20,21] various species of *Nicotiana* (tobacco) have become the most popular experimental material.[22-25] The popularity of tobacco is due to the availability of an extensive literature on the control of cell proliferation and organogenesis in diploid cell lines of tobacco, and because of the comparative ease with which tobacco can be used for many experimental purposes. The technique and nutrient media[26] used for this purpose are rather simple and it is, therefore, difficult to understand the failure of earlier attempts which resulted in proliferation and some organogenesis only from the diploid tissues of the anther,[27-29] mostly from the cut end of the anther filament or the connective region. Buds with anthers containing young uninucleate microspores to 2-celled immature pollen grains are surface sterilized by a quick dip in ethanol followed by 5 to 10 minutes in 5% sodium hypochlorite. Individual anthers are

[19] Modified White's nutrient medium as used by Ito and Stern[11] (g/liter): $Ca(NO_3)_2 \cdot 4\,H_2O$, 0.3; KNO_3, 0.08; KCl, 0.065; $MgSO_4 \cdot 7\,H_2O$, 0.75; Na_2SO_4, 0.2; $NaH_2PO_4 \cdot H_2O$, 0.019; $MnSO_4 \cdot 4\,H_2O$, 5×10^{-3}; $ZnSO_4 \cdot 7\,H_2O$, 3×10^{-3}; H_3BO_3, 15×10^{-4}; KI, 75×10^{-5}; $CuSO_4$, 1×10^{-5}; Na_2MoO_4, 1×10^{-6}; $Fe_2(SO_4)_3$, 0.001; glycine, 0.003; nicotinic acid, 5×10^{-4}; thiamine, 1×10^{-4}; pyridoxine, 1×10^{-4}. Sucrose was added to a concentration of 0.3 *M*, and pH was set at 5.6–5.8.

[20] S. Guha and S. C. Maheshwari, *Nature* (*London*) **204**, 497 (1964).

[21] S. Guha and S. C. Maheshwari, *Nature* (*London*) **212**, 97 (1966).

[22] J. P. Bourgin and J. P. Nitsch, *Ann. Physiol. Veg.* **9**, 377 (1967).

[23] K. Nakata and M. Tanaka, *Jap. J. Genet.* **43**, 65 (1968).

[24] J. P. Nitsch, *Phytomorphology* **19**, 389 (1966); *Z. Pflanzenzüechtg.* **67**, 3 (1972).

[25] N. Sunderland and F. M. Wicks, *Ann. Bot.* **22**, 213 (1971).

[26] Medium used by Nitsch,[24] and many others, for culture of tobacco anthers (mg/liter): KNO_3, 950; NH_4NO_3, 720; $MgSO_4 \cdot 7\,H_2O$, 185; $CaCl_2$, 166; KH_2PO_4, 68; 5 ml/liter of a solution obtained by dissolving in 1 liter of distilled water 5.57 g of $FeSO_4 \cdot 7\,H_2O$ and 7.45 g of Na_2–EDTA; $MnSO_4 \cdot 4\,H_2O$, 25; H_3BO_3, 10; $ZnSO_4 \cdot 7\,H_2O$, 10; $Na_2MoO_4 \cdot 2\,H_2O$, 0.25; $CuSO_4 \cdot 5\,H_2O$, 0.025; *myo*-inositol, nicotinic acid, 5; glycine, 2; pyridoxine hydrochloride, 0.5; biotin, 0.05; thiamine hydrochloride, 0.5; folic acid, 0.5; sucrose, 2%; Difco bacto-agar, 0.8%. pH of the medium was set at 5.5.

[27] I. K. Vasil, *In* "Plant Tissue and Organ Culture—A Symposium" (P. Maheshwari and N. S. Rangaswamy, eds.), Intern. Soc. Plant Morphol., Delhi (1963).

[28] R. N. Konar and K. Nataraja, *Phytomorphology* **15**, 245 (1965).

[29] M. Niizeki and W. F. Grant, *Can. J. Bot.* **49**, 2041 (1971).

then carefully dissected out and placed on agar nutrient media.[26] In order to avoid damaging the anthers during dissection and later handling, each anther is removed along with its filament, which serves as a convenient handle. Just before implantation on the medium, the filament is carefully removed. If the filament is retained, it may proliferate and regenerate diploid plants. Cultures are incubated in alternating periods of light (12–18 hours; 5000–10,000 lux/m^2) at 28°C and darkness (6–12 hours) at 22°C. Depending on the composition of the medium, callus or plantlets develop from the vegetative cell in the pollen grain within 3 to 6 weeks. After this period, plantlets are transferred to a low sugar (1 as compared to 2%) medium without indoleacetic acid. The plantlets can be transferred to pots in the greenhouse after they develop a sufficient root system.

Nutritional requirements for inducing cell proliferation in pollen grains vary greatly. Most species of tobacco which yield large numbers of haploid plants *in vitro* are exceptional in that these do not require any exogenous supply of plant growth substances for induction of supernumerary mitoses. Sucrose seems to be the only specifically required component of the nutrient medium for cell proliferation and callus formation. On such a medium embryoid formation proceeds up to the globular stage only. At least three factors seem necessary for the production of embryoids from pollen grains in tobacco: (a) sugar, (b) iron, and (c) other mineral salts. Vitamins and plant hormones like auxins are not indispensable; media containing kinetin and auxins produce a haploid callus only and do not differentiate embryoids. In *Datura,* kinetin is required to form haploid callus and embryoids from pollen, while auxins, yeast extract, and casein hydrolysate induce cell proliferation from the diploid somatic tissues of the anthers. Auxins are, however, quite effective in inducing pollen growth in plants like *Brassica oleracea, B. oleracea* × *B. alboglabra,* and *Oryza sativa.* Coconut milk and plum fruit juice, both rich in plant growth substances, are more effective than many synthetic plant hormones in anther cultures.

Anthers containing uninucleate microspores, which are devoid of any storage starch and have a large central vacuole, have been found to be most receptive for the production of haploid callus and/or plants. In some cases, anthers with 2-celled young pollen grains have also given rise to callus and plantlets. In both cases it is the vegetative nucleus which divides mitotically and goes through a series of definite morphogenetic stages simulating the development of a zygotic embryo to form the haploid plants.

Haploid plants produced by anther culture are sterile and do not set any seed. Homozygous diploids can be produced from these plants by colchicine treatment, or by culturing stem or petiole segments of the haploid plants in media containing kinetin and 2,4-dichlorophenoxyacetic acid. During the formation of callus, endomitosis takes place which doubles the number of chromosomes. Later by culturing such callus in appropriate nutrient media (mainly by using different concentrations of auxins and cytokinins; in some plants organogenesis will take place only if the auxins and cytokinins are eliminated entirely from the medium) homozygous diploid plants can be easily obtained in large numbers. Plants ranging in ploidy level from haploid to pentaploid (homozygous) have been obtained from pollen grains.[30,31] Variations in the ploidy levels of the callus tissue and/or the plants regenerated from proliferating microspores can be largely avoided by employing the nurse culture technique which makes it possible to isolate callus from individual microspores.[32] The amino acid analog, *p*-fluorophenylalanine, has also proved effective in selectively maintaining the growth of haploid tobacco cells with simultaneous inhibition of the growth of diploid cells.[33] Recently, haploid tissue obtained from anthers of tomato has been used to study phage-mediated transfer and the subsequent expression of the galactose and lactose operons of *Escherichia coli.*[34] The doubled haploid plants, obtained by culturing vegetative tissues of haploid plants, show high meiotic stability as well as a stable chromosome number.[35] Many uses of such plants can be made in studies encompassing the fields of genetics, plant breeding, plant physiology, and morphogenesis. Continuing haploid cell lines can also be established from proliferating pollen grains provided endomitosis, a relatively common phenomenon in callus cultures, can be effectively controlled.[36]

[30] M. Devreux, F. Saccardo, and A. Brunori, *Caryologia* **24,** 141 (1971).
[31] M. Zenkteler, *Experientia* **27,** 1087 (1971); and a personal communication.
[32] W. R. Sharp, R. S. Raskin, and H. E. Sommer, *Planta* **104,** 357 (1972).
[33] N. Gupta and P. S. Carlson, *Nature (London) New Biol.* **239,** 86 (1972).
[34] P. M. Gresshoff and C. H. Doy, *Planta* **107,** 161 (1972).
[35] G. B. Collins and R. S. Sadasivaiah, *Chromosoma* **38,** 387 (1972).
[36] For a detailed review of the biology of pollen, see I. K. Vasil, *Naturwissenschaften,* in press (1973).

CHAPTER 29

Aspen Callus

L. L. Winton and M. C. Mathes

Firm white callus from aspen trees has been initiated and maintained in an actively growing condition since 1961 at The Institute of Paper Chemistry.[1] Seven species of *Populus* have been cultured,[2] as well as numerous hybrids and polyploids.[3] One advantage of this technique is its simplicity over collecting cambial

[1] M. C. Mathes, *Forest Sci.* **10,** 35 (1964).
[2] M. C. Mathes, *Tappi* **47,** 710 (1964).
[3] L. L. Winton, *Forest Sci.* **17,** 348 (1971).

explants from mature trees.[4-6] By our method, callus is derived from the cambium of succulent juvenile plants and is composed of large, thin-walled cells which are chemically similar to xylem-initial cells normally produced in trees.[7] Callus initiated in 1961, on a modified White's Medium supplemented with coconut milk,[1] has been used to study antimicrobial secretions,[8,9] plantlet formation,[10] and chloroplast ultrastructure.[11] Callus initiated from 1966–1968,[12-14] on modifications of Wolter and Skoog Medium,[15] has been used for shake cultures of firm callus and rooting studies,[16] as well as for plant[17] and tree[3,18] production. Although several species have been cultured on a variety of media, most of our work has been with callus from one natural triploid clone (T-2-56) of normally diploid quaking aspen (*P. tremuloides* Michx.).

Root Sprout Production

This method was developed by Mathes[1] for internode explants from rooted root sprouts, although current-year branches from young trees, natural field suckers from roots, or seedlings can be used if the bark is relatively smooth. For the production of root sprouts, lateral roots 1 inch or less in diameter are collected from 2 to 3 inches beneath the soil surface, scrubbed in soap and water, cut into sections 10–15 cm long, then dipped in captan solution made with 1.5 tablespoons of captan 50W per gallon of water.[19] In order to reduce fungus contamination, the section ends are dipped in melted wax and the sections planted in a sterilized 1:1 mixture of sand and vermiculite and watered sparingly. Sprouts should be excised from the root sections when 3–5 cm tall, and planted in the same type of mixture in flat dishes covered with plastic bags. Rooted root sprouts are then transplanted to uncovered 3-inch pots and grown to 20–40 cm in height.

Collection of Explants and Sterilization

Bacteria are usually found in the lower stem, so the root sprouts are severed 5–10 cm above the soil surface and the leaves at the base of the petioles removed

[4] R. J. Gautheret, *Ann. Rev. Plant Physiol.* **6,** 433 (1955).
[5] C. Jacquiot, *J. Inst. Wood Sci.* **16,** 22 (1966).
[6] K. E. Wolter, *Nature (London)* **219,** 509 (1968).
[7] M. C. Mathes, D. W. Einspahr, and L. L. Winton, *Gen. Physiol. Notes* No. 8, 1–9 (1970).
[8] M. C. Mathes, *Science* **140,** 1101 (1963).
[9] M. C. Mathes, E. D. Helton, and K. D. Fisher, *Plant Cell Physiol.* **12,** 593 (1971).
[10] M. C. Mathes, *Phyton* **21,** 137 (1964).
[11] S. J. Blackwell, W. M. Laetsch, and B. B. Hyde, *Amer. J. Bot.* **56,** 457 (1969).
[12] L. L. Winton, *Phyton* **25,** 15 (1968).
[13] L. L. Winton, *Phyton* **25,** 23 (1968).
[14] L. L. Winton, *Phyton* **27,** 11 (1970).
[15] K. E. Wolter and F. Skoog, *Amer. J. Bot.* **53,** 263 (1966).
[16] L. L. Winton, *Amer. J. Bot.* **55,** 159 (1968).
[17] L. L. Winton, *Science* **160,** 1234 (1968).

with a razor blade. Discard the top 5–10 cm of the stem, to give a firm green section 10–20 cm long and ranging from 2 to 3 mm in diameter at the base to 1–2 mm at the top. Cut a dozen or more stems into sections 3–5 cm long and distribute them among three or four sterile 250-ml Erlenmeyer flasks. Cover the stem sections with undiluted household bleach containing 5–6% NaOCl, and add a few drops of a surfactant such as Tween-20 (polyoxyethylene sorbitan monolaurate).

Shake the flasks occasionally for 10 to 15 minutes, then decant the bleach, rinse the stem sections three times with autoclaved distilled water, decant all liquid, and shake the sections into sterile covered dishes 15 cm in diameter. For larger diameter sections with cracks in the bark, sterilization times of 20 to 40 minutes may be necessary. For each type of parental material, preliminary tests at 5-minute intervals should be run to determine the optimum time of sterilization. If bacteria have already invaded the outer cortical cells, the sections can be flamed with alcohol and the bark stripped off completely.

Callus Initiation

Internodal explants, 5–10 mm long, are aseptically excised with razor blades and both end sections discarded. The apical portion of each explant is then plunged vertically into agar medium, five segments per Petri dish. For larger segments, 3–7 mm in diameter and 10–15 mm long, explants may be laid horizontally on the agar surface and spun slightly to cover the entire surface of the segment with a film of moisture. For larger explants, or where bark fissures indicate possible fungus contamination, only one explant per container should be placed in test tubes, baby food bottles, 1-ounce French square bottles,[20] or 125-ml Erlenmeyer flasks.

Aspen callus is produced in the dark at 27° ± 1°C and at a relative humidity supplied by open pans of water in the incubator. Firm white callus develops slowly, but usually can be isolated from the explants after 4 to 8 weeks in incubation, depending on the medium used (Fig. 1). Soft friable callus can be induced by adding a tenfold increase in the auxin to the medium.[12,13]

Callus Isolation and Growth

Callus production may vary between explants, with respect to both the rate of growth and quality of the tissue. However, the final stock culture should be composed of uniformly firm white callus, preferably originating from one parental explant. For this reason, one-half dozen or fewer explants are selected having the greatest growth of the highest quality of firm white callus. The callus "button" from each explant is excised and quartered, and inocula from the same callus

---

[18] L. L. Winton, *Amer. J. Bot.* **57**, 904 (1970).
[19] M. K. Benson and D. E. Schwalbach, *Tree Planters' Notes* **21**, 12 (1970).
[20] P. R. White and P. G. Risser, *Physiol. Plant.* **17**, 600 (1964).

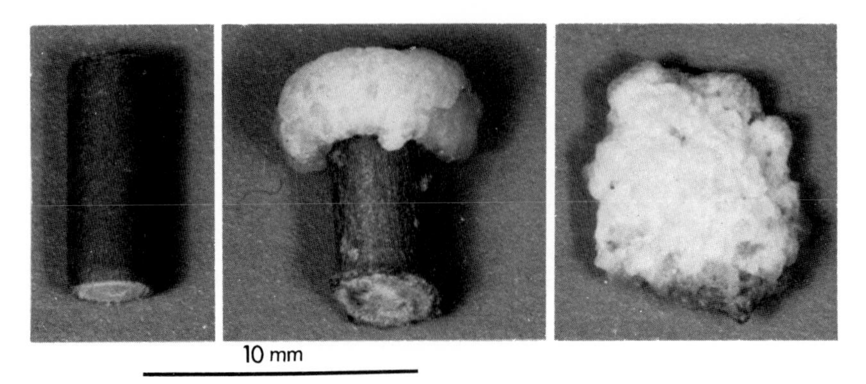

Fig. 1. An internodal stem segment from a rooted root sprout of quaking aspen (left), callus formation on the basal end of an inverted section after approximately 20 days of growth (center), and isolated callus tissue (right). Reprinted from M. C. Mathes,[2] with permission.

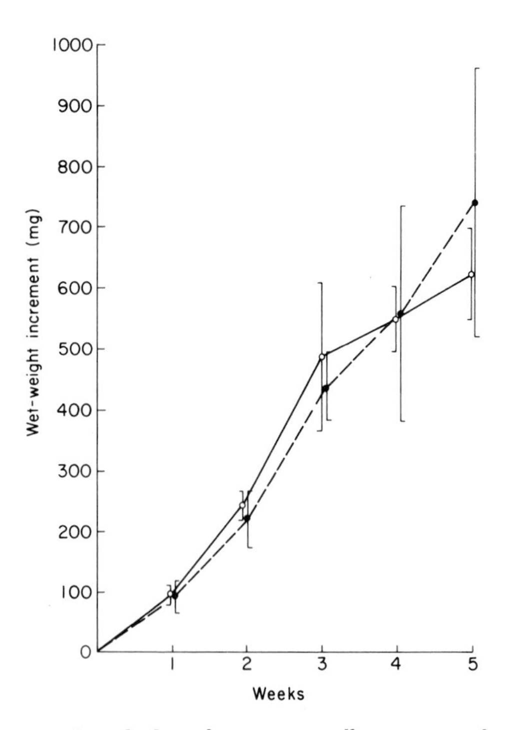

Fig. 2. Growth curves of triploid quaking aspen callus, 1 year after initiation, from the same root sprout source collected 12-7-1966 (solid line) and 1-23-1967 (dashed line) and subcultured monthly on modified Wolter and Skoog[15] Medium in the dark. Vertical lines indicate one standard deviation on either side of the average. Growth increments were not significantly different at the 5% level, but the 1-23-1967 callus (dashed line) had the better quality of firm white tissue and was retained as the stock culture.

piece are grown in the same dish. During succeeding monthly subcultures, the best clonal lines are selected until only the one best clone remains as the stock culture.

Inocula are cut 3–5 weeks after subculture, from near the edge of the callus piece, and should be approximately 3 mm on each side of a cube with the original callus surface left only on top. Callus grows about equally well on modified White's Medium with coconut milk[1] or on modified Wolter and Skoog Medium.[15] Growth curves are shown in Fig. 2 of two callus cultures initiated about 2 months apart on the latter medium (Winton, unpublished data). One year from initiation, the firm white callus doubled in wet-weight increment during both the second and third week after subculture. Only one of the cultures was maintained as the stock culture, but now, 5 years after callus initiation, its growth has slowed to about one-third the original. Supplements of amino acids and nucleotides significantly accelerated the growth of firm white callus during the second and third year after initiation.[18] However, these cultures are also slow growing. We recommend that new cultures be initiated every 2–4 years if the average growth declines.

Callus Types

In the dark, white callus indicates actively growing cells, but a decrease in growth is usually accompanied by an increase in brown pigmentation. Most cells are thin walled with large vacuoles (parenchymatous), but scattered areas of tracheid-like cells reflect a degree of differentiation.[2] The modified Wolter and Skoog Medium[15] is basically a chemically defined improvement of White's Medium, but aspen callus grown on either medium in the light turns red and usually dies. However, if the sucrose is omitted the callus turns green.[2] Green aspen callus also grows well in the light on Murashige and Skoog Medium,[21] modified for culturing callus isolated from conifer trees.[22] The formulation differs in many points from media derived from White's Medium, and permits the growth of firm green callus in light in the presence of sucrose.

[21] T. Murashige and F. Skoog, *Physiol. Plant.* **15**, 473 (1962).

[22] L. L. Winton, *Forest Sci.* **18**, 151 (1972).

CHAPTER 30

Plant Embryos

John G. Torrey

Until about 1960 plant embryos for experimental work *in vitro* were obtainable only by somewhat painstaking procedures of dissection from the ovular tissues of the maternal plant. This method is still useful and necessary and is outlined below. With the discovery that tissue culture manipulation will lead in a number of predictable cases to the initiation and development of large numbers of embryos in a somewhat synchronized developmental sequence, new approaches have been opened to experimentation with embryos. In a few special cases, that is, the culture of whole anthers of certain dicot genera, embryos develop quite predictably and are of particular interest since they are haploid plants, having been derived from the haploid pollen cells produced by meiosis. For an account of this procedure, see the article by I. K. Vasil, in this Section.

The Dissection and Culture of Plant Embryos Initiated *in Ovulo*

The fertilized egg or zygote is the single-celled stage with which one would like to begin to study plant embryogenesis. It would be useful if one could obtain easily and in large numbers unfertilized eggs, fertilize them *in vitro* at time zero, and follow their development from embryos into complex multicellular organisms. This type of study is possible in the brown alga, *Fucus*, growing in sea water in the light. No such accessibility to embryos is possible among the vascular plants where typically the eggs develop encompassed by multiple layers of tissues of the parent plant (the archegonium in the ferns, the ovule in the seed plants) and fertilization is achieved by complicated mechanisms which allow the fertilized egg to develop within the protective tissues of the mother plant.

To obtain the immature stages of developing embryos, one must dissect them from the protective surrounding tissues of the maternal plant. It has been proved impractical to dissect out single-celled zygotes in the case of flowering plants although this has been done for ferns.[1,2] In the flowering plants, one must achieve dissection of young flowers soon after pollination. A few particularly favorable species have been studied and are described below. In their review, Narayanaswami and Norstog[3] listed and gave references to genera of ferns,

[1] A. E. DeMaggio, *Phytomorphology* **11**, 64 (1961).

[2] A. E. DeMaggio and R. H. Wetmore, *Amer. J. Bot.* **48**, 551 (1961).

[3] S. Narayanaswami and K. Norstog, *Bot. Rev.* **30**, 587 (1964).

gymnosperms, monocots, and dicots whose embryos have been grown in culture. One must understand the timing of pollination and fertilization in each case if one is to achieve isolation of early embryonic stages for culturing.

Embryos from Dicotyledonous Plants

One of the most studied dicot embryos is the common weed, shepherd's purse *Capsella bursa-pastoris* L. which flowers readily and profusely; in the inflorescence, the embryonic stages are arranged along the axis with the younger stages at the top and the older stages toward the base. Rijven[4] described stages in detail, methods of isolation and culture, and presented a "how-to-do-it" account for this species. The following description of the isolation of embryos is his:

> The first thing to be done was to arrange all implements and appliances in working order on the table. Amongst these was a glass of water in which were the cut inflorescences of *Capsella*. A number of silicles, containing embryos in the desired stage, was selected and opened successively. Some practice was required for this operation, as the walls of the silicles had to be removed without severing the ovules from the placenta, in order to prevent their touching any unsterile parts. This was achieved by tearing the walls down from the ribs by means of a pair of coverglass tweezers after Kühne. The remainder of the silicle, including the ovules, was then placed in a sterilized watch glass; then the ovules were pulled off from the placenta with sterilized needles under a dissecting microscope. Another watch glass with 5 ml sterile medium, covered by a larger watch glass, was ready for the receipt of the ovules. The excision of the embryos followed by cutting a gap in the top of the ovule and exerting a slight pressure on its base near the micropyle. To prevent infection during these manipulations, the operator's head, except the eyes, was covered by a sterile cloth, as is usual in surgery.
>
> About 20 embryos could be excised within ten minutes. By means of a braking pipette we then brought them to a second, and from this to a third covered watch glass with medium in order to clean them from adhering remains of endosperm and from eventual contaminations.
>
> These operations having been repeated for a number of silicles, the culture was started.

The medium for successful culture of young heart-stage embryos of *Capsella* was described by Rijven and improved media were developed by Raghavan and Torrey[5] for culturing *Capsella* embryos at the globular stage and as small as 50 μm in diameter. Although less well developed embryos, even down to 4- or 8-celled stage could be isolated, no success was achieved in their culture.

Similar studies on embryos of *Datura stramonium* and hybrid species were described in some detail by Rietsema and Blondel,[6] who also gave in some detail the raison d'etre for work of this sort.

[4] A. H. G. C. Rijven, *Acta Botan. Neerl.* **1**, 157 (1952).

[5] V. Raghavan and J. G. Torrey, *Amer. J. Bot.* **50**, 540 (1963).

[6] J. Rietsema and B. Blondel, *In* "Blakeslee: The Genus Datura" (A. G. Avery, S. Satina, and J. Rietsema, eds.), pp. 196–219. Ronald Press, New York, 1959.

Embryos from Monocotyledonous Plants

Methods of isolation, culture, and nutrition of monocot embryos are not very different from those of dicots. A carefully studied genus is barley, *Hordeum vulgare* L. Norstog[7,8] presented details of isolation and successful culture of embryos down to 100 μm in size. He described his method as follows[8]:

> The exposed caryopses were placed on a flamed slide under a dissection microscope at 25–50× magnification. Using watchmaker's forceps and bacteriological inoculating needles ground to fine points, embryos as small as 0.2 mm were readily excised and transferred directly to the medium. Smaller embryos required different treatment because they tended to dry out in transit. The region of the ovule that contains the embryo is beaklike in barley. The beak was excised and transferred to a small drop of sterile paraffin oil. The embryo was then teased apart from the beak, lifted out in a film of oil using a microspatula, and transferred to the medium. The oil appeared to float free, and it was possible to push the embryo to an oil-free area. The oil did not appear to interfere with the embryonic growth.
>
> The synthetic medium[9] used in the cultures was a White's agar medium modified by greatly increasing the phosphate concentration, raising the sucrose level to 9%, adjusting pH to precisely 4.9 using malic acid as a buffer, and adding an amino acid mixture.

Orchid embryos have been as much studied as any monocots. In many orchid species, the mature seed is released from the mother plant when the embryo is still not fully developed as, for example, still in the globular stage. Orchid seeds must therefore be nurtured with great care. Hence, there exist the elaborate sterile technical procedures used routinely by commercial orchid growers who are in essence culturing immature orchid embryos through early embryogenesis *in vitro*. An introduction to embryogenesis in the orchids and to methods of culturing the seed may be found in the monograph by Withner.[10]

The Production of Embryos from Tissue Cultures

In 1959 Reinert[11] described the development of embryos of carrot in callus tissues which had been carried through a series of nutrient changes. The callus tissue was initiated in the first place from the secondary vascular tissues of the storage root of carrot. More recently several accounts of the initiation and development of what are termed "embryoids" or "adventive embryos" from cultured carrot callus tissues have been made.[12–14] The basic procedure was to

[7] K. Norstog, *Amer. J. Bot.* **48**, 876 (1961).
[8] K. Norstog, *Amer. J. Bot.* **52**, 538 (1965).
[9] K. Norstog and J. E. Smith, *Science* **142**, 1655 (1963).
[10] C. L. Withner, "The Orchids. A Scientific Survey." Ronald Press, New York, 1959.
[11] J. Reinert, *Planta* **53**, 318 (1959).
[12] W. Halperin and D. F. Wetherell, *Amer. J. Bot.* **51**, 274 (1964).
[13] H. Kato and M. Takeuchi, *Plant Cell Physiol.* **4**, 243 (1963).
[14] F. C. Steward, M. O. Mapes, A. E. Kent, and R. D. Holsten, *Science* **143**, 20 (1964).

develop a proliferating callus tissue on a nutrient medium, then change the medium (usually withholding auxin from the medium), and allowing organogenesis to begin. Kato and Takeuchi[13] gave the following account of their procedure:

> Throughout the present investigation, tap-roots of *Daucus carota* were used as starting material. A portion of the tap-root was excised aseptically by means of a cork-borer, and cut into disks 1 mm thick and 6 mm in diameter. These disks were transplanted onto a solid medium consisting of White's basic solution with addition of 0.5% agar supplemented with 0.1% Difco yeast extract and 1 or 10 ppm β-indoleacetic acid. The culture was grown in a dark room at $27 \pm 1°$, and transplanted every twenty days to a fresh medium of the above mentioned composition.
>
> The explants showed vigorous growth to produce a firm callus tissue, orange-yellow in color.
>
> After four months of culturing, single cells or small cell clumps were separated from the surface of the friable callus tissue, some of them floating on the wet surface of the agar and others penetrating into the agar medium. These cell colonies continued to grow on the agar surface and underwent a series of successive differentiation apparently similar to that of normal embryogenesis, finally to develop into a small but complete plantlet. Several stages of development were distinguished; single cell, globular, heart-shaped, torpedo-shaped bodies, and finally, young plantlet.

From callus tissue grown in liquid culture, cell suspensions could be separated by a mechanical filter and the small multicellular colonies isolated could be subcultured in synthetic medium to allow embryo development to occur. Embryoids developed singly from suspensions sieved at 45 μm and clumps of embryoids developed from the fraction 45–75 μm.[15] Although initial cultivation was facilitated with complex addenda to the media such as coconut milk or powdered yeast extract, the entire cultivation could be carried out in defined synthetic media. Removal of high concentrations of auxin was critical for the conversion from proliferation of unorganized callus to organized embryos. Normal embryonic stages were observed although no real synchrony in this type of development has yet been achieved.

Embryogenesis from cultured callus tissues has been observed in a large number of species. Callus initiated from immature or mature embryos appears to form embryoids more readily when exposed to sequential treatment on different media than callus derived from mature tissues. No explanation for this responsiveness is known. Embryoids arise from callus tissues cultured for long periods of time provided the proper nutrient medium is provided.[16] Reinert *et al.*[17] have conclusively demonstrated with photographic evidence that embryoids can arise from single isolated cells grown in culture. This conclusion had been reached earlier by Steward *et al.*[18] on indirect evidence based on the percentage embryoid formation from plated cell suspensions of carrot cells.

[15] W. Halperin, *Amer. J. Bot.* **53**, 443 (1966).

[16] I. K. Vasil and A. C. Hildebrandt, *Amer. J. Bot.* **53**, 869 (1966).

[17] J. Reinert, D. Backs-Husemann, and H. Zerban, *Colloq. Intern. Cent. Nat. Rech. Sci.* **193**, 261 (1971).

[18] F. C. Steward, L. M. Blakely, A. E. Kent, and M. O. Mapes, *Brookhaven Symp. Biol.* **16**, 73 (1964).

To date the use of embryos available in large numbers and at selected stages of development from cultured callus tissues or cell suspension cultures has been limited. Some abnormalities may be observed in embryonic development and must be selected out and discarded. Also there has been some debate as to the identity of embryos formed *in ovulo* with those formed *in vitro*. Yet here is a new source of embryolike structures for analysis and further study. "Perhaps further experimental study of these induced embryoids will contribute to a greater knowledge of the factors involved in normal embryogenesis."[19]

[19] E. G. Cutter, "Plant Anatomy: Experiment and Interpretation," Part 2. Edward Arnold, Ltd., London, 1971.

CHAPTER 31

Somatic Plant Cells

T. Murashige

Primary cultures of somatic plant cells have been established from virtually every organ or tissue and from diverse plants, from the moss to the monocotyledonous angiosperm. Usually the culture is initiated on an agar medium, with recultures being carried out either on agar or in liquid. Sometimes the culture is started directly in liquid nutrient. Unlike animal cell cultures, where often distinguishing characteristics of the progenitor tissue or organ are retained, plant cultures generally result in the development of a callus, the constituent cells of which rarely manifest characteristics identifiable with the original organ or tissue. The statement that the callus is composed of undifferentiated cells is also wrong; callus cells are not uniform morphologically or physiologically, and genetic variations in the form of endopolyploidy are common.

Some Basic Principles

In initiating callus cultures of a new plant it is helpful to consider certain characteristics with respect to (1) the explant, (2) the physical qualities of the nutrient medium, (3) the chemical composition of the medium, and (4) the temperature and light requirements. Explants from mature tissues and organs are expected to be disproportionately high in content of highly differentiated and endopolyploid cells. Such cells have repressed regenerative capabilities. Large explants generally proliferate faster than small ones; however, the probability of including viruses and mycoplasma, as well as certain bacteria and fungi, is also greater. It is possible to establish axenic cultures by selecting suitable tissues or organs as sources of explant, e.g., shoot apical meristem.

The major physical qualities of the nutrient medium include degree of gelation, rate of agitation of liquids, pH range, and relationship of medium volume to explant size. The type of culture vessel and vessel closure employed may also be important. With gelled media a concentration of agar equivalent to 0.6–1.0% Bacto-agar is generally used. Liquid media can be agitated very slowly by maintaining cultures on a rotating apparatus or very rapidly on gyrotory or reciprocating shakers. The pH adjustment is made during medium preparation and with many plant cultures the pH is set in the range 5.0–6.0. When agar is used the pH should be 5.5 or higher, otherwise, failure of gelation may result. Survival of extremely small explants requires the use of correspondingly small amounts of medium; on the other hand, growth rate of large explants is sometimes directly proportional to medium volume.

The nutrient medium composition can be evaluated by considering separately the inorganic salt portion and the organic addenda. For the salt requirement the formulation of White,[1] Heller,[2] or Murashige and Skoog,[3] each as is or with some modification, has been satisfactory for diverse plants and purposes. The essential organic addenda are sucrose (2–3%), thiamine·HCl (0.1–10 mg/liter), auxin (0.1–3 mg/liter of indole-3-acetic acid, indole-3-butyric acid, α-naphthaleneacetic acid, or 2,4-dichlorophenoxyacetic acid), and cytokinin (0.03–10 mg/liter kinetin). Additional vitamins, e.g., *myo*-inositol, nicotinic acid, pyridoxine, etc., as well as certain amino acids and amides (e.g., L-asparagine and L-glutamine) are also used sometimes.

With some plant cell cultures it has been necessary to include certain addenda of relatively undefined composition, such as endosperm fluids, protein hydrolysates, extracts of yeast and malt, and fruit juices.

A constant temperature in the range 25°–28°C has been standard. This temperature provision is not consistent with the conditions which exist in the plant's natural habitat and appropriate adjustments should be considered. For the purpose solely of multiplying plant cells, as callus or free suspensions, no light is necessary; illumination generally retards tissue growth. Primary cell cultures intended for organ initiation, however, may need light and the requirements with respect to illumination intensity, daily duration, and quality should be examined.

Culture of Tobacco Stem Pith

As illustration, steps toward establishing primary cultures from tobacco stem pith will be described. This example has been chosen because major require-

[1] P. R. White, *Growth* **7**, 53 (1943).
[2] R. Heller, *Ann. Sci. Nat. Bot. Biol. Veg.* **14**, 1 (1953).
[3] T. Murashige and F. Skoog, *Physiol. Plant.* **15**, 473 (1962).

ments have been established through systematic investigation. Furthermore, cultures can be obtained from nearly every tissue or organ of the tobacco plant, initiation of organs and reconstitution of complete plants is regulatable, totipotentiality is demonstrable in virtually all tobacco cells including microspores, and callus as well as liquid suspension of free-living cells is readily obtainable.

Nutrient Medium Preparation. The basal medium consists of the inorganic salts in kinds and amounts as described by Murashige and Skoog[3] and the organic addenda sucrose (3%), thiamine·HCl (0.4 mg/liter), and *myo*-inositol (100 mg/liter). A culture composed of callus and roots will result in the medium supplemented with 3 mg/liter indole-3-acetic acid (IAA) and 0.3 mg/liter 6-furfurylaminopurine (kinetin). The root initiation can be repressed and friability of the callus enhanced by using (2,4-dichlorophenoxy)acetic acid (2,4-D) in place of the IAA, and by an additional supplement of 1 mg/liter gibberellin A_3 (GA_3)

To obtain a culture composed of callus and shoots the supplements should be 2 mg/liter each of IAA and kinetin, 80 mg/liter adenine sulfate·dihydrate, 100 mg/liter L-tyrosine, and 170 mg/liter $NaH_2PO_4 \cdot H_2O$.

Bacto-agar is included at a rate of 1% and the pH of the medium is set at 5.7. The medium is distributed in 25-ml aliquots into 25 × 150 mm culture tubes and the tubes are capped with polypropylene closures (Kaputs). Sterilization of medium is accomplished by autoclaving 15 minutes at 15 psi. The nutrient agar is cooled as slants.

Disinfection and Excision. Young tobacco plants with stems about 250 cm should be used. Their leaves are removed and 5-cm sections of the stems are obtained. Swabbing the sections with 95% ethanol will remove much of the pubescence. The sections are immersed in dilute laundry bleach (Clorox or Purex diluted tenfold and containing a small amount of surfactant) for 10 minutes and rinsed three times with autoclaved water. Using a sterile No. 2 stainless steel cork borer, a cylinder of the central tissue is removed from each section and transferred into a sterile Petri dish. A 2-mm section from each end of the cylinder is discarded. The remainder of the cylinder is then sliced into 2-mm thick disks. The disks are finally transferred to culture tubes, 1 disk per tube.

There are variations along the length of the stem with respect to the incidence of endopolyploid cells and the morphogenetic behavior of explants. Preferably, explants should be confined to the upper 10 cm of the stem.

Culture Conditions. The cultures should be maintained in complete darkness if only unorganized callus is desired. For root and shoot initiation 16 hours of daily illumination with low intensity (1000 lux) fluorescent light may be desirable. A constant temperature of 27°C is satisfactory for tobacco cultures. Fresh weight yields may be used as an index of callus formation and growth. Quantitative data on organ initiation can be obtained by counting individual roots and shoots. Each of these measurements can be made after a 3- to 4-week culture period.

CHAPTER 32

Roots

H. E. Street

Root cultures are initiated by transfer of sterile root tips to an appropriate culture medium. The root tips develop as organ (root) cultures having the morphology and cellular structure characteristic of seedling roots. The root cultures show either only primary root anatomy or very limited secondary formation of vascular tissue (secondary formation of vascular tissue does not persist on subculture). Lateral roots develop in many cases; their development is essential for the building up of clonal cultures.

The history of the development of root cultures has been reviewed by White,[1] Street,[2] and Butcher and Street.[3] General techniques are described by Street and Henshaw.[4]

Root Culture Media

A medium which has been very widely used in root culture is a modified White's Medium. This medium has the composition shown in Table I. The medium is prepared by using a stock solution of the inorganic salts (minus the ferric chloride) at ten times the strength required in the culture medium. This stock solution is stored at 4°C and replaced at monthly intervals. The "vitamin solution" (containing the three B vitamins plus glycine) is prepared as a second stock at 100 times the final concentration and stored in measured aliquots in Pyrex tubes at −20°C. The ferric chloride is available as a concentrated "Specpure" solution (Johnson Matthey Ltd., Covent Garden, U. K.). To prepare a batch of medium the sugar is dissolved in distilled water. To this is added in turn the solution of ferric chloride, the vitamin solution, and the inorganic solution. The mixed solution is diluted almost to volume with distilled water and then its pH is adjusted to 4.9 to 5.0 with the aid of a small volume of NaOH or HCl. The medium is adjusted to final volume and distributed to the culture vessels for sterilization by autoclaving at 15 psi for 5 to 15 minutes.

The standard medium contains 2% sucrose but this may not be the optimal concentration for a particular clone, and for roots of monocotyledons the sucrose

[1] P. R. White, "A Handbook of Plant Culture." Cattell & Co., Lancaster, Pennsylvania, 1943.

[2] H. E. Street, *Biol. Rev. Cambridge Phil. Soc.* **32**, 117 (1957).

[3] D. N. Butcher and H. E. Street, *Bot. Rev.* **30**, 513 (1964).

[4] H. E. Street and G. G. Henshaw, *In* "Cells and Tissues in Culture" (E. N. Willmer, ed.), Vol. 3, p. 459. Academic Press, New York, 1966.

TABLE I
Modified White's Root Culture Medium[a,b]

Compound	Content per liter of medium expressed as weight of anhydrous compound
Calcium nitrate, $Ca(NO_3)\cdot 4\ H_2O$	200 mg
Potassium nitrate, KNO_3	80 mg
Potassium chloride, KCl	65 mg
Sodium dihydrogen phosphate, $NaH_2PO_4\cdot 4\ H_2O$	16.5 mg
Manganese chloride, $MnCl_2\cdot 4\ H_2O$	4.5 mg
Zinc sulfate, $ZnSO_4\cdot 7\ H_2O$	1.5 mg
Potassium iodide, KI	0.75 mg
Sodium sulfate, $NaSO_4\cdot 10\ H_2O$	200 mg
Magnesium sulfate, $MgSO_4\cdot 7\ H_2O$	360 mg
Boric acid, H_3BO_3	1.5 mg
Molybdic acid, H_2MoO_4	0.0017 mg
Copper sulfate, $CuSO_4\cdot 5\ H_2O$	0.013 mg
Ferric chloride, $FeCl_3$	2.5 mg
Thiamine hydrochloride	0.1 mg
Pyridoxine hydrochloride	0.1 mg
Nicotinic acid	0.5 mg
Glycine	3.0 mg
Sucrose	20 g

[a] P. R. White, "A Handbook of Plant Tissue Culture." Cattell & Co., Lancaster, Pennsylvania, 1943.
[b] H. E. Street and S. M. Gregor, *Ann. Bot.* **16,** 185 (1952).

should be replaced with an appropriate concentration of glucose. For some clones the glycine of the standard medium can, with advantage, be omitted.

Various workers have used different solutions of inorganic salts in preparing their root culture media and one or another of these may be superior, for a particular clone, to the inorganic salt mixture detailed in Table I. Alternative formulae are detailed in papers by Robbins,[5] Bonner and Devirian,[6] Almestrand,[7] and Heller.[8]

Growth of roots in modified White's Medium (Table I) results in a rise in pH; a single excised tomato root tip growing in 50 ml of this medium causes the pH to rise from the initial value of 4.8 to 4.9 to 5.8 to 6.0 during the 7-day incubation. Above pH 5.2, iron is rendered insoluble to the extent that its deficiency limits further growth. The simplest way to prevent this is to replace the ferric chloride of the standard medium by ferric sodium ethylenediaminetetraacetate (Fe-EDTA). A suitable Fe–EDTA preparation is as follows: 0.8 g disodium ethylenediaminetetraacetate is dissolved in water, 3.0 ml of a 10% w/v solution of ferric

[5] W. J. Robbins, *Bot. Gaz.* **73,** 376 (1922).
[6] J. Bonner and P. S. Devirian, *Amer. J. Bot.* **26,** 661 (1939).
[7] A. Almestrand, *Physiol. Plant.* **2,** 372 (1949).
[8] R. Heller, *Ann. Sci. Nat. Botan. Biol. Veg. Ser. II* p. 1 (1953).

chloride is added, and the volume adjusted to 1 liter; 6.5 ml of this solution per liter of medium gives the standard iron concentration. Fe–EDTA should always be used when culture is to be prolonged or the effect of pH on root growth is being examined. We have found it slightly inferior to ferric chloride for normal clonal maintenance and multiplication.

The modified White's Medium is very weakly buffered. No satisfactory method of increasing significantly the buffer capacity by adding soluble salts has been discovered. Sodium phosphates, for instance, have to be added in amounts which are markedly inhibitory to the growth of all root cultures examined. Studies of the relationship between pH and the growth of cultured roots have, however, been carried out by using as "solid buffer" the sparingly soluble salts, amorphous calcium dihydrogen orthophosphate $Ca(H_2PO_4)_2$, precipitated calcium phosphate prepared according to the British Pharmaceutical Codex, and calcium carbonate.[9] Appropriate mixtures of these compounds can be used to stabilize pH at any desired value within the range 4.2–7.5.

Root cultures are relatively insensitive to sodium and chloride ions, and culture media lacking particular mineral elements can be prepared by using purified salts (for general methods of purifying salts see Hewitt[10]) and substituting the corresponding sodium salts or chlorides. In view of the high sugar content of root culture media, purification of the sugar by suitable exchange resins may be essential to induce deficiency. Root cultures are very sensitive to residues from the reagents used in salt purifications, and it is essential to test for full restoration of the growth-promoting activity of the purified medium by addition of an effective concentration of the omitted element.

Many organic substances, including most natural sugars, suffer chemical change during autoclaving, particularly in the presence of the mixed salt solution of root culture media. Such substances should, therefore, either be autoclave sterilized in pure aqueous solution or sterilized by filtration and then added aseptically to the remainder of the autoclaved medium.

Culture Vessels

All the glassware should be very carefully cleaned and there should be a final rinse with distilled water. Traditionally, chromic acid–sulfuric acid has been used, but probably detergents are a satisfactory alternative provided they are very thoroughly removed by rinsing. All glassware should be of borosilicate glass (Pyrex) and the distilled water should be prepared from stills of this glass. "Analar" grade chemicals should be used to prepare culture media. Clean glassware should be protected from dust contamination. Culture vessels should be closed with nonabsorbent cotton plugs wrapped in dressing-free gauze or wide-open woven bandage and these plugs protected from drip in the autoclave by inverted glass or aluminum beakers.

[9] D. E. G. Sheat, B. H. Fletcher, and H. E. Street, *New Phytol.* **58**, 128 (1959).

[10] E. J. Hewitt, "Sand and Water Culture Methods Used in the Study of Plant Nutrition." Commonwealth Agricultural Bureau, Farnham Royal, U. K., 1966.

Origin of Root Tips

Normally root tips are obtained by surface sterilization of dry seeds followed by germination under aseptic conditions in Petri dishes containing moistened filter paper. The technique of seed sterilization must not cause any injury to the embryo. A 1% (w/v) solution of bromine is the sterilizant of choice if embryo injury does not result. Alternative sterilizants are 1% chlorine containing detergent or 0.1% aqueous mercuric chloride combined with an alcohol wash.

The 10-mm apex is excised from seminal roots 20–40 mm long and transferred carefully to the surface of the culture medium.

Technique of Subculture

Ten-millimeter apical tips of such seedling roots are excised with a sterile scalpel and transferred with a platinum loop singly to the surface of sterile culture medium. One-hundred milliliter Pyrex wide-mouthed Erlenmeyer flasks containing 50 ml of culture medium are most suitable for stock root cultures. The cultures are then incubated at 25°–27°C for a suitable period. With clones of tomato (*Lycopersicum esculentum*) this period is 7 days at 27°C. The root grows in length and lateral roots emerge from the main axis. A clone can be established from a single root culture of this kind by now setting up from it one or more "sector" cultures. Using a pair of fine iridectomy scissors, portions of the main root axis are cut out, each of which bears four or five young, lateral roots (with tomato these laterals should be 3–8 mm long). These sectors are then transferred singly to new flasks of culture medium and again incubated. During incubation the laterals grow in length and in turn come to bear laterals. From such a developed "sector" culture one can excise the 10 mm apical tips of the primary laterals and new sector pieces. The main lateral tips when cultured give roots similar to those developed from the initial seedling root tip; such cultures are often referred to as "tip" cultures and are the kind used in experiments. The sector pieces serve to propagate the clone and to yield further root apices from which to initiate experimental tip cultures. This procedure of clonal maintenance and multiplication is illustrated in Fig. 1.

This general technique is applicable when the root cultures develop a regular sequence of laterals and when such laterals are capable of rapid growth from sector initials. To maintain a high and uniform growth rate of any new clone, certain aspects of this basic technique must be approached experimentally. Aspects of the technique which should be varied toward this end are: length of root tips excised for tip cultures, size of laterals on the sector pieces, duration of the incubation periods (passage lengths) for both tip and sector cultures, incubation temperature, and composition of culture medium (particularly sugar concentration). There is no evidence that solidified media are preferable to liquid media or that aeration is likely to be a critical factor in the growth of cultures in the standard

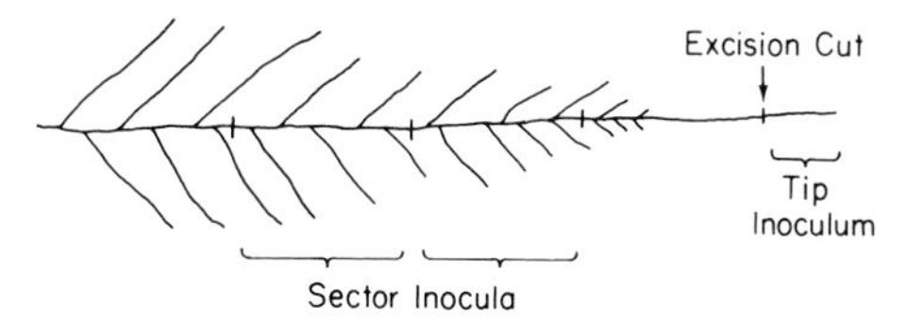

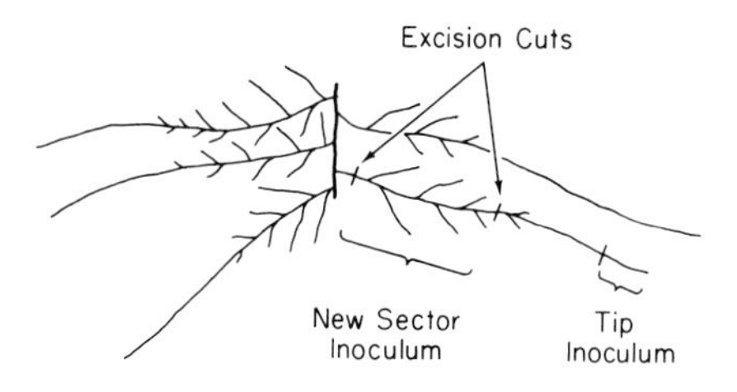

Fig. 1. "Tip" and "sector" cultures showing portions excised to initiate both sector and tip cultures. Drawings based on root cultures of tomato.

vessels described. Roots from the first list of species given later under *Range of Species* have been cultured as clones by this general technique.

When an actively growing clone cannot be established by the above technique, it may be possible, using an appropriate passage length and culture medium, to grow each individual root continuously by repeated excision and transfer to fresh culture medium of the apex of the main root axis. Roots from the second list of species given later (*Range of Species*) have been maintained in continuous culture by this technique.

Since many organisms grow only sluggishly in standard root culture media, it is important to periodically carry out a sterility test. A simple sterility test is to enrich the standard root culture medium (Table I) by incorporating 200 mg per liter acid-hydrolyzed casein. This medium is not inhibitory to the growth of most root cultures but promotes the growth of many microorganisms. The whole clone can periodically be screened by a passage in this enriched medium.

Measurements of Growth

Linear growth of the cultures is expressed by the following criteria: increase in length of main axis (millimeters); number of emergent laterals (lateral number); total length of laterals per root (millimeters). Fresh and dry weights are usually recorded by bulking five or ten roots. Reproducible fresh weights can be obtained by adopting a precise blotting technique and dry weights by gentle washing of the roots followed by drying to constant weight in small metal boats at 80°C. Measurements of cell expansion can be based upon the length and transverse diameter of exodermal cells in roots fixed in 70% ethanol and cleared with lactophenol. Usually ten cells are measured in each of five replicate roots at a point (5 mm or more from the extreme tip) where cell expansion is complete.

Measurement of the rate of production of new cells per culture per 24 hours can be carried out by the method of Butcher and Street.[11]

Range of Species

Clones of isolated cultured roots of the following species of higher plants have been successfully established in continuous culture: *Senecio vulgaris* L.; *Medicago sativa* L.; *Trifolium repens* L.; *Datura stramonium* L.; *Lycopersicum esculentum* Mill.; L. *pimpinellifolium* Mill.; *Solanum tuberosum* L.; *Secale cereale* L.; *Triticum vulgare* L. (var. Hilgerdorf); *Androcymbium gramineum* Cav.; *Pinus* spp. including *P. ponderosa* Dougl. and *P. serotina* Michx. Roots of the following species have been maintained in culture for prolonged periods although due to poor lateral root development, multiplication of clones from individual roots has not been achieved: *Callistephus hortensis* Cass; *Helianthus annuus* L; *Raphanus sativvus* L; *Brassica nigra* Koch; *Convolvulus arvensis* L; *Isatis tinctoria* L; *Acer melanoxylon* R.Br.; *Melilotus alba* Destr.; *Pisum sativum* L; *Linum usitatissimum* L; *Fagopyrum esculentum* Monch.; *Petunia violacea* Lindl. The roots of a number of other species have been cultured for limited periods. The data on the culturability of excised roots is summarized in tabular form in a recent review.[3] The difficulties encountered in attempting to culture the isolated roots of many species and the differences in culturability between varieties or strains within species have been discussed in some detail.[2]

Applications

Applications of the study of various aspects of root growth, physiology, and biochemistry are reviewed by Torrey[12] and Street.[13]

[11] D. N. Butcher and H. E. Street, *J. Exp. Bot.* **11**, 206 (1960).

[12] J. G. Torrey, *In* "Handbuch der Pflanzenphysiologie" (A. Lang, ed.), Vol. 15, Part 1, p. 1256. Springer Verlag, Berlin, 1965.

[13] H. E. Street, *In* "Plant Physiology" (F. C. Steward, ed.), Vol. 5B, p. 1. Academic Press, New York, 1966.

CHAPTER 33

Leaves

T. A. Steeves

In contrast to most other plant cultures, excised leaves do not undergo indefinite growth in culture, even when provided with adequate nutrients and transferred or subcultured at intervals. The leaf, in culture as on the intact plant, develops as a determinate organ, and its ability to develop in this way in isolation shows that it is self-differentiating. Indeed the objective of most leaf culture to date has been to study the total development of this organ under conditions which can be controlled in contrast to the largely unknown circumstances in which the leaf develops as part of an intact plant. It is even possible in certain cases to investigate the process of leaf determination by excising primordia of leaves prior to the completion of this process. In true leaf cultures, which will be described here, growth and differentiation occur at the expense of nutrients incorporated in the culture medium as well as those produced by the autotrophic leaf itself in the light. This is distinct from the study of detached, mature leaves which are often maintained for protracted periods for physiological or biochemical studies. In addition, disks or segments of leaf tissue are sometimes used to test the effects of physiologically active substances on plant tissue, and leaves, like other parts of the plant, can be used to initiate callus cultures.

Leaf Preparation

The most extensive leaf culture studies to date have been carried out on ferns, particularly the cinnamon fern (*Osmunda cinnamomea*).[1] The compact apical bud of this fern contains a large number of immature leaves ranging from newly formed primordia adjacent to the shoot apex to leaves essentially ready to undergo final expansion and maturation. Experience has shown that any of these can be cultured to maturity on a nutrient medium of extremely simple composition. After removal of outer leaf bases, including those of the current expanded fronds, the bud is thoroughly washed, then surface sterilized in a solution of calcium hypochlorite,[2] and finally rinsed in several changes of sterile water. A few damaged outer leaves are removed with sterile forceps and the inner leaves of whatever stage is required are removed sterilely either by break-

[1] T. A. Steeves and I. M. Sussex, *Amer. J. Bot.* 44, 665 (1957).

[2] In our laboratory, buds are sterilized for 20 minutes in a 7% w/v solution of Pittchlor, a commercial product containing 70% calcium hypochlorite manufactured by Columbia-Southern Chemical Co., Pittsburgh, Pennsylvania.

ing off with forceps or in the case of smaller primordia excising with a knife fashioned from a razor blade fragment. They may be placed directly on nutrient medium or may first be collected in a moistened Petri dish and planted later. In the case of younger primordia it has been found expedient to collect the leaves on the surface of sterile agar (ordinarily 2%) in a Petri dish prior to planting so that they may be counted and examined. Any fern or other plant with a sufficiently tight apical bud to permit surface sterilization should be amenable to this technique. However, difficulty has been encountered in the case of some ferns because the interior of the bud does not appear to be free of microorganisms. To date no technique has been perfected for the sterilization of individual leaf primordia. In certain cases[3] leaves have been excised from plants which had been grown sterilely in culture. Leaves of several flowering plants have also been cultured by the same general methods. In these plants it has been found unnecessary to carry out even a surface sterilization.[4,5] Outer exposed leaves are removed with sterilized forceps, and the inner primordia for culture are then excised with fresh forceps or a knife as previously described. These inner primordia and the apex itself, appear to be free of microorganisms which can multiply on the nutrient media employed.

Culture Method

A particularly advantageous aspect of the culture of leaf primordia is the extreme simplicity of the nutrient media required for their complete development. For both ferns and flowering plants, satisfactory growth has been achieved on a medium in which the nutrients consist solely of a balanced mineral salt solution and a carbohydrate source. In our laboratory we commonly use either Knop's solution or Knudson's solution of mineral salts supplemented with ferric citrate (0.4 ml of a 2.5% stock solution per liter) and a solution of microelements.[6] More vigorous growth has been noted on Knudson's solution in the case of angiosperm leaves and the younger primordia of several ferns,[7] and this has been attributed to the presence of reduced nitrogen in this solution. Other workers have used different mineral solutions. The carbohydrate source is ordinarily sucrose at a concentration of 2%. Usually, especially in the case of larger primordia, the medium is solidified with agar, and 0.8% has been found to be a satisfactory concentration. Leaves, however, may be cultivated on a liquid medium on which they ordinarily float; and our observations indicate that, where very young primordia are concerned, both survival and growth are favored by the liquid state. The medium is adjusted to pH 5.5 before autoclaving for 20 minutes. The type of vessel, whether test tube, flask, Petri dish, or bottle, has not been found to be critical except in the case of very small primordia. For

[3] I. M. Sussex and T. A. Steeves, *Bot. Gaz.* **119**, 203 (1958).
[4] T. A. Steeves, H. P. Gabriel, and M. W. Steeves, *Science* **126**, 350 (1957).
[5] L. J. Feldman and E. G. Cutter, *Bot. Gaz.* **131**, 39 (1970).
[6] D. P. Whittier and T. A. Steeves, *Can. J. Bot.* **38**, 925 (1960).
[7] T. A. Steeves, *Phytomorphology* **11**, 346 (1961).

these the highest survival rate has been achieved by the use of 1-ounce square tablet bottles with 10 ml of medium.[7] It is believed that the critical factor here is the volume of atmosphere in the vessel relative to the surface area of the medium, but it is not known why this is important. Leaves are ordinarily cultured in the light or in a 12-hour light–dark cycle, and morphogenetic abnormalities are noted when they are cultivated in the dark. Although in the light the leaves are green, it is doubtful that their photosynthetic product is an important energy source under ordinary conditions of culture.[8]

Morphology

Leaves grown from primordial stages to maturity in culture are relatively normal in form but they are greatly reduced in size and often show reduction in morphological complexity. Where reduced size has been investigated in detail in the case of some fern leaves, it has been found to be the result of decreased cell number rather than diminished cell size.[9] Attempts to increase the size of cultured leaves by the addition of complex substances to the medium have met with limited and inconsistent success. In the case of fern leaves, increases in the concentration of sugar in the medium to 6% or more have resulted in substantial size increases and have also increased morphological complexity. Concentrations above 3%, however, interfere with normal expansion of the leaves and it is necessary to transfer them from a high sucrose medium to a lower concentration after an initial phase of meristematic growth.

The leaf culture technique is an extremely simple one in which the only difficult task is the identification and excision of leaf primordia of the desired stage of development without excessive injury. It makes possible the investigation of development in an isolated organ under highly controlled conditions. For the developmental physiologist it provides the opportunity to assess the direct effects of nutrients, hormones, light, temperature, and other factors upon the organ itself without the mediation of the complex system of the intact plant.

[8] I. M. Sussex and M. E. Clutter, *Phytomorphology* **10**, 87 (1960).
[9] J. D. Caponetti and T. A. Steeves, *Can. J. Bot.* **48**, 1005 (1970).

SECTION III

Cell Harvesting

Editors' Comments

The method selected for cell harvest is dictated, for the most part, by the ultimate use the investigator has in mind for the cell culture. Suspension cultures can be harvested by simple centrifugation. Monolayer cultures must be treated severely enough to release them from their surface support, but gently enough, if they are to be subcultured, so that lethal damage is not inflicted. Hydrolytic enzymes fulfill the necessary criteria for releasing cells from their artificial matrix. With some cells, though all have not been tested, release can be achieved by use of low serum and cold temperature. Scraping of cells, as a rule, is not recommended. However, a specialized technique using a cellophane "wipe" is described in the next section.

Included in this section are procedures for harvesting cells in the mitotic stage of the cell cycle from monolayer cultures and cells in the various stages of the cell cycle from suspension cultures using Ficoll gradient centrifugation. For further discussion on the cell cycle, the reader is referred to Section VIII.

CHAPTER 1

Pronase

John F. Foley and Byron Th. Aftonomos

"Pronase" is the trade name of a crude preparation of proteolytic enzyme(s) obtained from a broth culture of *Streptomyces griseus*, used in the manufacture of streptomycin.[1] It has been found to more rapidly and efficiently disperse monolayer fibroblastic cell lines than any other available enzymatic method. It also disperses epithelial monolayers although it seldom provides single cell dispersion suitable for cloning experiments.

Properties

Five to seven components[2,3] have been demonstrated chromatographically, the major one being a neutral protease. Three elastaselike enzymes and one crude trypsinlike enzyme have been separated from the crude preparation and could partly account for the exceptionally broad substrate specificity of pronase.

Pronase can hydrolyze a large number of protein peptide bonds with the subsequent liberation of 70 to 90% of free amino acids, in comparison with acid hydrolysis. The enzyme exhibits maximum activity at pH 7 to 8 and high stability at pH 6 to 9 at room temperature. The stability markedly declines below pH 4 and above pH 10, and although inactivation occurs by heating to 80°C for 15 to 20 minutes, heat stability can be considerably increased by adding calcium ions.

Enzyme Preparation

1. Pronase B Grade (Calbiochem. Co., or equivalent).[3]
2. Make an isotonic solution of sodium and potassium salts, in a ratio of 16 to 1.
3. Mix 1 g Pronase powder with small amounts of the above solution so

[1] M. Nomoto, Y. Narahashi, and M. Murakami, *J. Biochem.* **48**, 593 (1960).
[2] A. Gertter and M. Trop, *Eur. J. Biochem.* **19**, 90 (1967).
[3] M. Trop, R. R. Artalion, Z. Malik, and A. Pinsky, *Biochem. J.* **119**, 339 (1970).

as to form a paste. Continue adding increasing amounts of the solution, with constant stirring, until a uniform suspension is obtained. Bring final volume to 100 ml and allow several hours in the refrigerator at 4°C. Centrifuge to remove coarse particulate matter.

4. Sterilize by filtration and dispense in suitable small aliquots. Store frozen at −20°C, at which temperature the enzyme is stable many months for dispersion purposes.

Once the enzyme preparation has been thawed it can be used for at least 4 weeks when kept at 4°C. For use on cell cultures, it is diluted to 0.1% with a balanced salt solution.

Cell Dispersion

The cell culture is drained of its growth medium, washed with balanced salt solution to remove the old medium and debris, and sufficient 0.1% enzyme solution to cover the cells is added for up to 20 minutes at 36.5°C. If dispersion for cloning purposes is not required, the cell culture can be kept at room temperature and periodically observed for separation of the cells which usually will occur within 3 to 10 minutes. The detached cells are further dispersed by pipetting several times and centrifuging at 60 g for 6 minutes. Following removal of the enzyme solution, the cells must be washed with a balanced salt solution twice before mixing with growth medium.

Using the 20-minute exposure period, we found that with five primary fibroblastic cell lines the highest percentage of clumps was 5.8% with a median of 1.5% (Table I) in contrast to a similar exposure to 0.2% trypsin (Nutritional Biochemical Company, 1:300) where the highest percentage of clumps was 8.5% with a median of 3.3%. Furthermore, a portion of the cell monolayer was frequently left after trypsin treatment and multi-cell clumps were observed with trypsin in contrast to the usual two-cell clumps with Pronase.

TABLE I

Cell Harvesting with Proteolytic Enzymes

	Average percentage clumps[a]	
Cell line	Trypsin	Pronase
Rat mesenchyme fibroblast	0.5	0.2
Human placenta fibroblast	4.5	0.0
Human placenta fibroblast	8.5	5.8
Human adult dermis fibroblast	7.0	4.3
Human fetal dermis fibroblast	3.3	1.5
HeLa epithelium	7.5	22.0
KB epithelium	7.5	22.0
Monkey kidney epithelium	1.0	4.0

[a] Based on counting 200 consecutive cells or clumps in a counting chamber for one to three experiments each.

Fig. 1. HeLa cell monolayer detached with 0.1% Pronase. ×5.

On the other hand, trypsin proved to be superior to Pronase in providing single-cell suspensions of HeLa and KB cells (Table I). With these two cell lines the monolayers separated from their substrate in large sheets with Pronase (Fig. 1). Both Pronase and trypsin provided 96% or better single-cell suspension of a monkey kidney epithelial cell line (Table I).

Discussion

Pronase is the single best proteolytic agent now available for the dispersion of fibroblastic cell lines. It is superior to trypsin not only in separating monolayers from their substrate but also in producing single cell suspensions. It must be remembered that Pronase is not inactivated by serum as is trypsin and therefore must be washed thoroughly from the cells before dispersing them in growth medium. By ellipsometry[4,5] it has been demonstrated that even with washing, some residual enzyme is left at the cell surface. Probably neglect to observe this precaution may account for the failure of one group's[6] cell lines to survive Pronase treatment. We have had no difficulty in maintaining cell lines serially dispersed with Pronase over months, and this experience is supported by other laboratories.[7,8] A detailed study[7] of the influence of Pronase as contrasted to trypsin on the chromosomes of human fibroblasts did not reveal any significant change in chromosome number, tetraploidy, breaks or achromatic gaps.

[4] J. F. Foley and B. Aftonomos, *J. Cell. Physiol.* **75**, 159 (1970).
[5] G. Poste, *Exp. Cell Res.* **65**, 359 (1971).
[6] J. Kahn, M. J. Ashwood, and D. M. Robinson, *Exp. Cell Res.* **40**, 445 (1965).
[7] D. Weinstein, *Exp. Cell Res.* **43**, 234 (1966).
[8] J. C. Sullivan and I. A. Schafer, *Exp. Cell Res.* **43**, 676 (1966).

CHAPTER 2

Trypsinization of Diploid Human Fibroblasts

Jack Litwin

At the present time human embryonic diploid fibroblasts can be grown only on suitable substrates from which they must be removed when they are passed into new cultures. They may be removed in a variety of ways: proteolytic enzymes such as trypsin or Pronase, chelating agents such as Versene, a combination of the two, or scraping. All of these techniques damage the cells to a greater or lesser extent causing the release of low molecular weight intracellular components.[1-4] In some cases the damage may be sufficient to reduce the subsequent growth of diploid cells and eventually reduce their *in vitro* life expectancy (longevity). Most of the cells treated with trypsin are capable of repairing this damage; however, scraping human diploid cells off a glass surface produces such extensive damage that only little net growth occurs. Versene also greatly damages these cells.[5]

[1] C. R. Rebb and M. Y. W. Chu, *Exp. Cell Res.* **20**, 453 (1960).
[2] S. Levine, *Exp. Cell Res.* **19**, 220 (1960).
[3] W. E. Magee, M. R. Sheek, and B. P. Sagik, *Proc. Soc. Exp. Biol. Med.* **99**, 390 (1958).
[4] H. J. Phillips and J. E. Terryberry, *Exp. Cell Res.* **13**, 341 (1957).
[5] J. Litwin, *Appl. Microbiol.* **20**, 899 (1970).

TRYPSIN

Although excellent results may be obtained with crystalline trypsin, even with concentrations down to 10 to 20 μg/ml, the trypsin preparation used most commonly is a relatively crude pancreatic extract. This type of preparation, along with the serum used in most cell culture media, is a completely undefined component to which cells are exposed. Variations in the general toxicity to cells from batch to batch have been observed. One batch of trypsin killed over 90% of the cells treated at 37°C for 15 minutes. As a general rule, when diploid cells suddenly stop growing before they should, it is best to check the trypsin as well as the serum batch.

BUFFERS

The buffers used for trypsinizing cells are frequently selected on the basis of subjective judgment or empirical observations. When the influence of various buffers on the growth potential and the subsequent attachment of human diploid fibroblasts was studied, the following observations were made.[6] (1) phosphate-buffered saline (PBS) without Ca^{2+} or Mg^{2+} produced good monodispersions of cells which were easy to count in a hemocytometer but the subsequent attachment was low. However, when very low concentrations of trypsin were used, such as 0.01% crude trypsin or 10 to 20 μg/ml crystalline trypsin, it became necessary to use a buffer without Ca^{2+} or Mg^{2+}; (2) Hanks' buffer produced the same growth pattern as PBS but better cell attachment was obtained in the subsequent passages, which indicated that less cell damage was produced; (3) 0.33 *M* sucrose *plus* 10^{-3} *M* Mg^{2+} produced as good results as the other buffers and also good attachment; (4) Eagle's medium without calf serum had a slight advantage over the other buffers. The cells had a slightly longer longevity and better attachment. Increasing the concentration of the amino acids and vitamins fourfold produced no beneficial effect; (5) 0.5% lactalbumin hydrolysate in Hanks' buffer produced the best growth and longevity.

The pH of these buffers was about 7.3 at the beginning of the process. Better results were obtained in all cases if the pH was raised to between 7.8 to 8.0 by the addition of 2% Tris (hydroxymethyl)aminomethane. During the course of trypsinization the pH usually drops and the higher pH at the beginning avoids too acid a solution at the end of the process as well as starting at a pH closer to the optimum for trypsin.

It is obvious that the type of buffer used is of little importance, although there may be an advantage in using some nutritional supplement with the trypsin. Nevertheless, it is of some importance to be consistent in the use of one or another buffer since it became apparent that poorer growth and attachment was obtained when a different buffer was used with the trypsin for each passage.

Other additives to the trypsin which have shown certain advantages are

[6] J. Litwin, *Appl. Microbiol.* **21**, 169 (1971).

antibiotics. In addition to 100 IU/ml penicillin and 100 μg/ml streptomycin one may add 1 μg/ml amphotericin B, 10 μg/ml mycostatin, and either 10 μg/ml aureomycin, gentamicin, or kanamycin. These mixtures will not only inhibit possible contaminating microorganisms in the trypsin, but will improve the growth and longevity of the cells, especially if these mixtures are also used in the growth medium.[6]

Temperature

The temperature of trypsinization has a marked influence on the attachment of cells onto glass (Fig. 1) in subsequent passages and even on the longevity (Fig. 2). Trypsinizing cells at 4°C produced less cellular damage which is reflected by a higher proportion of cells attaching to the glass in subsequent passages and a greater longevity than cells trypsinized at higher temperatures. The disadvantages to low temperature trypsinization are that the time required to loosen cells from the glass at 4°C is about 10 minutes longer than at 37°C and there is a greater tendency for cells to be suspended in clumps instead of being monodispersed.

Cell Harvest

After the cells are off the glass surface and are in suspension the general procedure is to centrifuge and resuspend them in a medium free of trypsin. Centrifugation is not absolutely necessary. One can add cells directly from the trypsin suspension to the new culture. The small amount of trypsin added does not usually affect the cells since the attachment of these cells is better than found with centrifuged cells. However, in our experiments the subsequent growth rate and longevity of the former cells has been less than those centrifuged possibly because we always changed the medium after

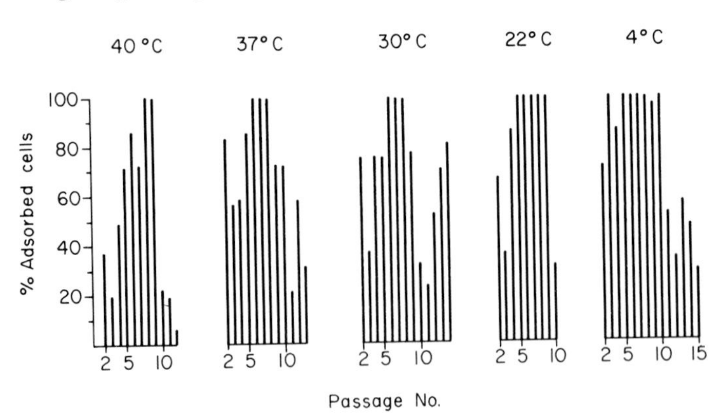

Fig. 1. The percentage of cells adsorbed to the glass surface after overnight incubation from the growth experiment shown in Fig. 2 (J. Litwin[6]).

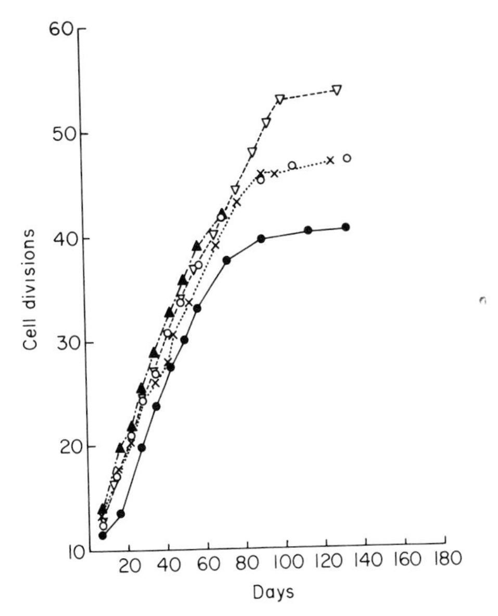

Fig. 2. The effect of temperature of trypsinization on the growth of human embryonic diploid lung fibroblast cells trypsinized in 0.25% trypsin in Hanks' buffer, pH 7.3, at the following temperatures: 40°C (●), 37°C (○), 30°C (×), 22°C (▲), and 4°C (▽) (J. Litwin[6]).

overnight incubation.[7,8] It has been observed empirically that the medium in a newly seeded culture of diploid cells should not be changed before at least 2 days incubation.

Centrifugation at forces of 120, 500, or 1500 *g* with each passage produced a slight reduction in longevity with increase in centrifugal force, but there was no influence on growth rate or attachment.[8] Although it is often recommended to centrifuge cells at as low a force as possible, it was found that with forces of 120 *g* a large proportion of cells were frequently lost when the supernatant was decanted. Therefore, a force no lower than 500 *g* for 20 minutes should be used to sediment the cells. Much higher forces can be used if the temperature is kept low.[9]

Serum should be present in the medium used to resuspend the cells after centrifugation because it slows down the rate of cell attachment to glass surfaces. If a medium without serum is used, cells fasten onto glass surfaces shortly after contact and considerable cells may be lost.

The trypsin preparations used to suspend cells require better standardization to ensure more reproducible results. It is recommended that samples of several batches be tested over four or five consecutive cell passages, and the

[7] J. Litwin, *Acta Pathol. Microbiol. Scand.* **78B**, 273 (1970).
[8] J. Litwin, *Appl. Microbiol.* **21**, 575 (1971).
[9] D. I. C. Wang, T. J. Sinskey, R. E. Gerner, and R. P. DeFilippi, *Biotechnol. Bioeng.* **X**, 641 (1968).

batch which gives the best growth and attachment results be selected. It is also possible that better reproducibility could be obtained with crystalline trypsin which could be used at lower concentrations than crude trypsin and thus the costs would be comparable.

CHAPTER 3

Transplantable Rat Tumors and Cold Temperature Release[1]

M. K. Patterson, Jr.

Transplanted tumors of rats have been used in this laboratory since 1956 as a source of cells for a variety of tissue culture studies. The procedure to be described is for the preparation of cultures of the Jensen sarcoma[2] (ATCC-CCL 45), but it has also been applied to Walker 256 carcinoma,[3] a 3′-methyl-4-dimethylaminoazobenzene (3′-MDAB)-induced hepatoma and transplantable "cell substrains" of the Jensen and Walker 256 tumors. A possible advantage of the method, other than its simplicity, is that *no lytic enzymes* are used either in the preparation of the cell inoculum or for the release of cells from the culture flask surface.

Cell Inoculum Preparation

The tumor is routinely carried as an intramuscular transplant in the *rectus femoris* muscle, but tumors carried at other sites can be used. The animal is killed by cervical fracture. The area of excision is shaved and painted with tincture of iodine. The animal is then strapped on a small surgical board and, except for the area of excision, covered with sterile towels. The tumor is removed and placed in a sterile Petri dish containing Earle's solution. Necrotic

[1] This procedure is included under cell harvesting to illustrate cold temperature release of cells and because cultures initiated from transplantable tumors are considered as continuations of the injected cell line or strain (cf. "Proposed Usage of Animal Tissue Culture Terms," *In Vitro* **2**, 155–159, 1966; Committee on Terminology, Tissue Culture Association, Inc., Chairman, S. Fedoroff).

[2] T. A. McCoy, M. Maxwell, and R. E. Neuman, *Cancer Res.* **16**, 979 (1956).

[3] T. A. McCoy and R. E. Neuman, *J. Nat. Cancer Inst.* **16**, 1221 (1956).

and connective tissues are removed and small pieces of the tumor tissue placed in a sterile 5-ml syringe fitted at the base with three circular pieces of a 24-mesh (0.014) stainless steel screen.[4] One gram wet weight tissue (8 to 9 days after transplant) yields approximately 10^8 cells, which are 60–70% viable. Viability increases with tumors of shorter transplant duration and, *in vitro*, with the time elapsed from moment of excision to planting in tissue culture medium. The tissue is pressed through the screen into a 15-ml centrifuge tube and suspended in 10 ml Earle's solution. The suspension is poured into a small glass funnel containing several layers of glass wool and the filtrate collected in a 15-ml centrifuge tube. The tube is plugged with gauze and centrifuged at 120 *g* for 5 minutes, and the supernatant is discarded. The cell button is resuspended in 10 ml Earle's solution and an aliquot removed for a cell count in a hemocytometer.

Cell inocula are prepared by diluting the cells to the desired number in McCoy's 5a "Acid" Medium[5] supplemented with whole, dialyzed, or Sephadex-treated 5% bovine serum. Acidification of medium is done by gassing the medium with 100% CO_2. A typical inoculum for a T-15 flask is 2 ml containing 2×10^5 Jensen sarcoma cells per milliliter. The flasks are stoppered with gauze plugs and placed in a 37°C gas-flow incubator using 5% CO_2 and 95% air saturated with water through use of a bubbler system. After 24 hours incubation, the medium is removed and 3 ml of McCoy's 5a Basic Medium (pH 7.4) added to the cultures. The flasks are plugged with silicone stoppers and placed in a conventional 37°C incubator. Because of the lactic acid production by these tumor cells, it is necessary to perform media changes at 24-hour intervals.

Cell Harvest

Saturation densities (2 to 4×10^6 cells per T-15 flask) for the Walker 256 carcinoma and the Jensen sarcoma are normally reached in 5 days. The cells are collected by removal of the spent medium, by addition of 3 ml Earle's solution, or complete replacement medium and by placing the flasks in a refrigerator at 4°C for 1 hour. By vigorously shaking the flask, the cells can then be dislodged as single cell suspensions.

Growth Characteristics

Growth curves of these tumors show a 24- to 48-hour lag period followed by exponential growth with a doubling time of approximately 24 hours (Fig. 1). After the first subculture the lag period is shortened. Because of the high

[4] Available from Ludlow-Saylor Wire Cloth Co., 8474 Delport Drive, St. Louis, Missouri 63114.

[5] H. G. Morton, *In Vitro* **6**, 89 (1970).

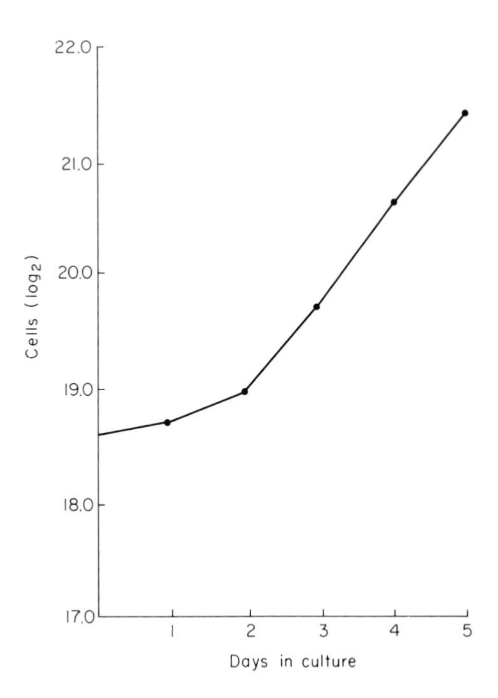

Fig. 1. Five-day growth curve of Jensen sarcoma cells initiated from transplantable tumor. Initial and final cell counts were 2.0×10^5 and 3.2×10^6 cells per milliliter respectively. McCoy's 5a Medium supplemented with 5% bovine serum changed at 24-hour intervals. At daily intervals cells were harvested by exposure to cold (see text) for either cell counting or subculturing.

lactic acid production of the Walker 256 carcinoma, it is necessary to reduce its cell inoculum to 1×10^5 cells per milliliter.

Notes on Cell Preparation

These procedures have been used to prepare cell inoculum for single cell isolation and cloning experiments,[6] for stationary culture flasks or Petri dishes of varying sizes,[7] as well as for perfusion-type systems.[8]

[6] M. K. Patterson, Jr., M. D. Maxwell, and E. Conway, *Cancer Res.* **29**, 296 (1969).
[7] M. K. Patterson, Jr. and M. D. Maxwell, *Cancer Res.* **30**, 1064 (1970).
[8] P. F. Kruse, Jr., E. Miedema, and H. C. Carter, *Biochemistry* **6**, 949 (1967).

CHAPTER 4

Cells at Defined Stages of the Cell Cycle

A. Selection of Mitotic Cells from Monolayer Culture

S. Shall

There are two broad categories of methods for obtaining cells at a given stage of the cell cycle: (1) by induction or coercion, and (2) by selection. The coercive methods rely on blocking the progress of the cells at a known, fixed point in the cell generation by the use of an appropriate inhibition. All the cells in the culture will then accumulate behind this barrier. When the inhibition is removed, the cells will all start from this fixed point together. Two processes are available for inhibition; DNA synthesis and mitosis. Inhibition of mitosis will permit the cells to progress through the cell generation but will block and accumulate them at mitosis. Upon release of the inhibition they should all complete mitosis together. In a similar manner, inhibition of DNA synthesis will allow cells not in DNA synthesis to proceed until they reach the S phase when they will be blocked and will, therefore, accumulate at the beginning of the S phase.

The compounds most commonly used to block DNA synthesis include high levels of thymidine, e.g., 2 m*M*. This appears to block deoxycytidine triphosphate synthesis by feedback inhibition. The inhibition may be released by washing out the thymidine or by adding deoxycytidine. Other compounds in use are fluorodeoxyuridine which inhibits TTP synthesis; hydroxyurea, amethopterin (methotrexate) or isoleucine and glutamine starvation. Mitosis may be inhibited by colchicine or Colcemid.

All available coercive methods induce gross distortions of cell metabolism. Those that inhibit DNA synthesis allow other metabolic processes to continue so that a different pattern of metabolism results. Consequently, all methods which coerce the cells into synchrony by the use of metabolic inhibition of one sort or another are best avoided unless specific indications suggest their use. Descriptions of these coercive methods may be found in recent reviews.[1,2]

We will confine ourselves here to methods which obtain a synchronous, homogeneous population of cells by selecting the desired fraction from an

[1] G. C. Mueller and K. Kajiwara, *In* "Fundamental Techniques in Virology" (K. Habel and N. P. Salzman, eds.), p. 21. Academic Press, New York, 1969.

[2] E. Stubblefield, *In* "Methods in Cell Physiology" (D. M. Prescott, ed.), Vol. 3, p. 25. Academic Press, New York, 1968.

asynchronous population. Mitotic cells may be harvested very conveniently and efficiently from monolayer cultures.[3,4]

Principle of the Method

The principle of the method is that some cell types when grown in a monolayer bind firmly to the glass or plastic, but during mitosis the cells round up and become more loosely attached. After mitosis is complete the cells again become firmly attached to the surface. During mitosis the more loosely attached cells may be removed selectively.

This method was originally devised by Terasima and Tolmach.[3,4] Subsequently, it was observed that calcium ions increased the firmness of the attachment of the cells to the glass.[5] Now media are used in which all the calcium in the salt solution is replaced by sodium, and usually the phosphate concentration is increased. In these media the sole source of calcium for the cells is the serum. It is very necessary that calcium-free media be used in this method. In addition, media low in magnesium may be used. This method will yield 10^4–10^6 cells with a mitotic index of 0.80 to 0.95. A complete discussion and description of this method has been given.[6,7]

Procedure

The cells are grown to about the middle of the logarithmic phase. It is helpful to have about 10^7 cells per bottle since the yield will be about 1 to 5%, depending on the growth rate. During the entire procedure take care to maintain pH by gassing with CO_2 if bicarbonate buffer is used and maintain the temperature at 37°C.

The loosely bound cells are first shaken free, holding the bottle horizontally and shaking twenty times in 3 seconds with an amplitude of about 2.5 inches. This is best done in a shaking water bath, but may be done manually. The bottle is held with the short axis in the direction of agitation and with the monolayer down so that the medium sweeps over the cells. The medium is decanted and kept on ice.

Fresh, prewarmed, and gassed medium is added to the culture bottles, taking care that the medium is not pipetted onto the cells. The cells are then allowed to grow for a further period, say 15, 30, 45, or 60 minutes. The harvesting is repeated and fresh medium is added to the culture bottle.

The first harvest will contain much cell debris and many dead cells. The second harvest yields a very small number of cells.

[3] T. Terasima and L. J. Tolmach, *Nature (London)* **190**, 1210 (1961).

[4] T. Terasima and L. J. Tolmach, *Exp. Cell Res.* **30**, 344 (1963).

[5] E. Robbins and P. I. Marcus, *Science* **144**, 1152 (1964).

[6] R. A. Tobey, E. C. Anderson, and D. F. Petersen, *J. Cell. Physiol.* **70**, 63 (1967).

[7] D. F. Petersen, E. C. Anderson, and R. A. Tobey, *In* "Methods in Cell Physiology" (D. M. Prescott, ed.), Vol. 3, p. 347. Academic Press, New York, 1968.

After a further time period a third harvest is collected as before. Subsequent harvests may be collected after allowing the cells to grow for the fixed time period between each harvest. The cells from the third harvest onward may be used directly for experimental purposes. Alternatively, all the harvests may be cooled immediately and kept on ice for several hours. It is possible to continue this cycle of harvesting, ten or more times, and each harvest after the first two may be used for experimental purposes.

The degree of synchrony is best estimated by determining the mitotic index of the harvested cells. The method is described in the next paper (Section III, Chapter 4B). The volume of the harvested cells may be estimated with a Coulter Counter and a cell volume (J) plotter.

Determination of Precise Conditions for a New Cell Type

Two factors are of importance in this technique. Different cell types will attach to the glass with different degrees of firmness. Thus the appropriate force required to dislodge the mitotic cells selectively must be experimentally determined with each cell type. The precise conditions described above should only be used as a guide. For each new cell type, the cell yield and mitotic index should be estimated as a function of the ampltitude, the frequency, and the duration of shaking. Too little force will give a small number of cells with a high mitotic index, the correct force will give a larger number of cells with a high mitotic index, and excessive force will give cell debris and a lowered mitotic index.

The second parameter is the interval between harvests. If this period is too short the yield of cells is very low; at the optimum interval the yield of cells is higher and the mitotic index is still high. If the growth interval is too long then the mitotic index will drop. It is necessary to consider which cell type is most appropriate; it is possible to select a subclone of a given cell type which shows more suitable attachment properties. In addition, both the medium used and the type of bottle affect the firmness with which interphase and mitotic cells adhere to the surface.

General Observations

The mitotic selection method is the best available technique for harvesting mitotic cells and for obtaining a synchronous population of cells. It has two basic limitations. (1) It is only suitable with monolayer cultures; and (2) it yields only small numbers of cells.

The yield is usually quite adequate for cytological methods but is generally too low for standard enzyme and biochemical measurements.

The low yield arises from two causes: (1) monolayer cultures give only relatively small numbers of cells, and (2) the method selects a small fraction of the total population.

There are three possible methods of increasing the yield.

1. You may use large bottles, say 20- or 50-ounce medical flats, and you may use many bottles and pool the harvest.

2. Once harvested, the cells may be rapidly cooled at 4°C. However, the cold shock may have undesirable metabolic effects in a few specific cases.

3. You may harvest five to ten bottles immediately one after another, pool the harvests, and collect all the cells together.

In general, the mitotic selection method will yield 10^4–10^6 cells with a mitotic index of 0.80 to 0.95 quite readily. It may be scaled-up and is usually applicable to most cells grown in monolayers.

CHAPTER 4

Cells at Defined Stages of the Cell Cycle

B. Sedimentation in Sucrose and Ficoll Gradients of Cells Grown in Suspension Culture

S. *Shall*

It is sometimes necessary to harvest cells at a specified part of the cell cycle, like mitosis, G_1, DNA synthetic period, or G_2. To select the desired cells one may use the mitotic selection procedure described in Section III, Chapter 4A. This method yields mitotic cells and is always the method of choice when applicable. It does have two limitations; (1) it can only be used with monolayer cultures and (2) it yields only a limited number of cells. If you use suspension cultures then the mitotic selection technique is not applicable. Alternatively, if you require a very large number of cells another procedure is required. The method to use then is to separate cells by velocity sedimentation in a gradient.[1]

Principle of the Method

The principle of the method is to sediment cells through a column of liquid so that the cells are distributed through the column. The rate at which cells sediment through a column of liquid is:

$$S = 2/9 \times g \times (\rho_c - \rho_s)/\eta \times r^2 \times 3.6 \times 10^{-4} \qquad (1)$$

[1] S. Shall and A. J. McClelland, *Nature N. Biol.* **229**, 59 (1971).

where S = rate of sedimentation (mm/hour), ρ_c = density of the cell (g/ml) (usually 1.05–1.10), ρ_s = density of medium, η = viscosity of medium (poises), r is radius of cell (μm), and g = gravitational acceleration (981 cm sec^{-2}). Equation (1) reduces to:

$$S = 0.0785\ (\rho_c - \rho_s)/\eta \times r^2 \tag{2}$$

If all the cells in the population have very similar densities (ρ_c), then the cells will separate on the basis of the radius (r). Cells with twice the radius will sediment four times faster than smaller cells. When you are dealing with a relatively homogeneous population of cells, the cell density does not vary much.[2] In this case the cells will separate according to their radius. If you have a mixed population of different types of cells, then their densities are often very different; this can be used to separate cell types one from another. With cultures of a single cell type the range of densities may be quite small. If, in addition, the density of the gradient (ρ_s) does not vary a lot, then $(\rho_c - \rho_s)/\eta$ will effectively be a constant, with a value of about 1 to 5 g/ml/poise. The rate at which the cells sediment will then depend almost entirely on their radius.

We assume that the volume of the cells increases as the cells progress through a cell generation. A twofold increase in volume is equivalent to about a 1.6-fold increase in r^2 and, therefore, in sedimentation rate. Thus, we use this technique to separate cells according to their volumes. Larger, older cells will sediment more rapidly to the bottom and younger, smaller cells will be found nearer the top of the gradient.

Preparation of Gradient

It is necessary to include a gradient of some kind in the column of liquid in order to stabilize it against convection. A variety of compounds may be used to make the gradient, including protein (bovine serum albumin), serum,[2] Ficoll[3,4] (a sucrose polymer), and sucrose.[1] We use sucrose because it is cheap and pure. Constant osmotic pressure must be maintained throughout the gradient.

The gradient is made by preparing two solutions, one of 10.0% (w/v) sucrose and the other of 2.72% (w/v) sucrose in complete medium. To maintain constant osmotic pressure, the concentration of sodium chloride must be decreased by 146 m*M* in the 10% sucrose solution and by 40 m*M* in the 2.72% sucrose solution. This means the preparation of special media containing decreased sodium chloride concentrations. The gradient is made with the simple glass apparatus shown in Figs. 1 and 3.

The apparatus to be used is determined by the scale of operations. Figure 1 shows the apparatus suitable for 30 to 50 $\times$ 10^6 cells; the apparatus shown in Fig. 3 should be used for up to about 10^9 cells.

[2] H. R. Macdonald and R. G. Miller, *Biophys. J.* **10**, 834 (1970).
[3] S. R. Ayad, M. Fox, and D. Winstanley, *Biochem. Biophys. Res. Commun.* **37**, 551 (1969).
[4] A. M. H. Warmsley and C. A. Pasternak, *Biochem. J.* **119**, 493 (1970).

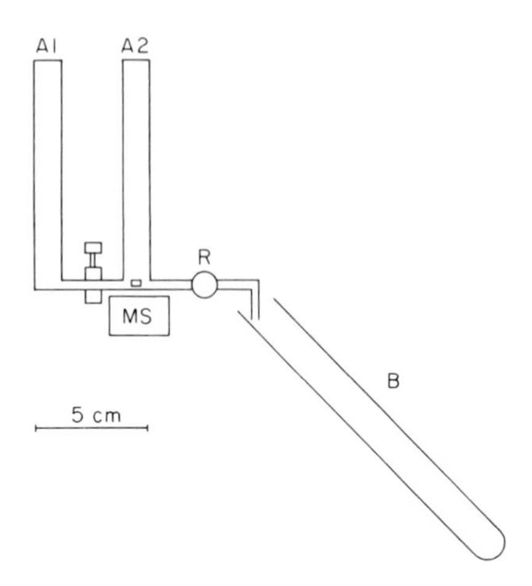

Fig. 1. Apparatus for generation of a gradient on a small scale. The actual separation is carried out in a glass test tube (B) 150 × 16 mm. The gradient is made in cylinders A1 and A2 connected by a short length of silicone tubing with a standard clip to open and close it. The outflow is controlled by a "Rotaflo" tap (R), which allows close control of flow rate. The dimensions of the cylinders A1 and A2 are 12 mm ID and 100 mm high. Cylinder A2 has a small bar magnet in it, controlled by a magnetic stirrer (MS) on which the apparatus is placed.

The gradient is formed by transferring heavy solution (10% sucrose) from the right-hand chamber (A2, Fig. 1) to the gradient tube (B, Fig. 1). As solution leaves the right-hand chamber, the lighter solution (2.72% sucrose) will flow from the left-hand (A1, Fig. 1) to the right-hand chamber; if this is being vigorously stirred, the concentration of sucrose in the right-hand chamber will decrease progressively as heavy liquid leaves and is replaced by lighter solution. If the two chambers are of the same dimensions then the gradient of sucrose will be linear from 10 down to 2.72%.

If growth experiments are done the gradient apparatus is sterilized before use.

Small-Scale Preparations (*Fig. 1*). A 7.5 ml solution containing 10% sucrose is placed in the gradient vessel (A2, Fig. 1) containing the bar magnet. Sucrose medium, 7.5 ml of the 2.72%, is placed in the other gradient vessel (A1, Fig. 1). The two vessels are kept level and the magnet is set to stir vigorously. The clip between the two reservoirs is opened and then the "Rotaflo" tap is carefully opened so that the solution drips out quite slowly into a 150 × 16 mm test tube (B, Fig. 1). The solution is allowed to run down the side of the test tube to the bottom.

Exponentially grown cells are collected by centrifugation at about 100 g for 5 minutes. The medium is pipetted off and retained. The cells are resuspended in 1.0 ml of this used medium and layered onto the gradient with a wide-mouthed pipette held against the wall of the tube. The cells run slowly down the side of the tube and across the surface of the gradient. There is some turbulence at the top of the gradient, but no streaming is observed. Usually

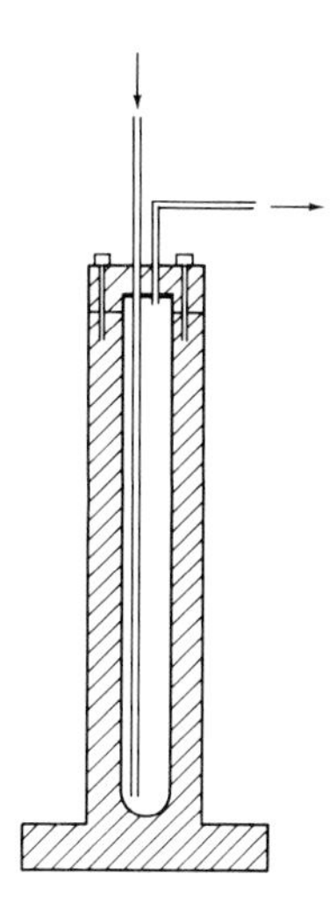

Fig. 2. Apparatus to fractionate gradient used in Fig. 1. The unit is made from Perspex and is hollowed out so that the test tube used (150 × 16 mm) fits in exactly. The top is held tightly to the holder with two screws. The rim of the test tube inserts into a depression in the top, where a rubber gasket is used to form a watertight seal. Two thin metal tubes pass through the top; one goes right to the base of the test tube, the other is flush with the inside of the top. The gradient is displaced from the test tube by pumping 25% (w/v) sucrose through the long tube to the bottom of the test tube. The gradient will be displaced upward through the short tube. Fractions of 1.0 ml may be collected.

about 30×10^6 cells are applied to the gradient. The tube is then left standing upright in the incubator at 37°C for 50 minutes while the cells sediment down the gradient. The topmost 1 ml of the gradient may then be removed with a pipette or the whole column may be fractionated using the device shown in Fig. 2. The topmost 1 ml contains the youngest and smallest cells. The sucrose should be washed off and the cells returned to normal growth medium. These cells will then show synchronous growth for one or two generations.

If the cells are quite large, say with an average volume of about 1500 μm^3 and have a long doubling time, say 24 hours, then the sedimentation may be done conveniently at unit gravity in an incubator at 37°C. If the cells are smaller or have shorter doubling times the separation may be done in the cold overnight (10–15 hours) or in a centrifuge at 37°C for 3 to 5 minutes at 100 to 200 *g*. From Eq. (2) you may obtain an approximate estimate of the sedimentation velocity. The cell density may be assumed to be 1.05–1.10, unless otherwise known. The sucrose gradient described has a density gradient from 1.01 g/ml at the top to 1.03 g/ml at the bottom.

If the sedimentation period is more than 5 to 10% of the doubling time, then the sedimentation should be done at 25° or 4°C to slow or stop cell growth during the separation period.

Large-Scale Preparations (Fig. 3). The 10% sucrose medium solution (150 ml) is placed in gradient reservoir A1, (Fig. 3) and the 2.72% solution in reser-

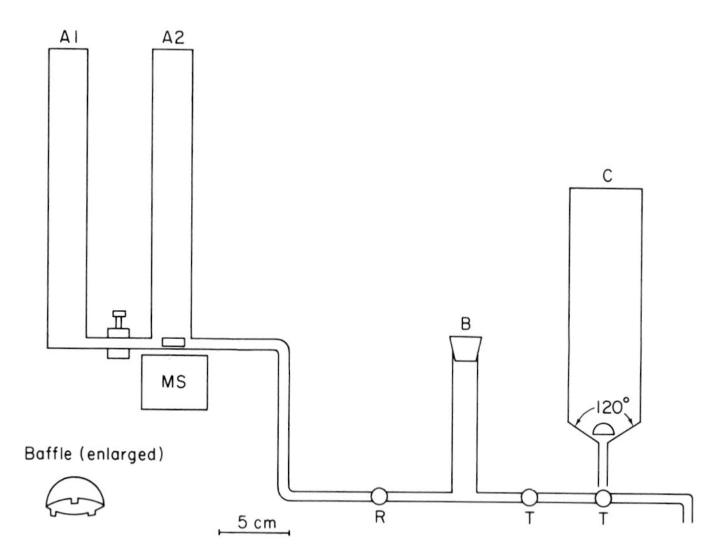

Fig. 3. Apparatus for sedimenation separation on a large scale. Gradient-forming vessels A1 and A2 are 27 mm ID and 200 mm high. The bottom of the gradient former should be higher than the top of the separating chamber (C). The "Rotaflo" tap (R) controls the rate of flow from the gradient. The all glass components are connected by sterilizable silicone tubing. The cell chamber (B) is 18 mm ID and 100 mm long and is tightly sealed with a No. 17 silicone bung. The separating chamber (C) is 51 mm ID and 150 mm on its straight edge. The baffle (see insert) is made from the base of a 16-mm diameter glass test tube. The bar magnet in cylinder A2 is motivated by the magnetic stirrer (MS). (T), two-way and three-way glass taps with wide-bore channels. Separating chamber C is covered to exclude dust. The entire apparatus may be steam-sterilized after assembly. "Rotaflo" taps (R) are sterilizable and permit fine control of flow rate; they are manufactured by Quickfit & Quartz Ltd.

voir A2 (Fig. 3). Ensure that the levels of the solutions in reservoirs A1 and A2 are the same, and then open the clip connecting the two reservoirs. Then, 10 ml of medium without serum is placed in the separating column (C, Fig. 3). The baffle is positioned, after air bubbles are removed, with the concave face downward. The baffle ensures that the gradient does not spurt into the separating column, but enters as a thin layer. Then 5–10 ml of concentrated cell suspension containing 10^8–10^9 cells, are placed in the cell chamber (B, Fig. 3) which is then tightly corked. The cell suspension is driven into the separating column (C, Fig. 3) by forcing air into the cell chamber (B, Fig. 3) by means of a needle and syringe. The cell suspension should form a narrow band underneath the layer of medium. Without admitting any air into the line, you now allow the gradient to flow into the separating column (C, Fig. 3) thus forcing the cells up into the column. It is important that gradient reservoir A2 is always vigorously stirred by the magnetic stirrer.

When the gradient is complete, the separation process may begin. This may be conducted at 37°, 25° or 4°C. The lower the temperature, the smaller the volume of the cells and the more viscous the medium; therefore the cells sediment more slowly. The appropriate time must be ascertained for each cell type.

With a volume of about 1500 μm^3, about 2 hours at 37°C is necessary. With smaller cells, it is necessary to use 4° or 25°C.

At the end of the separation the column is fractionated from the bottom into 5- or 10-ml fractions. The cells are gently centrifuged and resuspended. They may then be used either for growth or for analysis. The smallest and youngest cells are at the top and as you go down the gradient you pass G_1, then S and then G_2 cells. The G_1 cells are separated most cleanly, the others less so.

Instead of sucrose you may use a 5–20% (w/v) gradient of Ficoll[3,4] (Pharmacia Ltd.). In this case, the Ficoll may be added to the complete medium and the change in osmotic pressure ignored. Ficoll is toxic to some cell types, however. Bovine serum albumin or serum may also be used, but the first is sometimes toxic and the second expensive and rather variable.

This velocity sedimentation selection technique may be performed in smaller or larger tubes. To scale up still further, a zonal centrifuge to sediment the cells may be used.[4]

The velocity sedimentation technique separates on the basis of volume, not age. There is a dispersion of volumes at any given age and, therefore, there is a spread of ages at any given volume. The degree of synchrony is therefore less than with mitotic selection. Nonetheless, degrees of synchrony between 0.50 and 0.80 should readily be achieved. This technique may not work with heterogeneous cell populations because of the overlap. An important source of heterogeneity is aneuploidy which gives cells of widely different volumes at a fixed age.

The volumes of the cells through the gradient may be measured microscopically with phase-contrast microscopy and a graticule or with a Coulter Counter and a cell volume plotter. Alternatively, the asynchronous culture may be given a 30-minute pulse of thymidine (5.0 μCi/ml) before sedimentation. The S phase cells are then identified by measuring the radioactivity insoluble in 5% (w/v) trichloroacetic acid in each fraction. The mitotic cells may be identified by estimating the mitotic index of each fraction.

These methods may be used also to separate different cell types one from another, making use either of differences in volume or density.

Estimation of Mitotic Index

Spin the cells at about 100 *g* for 5 minutes in a bench centrifuge *in the cold.* Remove the supernatant with a Pasteur pipette and discard. Add several milliliters of precooled 0.6% (w/v) sodium citrate. Resuspend the cells by gentle shaking and leave for 20 minutes at 0°C. Centrifuge the cells *in the cold* as before and discard the supernatant. Add several milliliters of ice cold fixative (acetic acid:methanol, 3:1, v/v) dropwise down the side of the tube on to the cell pellet. Stand for 5 minutes at 0°C and centrifuge. Remove and discard the supernatant. Add 1.0 ml fixative and stand for 2 minutes at 0°C. Centrifuge and discard the supernatant. Add 0.2 ml of ice cold fixative and resuspend the cells by gentle shaking. Place several drops of this cell suspension on to a clean cover slip and allow to dry completely in air. Stain for 10 minutes in natural aceto-orcein, wash twice in methoxyethanol, and twice in Euparal essence. Leave to dry in air. Mount in Euparal on clean slides.

This preparation gives permanent slides. You should count several fields, scoring the number of mitotic cells. It is wise to count at least 400 cells but preferably 1000 cells should be counted.

CHAPTER 5

Processing Cells for Enzyme Assays

V. J. Cristofalo and Joan Kabakjian

The processing of cells for enzyme assay requires several steps: (1) growing the cells in large enough quantities and under conditions which allow convenient assay of the enzyme under study; (2) harvesting of the cells and intermediate processing such as washing residual medium and trypsin from the cell pack; (3) disruption of the cells under conditions which maintain the activity of the enzyme to be studied; (4) postdisruption processing such as isolation or removal of specific organelles.

The procedures described here have resulted from studies designed to evaluate the activities of a variety of enzymes in human cells grown in culture. The techniques have been worked out principally for human fibroblastlike cell lines, especially WI-38 (ATCC-CCL75). In most cases, they have been applied to the SV40 virus-transformed cell lines derived from WI-38[1] and occasionally to permanent cell lines such as HeLa (ATCC-CCL2). What we present here is a summary of techniques we have found to be successful.

With cell cultures, as with animal tissues, there is no single general procedure for cell preparation that is universally applicable. Differences in the

[1] A. J. Girardi, F. C. Jensen, and H. Koprowski, *J. Cell. Comp. Physiol.* **65**, 69 (1965).

[2] V. Allfrey, *In* "The Cell" (J. Brachet and A. E. Mirsky, eds.), p. 193. Academic Press, New York, 1959.

[3] O. Bodansky and M. K. Schwartz, *In* "Methods in Cancer Research" (H. Busch, ed.), Vol. II, p. 446. Academic Press, New York, 1967.

[4] R. K. Murray, R. Suss, and H. C. Pitot, *In* "Methods in Cancer Research" (H. Busch, ed.), Vol. II, p. 239. Academic Press, New York, 1967.

[5] V. R. Potter, *In* "Methods in Enzymology" (S. P. Colowick and N. O. Kaplan, eds.), Vol. I, p. 10. Academic Press, New York, 1955.

properties of the specific enzymes to be studied and the cell types under investigation will require modification of the procedures. For guidelines to such modifications the reader is referred to the excellent general review articles by Allfrey,[2] Bodansky and Schwartz,[3] Murray *et al.*,[4] Potter,[5] Shonk and Boxer,[6] and Umbreit *et al.*[7]

Cell Cultivation

It is desirable to assay enzyme activity from cells grown and harvested under uniform conditions of cell density, media composition, pH, and handling. Our routine methodology for the cultivation of human diploid cells in monolayer has been described elsewhere.[8] Typically, for enzyme assays appropriate vessels are seeded with 1–3 × 10^4 cells/cm^2 of growing surface. After seeding, the cultures are permitted to grow at 37°C until they are approximately 30% confluent. One-half of the growth medium is then removed and replaced with fresh medium. The cells are harvested 24 hours after this refeeding. Under these conditions the cells are in the log phase of growth when harvested.

Saturation density changes with passage number during the life cycle of diploid cells, but as a rough estimate, a monolayer of WI-38 cells harvested at confluency will yield approximately 0.25 μl of packed cells/cm^2 of growing surface. With permanently proliferating cell lines, the yields will be higher.

Cell Harvesting

In preparation for harvesting, the growth medium is decanted and the cell monolayer is washed twice with 0.25% trypsin solution prepared in Ca^{2+}- and Mg^{2+}-free phosphate-buffered saline. A thin film of trypsin solution is allowed to remain on the cell sheet after the second washing. The culture vessels are incubated at 37°C until the cell monolayer detaches from the growing surface. The cells are then harvested into a centrifuge tube by repeating washings of the culture vessels with growth medium containing 10% bovine or fetal bovine serum. Cell clumps are dispersed by aspiration either with a syringe fitted with a No. 15 needle or with a pipette. An aliquot of the dispersed cell suspension can be removed at this point for counting the cells.

The cell suspension is then centrifuged at 500 *g* for 2 to 3 minutes at 4°C. The supernatant medium is decanted and the cells are washed three times by centrifugation as above with a cold, buffered balanced salt solution. The resulting cell pack is placed on ice until homogenization. We have always homogenized as soon as possible after harvesting.

[6] C. E. Shonk and G. E. Boxer, *In* "Methods in Cancer Research" (H. Busch, ed.), Vol. II, p. 581. Academic Press, New York, 1967.

[7] W. W. Umbreit, R. H. Burris, and J. F. Stauffer, "Manometric Techniques," pp. 114–192. Burgess, Minneapolis, Minnesota, 1964.

[8] V. J. Cristofalo and B. Sharf, *Exp. Cell Res.*, in press, 1973.

Cell Disruption

The methods of cell disruption used for animal cells *in vitro* are simply adaptations of methods developed for excised animal tissues.

In preparing cells for enzyme assay we have used mechanical, sonic, and explosive decompression methods of homogenization. Important variables for each method are (1) the time and intensity of the treatment and (2) the suspending medium. In all cases, manipulations were carried out at or near 4°C and cell breakage was always monitored by phase microscopy.

Mechanical Disruption

1. Potter-Elvejhem Homogenization Apparatus. In this method we employ a glass homogenization tube and a motor driven Teflon pestle (Arthur H. Thomas, Inc., Philadelphia, Pennsylvania). The cells are forced between the wall of the tube and the pestle and the resulting shear forces disrupt the cells. The packed cells are washed into a prechilled homogenizing tube with nine volumes of an extraction medium which contains 0.15 *M* KCl, 0.05 *M* $KHCO_3$, and 0.006 *M* EDTA (Shonk and Boxer).[6] The homogenization tube is placed in a beaker of ice to offset local heating. With the stirring motor rotating at 9000–10,000 rpm, the pestle is forced through the suspension by rapidly moving the tube up and down. About 12 such cycles have been found to be sufficient for breaking essentially 100% of WI-38 cells. More are required for some of the permanent cell lines studied. This procedure has been used extensively in our laboratory for studies in which total enzyme activities were desired and preservation of intact cellular organelles was not required. The crude homogenate obtained has been used, after appropriate dilution, for the assay of acid phosphatase (EC 3.1.3.2) alkaline phosphatase (EC 3.1.3.1), and β-glucuronidase (EC 3.2.1.31).[9] After centrifugation of the homogenate at 105,000 *g* (4°C) to remove the particulate matter, the resulting extract has been stabilized with 1 mg/ml dithiothreitol and used for the assay of a variety of enzymes of glucose metabolism.[10]

2. Dounce Homogenization. The Dounce homogenizing apparatus (Kontes Glass Inc., Vineland, New Jersey) consists of a tube and two glass pestles, one loose fitting and one tight fitting. The principles of homogenization in this method are essentially the same as those described above. Here however, disruption is more gentle, and the pestle is driven by hand rather than mechanically. In our procedure the packed cells are washed into a prechilled Dounce homogenization tube with nine volumes of cold 10^{-3} *M* phosphate buffer, pH 7.4. Other hypotonic solutions have also been used successfully. The cell suspension is allowed to remain at room temperature for 10 minutes with occasional stirring to keep the cells from settling. The tube is then placed in ice and the cells homogenized. We have found that, for WI-38 cells, thirty thrusts with the loose fitting pestle and four thrusts with the tight fitting pestle are adequate to disrupt 90–100% of the cells. The permanent cell lines we used required approxi-

[9] V. J. Cristofalo, N. Parris, and D. Kritchevsky, *J. Cell. Physiol.* **69**, 263 (1967).

[10] V. J. Cristofalo, *In* "Aging in Cell and Tissue Culture" (E. Holeckova and V. J. Cristofalo, eds.), p. 83. Plenum, New York, 1970.

mately twice as many thrusts with the tight fitting pestle. As soon as breakage is complete, a volume of 0.5 *M* sucrose equal to the volume of the broken cell suspension is rapidly added to the homogenate to restore isotonicity. Any further dilutions are made with 0.25 *M* sucrose. This procedure has been used in our laboratory for a variety of enzyme assays including acid phosphatase (EC 3.1.3.2) and β-glucuronidase (EC 3.2.1.31). This is a relatively gentle technique and is particularly useful when specific subcellular fractions are required. Using modifications of standard techniques for centrifugal separation of cell organelles, we have obtained good preparations of nuclei, mitochondria, and lysosomes. We have found that with both mechanical methods, the smallest commercially available homogenizers require a minimum of 2 to 3 ml of cell suspension for adequate breakage.

Disruption by Sonication. The energy of sound waves transmitted through the suspension causes disruption of the cells. We use sonication routinely for the assay of thymidine kinase from WI-38 cells. The cell pack is suspended in 0.05 *M* Tris buffer (pH 7.8) containing 0.01 *M* dithiothreitol to yield a cell concentration of approximately 10^6/ml of suspension. The tube containing the cell suspension is placed in an ice bath and the standard microtip probe of the sonic oscillator (Branson Sonifier-Cell Disruptor, Heat Systems-Ultrasonics Inc., Plainview, New York) is placed in the center of the cell suspension. The sonicator is turned on, "matched" for maximum energy and sonicated at setting 2 for 1 minute. This treatment breaks 100% of the cells. Although wide variation exists in sensitivity to breakage, a general starting point for use of this instrument to disrupt animal cells is on the order of 5 to 10 seconds of sonication per milliliter cell suspension. One practical advantage of this method is that fewer cells are required than for the mechanical methods described above.

Explosive Decompression. Cell disruption by explosive decompression (Parr Bomb, Parr Instrument Co., Moline, Illinois) is based on the fact that nitrogen taken into the cells under high pressure will expand rapidly when the pressure is suddenly reduced to atmospheric levels. The rapid expansion of the gas causes the cells to rupture. For total cell breakage of L-5178Y cells,[11] Manson *et al.*[12] have described the following procedure. The cell pack is washed into a cellulose nitrate tube with 0.25 *M* sucrose solution containing 2×10^{-4} *M* $CaCl_2$ to a concentration of approximately 5×10^8 cells/ml. For total breakage the pressure is raised to 1500 psi and maintained for 15 minutes. At the end of this exposure the cells are allowed to escape through an outlet valve, and the rapid reduction in pressure causes the cells to lyse. We have had only limited experience using this method with diploid cells. A minimum quantity of 5 ml of suspension and, in general, somewhat lower pressures and shorter times are required for breakage of the diploid fibroblastlike cells. A modification of this method appears to be especially useful for the preparation of nuclei as pressure of 500–800 psi for 10 minutes will disrupt the cells but not the nuclei.

[11] G. A. Fisher, *Ann. N. Y. Acad. Sci.* **76**, 673 (1957).
[12] L. A. Manson, G. V. Foshi, and J. Palm, *J. Cell. Comp. Physiol.* **61**, 109 (1963).

SECTION IV

Replicate Culture Methods

Editors' Comments

Early work with animal cells *in vitro* was limited mostly to cultures initiated from explants of tissues. Over 20 years ago two developments occurred which changed the course of tissue culture experimentation. One was the establishment of routine methods for preparing single cell suspensions from excised tissues (see Sections I and III). The other was the observation that certain animal cell types could proliferate while suspended in nutrient medium. Thus, animal cell culture methodology could be developed according to time-tested microbiological procedures. Soon, means were at hand to prepare and maintain quantities of replicate cultures, all established from a common pool of free cells. This opened possibilities for research projects in nutritional requirements, nutrient metabolism, virus propagation, and vaccine production (e.g., polio vaccine), and many life scientists were attracted to the animal world *in vitro*.

Several procedures in this section are representative of replicate culture methodology and are applicable to a wide variety of cell types. The section also contains a novel and recent procedure for "subculturing" without prior cell harvest and a procedure for working with replicate cultures of nonmitotic cell populations, which have received relatively little study to date.

The reader will find aspects of the subject of this section discussed in other sections of this volume, a dichotomous situation which we have attempted to reconcile with extensive cross indexing.

CHAPTER 1

Monolayer Cultures

Virginia J. Evans and William G. Taylor

Whether analyzing and quantitating the *in vitro* influence of a metabolite, drug, microbe, or culture technique, the experimental design must include elements of *replicacy* and *randomness.* First, large numbers of replicate tissue cultures must be prepared with a minimum trauma to the cell population. These cultures then must be distributed randomly into sets of control and experimental cultures; in this way the inherent sampling error of the planting procedure in any one set of cultures is diminished. Next, the variable(s) under study can be introduced. If fresh growth medium is required during the incubation period, the old medium must be removed with a minimum loss of cells.

The methods outlined[1–3] describe the preparation of replicate monolayer cultures as used in this laboratory. These methods minimize sample variation in replicate cell cultures ($\lesssim$5–7% deviation), changes in pH during planting ($<$0.1 pH unit), and loss of nonadherent cells during fluid renewals. Techniques for replicate microtest plate[4,5] and suspension cultures[6] are also available.

Preparation of the Cell Suspension

Several parent cultures are incubated in a medium of choice until a late logarithmic phase population is obtained. For rapidly proliferating continuous cell lines, such as strain L, this may require 4–5 days with a final cell population of 4 to 5 $\times$ 10^6 cells per T-15 flask. Prior to planting replicate cultures, it may be useful to incubate the parent flasks for 18 to 24 hours in a maintenance medium free of the variable(s) under test.

Cell suspensions can be prepared as previously described (see Section I).

[1] V. J. Evans, W. R. Earle, K. K. Sanford, J. E. Shannon, and H. K. Waltz, *J. Nat. Cancer Inst.* **11**, 907 (1951).

[2] V. J. Evans, *In* "Methods in Medical Research" (M. B. Visscher, ed.), Vol. 4. Year Book Medical Publishers, Chicago, Illinois, 1951.

[3] W. G. Taylor, F. M. Price, R. A. Dworkin, and V. J. Evans, *In Vitro* **7**, 295 (1972).

[4] R. A. Goldsby and E. Zipser, *Exp. Cell Res.* **54**, 271 (1969).

[5] F. Suzuki, M. Kashimoto, and M. Horikawa, *Exp. Cell Res.* **68**, 476 (1971).

[6] F. J. Weirether, J. S. Walker, and R. E. Lincoln, *Appl. Microbiol.* **16**, 841 (1968).

In the case of cell lines continuously subcultured in serum- or protein-free growth medium, or in instances where enzymatic or chemical cell disaggregating agents are undesirable, the following may be used. Cells are removed from the floor of the parent vessel by: (1) dispersing the cell sheet with a piece of sterile perforated cellophane[7] mopped over the flask floor with a sterile, unpolished, bent tip glass rod, or (2) disrupting the cell sheet with a sterile, platinum-iridium bacteriological loop. Clusters of cells are disrupted by gently aspirating the suspension; a sieving pipette containing 80 mesh platinum gauze[8] is useful if working with cell lines, such as MK_2, which form multicellular clumps when dispersed.

The pooled cell suspension is centrifuged gently (110–140 g, 10–15 minutes, 25°C), then resuspended in a volume calculated to yield the cell density desired. Alternatively, the suspension from individual flasks can be added to a pool of planting medium until the final inoculum size is reached. For our purposes, a frequently used inoculum level is $1.2–1.5 \times 10^5$ cells per T-15 flask ($8–10 \times 10^4$ cells/cm^2), as determined by hemocytometer or automated counting methods (see Section VIII). NOTE: Maintain the cell suspension at a physiological pH (7.2–7.4) at all times; dilute to the desired inoculum level, and plant immediately!

Planting Procedures

Method A.[1,2] The monodispersed cell suspension is introduced into a reservoir containing an impeller[9] (Fig. 1A) where it is stirred at 450 rpm during sampling and for 15 minutes before sampling. The number of cells in the sample, determined by counting, is used in calculating the dilution of the stirred suspension necessary to yield the desired number of cells for inoculation of culture flasks. The side arms fitted with cotton filters permit continuous aeration with humidified CO_2 (10% CO_2/90% air) to adjust and maintain the pH at 7.2 to 7.3.

By means of a three-way stopcock, a 1-ml sample is discharged from the calibrated delivery tip of the reservoir by a jet humidified CO_2/air at about 2 psi. The gas mixture is allowed to flow for an additional 7–10 seconds to replace the ambient air with humidified 10% CO_2. Flasks are closed immediately, randomized, and then reopened so that experimental variables can be introduced in that volume of culture medium not used in the planting procedures. All cultures are gassed again (10% CO_2/air) and incubated.

To replace old culture medium with fresh fluid, flasks are centrifuged at 110 g for 20 minutes, then placed upright in a suitable rack. An assembly design to assure the removal of the used fluid in the axis of the cone-shaped tip of the flask is shown in Fig. 1B. This is particularly important for cell lines which adhere poorly to the floor of the culture flask.

[7] Available from Microbiological Associates, Inc., 4733 Bethesda Avenue, Bethesda, Maryland 20014.

[8] Available from the Kontes Glass Co., Vineland, New Jersey 08360.

[9] PGC Scientifics, 12111 Parklawn Drive, Rockville, Maryland 20852.

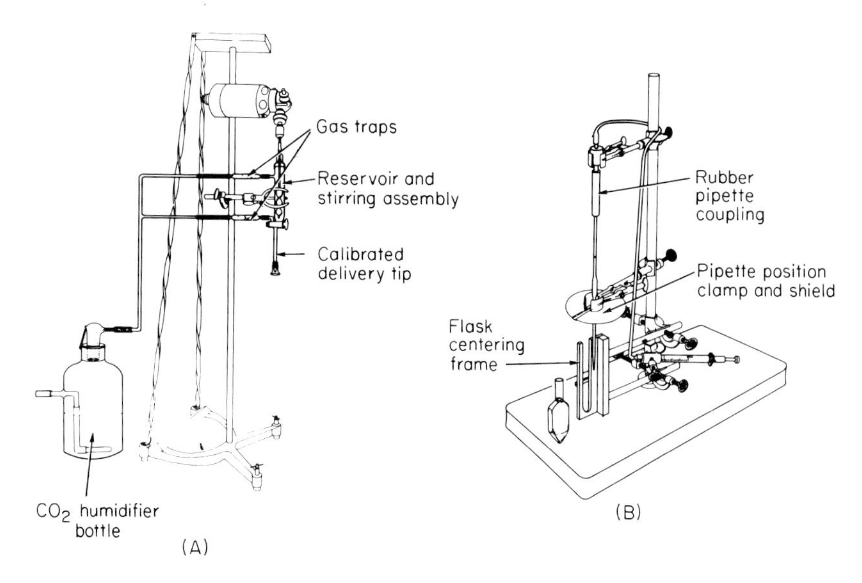

Fig. 1. (A) Apparatus for dispensing cell suspensions. (B) Fluid renewal assembly for withdrawing supernatant culture medium from T-15 flasks.

Method B.[3] A monodispersed suspension—75–100 ml in excess of that required—is prepared in a sterile flat-bottom boiling flask (or any other sterile vessel), positioned eccentrically on a magnetic stirrer, and stirred gently with two sterile Teflon-coated 15 × 1.5 mm magnetic stirring bars, as shown in Fig. 2A. The silicone cap and fittings are aseptically placed over the mouth and neck of the reservoir flask; the glass withdrawal tube should extend to within 3 to 5

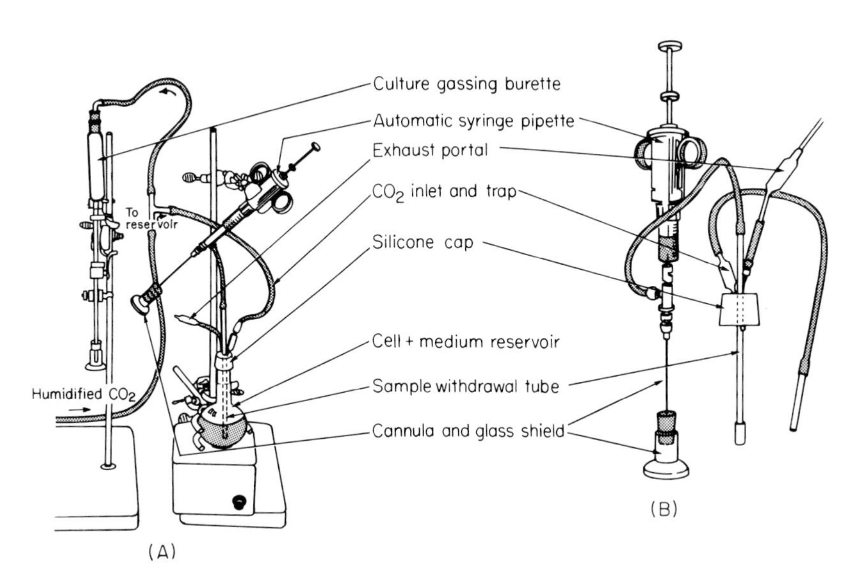

Fig. 2. (A) Replicate culture planting unit with stirred cell suspension and culture gassing burette. (B) Replicate culture planting device.

mm of the floor of the reservoir. The cell suspension is continuously aerated with sterile humidified 10% CO_2 and 90% air entering through the CO_2 inlet. The exhaust portal can be eliminated if the silicone cap is bored to fit loosely over the mouth and neck of the reservoir flask. An aliquot of the cell suspension is drawn into the automatic syringe pipette, air bubbles removed, and the first five volumes discarded. As the mouth of each culture flask is centered beneath the glass shield and cannula, a constant preset volume of suspension is dispensed rapidly by aspirating the syringe. Ambient air is flushed from the inoculated flask with sterile humidified CO_2 (7–10 seconds) through the gassing burette.[9] Flasks are closed immediately. Randomization and addition of the experimental variables are performed as described in Method A.

Immediately before cell planting, the replicate culture planting device (Fig. 2B) is flushed with triple glass-distilled water, wrapped in Patapar paper,[10] and autoclave sterilized. Distilled water must be present in the syringe and valve assembly during sterilization to avoid drying and adherence of the syringe barrel and valves to the surrounding assembly. Immediately after use, the entire unit is rinsed with triple glass-distilled water so that neither cells nor medium collect or dry within the device.

Randomization of Cultures

To minimize the variability inherent in replicate planting procedures, flasks are numbered in the sequence planted, then distributed into sets with a random numbers table. This can be done as soon as the cells have attached; ideally, the cultures should be kept at the desired incubation temperature during randomization. A typical table for randomizing 48 replicate cultures, e.g., eight variables with six cultures per variable, is shown in Table I, and any random numbers table can be used for larger or smaller experiments.

TABLE I

Randomization Table for Preparing Sets of Control and Experimental Cultures[a]

Variable							
A	B	C	D	E	F	G	H
42	19	14	3	46	25	41	16
26	27	13	22	4	36	43	8
44	18	33	6	30	17	10	24
34	11	23	48	12	40	35	2
31	39	5	9	32	1	21	37
38	47	28	15	29	20	45	7

[a] Numbers shown are flask nos.

[10] Available from A. J. Buck & Son, Inc., 10534 York Road, Cockeysville, Maryland 21030.

Harvesting and Evaluating Cell Populations

After appropriate incubation period (5–6 days), cell populations are dispersed with 0.125% trypsin. If viable cell counts are desired, a dye exclusion method is used (see Section VIII); for a total count, a formalin-isotonic saline solution is used to preserve the cells[11] following trypsinization, and the cell numbers are determined by automatic counting procedures[12] (see Section VIII). A small volume (0.1–0.3 ml) of serum may be useful in preventing cell clumping; if clusters cannot be dispersed, total protein[13] and nucleic acid[14] determinations have been found to be a useful measure of growth.

[11] E. L. Pruden and M. E. Winstead, *Amer. J. Med. Technol.* **30**, 1 (1964).
[12] W. G. Taylor, R. A. Dworkin, R. W. Pumper, and V. J. Evans, *Exp. Cell Res.* **74**, 275 (1972).
[13] V. I. Oyama and H. Eagle, *Proc. Soc. Exp. Biol. Med.* **91**, 305 (1956).
[14] L. Siminovitch, A. F. Graham, S. M. Lesley, and A. Nevill, *Exp. Cell Res.* **12**, 299 (1957).

CHAPTER 2

Plant Cell Suspension Cultures

A. C. Hildebrandt

Plant cell cultures have been used in a variety of ways to clarify normal and diseased growth and differentiation of plants. The idea of isolating and growing cells and tissues under controlled environments on a synthetic medium was considered by many workers.[1–3]

Procedures

Cultures can be established and maintained either on agar or in liquid media. In many cases, agar medium is preferred for the original isolation of

[1] R. J. Gautheret, "Culture des Tissue Vegetaux." Masson et Cie, Paris, 1959.
[2] D. R. White, *Cancer Res.* **5**, 302 (1945).
[3] L. Hirth and G. Morel, eds. *Colloq. Intern. Cent. Nat. Rech. Sci.* **193**, (1971).

plant materials as a source of cell and tissue cultures. Diseased or normal tissues can be removed aseptically from the plant and explanted to the medium. Surface sterilization with sodium hypochlorite is used, and pieces of stem, leaf, tuber, or gall, or whole seeds, excised embryos, or anthers, for example, are removed with a sterile scalpel. If many explanted pieces are used, chances are that at least one tissue explant will be sterile and will proliferate to produce callus cells. Callus cells, once produced, should be divided and transferred to fresh medium at regular intervals while the tissue mass is in an active, vigorous condition.

Culture Media. Many different types of media have been developed for culture of plant cells, tissues, and organs *in vitro*. Tables I and II give the composition of several media that have been used extensively for the original isolation and for continued maintenance of cell cultures.[4] Another medium (SH) has recently been described[5] that has been useful in providing a friable callus cell culture of many species. Coconut milk used for the C and D media is secured from fresh coconuts. The milk is drained through cheesecloth to remove any debris. Care is exercised to ensure that no fermented milk is used.

TABLE I
Constitution of Various Nutrient Media with T Medium Serving as the Basic Medium[a]

Medium	Amount (mg/liter)	Medium	Amount (mg/liter)
T Medium		THS Medium[b]	
Na_2SO_4	800.0	KCl	845.0
$Ca(NO_3)_2 \cdot 4\ H_2O$	400.0	$NaNO_3$	1800.0
$MgSO_4 \cdot 7\ H_2O$	180.0	$NaH_2PO_4 \cdot H_2O$	300.0
KNO_3	80.0	$(NH_4)_2SO_4$	790.0
KCl	65.0	Kinetin	0.5
$NaH_2PO_4 \cdot H_2O$	33.0	*myo*-Inositol	100.0
$MnSO_4 \cdot 4\ H_2O$	0.45	Calcium pantothenate	2.5
$ZnSO_4 \cdot 7\ H_2O$	0.6	C Medium[b]	
H_3BO_3	0.00375		
KI	0.03	Coconut milk	150[c]
$Fe_2(C_4H_4O_6)_3$	40.0	Calcium pantothenate	2.5
Glycine	3.0	α-Naphthaleneacetic acid	0.1
Thiamine·HCl	0.1	D Medium[d]	
Sucrose	20000.0		
Agar	6000.0	2,4-Dichlorophenoxyacetic acid	6

[a] A. C. Hildebrandt, *In* "Modern Methods of Plant Analysis," Vol. 5, p. 383. Springer-Verlag, Berlin, 1962.
[b] Complete T medium plus the additions listed.
[c] Amount in ml/liter.
[d] Complete C medium plus the addition indicated.

[4] I. K. Vasil and A. C. Hildebrandt, *Planta* **68**, 69 (1966).
[5] R. U. Schenk and A. C. Hildebrandt, *Can. J. Bot.* **50**, 199 (1972).

TABLE II

Preparation[a] and Composition of Murashige and Skoog's[b,c] Medium (MS) as Used in the Present Experiments

Stock solution	Constituents	Conc. in stock solution (g/liter)	Volume of stock solution in final medium (ml/liter)	Final conc. in medium (mg/liter)
A	NH_4NO_3	82.5	20	1650.0
B	KNO_3	95.0	20	1900.0
C	H_3BO_3	1.24	5	6.2
	KH_2PO_4	34.00		170.0
	KI	0.166		0.83
	$Na_2MoO_4 \cdot 2\ H_2O$	0.05		0.25
	$CoCl_2 \cdot 6\ H_2O$	0.005		0.025
D	$CaCl_2 \cdot 2\ H_2O$	88.0	5	440.0
E	$MgSO_4 \cdot 7\ H_2O$	74.0	5	370.0
	$MnSO_4 \cdot 4H_2O$	4.46		22.3
	$ZnSO_4 \cdot 7\ H_2O$	1.72		8.6
	$CuSO_4 \cdot 5\ H_2O$	0.005		0.025
F[d]	$Na_2 \cdot EDTA$	7.45	5	37.35
	$FeSO_4 \cdot 7\ H_2O$	5.57		27.85
G	Thiamine·HCl	0.02	5	0.1
	Nicotinic acid	0.1		0.5
	Pyridoxine·HCl	0.1		0.5
	Glycine	0.4		2.0

[a] The stock solutions A–G prepared and stored in a refrigerator (never more than 4–6 weeks) and mixed just before preparing the final medium.

[b] T. Murashige and F. Skoog, *Physiol. Plant.* **15**, 473 (1962).

[c] Addendum: sucrose 30 g/liter, *myo*-inositol 100 mg/liter, indole-3-acetic acid 10 mg/liter, kinetin 0.04 mg/liter, agar 10 g/liter.

[d] The $FeSO_4 \cdot 7\ H_2O$ is dissolved in ca. 200 ml double-distilled water. The $Na_2 \cdot EDTA$ is dissolved in ca. 200 ml double-distilled water, heated, and mixed (under continuous stirring) with the $FeSO_4 \cdot 7\ H_2O$ solution. After cooling, the volume is adjusted to 1000 ml. Heating and stirring result in a more stable Fe-EDTA complex.

The coconut milk is added directly to the medium and the medium sterilized in the autoclave. Kinetin (2,4-dichlorophenoxyacetic acid) and indoleacetic acid may be dissolved in a few drops of 95% ethyl alcohol before adding to the media.

Plant cell and tissue cultures are also often originated and maintained in liquid media. Liquid medium for suspension cultures is of the same composition as the solid medium, but with the agar omitted. Liquid media also have the advantage of avoiding the addition of unknown compounds present in and comprising the agar.

Once the cell culture is established, whether on agar or in liquid medium, it is maintained by subculture at regular intervals when the culture is at the active stage of growth. Cultures permitted to grow long periods and until they turn brown before transfer may require a long period of incubation to assume a normal growth rate and appearance.

Types of Culture Vessels. Various types of culture vessels have been used for the liquid suspension cultures of plant cells, tissues, and organs. These include Erlenmeyer flasks, test tubes, screw-cap vials, prescription bottles, and flasks of various shapes and sizes. In the case of most liquid suspension cultures, some type of aeration is provided to the cells.

Test tube cultures,[6] special culture tubes,[7] and bottles[8] were rotated on a roller tube drum arrangement. Other apparatus[9] permitted renewal and circulation of the medium. Comparisons of various culture vessels and aeration procedures[8] showed that excellent growth in liquid culture develops in 6-ounce prescription bottles in an upright position with 40 ml of liquid medium on a reciprocating shaker. The reciprocating shaker made sixty complete cycles per minute with a 4-inch swing. Comparable results were seen with a rotary shaker sufficiently fast to keep the cells well suspended in the liquid. More sophisticated and large volume cultures were agitated and aerated by using a shaker or by bubbling sterile air through the medium.[10–13]

Effect of Light and Temperature

Cell cultures may be incubated in the light or in the dark. Stem tip or embryo cultures often grow best in the light. Callus and root tip cultures grow well in light or dark. Chlorophyll and other pigments, if desired, may develop with cultures grown in light.[3,4] The quality and quantity of light often influences the amount of pigment formation as well as the differentiation of roots, stems, leaves, and plants from undifferentiated cell cultures.[4] A combination of cool white and Gro-Lux fluorescent and incandescent lamps is useful for this purpose. Most cultures grow well at 26–30°C. When extra light is used, it is essential to maintain the optimum temperature at the culture level. Excessive drying of cultures may be reduced by maintaining a relative humidity of 35–50% in the culture room.

Subcultures

Stock cultures of plant cells may be maintained indefinitely by regular subculture. Optimum rates of growth are maintained by transferring a portion of the stock or mother tissue when it is in its most active phase of growth. The optimum phase of growth varies with the plant species and with the strain

[6] P. R. White, "The Cultivation of Animal and Plant Cells." Ronald Press, New York, 1954.

[7] S. M. Caplin and F. C. Steward, *Nature (London)* **163**, 920 (1949).

[8] W. H. Muir, A. C. Hildebrandt, and A. J. Riker, *Amer. J. Bot.* **45**, 589 (1958).

[9] R. S. DeRopp, *Science* **104**, 371 (1946).

[10] L. G. Nickell and W. Tulecke, *J. Biochem. Microbiol. Technol. Eng.* **2**, 287 (1960).

[11] W. Tulecke and L. G. Nickell, *Trans. N. Y. Acad. Sci.* **22**, 196 (1960).

[12] W. Tulecke, R. Taggart, and L. Colavita, *Contrib. Boyce Thompson Inst.* **23**, 33 (1965).

[13] H. E. Street, P. J. King, and K. J. Mansfield, *Colloq. Intern. Cent. Nat. Rech. Sci.* **193**, 17 (1971).

of tissue. Usually best results are obtained if actively growing tissue is transferred at 2 to 4 week intervals to fresh medium. Seed tissue pieces of agar-grown stock weighing 50–500 mg are usually sufficient for transfer to liquid suspension cultures. If the agar-grown tissue is firm, it may be advisable to asceptically divide the mother stock piece in a sterile Petri dish with a sterile scalpel into 50-mg pieces and to transfer with a sterile transfer loop 10–30 pieces to 40 ml liquid in a 6-ounce prescription bottle. If larger volumes of liquid suspension cultures are desirable, increasing the amount of stock required may be obtained by preparing a series of cultures using progressively larger amounts of liquid and tissue in succeeding cultures. On the other hand, if the cell cultures consist largely of single cells and small colonies of cells, as opposed to large, firm clumps of tissue, then subcultures may be propagated by transferring an aliquot of the suspension culture with a sterile pipette. One-half to one milliliter of such a thick cell suspension is usually adequate to inoculate a 40-ml liquid medium shake culture, but a much smaller amount of inoculum may also suffice. Proportionately larger aliquots of the cell suspension may be used to inoculate larger volumes of liquid. Sterile pipettes with sterile cotton in the mouth end may be used to transfer the cell inoculum, or it may be poured directly from established cultures to fresh culture medium following a flaming of the lip of the mother culture vessel.

Contamination of Cultures

Contamination of cultures by microorganisms is always a hazard, but with careful aseptic technique it may be kept to a minimum. Antibiotics are seldom used with plant cell cultures for this purpose. A special transfer room or chamber is desirable for extensive culture work. A desk top transfer chamber is adequate for small operations. During subculture, only one mother stock piece should be transferred to each Petri dish for dividing with a sterile scalpel. After transferring most of the seed pieces to fresh medium, a few representative pieces may be macerated in the dish and a plate poured with a medium, such as one containing yeast extract, or peptone or a similarly rich medium. All stock cultures from seed pieces that subsequently show contamination should be discarded. If a valuable stock culture becomes contaminated, it may sometimes be possible to save it by carefully removing very small portions of the tissue from areas of the culture apparently free of contamination or single cells for transfer to fresh medium.

Additional details for plant cell suspension culture work have also been described in Sections V and XI.

CHAPTER 3

Subculturing Human Diploid Fibroblast Cultures[1]

Leonard Hayflick

The following methods are those currently employed in our laboratories for the cultivation of human diploid cell strains.[2,3] Several common pitfalls encountered in the cultivation of these cells are identified, and the theory of the finite lifetime of human diploid cell strains is briefly described.[4-6]

Reagents

Growth Medium. Eagle's Basal Medium; 10% calf serum; 50 μg aureomycin/ml. The medium should be prewarmed to 37°C before use. The final pH should be brought to 7.2 before serum addition. The pH of the medium after equilibration of the culture at 37°C must be less than 7.4.

Aureomycin (Lederle Product No. 4691-96, intravenous) is bottled in 500-mg amounts. Reconstitute in 50 ml of warm (37°C) sterile, distilled water. Agitate to ensure a clear amber solution. Prepare 5-ml aliquots and store at −20°C. Use one 5-ml aliquot per liter of medium. Final concentration is, therefore, 50 μg/ml. Nine thousand cell cultures grown in aureomycin in lieu of penicillin and streptomycin have been found, upon testing, to be free of mycoplasmas over a 10-year period.

Trypsin. Add as 0.25% in phosphate buffered saline (PBS) or any BSS plus aureomycin as above. Final pH of trypsin solution *must be at least 7.4.*

Subcultivations

1. Decant all of the medium.
2. Add enough prewarmed trypsin to cover the sheet. Allow the culture

[1] Supported, in part, by research Grant HD 04004 from the National Institute of Child Health and Human Development, National Institutes of Health, Bethesda, Maryland.

[2] L. Hayflick and P. S. Moorhead, *Exp. Cell Res.* **25**, 585 (1961).

[3] L. Hayflick, *Exp. Cell Res.* **37**, 614 (1965).

[4] L. Hayflick, *In* "Perspectives in Experimental Gerontology" (N. W. Shock, ed.), Chapter 14, p. 195. Thomas, Springfield, Illinois, 1966.

[5] L. Hayflick, *Sci. Amer.* **218**, 32 (1968).

[6] L. Hayflick, "Aging and Development." Academy of Science and Literature, Mainz, Germany, F. K. Schattauer, Verlag, Stuttgart, in press, 1973.

to stand at room temperature for 1 minute. Decant all of the trypsin. Allow the culture to stand for 5 to 20 minutes more at room temperature or at 37°C. Incubation at 37°C during this period of time will hasten the trypsinization process.

3. The trypsinization process will be completed when the cell sheet appears to be loosened from the glass surface. This can be seen macroscopically by holding the bottle up to the light in a vertical position and observing the cell sheet sloughing off of the glass surface.

4. After this has occurred, add a small amount of fresh medium and splash it over the cell sheet. All of the cells should be removed. Aspirate the medium plus the cells with a pipette onto the glass surface to remove all remaining cells. It is essential that this aspiration be done as completely as possible with a small-bore 5- or 10-ml pipette so as to obtain single dispersed cells. This is one of the most crucial steps, for if the cells are not broken up, the new culture will contain numerous microcolonies or "explants" that will ultimately lead to early senescence of the culture.

5. Add sufficient fresh medium to the aspirated suspension so that the total volume will cover the glass surface of two bottles, each having the same surface area as the original bottle (or use a single bottle having twice the floor area of the original bottle). This is called a 1:2 split. (The old bottle can be reused without washing.)

6. Incubate the bottle (or bottles) at 37°C. No intervening culture feedings are necessary.

7. When making 1:2 splits, subcultivation of these cultures should be done on a rigid 3- or 4-day schedule, at which time confluent cell sheets should occur. Surplus cells can be stored at −70°C. It is important to point out that the pH of the medium is of extreme importance. The final pH of the medium must not exceed 7.4 after equilibration of the culture at 37°C. A higher pH may result if too few cells are contained in the culture or if the original pH of the medium is too high, or, and most importantly, if the gas phase of the culture vessel is too large. If the latter is the case, it is then essential that the Eagle's Medium be prepared in Hanks' balanced salt solution or in 28 μM HEPES buffer.

8. The passage number of the strain is indicated by an encircled number. Increase this number by one at each 1:2 subcultivation.

9. By making repeated 1:2 splits (twice a week) it can be seen that the number of culture bottles can be increased geometrically (1, 2, 4, 8, 16, 32, 64, etc.) in a short period of time for the production of large quantities of cells for various purposes.

10. A human diploid embryonic cell strain has a "passage potential" of about fifty 1:2 subcultivations at which time the cells will cease to divide and eventually die.[2,3] Cell populations derived from adult tissues have a lower passage potential.[3]

11. Although the strain will be lost as a continuously passaged population, it will not be lost for use since frozen ampoules can be obtained at almost every passage and thus the strain can be restored to continuous passage again, up

to a cumulative total of 50. By repeating this procedure, the number of cells that can be obtained is almost unlimited for all practical purposes.

12. Using split ratios higher than 1:2 results in the advantage of minimizing the number of manipulations necessary to obtain a specific cell density or the number of bottle cultures. Since human diploid cell strains pass through a finite number of population doublings *in vitro*, it is necessary to keep a record of the number of population doublings that have elapsed. With a 1:2 split ratio this is achieved by simply adding "1" to each split since this ratio yields one population doubling. Larger split ratios can be used. For example, a split ratio of 1:4 would yield two doublings per 1:4 split; a 1:10 split ratio would yield a 3.25 doublings per 1:10 split. In order to have knowledge of the approach of phase III it is essential to keep records of the number of elapsed population doublings.

Theory of Population Increase by Subcultivation

Since human diploid cells multiply by fission, the increase in population may be expressed per cell as follows:

$$\begin{array}{cccccl} 1 & 2 & 4 & 8 & 16 & \ldots \text{ number of cells} \\ 0 & 1 & 2 & 3 & 4 & \ldots\ n \text{ (number of generations)} \end{array}$$

one generation, one generation

Expressed exponentially, the population after n generations is 2^n per cell in the inoculum, or total population N is the initial population, X_0, multiplied by 2^n or

$$N = X_0 2^n \tag{1}$$

The data needed to determine the number of generations, n, will be the number of cells per unit volume in the inoculum, X_0, at time t_0, and the final population, N, at time $= t_2$. The number of generations, n, can be most readily evaluated by expressing Eq. (1) in logarithmic form. Using logarithms to the base 10, this equation becomes

$$\log N = \log X_0 + n \log 2 \tag{2}$$

or rearranging

$$n = (\log N - \log X_0)/\log 2 \tag{3}$$

since $\log 2 = 0.301$:

$$n = 3.32\,(\log N - \log X_0) \tag{4}$$

Logarithms to the base 2 should be used for biological systems because an increase of one logarithmic unit would correspond to one doubling or one generation. If this is done, the log 2 drops out of the denominator of Eq. (3). Natural logarithms (base e, written ln) may also be used. Regardless of the base of the logarithms used, the equation will take the same form, and conversion from one form to another can be made by multiplying by a constant, i.e., log 2 = 0.301 or 1/log 2 = 3.32, so $\log_2 = 3.32 \log_{10}$.

The multiplication rate, *r*, or number of generations per unit time can be obtained for Eq. (4) by dividing by the time interval between inoculation, t_0, and the time at which the final population, *N*, was taken, i.e., t_2. Therefore, the multiplication rate, *r*, is

$$r = n/(t_2 - t_0) \tag{5}$$

or

$$r = [3.32 (\log N - \log X_0)]/t_2 - t_0 \tag{6}$$

One must specify the units of *r*, both the base of the logarithms used, and the units of time—usually in days (units of 24 hours) for tissue culture systems, i.e., doublings in population per 24 hours. To write formulas (2) and (5) in general form, one may determine the number of generations or multiplication rate over any interval in which the initial count is X_1, at any selected time t_1, and the final count X_2, at time t_2. Thus

$$n = (\log X_2 - \log X_1)/\log 2 = 3.32 (\log X_2 - \log X_1) \tag{7}$$

or

$$n = 3.32 \log X_2/X_1 \quad \text{or} \quad n = (\log X_2/\log X_1) \tag{8}$$

and

$$r = (3.32 \log X_2/X_1)/t_2 - t_1 \tag{9}$$

Since the generation time, *g*, is the time for the population to double, it is the reciprocal of the doubling, per unit time,

$$g = 1/r \tag{10}$$

generation time = time elapsed per doubling in number.

CHAPTER 4

Apparatus for Changing Tissue Culture Media

D. O. Cliver

Many types of tissue cultures require one or more changes of medium between the time of planting and experimental use. We mechanize these changes as much as possible, so as to reduce the amount of labor.

A medium change simply involves removing the spent medium and adding fresh medium. Either or both of these steps might be mechanized, depending on the number of culture units to be changed, the volume of medium per culture, the configuration of the culture vessel, and the type of closure. One would prefer a residue of $<5\%$ of the volume of spent medium and a volume of fresh medium dispensed within $\pm 5\%$ of that selected, though these are seldom critical requirements.

The spent medium may be removed by pouring or by aspiration. Pouring is satisfactory if the number of culture units is small, if the unit volume of medium is quite large, or if several culture vessels share a single closure (as with some types of tube racks). For large numbers of culture units, and especially where the vessel has a constricted opening (as with small plastic flasks), aspiration is preferred. The use of a pipetting machine as an aspirator provides control of the timing and volume of intake. This limits unnecessary drawing of room air into the needle or into the culture vessel. We set the stroke volume at approximately three times the volume of fluid in the cultures; this leaves little residual spent medium, yet the total excess aspirated volume is much less than the free air space in the culture vessel.

APPARATUS

If a double pipetting machine with a single-cycle attachment (which stops the machine after a single complete pipetting cycle) is used, one can dispense fresh medium into the cultures in the same operation.[1] Events during a single cycle are diagrammed in Fig. 1. During the first $\sim 180°$ of rotation, the machine's left syringe fills with fresh medium from a reservoir, while the right syringe aspirates the spent medium from the culture. During the second half-cycle, the left syringe dispenses the fresh medium into the culture, and the right syringe discards the spent medium. The machine turns itself off at 360°.

This combination of functions has been made possible by a double, 14-gauge

[1] D. O. Cliver and R. M. Herrmann, *Health Lab. Sci.* 6, 5 (1969).

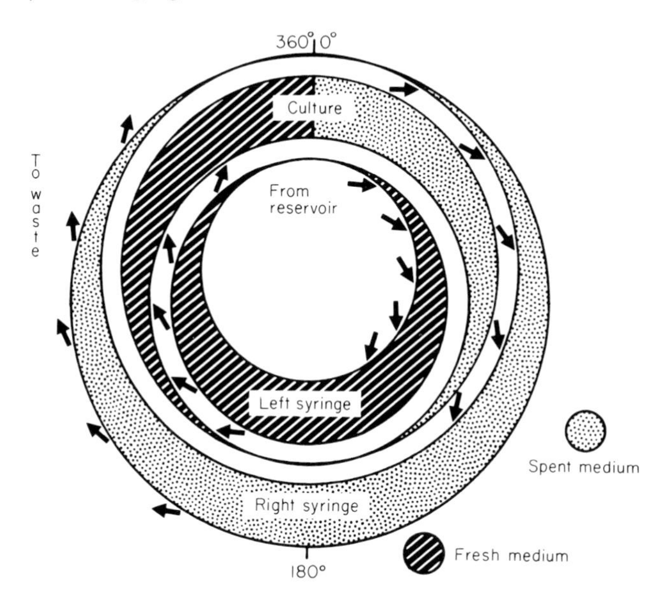

Fig. 1. Cycle used in changing culture medium with a double-syringe pipetting machine. During a single rotation of the machine, spent medium (dotted area) is aspirated from the culture and discarded by the right syringe, while the left syringe fills with fresh medium (barred area) and then dispenses it into the culture.

needle produced to our specifications (SH-1085) by the CSI Division of Becton, Dickinson and Company, Rutherford, New Jersey. A detail of the tip (longitudinal section) appears in Fig. 2. Spent medium is aspirated with the lower needle, through orifice "A." Fresh medium is dispensed from the upper needle, during the other half-cycle, through orifice "B." The free length of this needle, from mounting point to tip, is ~16 cm. This permits it to reach the bottom of any of the culture vessels we use.

An experienced operator can change 200 screw-cap flask cultures (5-ml fluid volume) per hour using this apparatus. Manipulation of flasks and caps is the rate-limiting factor; the machine cycle takes only ~1 second per culture.

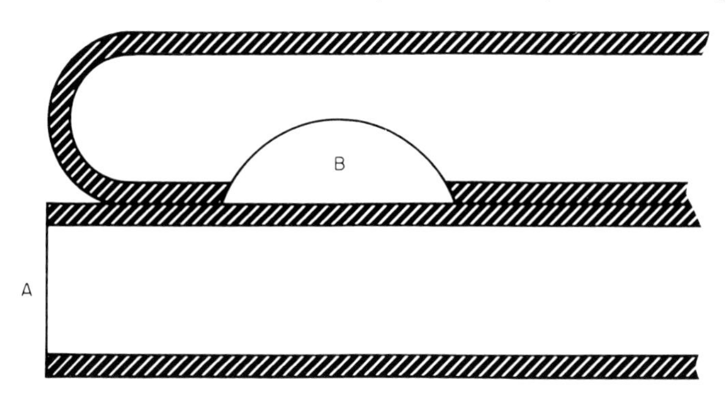

Fig. 2. Detail (longitudinal section) of the tip of the 14-gauge double needle. Spent medium is aspirated through the end orifice A, after which the fresh medium is dispensed through the side orifice B.

Contamination has been no problem in practice. The inherent risk of airborne or manual contamination, using this apparatus, is evidently very low. Neither has transfer of contamination from one culture to another on the needle been a problem, apparently because the first culture has not become contaminated. Our 8 years' experience with this apparatus indicates that it provides a rapid, safe, and economical means of changing medium in large numbers of tissue culture units.

CHAPTER 5

Single "Subculture" in Situ Using Glass Beads

G. M. Kolodny

Investigations of intracellular changes have been conducted on cells released from contact inhibition by reseeding cells in a larger growth area after trypsinization, by addition of fresh serum,[1,2] or by transforming these cells.[3-6] These methods of releasing contact inhibition pose some disadvantages for the interpretation of experimental results. It is uncertain, for example, that early biochemical changes seen after reseeding trypsinized cells are due solely to loss of contact inhibition. Some may represent changes necessary to repair cellular damage caused by the trypsinization.

Serum stimulation in contact-inhibited cells does not cause physical cell separation, and biochemical alterations seen after addition of fresh serum to a confluent culture might be due to stimuli other than those operating in an authentic release of contact inhibition. Similarly, virus infection of contact-inhibited cells can and does activate processes other than those simply concerned with cell to cell contact.

A method of releasing contact inhibition without disturbing cell substrate attachment, or the chemical composition of the medium, should provide for an

[1] G. J. Todaro, G. K. Lazar, and H. Green, *J. Cell. Comp. Physiol.* **66**, 325 (1965).

[2] H. Yoshikura and Y. Hirokawa, *Exp. Cell Res.* **52**, 439 (1968).

[3] J. Kara and R. Weil, *Proc. Nat. Acad. Sci. U. S.* **57**, 63 (1967).

[4] H. Green and G. J. Todaro, *Symp. Fundamental Cancer Res.*, University of Texas, M. D. Anderson Hospital and Tumor Institute, Houston, Texas. Williams & Wilkins, Baltimore, Maryland, 1967.

[5] R. Dulbecco, *Perspec. Biol. Med.* **9**, 298 (1966).

[6] L. Sachs, *Curr. Top. Develop. Biol.* **2**, 129 (1967).

increase in the space available for growth of a confluent culture. Ideally, such an increase in growth area should be uniformly available to every cell within the culture. Increasing the growth area along a border of the culture leads to stimulation of cell division in only a very small and insignificant fraction of cells immediately adjacent to the border. Cells within the center of the culture are not released from contact inhibition.[7]

In this laboratory we have studied the macromolecular changes accompanying release of contact inhibition by using a method by means of which a large fraction of cells in a confluent monolayer are released from contact inhibition of growth and division.[8,9] This is accomplished without chemical treatment of the cells, without change of medium, and in such a way that most cells are provided with free growth area. The procedure involves growing the cells to confluence on surfaces uniformly covered with glass beads 200 μm in diameter. When confluence has been attained, contact inhibition may be released by discarding the beads, leaving behind numerous spaces throughout the culture. Removal of the beads dislodges few if any of the cells. After release of contact inhibition by removing the beads, the cultures double in cell number following the first round of DNA replications, and continue to grow until they are again contact inhibited. Cell types used have included several cell lines and strains including both primary cell cultures of neonatal rat heart cells and established 3T3 and 3T6 mouse fibroblast cell lines.

Cell Cultures

Cells are cultured in Dulbecco's medium with the addition of 10% calf serum, 75 units/ml penicillin, and 50 μg/ml of streptomycin. Serum and medium are supplied by Grand Island Biological Co., Grand Island, New York. The cultures are incubated at 37°C in a humid atmosphere of 5% CO_2 in air. Cultures are checked periodically for contamination by microorganisms.

Maintenance of Cultures with Glass Beads

Cultures are seeded in the presence of glass beads. One gram of glass beads (200 μm average diam., 3M Company, Reflective Products Division, St. Paul, Minnesota), is placed in a 60-mm plastic tissue culture Petri dish (Falcon Plastics, Los Angeles, Calif.) after dry heat sterilization of the beads for 6 to 12 hours at 170°C. The cells in 5 ml of medium are then seeded into the culture dish with the glass beads. By gentle tipping and swirling motions an even layer of beads is made to cover the bottom of the dish. The dish is carefully placed in the incubator and after a short time the cells settle between the beads. The cells attach on the surface of the dish between the beads and multiply until the spaces between the beads are completely occupied by adherent contact-inhibited cells.

[7] R. Dulbecco and M. G. P. Stoker, *Proc. Nat. Acad. Sci. U. S.* **66**, 204 (1970).
[8] G. M. Kolodny and P. R. Gross, *Exp. Cell Res.* **57**, 423 (1969).
[9] G. M. Kolodny, *Amer. J. Roentgenol. Radium Ther. Nuc. Med.* **108**, 736 (1970).

Medium in the Petri dish is changed every 3 days. The spent medium is removed with a curved-tip Pasteur pipette fitted with a rubber bulb, slowly and carefully so as not to disturb the beads. Only about 85% of the spent medium is removed. If the dish is drained dry, some beads float off the surface when fresh medium is returned to the dish. The fresh medium is added slowly to the dish through a curved-tip pipette lying against the side of the dish.

Removal of Beads

When the surface of the Petri dish between the glass beads is fully covered with cells, and at least 48 hours after a change of medium, contact inhibition of cell division may be released by discarding the glass beads. The culture dish is tipped slightly to one side. The medium is sucked up into a wide-bore pipette and then allowed to drip back onto the dish so as to wash all the glass beads to the low side of the tipped dish. The beads and medium are then pipetted together into a centrifuge tube which is spun at 1600 g for 1 minute to pellet detached cells, dead cells, and large debris with the beads. The cell-free supernatant medium is then returned to the tissue culture dish, which now contains a cell layer interspersed throughout with holes left by the discarded beads.

Release of Contact Inhibition

When the beads are discarded, the cells in the dish start to multiply and fill up the holes. Within 12 to 36 hours, depending on the cell type, mitotic cells in large numbers can be seen within and at the periphery of the holes. In 72 to 96 hours a confluent monolayer of contact-inhibited cells has been formed.

Figure 1 shows a plot of DNA synthesis (thymidine incorporation) and mitotic index in the hours following release of contact inhibition by the glass bead method. There is a semisynchronous burst of DNA synthesis with a peak at about 30 hours after release followed by a semisynchronous peak in the mitotic index at about 40 hours.

Cell Growth

Normal cell growth characteristics are preserved in the presence of the beads and after their removal. Figure 2 is a plot of daily cell counts in replicate 3T3 cultures, before and after removal beads. Media are changed typically 2 days after inoculating the cultures and every 3 days thereafter. The cell cultures grow logarithmically until confluent at about 10^5 cells per plate and remain contact inhibited until the beads are removed. After removal of the beads, the cells once again multiply logarithmically as shown in the figure. No change of media at times prior to removal of the beads is necessary for release of contact inhibition since growth curves similar to that seen here can be obtained without any change

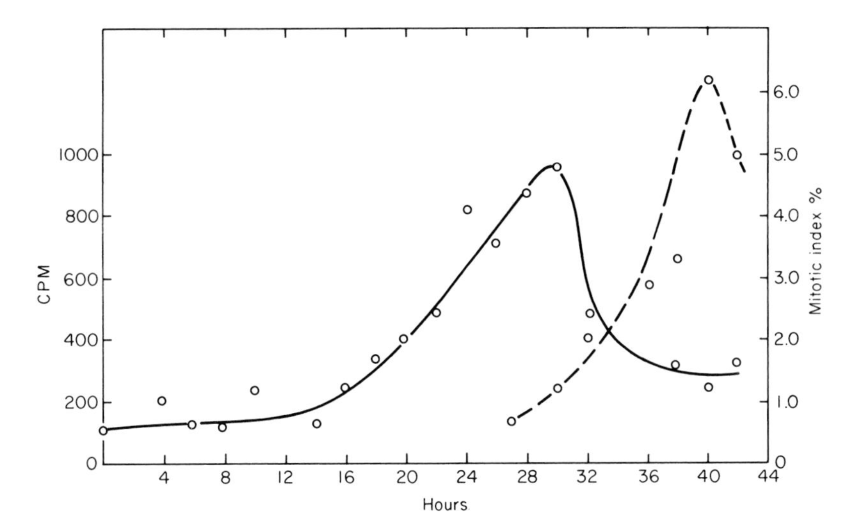

Fig. 1. DNA synthesis (thymidine incorporation) and mitoses in 3T3 cells following removal of beads. [^{3}H]thymidine (2.5 μCi) was added to each dish and incorporation during 20-minute pulses measured as TCA precipitable cpm. These are plotted over the 42-hour period following release of contact inhibition. Mitotic index in the same period is also plotted.

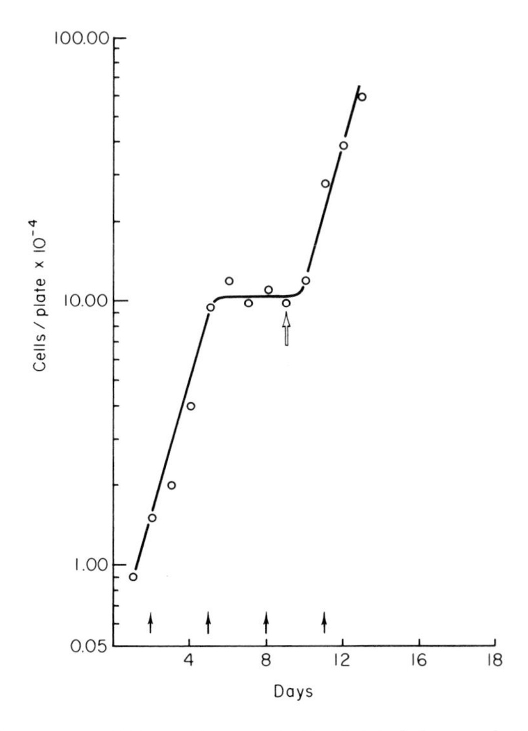

Fig. 2. 3T3 cells were seeded on 60-mm plastic Petri dishes with 1 g of 200 μm glass beads in each dish. Log counts of trypsinized cells before and after removing beads are plotted against days after seeding cultures. Filled arrows indicate time of medium change; unfilled arrow shows time of removal of beads.

of medium throughout the period of growth with beads, removal of the beads, and reestablishment of a confluent culture at a higher cell density.

Comments

Certain considerations and precautions should influence use of the glass bead technique to maintain and release contact inhibition of cell division.

The beads must be sterilized with dry heat for at least 6 hours at 170°C. Sterilization in steam leads to caking of the beads so that they cannot be dispersed as a uniform layer. Sterilization for less than 6 hours occasionally leads to growth of resistant microorganisms.

The use of beads smaller or larger than 200 μm has given less desirable behavior of the contact released monolayers. If the beads are smaller than 200 μm they do not settle well in the culture dish and do not provide sufficient space for growth between the beads. If the beads are larger than 200 μm there are many more cells between beads and a lower percentage is therefore released from contact inhibition when the beads are removed.

The presence of glass beads in intimate association with the cells has no apparent effect on their characteristic morphology, as determined by phase microscopy before and after removal of the beads. The glass beads used are composed of a soda-lime glass which does not provide a good substrate for growth of tissue culture cells. Therefore cells grown in the presence of the beads do not adhere to the beads and are apparently not damaged physically to any significant extent when the beads are removed.

Normally, beads are discarded at least 48 hours after medium change and the same medium returned to the cultures to exclude the effect of releasing contact inhibition by medium change and fresh serum addition.[7]

Mitotic indices, in our experience, are best obtained from mitotic cell counts using ×150 phase microscopy. The use of fixatives and stains in the counting of mitotic cells leads to an underestimation of mitotic cells. Mitotic cells round up and show decreased adhesiveness to the plastic substrate upon which they are growing.[10] Agitation caused by adding fixatives and stains dislodges a great many of these. Observation of dislodged cells show that greater than 50% are in mitosis.

[10] T. Terasima and L. J. Tolmach, *Exp. Cell Res.* **30**, 344 (1963).

CHAPTER 6

Maintenance of Diploid Fibroblast Cultures as Nonmitotic Populations

*R. T. Dell'Orco, J. G. Mertens, and P. F. Kruse, Jr.**

The use of normal diploid fibroblasts for *in vitro* aging studies has become well established in recent years. The majority of these investigations have employed actively proliferating populations; however, *in situ* fibroblasts of connective tissue have little or no mitotic activity except during wound healing.[1-3] Therefore, a closer approximation to the *in vivo* situation might be the use of nonmitotic cell populations for aging studies *in vitro*.

Several attempts have been made to establish conditions for handling replicate cultures of nondividing populations; however, they have met with only limited success.[4-6] Perhaps the most acceptable and the most convenient of the methods that have been employed to date is that of reducing the serum concentration of the incubation medium.[6-8] A procedure for this and some data characteristic of the cell populations so treated are described below.

Induction of the Nonmitotic State

Three human diploid fibroblast cell strains have been successfully maintained as nonmitotic cell populations for extended periods of time in this laboratory. These cell strains include two that were established from foreskin tissue—the NF-HFO-1[9] strain which was derived from newborn tissue and characteristically enters phase III after 35 ± 5 population doublings *in vitro* and the NF-HFO-3 strain which was established from adult tissue and enters phase III after 15 to 20 population doublings. The third cell strain is the WI-38 which was derived from embryonic lung and characterized by Hayflick.[10]

Confluent cultures, derived from stock lines carried in antibiotic-free me-

* Deceased.
[1] C. P. Leblond and B. E. Walker, *Physiol. Rev.* **36**, 255 (1956).
[2] B. E. Walker and C. P. Leblond, *Exp. Cell Res.* **14**, 510 (1958).
[3] W. S. Bullough and E. B. Laurance, *Exp. Cell Res.* **21**, 394 (1960).
[4] G. C. Yuan, R. S. Chang, J. B. Little, and G. Cornil, *J. Gerontol.* **22**, 174 (1967).
[5] R. J. Hay, R. A. Menzies, H. P. Morgan, and B. L. Strehler, *Exp. Gerontol.* **3**, 35 (1968).
[6] P. F. Kruse, Jr., W. Whittle, and E. Miedema, *J. Cell Biol.* **42**, 113 (1969).
[7] J. S. McHale, M. L. Mouton, and J. T. McHale, *Exp. Gerontol.* **6**, 89 (1971).
[8] R. T. Dell'Orco, J. G. Mertens, and P. F. Kruse, Jr., *Exp. Cell Res.*, in press. 1973.
[9] Referred to as cell line CF-1 in reference of footnote 8.
[10] L. Hayflick, *Exp. Cell Res.* **37**, 614 (1965).

dium, are subcultured at a 1:4 split ratio and grown to confluency in McCoy's Medium 7a supplemented with 10% fetal bovine or whole calf serum and 25 μg/ml of both penicillin and streptomycin. The incubation medium is changed every 48 hours until confluency is reached. At this time the growth medium is decanted and replaced with McCoy's Medium 7a supplemented with 0.5% serum and antibiotics. The cultures are incubated in a closed system under an atmosphere of 8% CO_2. The incubation medium is changed twice weekly for the duration of the maintenance period.

Cultures have been arrested in this manner for as long as 6 months with complete recovery of proliferative capacity; however, for practical purposes, experimental times are generally limited to 3 weeks, during which time the population has become stabilized. Cells are routinely recovered from this stabilized, nonmitotic state by subculturing at a 1:2 split ratio into medium containing 10% serum.

Characteristics of Nonmitotic Populations

Cells from any passage number before the entrance into phase III can be arrested and recovered to complete their characteristic number of population doublings. The percentage of mitotic cells in arrested cultures is less than 0.1 after 7 days of incubation and is 0 at day 14. DNA synthesis, as determined by incorporation of [^{3}H]thymidine and autoradiography, decreases rapidly during the first 7 days of incubation and less than 1% of the cells show any incorporation through day 21. There is approximately a 15% cell loss during 21 days of exposure to low serum medium which is attributed to culture manipulations, i.e., medium replacement and sample preparation. While protein loss on a per cell basis varies between cell strains, it occurs during the first 7 days of exposure and thereafter remains constant. Supplementation of the low serum maintenance medium with hydrocortisone, prednisolone, human serum, ascorbic acid, cholesterol, or vitamin E does not appear to affect significantly maintenance of cell numbers or cell protein content of these cells as nonmitotic populations. Addition of 1 IU/ml insulin to the maintenance media results in an average cell protein content ca. 20% higher than in its absence, an effect of insulin similar to that observed by Griffiths.[11]

Population Doublings and Metabolic Time

The process described for arresting cells in a nonmitotic state has been employed to determine that it is the number of population doublings which human diploid cells undergo *in vitro* and not the length of time maintained in culture that is the predominant factor in determining their *in vitro* life-span.[8] Table I contains data from two experiments with NF-HFO-1 cells that show this. NF-HFO-1 cultures exposed to low serum medium at passage numbers ranging from

[11] J. B. Griffiths, *J. Cell Sci.* **7**, 575 (1970).

TABLE I
Extension of in Vitro Life-span of NF-HFO-1 Cells after Incubation with Low Serum Medium[a]

Experiment	P[b]	M-time[c] (days)	P at phase III[d]		Days in culture[e]	
			Maintenance	Control	Maintenance	Control
1	19	21	40	36	108	69
	23	21	38		90	54
	29	21	39		73	34
2	13	21	34	32	116	77
	23	21	32		80	41
	27	21	37		66	27
	15	77	37		161	70
	13	177	36		287	77

[a] From R. T. Dell'Orco *et al.*[8]
[b] Passage number of cells when exposed to medium containing 0.5% serum.
[c] Days of incubation with medium containing 0.5% serum.
[d] Passage number at entrance into phase III.
[e] Days in culture from indicated passage number (P) until entrance into phase III.

13 to 29 and for durations of up to 177 days are all capable of achieving passage levels equivalent to or greater than the controls which have been continuously grown in medium containing 10% serum. It is noted that the arrested cells take a proportionately longer calendar time to achieve these equivalent passage numbers.

SECTION V

Single Cell Isolations and Cloning

Editors' Comments

The homogeneity of a cell culture population is of general concern to the investigator. Thus, techniques have been developed for single cell isolation and for the growth of the "isolates" to yield cell clones and ultimately populations sufficient in size for study. Numerous techniques have been developed for isolating single cells, each with its merits of applicability. The methods described in this section are the most commonly used.

Cloning techniques have wide application. For example, they have been used (a) to provide populations derived from a single cell to establish cell lines (see also Section II); (b) to test the genetic consistency of a cell population; (c) to isolate variants (see also Section XII); and (d) to evaluate the effect of materials on the growth potential of cells, e.g., the relative "plating efficiency" after exposure to a particular treatment. Soft agar cloning techniques have provided a means for separating viral transformed cells from their untransformed counterparts.

It became apparent early in the study of single cell "isolates" and clones that the "environmental" requirements of these low density populations differed from high density cultures. Thus, the nutritional considerations and the gas phase of these cultures are also discussed in this section.

CHAPTER 1

Animal Cells

A. Capillary Techniques

K. K. Sanford

A capillary cloning technique was first developed in 1948[1] before the successful growth of any single, isolated tissue cell into a clone and before the introduction of improved chemically characterized culture media. The development of this technique was based on the hypothesis that a single cell failed to grow because even our best culture media were so inadequate as to need extensive modification by the cells before utilization. Whereas a large clump of cells could adapt a reasonable volume of the usual culture medium, one cell grown in an identical volume was unable to adjust the fluid sufficiently for survival and proliferation. On the basis of this concept, an attempt was made (1) to reduce the amount of culture medium bathing the single cell to that volume which the cell could adjust, and (2) to supply the cell with a culture medium already adjusted by the growth of large cultures of living cells. The technique so developed had certain advantages in that one cell was physically separated from all other cells; the cell could be readily observed; pH shifts and evaporation of culture fluid were negligible; and culture fluids could be renewed easily.

The original technique, although successfully applied to a number of cell types,[2–5] was arduous, and a simplified procedure described in 1961[6] was successfully applied to additional cell types.[7–9] Subsequently, further changes were introduced,[10] which will be described herein.

[1] K. K. Sanford, W. R. Earle, and G. D. Likely, *J. Nat. Cancer Inst.* **9**, 229 (1948).

[2] G. D. Likely, K. K. Sanford, and W. R. Earle, *J. Nat. Cancer Inst.* **13**, 177 (1952).

[3] G. D. Likely, K. K. Sanford, V. J. Evans, and W. R. Earle, *J. Nat. Cancer Inst.* **18**, 701 (1957).

[4] V. P. Perry, K. K. Sanford, V. J. Evans, G. W. Hyatt, and W. R. Earle, *J. Nat. Cancer Inst.* **18**, 709 (1957).

[5] K. K. Sanford, R. M. Merwin, G. L. Hobbs, J. M. Young, and W. R. Earle, *J. Nat. Cancer Inst.* **23**, 1035 (1959).

[6] K. K. Sanford, A. B. Covalesky, L. T. Dupree, and W. R. Earle, *Exp. Cell Res.* **23**, 361, (1961).

[7] K. K. Sanford, T. B. Dunn, A. B. Covalesky, L. T. Dupree, and W. R. Earle, *J. Nat. Cancer Inst.* **26**, 331 (1961).

[8] K. K. Sanford, T. B. Dunn, B. B. Westfall, A. B. Covalesky, L. T. Dupree, and W. R. Earle, *J. Nat. Cancer Inst.* **26**, 1139 (1961).

PREPARATORY STUDIES

Before applying the capillary cloning technique, it is advisable to determine the plating efficiency of the cells to estimate the probability of success in establishing a clone. If the plating efficiency is low, various culture conditions such as different O_2 concentrations in the gas phase,[11] sources of serum and culture fluid, methods of cell dispersion, etc., should be explored for selection of more optimal conditions to support single cell growth.

MATERIALS

Capillary tubes should be approximately 8 mm long with a 0.1-mm bore. These may be drawn from clean Pyrex glass tubing (3 mm OD, 0.6 mm wall thickness) by using a Hoke-Jewel oxy-gas torch No. 71 with No. 5 tip. In addition, a cylinder of oxygen equipped with a Hoke regulator 901B10 with appropriate CGA connection is needed (Hoke, Inc., Tenakill Park, Cresskill, New Jersey 07626). Short 2-inch lengths of the glass tubing should first be thoroughly cleaned, preferably in hot reagent grade sulfuric acid and about 0.5% nitric acid, rinsed in tap water, boiled in dilute Calgolac solution, 1 teaspoon Calgolac per liter distilled water (Calgon Corporation, 7501 Page Avenue, St. Louis, Missouri 63166), and repeatedly rinsed in hot tap water followed by distilled water. Each capillary tube should be checked to be certain it will fill when one end is immersed in triple glass-distilled water. The capillary tubes are then placed in a 60-mm glass Petri dish and autoclaved.

Two or three stainless steel watchmaker's forceps (4⅛ inch, fine tip; obtainable from Clay-Adams, Inc., 141 E. 25th Street, New York 10010) should be used. Sterilize by dry heat.

Fifty small culture vessels, preferably of Pyrex should be obtained. These can be Carrel D3.5, T-9 flasks, or Leighton tubes (Kontes Glass Company, Vineland, New Jersey 08360).

Perforated cellophane (Beckley design OC obtainable from Microbiological Associates, Inc., Bethesda, Maryland 20014) should be cut to fit the floor of the culture vessel. For cleaning, the cellophane should be soaked in two changes each of reagent quality ether, alcohol, acetone, and triple-distilled water, allowing 15 minutes for each change. The cellophane sheets should then be dried on filter paper and each sheet inserted into the culture flask by means of forceps so that the fingers do not touch the sheet. The flasks with the inserted cellophane sheet may be sterilized by autoclave. Also necessary are silicone stoppers (X9711

---

[9] K. K. Sanford, B. B. Westfall, E. H. Y. Chu, E. L. Kuff, A. B. Covalesky, L. T. Dupree, G. L. Hobbs, and W. R. Earle, *J. Nat. Cancer Inst.* **26**, 1193 (1961).

[10] The use of short capillary segments rather than capillary pipettes was a modification introduced by Mr. F. M. Price of our laboratory.

[11] A. Richter, K. K. Sanford, V. J. Evans, and W. G. Taylor, *J. Nat. Cancer Inst.* **49**, 1705 (1972).

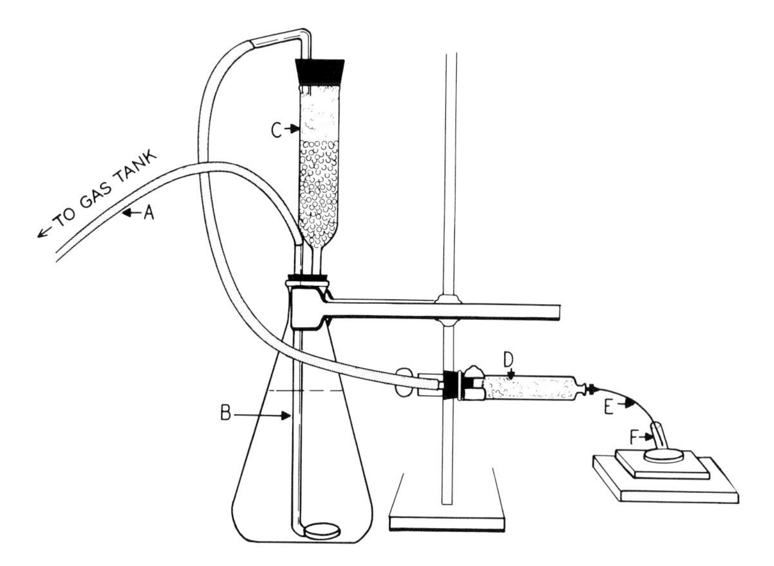

Fig. 1. Humidifier and gas dispersion apparatus. Further description in text.

silicone) to fit culture vessels (obtainable from The West Company, Phoenixville, Pennsylvania 19460), a needle or spatula such as a platinum alloy spatula,[12] that can be repeatedly flamed without producing toxic products, and a cylinder of 10% CO_2 in air or 10% CO_2:1% O_2:89% N_2 of high purity, equipped with a Hoke regulator as above.

A humidifier and gas dispersion apparatus (Fig. 1) is used. The gas is dispersed through a medium porosity fritted glass disk (Fig. 1B) (Arthur H. Thomas Company, Philadelphia, Pennsylvania 19105) inserted in 1-liter Erlenmeyer flask containing 0.5% sulfuric acid (isotonic and acid to prevent fungus growth). After bubbling through the solution, the gas is freed of droplets by passing consecutively through glass and cotton in a tube (Fig. 1C). The gas is then sterilized by filtering through a sterile (autoclaved) 10-ml syringe barrel packed with cotton (Fig. 1D). The syringe is fitted with a sterile 3½ inch, 18-gauge needle, preferably platinum-iridium (Fig. 1E) which can be flamed. The needle is bent to jet the gas into the Petri dish or culture vessel (Fig. 1F).

Two syringe barrels with needles are needed and can be connected by rubber hose and Y-tube to the gas dispersion apparatus with pinch clamps to cut off flow.

Petri dishes (60-mm diameter) and an inverted microscope such as the Zeiss standard UPL (Carl Zeiss, Inc., 444 Fifth Avenue, New York 10018), fitted with 10x oculars and 25x objective, are also requisite.

Preparation of Single Cell Suspension. The method of preparing a uniform single cell suspension, whether by mechanical sieving, enzymatic digestion, vig-

[12] V. J. Evans, W. R. Earle, K. K. Sanford, J. E. Shannon, and H. K. Waltz, *J. Nat. Cancer Inst.* **11**, 907 (1951).

orous pipetting, or removal of cell aggregates by sedimentation procedures, will depend on the cell type used. The technique selected should yield a single cell suspension free of cell aggregates and debris and yet provide minimal injury to the cells. The suspension should be adjusted to approximately 1.6×10^4 cells/ml and 5 ml transferred to a 60-mm Petri dish. The dish should be provided at all times with a slow stream of humidified gas, either 10% Co_2:90% air or 10% CO_2:1% O_2:89% N_2, to maintain the pH at approximately 7.3. The gassing can be eliminated only if an organic buffer can be used effectively without reducing plating efficiency.

Isolation of Single Cells. To a second Petri dish or culture flask without cellophane, add sufficient culture fluid to cover the floor. Place on the stage of an inverted microscope and gas continuously from the second gassing outlet. With sterile forceps pick up a capillary tube and insert one end in the Petri dish of cell suspension, while agitating the suspension. Place the capillary in the Petri dish or flask under the microscope. To prevent emptying of the capillary tube, always hold it in a horizontal position during immersion in culture fluid. Either examine each tube immediately or prepare all tubes and incubate at 37°C for several hours to allow cell attachment. If the tubes are in a Petri dish instead of a stoppered flask, the dish should be maintained in a CO_2-humidity incubator.

Planting of Capillary Tubes. To the fifty small culture vessels, add 2 to 3 ml of culture fluid, straighten and smooth the cellophane sheet against the floor of the flask with the needle or spatula, gas to adjust pH if necessary, and stopper. Examine each capillary tube under the microscope. One advantage of the capillary tube over the coverslip method is that the tube can be rolled over for ready examination of all surfaces of the interior. Plant any capillary tubes containing one or two cells in the culture vessel by first immersing the capillary by forceps in the culture fluid and then inserting it under the sheet of perforated cellophane. Position the capillary in the center of the flask, withdraw most of the culture fluid, gas, and stopper the flask. With a glass marking pencil, indicate the position of the capillary on the under surface of the flask and label the culture. If it is necessary to resterilize forceps during the operations, they can be dipped in a tube of reagent quality chloroform and placed horizontally in a sterile test tube to dry. Incubate capillary tube cultures at 37.5°C.

Observations

Capillary tube cultures should be carefully examined under the microscope once during the first 12 hours after planting and twice a day thereafter, to be certain only one cell survives and divides. Fluids can be renewed once or twice a week by adding 2–3 ml and then withdrawing most of this to prevent displacement of the capillary tube. The pH should be adjusted by gassing.

Although it is essential to examine and record observations on each culture twice daily, additional exposures to light or reduced temperature may be damaging to the cells. After the cells emerge from the open ends of the capillary and cover the floor of the flask, more frequent renewals of culture fluid and a larger volume of fluid may be required.

CHAPTER 1

Animal Cells

B. Microdrop Techniques

Ian Macpherson

A clone is a population derived from a single cell. It provides a genetically purified stock. Colonies developing in cultures seeded with small numbers of well dispersed cell suspensions have a high probability of being true clones but some of the colonies may have been derived from small clumps of cells. Cloning methods in which a cell is separated visually greatly increase the certainty that a clone is truly derived from a single cell.

A method in which single cells were isolated with finely drawn pipettes into droplets of medium under paraffin[1-3] has been superseded in this laboratory by the method described below.

Equipment

An inverted microscope or plate microscope with a total magnification of about ×50 (e.g., Olympus CK, Tokyo).

Sterile multicup (6 mm) polystyrene tissue culture dishes (e.g., Linbro Chemical Co. Inc. "Multi-Dish Disposo-Trays" Cat. No. 15-FB-96-TC). These have 96 cups of about 0.2-ml capacity.

Sterile polystyrene Petri dishes of tissue culture grade (NUNC or Falcon).

Sterile polystyrene Petri dishes of bacteriological grade (e.g., Sterilin 9-cm dishes).

[1] A. Lwoff, R. Dulbecco, M. Vogt, and M. Lwoff, *Virology* **1**, 128 (1955).

[2] P. Wildy and M. Stoker, *Nature* (*London*) **181**, 1407 (1958).

[3] I. Macpherson, *In* "Fundamental Techniques in Virology," (K. Habel and N. Salzman, eds.), p. 17. Academic Press, New York, 1969.

Sterile Pasteur, 5- and 10-ml pipettes.

Tissue culture medium to suit cells. The more complex media such as Waymouth's or Ham's supplemented with nonessential amino acids, Krebs cycle intermediates, and 15–20% fetal calf serum are best.

Sterile trypsin (Difco 1:300) solution, 0.01% in Dulbecco's phosphate-buffered saline.

Mouth piece for micropipettes. A short length of glass tubing with about 50 cm of rubber tubing to fit the micropipettes.

Micropipettes drawn from thin tubing. These are made in two stages from clean, sterile plugged soft glass tubing (Fig. 1a–c). In the first stage the tubing is pulled into two fine pipettes with an external diameter tapering to about 1 mm. When cool they are softened again at the edge of Bunsen flame or over the pilot light of a burner. Pipettes of fine bore are made by pulling the softened glass immediately after it has been removed from the flame. The most suitable pipette bore can be found by trial and error. Generally speaking the finer they can be drawn the better. An internal diameter of about 50 to 100 μm or about five to ten cell diameters is suitable.

Method

The technique described here is suitable for cell lines with a high degree of autonomy in culture, i.e., a colony-forming efficiency in a large volume of medium (plating efficiency) of more than 3%, e.g., cell lines such as HeLa, L, BHK21, and 3T3.

Selecting the Cells. If the cell suspensions for cloning are derived from colonies in Petri dishes suitable cell types may be preselected. A preliminary selection may be made with the aid of a beam of light from a microscope lamp directed on the underside of a dish containing 7- to 10-day old colonies. Rare colonies with distinct morphological characteristics can be found and

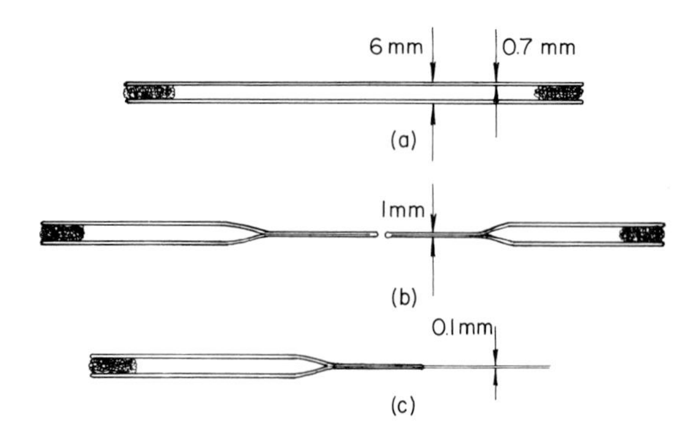

Fig. 1. Preparation of micropipette: (a) sterile, plugged glass tube about 12 cm long; (b) first stage in which tube is drawn to about 1 mm; and (c) second stage in which pipette is drawn out again to fine bore and then broken off.

marked with a circle of grease pencil or marker on the underside of the dish. Virus-transformed colonies which are refractile and piled up are easily detected in this way. Large colonies are presumably derived from rapidly growing cells well adapted to the medium used for plating. The cell content of marked colonies can be checked with an inverted microscope.

Harvesting the Cells for Cloning. The medium is removed from a dish with marked colonies and about 5 ml of trypsin solution added to the dish, left for about 10 seconds, and then sucked off. The cells in the colonies round up after about 5 to 10 minutes at room temperature but remain loosely attached to the dish. Cells from the area of a marked colony are then sucked into a fine pipette containing medium and transferred as a localized pool to 9-cm bacteriological Petri dishes containing about 5 ml medium. A bacteriological grade dish is used because the cells do not attach to its surface and are available floating freely for easy isolation.

Picking Single Cells. A finely drawn pipette is attached to the mouth piece and partially filled with medium. The 9-cm dish containing the cells is placed on an inverted microscope and a part of the cell pool with well-separated cells brought into focus. It is advisable to blow medium slowly from the pipette until the tip comes into focus. This will ensure that cells do not enter the pipette before a visible selection has been made. The tip of the pipette is brought into the field of vision and one cell is sucked into the pipette.

Growing Up the Isolated Cell. The micropipette is withdrawn from the dish and a small bead of medium containing the isolated cell dropped into a cup in the multiple culture dish. Isolations are continued until a row is completed. Each drop is then examined on the inverted microscope to check that it contains only one cell. About 0.1 ml of medium (2 drops from a standard Pasteur pipette) is added to each cup in the row and the procedure is repeated until sufficient isolations are made to produce the number of clones required, e.g., if the colony-forming efficiency is 10% and three clones are required then a whole dish (96 isolations) should be used. The dishes are incubated in a 5–10% CO_2 in air incubator until large colonies have grown. This usually takes about 7 to 10 days. Colonies may then be transferred to 5-cm Petri dishes or bottles following their resuspension by trypsinization.

This method may be applied to cells with low colony-forming efficiencies (i.e., $<1\%$) if the isolates are supported by "feeders," i.e., cells that have been X-irradiated[4,5] or treated with mitomycin C.[6] Such cells are incapable of division but they retain the ability to help the growth of sparse viable cells either by "conditioning" the medium or by making direct contacts with the isolated viable cells and supporting them by the process of "metabolic coopera-

[4] H. W. Fisher and T. T. Puck, *Proc. Nat. Acad. Sci. U. S.* **42**, 900 (1956).

[5] K. H. Rothfels, E. B. Kupelwieser, and R. C. Parker, *Proc. Can. Cancer Res. Conf. No.* **5**, 191 (1963).

[6] I. Macpherson and A. Bryden, *Exp. Cell Res.* **69**, 240 (1971).

tion" described by Subak-Sharpe *et al.*[7] in which a cell may share the metabolic activity of a neighboring cell. The actual mode of action is not known.

Mitomycin C-treated cells are prepared by incubating semiconfluent cultures overnight in medium containing 2 μg of the drug per 10^6 cells. The following day the cells are dispersed with trypsin, resuspended in medium, and 10^3 added in 0.1 ml to each multicup. The same cells that are being cloned can be used as feeders. This reduces any problems that may arise as a result of the transfer of genetic information or viruses from a heterologous cell. The feeder cells become giants and come off the surface of the dish after about 7 to 10 days. If the medium is changed at 3- to 4-day intervals the cups containing proliferating clones will rapidly become obvious as the pH of the medium drops rapidly. Cups without viable cells will maintain their initial pH.

[7] J. H. Subak-Sharpe, R. R. Bürk, and J. D. Pitts, *Heredity* **21**, 342 (1966).

CHAPTER 2

Plant Cells

A. C. Hildebrandt

Single cell clones of bacteria and fungi have been used to clarify much of modern microbiology including genetics and biochemical mutations. Isolation and growth of single higher plant cells of normal and diseased origins have long been desired. Pure cultures of higher plant cells, tissues, and organs have been of interest to biologists for understanding cellular growth and development and interactions between host cells and a variety of pathogens and pathogenic agents. It was important to know how cells in tissue masses were alike or different. Are all the cells in diseased growth alike or are only certain cells diseased and others normal?

Single cell clones of plant pathogenic bacteria were established in 1930.[1] More recently single fungus, animal, and higher plant cells have been isolated and cloned. In certain cases cultures have been established by selecting colonies of cells derived by plating out on agar and from suspensions of single cells and small colonies. These colonies were not necessarily derived from single cells and the cultures thus derived were, therefore, not necessarily single cell clones.

[1] W. H. Wright, A. A. Hendrickson, and A. J. Riker, *J. Agr. Res.* **41**, 541 (1930).

In this discussion a single cell culture or single cell clone refers only to one that originated from a single, mechanically picked, individual, plant cell. These methods have opened the way to work with pure culture techniques applied to tissue cultures of higher plants in ways similar to those used for microorganisms.

Early attempts by Haberlandt[2] (and by others later) to isolate and grow single higher plant cells were unsuccessful, perhaps, because of the mature type of cells selected and the lack of favorable culture media. More recently it has been possible to isolate and to grow single higher plant cells and to establish clones of tissue. Such single cell cultures and clones have been made possible by several methods.

Procedures

Filter Paper Raft-Nurse Cultures. Clones of plant tissue were first established from hand-picked single cells of tobacco and marigold callus cultures by Muir *et al.*[3] using a nurse culture method (Fig. 1). Established cultures of plant cells of many species when grown in a liquid medium on a variety of shaking devices often produce a suspension of single cells and small colonies of cells. The liquid medium is of the same composition as that of the semisolid medium except that the agar is omitted. The shaker or roller wheel type of agitator thus provides a means for additional aeration and may also loosen the colonies of cells to free single cells. The suspension of cells is then spread over an agar medium, since it is more difficult to pick up single cells from agar grown cultures. Single cells, spread over the agar surface, are located under a dissecting microscope and picked up with a microneedle or a micropipette under aseptic conditions. With the nurse culture method, the single cell is transferred with a sterile needle, loop, or micropipette to an 8 × 8 mm square of sterile filter paper (Reeve Angel, crepe surface, No. 202). These filter paper squares have been previously placed

Fig. 1. The filter paper raft-nurse culture method of culturing single isolated higher plant cells and clones. Single tobacco cell placed on filter paper on nurse tissue at left. Subsequent growth of cell produced colonies of cells seen at center. Far right shows established clone of tissue transferred to agar medium that may be subcultured for unlimited periods. A. C. Hildebrandt and A. J. Riker.[10]

[2] G. Haberlandt, *Sitzungsber Akad. Wiss. Wien, Math. Naturwiss. Kl* III, 69 (1902).
[3] W. H. Muir, A. C. Hildebrandt, and A. J. Riker, *Science* **119**, 877 (1954).

for 2 to 3 days on top of established young cultured nurse tissue pieces of similar or different species growing on agar medium. This pretreatment of the filter paper squares permits the filter paper to absorb the liquids and nutrients from the mother host cell culture. After the single plant cell is placed on the filter paper raft, the filter paper raft and cell are returned to the nurse host tissue piece in the culture vessel and incubated. The various steps are best done quickly to avoid excessive desiccation of cell and moist filter paper raft. The isolated cell and progeny cells remain on the top of the raft away from direct contact with the nurse culture. The nurse tissue piece, however, continues to provide liquid and nutrients to the isolated cell. The nurse culture may lose vigor, and it may be necessary to transfer the raft with the isolated single cell or developing colony of cells to fresh nurse pieces several times during the 6–10 weeks required to develop the small colony of cells. At any time the raft may be removed and examined under the microscope to verify the single cell nature of the isolation or colony development and that the colony growth is from division of the single cell and not from cells of the nurse pieces below the raft. Colonies of single cell origin after reaching approximately 4 mm diameter are then transferred directly to the surface of the agar medium with a sterile transfer loop, where they grow independently. Such colonies after 2 to 4 weeks, depending on the rate of growth, may be divided with a sterile scalpel and subcultured to fresh medium and maintained as clones of tissue of single cell origin. Single cell clones of many species have been established with the nurse culture-filter paper raft method.

Efforts have also been made to isolate and to grow single cells in hanging drop cultures,[4] but only limited cell growth was sustained. The cell plating technique of Bergmann[5] and the cell sieving method have been useful to study the requirements for growth and differentiation of cells, tissues, and plants.[6,7]

Microscope Slide Cultures. A microculture chamber devised by Jones *et al.*[8] has provided a means for culturing single cells which permits cytological observations of the living, unstained cells while dividing, enlarging, differentiating, maturing, and dying. Although originally developed to examine tobacco cells, it has now been used for the isolation and growth of single cells of many species.

The microculture chamber (Fig. 2) is prepared aseptically using a standard microscope slide, U. S. P. heavy mineral oil, and 22-mm square No. 1 coverslips. Each of two sterile coverslips is lowered onto a drop of sterile mineral oil at each end of the slide. A rectangle of mineral oil is placed on the slide to connect the two coverslip risers. One cell, as desired, along with the medium, is picked up with a sterile micropipette and placed within the rectangle of mineral oil on the slide. A third coverslip is then lowered over the drop and cell so as to straddle the two coverslip risers and to enclose the cell within the mineral oil

[4] J. G. Torrey, *Proc. Nat. Acad. Sci. U. S.* **43**, 887 (1957).

[5] L. Bergmann, *J. Gen. Physiol.* **43**, 841 (1960).

[6] L. M. Blakely and F. C. Steward, *Amer. J. Bot.* **51**, 780 (1964).

[7] W. Halperin, *Symp. Int. Soc. Cell Biol.* **9**, 169 (1970).

[8] L. E. Jones, A. C. Hildebrandt, A. J. Riker, and J. H. Wu, *Amer. J. Bot.* **47**, 468 (1960).

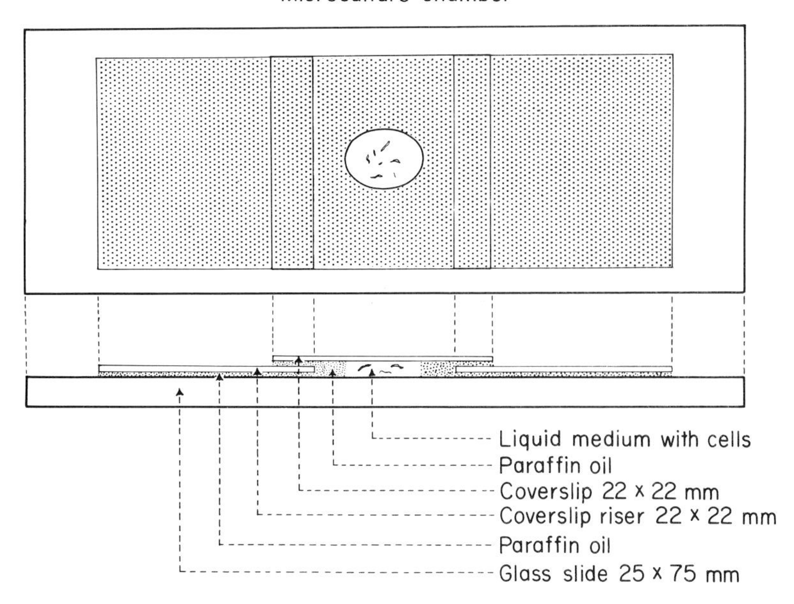

Fig. 2. The microculture method of culturing single cells of higher plants. Single cells or suspensions of cells may be kept alive under aseptic conditions in a drop of medium surrounded by mineral oil. Observation of growth and division of the cell in a living, unstained condition is possible for weeks with phase microscopy. L. E. Jones, A. C. Hildebrandt, A. J. Riker, and J. H. Wu.[8]

barrier, providing a microchamber for the cell and medium. An alternative method is to place the cell and medium on the third coverslip and then to invert this on the slide so the cells and medium are in the chamber within the mineral oil barrier. The cells and medium enclosed with the small chamber have routinely remained alive for days but, in many instances for as long as several months, without adding fresh nutrient. In preparing the microculture chamber, it is important to avoid desiccation, tearing, or overheating of the cells. The cellular activities can be observed with phase and interference microscopy and recorded with still and time-lapse photography. Fresh nutrient may also be added, if desired, to the microculture drop to maintain the culture and any subsequent colony development. Between observations the cultures are conveniently stored in sterile Petri dishes in a dark room at 26°C.

Applications

The isolation and growth of single cells and single cell clones of higher plants thus provide many opportunities to examine cellular details, growth, differentiation, and host cell–environment interactions. Such single cell clones have been useful to clarify biochemical, morphological, and genetic similarities and differences between normal and diseased plant cells, and to verify totipotency of somatic cells. Many different sizes and shapes of cells may be observed in

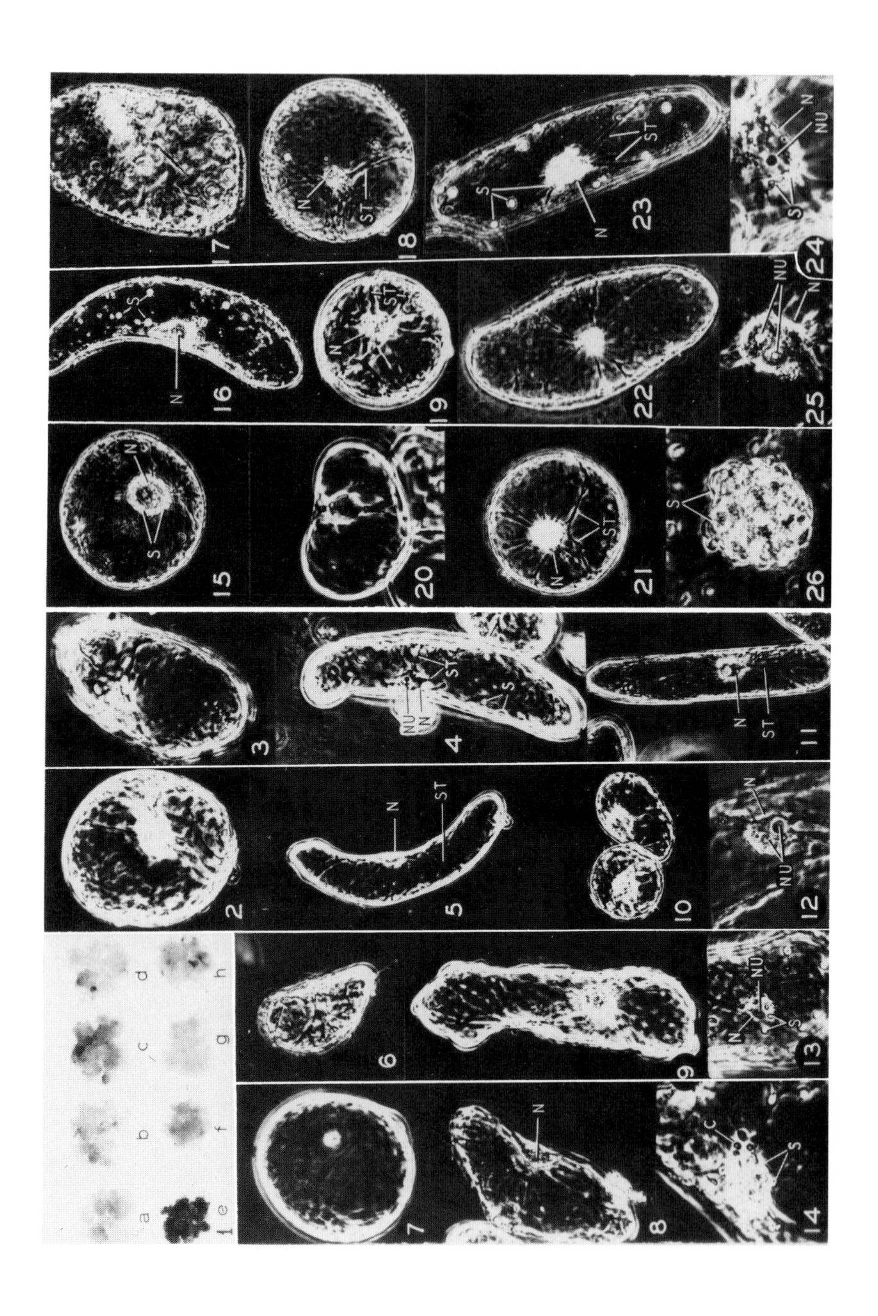

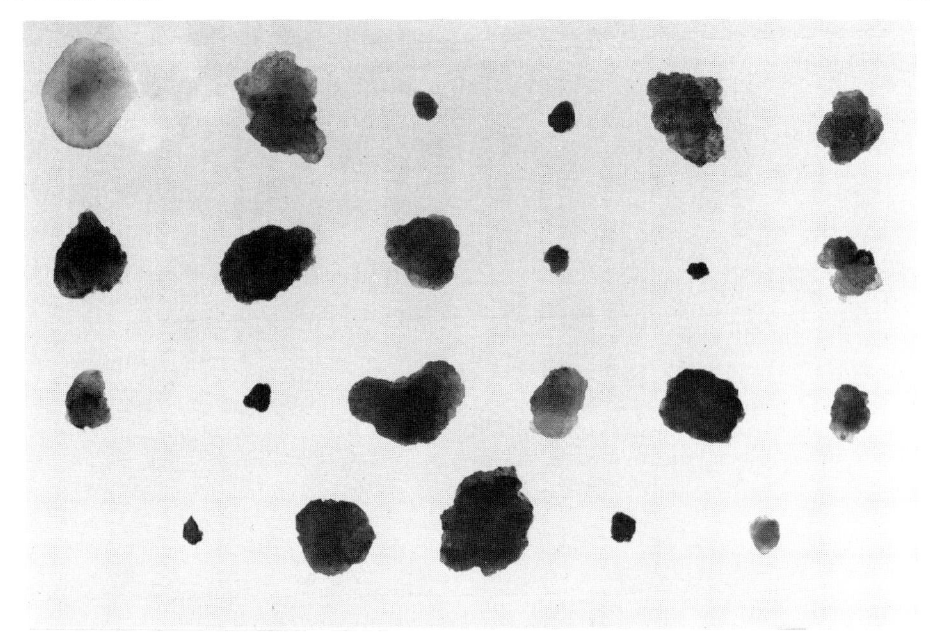

Fig. 4. Appearance and relative size of a representative tissue piece from the parent clone and each secondary single cell clone of H 196 after 53 days on casein hydrolysate medium with 2% lactose added. The clones are arranged as follows (from left to right): Top row, clones 1–6; second row, clones 7–12; third row, clones 13–19 (H 196-15 has been omitted); bottom row, clones 20–23 and the parent. Some clones are light colored, some dark, and some mixed. An indication of the range of colors is seen in the contrasts in the tissue pieces. R. C. Sievert and A. C. Hildebrandt.[12]

the liquid suspension cultures (Fig. 3). The filter paper raft-nurse culture method was used originally to study cells of normal and crown gall origins[3] and Braum[9] used the method to produce plants to clarify the teratomaceous crown galls. Single cell clones derived by this method varied in growth rate and susceptibility

[9] A. C. Braum, *Proc. Nat. Acad. Sci. U. S.* **45**, 932 (1959).

Fig. 3. (1) Calli of 7 species of edible plant tissues grown for 5 weeks on C or D medium (a, carrot; b, endive; c, lettuce leaf petiole; d, lettuce stem; e, spinach; f, navy bean; g, parsley; h, red kidney bean). (2–14) Phase-contrast photomicrographs of living representative cells of edible plant callus tissues; 2–4: spherical, oval, and elongated cells of carrot; 5 and 6: elongated and oval cells of endive; 7–9; spherical, oval, and elongated cells of lettuce leaf petiole; 10 and 11: spherical, oval, and elongated cells of lettuce stem; 12–14: nuclear areas of single cells of lettuce stem, endive, and lettuce leaf petiole, respectively. C, Chloroplast; N, nucleus; NU, nucleolus; S, starch granule; ST, cytoplasmic strands. Magnification: 2–9 and 11, ×184; 10, ×115; 12 and 13, ×459; 14, ×918. (15–26) Phase-contrast photomicrographs of living representative cells of edible plant callus tissues; 15: spherical cell of spinach; 16–18: elongated, oval, and spherical cells of navy bean; 19 and 20: spherical and oval cells of parsley; 21–23: spherical, oval, and elongated cells of red kidney bean; 24–26: nuclear area of parsley, navy bean, and spinach, respectively. Note in 26, the nucleus is completely covered by starch granules. N, Nucleus; NU, nucleolus; S, starch granule; ST, cytoplasmic strands. Magnification: 15–23, ×214; 24–26, ×534. U. Kant and A. C. Hildebrandt, *Can. J. Bot.* **47**, 849 (1969).

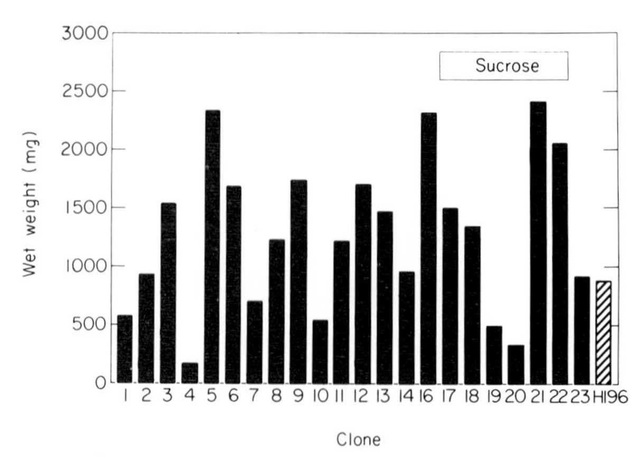

Fig. 5. Average weight per tissue piece of the parent and secondary single cell clones of H 196 after 5 weeks on casein hydrolysate medium with 2% sucrose added. The parent clone is indicated by diagonal lines. R. C. Sievert and A. C. Hildebrandt.[12]

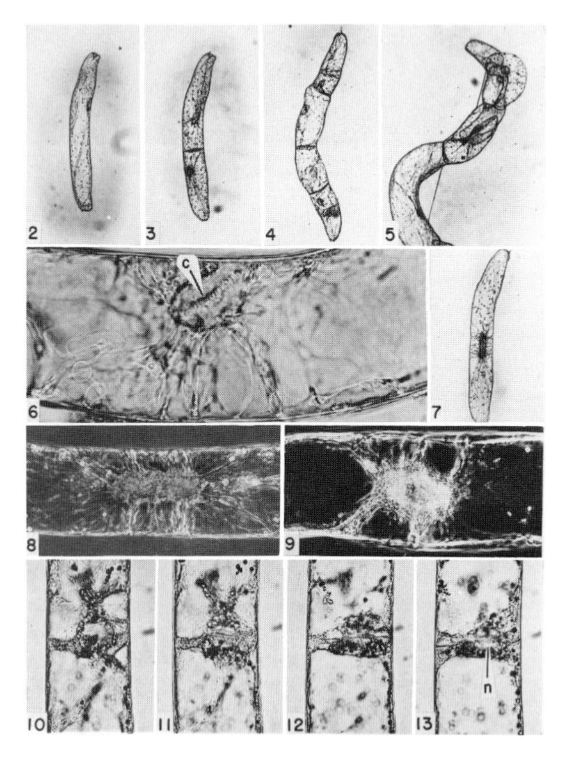

Fig. 6. (2–13). Growth of isolated single cell in microculture. 2: Initial single cell; 3: 2 days later; 4: 4 days later. 5: 7 days later. 6–13: Mitosis and cell division. All except 6 and 9 are of the first division of the single cell seen in 2. 6: Mitosis in vacuolate cell ×510; c, chromosomes at metaphase. 7: Early prophase at 2:18 PM. ×85. 8: Early prophase at 2:20 PM; phase contrast. Numerous thin strands of cytoplasm surround the nucleus. 9: Late prophase; phase contrast. Cytoplasmic strands are coalescing. 10: Late metaphase at 4:47 PM. Phragmosome of cytoplasm already formed. 11: Early telophase at 4:57 PM. Cell plate has started to form. 12: Midtelophase at 5:10 PM. Cell plate is extending along the phragmosome. Cytoplasm is amassing near telophase nuclei and moving into the parietal sheath. 13: Cell plate is nearly completed at 5:22 PM. Daughter nuclei (n) still are adjacent to the new cell wall. 2–5, ×75; 8–13, ×305; others, as indicated. L. E. Jones, A. C. Hildebrandt, A. J. Riker, and J. H. Wu.[8]

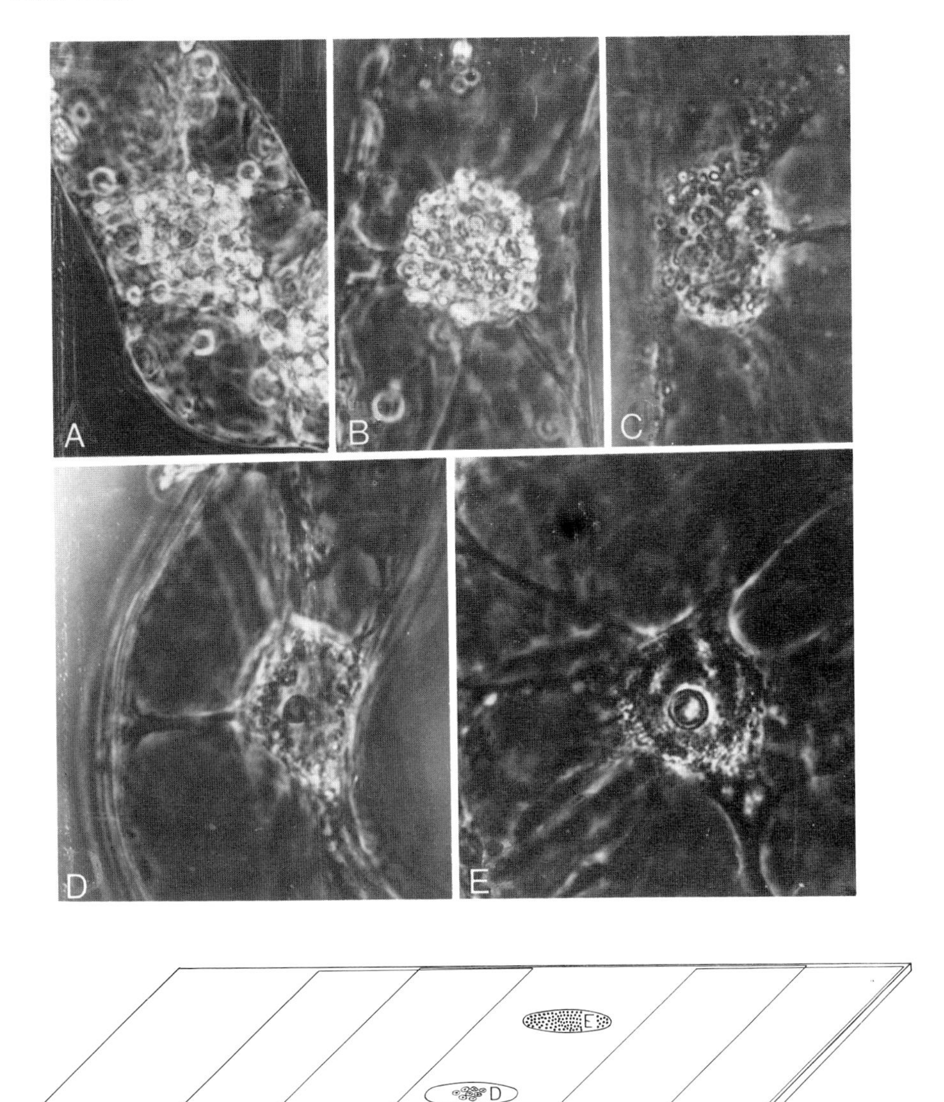

Fig. 7. (A–E) Living tomato tissue culture cells in microculture illustrating rating of starch content on a scale of 0–4. A, 4; B, 3; C, 2; D, 1; and E, 0. Phase contrast. (A) ×700; (B) ×730; (C) ×800; (D) ×560; (E) ×490. (Bottom) Diagrammatic representation of the microculture. (A) Microscope slide; (B) 22-mm^2 coverslip; (C) 22 × 40 mm coverslip; (D) drop of medium containing tissue culture cells; (E) drop of medium containing bacterial cells (or drop of sterile medium for control cultures). The microculture was prepared aseptically. M. Kalil and A. C. Hildebrandt.[11]

to tobacco mosaic virus infection.[10,11] The comparative growth rates of single cell clones and secondary single cell clones derived from the original mother

[10] A. C. Hildebrandt and A. J. Riker, *Fed. Proc. Fed. Amer. Soc. Exp. Biol.* **17**, 986 (1958).

[11] A. C. Hildebrandt, *Proc. Nat. Acad. Sci. U. S.* **44**, 354 (1958).

Fig. 8. Differentiation and plantlet formation in a single cell clone of tobacco. A, Undifferentiated mass of callus; B, differentiation of a large number of shoots with normal leaves; C, root formation in plantlets; D, plantlets with roots and shoots (A in D medium; B and C in MS medium; D in White's Medium; all two-thirds natural size); E, plantlet transferred to soil in greenhouse; inverted culture bottle retains high humidity around the young plant (⅓ natural size); F–H, Various stages in the further development and eventual flowering of the tobacco plants initially transferred from the culture bottles (F, one-half natural size; G, one-third; H, one-sixth). V. Vasil and A. C. Hildebrandt.[15]

single cell clone have been compared on various common sugars[12] (Figs. 4 and 5).

The microculture method was subsequently used in many ways to clarify normal and diseased cell growth and host cell–pathogen interactions. The cytology of the living, unstained cells was seen with phase microscopy as cells grew and divided and matured[8] (Fig. 6). The relation of tobacco mosaic virus to the host tobacco cells has been examined in detail in microculture and recorded with still and motion pictures.[13] A double-chambered microculture[14] was used to examine the effects of crown gall bacteria on degradation of starch grains of host cells of tomato (Fig. 7). The totipotency of the single somatic tobacco cells was also demonstrated.[15] A single tobacco cell in microculture produced a colony of cells, and this colony was subsequently removed to agar medium and induced to differentiate into a complete tobacco plant (Fig. 8). The totipotency of the single somatic cells was thus demonstrated in the absence of nurse tissue, neighboring cells, and conditioned medium. More recently the microculture

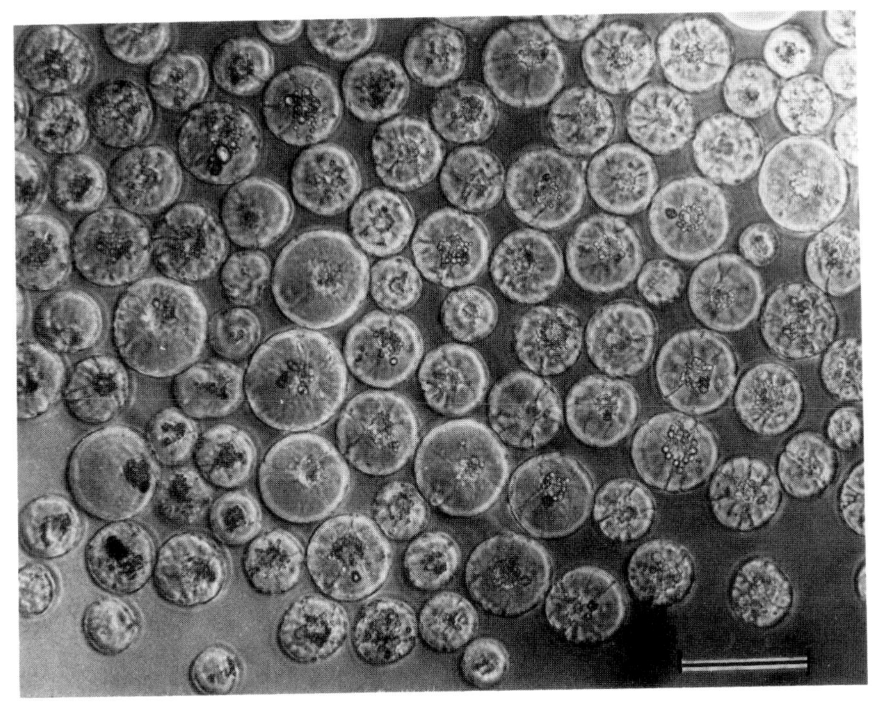

Peanut (VB) Protoplasts

Fig. 9. Healthy, naked protoplasts (cell walls removed enzymatically) floating at the top of a hanging drop, *Arachis hypogaea* (peanut), cultivar Virginia Bunch 67. R. U. Schenk and A. C. Hildebrandt.[16]

[12] R. C. Sievert and A. C. Hildebrandt, *Amer. J. Bot.* **52**, 742 (1965).
[13] M. Singh and A. C. Hildebrandt, *Virology* **30**, 134 (1966).
[14] M. Kalil and A. C. Hildebrandt, *Phyton* **28**, 177 (1971).
[15] V. Vasil and A. C. Hildebrandt, *Science* **150**, 889 (1965).

method has been useful to follow enzymatic plant cell wall removal, and protoplast fusions in studies to induce somatic hybrids in plants[16] (Fig. 9). Further details on single cell clones and their applications to clarifying normal and diseased cell growth and differentiation have recently been elaborated.[17,18]

[16] R. U. Schenk and A. C. Hildebrandt, *Colloq. Int. Cent. Nat. Rech. Sci.* **193**, 319 (1971).
[17] A. C. Hildebrandt, *Symp. Int. Soc. Cell. Biol.* **9**, 147 (1970).
[18] A. C. Hildebrand, *Colloq. Int. Cent. Nat. Rech. Sci.* **193**, 71 (1971).

CHAPTER 3

Dilution Plating and Nutritional Considerations

A. Animal Cells

Richard G. Ham

Dilution plating is a classical microbiological procedure in which an inoculum is diluted sufficiently so that each viable cell forms a well isolated colony on an agar surface. This technique was adapted to allow the formation of colonies by animal cells attached to solid surfaces under liquid media by Puck and Marcus[1] and has since been employed extensively in many areas of investigation.

The essential features of dilution plating (or "clonal growth" as it is often called) are (1) preparation of a suspension of viable single cells; (2) determination of the number of cells in the suspension; (3) dilution to the desired number of cells per Petri dish or culture bottle; and (4) growth of colonies from the individual cells.

This chapter begins with a brief analysis of the special nutritional requirements for growth of isolated cells. Detailed procedures for dilution plating of established cell lines are then presented, followed by modifications necessary for primary cultures, and finally by a discussion of the practical applications of dilution plating. Additional procedures and a more complete discussion of the principles that are involved in clonal growth have been published elsewhere.[2,3]

[1] T. T. Puck and P. I. Marcus, *Proc. Nat. Acad. Sci. U. S.* **41**, 432 (1955).
[2] R. G. Ham and T. T. Puck, *In* "Methods in Enzymology" (S. P. Colowick and N. O. Kaplan, eds.), Vol. 5, p. 90. Academic Press, New York, 1962.
[3] R. G. Ham, *In* "Methods in Cell Physiology" (D. M. Prescott, ed.), Vol. V, p. 37. Academic Press, New York, 1972.

NUTRITIONAL CONSIDERATIONS

The principal feature that distinguishes dilution plating from other forms of cell culture is the use of a small number of cells in a relatively large volume of medium. The cells are therefore unable to "condition" their medium and can survive and multiply only if the medium is both free from toxicity and nutritionally complete as formulated.

All traces of toxic materials must be carefully avoided, particularly if they can be taken up by the cells, since each cell in a clonal inoculum has the opportunity to collect such materials from a large volume of medium. Precise nutrient balance and the avoidance of toxic excesses of required nutrients are also critical considerations for clonal growth.

Another major requirement is that the medium must provide products of cell metabolism which, under clonal conditions, escape from the cells before they can be utilized in subsequent biosynthetic reactions. Similar requirements are not observed in crowded cultures since the metabolic products accumulate in the extracellular fluid and adequate intracellular pools are maintained by equilibration. Common examples of such "population-dependent" requirements include pyruvate, nonessential amino acids, and carbon dioxide.

The current availability of special organic buffers for cell culture[4] makes it particularly important to stress the fact that carbon dioxide is an essential nutrient for animal cells, even though it is normally produced in greater amounts than are needed. If no exogenous supply of carbon dioxide is supplied under clonal conditions, the carbon dioxide that is produced by the cells will be very rapidly diluted below the concentration needed for essential biosynthetic reactions and the cells will be unable to multiply. Thus, if organic buffers are used for clonal growth, they must be used *in addition to* the bicarbonate-carbon dioxide system and not in place of it.

The selection of a medium for clonal growth must ultimately be based on what works best for the cell population in question. Medium F12M[3,5] supplemented with 10% fetal bovine serum is recommended in the procedures described below, both because F12M is designed specifically for clonal growth[5] and because this particular combination of medium and serum has proved effective for dilution plating of many different types of primary and established cultures. However, in many cases, other media such as Eagle's MEM with optional supplements for clonal growth[6] can also be utilized. In cases where clonal growth has previously been described in the literature, it is generally best to start with the medium used in the original experiment before attempting variations. A detailed analysis of vertebrate cell nutrition including the requirements for clonal growth is being published elsewhere.[7]

Although clonal growth of some mammalian cells has been achieved in defined

[4] H. Eagle, *Science* **174**, 500 (1971).
[5] R. G. Ham, *Proc. Nat. Acad. Sci. U. S.* 53, 288 (1965).
[6] H. Eagle, *Science* **130**, 432 (1959).
[7] R. G. Ham, to be published.

or semidefined media, it is generally better to add serum or serum proteins if they will not interfere with the desired experiment. Fetal bovine serum at a concentration of 10% is recommended for general purpose media; however, for many established cell lines other types of serum and/or lower concentrations are equally satisfactory.

Adequate humidification of the culture incubator is essential for good dilution plating. The experiments are generally done in Petri dishes with loose fitting tops and require relatively long incubation periods. Since the range of osmotic pressures that can be tolerated by cultured cells is small, it is very important to prevent evaporation by keeping the humidity inside the incubator close to saturation at all times. Details of incubator design and humidification have been discussed previously in greater detail.[2,3,8]

The nature of the culture surface is also very important. Plastic Petri dishes with special surfaces for tissue culture (e.g., Falcon or Lux) are generally used. Adequately cleaned glass dishes are also acceptable, but their optical properties are generally not as good as those of the plastic dishes.

Procedures for Dilution Plating

The following procedures will result in colony formation with a good plating efficiency for cells of many established lines, including HeLa, mouse L, and Chinese hamster ovary. (These procedures assume the use of a 75-cm^2 plastic culture flask–e.g., Falcon No. 3024, containing 20 ml of culture medium for the monolayer culture. However, with appropriate adjustment of the volume of solutions used they are equally applicable to other culture vessels.)

1. Two or three days before the dilution plating experiment, subculture the parental monolayer culture at a population density such that the cells will be in a state of active multiplication and the culture surface nearly covered at the time of the experiment.

2. One or two hours before the experiment, prepare the culture medium (90% medium F12M,[3,5] 10% fetal bovine serum), place 5 ml in each of several 60 × 15 mm plastic, tissue culture Petri dishes (e.g., Falcon No. 3002), and set the dishes of medium in the tissue culture incubator to equilibrate with the 5% CO_2 atmosphere. (If the incubator has removable shelves, the dishes can be kept on a shelf throughout the inoculation procedures described below.)

3. Examine the monolayer cultures with an inverted microscope, select one with a uniform population of healthy appearing cells, and remove the used culture medium from the flask with a 25-ml pipette.

4. With a 10-ml pipette add 5 ml of 0.05% trypsin in Ca- and Mg-free saline (prewarmed to 37°C) to the culture flask. Keeping the pipette in the flask, gently rock the flask so that the trypsin flows over the cells. Then remove as much of the trypsin as possible with the pipette. Using a fresh pipette repeat the entire rinse a second time.

5. Add 5 ml of the trypsin solution to the culture flask, cap it tightly, and

[8] R. G. Ham and T. T. Puck, *Proc. Soc. Exp. Biol. Med.* **111,** 67 (1962).

incubate at 37.5°C with the trypsin in contact with the cells until they are adequately loosened.

6. While waiting for the cells to be released, place 9 ml of neutralized (pH 7.3) culture medium (or medium prepared without sodium bicarbonate) in a sterile, metal-capped test tube. Also have additional tubes and neutralized medium ready for diluting the cell suspension.

7. As soon as the majority of the cells can be shaken loose from the culture surface (usually 5–15 minutes), draw them up and down with a 5-ml pipette a few times to break up the clumps and obtain a smooth single cell suspension.

8. Quickly add 1.0 ml of the cell suspension to the 9.0 of neutralized medium previously placed in a capped test tube. Mix the diluted cell suspension by drawing it up and down with a 10-ml pipette. (The serum in the neutralized medium stops further tryptic action. This step should be done as soon as possible after the cells are loose. If serum-free or low serum media are being used, it may be necessary to centrifuge the cells out of the trypsin and resuspend them in the medium.)

9. If monolayer subcultures are to be made, add appropriate amounts of the concentrated cell suspension from the original flask to the new flasks containing culture medium at this time.

10. Mix the concentrated cell suspension in the original culture flask thoroughly by drawing it up and down with a 10-ml pipette. Then, using a 0.1-ml pipette held as nearly horizontal as possible, quickly place one drop of the cell suspension from the culture flask in each side of a hemocytometer (American Optical No. 1492). Count all cells in one large square (nine large squares per side). This count multiplied by 10^4 yields the number of cells per milliliter in the suspension. Count at least one square on each side of the hemocytometer. If the number of cells present is low, count enough squares to obtain a total count of at least 200 cells. (Cf. Section VIII for additional details on the use of hemocytometers.)

11. Using neutralized medium and sterile, capped test tubes, dilute the cell suspension from the first dilution tube (prepared in step 8) to a final concentration of 100 cells per milliliter.

12. Add 0.1 ml of the final cell suspension (100 cells) to each Petri dish previously prepared. Stir the suspension regularly to avoid settling of the cells, and use a separate 0.1-ml pipette for each dish. Gently swirl each dish immediately after the cells are added to be certain that they are uniformly distributed throughout the medium.

13. Incubate at a temperature of 37.5°C in an atmosphere of 5% carbon dioxide in air saturated with water vapor. Do not disturb the dishes during the incubation period as this may shake mitotic cells loose from the developing colonies and lead to the formation of smaller satellite colonies as the incubation continues. If it is desirable to monitor growth, prepare an additional dish and keep it on a separate shelf in the incubator so that the clonal cultures will not be disturbed when it is examined.

14. For most cell lines, the colonies will be large enough to subculture, or to fix and stain after 7 to 10 days. For determination of plating efficiency, the

colonies can be fixed with 10% formalin and stained with 0.1% crystal violet in water for approximately 10 minutes.

The above procedures are based on the premise that the cells to be cultured are already in a monolayer, either as an established culture or as an outgrowth from a primary explant. If the source of cells is an actively growing suspension culture, it is possible to start at step 11 although it is often better to adapt the cells to attached growth through one or more monolayer passages before attempting dilution plating.

The preparation of viable cell suspensions from tissue fragments or organ cultures is still very much of an art with few firmly established rules. Generally, however, it involves some sort of enzymatic digestion (trypsin, collagenase, elastase, ficin, Pronase, etc.) often accompanied by mechanical agitation. The following procedure is employed in this laboratory for the preparation of viable cell suspensions from rabbit ear cartilage.[3,9]

1. Wash the ear of a freshly killed rabbit with 70% ethanol and cut out a portion approximately 2.5 × 2.5 cm from near the center of the ear where the cartilage is relatively flat and of uniform thickness.

2. Remove the skin with sterilized dissecting instruments, and place the tissue in a glass Petri dish containing neutralized culture medium (90% F12M, 10% fetal bovine serum.[3,5] The dissecting instruments can be sterilized by dipping in 95% ethanol and burning it off.

3. Carefully dissect away the thin layer of connective tissue and remove any adhering blood vessels, leaving a piece of clean shiny white cartilage.

4. Transfer a portion of the cartilage approximately 1 cm square to a Petri dish containing 5 ml of dissociating medium FBC (0.25% collagenase plus 10% fetal bovine serum in saline—cf. Appendix B in footnote 3). Cut the cartilage into 1–2 mm cubes with a scalpel.

5. Incubate at 37.5°C for 15 minutes, then draw the fragments vigorously up and down a wide-mouth 10-ml pipette.

6. Draw off the dissociating medium with a pipette, allowing the tissue fragments to remain in the Petri dish. Add 5 ml of fresh FBC and continue the incubation at 37.5°C until a sufficient number of cells have been liberated. (This will normally require several hours; the exact time is dependent both on the age of the rabbit and the concentration of the collagenase.)

7. When a sufficient number of loose cells appear in the dissociation mixture, add an equal volume of neutralized serum-containing medium, and draw the tissue fragments up and down in a wide-mouth 10-ml pipette.

8. Transfer the suspension with the fragments to a sterile 15-ml conical centrifuge tube and centrifuge gently to remove the fragments. (In this laboratory a Model CL International Clinical centrifuge with a No. 221 head is run for 10 seconds at speed setting 3; equilibrium speed is not reached in that time.)

9. Transfer the remaining suspension to another sterile 15-ml conical centrifuge tube and centrifuge gently until the cells form a pellet at the bottom of

[9] R. G. Ham and G. L. Sattler, *J. Cell. Physiol.* **72**, 109 (1968).

the tube (3 minutes at speed setting 3, International Clinical Centrifuge, Model CL with No. 221 head).

10. Discard the supernatant solution and resuspend the cell pellet in 1 ml of neutralized growth medium. Count, dilute, inoculate, and incubate as described in steps 11–15 above.

Procedures similar to the above, utilizing collagenase, Pronase, and trypsin digestions have been employed for clonal growth of many types of avian and mammalian cells in primary culture.[3,9,10]

Establishment of Clonal Populations

Dilution plating leads to the formation of colonies from single, isolated cells. For the establishment of clonal populations, it is only necessary to transfer one colony to a separate culture vessel without contamination by other cells. The following procedure has been used extensively.[2,3]

1. Place stainless steel cloning cylinders (about 6 mm in diameter and 12 mm high, with 1 mm walls) in a glass Petri dish and sterilize by autoclaving. In a separate glass Petri dish, autoclave some Dow-Corning silicone stopcock grease.
2. Examine a Petri dish containing living colonies with an inverted microscope. Select an average-sized colony that is well separated from its neighbors. (Extra large colonies should be avoided because of the possibility that they might have arisen from clumps of cells.) Mark the position of the colony on the outside of the Petri dish with a sharpened wax pencil.
3. Remove the medium and rinse the colonies twice with sterile saline, leaving only enough to keep the cells moist.
4. Using forceps sterilized by dipping in alcohol and flaming, touch the end of a cloning cylinder to the sterile silicone grease and then place it over the colony, verifying its location with the inverted microscope.
5. Add 0.05% trypsin to the cylinder with a sterile hypodermic syringe, incubate until loose cells can be seen with the inverted microscope, agitate gently with a 20-gauge needle on a syringe, withdraw the cell suspension into the syringe, and transfer it to a new culture vessel containing growth medium.

For greater assurance that the selected colony has arisen from a single cell it is possible to place the cylinder around a single cell shortly after it has attached and allow the colony to develop within the cylinder.

Replicate Clonal Cultures

Dilution plating is a powerful tool for precise quantitative evaluation of cellular responses to factors influencing their survival and growth. An investigator

[10] R. D. Cahn, H. G. Coon, and M. B. Cahn, *In* "Methods in Developmental Biology" (F. H. Wilt and N. Wessels, eds.), p. 493. Crowell, New York, 1967.

can easily prepare within a short time a hundred or more replicate Petri dish cultures all in identical background media and all from the same concentrated cell suspension. By systematically varying the treatment received by the cells or the composition of the culture medium, he can analyze in detail the responses of the cells over a wide range of the variable being studied. The following are a few of the areas where this approach is frequently utilized.

Cell Survival. For the determination of cell survival the cell inoculum is divided into a series of aliquots and exposed to graded amounts of a deleterious agent (e.g., X rays, ultraviolet radiation, viruses, antibiotics, toxic chemicals, or adverse culture conditions). Each aliquot is then cultured in the normal manner in media identical to that used for an untreated control aliquot. At the end of the culture period, plating efficiencies are determined and percentage survival is plotted on a log scale against the amount of treatment to determine the mean lethal dose and killing kinetics.[11]

Nutrient Requirements. For the demonstration of responses to nutrient factors, both the background medium for the experiment and the neutralized medium used to dilute the cell suspension are prepared *without* the nutrient being studied. That substance is then added to the culture dishes in graded amounts over a wide range of concentrations and all dishes are given identical cell inocula. The extensive cell multiplication involved in colony formation (normally greater than 1000×) virtually eliminates problems due to nutrient carry-over in the cells of the inoculum and a very sensitive response to the composition of the culture medium is obtained. A plot of colony size and plating efficiency versus the logarithm of the nutrient concentration is prepared and used to determine the minimum amount of the nutrient needed for growth, the maximum amount tolerated without toxicity, and an "optimum" concentration that is as far removed from both as possible.[5]

Genetic Analysis. Dilution plating permits each cell in an inoculum to express its genetic potential independently and provides many opportunities for the selection of variant cell populations with desired properties. One example is the selection of auxotrophic mutants by plating a mutagenized cell population in a minimal medium containing 5-bromodeoxyuridine, irradiating with short wavelength visible light, and then transferring the surviving cells (which did not replicate their DNA in the BUdR-containing minimal medium) to a more complete medium. Variant cells that are able to replicate their DNA only in the more complex medium are selected by this procedure.[12]

Cellular Differentiation. The ability to express differentiated properties can be determined individually for each cell in a clonal inoculum through the use of dilution plating. For example, when rabbit ear chondrocytes are cultured as described earlier in this chapter, most of the colonies that are formed produce

[11] T. T. Puck, *Rev. Mod. Phys.* **31**, 433 (1959).
[12] T. T. Puck, *Symp. Int. Soc. Cell Biol.* **9**, 135 (1971).

large amounts of cartilage like matrix.[9] Variation of the composition of the culture medium permits a precise evaluation of environmental controls over the expression of differentiation. Comparison of expression by cells from different sources plated in identical media permits an analysis of their potential for differentiation. The inheritance of tissue-specific determination over many generations in culture can be demonstrated by subculturing cells from differentiated colonies.[9,10]

In summary, dilution plating is a powerful culture technique that is valuable not only for the establishment of clonal populations, but also for the precise evaluation of many types of cellular responses. Its special merits lie in the ability to consider each cell as an individual, in the large number of cell multiplications that are involved in a single clonal passage, in the absence of interfering effects from neighboring cells, and in the fact that the culture environment is not appreciably altered or "conditioned" by the cells. These factors facilitate many types of quantitative measurement that would be much more difficult in the presence of larger numbers of cells.

CHAPTER 3

Dilution Plating and Nutritional Considerations

B. Plant Cells

D. K. Dougall

Dilution plating of plant cells can be used to obtain lines of cells or to screen large populations of cells for desired characteristics. The latter use has been least explored. Its potential is shown in the demonstrations of biochemical variants[1,2] and of nutritional variants[3,4] in plant cell cultures. The method may also be used to obtain clones if the specific single cell origin of a colony is established by microscopic observation of the inoculated plate and marking of isolated single cells. Dilution plating of tobacco protoplasts to yield colonies has been recently described.[5] This demonstration increases the range

[1] Y. M. Heimer and P. Filner, *Biochim. Biophys. Acta* **215**, 152 (1970).
[2] J. M. Widholm, *Biochim. Biophys. Acta* **261**, 52 (1972).
[3] J. L. Gibbs and D. K. Dougall, *Exp. Cell Res.* **40**, 85 (1965).
[4] R. C. Sievert and A. C. Hildebrandt, *Amer. J. Bot.* **52**, 742 (1965).
[5] T. Nagata and I. Takebe, *Planta* **99**, 12 (1971).

of methodology available significantly. Mutant lines have been isolated from mutagen treated haploid tobacco cell cultures by dilution plating.[6]

Preparation of Single Cell Suspensions

Suspension cultures of plant cells contain single cells. The proportion of single cells is likely to rise in the latter portion of the growth curve. The single cells are freed from clumps, etc., by passage through either a stainless steel screen over a beaker or through bolting cloth[7] fixed over the end of a glass tube[8] using a flask or large tube as a receiver. The mesh size required to give an acceptable proportion of single cells should be chosen experimentally.

The cell density of the filtered suspension is determined by counting in a hemocytometer (0.2 mm clearance) or in a Sedgewick-Rafter plankton counting chamber. The density of the cell suspension may be adjusted to the desired value by dilution or by centrifugation (150 *g* for 5 minutes) and reconstitution to the required volume. The medium from the original culture may be replaced by fresh medium after centrifugation and removal of the old medium.

Plating

Aliquots (1.0 ml) of the single cell suspension are pipetted into Petri dishes (90 mm). The pipetting requires care because the cells settle very rapidly. It is preferable to swirl the suspension in a circular manner by hand (magnetic stirrers lead to loss of viability[9]) and to fill the pipette so that it contains the desired volume, then delivering the contents to the Petri dish. The media containing agar is melted by steaming at 100°–105°C then held in a water bath at 40°–45°C. At this temperature the agar does not gel prior to mixing with the cell suspension and is not hot enough to influence the yield of colonies. The required amount of (10 ml/90-mm Petri dish is reasonable) is added to the dish and mixed with the aliquot of the cell suspension. This is best done by tilting the plate so that the medium runs over the surface of the plate until thoroughly mixed and evenly distributed. Keeping the plates warm (35°C) during mixing prevents the agar from becoming too viscous for mixing or from setting prematurely. After mixing, the plates need to stand until the agar has set. During incubation of the plates, provision needs to be made to prevent extensive desiccation. This can be done either by taping the lids all around[9] or by incubation in a closed container.[3] Plates should be incubated with the medium down.

The plates are inspected at intervals for visible colonies. These should appear

[6] P. S. Carlson, *Science* **168**, 487 (1970).

[7] Available from Tolber, Ernst and Traber, Inc., 420 Sawmill River Road, Elmsford, New York 10523.

[8] L. Bergmann, *J. Gen. Physiol.* **43**, 841 (1960).

[9] L. M. Blakely and F. C. Steward, *Amer. J. Bot.* **51**, 780 (1964).

between 10 and 20 days and can be counted. The number of colonies may increase with further incubation.

Plating efficiency is the number of visible colonies obtained relative to the number of cellular units put into a plate. The expression cellular units is used here because filtration does not yield a population containing exclusively single cells; there are groups of two cells and larger. Because plant cells do not migrate, it is probable that cells of a group are the progeny of a single cell. Thus each group may be regarded as a unit in evaluating the efficiency of a particular plating experiment. Plating efficiency may therefore be expressed as

$$\frac{\text{Visible colonies obtained}}{\text{Number of single cells} + \text{number of groups of cells}} \times 100$$

The observed plating efficiencies for various tissues are given in the following tabulation.

Tissue	Plating efficiency (%)
Nicotiana tobaccum var. Hickory Prior[3]	100
Daucus carota L.[9]	6
Phaseolus vulgaris[8]	13
Convolvulus arvensis[10]	35

Media

The media used are generally solidified using 0.7–1.0% agar. Cells can be plated in the absence of agar because there is a slight tendency for them to adhere to glass, but care must be taken during transportation of the plates.[3] The media composition is usually based on that required for cells to grow in liquid medium. In difficult cases media for plating may be supplemented with conditioned media, coconut water, yeast extract, malt extract or casein hydrolysate to improve the plating efficiency. Pieces of tissue growing concurrently on the plate stimulate the growth of single cells.[9] Where optimum plating efficiency is required, all components of the media including pH and agar need to be checked for optimum concentration. Other factors which have been shown to affect plating efficiency are (1) period of previous subculture,[3,10] (2) medium of previous subculture,[3] (3) extent of washing on filtration,[9] (4) population density on inoculation,[3,5,9] (5) temperature of incubation, (6) light intensity,[5,9] and (7) period of subculture prior to establishment of the suspension culture.[3]

In addition to these factors in plating efficiency, one can anticipate the appearance of additional nutritional requirements, particularly in defined media and at low cell densities. Although the observations of Stuart and Street[11,12] were

[10] E. D. Earle and J. G. Torrey, *Plant Physiol.* **40**, 520 (1965).
[11] R. Stuart and H. E. Street, *J. Exp. Bot.* **20**, 556 (1969).
[12] R. Stuart and H. E. Street, *J. Exp. Bot.* **22**, 96 (1971).

not made in dilution plating experiments, they do bear on nutritional requirements at low inoculum densities in defined media. The factors which Stuart and Street[11,12] found significantly different from the requirements for mass culture of *Acer pseudoplatanus* were (1) initial pH of medium, (2) requirement for gibberellic acid, (3) requirement for amino acids, and (4) specific requirement for sucrose in the medium. The appearance of additional requirements for growth at low cell densities is a probable reason for the observed decrease in plating efficiency with decreased inoculum size observed by a number of investigators. Another probable reason is the decreased ability of cells to cross-feed one another at low population densities.

CHAPTER 4

Dilution Plating on Coverslip Fragments

G. M. Martin

In typical applications of dilute plating techniques, one is never certain that a given colony consists entirely of progeny of a single parental cell. In situations where this must be unequivocally established for each colony, there is no substitute for the direct visual isolation of a single cell. The technique which we have found most convenient for this purpose involves the isolation of single cells on coverslip fragments.[1-3] About 1000 clones of human fibroblasts (including many tetraploid clones) have been successfully established in our laboratory with such methods. As with dilute plating studies, cloning efficiency varies dramatically as a function of cell strain, passage number, and type of cloning medium, including the batch of fetal calf serum and its concentration. As a general rule, cloning should be carried out in the first few passages with pretested lots of sera; the latter is best carried out with dilute plating methods in one's own laboratory using reference human diploid fibroblasts and a range of serum concentration.

[1] D. M. Schenck and M. Moskowitz, *Proc. Soc. Exp. Biol. Med.* **99**, 30 (1958).
[2] A. E. Freeman, T. G. Ward, and R. G. Wolford, *Proc. Soc. Exp. Biol. Med.* **116**, 339 (1964).
[3] G. M. Martin and A. Tuan, *Proc. Soc. Exp. Biol. Med.* **123**, 138 (1966).

Coverslip Preparation

Coverslip fragments can be homemade or obtained commercially, although not all manufacturers are willing to provide coverslips small enough to be of much use. We have found number 3 thickness, 3-mm square coverslips to be the most convenient type consistent with commercial production and easy manipulation. At the present time, these are provided by Bellco Co., Vineland, New Jersey. The coverslips are washed in warm 50% nitric acid, exhaustively rinsed in cold tap water and with at least six changes of single distilled water, after which they are blotted dry with Whatman No. 1 filter paper and sterilized in glass Petri dishes at 165°C for 2 hours. In preparation for the cloning experiments, a series of 35-mm plastic tissue culture dishes (Falcon Co., Los Angeles) are preloaded with approximately thirty coverslips and placed in square plastic phage dishes (Falcon Co., Los Angeles, California). Such dishes are convenient for microscopic examination; they also cut down on the possibility of contamination and serve as mini-incubators.

Cell Inoculation

Well-fed subconfluent cultures are gently but rapidly trypsinized[4] and plated in concentrations varying from about 150 to 1500 cells per dish in 2–2.5 ml of pretested cloning medium. Our presently preferred formulation is a modified Dulbecco-Vogt medium[5] fortified with 1 m*M* sodium pyruvate and 0.1 m*M* nonessential amino acids (L-alanine, L-asparagine, L-aspartic acid, L-glutamic acid, L-proline, L-serine, and glycine) in addition to the amounts already present in Dulbecco-Vogt and in fetal calf serum (usually used at concentrations of 20% and inactivated by heating at 56°C for 30 minutes). In most of our experiments, 27 m*M* bicarbonate has been used in an atmosphere of 5% CO_2 in air (pH 7.4) but recent experiments suggest that 40 m*M* HEPES buffer[6] (pH 7.4) in the presence of an initial concentration of 9 m*M* bicarbonate gives acceptable results with reduced CO_2 levels. Penicillin (100 units/ml) and streptomycin (50 μg/ml) are usually added. After an incubation period of several hours to overnight at 37°C to allow for the attachment of cells, the dishes are examined with an inverted microscope (preferably with phase optics) in order to select those containing coverslips with only one or a few cells. One by one, the coverslips are transferred with a sterile watchmaker's forceps to a 35-mm Petri dish containing 2 ml of prewarmed medium buffered with 0.2 *M* Tris–HCl or 0.04 m*M* HEPES (pH 7.4). A systematic search is made for cells along the surfaces and margins of the coverslip and all but one cell is "brushed" off with a sterile No. 27-gauge needle (attached to a syringe) under

[4] G. M. Martin, *Proc. Soc. Exp. Biol. Med.* **116**, 167 (1964).
[5] H. Ginsburg and D. Lagunoff, *J. Cell Biol.* **35**, 685 (1967).
[6] H. Eagle, *Science* **174**, 500 (1971).

direct visualization with the inverted microscope. It is important to avoid coverslips with more than two or three cells, since there is an increased probability of missing cells (especially along edges) with increased cell density. The ideal procedure is to select only those coverslips with a single cell. This can usually be achieved by plating about 250 cells per dish and isolating them 5–6 hours later. Each coverslip is then placed in a new 35-mm dish containing 2–2.5 ml of fresh medium and incubated at 37°C undisturbed for approximately 7 days, at which time medium changes can begin, the frequency depending upon the rate of cell proliferation. The cells will eventually spread from the coverslip to the surrounding plastic surface and the clones can usually be transferred to 4-ounce prescription bottles in 2 to 3 weeks. Temperature maintenance of 37°C is essential and is best achieved by the use of an incubator reserved for cloning purposes only.

CHAPTER 5

Microtest Plates

A. Single Cell Clones

John E. K. Cooper

A quick, efficient method for isolating large numbers of cell clones, each unquestionably derived from a single cell, is potentially a valuable tool for application to genetic studies, for purifying mixed cell populations, and for isolating slow-growing tumor cells from the faster-growing stromal elements. This method[1] utilizes sterile MicroTest II tissue culture plates (microplates: Falcon Plastics, Los Angeles, California).

Each microplate contains 96 wells, each 6 mm in diameter with a straight side and an optically clear, flat bottom, capable of holding 0.4 ml of fluid. A sterile lid is provided to cover the entire microplate. The principal advantage in utilizing these plates as cloning vessels derives from the fact that each well is, in effect, a separate culture vessel. Thus, these plates avoid contamination of a clone by dislodged or migratory cells from adjacent developing colonies which can happen with cloning in Petri plates.[2] Moreover, the individual wells

[1] J. E. K. Cooper, *Tex. Rep. Biol. Med.* **28**, 29 (1970).
[2] T. T. Puck, P. I. Marcus, and S. J. Cieciura, *J. Exp. Med.* **103**, 273 (1956).

can be inspected visually to ensure that only those initially containing one cell are utilized.

Inoculation of Microplates

The cells to be cloned must first be dispersed as a dilute single cell suspension. Tissues or cell cultures are trypsinized, single cell suspensions are prepared, and the cells are resuspended in a small volume of complete growth medium (cf. Sections I and III, this volume). A portion of this suspension is removed for cell counts. The suspension is then diluted appropriately with complete growth medium to yield a final density of 10 cells per milliliter. The diluted cell suspension is dispensed under aseptic conditions into the wells of the microplates with a tuberculin syringe fitted with a 19- or 20-gauge needle. Each well is inoculated with 0.1 ml of cell suspension, yielding, on the average, one cell per well. While many wells will actually contain only one cell, some will contain more than one and others none. The microplate must therefore be inspected visually with an inverted microscope fitted with a low-power objective, 2.5× or 3× being the most suitable. Inspection can be performed either immediately after seeding the microplate or 24 hours later. In the former case, the cells appear as round, brightly refractile objects on the flat bottom of the wells, and usually are readily distinguished from debris. In the latter case, most cells will tend to be flattened and attached to the well bottom and may be difficult to visualize. In that event, trial and error adjustment of the incident light will enable better visual inspection.

Those wells containing only one cell as ascertained by microscopic inspection are identified with a mark on the overlying lid for subsequent feeding and further examination. In general, it is recommended that visual inspection be performed immediately after inoculation. Examination of wells 24 hours after seeding is recommended only for slow-growing cells, because fast-growing cells may already have divided one or more times, giving a falsely high estimate of wells with more than one cell initially. Similarly, care must be exercised in the event that two separate cells may have settled close together in one well; this would be detected only immediately after seeding the microplate.

Incubation of Microplates

Because the microplates are not sealed against exchange of gases, they therefore must be incubated in a moist atmosphere containing 5% CO_2. This presents no problem in laboratories equipped with gas-flow incubators. For laboratories without such equipment, an alternative is to stack the microplates in a large desiccator, fill the bottom of the desiccator with water, and gas the desiccator with 5% CO_2 and 95% air before sealing.

After several days incubation, the medium is changed by inverting the uncovered microplate onto a sterile absorbent pad, and blotting the excess medium from the well edges. Fresh growth medium, 0.2 ml per well, is dispensed only into the marked wells (those initially containing one cell), and the remaining wells are not fed. For relatively fast-growing cells, the medium is first changed 4 or 5 days after inoculation; for slow-growing cells, change the medium 7 to 8 days after seeding. The microplates are then incubated further until examination with the inverted microscope shows that the developing colonies are of suitable size for subculturing into larger, individual culture vessels. Generally, colonies can be successfully subcultured 8 to 10 days after inoculation or, if cell growth is very slow, about 14 days after seeding.

Utilization of microplates offers distinct savings in time, effort, and space for isolating large numbers of clones. Each microplate occupies only the space necessary for a single 60-cm^2 culture flask, yet each plate contains 96 individual vessels. The effort of handling one microplate with the potentially large number of clones is obviously much less than that required for the same number of separate culture vessels. Each microplate can be seeded with a cell suspension in a few minutes, the useful wells containing only one cell each are then marked, and subsequent feeding again can be accomplished in less than 2 minutes.

Subculturing

Subculturing of clones in microplates can be achieved with no danger of cross-contamination from other clones, because trypsinization is performed in the individual wells. The medium is decanted from the microplates as before, and the wells to be subcultured are rinsed with 0.2 ml of Hanks' balanced salt solution without Ca^{2+} or Mg^{2+} (rinsing solution). To each well to be subcultured, add two or three drops of trypsin solution used for routine subculturing of cells (cf. Section III, this volume). Incubate the covered plates for several minutes and inspect with the inverted microscope. When the cells are rounded and have loosened somewhat from the well bottom, 0.2 ml of fresh growth medium is added to the well with a sterile Pasteur pipette. The cells are dislodged by careful, but vigorous pipetting. The suspended cells from each well are generally seeded into separate culture flasks, usually 25-cm^2 size, together with 3 ml of fresh growth medium. After 3 days incubation, one-half volume of the medium in the culture flasks is replaced with fresh growth medium. Within a few days, the cells will have formed many new colonies in the flasks, and may then be further subcultured, subjected to subcloning if desired, or harvested.

Staining and Photographing Clones *in Situ*

Colonies grown from single cells in microplate wells may readily be stained *in situ* with many standard cytological stains, including Giemsa's blood stain, toluidine blue, acetic orcein, and Feulgen, after suitable fixation. Fixation with methanol:acetic acid (3:1) mixture is rapid and satisfactory. Three drops of fixative are added to the medium in each well and the plate is drained on an absorbent pad. Two changes of fresh fixative (0.2 ml per well) are dispensed and drained. After the last change the plates are air-dried and are then ready for staining.

All colonies in a microplate may be treated with the same staining procedure, or, because each colony is effectively in a separate staining dish, different stains may be applied to different colonies in the same microplate. Three drops of the desired staining solution are placed in each well, using a Pasteur pipette. After a suitable interval for staining, the excess stain is removed from the wells through repeated gentle washings, consisting of alternately flooding the wells with wash solution and decanting. The necessary time for staining and subsequent washing depends upon the staining procedure employed. The stained microplates are again allowed to air-dry.

Because of the optically clear, flat bottoms of the wells, the stained colonies may be photographed through a microscope with high dry objectives, yielding good definition at moderately high magnifications. Oil immersion objectives are not suitable, because of the thickness of the well bottoms. The wells are slightly overfilled with immersion oil, and the lid is replaced and taped in place, care being taken to exclude air bubbles. The microplate is inverted over a standard microscope fitted for photomicrography, and the substage condenser is adjusted to yield the brightest image without dark halos at the well edges.

Live, unstained colonies may also be photographed *in situ*, using phase-contrast optics. Under aseptic conditions, the wells are filled with medium, and the sterile lid taped in place. Again, care is taken to avoid air bubbles as much as possible. The microplate is inverted over the microscope stage, and the condenser is adjusted to give the greatest contrast between the cells and the well bottom. After colonies have been photographed, the microplate is turned upright and the excess medium is decanted as before. The clones may be fed with 0.2 ml fresh medium per well or may be subcultured as before. In this way, a record of clonal growth pattern is obtained while the clone itself can be subcultured for further study.

CHAPTER 5

Microtest Plates

B. Replica Plating

James A. Robb

Mammalian cells can be successfully cultured in volumes of 0.010 ml and less. Combined with replica plating techniques these microcultures can be used for screening purposes for various objectives, including virus titers and genetic analyses. An important advantage is that an investigator can with limited budget, space, and personnel perform research using cell culture systems.

Inoculation of Microtest Plates

Cells can be inoculated into Microtest plates (Falcon Plastics, No. 3034, 2 × 3 inches, 60 wells per plate) as either single cells[1] or at a concentration sufficient to form a confluent monolayer.[2] All cell inoculations are made (Fig. 1) with a Hamilton Repeating Dispenser (Hamilton Co., Whittier, California, No. PB600-1) fitted with various sized Hamilton syringes without fixed needles (Luer tip syringes). The repeating dispenser delivers 1/50th of the syringe volume per squirt. The Teflon disk used in the dispenser should be removed and cut so that its thickness is reduced by about 50%. This procedure permits a smoother inoculation of fluid into the wells and helps eliminate splashing. The dispenser with attached syringe can be sterilized by 100°C dry heat for 24 hours or by autoclaving for 20 minutes. Repeated autoclaving of the dispenser is not suggested. A 3-inch Luer tip stainless steel needle is placed on the syringe immediately prior to use. Syringes with fixed needles will break under the sterilizing conditions. The stainless steel needle is sterilized separately.

For single cell culturing or cloning, 0.001 ml of a 1000 cells/ml suspension is placed into each well using a syringe with a 0.050-ml capacity (No. 705-LT). After the desired number of plates have been inoculated, they are immediately observed with an inverted or upright microscope, and the specific wells containing only one cell are recorded. Because the 0.001-ml volume is just sufficient to cover the bottom of the well, the rounded-up, refractile cells are easily counted on the bottom and lower sides of the well. After the plates are scored, 0.010 ml of medium is placed in each well using a repeating dispenser fitted

[1] J. A. Robb, *Science* **170**, 857 (1970).
[2] J. A. Robb and R. G. Martin, *Virology* **41**, 751 (1970).

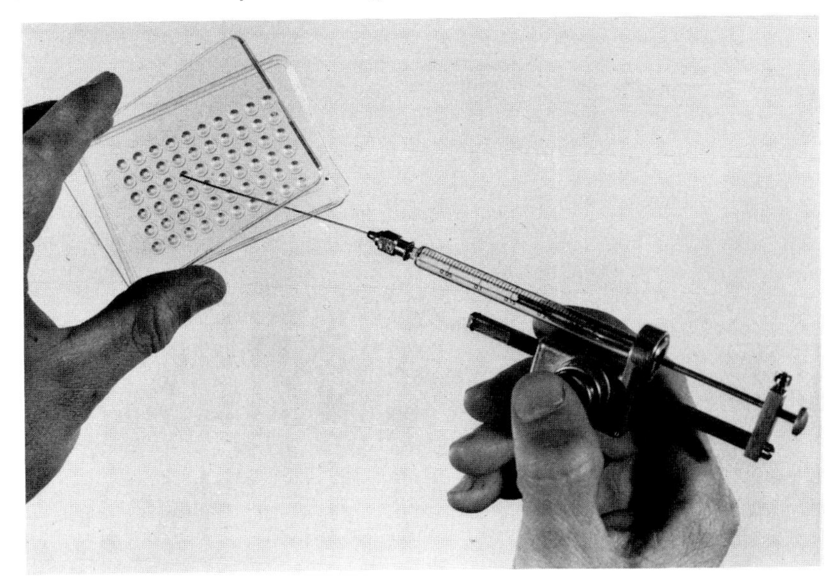

Fig. 1. Inoculation of Microtest plates with a Hamilton repeating dispenser.

with a 0.50-ml capacity syringe (No. 750-LT). In order to prevent evaporation, 0.3–0.4 ml sterile saline should be placed in the corner of each Microtest plate and moved around the inside edge of the plate by tilting. This step can be performed prior to or after cell inoculation. Care must be taken to prevent saline contamination of the wells. After cell inoculation, the Microtest plates are placed into plastic bags along with a saline-moistened piece of absorbent paper. The bags are tightly sealed with wire twists and placed into the incubator. If bicarbonate buffering is used, the bags should be left open in the incubator for a few minutes before sealing.

A single cell will form a 100–500 cell clone in about 2 weeks at 37°C. "Instant" monolayers can be achieved by placing 0.010 ml of a $1\text{–}5 \times 10^5$ cells/ml suspension into each well, the actual cell concentration depending upon the cell type and the percentage confluency desired. Powdered Ham's F-12 Medium (Schwarz Mann BioResearch, Orangeburg, New York) has proved to be an excellent medium for cloning single cells in Microtest plates. The medium is buffered with organic buffers[3] such as Tricine (Sigma Chemical Co., St. Louis, Missouri). Some success has also been achieved using bicarbonate buffering if the plates are processed and incubated rapidly in the appropriate CO_2 atmosphere.

Medium Change, Cell Removal, and Freezing

Medium can be changed in the Microtest wells by the following procedure. A sterile 9-inch Pasteur pipette attached to low vacuum is gently touched to the side of each well, not the bottom. Care must be taken not to aspirate all

[3] H. Eagle, *Science* **174**, 500 (1971).

the medium from the well (leave about 0.002 ml) or the cells in the center of the well will be rapidly killed by drying. Residual medium can be diluted by adding 0.010 ml saline per well. After aspirating the saline, growth or test medium is then added at 0.010 ml per well.

Cells can be removed from the wells by the following procedure. The medium is removed and the cells are rinsed once with saline as above. The cell detachment solution [e.g., 0.05% trypsin–0.02% ethylene glocolbis (β-aminoethyl ether)*N*,*N'*-tetraacetic acid (EGTA, Sigma)] in Puck's Saline A (Grand Island Biological Co., Grand Island, New York) is added at 0.005 ml per well (syringe No. 725-LT). When the cells have detached from the surface of the well (at either 23° or 37°C), 0.010 ml per well medium with serum is added. A 1-ml disposable tuberculin syringe with attached disposable needle is used to suspend the cells, to remove the suspension from the well, and to transfer the suspension into additional growth medium for culture.

Cells are frozen directly in the Microtest plates by either of the following procedures. (1) Medium containing 50% glycerin or 50% dimethyl sulfoxide (DMSO) is added at 0.002 ml per well (syringe No. 710-LT) giving a final concentration of 10%, and the plates are frozen at −70°C. (2) The medium is aspirated and replaced with 0.010 ml medium containing 10% glycerin or 10% DMSO prior to freezing. After thawing at room temperature, the medium is aspirated and fresh growth medium is added at 0.010 ml per well. A second aspiration and addition of medium is sometimes helpful. The plates are incubated until confluent monolayers are established.

Replica Plating of Cells

Replica plating in Microtest plates has been developed for both cells[1] and viruses[2] using the same replicator device (Fig. 2). The replicator consists of an aluminum plate (7.6 × 5.0 × 0.6 cm) in which two sets of holes are drilled. One set consists of sixty vertical holes that match exactly the sixty wells in the Microtest plate. Each of these holes contains a 1.3 cm 2-56-NC stainless steel bolt secured by a 2-56-NC stainless steel hexagonal nut. Each bolt has a 0.4 × 0.4 mm groove at its tip to increase the capillary action of the bolt. The other set of eleven horizontal 3.0 mm diameter holes are placed between and at each end of the rows of bolts. These holes are necessary for rapid cooling after the replicator is flame sterilized (see below). The aluminum handle is secured to the 1.2-cm wide aluminum arc by a stainless steel bolt, and the arc is welded to the plate. When the replicator is placed in a Microtest plate, the tip of each bolt should rest at about three-fourths the depth of the well. The replicator can be ordered from Mr. William C. Stouffer, 10131 Glen Road, Rockville, Maryland 20854.

The majority of cells in the wells to be replica plated must have no attachments to the well surface. Cell lines that attach to the well surface must be freed of their attachments by the cell detachment procedure described above. Some types of cell lines (e.g., lymphoblasts and some insect cell lines)

Fig. 2. The replicator used for replica plating animal cells and virus.

do not need any treatment. Replica plating is performed by the following procedure. Two 1000-ml beakers are placed in ice: 1 beaker filled with approximately 100 ml of glass-distilled water, the other filled with about 100 ml absolute (100%) ethanol. The replicator is placed in the ethanol, tapped once to remove excess ethanol, and flamed. After the flame has disappeared, the replicator is allowed to air-cool for 1 minute. It is then dipped into a Microtest plate filled with about 5 ml sterile saline. The sterile saline further cools the replicator, fills the grooves in the bolts, and dilutes any residual ethanol. One saline filled plate can be used two to three times. The replicator is then placed into the plate to be replica plated (the master plate) and rocked several times to suspend the cells. Following this, the replicator is placed into a recipient plate containing 0.010 ml per well growth or test medium and is rocked several times to ensure adequate release of cells into the recipient wells. Inoculation of the recipient plate can be performed as many times as is desired to ensure that an adequate number of cells are transferred. About 0.001 ml is transferred per well per inoculation. After one master plate has been replica plated, the replicator is dipped into the glass-distilled water to remove cells and medium and is then placed into the ethanol until the next master plate is replicated. The recipient plates are handled as described in the above section (Inoculation of Microtest Plates).

Replica Plating of Viruses

Exactly the same procedure described above for replica plating cells is used for viruses. The techniques used for growing viruses in Microtest plates are described in Section XI (Microculture Procedures: Simian Virus 40). The wells containing virus are replicated when a sufficient amount of virus has been released into the medium. Sometimes the release of virus into the medium is facilitated by freezing and thawing the plates at −20°C (e.g., simian virus 40). The recipient wells contain the host cells in 0.010 ml medium per well. These recipient cells can be placed into the wells immediately prior to the viral replica plating or can be attached as a monolayer culture prior to virus inoculation.

CHAPTER 6

Low Oxygen Tension Technique

Alan Richter

Standard conditions for the optimum growth of clones of mammalian cells thus far have included the perfusion of 5 to 10% carbon dioxide in air in so-called continuous flow, humidified, carbon dioxide incubators. The experiments previously reported[1] made use of the Stulberg chamber[2] which is a box[3] constructed of plexiglass (acrylic). Although Stulberg used this box as a continuous flow system, it was used in the experiments described here and previously

[1] A. Richter, K. K. Sanford, and V. J. Evans, *In Vitro* **6**, 378 (1971) (Abstr.).

[2] C. S. Stulberg, W. D. Peterson, Jr., and L. Berman, *Natl. Cancer Inst. Monogr.* **7**, 17 (1962). In order to reduce mold contamination it is recommended to install a Millipore gas-line filter holder with 0.22 μm filter between the source of the compressed gas and the inlet of the chamber.

[3] Available from Emmitt Scientific, Inc., Farmington, Michigan 48024.

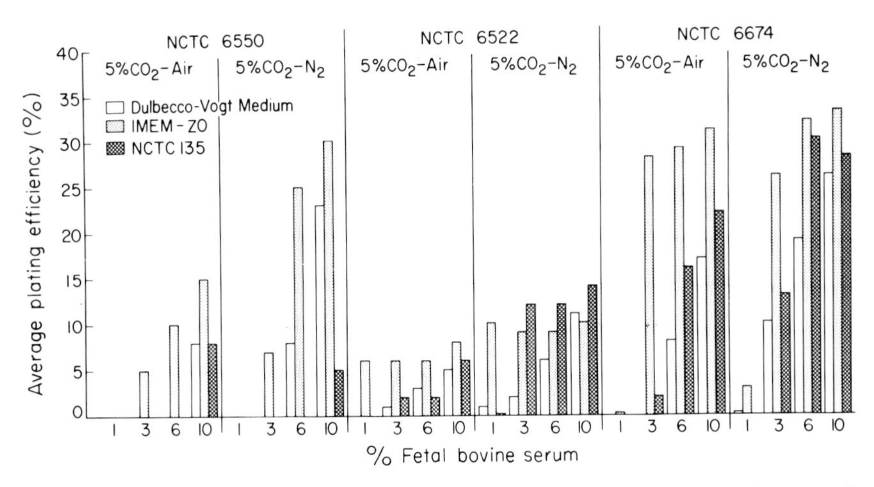

Fig. 1. Influence of culture medium, serum concentration, and gas atmosphere on plating efficiency of three rodent embryo cell lines. Each determination was based on four cultures. Cell lines 6550 and 6522 were derived from 12- and 13-day $C3H_f$-H_e mouse embryos, respectively, and line 6674 from 11-day Albany-N rat embryo.

as a closed system in which the box is flushed for 10 minutes with the desired gas phase and then sealed to the atmosphere for the duration of the experiment, usually a period of 10 days at 37°C. Plating was carried out in Falcon disposable culture dishes with a 50-mm diameter floor surface (Catalog No. 3002). Two hundred cells per dish were dispensed in 5 ml of media. When various gas phases were compared, it was found that all cells tested grew best when the box had been flushed with 10% carbon dioxide: 90% nitrogen compared to 10% carbon dioxide: 90% air. A sample experiment using 5% carbon dioxide is shown in Fig. 1. The plating efficiency of each cell line was better in the nitrogen atmosphere compared to air. In some instances, plating efficiency was nearly equal in the two atmospheres in a specific medium (NCTC 6674 in Richter's Improved MEM Zinc Option[4]) but the effect could be demonstrated in another medium (same cell in Dulbecco Vogt medium). Each medium gave either improvement or no effect when tested in the nitrogen atmosphere, depending on the particular cell line used. Other cells tested included primary and long-term mouse embryo cell cultures, human embryo lung WI-38, and Coon's differentiated rat liver. A sample of the gas phase inside the chamber was taken halfway during the 10-day incubation period and analyzed by mass spectrometry[5] and shown to contain 3.3% oxygen. It is believed that this is an overestimation due to the difficulty of retrieving a sample unexposed to air[6]; the covered dishes contained an air phase when placed in the box.

The superior growth of cells in environments of low oxygen is unexplained,

[4] A. Richter, *In Vitro* **6**, 220 (1970).

[5] Gollob Analytical Service Corporation, Berkeley Heights, New Jersey 07922.

[6] Additional details may be found in A. Richter, K. K. Sanford, and V. J. Evans, *J. Nat. Cancer Inst.* **49**, 1705 (1972).

but nevertheless the facts remain that under the conditions of these experiments oxygen concentrations of 20% as found in air were inhibitory to the growth of clones. The possibility arises that many basic phenomena of tissue culture, including the production of viruses and hormones in cell cultures, may be enhanced by the use of low oxygen tension gas phases.

CHAPTER 7

Soft Agar Techniques

Ian Macpherson

Cells derived directly from trypsinized normal animal tissues or from primary or low passage culture do not multiply in liquid or semisolid suspension culture. This characteristic is retained to some extent by certain cell lines that have acquired a higher degree of autonomy in culture (e.g., BHK21, 3T3). These cells also respond to growth controls *in vitro* including contact inhibition of movement,[1] density-dependent inhibition of growth,[2] topoinhibition,[3] and serum dependence.[4] Cell lines that have been carried *in vitro* through many generations, cancer cells, and cells transformed by oncogenic viruses all multiply to a greater or lesser extent in suspension culture.

The acquisition of the ability to grow in suspension has been utilized in a selective assay for virus-transformed cells.[5] If cells that do not multiply in suspension are infected with a tumor virus and cultured in soft agar medium only transformed cells grow progressively to form colonies. This technique may also be used as a convenient method for cloning. Colonies may be picked from the agar with pipettes and used to initiate conventional cultures.

The selectivity of agar suspension culture is due to the inhibitory effect of acidic and sulfated polysaccharides in most agars on normal cell multiplication.[6] Normal cells also require extension on a solid substrate in order to undergo cell division. Agar medium in which the polyanions have been complexed with DEAE–dextran is capable of supporting at least limited multiplication of some

[1] M. Abercrombie and J. E. M. Heaysman, *Exp. Cell Res.* **6**, 293 (1954).
[2] M. G. P. Stoker and H. Rubin, *Nature* (*London*) **215**, 171 (1967).
[3] R. Dulbecco, *Nature* (*London*) **227**, 802 (1970).
[4] R. R. Bürk, *Nature* (*London*) **212**, 1261 (1966).
[5] I. Macpherson and L. Montagnier, *Virology* **23**, 291 (1964).
[6] L. Montagnier, *Ciba Found. Symp. on Growth Control in Cell Cultures* p. 33 (1971).

cell lines unable to grow in untreated agar.[6,7] Agarose, which is agar with most of the large charged molecules removed also supports the growth of some cells. Conversely an increase in the concentration of polyanions by the addition of dextran sulfate reinforces the inhibitory effect of the agar gel.[6,8]

A basic technique for the preparation of soft agar cultures suitable for the assay of virus transformation in cell lines such as BHK21 is described here. Factors affecting the growth of cells in agar are listed in Table I.[5–7,9–24] Modifications for special purposes can be achieved, e.g., by the substitution of agarose for agar. Additional promotors or inhibitors can be applied within these general procedures. If the technique is to be used for the production of clones, then a number of promoters should be tested in the medium to find the optimum conditions of culture, e.g., the basic method with the addition of DEAE-dextran to a level that promotes the maximum number and size of colonies in medium plus feeder cells would be an appropriate starting point. If this was unsatisfactory then other additions such as insulin, purines, or collagen could be tested. When the medium is to be used for selective culture, the "background" growth of the untransformed or untreated cells can be suppressed to a suitable level by the addition of dextran sulfate. Since mycoplasma infection of cells may promote their growth in soft agar, care should be taken to control for this factor by testing the effect of broad-spectrum antibiotics on growth in agar.

An additional advantage of the soft agar method is its use for producing diffusion gradients in the medium. This can be achieved by placing filter paper disks soaked in the test material in the dishes before adding the agar base.

Growth in agar medium is correlated with malignant potential in animal cells.[25] Variants of BHK21/13 cells resistant to dextran sulfate are highly tumori-

[7] I. Macpherson, unpublished data.

[8] L. Montagnier, *Bull. Cancer* **57**, 13 (1970).

[9] I. Macpherson, *J. Cell Sci.* **1**, 145 (1966).

[10] R. M. McAllister, M. O. Nicolson, A. M. Lewis, I. Macpherson, and R. J. Huebner, *J. Gen. Virol.* **4**, 29 (1969).

[11] P. H. Black, *Virology* **28**, 760 (1966).

[12] R. M. McAllister and I. Macpherson, *J. Gen. Virol.* **4**, 29 (1969).

[13] I. Macpherson, *Science* **148**, 1731 (1965).

[14] H. Rubin, *Exp. Cell Res.* **41**, 138 (1966).

[15] J. Zavada and I. Macpherson, *Nature (London)* **225**, 24 (1970).

[16] M. G. P. Stoker, *Nature (London)* **218**, 234 (1968).

[17] I. Macpherson and W. Russell, *Nature (London)* **210**, 1343 (1966).

[18] I. Macpherson and A. Bryden, *Exp. Cell Res.* **69**, 240 (1971).

[19] I. Macpherson, this volume, Section V, Chapter 1.B.

[20] G. D. Clarke, M. G. P. Stoker, A. Ludlow, and M. Thornton, *Nature (London)* **227**, 798 (1970).

[21] G. D. Clarke and M. G. P. Stoker, *Ciba Found. Symp. on Growth Control in Cell Cultures.* p. 17 (1971).

[22] H. Otsuka, *J. Cell Sci.* **10**, 137 (1972).

[23] E. Tjötta, M. Flikke, and O. Lahelle, *Arch. Ges. Virusforsch.* **23**, 288 (1968).

[24] F. K. Sanders and J. D. Smith, *Nature (London)* **227**, 513 (1970).

[25] T. Kakunaga and J. Kamahora, *Biken J.* **11**, 313 (1968).

TABLE I
Factors Affecting Growth in Soft Agar Medium

Factors	References
Promotors	
1. Transformation by oncogenic viruses	
(a) Polyoma	5
(b) SV40	11
(c) Adenoviruses	10, 12
(d) Rous sarcoma virus	13, 14
(e) Murine sarcoma virus	15
2. Abortive transformation by polyoma virus	16
3. Infection with mycoplasma	9, 17
4. Addition of X-irradiated or mitomycin C-treated cells as feeders (5×10^4/culture in top layer)	18, 19
5. DEAE-dextran (mol. wt. 2×10^6) added to complex acidic and sulfated polysaccharide in Difco Bacto-Agar (approx. 5 μg/ml)	6, 7
6. Insulin about 0.1 to 1.0 μg/ml	7, 20
7. Serum	
(a) Increased levels up to 15%	21
(b) Pig	22
8. Conditioned medium	23
9. Collagen	6, 24
10. Purine derivatives	6
Inhibitors	
1. Dextran sulfate (mol. wt. 5×10^5–4×10^7) up to 50 μg/ml	6, 21
2. Heparin, up to 50 μg/ml	21

genic and also have a reduced serum requirement for growth.[26] However, the correlation is not absolute and some lines of cells that grow well in agar containing sulfated polysaccharides are nontumorigenic.[7,12] Agar suspension culture may be used to grow tumor cells from trypsinized tumors and suppress the growth of stromal fibroblasts.

Methods

Preparation of 1.25% Agar Stock. Add 12.5 g of Difco Bacto-agar to about 800 ml of cold water (tissue culture quality). Boil to dissolve the agar and make up to 1 liter with hot water. Dispense 80-ml amounts (±1 or 2 ml) of hot agar in screw-capped bottles (approx. 250-ml capacity). Sterilize by autoclaving at low pressure with the screw caps loosened. Cool to room temperature, tighten caps, and store at room temperature. For use, melt agar in boiling water checking to see that it does not contain any residual pieces of unmelted agar. Cool bottles briefly at room temperature (to prevent cracking) before transferring them to a water bath at 44°C. Ensure that the level of the water

[26] G. D. Clarke, unpublished data.

bath is above the level of the agar in the bottle otherwise it may gel at the meniscus.

Synthetic Medium (2×). Prepare double-strength synthetic medium (i.e., with twice the normal concentration of all components). Sterilize the medium by filtration through Millipore membranes. Dispense in 80-ml amounts in screw-capped bottles. Store at 4°C for periods of up to 1 week.

Serum. Millipore-filtered calf or fetal calf serum. Store frozen and heat at 56°C for 30 minutes before use.

Tryptose Phosphate Broth. Autoclaved Difco Bacto tryptose phosphate broth. Store at room temperature.

Preparation of Complete Agar Medium. Add 20 ml of serum and 20 ml of tryptose phosphate broth to the 2× synthetic medium, warm to 44°C, and then add to the melted agar. Mix by swirling but avoid frothing and trapping air bubbles in the medium.

Preparation of Agar Base Layers. Pipette 7 ml of agar medium into 5-cm Petri dishes and leave to set at room temperature. The dishes may be glass or plastic. The latter need not be of tissue culture grade since the cells do not come into contact with the surface of the dish. It is inadvisable to prepare the base layers more than an hour or so before use or to store them in the refrigerator, since fluid is exuded from the base layer and this prevents the adhesion of the top layer. The base layer permits the concentration of the cells in a thin top layer which facilitates the inspection of the plates by low-power microscopy. The base layer also provides the bulk of the nutrients for growth.

Preparation of the Cell Containing Top Layer. When the agar suspension method is applied to the assay of virus-transformed colonies the cells may be infected in suspension or as monolayers which are subsequently resuspended with trypsin. In either case care must be taken to ensure that the suspension is free of cell aggregates. Some strains of small-plaque polyoma virus cause hamster cells to agglutinate when they are infected in suspension with high concentrations of virus. If this occurs, the virus should be adsorbed to monolayers. Following adsorption the cultures are fed with fluid medium and left for 1 or 2 hours to allow the virus to penetrate the cells. They may then be trypsinized.

One volume of cell suspension in medium at 37°C is mixed with two volumes of 0.5% agar medium (as prepared above) at 44°C. Mixing is achieved by gentle swirling or by gentle pipetting once or twice. Large volumes should not be mixed at one time since it is difficult to maintain the agar medium fluid during pipettings. A 30-ml screw-capped bottle with a total of 15 ml of agar medium and cell suspension can be easily handled. The volume of the top layer is 1.5 ml. The optimum number of cells per culture depends on the type of cells used and the number of transformed colonies expected. If large well-separated colonies are required, fewer cells should be plated, e.g., for BHK21/13 cells and polyoma virus the optimum number of cells plated for transformation assays is 5×10^4–10^5 cells per dish and 10^3 cells per dish when the method is used to isolate transformed colonies.

Cultures are incubated at 37°C in a well humidified atmosphere of 5 to 10% CO_2 in air. If a suitable CO_2 incubator is not available the dishes may be enclosed in a gas-tight box or plastic bag containing a suitable CO_2/air mixture. Cells capable of dividing do so shortly after incubation has commenced and grow progressively to form colonies 0.1–0.2 mm in diameter in 7 to 10 days. If cultures are to be incubated for more than 10 days they should be fed with 1 to 2 ml medium on the top layer. This can be changed by careful pipetting.

Counting Colonies

Colonies are examined and counted unfixed and unstained with the aid of a plate microscope (25–50× magnification). Cultures may be left for several days in the refrigerator before being examined. Occasionally normal cells undergo a few divisions and form small colonies. By examining control cultures it is possible to determine the size of colony that should be considered as "background." The longer cultures are incubated the more obvious this distinction becomes. Transformed cell colonies increase in size more rapidly than the "abortive" normal cell colonies.

A more certain method of classifying colonies is to use an image-shearing eyepiece when counting (manufactured by W. Watson & Sons, Barnet, Herts., England). This splits the image of a colony into green and red components. The eyepiece may be calibrated for a particular colony size, say 0.1 mm. Colonies of this size present two images, just touching. Colonies larger than 0.1 mm in diameter will have overlapping images and colonies smaller than 0.1 mm will have separated images.

In cultures with many colonies a convenient method of counting is to lay a small coverglass (e.g., 5 × 20 mm) on the top layer surface and make a sample count of the colonies below the coverglass. A grid may be ruled on the coverglass with a ball-point pen.

Subculturing Colonies

Colonies may be removed from the agar with finely drawn Pasteur pipettes. The whole colony is transferred to a tube containing 1 ml of medium and pipetted until it breaks into several fragments. It is essential to remove the agar from the colony, otherwise the cells will be prevented from migrating onto the glass and dividing. A true assessment of the type of cells in colony isolates can only be obtained if the initial growth is trypsinized and subcultured.

The whole cell growth from an agar suspension culture can be harvested by pipetting the top layer off and pipetting in medium to free the colonies from the agar. The colonies may then be cultured directly on glass or trypsinized.

SECTION VI

Perfusion and Mass Culture Techniques

Editors' Comments

There are a number of reasons why one may wish to work with perfusion cell and tissue cultures. One of the principal reasons is that fluctuations in the culture environment can be minimized, and this feature can be of fundamental importance in nutritional and metabolic studies as well as in other areas of investigation. More specific reasons can be cited. For example, in the study of pharmacodynamics *in vitro* the time-course of drug exposure to cells can be controlled by adjusting the rate of perfusion following injection of a drug, thus duplicating to some extent the "clearance" mechanisms following drug injection *in vivo*. This cannot be done satisfactorily in any other type of culture system.

An ancillary benefit to be derived from perfusion cultures is that one can usually work with multilayered—and therefore more "tissuelike"—cell systems, as well as with greater cell production. The latter is usually encountered because with nonperfused cultures it is inconvenient to supply fresh media more than once daily during the daylight hours of the work week; whereas, this limitation is not present in perfusion systems which contain automated means for feeding and elimination.

One may wish to perfuse small size cultures mounted on microscope stages, with or without provision for photography, or larger single or replicate flask cultures, or rolling bottle cultures, or cultures containing cells in suspension. Protocols are given in this section for all of these possibilities, plus more specialized situations such as those for cell production and function in glass or artificial capillary systems. In addition the reader is referred to Section VII which describes a method for propagating cells, with and without perfusion, on "microcarriers" in suspension.

This section also includes methods specifically designed for production of large quantities of cells, with or without perfusion of nutrient. The need for more and better means for production of cells (and cell products) is readily apparent. Some of the newest and most promising developments are presented here.

CHAPTER 1

Dual-Rotary Circumfusion System[1]

George G. Rose

The dual-rotary circumfusion system[2,3] is a self-contained tissue culturing device which circulates oxygen-saturated fluid nutrient through multiple units of interconnected culture chambers residing in a rotating wheel (Fig. 1). Each system accommodates twenty-four multipurpose culture chambers usually as two units of twelve chambers each, although four units of six chambers or one unit of twenty-four chambers may also be constructed. Occasionally, units within one system are prepared with switching devices, so that the fluid of one group of chambers may be combined with or separated from another group (Fig. 2). The gas permeability of the interconnecting Teflon[4] and Tygon[5] (pvc) tubings effects an oxygen-saturated circulating fluid nutrient. Conversely, the loss of CO_2 through the tubings necessitates its constant replacement. The rotation of the chamber wheel (3.5 rpm) acts as a temperature stabilizer and assists in the distribution of the circulating nutrient to the culture microenvironments. Additionally, the hydrostatic pressure of the nutrient induced by the pump is transmitted through the tubings to each chamber and gradually changes from 18 to 32 mm Hg with each revolution of the wheel; the chamber in the 6 o'clock position contains fluid at 32 mm Hg and the chamber in the 12 o'clock position 18 mm Hg. The pulse pressure induced by the 3 rollers of the pump may be transmitted to each chamber through a controllable range of 0 to 5 mm Hg.

HARDWARE (NONSTERILE)

The entire circumfusion system is housed in a Plexiglas[6] incubator measuring 14½ × 14½ × 17½ inches (Fig. 3).[7] Two peristaltic pumps are on either side of

[1] This work was supported by U. S. Public Health Service, National Institutes of Health, Grant DE 01547 from the National Institute of Dental Research and Grant CA 10790 from the National Cancer Institute.

[2] G. G. Rose, M. Kumegawa, H. Nikai, M. Bracho, and M. Cattoni, *Microvasc. Res.* **2**, 24 (1970).

[3] G. G. Rose, "Atlas of Vertebrate Cells in Tissue Culture." Academic Press, New York, 1970.

[4] E. I. du Pont de Nemours & Co., Wilmington, Delaware.

[5] Norton Plastics and Synthetics Division, Akron, Ohio.

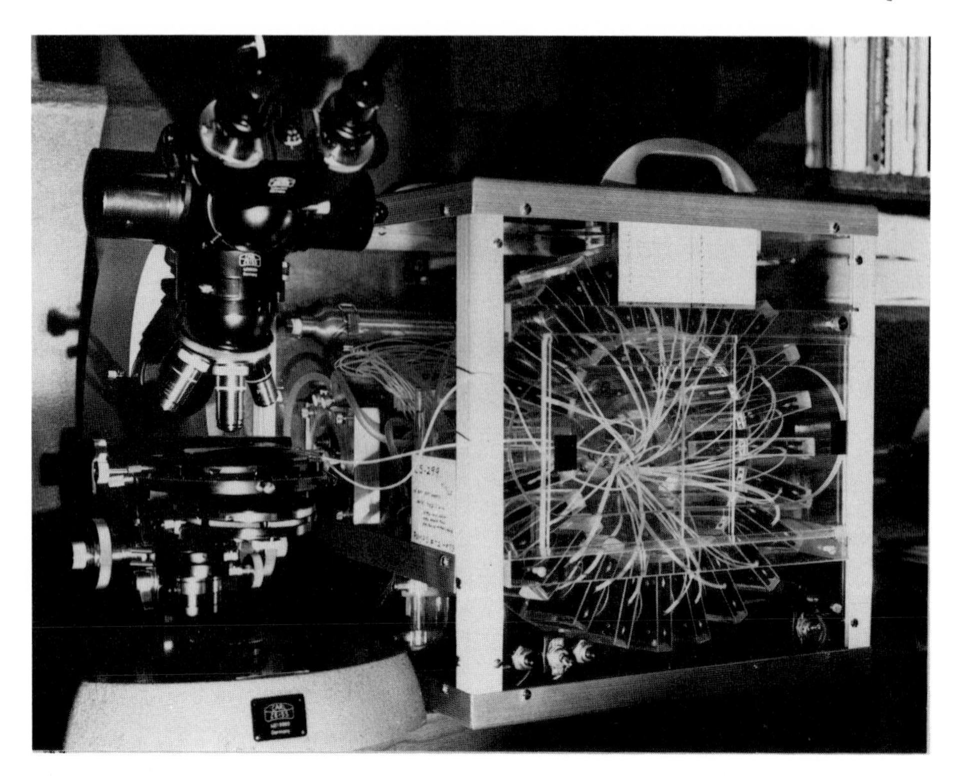

Fig. 1. Dual-rotary circumfusion system with one chamber in position for observation on the microscope stage. The front sliding panels open from either side, so the circumfusion system may be placed on either side of the microscope.

a Plexiglas pump hotbox located in the back of the incubator. A common shaft geared to a fan-cooled electric motor (torque: 65 in lb) rotates the two pumps at 22 rpm. Also, in the hotbox are two additional fans; one blows air heated by the pump into the incubator and the other exhausts hot air through a flap-protected orifice in the back of the incubator. The fans are governed reciprocally by a liquid thermostat placed in the baffled warm air stream coming into the incubator from the pump hotbox and adjusted so that the chamber temperature monitored by a thermistor attached to a cover slip is 37°C. A second "fail-safe" thermostat in the pump motor circuit is set at a slightly higher temperature.

A Plexiglas rotating chamber wheel is located in the front of the incubator and contains twenty-four chamber pockets. The center of the wheel is a 3-inch tube through which passes the Teflon tubing interconnecting the chambers with the reservoir bottles. This tube is linked to the drive motor with a rubber belt, so that after each 17-second revolution, a reversing switch is tripped. A slight tilt to the wheel prevents the chambers from being inadvertently dislodged during rotation, The 4 reservoir bottles (two for each twelve-chamber unit) are placed in receptacles between the chamber wheel and the pump hotbox. A small 150-ml gas

[6] Rohm & Haas Co., Philadelphia, Pennsylvania.
[7] R. J. Matthias & Associates, Houston, Texas.

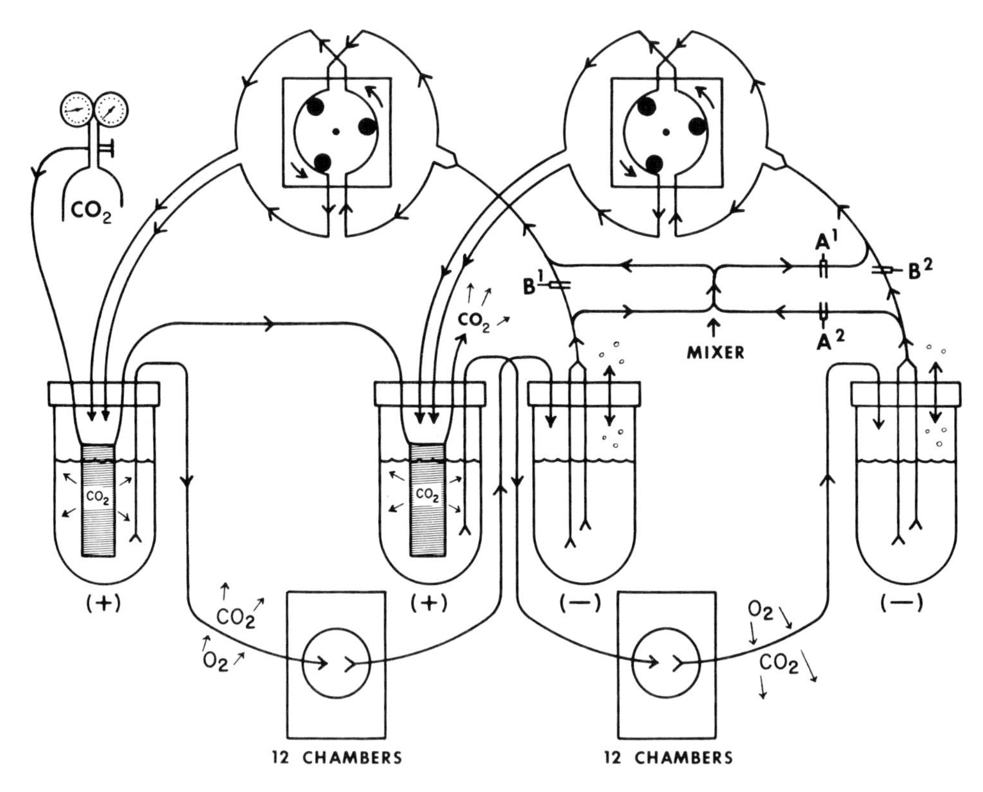

Fig. 2. Diagram of the fluid flow through two twelve-chamber units interconnected with a Tygon mixer. When clamps are placed at A^1 and A^2 but not at B^1 and B^2, the two twelve-chamber units are independent systems. Clamping at B^1 and B^2 but not at A^1 and A^2 effects a mixing of the nutrient from the two negative (−) pressure reservoir bottles through the mixer segment of Tygon. Only one CO_2 cylinder is required, as the two CO_2 coils are easily connected. The double-ended arrows in the negative pressure bottles represent air vent needles inserted through the rubber gasket in the tops of the bottles.

cylinder[8] fitted with a miniature regulator[8] and containing Coleman Grade CO_2 (99.99%) is located in clamps over the hotbox. A swinging panel at the top of the incubator provides access to the CO_2 regulator, and a bubble flowmeter for monitoring the CO_2 flow (one bubble per 5–10 seconds) is located in the front left corner of the incubator. Sliding removable panels on either side provide easy access to the incubator for installing the softwear (culture harness) or for periodic exchanging of the nutrient in the reservoir bottles. Sliding panels at the front of the incubator provide ports of exit for chambers during microscopic observation or cinematographic recording (Figs. 1 and 4). On the front panel, there are switches for the pump and chamber wheel motors (Figs. 1 and 3). Pushbutton controls for the chamber wheel are also provided for specific chamber selection. The presently used Mark II multipurpose culture chambers[9] have holes set with 42-mm spacing (edge-to-edge) for coverslips measuring 40 × 40 mm (Fig. 5).

[8] The Matheson Company, Inc., East Rutherford, New Jersey.
[9] Wahlberg-McCreary, Inc., Houston, Texas.

Fig. 3. Dual-rotary circumfusion system with side panel removed before insertion of the two culture harnesses and twenty-four chambers. The power switch for the chamber wheel is on the left lower corner of the front panel. When turned off, the two pushbuttons may be used to turn the wheel in either direction for specific chamber selection. The pump hotbox is seen in the rear of the incubator with one of its two three-roller peristaltic pumps. On the front surface of the hotbox is the fan for blowing baffled heated air into the incubator. The 150-ml CO_2 gas cylinder and delivery gauge are clamped above the hotbox. The holder for two reservoir bottles is seen between the chamber wheel and hotbox.

This spacing prevents the objectionable bowing of the older style chambers which had holes spaced ½-inch farther apart.

Software (Sterile)

The culture harness is the basic unit of the circumfusion system (Fig. 6). It is constructed by the laboratory technician from the following: (1) twelve rubber chamber gaskets,[10] (2) twenty-four 3-foot lengths of 0.034-inch diameter Teflon spaghetti tubing, (3) two Nalgene[11] polycarbonate centrifuge bottles (250 ml), (4) two special metal caps[7] for the bottles (two pieces each), (5) two round rubber gaskets[10] for insertion in the special bottle caps, (6) two pieces of Teflon tubing ¼-inch × 5 inches, (7) two pieces of Teflon tubing ¼-inch by 2 inches, (8) two lengths of Tygon (R3603) tubing ¼-inch OD (⅛ inch ID) × 36 inches, (9)

[10] American Packing and Gasket Co., Houston, Texas.
[11] The Nalge Co., Inc., Rochester, New York.

Fig. 4. Mark II multipurpose culture chamber being removed for microscopic observation. Note the way the Teflon tubing passes through the center hole in the chamber wheel (see also Fig. 1).

four pieces of amber rubber laboratory tubing ¼-inch ID × ¾-inch, (10) one piece of Teflon tubing ½ × 2¼ inches, and (11) 11 feet of 0.022-inch Teflon spaghetti tubing.

The chamber gaskets[10] for the Mark II multipurpose culture chamber are octagonal (1¾ inches side-to-side × ⅛-inch thick and a center hole 1⅛ inches diameter; Fig. 5), and made of pure gum rubber[12] (floating stock No. 107). They are pretreated by boiling in a biodetergent, reboiling in water, and rinsing overnight in alcohol. After pretreatment, they are reused after cleansing by routine washing procedures. Two lengths of 0.034-inch Teflon tubing are inserted through opposite edges of a gasket by first carefully placing a 14-gauge thin-walled needle with stylus through the edge of the gasket, withdrawing the stylus, inserting the Teflon tubing through the needle, and then withdrawing the needle. This effects a snug, watertight union between the gasket and tubing. The gasket supported on either

[12] Acme-Hamilton Mfg. Corp., Trenton, New Jersey.

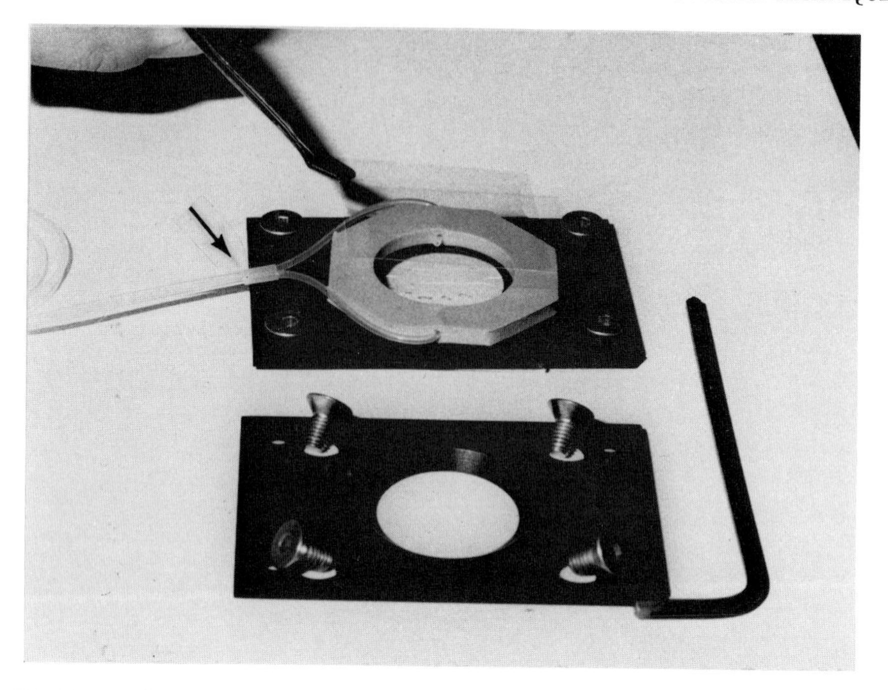

Fig. 5. Partially assembled Mark II multipurpose culture chamber. A short length of 0.076-inch Teflon spaghetti tubing (arrow) unites the two pieces of 0.034-inch tubing as they pass out of the rubber gasket. This consolidates the tubing and effects smoother withdrawal and insertion of individual chambers into the chamber wheel.

side by two pieces of Plexiglas, is placed in a desk vice for insertion of the 14-gauge needle and Teflon tubing installation procedure. The two round rubber gaskets (1⅝ inches × 3⁄16-inch) used in the metal caps are of the same grade of rubber. They are each drilled in their centers with two ⅛-inch holes for a snug insertion of the two pieces of ¼-inch Teflon tubing attached to the two pieces of Tygon. Short lengths of amber tubing are used as retaining sleeves to strengthen the union of the Tygon with the ¼-inch Teflon. The 2-inch lengths of Teflon are inserted into the round gasket in the top of the positive pressure reservoir bottle, and the 5-inch lengths in the gasket of the negative pressure reservoir bottle (Fig. 6). Using the 14-gauge needle, the two free ends of 0.034-inch Teflon attached to the chamber gaskets are inserted through the two round gaskets, one into one bottle gasket and one into the other. An 8-foot coil of 0.022-inch Teflon spaghetti tubing is wound around a 2¼-inch length of ½-inch diameter Teflon tube and after locking the coil by passing the tubing through holes in the ½-inch tube, the two ends are inserted up through the round rubber gasket of the positive pressure reservoir bottle. One end is connected to the CO_2 regulator and the other end to the CO_2 flowmeter. Each gasket is enclosed in a paper syringe bag (Fig. 6) and the entire assembly clothwrapped for gas sterilization. Autoclaving may be used, but special wrapping precautions must be taken to prevent occlusive deformities from occurring in the Tygon tubing. Construction of a harness requires about 4 hours.

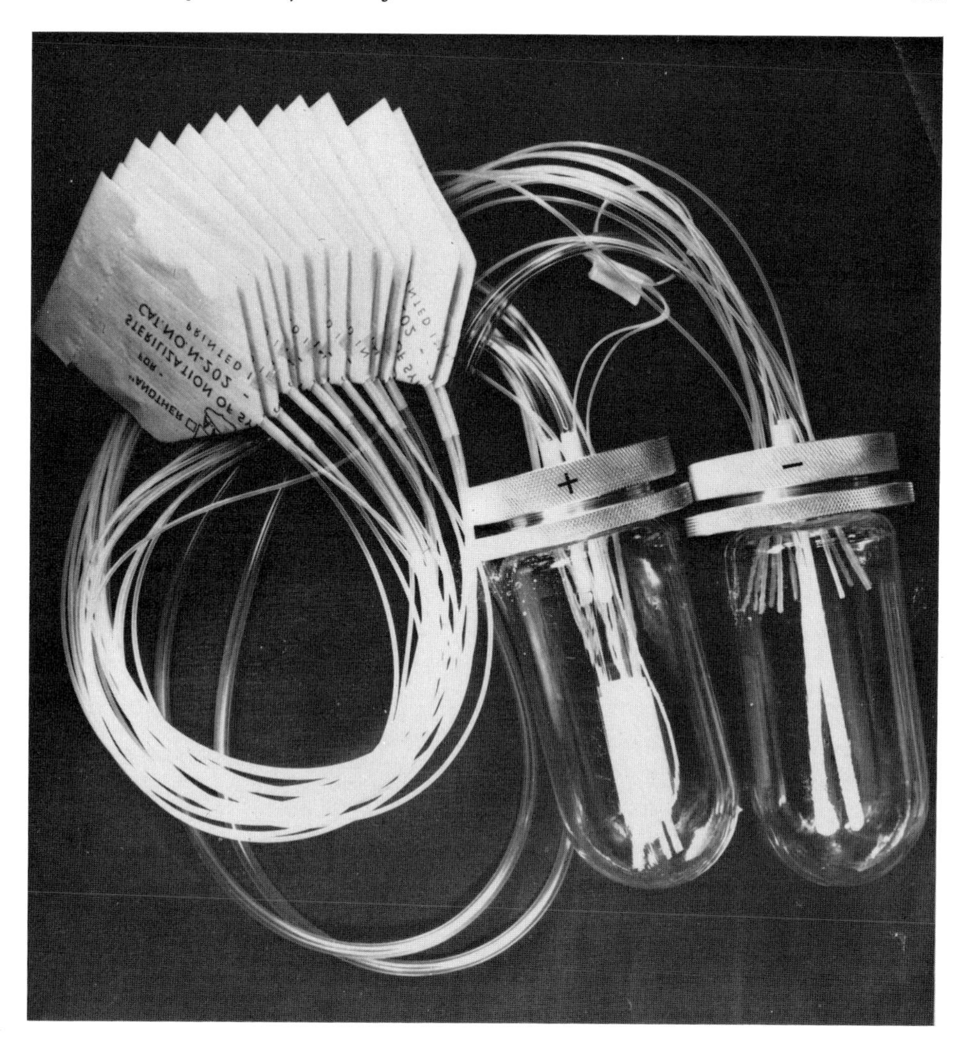

Fig. 6. A twelve-chamber culture harness after sterilization. The gaskets are individually enclosed in syringe paper bags. These are removed one at a time as individual chambers are fabricated (Fig. 5). After all of the chambers have been fabricated, the empty reservoir bottles are replaced with ones containing nutrient. The short insertions of 0.034-inch Teflon in the negative (−) pressure bottle permit a visualization of the dripping nutrient flowing from each chamber. Nutrients pass up the longer ¼-inch tubing in the negative bottle, through the Tygon inserted in the pump, and then into the positive (+) pressure bottle through the short lengths of ¼-inch Teflon tubing. The twelve long lengths of 0.034-inch Teflon which return the fluid to the chambers are encased in the ½-inch Teflon tubing used in constructing the CO_2 coil.

Switching Devices. An alternate construction of the Tygon tubing is to unite four pieces of tubing for two pumps by a mixing unit, as shown in Fig. 2. This may be done by using polypropylene Y- and T-connectors with the extra pieces of Tygon.

Coverslips. Culture coverslips are Corning[13] glass 40 × 40 mm, No. 1 thin. Closing coverslips are made in the laboratory from thin (0.010-inch) sheets of polycarbonate (Lexan).[14] The culture coverslip is covered on its outer surface with a thin sheet (0.0015-inch) of sterile Kodacel.[15] If coverslip cracking should occur during the first 24 hours after chamber construction, the Kodacel will protect the system against contamination. Afterward, windows may be cut in the Kodacel with a scalpel for high-power (100×) phase-contrast observations. Similarly, the polycarbonate coverslips are used to prevent breakage and resultant collapse of the system due to leakage of fluid nutrients and/or contamination.

Cultivation and Maintenance Procedures

Using one of several cellophane techniques,[16,17] the chambers are constructed in routine fashion; the bags covering the gaskets are opened one at a time as each chamber is fabricated. The rest of the culture harness does not require any special protection after sterilization. After the twelve chambers of one system have been assembled, the two empty centrifuge bottles of the culture harness are replaced by two nutrient-containing bottles, 250 ml in the positive pressure bottle, and 150 ml in the negative pressure bottle. The bottles are placed in their respective receptacles in the incubator (Fig. 3) and the chambers passed forward through the center hole in the chamber wheel (Figs. 1, 3, and 4). They are then installed in individual chamber pockets in the front of the wheel. The two Tygon tubings are next installed in the peristaltic pump on one side of the hotbox and the pump motor and chamber wheel are switched on. The nutrient flow is immediate, although occasionally a few bubbles will lodge in some of the tubings. These are removed simply by flicking them with the fingers. One or two chambers may not fill immediately but can be induced to do so by pressing the thumb on the plastic coverslip a few times. A needle is inserted into the top of the negative pressure bottle to act as an air vent. This prevents a slow positive pressure buildup from occurring in this bottle due to an escape of dissolved gases. If the air-vent needle is not used, the positive pressure induces a resistance against the flow of the nutrient and effects a bulging of the chamber coverslips. This not only produces a lens effect but may crack the glass coverslips, as well. The air-vent needle is capped with a portion of its own paper-cellophane container to act as a dust protector.

Nutrients are most easily exchanged by unscrewing the positive pressure bottle and quickly replacing it with one containing fresh nutrient. The hydrostatic and pulse pressure within individual chambers may be monitored with miniature

[13] Corning Glass Works, Corning, New York.

[14] Cadillac Plastic and Chemical Co., Detroit, Michigan.

[15] Eastman Kodak Co., Rochester, New York.

[16] G. G. Rose, C. M. Pomerat, T. O. Shindler, and J. B. Trunnell, *J. Biophys. Biochem. Cytol.* 4, 761 (1958).

[17] G. G. Rose, *Int. Rev. Exp. Pathol.* 5, 111 (1966).

pressure transducers and recording equipment,[18] as detailed in an earlier publication.[2] The key to regulating the pulse pressure is the potential air space in the positive pressure bottles. If this is eliminated by allowing the bottle to fill, the pulse pressure will be transmitted from the positive pressure bottle and will rise to 5 mm Hg. With a 2-inch air space over the fluid nutrient in the positive pressure bottle, the pulse pressure in a chamber is zero. To fill the positive pressure bottle one simply inverts it and the air flows out the 0.034-inch tubings to the chambers and then to the negative pressure bottle. To add air to the positive pressure bottle, one inverts the negative pressure bottle, so that air is pumped from the negative pressure bottle to the positive pressure bottle. The pulse pressure must be zero when a chamber is selected for cinematographic recording as a pulse pressure of any magnitude will effect a rise and fall of the coverslip.

A thermistor is attached to the coverslip of one of the chambers to monitor the chamber temperature on a chart recorder.[19] Samples of fluid may be removed for PH, pO_2 and pCO_2 evaluation.[2]

[18] Statham Instruments, Inc., Oxnard, California.
[19] Rustrak Instrument Co., Inc., Manchester, New Hampshire.

CHAPTER 2

Automated Single Rose Chamber for Cardiac and Other Cells[1]

Frederick H. Kasten

Studies of living cells *in vitro* have been facilitated by the development of the Rose multipurpose chamber.[2] This chamber is essentially an enclosed system for growing and maintaining cells under controlled conditions for microscopic observation and manipulation. Extremely high resolution is attainable with

[1] This research was supported in part by USPHS Research Grants NS-09524 from the National Institute of Neurologic Diseases and Stroke and CA-12067 from the National Cancer Institute. It has also benefited from USPHS training Grant 5-T01-DE-00241-04 from the National Institute of Dental Research.
[2] G. G. Rose, *Tex. Rep. Biol. Med.* 12, 1074 (1954).

phase-contrast optics using the Rose chamber, since cells are separated from the microscope objective only by the thickness of a No. 1 or No. 1½ coverslip (0.13–0.16 mm or 0.16–0.19 mm), and long working condensers are available. The chamber may be employed for routine observations of cells, photomicrography, and cinematography. As will be described here, it may also be used simultaneously for continuous and automatic perfusion.

The chamber consists of two stainless steel or aluminum plates, enveloping—sandwich fashion—a coverslip on each surface, and a single rubber or silicone gasket between them. Both plates and the gasket have central holes which form a well, holding approximately 1.8–2.0 ml of fluid. The design of the chamber is shown in Fig. 1. The steps involved in putting together the chamber are shown in Fig. 2. Explants may be placed on one of the coverslips (plain or collagen-coated, with or without plasma clots) under dialysis cellophane prior to assembly. Alternatively, cell suspensions may be injected into the assembled chamber so as to fall on one glass surface. The central well has an area of approximately 7.5 cm^2, which permits a large viewing surface. It also has the

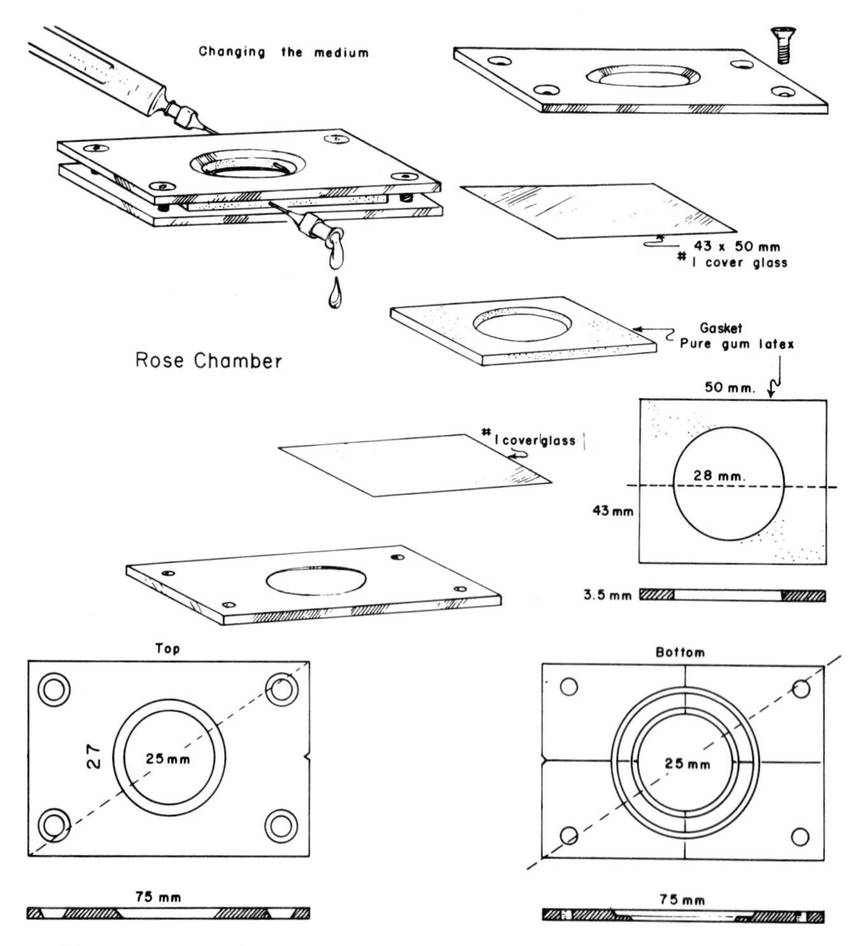

Fig. 1. Drawing of a Rose multipurpose chamber with its component parts.

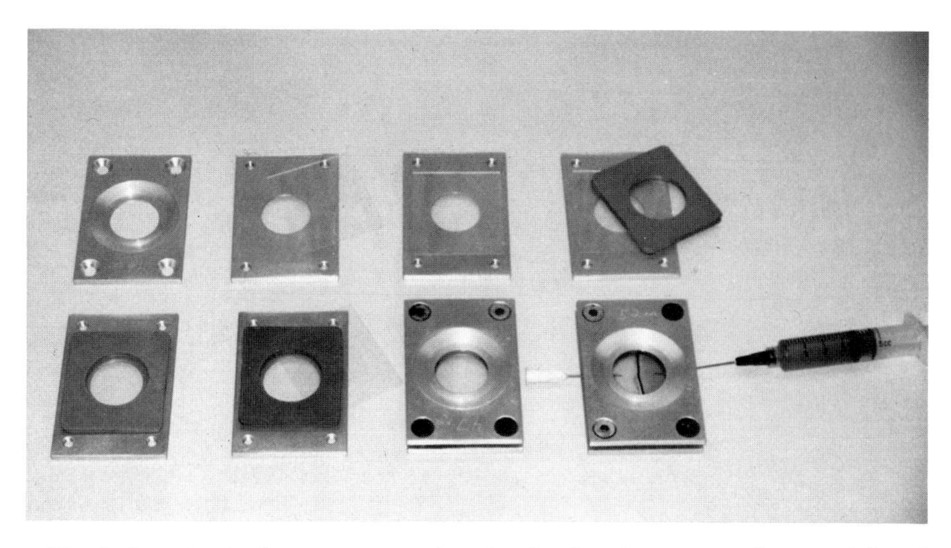

Fig. 2. Steps involved in putting together the chamber, beginning with upper left and proceeding to right, then lower left proceeding to right.

advantage of not requiring as frequent a medium change as in smaller chambers patterned after the Rose chamber. Medium is removed and added with the aid of sterile needles injected on both sides of the gasket after first swabbing the rubber surface with ethanol. We routinely use a No. 25-gauge needle to withdraw or add fluid and a No. 22-gauge needle inserted through the gasket on the opposite side to relieve air pressure and permit fluid to be removed (Fig. 3). The Rose chamber has a length equal to that of a standard microscope slide (7.5 cm) and generally fits onto the mechanical stage of most microscopes without great difficulty. Because of the thickness of the chamber (0.9 cm), it sometimes has a tendency to slip over the edge of the retaining bar on the stage; in this case, it is a simple matter to tape down the chamber. To avoid this minor complication, all the microscopes in our laboratory have had spring-loaded aluminum adapters fabricated and placed on the mechanical stages (Fig. 4). These adapters are specially designed to lock the Rose chamber (or standard microscope slides) on three sides under tension.

Preperfusion Methods Employing the Rose Chamber

For observing the effects of drugs and other agents on cultures in Rose chambers, several experimental approaches are possible. The simplest, although far from ideal method (which I call the frontal approach), is to simply mark the area under observation with ink, photographing or filming the cells as desired; then the chamber is carefully removed from the microscope without disturbing the position on the stage. The drug or agent is injected into the chamber, which is then placed on the stage for observation. A variation of this approach is to inject the drug while the chamber is on the stage and under observation. This is fraught with difficulty because it is hard to keep the cells

Fig. 3. Detailed view of chamber during injection of medium. Note the three explants and the cellophane strip overlying these tissues. Center well is sealed by No. 1 coverslips on both sides.

in focus, especially at high magnification while fluid is being added. In both cases, the fluid exchange time is difficult to control and reproduce. For long-term experiments, where cellular changes are expected a few hours or longer after the drug is added in full concentration, this approach may be adequate. For example, we employed it in studying the sequential effects of excess thymidine and its removal in CMP-human tumor cells.[3] The initial cytological effects took place about 3 hours after the drug was added—a period long enough to permit film recording.

A second general approach involves the gravity flow technique. A vessel containing the drug is placed at a suitable height above the chamber, to which it is connected by tubing and needle. An outflow needle is required as before. By adjusting the height of the vessel and choosing an appropriate diameter of tubing, a convenient flow rate can be obtained. Before perfusion pumps became available, this technique was especially useful for altering the biochemical environment. For example, Pomerat[4] employed a modification of this method in which a reservoir was attached to a hanging drop preparation. The apparatus consisted of a metal frame to which a large coverslip was attached underneath in a depression. Inlet and outlet tubes were affixed in grooves, allowing nutrients to pass through the field under observation. The method was used to study the effect of chlorpromazine upon the contraction rate of oligodendrocytes.[5]

[3] F. H. Kasten and F. F. Strasser, *J. Nat. Cancer Inst. Monogr.* **23**, 353 (1966).
[4] C. M. Pomerat, *15th Congr. Soc. Intern. Chirugie, Lisbon,* p. 236 (1954).
[5] C. M. Pomerat, *Z. Zellforsch.* **45**, 2 (1956).

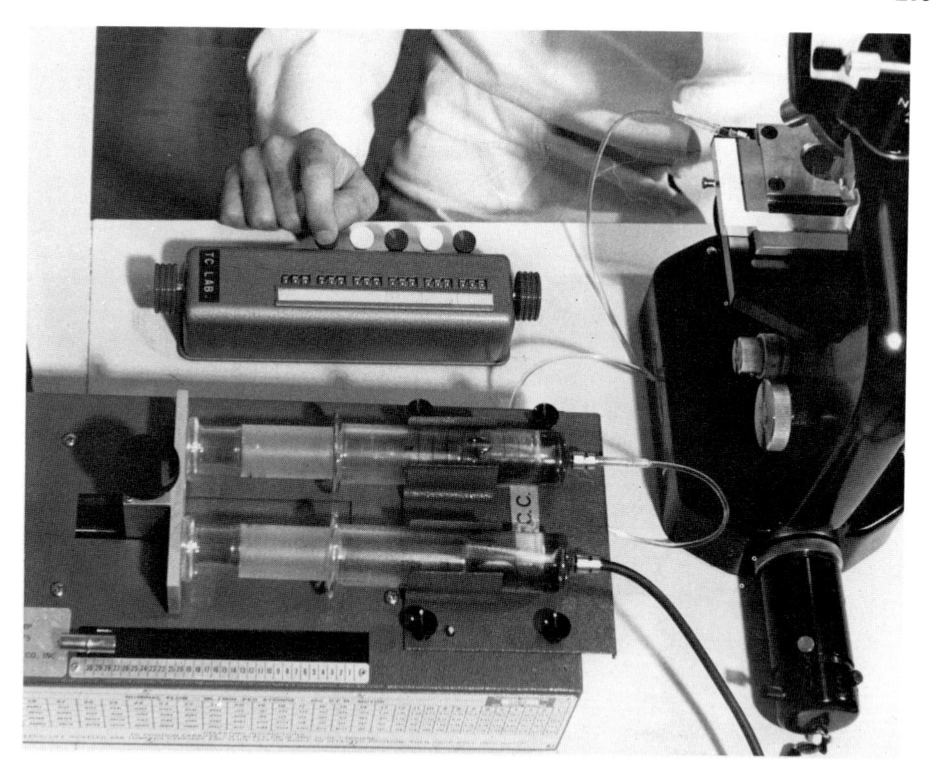

Fig. 4. View of perfusion pump with two syringes, one of which is connected by capillary tubing and needle to a Rose chamber on the microscope stage. A selector control on the pump is set on position 30. Immediately below it is a table listing the 120 possible flow rates attainable with four sizes of syringes. A hand tally counter is used for tabulating beat rates. A spring-loaded aluminum adapter is fastened to the mechanical stage of the microscope and locks the Rose chamber securely.

Automated Perfusion of the Rose Chamber

With the availability of various commercial perfusion (or infusion) pumps with variable speeds, it is relatively easy to employ one in conjunction with the Rose chamber. The setup to be described has been employed in this laboratory for 5 years, particularly for studies on beating myocardial cells, and can be used without modification on other cell and tissue systems as well.

We employ a Harvard variable speed infusion pump, Model No. 975, which may be placed close to the microscope. The pump contains a 100 rpm motor with switch, a selector control for choosing any one of thirty different gear combinations, and a carriage device for the syringe(s). The unit will accommodate one or more 5-, 20-, 50-, or 100-ml syringes of the ground glass or disposable type, with each size having its own spectrum of thirty flow rates, or 120 different rates in all. The precalibrated rates range from 0.000465 ml/minute to 77.4 ml/minute per syringe. For relatively short-term experiments where a drug solution is to be perfused into a chamber at a very slow rate, e.g., up to about 6 hours, we use a 5-ml syringe. However, it is feasible to use a 20-ml syringe for longer experiments, since there is some overlap in flow rates with the two sizes of syringes. Since the Rose chamber holds 2 ml of fluid, a flow rate is chosen which will produce either a rapid exchange of medium, e.g., 0.133 ml/minute with 2 ml replaced in 15 minutes, or a slow exchange, e.g., 0.0011 ml/minute, with 2 ml replaced in 3 hours. The choice is based upon whether one wishes to begin recording observations immediately or after some time.

Another consideration is the sensitivity of cells to rate of perfusion. This problem has not been considered adequately by most workers who use contracting myocardial cells. In our studies, which have been reported briefly,[6] it was found that relatively fast perfusion *with control medium* induces a rapid increase in beat rates and eventually a complete inhibition of contraction. Very slow perfusion (0.00465–0.0135 ml/minute) replaces the old medium in a minimum of 2.5 to 7 hours and does not generally disturb the normal contractile rhythm. It is apparent that the myocardial cell surface is sensitive to fluid flow. The effect reported is apparently not due to hydrostatic pressure, since it is seen as well when extra-large or multiple needles are used on the outflow. Because of this unusual phenomenon, which I call "fluid flow response," all our drug perfusion experiments are carried out routinely at a flow rate of approximately 0.005 ml/minute. In addition, normal medium is always employed as a control in parallel tests done at the same perfusion rate. It is very possible that some results reported in the literature are in error, since they are based on drug injections or rapid addition of drug solutions to beating myocardial cells. These cells are likewise sensitized to fluid movements by a prior medium change, as is demonstrated in an experiment described below.

Another important variable to be controlled is the temperature. Perfusion experiments are done at the microscope in a warm room maintained at 36°–37°C, by using an air curtain with a thermistor probe taped to the surface of the chamber, or by enclosing the microscope in an incubator. I have found it most convenient to use an air curtain. A warm room is not always available and can be uncomfortable during long periods of work. Years ago, it was customary to build incubators around microscopes, especially for cinematographic setups. However, they are not only bulky and awkward but almost impossible to get close to with other equipment, such as a perfusion pump. The air curtain is better, except for the cost when obtained commercially.

Description of a Typical Perfusion Experiment

This experiment was done to study the sensitivity of beating myocardial cells to a change in medium, followed by perfusion. The results are summarized in Fig. 5. Trypsinized suspensions of 3-day-old rat heart cells of ventricular origin were set up in Rose chambers, as described elsewhere in this volume.[7] At 2 days *in vitro*, chambers were examined by phase-contrast optics at 37°C. Two chambers were selected which contained healthy, rhythmically contracting cells. One

[6] A. N. Stroud, F. F. Strasser, K. Finck, and F. H. Kasten, *In Vitro* **4**, 68 (1969).
[7] F. H. Kasten, Section II, Chapter 10.

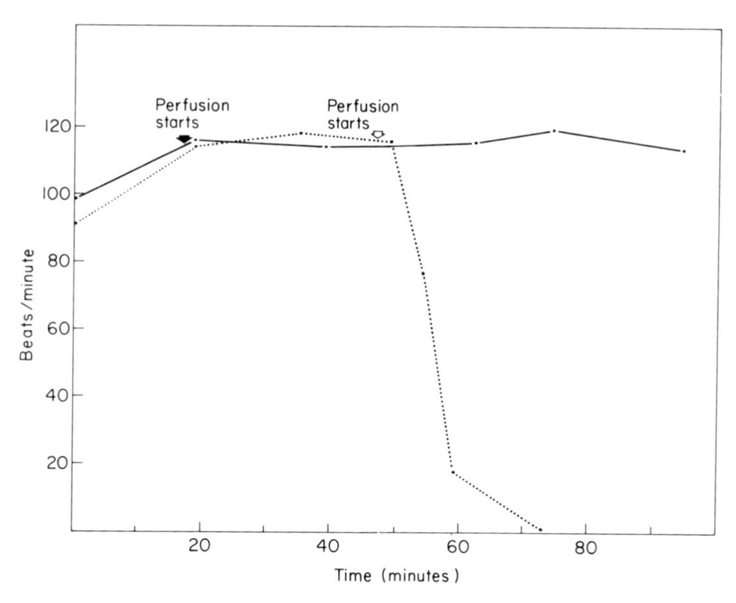

Fig. 5. Results of a perfusion experiment on beating rat heart cells *in vitro*. The results indicate that cells are sensitive to perfusion immediately after medium is changed. Each point represents the average beat rate for three cells. . . . ., Medium changed; ——, medium not changed. See text for further description of experiment.

chamber was given a change of medium and then placed on the microscope stage at 37°C. Fifteen minutes later, when temperature equilibrium was reached, the beating rates were determined for three different cells in a field. Counts were made with a hand tally counter at 15- to 20-minute intervals thereafter until a total of 45 minutes elapsed. At this time, contractions were relatively constant and averaged about 116 beats/minute. Perfusion was started with fresh medium at 37°C in a 20-ml syringe. The inflow needle in this experiment was No. 20 gauge and the outflow needle No. 22 gauge. Medium had been forced to the tip of the inlet needle prior to being inserted into the chamber. Sterile capillary tubing was used between the syringe and the needle. Perfusion was done at the rate of 0.077 ml/minute and beat rates were monitored at regular intervals, beginning 2 minutes after perfusion was started. As can be seen from Fig. 5, beating rates began falling almost immediately after perfusion started. Contractions ceased altogether in the three cells (and the other cells in the chamber), beginning 7 minutes after the start of perfusion. In a control perfusion experiment, successive counts were made on three cells in the other chamber which had not had its medium changed. The beat rates were approximately at the same level as in the other chamber at the time perfusion was started, i.e., 115 beats/minute. Contractions remained at this level for the entire period of perfusion, 70 minutes. According to the experiment described above and other data,[8] myocardial cells are very sensitive to changes in their physical and chemical environment. The results of experiments using perfused drugs and other agents on cultured myocardial cells need to be carefully interpreted.

[8] F. H. Kasten, unpublished studies.

CHAPTER 3

Culture Dish Perfusion with Cinemicrography[1]

LaRoy N. Castor

Growth of a cell culture depends on the interaction of individual cells with the extracellular environment and with each other. In conventional culture methods, even if the medium is changed as often as once a day, the environment does not remain constant. Consequently, rates of metabolism and synthesis fluctuate widely, as do parameters of population kinetics such as the rates of cell division and movement. Perfusion of the culture medium maintains it at an approximately constant composition and permits cultures to grow steadily to high population densities. The perfusion and cinemicrographic equipment to be described is suitable for obtaining accurate measurements of rates of cell division and movement, using bright-field microscopy with low power objectives. It is especially useful for determining changes in these parameters associated with contact interactions among cells.[2,3]

Perfusion of the Culture Dish

The basic components of the perfusion apparatus are a 35-mm plastic or glass culture dish, an aeration chamber, and a means for input, output, and stirring of the culture medium (Figs. 1 and 2). The upper and lower sections of the aeration chamber (Fig. 3) are made from 3-inch diameter stainless steel bar stock, with holes drilled to support 2.5-mm OD silicone rubber tubing. A top plate is cut from lantern-slide glass. The incoming CO_2-air mixture is humidified by bubbling through water at incubator temperature. The rate of flow of gas is maintained at about 50 ml/minute, measured by a flowmeter—a rate rapid enough to prevent any inward diffusion of atmospheric air between the loose-fitting components of the chamber. For ensuring sterility a 13-mm "hydrophobic" Millipore filter, contained in a Swinney holder in the gas line, is a decided improvement over the traditional piece of cotton, since it does not clog with moisture.

The input reservoir holding the fresh culture medium is kept in a small refrigerator, so that changes in the medium during long-term experiments are

[1] This equipment was developed with the support of USPHS Research Grants CA-07846, CA-10367, CA-06927, and RR-05539 from the National Institutes of Health, and by an appropriation from the Commonwealth of Pennsylvania.

[2] L. N. Castor, *J. Cell. Physiol.* **72**, 161 (1968).

[3] L. N. Castor, *Exp. Cell Res.* **68**, 17 (1971).

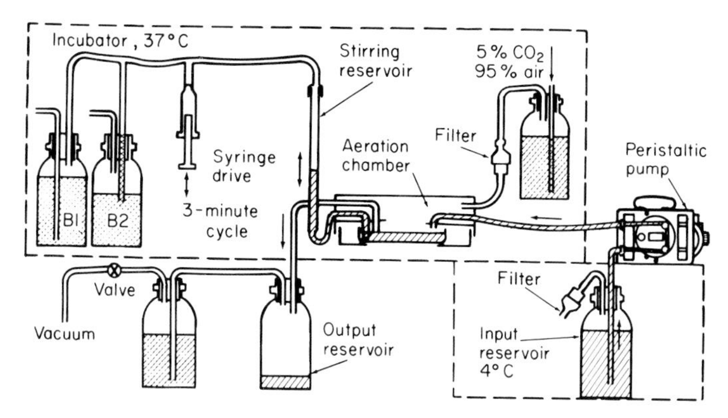

Fig. 1. Diagram of perfusion apparatus. Culture medium contained in a refrigerated input reservoir is fed to the culture dish by a peristaltic pump and removed by vacuum aspiration. Culture medium is indicated by single hatching. Water for bubblers, with 1% $CuSO_4$ to inhibit growth of microorganisms, is indicated by cross-hatching. Gentle stirring of the medium in the dish is provided (if desired) by the syringe with reciprocating drive and the pressure-regulating bubblers, B1 and B2. On the outward excursion of the syringe barrel, medium rises in the stirring reservoir and water rises in the tubing of B2. Water falls in the tubing of B1 until the tip is reached, when bubbling of air prevents any further change in any of the fluid levels. On the inward excursion of the syringe barrel, the medium in the stirring reservoir falls; the lowest level reached is only slightly below the level of medium in the dish, due to bubbling of air in B2.

minimized. We previously used a motor-driven syringe to feed the medium to the culture.[2,4] A newer and more flexible arrangement uses a bottle as the input reservoir, and the flow of medium is controlled by a peristaltic pump.[5] All tubing that contacts the culture medium is of silicone rubber[6] with glass connections, to avoid toxicity problems and to permit oven sterilization. A hydrophobic filter on the input reservoir supplies air to replace the medium as it is withdrawn. This filter can be connected to the CO_2-air mixture, as has been done in Fig. 2, so as to maintain the medium at the correct pH. The pH will rise, however, as the medium travels through the silicone tubing, but the rate of gas exchange in the culture dish is sufficiently rapid to reestablish the correct value.

The smallest pumping tube available for the pump permits a maximum perfusion rate of 33 ml/day, which is sufficiently rapid for almost all experiments with a culture dish of 9-cm^2 area. However, the lowest pumping rate attainable with the speed control on the pump is 8 ml/day—too rapid to avoid necrosis in nonconfluent cultures of many cell lines. The pumping rate is reduced by a cam-switch arrangement that operates the pump intermittently for any desired fraction of a 12-minute cycle. The pump as normally supplied contains a standby

[4] P. F. Kruse, Jr., *In* "Nutrition and Metabolism of Cells in Culture" (G. Rothblat and V. Cristofalo, eds.), p. 11. Academic Press, New York, 1972.

[5] Holter Pump, Model 903, Extracorporeal Medical Specialties, Inc., King of Prussia, Pennsylvania 19406.

[6] Medical Grade, Arthur H. Thomas Co., Philadelphia, Pennsylvania 19105, and Extracorporeal Medical Specialties, Inc., King of Prussia, Pennsylvania 19406.

Fig. 2. Perfusion and cinemicrographic apparatus (incubator cover removed). The aeration chamber at the left has a stirring tube installed. On the far right are flow meters for controlling gas flow, peristaltic pump (one pump accommodates two units), and refrigerator with bottles of culture medium serving as input reservoirs. Bubblers for humidifying the gas and for the stirring systems are at the rear.

battery for continued operation during a power failure, so that for this method of flow control the battery leads must be disconnected.

The output system begins with a length of 2.5-mm OD silicone tubing, supported with its tip about 3 mm above the bottom of the culture dish. When the tip is contacted by the rising medium in the dish, medium is aspirated to the output reservoir. About 0.5 ml is removed at a time, after which the CO_2-air mixture in the dish is pulled through the output system. The rate of fluid or gas flow to the output reservoir is monitored by a bubbler and controlled by a metering valve[7] at about 1 ml/minute. This rate is by no means critical, but it must be only a small fraction of the rate of flow of gas to the aeration chamber, lest some atmospheric air be pulled into the chamber.

[7] Type 1B-2SA, Nupro Company, Cleveland, Ohio 44110. This valve is set at a fixed flow rate and a second valve, Nupro B-2J2, used for turning the system on and off.

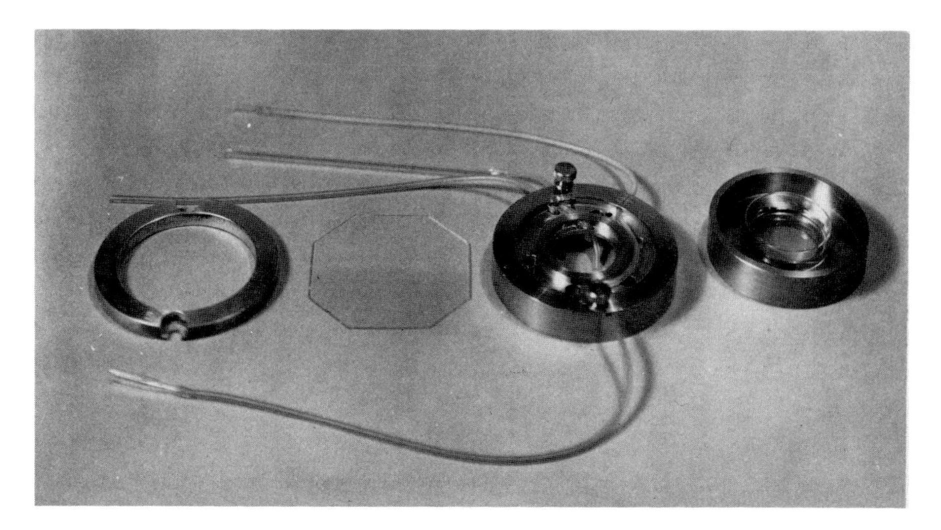

Fig. 3. Aeration chamber, disassembled. From left: clamping ring for top plate; top plate; upper section with silicone rubber tubing; lower section with 35-mm culture dish.

Gentle stirring of the culture medium in the dish is sometimes desirable to minimize concentration gradients that might be set up in the vicinity of the cells, and a means of accomplishing this is included in Fig. 1. About 0.5 ml of medium is withdrawn from the dish and returned in a 3-minute cycle. The bubblers act as pressure regulators to control the upper and lower levels attained by the medium in the stirring tube.

For the initiation of perfusion cultures, cells are dispersed by trypsin and seeded either directly in the culture dishes or in a larger dish containing 25 mm coverslips, one of which is later transferred to each culture dish in an aeration chamber. Cells in heavy suspensions tend to concentrate at the center of a small culture dish, so that the coverslip method provides greater uniformity of the population density.

How rapidly a culture can be perfused, and whether or not the medium can be stirred, depend on the conditions of culture and on the cell line. Evidently "conditioning factors"[8] produced by the cells and required by them for survival are present in the medium close to the cells, and must not be washed away too rapidly. Serum protein is a partial substitute for conditioning factors, since cells in 10% dialyzed calf serum will survive and continue rapid division at higher perfusion rates than will the same cells in lower concentrations of serum. In 10% serum, a perfusion rate of 12 ml/day in a 35-mm culture dish (1.3 ml/cm^2 per day) usually permits good growth of a confluent culture.[3] As the population density increases beyond confluence, a higher perfusion rate may be needed to maintain the constancy of the pH of the medium.[9] For nonconfluent cultures a lower rate, in proportion to the fraction of the dish covered by cells, may be

[8] A. Rein and H. Rubin, *Exp. Cell Res.* **49**, 666 (1968).
[9] P. F. Kruse, Jr., W. Whittle, and E. Miedema, *J. Cell Biol.* **42**, 113 (1969).

required to avoid necrosis and maintain a maximum rate of cell division. One expedient for obtaining good culture growth is to seed the cells on a small piece of coverslip, 0.5 cm^2 or less in area.[2] Perfusion can then be continuous at a slow rate (2 ml/day or less), or at a higher rate for only an hour or two each day. The success of this method probably results from conditioning factors diffusing away from the cells only slowly, although the rate of renewal of the medium is high in relation to the total number of cells.

Cinemicrographic Apparatus

Perfusion units assembled on two microscopes are shown in Fig. 2. The American Optical microscopes are mounted inverted, with their stages modified so as to be right side up. The height of the aeration chambers requires that the working distances of the condensers be 2 cm or more. The regular eyepiece section of each microscope has been replaced by a beam splitter,[10] which focuses images on both a focusing eyepiece with reticle[11] and the film plane of a Bolex 16-mm movie camera. No separate camera lens is required, since the flat field "infinity corrected" objective provides an image which is in focus over the 7 × 10 mm frame. A rotating-disk timing circuit turns on the microscope lamp and actuates the motor[12] that drives the shutter, so as to expose single frames periodically. The lamp is turned on for viewing purposes by a switch which turns off after a preset interval, so that the field of view is not inadvertently exposed to a long period of illumination. A variable transformer adjusts the intensity of illumination, which is measured by an exposure meter placed at the eyepiece. Although the cinemicrographic apparatus discussed here was constructed in our machine shop, components performing the same functions are available commercially. However, some special construction is usually necessary to fit the equipment to the microscope and incubator.[13]

The incubator consists of the base upon which the two microscopes are mounted and a clear Plexiglas cover (not in place in Fig. 2), which is covered by an opaque cloth to exclude most outside light from the culture. Inside the base are a heating element, controlled by a thermoregulator and mercury relay, and a centrifugal blower. Warm air from the blower is directed over the top of each aeration chamber, so as to prevent condensation of moisture from the humid atmosphere inside.

A suitable film is Kodak Plus X Reversal Film, No. 7276. Reversal processing service is available in large cities and by mail. Processing in the laboratory is probably more costly, but it has the advantage of being completed in an hour or two and of permitting the analysis of small lengths of film while an experiment is still in progress. The simple varieties of processing equipment sold for

[10] Component parts are available from Edmund Scientific Co., Barrington, New Jersey 08007.

[11] Unitron Instrument Co., Newton Highlands, Massachusetts 02161.

[12] Slo-Syn, Type SS25, Superior Electric Co., Bristol, Connecticut 06010.

[13] G. G. Rose (ed.), "Cinemicrography in Cell Biology." Academic Press, New York, 1963.

amateur use are too cumbersome for a busy laboratory, so that an automatic processor is highly desirable.[14]

Films are viewed using a projector capable of both single-frame and continuous projection in either forward or reverse direction, and having a digital frame counter.[15] The projector is mounted in an optical system, constructed from a mirror and prisms,[10] which projects the image on a horizontal surface. Thus mitoses can be counted and cell movements traced on sheets of ordinary paper.

[14] Processing by mail and also a satisfactory automatic processor (L & F Portable) are available from Superior Bulk Film Co., Chicago, Illinois 60610.

[15] Model AAP-200, Lafayette Instrument Co., Lafayette, Indiana 47902.

CHAPTER 4

Sykes-Moore Chamber

J. A. Sykes

The availability of phase-contrast optics and of improved methods and media for the cultivation of cells *in vitro* led to the requirement for a container in which cells could be grown and studied, while living, by phase-contrast microscopy. To meet these requirements, the specifications for such a container had to include not only the growing of the cells and the use of high magnifications for their study, but also a simple and reliable means for changing their liquid/gas environment. In addition, the container had to be small, easily assembled, sterilized, and manipulated. The first chamber to approach these ideals was the Rose Chamber[1] which fulfilled most of the requirements, but had the disadvantages of large size and was somewhat difficult to assemble. The Sykes-Moore Chamber[2] was an attempt to overcome the difficulties associated with the Rose Chamber, while preserving its desirable optical and cultural qualities.

Construction

The chamber consists of five parts: Two of metal, two of glass, and one of silicone rubber. The metal parts are readily machined from standard stainless steel

[1] G. Rose, *Tex. Rep. Biol. Med.* **12**, 1074 (1954).

[2] J. A. Sykes and E. Bailey Moore, *Tex. Rep. Biol. Med.* **18**, 288 (1960).

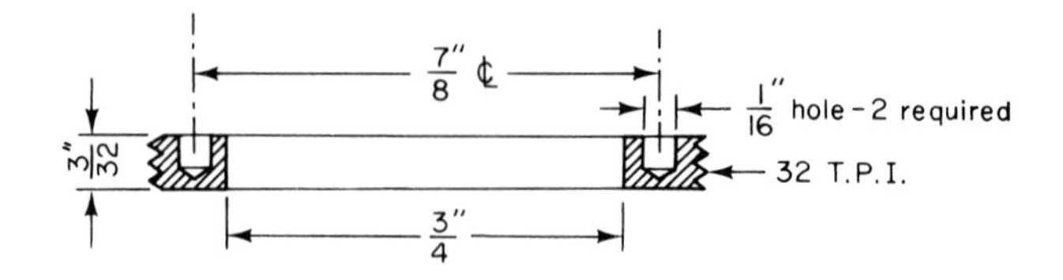

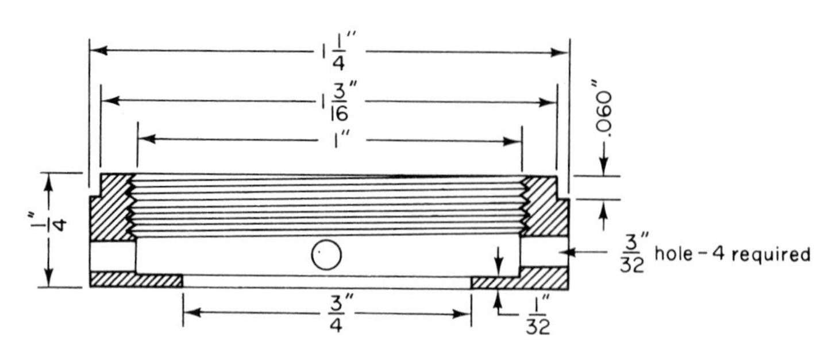

Fig. 1. Mechanical drawing of Sykes-Moore chamber.

or brass stock. When brass is used, it should be heavily chrome-plated to prevent tarnishing, which would lead to binding of the threads following repeated washings and sterilizations. The metal parts of the chamber are designed to be interchangeable to simplify the use and reuse of large numbers of chambers. If these requirements are to be satisfied, the dimensions shown in Fig. 1 must be maintained. No alignment of parts is necessary for introduction of the hypodermic needles used in the filling operation. The relationship of the parts of the

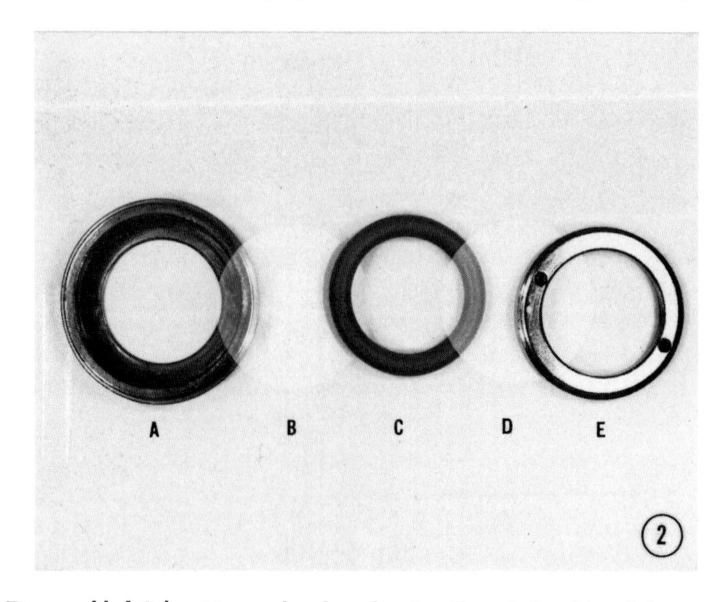

Fig. 2. Disassembled Sykes-Moore chamber showing the relationship of the various parts. The sequence of assembly is: Bottom (A), lower coverglass (B), silicone rubber "O" ring (C), upper coverglass (D), and top (E).

chamber is shown in Fig. 2. The top is the smaller metal ring and has two blind holes for use with a simple annular wrench (Fig. 3B). If necessary, the chamber can be held in the metal block (Fig. 3A), with the locating pin (arrow) fitting into any one of the four ports of the chamber base, thus preventing rotation of the chamber during final tightening of the top. The simple carrier (Fig. 3C) used to hold the chamber for microscopy is 3 × 1½ × 0.125 inch and fits most mechanical stages. The box structure (Fig. 4) is constructed from perforated aluminum sheeting and fitted with aluminum ledges to hold Plexiglas shelves.

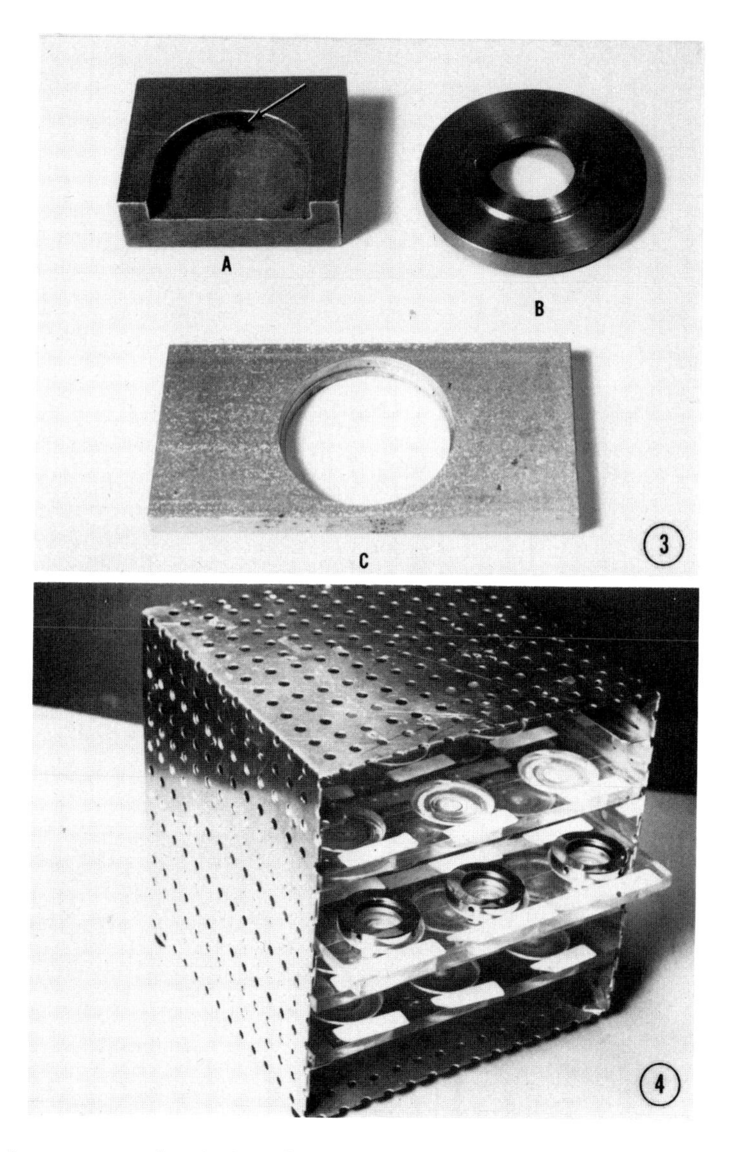

Fig. 3. Accessories used with the Sykes-Moore chamber. (A) The metal block for holding the chamber during final tightening—the arrow points to the locating pin; (B) the annular wrench for tightening the top; and (C) the carrier used for microscopy.

Fig. 4. A simple box structure for holding and incubating forty-five chambers.

It is used for incubating large numbers of chambers. Each shelf has nine (9) inset holes (0.125 inch deep) to prevent the chambers from sliding when moving the box to and from the incubator.

Assembly and Sterilization

The parts of the chamber—bottom, bottom coverglass, silicone rubber "O" ring, top coverglass, and top ring—are shown in the order in which they are assembled (from left to right) in Fig. 2. A silicone rubber "O" ring, which has been found nontoxic, is a red silicone rubber (American Packing and Gasket Co., Style DC-240, Size 6277-15), 0.125 inch thick and 1.0 inch OD (25 mm). The coverglasses are of Pyrex glass No. 2 thickness and 1.0 inch (25 mm) in diameter.

Because of the materials from which the chamber is constructed, sterilization is by autoclaving, although it may also be accomplished by flaming the individual parts after dipping them in alcohol. Autoclaving is the method of choice; disassembled single chambers contained in 60-mm Petri dishes can be sterilized using the fast exhaust and dry cycle with an exposure time of 20 minutes to steam at a pressure of 15 to 20 psi. If the top is tightened, the coverglasses are liable to rupture during autoclaving. The chamber is assembled for autoclaving by placing the bottom, threaded side uppermost, on a flat surface, dropping into place an alcohol-cleaned coverglass, followed by an "O" ring and another coverglass. The top is then placed in position and screwed in about one-half turn. Five loosely assembled chambers may conveniently be sterilized in a 100-mm diameter Petri dish. Slow exhaust must be used when autoclaving assembled chambers for a recommended 30 minutes at a steam pressure of 15–20 psi. Chambers autoclaved in this fashion will maintain their sterility for several months under normal laboratory conditions. It is not advisable to use ethylene oxide sterilization, as many sensitive cells will not grow in chambers that have been sterilized by this chemical.

Procedure

Disassembled Single Chambers. When using plasma clot, Earle's cellophane, or dialysis membrane to hold tissue fragments against the glass surface, a disassembled single chamber is used. The following procedure has been found to be satisfactory. Circles of the cellophane or dialysis membrane, 25 mm in diameter, are soaked overnight in glass-distilled water; the next day they are transferred to fresh double-distilled water in a suitable screw-capped jar and sterilized by boiling for 30 minutes or by autoclaving for 15 minutes at a steam pressure of 15 psi, using the slow exhaust cycle. The sterilized circles then may conveniently be kept at 4°C in the jar in which they were sterilized.

Before using the freshly sterilized or stored sterile circles of cellophane/dialysis membrane, they should be equilibrated for 5 to 10 minutes in the culture medium to be used. The assembly of the chamber is carried out as follows: Place the first coverglass in the bottom part of the chamber; place the pieces of tissue on the glass; cover with the medium-equilibrated membrane; place the "O" ring in position, then the top coverglass and the top ring. Screw down the top ring until the "O" ring can be seen to be compressed; insert a sterile No. 22 hypodermic needle through the "O" ring at one of the four ports. Complete tightening of the top until there is a good seal between the coverglasses and the "O" ring. Using a 2-ml syringe fitted with a No. 22 needle and filled with medium, displace the air through the first needle, taking care to orient the bevels of the needle as shown (Fig. 5); withdraw the syringe, then the vent needle, and incubate the chamber with the bottom (cellophane side) up.

Assembled Chambers. Assembled chambers are easily prepared to receive cells in suspension. First, tighten the top ring until the "O" ring is seen or felt to be slightly compressed; insert a No. 22 needle through the "O" ring via one of the ports by gentle pressure and repeated semi-rotation. This needle will serve as an air vent. The chamber top is then tightened and a 2-ml syringe, fitted with a No. 22 needle and containing the cell suspension, is inserted through one of the other ports and the chamber filled with the cell suspension. The number of cells per milliliter will depend on the cell line and the purpose of the experiment. With fast-growing cell lines, 1.0×10^6–5.0×10^6 cells per milliliter will usually give a monolayer of cells in 48 hours. After filling the chamber, it is incubated bottom down for the first 24 hours to permit cell attachment; thereafter, it is incubated bottom up, which permits cell debris to fall away from the viewing area.

Medium Changes. Changes of medium in chambers with or without cello-

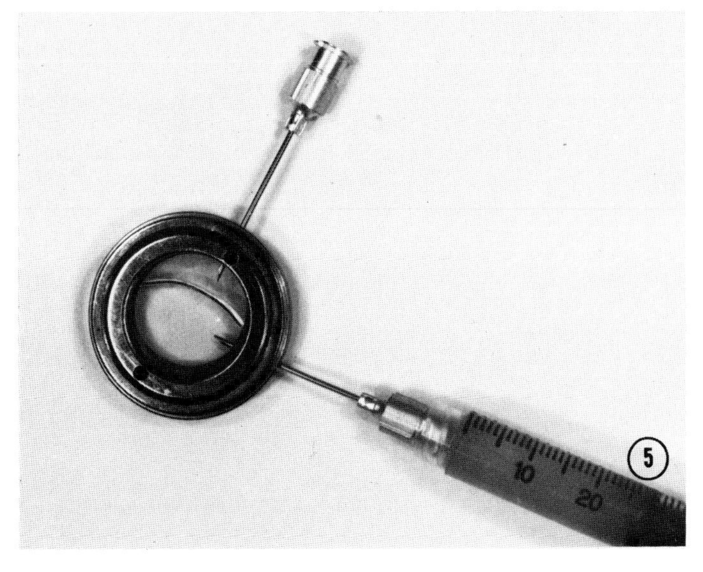

Fig. 5. A Sykes-Moore chamber partially filled with cell suspension, showing the position of the bevels of the vent and filling needles. The chamber is in approximately the same position as when it is filled.

phane are made in the same way. A sterile No. 22 needle for venting is pushed through one port, and a suitable size syringe (2–10 ml), containing medium and fitted with a No. 22 needle, is used to introduce medium through one of the other ports. As the approximate volume of an assembled chamber is 0.7 ml, a slow, pulsing, flushing of the chamber with 2 ml of medium will effect a complete change (based on experimental studies using dye solutions). Using this technique for medium changes permits the maintenance and study of the cultures over long periods of time without fear of contaminations. It is advisable to put a drop of 75% ethyl alcohol into each of the two ports to be used for the medium change as an insurance against bacterial contamination, particularly if the "O" ring has been punctured many times. Whenever possible, a fresh needle track is made at each medium change to ensure adequate sealing of the track on withdrawal of the needle. Very long-term cultures are possible by changing the "O" ring, top coverglass, and metal parts of the chamber, using strict aseptic techniques.

Staining Procedure. The same technique as used for medium changing can also be used for washing, fixing, and staining the cells *in situ.* An alternative procedure for fixing and staining culture is to insert a vent needle, remove the medium by suction, allowing the chamber to fill with air; remove the syringe but leave the vent needle *in situ* to prevent rupture of the coverglasses as the top ring is unscrewed. The coverglass carrying the cells is then readily handled for fixation, staining, etc.

Precautions with Infectious Agents. When dealing with hazardous viruses, the fluid contained in the chamber can be displaced, via a length of tubing, into a suitable vessel containing a disinfectant. Alternatively, the vent needle can be fitted with a Swinney-type air filter and the contents of the chamber drawn into a syringe for further study, as, for example, titration of a virus.

Perfusion. The chambers are easily used for perfusion studies by using a slightly modified microscope holder which permits introduction of the two needles needed for changing the liquid within the chamber. Best results are obtained by sealing the needles in place with a drop of RTV silicone rubber in the ports traversed by the needles. Suitable lengths of sterile tubing are led from the needles to the medium source and to a container to receive the spent medium. Under perfusion conditions, the contents of the chamber are at atmospheric pressure and thermal expansion of the liquid does not affect the flatness of the coverglass to which the cells are attached.

CHAPTER 5

Culture of the Lens (33°)[1]

Bernard Schwartz

The lens has long been recognized as an ideal experimental material for the study of the metabolism of the permeability of an organ *in vitro*.[2] Because of its avascularity and lack of nerve supply, as well as the fact that it is surrounded by a capsule and therefore can be easily removed and manipulated, the lens has been used increasingly by many investigators. Particular interest has been centered on the technique of lens culture whereby the lens is studied *in vitro* for considerable periods of time up to several days.[3-6]

The types of systems used to culture the lens have been classified into perfusion or open systems, batch systems, and closed systems.[7] The terminology has been taken from the physicochemical literature[8] and the terms are defined as follows: an open or perfusion system is one that exchanges both energy and matter with the environment; a closed system is one that exchanges energy but not matter with the environment. The batch system is a modification of the closed system whereby the culture fluid is removed entirely at specific time intervals and replaced with a volume of fresh fluid. For all practical purposes it can be considered a closed system with repeated smaller time intervals of culture.

Analysis of the technique of closed or batch systems indicates that throughout the entire period of culture there is a continual change in concentrations of metabolic substances within the medium.[7] A perfusion system, however, by using a constant flow of medium of constant composition past a tissue, circumvents the changing concentrations found in a closed or batch system. Thus the perfusion system imitates more closely the physiological conditions found *in vivo* than the closed or batch system. The problem of concentration changes in closed systems is not only fundamental to the study of lens metabolism but is a general biological problem wherever the study of tissue *in vitro* for prolonged periods of time is carried out.

[1] Supported in part by NIH Grant No. 5-R01-NB-01820.

[2] Kunde in Z. *Wiss. Zool.* VIII (1857) cited by R. Deutschmann, *Graefe's Arch. Ophthalmol.* **23**, 112 (1877).

[3] A. Bakker, *Graefe's Arch. Ophthalmol.* **135**, 581 (1936).

[4] F. C. Merriam and V. E. Kinsey, *Arch. Ophthalmol.* **43**, 979 (1950).

[5] J. E. Harris and L. B. Gehrsitz, *Amer. J. Ophthalmol.* **34**, 131 (Pt. 2) (1951).

[6] B. Schwartz, B. Danes, and P. J. Leinfelder, *Amer. J. Ophthalmol.* **38**, 182 (Pt. 2) (1954).

[7] B. Schwartz, *Arch. Ophthalmol.* **63**, 593 (1960).

[8] I. Prigogine, "Introduction to Thermodynamics of Irreversible Process." Thomas, Springfield, Illinois, 1955.

Since a primary aim of the culture of the lens was to imitate the physiological situation as closely as possible, it is important to point out that thermodynamically the closed or batch system approximates a chemical system approaching equilibrium while the perfusion system is much closer to the physiological condition of steady state. Since in a closed system the availability of metabolic substrate is limited and there is an increasing buildup of metabolic end products, the lens is gradually approaching an equilibrium state. However, in the perfusion system where metabolic substrate is unlimited and there is no buildup of a metabolic end product, the lens can theoretically approach and reach a steady state.

For the reasons mentioned above and also for these fundamental thermodynamic considerations which have been previously reviewed by the author,[7] the development of a perfusion system for the culture of the lens was undertaken. The perfusion system was then primarily used to determine those conditions of culture which would maintain the lens in the physiological steady state. Achievement of these conditions would lead to a greater understanding of the forces which maintain the lens *in vivo*.

Since a great deal of information was readily available regarding the rabbit lens including the composition of the aqueous humor[9] surrounding the lens, it was decided that this species be used to initiate the perfusion system of lens culture.

General Description of the Perfusion System[10]

Figure 1 is a schematic drawing of the lens perfusion system as finally evolved. The essential feature of the system indicated in the center of the drawing is the lens mounted in a small glass chamber on a microscopic stage (Fig. 2). The lens is observed through a dissecting microscope with appropriate photographical attachments. Flowing into the lens chamber from the left is the synthetic perfusion medium, which flows through other chambers containing a reference silver-silver chloride and a platinum redox electrode to an outflow sampling system. Simultaneously, as perfusion fluid is passing through the lens chamber, another stream of perfusion fluid bypasses the lens to flow through a similar reference-redox electrode chamber to another outflow system for collecting a sample of control medium. Thus, the total system is designed so that similar medium is used to nourish the lens and also to bypass the lens so that analyses on both sets of media can determine the contribution of the lens to the medium.

For each perfusion run the medium is mixed to final volume from frozen concentrated samples and is placed in appropriate flasks for gassing. The solution is then drawn through a Pyrex sintered-glass filter, regassed under sterile conditions, and withdrawn directly into the reservoir inflow unit. This is a large multisyringe pump with a synchronous motor with a capacity for holding five 100-ml syringes. All 5 syringes can be emptied simultaneously at flow rates varying from 3 to 200

[9] D. V. N. Reddy, *In* "Biochemistry of the Eye," XXth International Congress of Ophthalmology Symposium at Tutzing Castle (M. U. Dardenne and J. Nordmann, eds.), p. 167. Karger, New York, 1966.

[10] B. Schwartz, *Arch. Ophthalmol.* 63, 607 (1960).

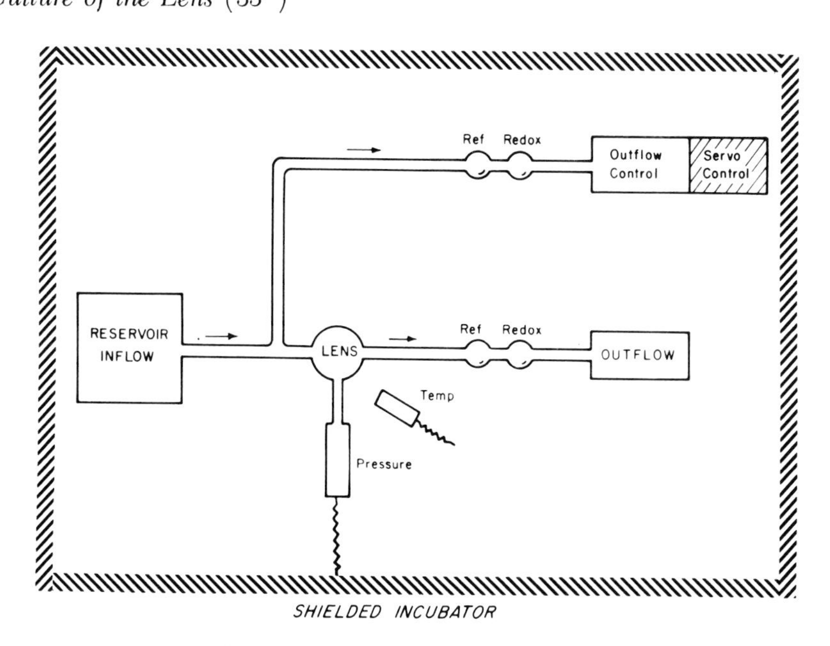

Fig. 1. Diagram of lens perfusion system.

Fig. 2. Rabbit lens in perfusion chamber. (From B. Schwartz.[10])

ml per 24 hours. Since the lens and medium are placed into a totally enclosed rigid system, and since it is necessary to maintain the pressure within the system at normal ocular pressure of about 15 mm Hg, appropriate controls are utilized to adjust the rate of fluid flow through the system so that the pressure is maintained at a constant level. This is done by using withdrawal syringe pumps on both outflow channels and a servo control attached to the outflow control side which is triggered by a pressure transducer attached to the lens chamber. The outflow syringe pump, collecting fluid flowing past the lens, functions at a constant rate. Therefore, appropriate changes for adjusting pressure in the system are produced by the servo control on the other outflow channel. By means of this mechanism, constant small rates of flow of fluid past the lens can be maintained for long periods of time such as 5 days under a steady pressure level.

The materials of the perfusion system consisting of Pyrex glass, including ball and socket joints coated with silicone grease, as well as Teflon sleeves are sterilized by heat. Only the pressure transducer which is of the Statham type and is commonly used in physiological laboratories for venous pressure measurement, and the reference and platinum electrodes are sterilized chemically by benzalkonium chloride. The whole system is placed in a large shielded incubator with a multichannel recorder outside the incubator to record constantly pressure within the lens system, temperature, the redox potentials in both the outflow and the outflow control channels and the function of the servo system.

Temperature of the System

The first perfusion experiments were run with the incubator, adjusted so that the temperature of the microscope stage was at 37.0°C.[11] The assumption that the lens of the rabbit eye was at 37°C *in vivo* was erroneous. Appropriate studies were made to determine the true temperature of the lens within the eye by a series of experiments which demonstrated a steep ocular temperature gradient[12] with the front of the eye being much cooler than the posterior segment. The temperature gradient was found to be affected by the environmental temperature.[13] Therefore, a temperature of 33°C was used for the culture of the lens which was appropriate for an environmental temperature of 20°C.

The Synthetic Medium for Lens Culture[14]

The aqueous humor which nourishes the lens *in vivo* has an extremely low concentration of protein in the order of 40 mg per 100 ml.[15] Thus aqueous humor

[11] B. Schwartz, *Arch. Ophthalmol.* **63**, 643 (1960).
[12] B. Schwartz and M. R. Feller, *Invest. Ophthalmol.* **1**, 513 (1962).
[13] B. Schwartz, *Arch. Ophthalmol.* **74**, 237 (1965).
[14] B. Schwartz, *Arch. Ophthalmol.* **63**, 625 (1960).
[15] F. H. Adler, "Physiology of the Eye—Clinical Application," 4th ed., p. 99. Mosby, St. Louis, Missouri, 1965.

is almost a balanced salt solution except for unknown quantities of hormones, vitamins, and enzymes. Once an adequate analysis of aqueous humor is known, it is fairly simple to construct a synthetic perfusion medium.

Using this principle as a guide, a synthetic medium was developed which contained no proteins or antibiotics and which closely simulated the composition of posterior chamber aqueous. The determination of the final composition was based primarily on the analyses of posterior chamber aqueous by Kinsey and his colleagues.[9,16] The medium closely imitated the posterior chamber aqueous in all constituents including salts, nonelectrolytes, and amino acids as well as pH, oxygen, and carbon dioxide content except for the content of vitamins, especially ascorbic acid.

For vitamins, concentrations were used which were commonly used in other tissue culture media[17] except for inositol which is known to be in high concentration in the aqueous humor of rabbit eyes.[18]

Ascorbic acid is a major constituent of posterior chamber aqueous humor and appears to be secreted by the ciliary body so that it is present in the rabbit aqueous humor at approximately twenty times its level in plasma. Unfortunately ascorbic acid in the synthetic culture medium is extremely susceptible to oxidation and within several hours drops precipitously to extremely low levels and at 24 hours is entirely oxidized.[14] Initially an attempt was made to maintain the level of ascorbic acid of the medium by increasing the reduced glutathione content of the medium. This resulted in severe opacification of the lens in culture[11] although analyses of the medium indicated that reduced glutathione did stabilize the content of ascorbic acid.[14] Furthermore, it was determined that with the oxidation of ascorbic acid in the medium there was a decrease in oxygen content.[19] In order to avoid these difficulties, ascorbic acid was omitted from the medium.

The oxygen concentration chosen for the perfusion medium was based on known oxygen determinations of anterior chamber aqueous, since no known data were available for posterior chamber aqueous.

Also, it was reasoned that the posterior chamber aqueous probably has an oxygen concentration equivalent to plasma since it is separated from the capillaries in the ciliary processes by a layer of two cells. Therefore, a partial pressure of oxygen of 100 mm Hg was chosen to be the maximum concentration of oxygen. A Bunsen solubility coefficient of 0.2413 for oxygen was determined from a 0.158 M sodium chloride solution at 33°C. The concentration of oxygen was therefore calculated at being 0.1228 μM/ml at 33°C and this is equivalent to a 12% oxygen concentration in a gas mixture.[20]

The main buffer constituent of aqueous humor is bicarbonate. The bicarbonate and levels of carbon dioxide were adjusted in the medium so that an

[16] D. V. N. Reddy, C. Rosenberg, and V. E. Kinsey, *Exp. Eye Res.* **1**, 175 (1961).

[17] J. F. Morgan, H. J. Morton, M. Campbell, and L. R. Guerin, *J. Nat. Cancer Inst.* **16**, 1405 (1956).

[18] R. Van Heyningen, *Biochem. J.* **65**, 24 (1956).

[19] B. Schwartz, unpublished data, 1965.

[20] B. Schwartz, unpublished data, 1965.

average pH of 7.56 was obtained at 33°C which is equivalent to posterior chamber aqueous pH.[21]

Problems of Maintaining pH of the Culture Medium in Relation to Rate of Flow past the Lens

The initial perfusion experiments indicated that a certain flow rate had to be maintained past the lens in order to avoid significant decreases in pH of the medium due to the production of large quantities of metabolic lactic acid.[11] The medium has a poor buffer capacity since it is essentially a balanced salt solution with a bicarbonate buffer system. Most of the buffer capacity of plasma and tissues is a function of their concentration of proteins. Since the medium is deficient in proteins, the buffer capacity was extremely low. Thus, a major problem for the lens perfusion system was to determine a flow rate which maintained pH, and was not so large as to cause washout of lens' constituents by diffusion out of the lens of substances which are not present in the medium. It was determined initially that a flow rate of 2 ml/hour would amply maintain the pH constant. This then was the main criterion for determining the flow rate to be used in the studies of lens in a culture system.

Placement of the Lens from the Rabbit Eye into the Perfusion System

In order to avoid any toxic effect of residual drugs, rabbits were killed only with intravenous air. The eyes were then enucleated and the lenses removed in a closed room under appropriate sterile conditions. The posterior portion of the eye was opened first with an aseptic surgical technique. The vitreous was removed so that the posterior surface of the lens was exposed upward and the anterior surface of the lens remained continuously covered by aqueous humor in the posterior chamber. With microscopic control, the lens was then removed by continuously cutting the zonules with scissors in a circumferential manner. Unless this is done with the utmost care, large tears in the lens capsule can occur since the zonules are extremely tough in the rabbit eye and slight tugging on them easily produces damage both to the lens and capsule. The lens was then immediately placed in the previously set up culture system filled with the synthetic medium. The lens was observed through the dissecting microscope in the lens chamber. If there was any question regarding the intactness of the capsule or its transparency, the lens was removed and a second lens was then taken for culture.

Light

Since the effect of light on culture of the lens is unknown, it was decided to initially perform all lens cultures under complete darkness. The lens in the chamber was appropriately covered so that the chamber was lightproof.

[21] B. Schwartz, *Invest. Ophthalmol.* 3, 96 (1964).

Pressure

The normal rabbit ocular pressure varies from approximately 10–20 mm Hg. Therefore, an attempt was made to maintain the pressure in the system in this range.

Support of the Lens

In the eye the lens is suspended in the aqueous of the posterior chamber by a ring of zonular fibers primarily composed of collagen. A similar device to support the lens in the perfusion system was not feasible. A compromise was devised whereby the lens rested on its posterior surface in a concave optical flat which allowed adequate optical resolution for observation and photography (Fig. 2). The lens' support was so designed to minimize the contact of the posterior surface and the resultant loss of surface area for nutrition of the lens. Since the main metabolic activity of the lens is localized in the anterior surface which is covered by epithelium, this support appeared to be adequate for maintaining the functon of transparency.

Observation of the Lens in the Culture System

The lens was observed directly through a dissecting microscope. Attached to this microscope were cameras to take 35 mm photographs as well as Polaroid photographs. Light was used to transilluminate the lens. In this way the smallest vacuoles or opacities of the lens could be observed and photographed.

Criteria for Maintenance of the Lens in the System

One of the major objectives of the perfusion system for lens culture is to set up a system which maintains the lens in its physiological state. Therefore, criteria for maintenance are required. These can be directly related to observations of the lens in the culture system or indirectly to the lens of the fellow eye.

Such observations as the gross transparency of the lens and its appearance within the system can be used. However, with detailed microscopic observations of the lens it was apparent that the criterion of gross transparency was inadequate. Vacuoles developed within the lens suture lines only seen under the microscope, some of which were reversible and others developed into permanent opacities. Therefore, the criterion of transparency has to be defined operationally.

Another sensitive criterion for lens maintenance is its weight. Any slight change in the permeability of the lens results in shifts of cations, anions, and water content of the lens. This is especially apparent if there is a slight rupture or tear in the capsule. Under optimal conditions of culture it is usually found that lenses gained weight of 3% of initial weight. Any tears in the capsule resulted in loss of weight probably by the leakage of proteins from the lens. Relative criteria can also be used to determine maintenance of the lens in the system by comparing the analyses of the cultured lens to a similar analysis of the lens in the fellow eye of the rabbit which is immediately frozen on removal from the eye. Very little difference exists between the lenses of two eyes when they are removed sequentially and analyzed in a similar fashion. A series of analyses have been done to exactly determine such small differences and whether they depend on the sequence in which the lenses are removed.

Operational criteria have been developed for lens concentrations of glucose, lactic acid, adenosine triphosphate (ATP), ascorbic acid, reduced glutathione, total amino acids, and soluble proteins among others. In addition, cytological criteria of the mitotic index of the lens epithelium have been formulated.

Besides an analysis of lens constituents to determine maintenance, the metabolic behavior of the lens can be studied by sequential analyses of the medium. A comparison of the constituents of the medium which has flowed past the lens in relation to the control sample of medium provides the means to determine the metabolic activity of the lens in relation to time.

After a short lag period of approximately 12 hours which allows for washout of the system, the lens can be shown to be in a metabolic steady state in relation to such parameters as utilization of glucose and production of lactic acid over a culture period of 4 days.

CHAPTER 6

Cardiac Cells in a Capillary Tube

M. A. Savageau

Perfusion has long been used in organ systems (especially heart and liver) as an experimental tool in establishing well-controlled conditions for kinetic studies. There are two essential features a system for perfusion must provide for kinetic studies: the ability to maintain distinguishable environmental steady states, and the ability to change abruptly from one to another. To meet these requirements a variety of perfusion techniques have been developed (see other articles in Section VI), each with their own particular advantages. Capillary tubing has proved to be an excellent perfusion chamber for continuous culture of tumor cells,[1] spectrophotometric observation of metabolic transients in isolated

[1] P. H. Dirstine, D. B. MacCallum, J. H. Anson, and A. Mohammed, *Exp. Cell Res.* **30**, 426 (1963).

cells,[2] and kinetic studies of the contraction rate for individual cardiac cells.[3] There are several reasons for its utility: (1) Capillary tubing can be used to isolate and maintain single cells[4]; (2) the simple geometry allows for accurate quantitative estimation of the flow rate at the surface of the cell; (3) the small volume allows for a rapid change of environmental conditions and permits prolonged perfusion tests even with small volumes of scarce substances; and (4) direct microscopic observation of the cells under controlled conditions is possible with relatively simple apparatus.

Preparation of Isolated Heart Cells for Perfusion

Hearts are obtained aseptically from six 2-day-old Wistar rats and pooled in a Stender dish containing 5 ml of culture medium. This culture medium consists of Eagle's Minimal Essential Medium, 10% fetal calf serum, 10% human serum, penicillin (100 units/ml), and streptomycin (50 μg/ml). The noncardiac tissue is teased away with forceps and the remaining tissue is washed free of blood with culture medium. This tissue is then transferred to a second sterile Stender dish containing 5 ml of culture medium and a magnetic stirring bar. The hearts are cut into pieces approximately 1 mm^3 with sterile Bard-Parker knife blades. After washing the pieces of tissue free of blood with culture medium, they are resuspended in 5 ml of a 0.5% trypsin solution made up of Difco (1:250) trypsin, Hanks' saline (pH 7.4) without Mg or Ca ions, and 4–6% chicken serum. Chicken serum is believed to protect the heart cells from damage by trypsin. However, very similar results are obtained by eliminating the serum and lowering the trypsin concentration to 0.125%. The dish is covered, placed on a magnetic stirrer, maintained at 37°C, and stirred at one revolution per second.

After 30 minutes the trypsin solution is discarded, 5 ml of fresh trypsin solution is added back to the tissue in the dish, and the dish returned to the incubator. Following both the second and the third incubation periods of 30 minutes, the trypsin solution is removed to a 15-ml conical centrifuge tube, cooled on ice for 5 minutes, and centrifuged at approximately 800 g for 4 minutes in a clinical centrifuge. The pellet of cells is resuspended in 2 ml of fresh culture medium. Cell counts are made in a hemocytometer (cf. Section VIII) and the final density of cells is adjusted to 5×10^5 cells per milliliter of medium for plating.

The suspension of cells is drawn by capillary action into sterile capillary tubes 2 cm in length, 0.4 mm ID, and pre-bent to eliminate rolling. The tubes, filled with the suspension of cells, are placed in Petri dishes (two or three per dish) with 5 ml of fresh culture medium. These Petri dishes are placed in jars which are gassed with a mixture of 5% CO_2 and 95% air, tightly sealed, and maintained at 37°C.

[2] C. Ritter, *Exp. Cell Res.* **40**, 169 (1965).

[3] M. A. Savageau and J. P. Steward, *Curr. Modern Biol.* **1**, 159 (1967).

[4] K. K. Sanford, A. B. Covalesky, L. T. Dupree, and W. R. Earle, *Exp. Cell Res.* **23**, 361 (1961).

The Appearance of Cardiac Cells in Capillary Cultures

After a few hours of incubation at 37°C the cells, which are spherical in suspension, become fastened to the glass on which they grow. Their flatness permits good observation of contraction using an inverted phase-contrast microscope. After 2 to 3 days in culture, attached cells are found only in the terminal portion of the bore of the capillary tubes. This terminal portion is equal in length to fifteen times the internal tube diameter. This distribution of cells undoubtedly occurs because diffusion is a limiting factor in the exchange of chemical substances between media in the center of the tubes and that in the Petri dish. During the first week in culture the cells exist as single cells or small clusters of cells; during the second week there are predominantly large clusters or monolayers of cells. The contraction of cells in clusters or monolayers is synchronous.

Perfusion of Cardiac Cells in Capillary Tubes

Sterile polyethylene tubing is used to connect one end of the capillary tube to a sterile collection flask and the other end to a syringe filled with sterile, aerated (5% CO_2, 95% air) medium for perfusion. The capillary tube is mounted on a microscope slide and inserted into a warm stage (Lab-Line/CS&E "Incustage") on the microscope. For the purposes of observation it has been found useful to maintain the cells at 30°C. The lower temperature decreases the contraction rate to allow increased resolution of the pulses. The syringe is driven at constant speed by a Harvard infusion pump. The experimental setup is shown schematically in Fig. 1. In changing from one medium to another a small air bubble is inserted in the tubing between the two media by briefly disconnecting the tubing from the syringe. (Air bubbles have also been inserted with a 27-gauge hypodermic needle.) This is suggested to provide a sharp interface. The bubble has been found not to affect the contraction rate of the cells. In order to increase the efficiency of the optical coupling the capillary tube is surrounded by a few drops of oil. The position of a number of cells in the capillary tube is recorded with reference to the settings on the mechanical stage of the microscope. Ten to thirty contractions for each cell are monitored every 5 or 10 minutes. Each pulsation is recorded by momentarily closing an electrical circuit which activates the writing pen of a strip-chart recorder, and the results are thus retained for future analysis. The rate of contraction in any sampling period is determined from the time required for a fixed number of contractions.

Temporal Response of Contraction Rate to Perfusion

Before describing various temporal responses, it is relevant to discuss two general features which are found in all cases I have observed. First, when cells are taken from the Petri dishes and subjected to perfusion, the rate of contraction is changed little when the capillary tube is first connected to the polyethylene

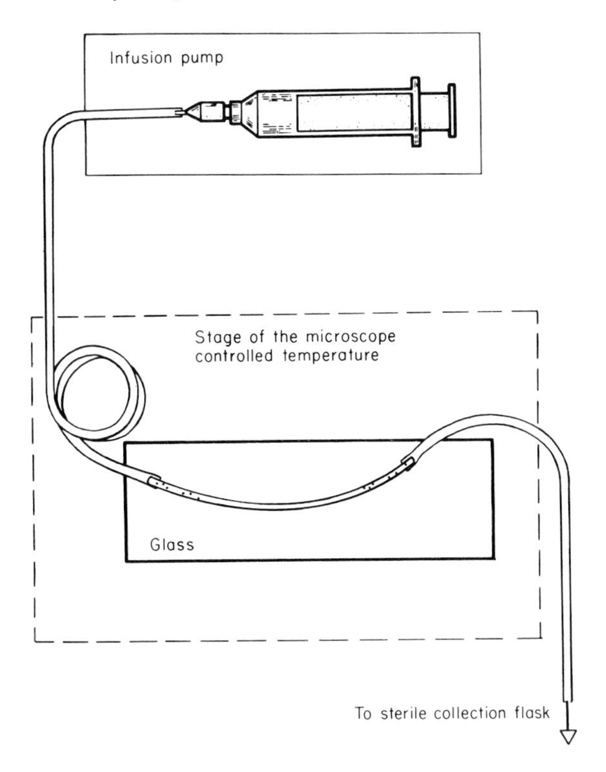

Fig. 1. Schematic diagram of the perfusion of cardiac cells in capillary tubing.

tubing containing fresh culture medium, but suddenly increases when perfusion is commenced. The new, elevated rate of contraction is established in less than 1 minute. Second, the responses appear to have a well-defined component upon which is superimposed random fluctuations with a standard deviation of 5 to 15 contractions per minute.

Two types of temporal responses to prolonged perfusion with fresh culture medium are observed and these depend upon the age of the cells in the culture. The response of a cell perfused for 500 minutes during the first week in culture is shown in Fig. 2a. This response is of short duration, the cell having stopped contracting after 40 minutes of perfusion. This type of response is different from that observed in cells perfused during the second week in culture as illustrated in Fig. 2b. In the latter case, the rate of contraction is seen to decrease with time, as before, but the response is of much longer duration and the cell does not stop contracting.

Another type of response is shown in Fig. 3. After prolonged perfusion with fresh culture medium, there appears to be a characteristic alteration in the rate of contraction when the perfusion medium is abruptly changed from fresh medium to medium that has been used to support the growth of a dense culture of cardiac cells for 5 days (conditioned medium). In seven of eight experiments the rate of contraction was found to increase, reach a peak, and decrease to a rate approximating the value before the change.

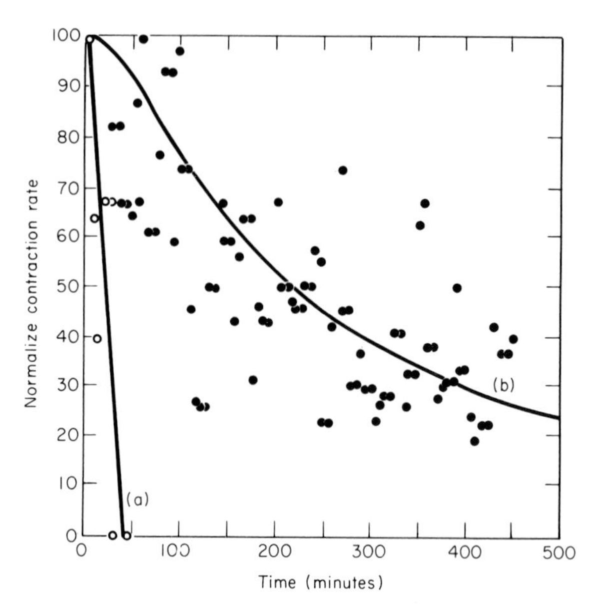

Fig. 2. Responses to perfusion at 900 μm/second with fresh medium. (a) Cell 2 days in culture; (b) another cell 10 days in culture. (From Savageau and Steward.[3])

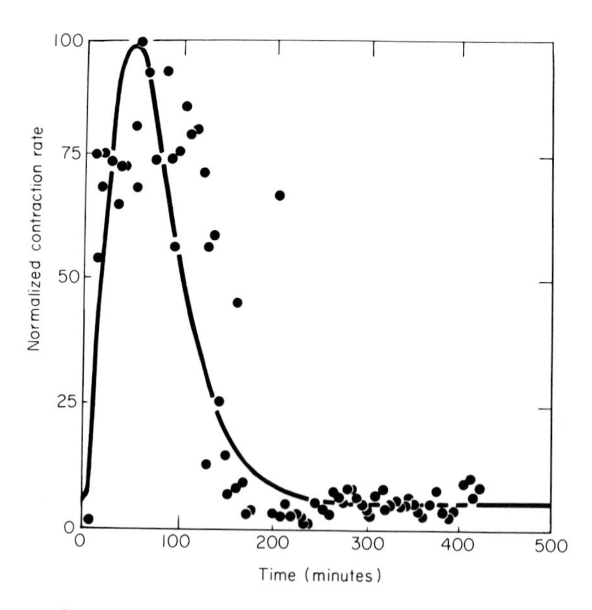

Fig. 3. Response of contraction rate of perfusion with conditioned medium. After prolonged perfusion with fresh culture medium at 900 μm/second the perfusion medium was changed. At time zero the fresh medium was changed to medium that had been used to support the growth of a dense culture of cardiac cells for 5 days. The cell was then perfused for 500 minutes at 900 μm/second. The cell in this example had been in culture for 12 days. (From Savageau and Steward.[3])

SUMMARY

Capillary tubing provides a superior perfusion chamber for many applications. Cells can be cultured and observed conveniently under well-defined environmental conditions with simple equipment. The observations using cardiac cells described above illustrate the utility of this technique for examining the interaction between cellular physiology and environmental factors.

CHAPTER 7

Artificial Capillaries: An Approach to Tissue Growth in Vitro

R. A. Knazek and P. M. Gullino

The ability to grow cells to tissuelike densities *in vitro* is desirable since certain characteristics of many cell lines change as a function of cell density. Diffusion of cell nutrients and products within tissue, however, is a major obstacle to attaining such high densities. Periodic or continuous replacement of nutrient medium to enhance diffusion into the cell mass causes a sudden change in fluid composition adjacent to the cells or prevents the accumulation of essential cell products. A method which does not permit such variations within the cellular microenvironment but still provides continual nutrient supply and product removal may, therefore, allow cells to grow to tissue densities *in vitro*. Cell culture within a perfused artificial capillary bed, as described in this chapter, provides such a technique.[1]

APPARATUS

Several types of capillaries are commercially available (Table I). These are tube-shaped, semipermeable membranes that limit diffusion through their walls to substances having molecular weights less than a nominal maximum.

[1] R. A. Knazek, P. M. Gullino, P. O. Kohler, and R. L. Dedrick, *Science* **178**, 65 (1972).

TABLE I
Commercial Sources of Artificial Capillaries

Composition	Manufacturer	Limiting mol. wt.	Approximate OD/ID (μm)
Cellulose acetate (HFU)	Dow Chemical Corp.	30,000	250/200
Polymeric (PM-30)	Amicon Corp.	30,000	340/190
Polymeric (XM-50)	Amicon Corp.	50,000	340/190
Silicone polycarbonate	Dow Chemical Corp.	Permeable to gases	260/180

Construction of Artificial Capillary Bed. Thirty XM-50 capillaries mixed with 30 silicone polycarbonate capillaries comprise an artificial capillary bed. The former capillaries provide nutrient supply and product removal, whereas, the latter increase gas transport within the bed. The mixed capillary bundle is pulled into a 90 × 8 mm glass shell having a ground glass male adaptor at each end and two glass side ports (Fig. 1a). Both bundle ends are then closed with a suture tie and immersed in a mixture of 12 g liquid silicone rubber (General Electric RTV-11) and 8 g 360 Medical Fluid (Dow Corning Corporation) that has been catalyzed with 1 drop or approximately 40 mg of stannous octoate (Tenneco,

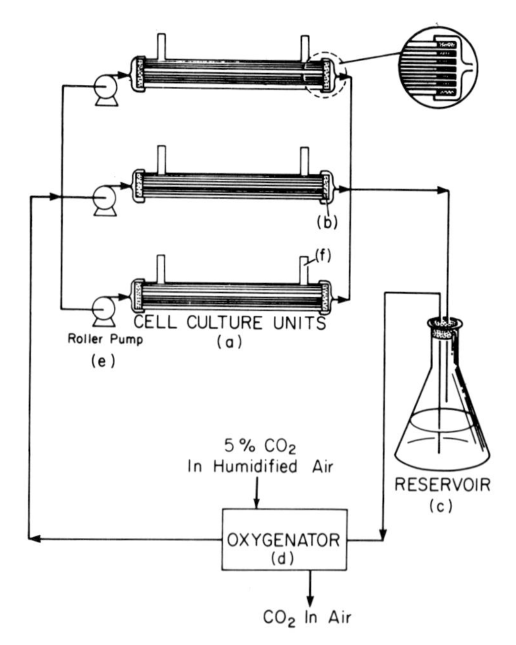

Fig. 1. Perfusion circuit. A cell culture unit (a) consists of a mixture of XM-50 and silicone polycarbonate capillaries sealed into each end of a 90 × 8 mm glass shell by silicone rubber (b). Units may be arranged in parallel as shown or in series. Nutrient medium stored in a 125-ml reservoir flask (c) was oxygenated and brought to the appropriate pH by exposure to a humidified 5% CO_2 in air mixture in a Dow Corning Mini-lung (d) prior to being pumped (Holter Co., pump RL-175) (e) through each capillary bundle. Cells were inoculated onto the capillary bundles using syringes connected to the two shell side ports (f).

Nuocure 28). After solidifying overnight, the excess silicone rubber is then trimmed flush to the shell to expose the patent capillary ends.

Assembly of Perfusion Circuit. The roller pump chambers (Holter Co., PC 7250), oxygenator (Dow Corning Corp., Mini-lung)[2] and 125-ml reservoir flask are connected as shown in Fig. 1 by 3.2 mm OD tubing (Dow Corning Corp., Silastic). Tapered female ground glass adaptors are then fitted to both ends of the shell and inserted into the perfusion circuit.

The free end of a Silastic tubing section fitted with a metal female Luer adaptor is attached to each side port.

Sterilization. The perfusion circuit should then be sterilized in ethylene oxide for 6 hours and subsequently aired sterilely for 2 days. Disposable sterile syringes, one containing sterile culture medium are inserted into the side port adaptors and the perfusate reservoir then filled with 100 ml of culture medium. The equipment is placed in a 37°C incubator at approximately 90% humidity where both the perfusion circuit and the shell surrounding the capillary bundle are flushed with culture medium for 2 days to remove remaining traces of ethylene oxide.

Inoculation of Artificial Capillary Bed. The perfusate reservoir should then be emptied and refilled with fresh medium whose pH may be maintained by exposure to a humidified 5% CO_2 in air mixture flowing through the oxygenator. The previously emptied side port syringes from each of the two parallel units are then discarded and the metal adaptors flamed. A suspension of 0.2–2 $\times 10^6$ cells in medium is then injected through the port onto each capillary bundle. The cells then settle and should be permitted to adhere to the capillaries for 1 day, after which, the extracapillary medium within the shell is replaced with fresh medium to remove unattached cells and cellular debris. Both Silastic tubing sections are then clamped close to each loading port. The perfusion medium can subsequently be replaced and/or analyzed as desired.

Histological Study. Cellular morphology may be observed when the experiment is terminated by replacing the extracapillary medium with a warm 4% agarose suspension in normal saline through a side port. This will hold the cells and capillaries in place when cooled. The glass ends of the shell are then fractured, the exposed ends of the bundle cut, and the agarose-encased capillary bundle removed for storage in cold 10% formalin. Sections perpendicular to the long axis of the bundle can be made 2 days later and stained for histological study.

Experimental Example

This technique has been used to culture several types of cells to tissuelike densities as described in the following example.

[2] P. M. Gullino, *In* "Organ Perfusion and Preservation" (J. C. Norman, ed.), p. 45. Appleton-Century-Crofts, New York, 1968.

Approximately 1.5×10^6 JEG-7 human choriocarcinoma cells[3] suspended in modified Ham's F-10 medium[4] were injected onto each of three capillary bundles arranged in parallel. The bundles were perfused at a rate of 0.7 ml per minute by the same type of medium maintained at pH 7.3. The extracapillary and perfusion media were replaced simultaneously every 1–2 days. Periodic analyses[5] of the perfusate human chorionic gonadotropin (HCG) showed an initial lag in HCG production with the subsequent production rate doubling every 1.6 days (Fig. 2b). The HCG production rate after 28 days of culture was equivalent to that produced by 217×10^6 cells in monolayer culture[3] but required less than

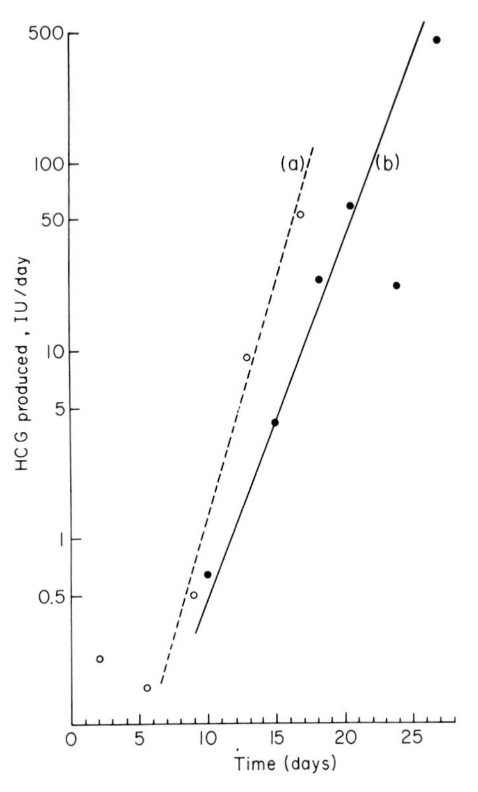

Fig. 2. Rate of HCG production versus duration of JEG-7 choriocarcinoma cell culture on mixed capillary bundles. (a) Perfused with 5 ml Medium A per minute without extracapillary medium replacement. (b) Perfused with 0.7 ml Medium B per minute with periodic extracapillary medium replacement. Medium A: 83.3% Ham's F-10, 13.5% horse serum, 3.2% fetal calf serum, 5000 units % aqueous penicillin, and 5 mg % streptomycin. Medium B: 83.3% Ham's F-10, 13.5% horse serum, 3.2% fetal calf serum, 5000 unit % aqueous penicillin, 5 mg % streptomycin, 0.5 mg % insulin, and 0.62 mg % cortisone acetate.

[3] P. O. Kohler, W. E. Bridson, J. M. Hammond, B. Weintraub, M. A. Kirschner, and D. H. Van Thiel, *Acta Endocrinol. Suppl.* **153**, 137 (1971).

[4] 83.3% Ham's F-10, 13.5% horse serum, 3.2% fetal calf serum, 5000 unit % aqueous penicillin, 5 mg % streptomycin, 0.5 mg % insulin, and 0.62 mg % cortisone acetate.

[5] W. D. Odell, P. L. Rayford, and G. T. Ross, *J. Lab. Clin. Med.* **70**, 973 (1967).

3 ml of actual culture volume. The gross appearance of one unit after 4 weeks of culture is shown in Fig. 3a. A histological section through the same bundle is shown in Fig. 3b. Multilayer cell growth with invasion of the porous capillary surfaces and bridging between adjacent capillaries is apparent. Spaces between the cell mass and capillaries are artifacts of fixation.

A subsequent experiment using two parallel units perfused by 5 ml of medium[6] per minute but without periodic extracapillary medium replacement, produced HCG at a rate that doubled every 1.2 days (Fig. 2a). This is in accord with cell doubling times of 1.1 to 1.3 days observed in the log phase of growth using standard monolayer cultures in our laboratory.

Discussion

The artificial capillary system requires much less volume than standard monolayer flasks that provide the same amount of surface area. This allows large numbers of cells to be cultured in a relatively simple apparatus with considerably less space and equipment than are required for other commercial culture techniques.

The formation of cell multilayers as described indicates that this technique permits cells to grow to solid tissue densities *in vitro* and may even permit growth of cells that had previously defied culture by other *in vitro* techniques. This may result from the nearly constant favorable conditions within the cellular microenvironment and/or the accumulation of essential cell products.

The formation of a tissuelike mass within a small volume lends itself to convenient study of cell behavior under conditions that more nearly approximate the *in vivo* state. Metabolic pathways, cell cycles, drug responses, and other phenomena may be investigated as functions of variables that are more easily controlled than those of the *in vivo* studies.

The semipermeable nature of the capillary wall will concentrate less diffusible cell products within the extracapillary medium. The study of such products will be informative and may reveal that they possess the capacity for stimulatory or inhibitory control of cell behavior. Diffusible cell products, however, may be monitored in the perfusate as indicators of cell function under various controlled conditions without the necessity of cell culture manipulation. Valuable cell products, e.g., hormones, can be retrieved from the perfusate or extracapillary medium for additional study or use.

Humoral symbiotic relationships between various cell types may also be studied under conditions of continual stimulation without cell to cell contact or cumbersome medium changes. The use of units in parallel or series arrangements will allow study of hormone-producing cell lines grown in one unit and their effect upon target cells present in the other units. Ontogenesis of normally juxtaposed, but different, embryonic tissues may also be studied in a similar fashion.

These suggestions undoubtably do not comprise a complete listing of the

[6] 83.3% Ham's F-10, 13.5% horse serum, 3.2% fetal calf serum, 5000 unit % aqueous penicillin, and 5 mg % streptomycin.

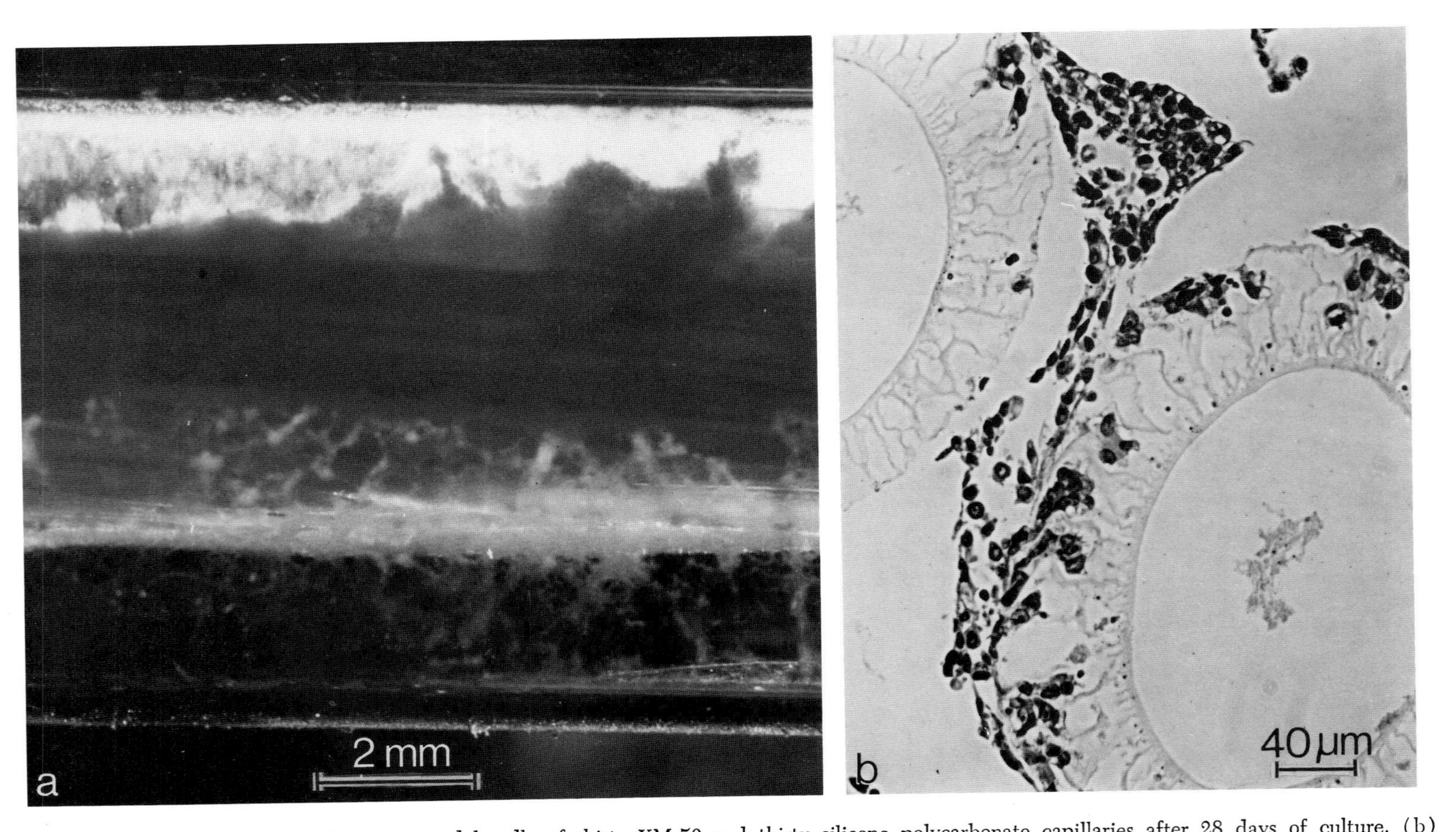

Fig. 3. (a) JEG-7 cell growth on a mixed bundle of thirty XM-50 and thirty silicone polycarbonate capillaries after 28 days of culture. (b) Histological section through the capillary bundle shown in (a). Note cells invading and bridging the porous outer surfaces of two XM-50 capillaries. Phase contrast used to display the capillaries makes the nuclei appear very dark.

possible applications for the artificial capillary system of cell culture. The authors believe that this technique will provide a more accurate and convenient method of elucidating the properties and behavior of cells and tissues *in vitro*.

CHAPTER 8

Replicate Roller Bottles

W. L. Whittle and P. F. Kruse, Jr.

Production of large numbers of "anchorage-dependent"[1] cells in tissue culture for the study of enzymes, virus production, etc., has long been a problem. Primarily, flasks have been used with cells sticking to one side of the flasks and overlaid with medium; thus, most of the surface areas of the flasks have not been used to support cell growth. Cultures in rolling bottles have the advantage of using almost all of the surface area of the culture vessels; in addition, medium and gas exchange is more easily made effective in roller bottles than in stationary flasks.[2]

These aspects of cell culture in rolling bottles can be enhanced by providing the bottles with some means of semiautomatic feeding and elimination of nutrient medium, as described previously.[2,3] An apparatus for perfusion of replicate roller bottle cultures is produced commercially[4]; an early model of this apparatus is illustrated in Fig. 1. The following procedure and cell yield data illustrate its use. The procedure for more current versions of the apparatus, including an eight-bottle, two-tier unit is essentially the same. A procedure for perfusion of T-flask type cultures is presented elsewhere.[5]

Bottle Cultures

Autoclave bottles and swivel caps[6] (S in Fig. 1) separately, protecting openings with small rubber caps or cotton-plugged tubing. Inoculate bottles with cell

[1] M. Stoker, C. O'Neill, S. Berryman, and V. Waxman, *Intern. J. Cancer* 3, 683 (1968).
[2] R. Elsworth, *Process Biochem.*, March, 1970.
[3] P. F. Kruse, Jr., L. N. Keen, and W. L. Whittle, *In Vitro* **6**, 75 (1970).
[4] New Brunswick Scientific Co., New Brunswick, New Jersey 08903.
[5] P. F. Kruse, Jr., *In* "Methods in Enzymology," Methodology of Biomembranes (S. Fleischer, L. Packer, and R. Estabrook, eds.), Academic Press, New York, in press.
[6] See Fig. 2 in reference given in footnote 3.

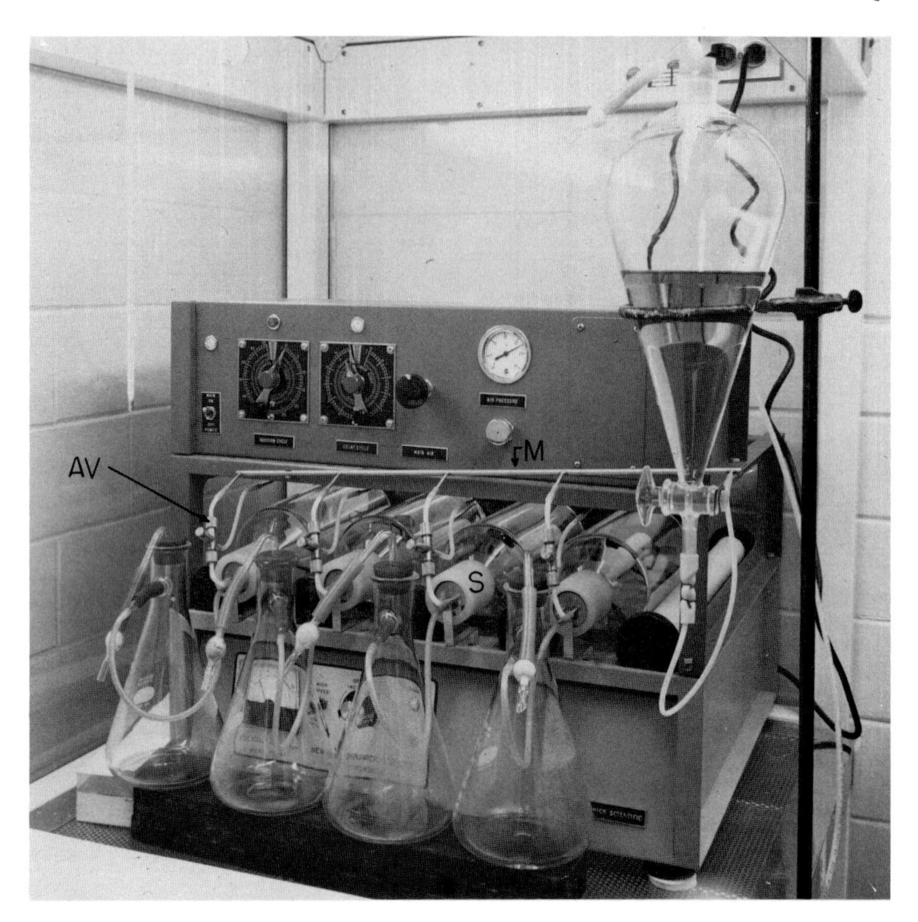

Fig. 1. Roller Bottle-Perfusion Unit (early model; New Brunswick Scientific Co.). S, Roller bottle swivel caps; M, influent medium manifold; AV, compressed-air-powered pinch valves.

suspension (suggest 1–2 × 10^7 diploid cells or 5–10 × 10^6 heteroploid cells) and 200 ml of medium per 290-mm bottle. Gas with 5–8% CO_2 in air (if using bicarbonate buffer), attach swivel caps with protected outside openings, and rotate overnight or longer at about 12 rph (dial setting = 8)[7] to establish uniform dispersion of cells on the glass.

Perfusion Train

Autoclave fresh medium reservoir (separatory funnel or other container with opening at top protected with cotton-plug tubing bent downward) containing flexible Teflon tubing from bottom opening of reservoir, protected at end of the tubing with cotton-plug which is removed subsequent to connection later with influent manifold (M in Fig. 1).

[7] From calibration chart of dial settings versus bottle rotation speeds.

Fit each end of the compressed air-powered pinch valves (AV in Fig. 1) with pieces of Teflon tubing, 1½ × ⅛ inch ID, and slip upper tubings over the manifold outlets using ¾ inch lengths of silicone tubing as additional sleeves; the Teflon tubing is slipped over the metal outlets as well as the silicone sleeves. Protect the four bottom Teflon-tubing air valve outlets with cotton-plug tubings which are removed and discarded later to connect valves to inlets of swivel caps. Protect inlet tube of manifold with cotton-plug tube, and autoclave entire inlet manifold-air valve assembly.

Using Effluent Manifold.[8] Attach U-shaped stainless steel tubings to manifold with 2 inch pieces of silicone tubing and fit other (short) arm of the U's with 1½ inch silicone tubings, which are later connected to swivel cap effluent tubes. Protect end of manifold with cotton plug; autoclave manifold assembly and 3- to 4-liter effluent bottle separately.

Using Individual Effluent Bottles. Autoclave four 1-liter suction flasks fitted with silicone tubing on side arms and long enough for eventual connection directly to each swivel cap effluent tube (Fig. 1). Flask openings are protected with cotton-plug tubing bent downward. (The effluent manifold is not used in this procedure.)

Connection of Bottle Cultures and Perfusion Train

1. Mount reservoir bottle, close stopcock, and fill two-thirds full with fresh medium. Gas bottle lightly with 5% CO_2 in air; pH of medium should be about 7.4–7.6.
2. Place inlet manifold-air valve assembly in clamps provided.
3. Attach line from reservoir to inlet manifold.
4. Attach the four air line tubings from unit to compressed air valves.
5. Turn on power switch and set air pressure gauge at 20–25 pounds (house air line to unit should contain an air purifier or filter).
6. Remove protective cap from inlet line on swivel caps and slip Teflon tubing with silicone sleeve over the stainless steel inlets. (Influent lines are fitted before effluent lines so that any slight pressure buildup in bottle cultures is released when effluent lines are attached.)
7. Attach effluent lines; position effluent bottles so that their inlet side arms are at a level equal to or slightly above the level of the effluent lines in the swivel caps.
8. Set clock No. 1 on 3 hours and clock No. 2 on 30 seconds. Open stopcock on reservoir and manually activate clock No. 1 two or three times until inlet lines are full. During this operation effluent lines to the suction flasks are clamped off with pinch clamps and breather outlets in swivel caps are open. After about ½ to

[8] Not shown in Fig. 1, but supplied with instrument (cf., footnote 4) for use when it is desired to collect a common effluent from all four bottle cultures.

1 hour (for temperature equilibration in warm room), remove pinch clamps from effluent lines and use them to clamp off the breather tubes in the swivel caps. Decrease the setting of clock No. 2 from 30 to 5–10 seconds. Clock No. 1 regulates the interval of time between feedings while clock No. 2 determines the amount of each feeding.

9. Increase bottle rotation speed to about 30 rph (dial setting = 16).[7] NOTE: Optimum rotation speed may be determined for each cell type, but 30 rph has been satisfactory for several cell types including human diploid cells.

10. On succeeding days shorten time interval on clock No. 1 and lengthen interval on clock No. 2 as necessary to maintain pH at 7.1 ± 0.1 and to provide adequate nutrition to increasing cell numbers and formation of multilayers. NOTE: Multilayered cultures can be held in a stationary phase (i.e., no DNA synthesis) by slower perfusion with medium containing less than 1% serum.

11. If one or more cultures are terminated prior to others, use the knurled thumb screw on the air valves to shut off the vacant lines during subsequent feeding of remaining cultures.

12. Gas effluent bottles to maintain satisfactory pH of effluent media if it is to be subjected to later analyses.

TABLE I

NF-HFO-2 Human Diploid Fibroblast Cells[a] Roller Bottle Perfusion Experiments

Days	Cell count (per bottle × 10⁶)	Increase (×)	M.E.[b]	pH	Perfusion rate (ml/day/bottle)
Experiment 1 (10th passage cells)					
0	8.94	—	0.09	—	—
3.1[c]	24.2	2.7	0.25	7.15	—
6.0	87.1	9.7	0.89	7.10	117
11.1	131	14.7	1.34	7.13	221
17.1	188[d]	21.0	1.92	7.04	277
Experiment 2 (24th passage cells)					
0	13.2	—	0.13	—	—
2.0[c]	35.7	2.7	0.36	7.40	—
6.8	102	7.7	1.04	7.25	118
9.8	135	10.2	1.38	7.20	201
13.8	199[d]	15.1	2.03	7.03	211

[a] Cells originally established from human foreskin tissue; these enter phase III *in vitro* after about 35 to 40 population doublings.

[b] M.E., monolayer equivalents; M.E. = 1.0 = a *tightly* packed confluent sheet one cell thick; confluency of NFO-HFO-2 cells in the 290 × 110 mm diameter bottles was reached at 98 × 10^6 cells/bottle, or 134,000 cells/cm^2 (same as WI-38 cells). Inside surface area of bottles = 730 cm^2.

[c] Perfusion was started on third and second days in experiments 1 and 2, respectively. Experiment 2 was run with New Brunswick Scientific Co. roller bottle perfusion system (Fig. 1, text).

[d] Cell yields are ca. fourfold those obtained in comparable experiments in stoppered, roller bottles with media changes every 2–3 days. In T-flask perfusion experiments cell yields of diploid human fibroblasts usually amount to four- to eightfold greater than those in concurrently run cultures in stoppered T-flasks.

Performance

Table I shows data obtained using NF-HFO-2 human diploid fibroblast cells; another human diploid line, WI-38,[9] gave similar results. However, using heteroploid cells much higher cell numbers are obtainable. Jensen rat sarcoma cells increased from 8.5×10^6 cells per bottle to 2.0×10^9 cells per bottle in 10 days, and WI-38VA13A[10] from 3.3×10^6 to 1.27×10^9 cells per bottle in 11 days incubation. These yields amounted to cell packs of 2.95 ml and 3.05 g wet weight (Jensen) and 1.50 ml and 1.55 g wet weight (VA13A).

[9] Embryonic human lung line, furnished by courtesy of Dr. L. Hayflick.
[10] SV-40 transformed WI-38, furnished by courtesy of Dr. V. Cristofalo.

CHAPTER 9

Replicate Tube Chambers for Bones

J. Bornstein, P. L. Schwartz, and R. E. H. Wettenhall

The system described here was devised for the study of the time course of growth, calcium deposition, and synthesis of collagen and proteoglycans in neonatal bones under strictly comparable conditions.

The use of a flow-through system enables such studies to be carried out under conditions of constant substrate concentration and of effective removal of metabolites. Multiple handling of the bone chambers is avoided and it is thus possible, with initial strict asepsis, to carry out the study in the absence of antibiotics.

Replicate Tube Chambers

The key to the technique is the organ chamber. It consists of a glass vessel 4 cm long and 1.5 cm diameter with an exit tube closed at one end by a Teflon stopper through which passes a stainless steel entry tube (Fig. 1a). The bone holder is a stainless steel plate 3.5 cm × 1.25 cm drilled along the long axis with six 3-mm holes, in which are placed small stainless steel springs which are the actual bone mounts (Fig. 1b). The flow-through system consists of one or more medium storage reservoirs, the medium being constantly oxygenated by gas

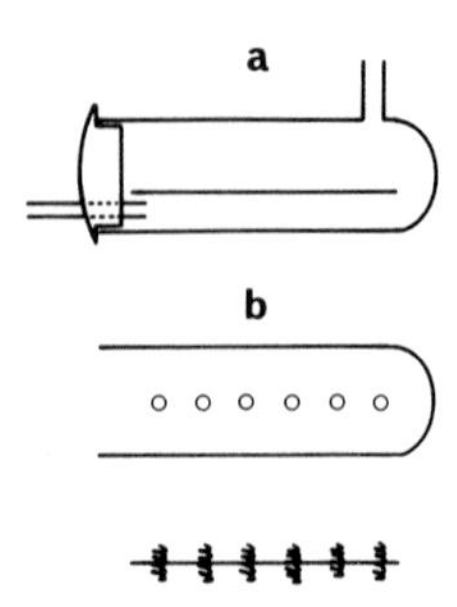

Fig. 1. Organ chambers (see text).

passed through a Millipore filter. The medium is pumped by a peristaltic pump from the reservoir through the organ chamber to a collection flask. The most suitable pump has been found to be a Technicon Autoanalyzer proportioning pump as, by the selection of Technicon coded plastic tubing, it is possible to vary the rate of flow through the system and to maintain a constant rate of flow in the experiment. The system is shown schematically in Fig. 2.

By the use of manifolds it is possible to maintain groups of bones under identical conditions and to remove any group of bones without disturbing the system.

Culture Technique

1. All glassware and system components are sterilized by autoclaving.
2. Medium components and additives are sterilized by passage through suitable filters.
3. All procedures are carried out with full aseptic technique; the culture and operating rooms are sterilized by ultraviolet radiation.
4. The neonatal animals are sacrificed by decapitation, washed with disinfectant, and the bones to be cultured dissected.
5. The bones are placed in the coils of the holders immersed in the medium used, and, when filled, transferred to the organ chambers containing medium.
6. When all chambers are filled, pumping with sterile, oxygenated medium begins, and is continued to the end of the experiment.

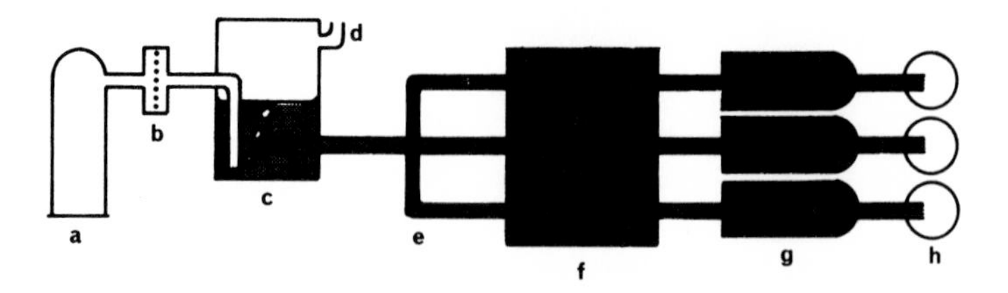

Fig. 2. Schematic of replicate bone chamber perfusion. a, Gas supply; b, Millipore filter; c, reservoir; d, gas outflow; e, manifold; f, pump; g, organ chambers; h, outflow reservoir.

It was found necessary to use the type of holder described, as even at pumping rates as low as 0.053 ml/minute there was sufficient turbulence in the chambers to produce bone damage.

This technique was used, by paired comparison, to study the effects of insulin and other hormones, and medium additives such as vitamins, based on the parameters mentioned above.

The system was capable of sustaining lineal growth (18% increase in length of paired tibiae in 48 hours), calcium deposition, collagen synthesis as measured by hydroxyproline content, and proteoglycan synthesis as measured by galactosamine content for at least 48 hours.

All parameters were accelerated by the addition of vitamin A (100 μg), vitamin B_2 (1 μg), and vitamin D (20 μg) to each liter of medium.

In purely artificial medium the increase in length is the same as *in vivo* for the first 24 hours, but thereafter is appreciably slower.[1]

The addition of insulin produced an acceleration of collagen and proteoglycan synthesis, but other hormones tested had no effect.[2]

On the work so far carried out, the system provides a technique for the comparison of medium changes and the effect of additives under strictly comparable conditions, and we would suggest that the principal problem encountered, i.e., loss of bone matrix,[1] may be overcome by the addition of crystalline serum albumin to the medium.

[1] P. L. Schwartz, R. E. H. Wettenhall, and J. Bornstein, *J. Exp. Zool.* **168**, 517 (1968).
[2] R. E. H. Wettenhall, P. L. Schwartz, and J. Bornstein, *Diabetes* **18**, 280 (1969).

CHAPTER 10

Multisurface Stacked Plate Propagators

Joseph B. Schleicher

The growth of large-scale quantities of tissue culture monolayers is desirable in view of the large quantities of living cells required in the research, development, and production of virus vaccines and biochemicals. Although the growth of tissue culture cells in suspension cultures on a large scale has been well established, this method of cell production is not adaptable to the production of vaccines and biochemicals for use in humans. The production of vaccines and biochemicals for humans is restricted to the use of tissue cultures of "normal" noncancerous tissue cells, such as, primary explants, diploid cell strains, and finite cell lines. These cells require a surface on which the cells can attach and proliferate.

The conventional techniques and equipment for growing tissue cultures on surfaces are not adaptable for large-scale production, particularly in a single unit. Conventional techniques usually consist of growing cells on the walls of various size glass or plastic containers. A typical process entails cell growth on the submerged glass surface of 1-liter rectangular, stationary bottles. In order to overcome the limitations of the bottle tissue culture system, a Mass Tissue Culture Propagator (MTCP) was invented at Abbott Laboratories.[1-3] Any large culture system must have a constant turnover of media to dilute waste products, replenish nutrients, and maintain a viable environment for the cell population. This is supplied by various perfusion techniques. In the MTCP this is performed by an airlift pump system. A perfusion technique was applied to roller bottles by Kruse and Miedema.[4] Molin and Heden used a rotating stack of titanium disks,[5] while Parisius *et al.* recently reported using the same technique but with glass plates.[6] McCoy *et al.* circulated medium through glass columns packed with glass helices.[7] This chapter will be concerned with a discussion of the Abbott MTCP.

Materials

In these studies we utilized established tissue culture techniques such as trypsinization, cell concentrations, growth media, and maintenance media routinely used for planting primary or other tissue culture cells in bottles. Bovine kidney cells were grown in Earle's BSS with 0.5% lactalbumin hydrolysate and 4% inactivated calf serum. WI-38 diploid cell cultures were grown in Eagle's BME with 10% fetal calf serum. Human embryonic kidney cells were grown in Medium 199 with 5% or 10% fetal calf serum. Chanock's Medium was used for the growth of *Mycoplasma pneumoniae*. Medium 199 with or without additional Eagle's vitamins and amino acids was used for maintenance.

The virus cultures were those adapted to bovine tissue culture. Newcastle disease virus was used to stimulate interferon in CCL1 and human embryonic kidney cells with Eagle's BME, with either 2% fetal calf serum or 0.1% human serum albumin.

[1] R. E. Weiss and J. B. Schleicher, United States Patent Office, Patent No. 3,407,120, (1968).

[2] R. E. Weiss and J. B. Schleicher, *Biotechnol. Bioeng.* **10**, 601 (1968).

[3] J. B. Schleicher and R. E. Weiss, *Biotechnol. Bioeng.* **10**, 617 (1968).

[4] P. F. Kruse, Jr. and E. Miedema, *J. Cell Biol.* **27**, 273 (1965).

[5] O. Molin and C. G. Hedén, *Progr. Immunobiol. Standardization* **3**, 106 (1968).

[6] J. W. Parisius, N. Cucakovich, and H. G. MacMorine, Paper read at *23rd Tissue Culture Assoc. Meeting, Los Angeles,* 1972.

[7] T. A. McCoy, W. L. Whittle, and E. Conway, *Proc. Soc. Exp. Biol. Med.* **109**, 235 (1962).

Procedure

Routine production of virus vaccines consists of planting suitable cell concentrations into flasks and incubating these until a confluent monolayer forms on the submerged surface. The growth medium is then removed and the cell sheet washed to remove traces of media containing serum. Maintenance medium containing the virus is then added. The virus replicates in the cell and usually is released into the production or maintenance medium after suitable incubation time and temperature. The medium is then processed for vaccine production.

During growth and maintenance, the cells consume oxygen and produce carbon dioxide. Thus, variations of pH occur in the medium due to changes in the content of gas in equilibrium with the bicarbonate buffered media in a bottle. Aseptic procedures must be used to prevent contamination. The pooling of harvests from a large number of bottles greatly increases the possibility of contamination. The MTCP was developed to overcome many of the limitations of the bottle system. This invention provides a suitable apparatus for the large-scale propagation of cells in tissue culture in a single unit. Various size MTCP units have been developed ranging from 1 to 200 liters.

Description of Apparatus

Figure 1 shows a flow sheet of the MTCP system. A Pyrex or stainless steel jar is used to house the stacked glass plates which are made from Pennvernon single strength grade B window glass. The glass is $\frac{3}{32}$ inches thick. The cells grow upon these glass surfaces. Notched type 316 stainless steel rods are used to support the plates in a $\frac{1}{4}$ inch center-to-center distance. This results in a $\frac{1}{8}$ inch space between glass surfaces.

In the bottle process, a large volume of gas phase is necessary to provide sufficient oxygen and to act as a reservoir for the carbon dioxide produced by the cells. In the MTCP, a sterile air-carbon dioxide mixture is introduced through the gassing inlet. By permitting the gas to flow from the sparge tube into the airlift pump, the rising gas bubbles create a hydrostatic head difference due to the decreased density in the air lift pump. As a result, medium is forced into the bottom of the pump and circulated upward where it is discharged gently at the surface of the fluid. Thus aeration of the medium is maintained while it is circulated. Ports are provided for sampling, air exhaust, feeding, and harvesting. The process for growing tissue culture cells and using them in production of virus or biochemicals follows below.

The apparatus is inoculated with cells in sufficient medium to immerse the plates, and the culture is aerated for 15 minutes to distribute the cells and equilibrate the medium. The air is stopped to permit the cells to settle on the top of the glass plates and attach to the surface. This may be different for various cells. However, most cells will settle out in 4 hours. Since circulation is so gentle the cells are not disturbed when the circulation is begun and they will attach

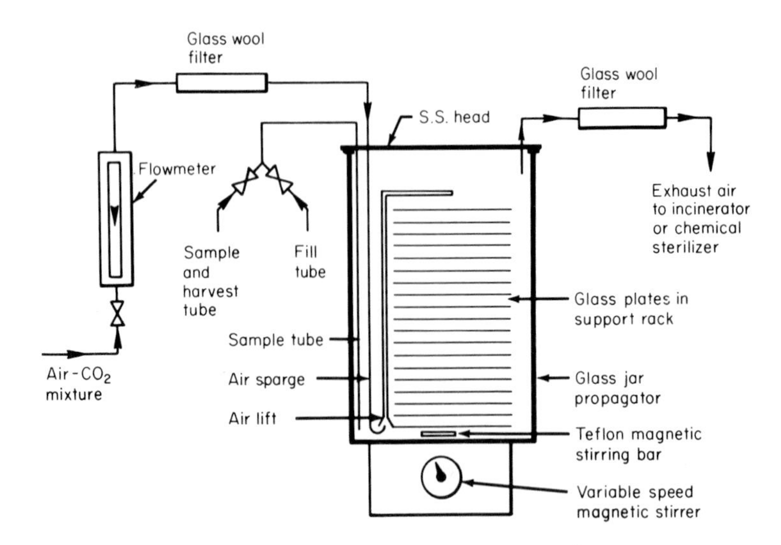

Fig. 1. Flow sheet: multisurface mass scale tissue culture propagator.

and begin to proliferate. Control of pH, dissolved oxygen, and carbon dioxide is attained by controlling the rates of gassing and by the composition of the gas mixture.

Once confluent monolayers are formed, the growth medium is removed and the cells are either harvested by trypsinization or washed and left for virus production or biochemical production. Tables I and II list the various types of cells and organisms produced in the MTCP. The MTCP has also been used to produce large quantities of the biochemicals interferon and urokinase.[8]

In order to facilitate contact between virus and cells in the MTCP a Teflon magnetic stirring bar may be added. Additional circulation may be attained by suspending the airlift pump in a central well by cutting 1-inch diameter holes in the center of the plates. Another method for increasing circulation, useful for removing cells from the glass surfaces, is to attach a thin Teflon paddle to a

TABLE I

Types of Cells Grown in the MTCP

Primary
Newborn bovine kidney (BK)
African green monkey kidney
Duck embryo
Chick embryo
Human embryonic kidney
Human diploid
Embryonic lung (WI-38)
Neonate foreskin (HF)
Serial cell lines
Baby hamster kidney (BHK-21)
Mouse (CCLI)

[8] G. H. Barlow and L. Lazer, *Thromb. Res.* 1, 201 (1972).

TABLE II
Types of Agents Produced in the MTCP

Influenza viruses
Mumps
Parainfluenza viruses
PPLO
Rubella virus
Rhinoviruses
Respiratory syncytial viruses

magnetic stirring bar in the central well. During virus production, the exhaust gas is usually discharged to an incinerator or chemical trap to kill any virus that it may contain.

The usual fermentation instrumentation such as pH and dissolved oxygen electrodes and automatic monitoring and controlling devices can be applied to the MTCP. Equipment for continuous feeding and sampling can also be provided.

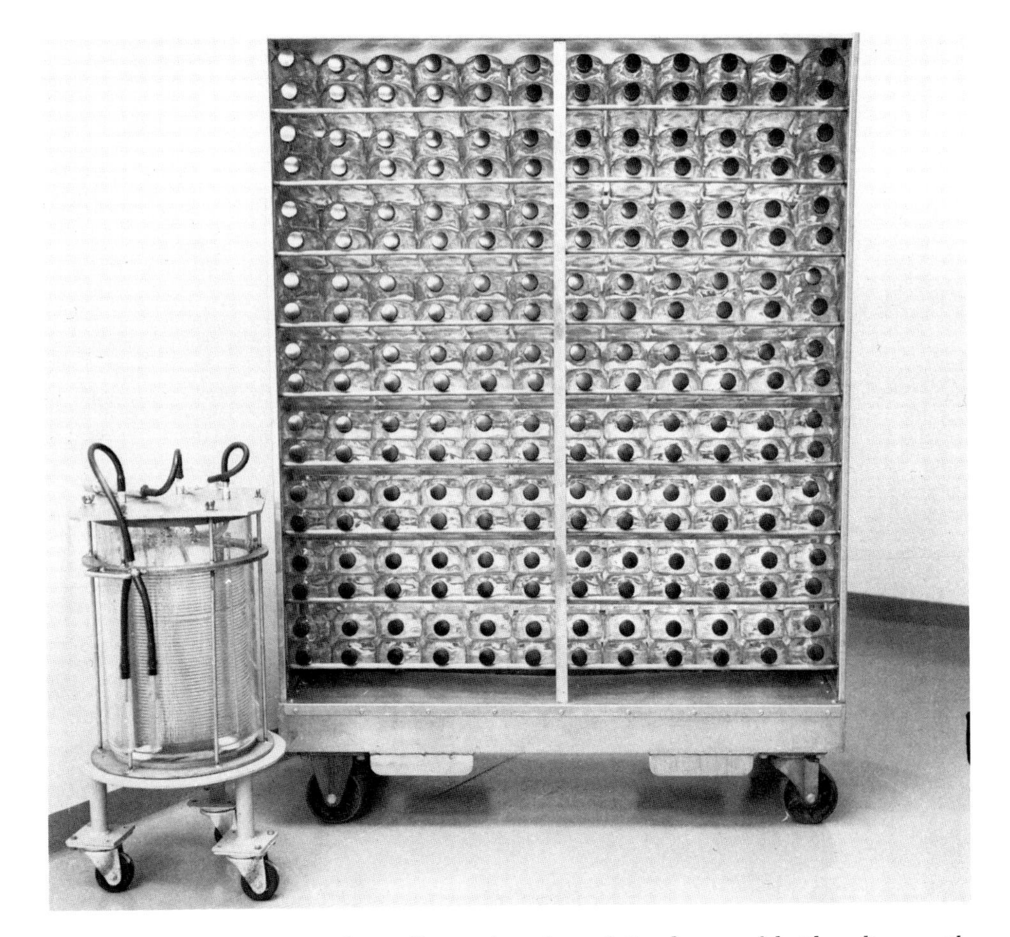

Fig. 2. Illustrating equivalent cell growth surface of tiered rows of bottle cultures with the multisurface propagator.

Surfaces other than glass can be used. However, ordinary single strength window glass is a very economical surface for cell growth since it can be reused many times.

Monitoring of cell growth was successfully followed by measuring glucose utilization which is an easy and accurate procedure. New microscopic systems are now available with long working distance lenses to directly observe cell growth in the MTCP.

The advantages of the propagator are basically operational and result because the MTCP provides a large surface area on which cell growth can occur in a relatively small total volume when compared with the large total volume required for equivalent cell growth surface with bottles and flasks (Fig. 2).

This results in less handling, reduced labor, reduced floor space, and less contamination. It also results in increased plant capacity. There is a fourfold reduction in labor and a two- to threefold reduction in space requirements. Other advantages in the use of the propagator include greater purity of product. This results from better washing of the cells prior to use. There is better control of environmental conditions during cell growth and production, as well as lower costs for tissue culture products.

CHAPTER 11

Bulk Culture of Cell Monolayers

William House

Increasing demand for cells requires increased production and improvements in production techniques.

Over the past 10 years 10-g quantities of cells which multiply only when they are attached to a surface have been grown in roller bottles,[1,2] bottles containing particles such as Sephadex[3] or, in culture vessels containing disks stacked on a spindle.[4] Roller bottles have given excellent service and they are cheap enough to be disposable. By contrast I have not found the other two methods satisfactory

[1] W. House and P. Wildy, *Lab. Pract.* **14,** 594 (1965).
[2] P. F. Kruse, L. N. Keen, and W. L. Whittle, *In Vitro* **6,** 75 (1970).
[3] A. L. Van Wezel, *Nature* (*London*) **216,** 64 (1967).
[4] R. E. Weiss and J. B. Schleicher, *Biotech. Bioeng.* **10,** 601 (1968).

for routine use, not least because they are difficult to automate and the preparation of the surfaces poses problems.

The disadvantage of roller bottles is that their surface-to-volume ratio is low, and, they cannot be used economically to grow 100-g quantities of cells because the number of bottles required takes up much warm room space and, of course, increases labor costs. When it became clear that cells would be required in 100-g quantities we decided to devise culture vessels which have a high area-to-volume ratio, are inexpensive, and are made of commercially available, sterile and disposable components. An additional requirement was that it should be possible to remove the cells without using enzymes.

Having tested various methods and materials cited in the literature we decided to pursue the use of Melinex, a polyester plastic film which is commercially available[5] in various thicknesses and widths, which can be autoclaved, and which had been used with good results by Firket.[6] (See also Section VII.) We found cells grew best on Melinex if it were washed by the standard method used for tissue culture glassware before being autoclaved.

The simplest way to obtain a high surface area-to-volume ratio was to insert a spiral of Melinex in a container. Preliminary experiments were done with a spiral of Melinex wound onto the film spool of a stainless steel photographic developing tank. The results were most encouraging; a good yield of cells was obtained. Obviously, the principle of the method was sound but we had to devise cheap plastic vessels to replace the stainless steel tanks. We experimented, therefore, with plastic tanks in which we fitted a spiral of Melinex, the surface of which was spaced (4-mm separation) by attaching corrugated strips of Melinex, 1 cm wide, to the sheet before it was wound into a spiral. The vessel was then filled with medium, seeded with cells, and gassed with 5% CO_2 in air from a cylinder via an airlift pump inside the vessel.[7] After various modifications were made to prototype vessels of this sort the system was passed to a commercial manufacturer of plastic products experienced in production of tissue culture dishes.

The commercially produced vessel[8] (Fig. 1) is a 2-liter clear polystyrene bottle molded in two halves to allow the insertion of Melinex spiral and, once assembled, the joint is sealed by the makers. The Melinex spiral is 200 cm $\times$ 20 cm $\times$ 125 μm with a 4-mm spacing. The vessel has three screw-capped inlets/outlets. The small central inlet contains a tube which passes through the center of the spiral to the bottom of the vessel for aeration. The other small opening serves as an outlet for gas; cells and medium are added through the peripheral larger opening. The vessel and airfilter tubes, which have screw attachments to fit the small outlets, are supplied sterile in a plastic bag.

[5] Melinex: Manufacturer, Imperial Chemical Industries Ltd., Plastics Division, Welwyn Garden City, Herts., England; Supplier, Boyden Data Papers Ltd., Parkhouse Street, Camberwell, London, England.

[6] H. Firket, *Stain Technol.* **41**, 189 (1966).

[7] W. House, M. Shearer, and N. G. Maroudas, *Exp. Cell Res.* **71**, 293 (1972).

[8] I.C.R.F. Cell Culture Vessel supplied by Sterilin Ltd., 12 Still Rise, Richmond, Surrey, England or Cooke Engineering Co., 900 Slaters Lane, Alexandria, Virginia 22314.

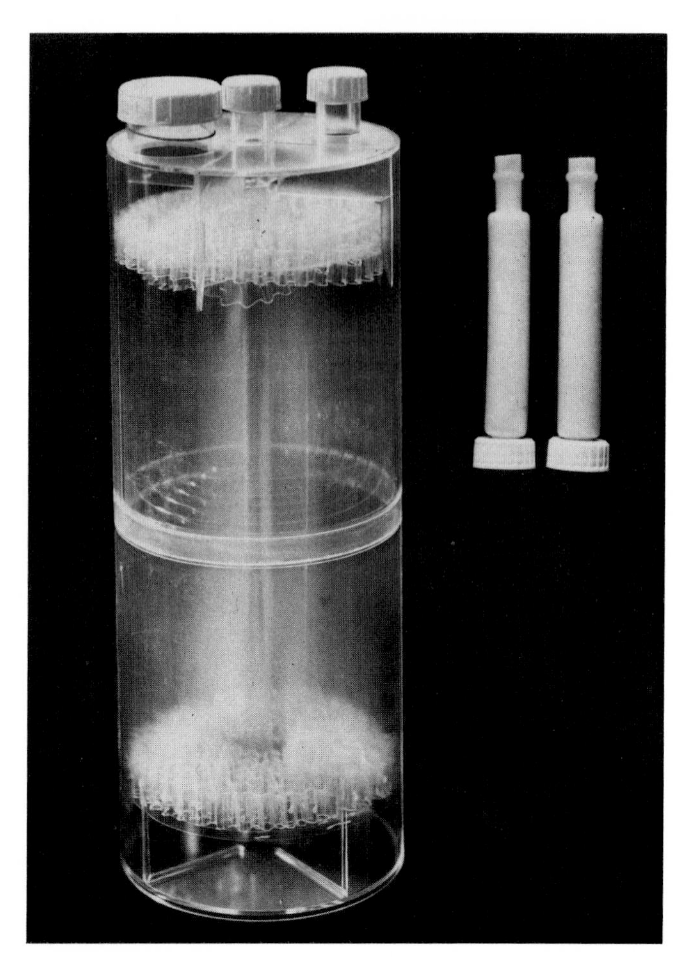

Fig. 1. Two-liter roller bottle assembly containing spiral of Melinex, 200 cm × 20 cm × 125 μm with a 4-mm spacing.

From our experience with these commercially made vessels we are convinced that the spiral system of cell culture is satisfactory for large-scale cell production. Table I shows typical yields of primary cell cultures of whole mouse embryos, $BHK_{21}C_{13}$, and BALB/c 3T3 cell lines grown on Melinex coils. These cells, and others, grow at least as well on Melinex as on glass and the Melinex spiral vessels are easier to handle, and save space and labor.

Mowatt and Radlett, at the Animal Virus Research Institute at Pirbright, have used basically similar Melinex spiral culture vessels and obtained similar yields of $BHK_{21}C_{13}$ cells. They have also shown that these vessels are suitable for the production of large amounts of FMD virus; indeed, the yield of FMD per cell is the same irrespective of whether the cells are grown on Melinex spirals or the surfaces of Roux bottles. Mowatt and Radlett find, as we do, that the Melinex sheet can be washed, sterilized, and repeatedly reused. They have used one spiral twenty times without any decline in cell yield.

TABLE I
Comparison of Cell Yields from Equal Areas of Melinex and Glass

Cells	Incubation (days)	Cell inoculum	Cell yield
Melinex (8000 cm²)			
$BHK_{21}C_{13}$	4	1×10^8	2×10^9
Whole mouse embryo	5	2×10^9	1×10^9
BALB/c 3T3	3	5×10^7	7×10^8
Glass (10 bottles, total area 8000 cm²)			
$BHK_{21}C_{13}$	4	1×10^8	2×10^9
Whole mouse embryo	5	2×10^9	1×10^9
BALB/c 3T3	3	1×10^8	5×10^8

Method of Using Culture Vessel

1. Add 1600 ml of suitable growth medium containing 0.4 ml of antifoam (MS Antifoam Emulsion RD)[9] to the bottle. A suspension of cells in 10 ml of medium is then added to the bottle.

Fig. 2. Assembled bottles resting on roller machine during cell dispersement and attachment at 2 to 4 rph.

[9] MS Antifoam Emulsion RD supplied by Hopkins and Williams, Freshwater Road, Chadwell Heath, Essex, England.

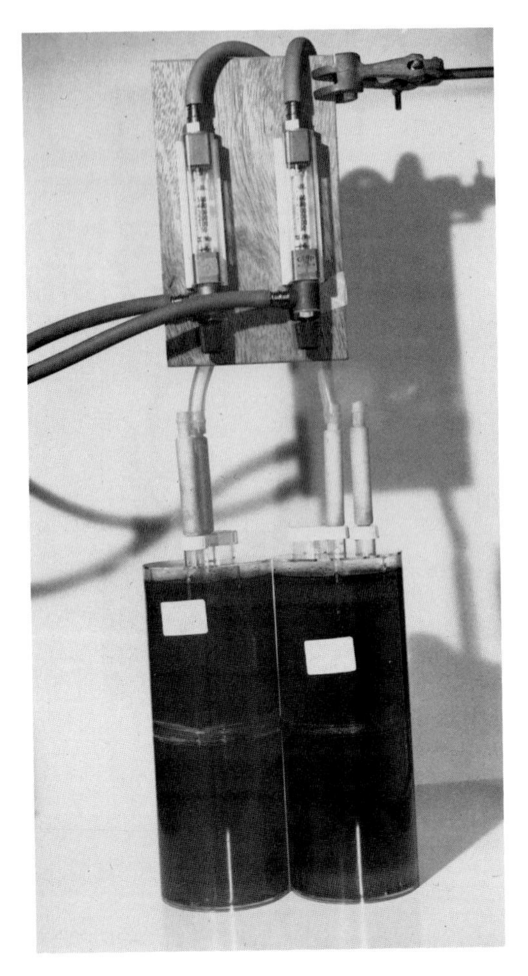

Fig. 3. Final assembly of bottle during cell growth period.

2. The cells are dispersed evenly throughout the bottle by gently inverting it five to ten times.

3. The bottle is then placed on a roller machine[10] (Fig. 2) and is rotated at two to four rph (faster speeds cause aggregation of the cells and uneven growth).

Cells of most lines adhere to the plastic within 3 hours but a few lines take longer and sometimes it is more convenient to leave the bottle rolling overnight. It is a simple matter to place the bottle under an inverted microscope and examine the outer layer of the spiral to check attachment of the cells.

4. When the cells have attached, the caps on the small diameter openings are replaced by the two air filters and a supply of 5% CO_2 in air is connected to the inlet filter. The flow rate of about ten bubbles per minute is regulated by a flowmeter (Fig. 3). The cultures are allowed to incubate for the time necessary for

[10] Roller Machine manufactured by Luckman Ltd., Victoria Gardens, Burgess Hill, Sussex, England.

Fig. 4. Partially assembled 20-liter roller bottle system.

the particular cells. If necessary during this incubation period the cultures can be examined with an inverted microscope.

5. To harvest the cells, the gas supply is disconnected from the inlet filter and the filters replaced by sterile screw caps. The medium is discarded and the cell sheets washed by adding 400 ml of phosphate-buffered saline and inverting the bottle gently three to five times. The saline is replaced by 200 ml of 0.25% of trypsin at 37°C, and the cells are washed in trypsin by inverting the bottle ten to twenty times over a period of 5 minutes, or, alternatively by placing the bottle with trypsin on a fast roller (2 rpm) for 10 minutes.

6. Once all the cells are in suspension they are removed and centrifuged in the normal way. The transparency of the Melinex makes it easy to see when the cells have been removed.

We have not yet devised a simple way of removing the cells from the Melinex spiral without using enzymes. At present we crack open the vessels, remove the spiral, and scrape off the cells with a silicone rubber blade, collecting them in a stainless steel tray.

We, and the Pirbright team, are now trying to increase production per unit space by using larger, 10- to 20-liter instead of 2-liter, vessels.

Preliminary results at Pirbright suggest that the 10-liter vessels only produce a fivefold increase in cells instead of the eightfold increase they obtain with the 2-liter vessels. So far, the reason for this lower yield has not been found.

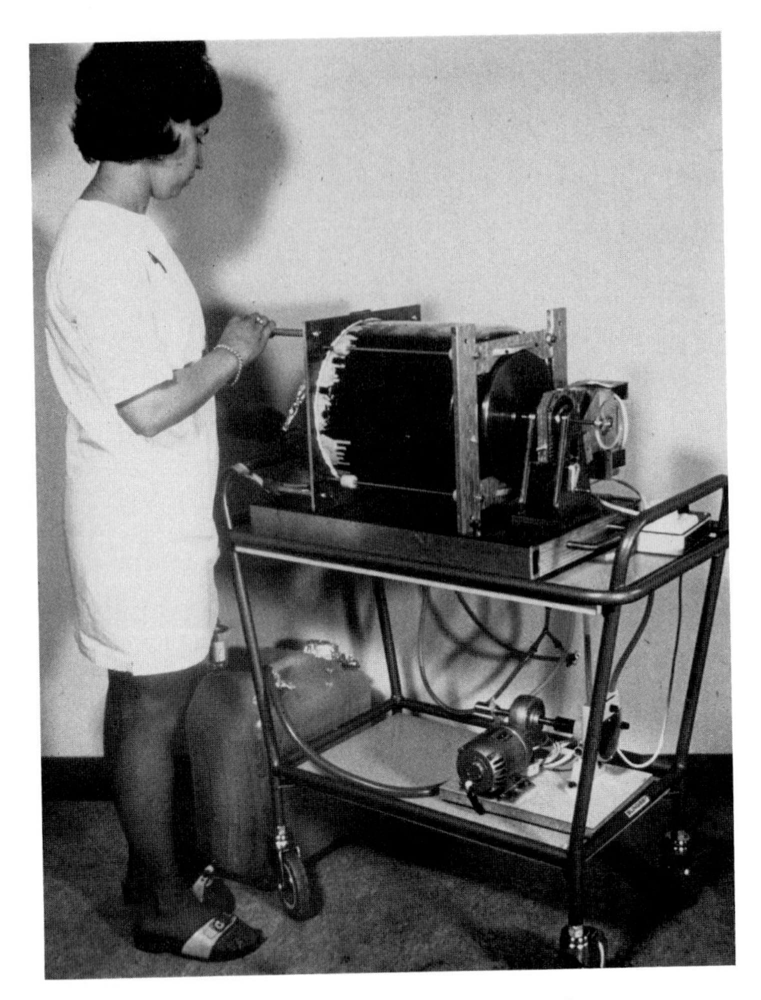

Fig. 5. Final assembly of 20-liter roller bottle system.

In this laboratory we have reached the stage where we are using an average of twenty-five 1-liter vessels per week, producing approximately 100 g cells per week. There is no doubt in my mind that we will require a 20-liter machine capable of producing 50 g cells per run, and this is now being developed. (Figs. 4 and 5.)

CHAPTER 12

Spin Filter Device for Suspension Cultures

Philip S. Thayer

The spin filter culture device[1] was developed for use in bacterial and cell culture studies to serve two purposes: to grow cells to high population densities, and to provide a system for the exposure of cells in suspension culture to cytotoxic drugs with provision for removal of drug at accelerated rates, exponentially or otherwise. It has been used in our laboratories since 1967 with cultures of murine tumor cells including L1210 and P388. The system should be adaptable to any suspension growing cell which can tolerate the conditions of spinner culture. Some suspension growing cells either will not tolerate these conditions or may require a period of selection or adaptation. The device has been used in other laboratories for the growth of bacteria and phytoflagellates. The present description is based on the commercially available device[2] which differs in details of construction from the original device described from this laboratory.[3]

Principles of Operation

The principle of the spin filter device is that rotating a filter of appropriate construction and geometry will prevent the accumulation of cells (or other particles) on the surface or within the filter. In aqueous media, this is effected not through centrifugal action but through turbulence of the boundary layer which has the effect of sweeping the surface of the filter clean, except for perhaps a thin layer of cells in the laminar flow region at the very surface of the filter. Since all removal of culture medium is effected through the filter, the culture has the aspect of a batch culture system. Provision could be made for the separate removal of some total culture (including cells) so that the device could be used for a high cell density chemostat.

Description

A partially schematic diagram of the spin filter device is shown in Fig. 1. The path of withdrawal of the culture fluid is through the filter (B), into the annular space within it, and through the diagonal hole (D) in the top support,

[1] U. S. Patent, 3,647,632, March 7, 1972, D. E. Johnson, L. R. Woodland, C. J. Kensler, and P. Himmelfarb.

[2] ADL Virglas Biospin Filter, The Virtis Company, Gardiner, New York 12525.

[3] P. Himmelfarb, P. S. Thayer, and H. E. Martin, *Science* **164**, 555 (1969).

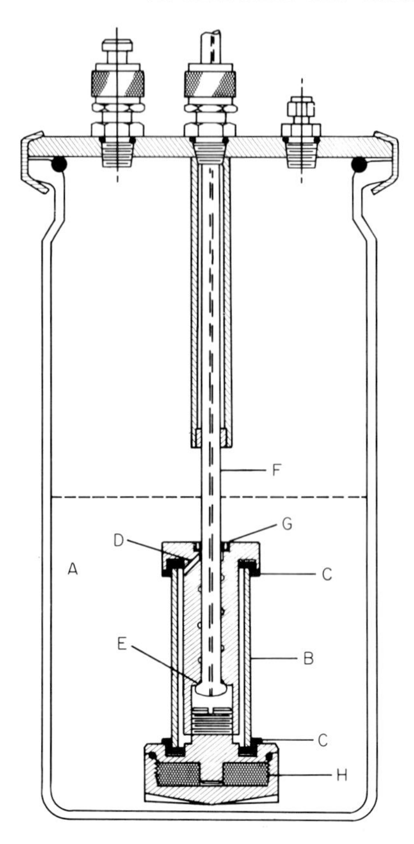

Fig. 1. Spin filter culture vessel. (A) Culture; (B) filter; (C) filter gasket; (D) diagonal flow channel; (E) bearing surface; (F) withdrawal tube; (G) ring seal; (H) magnet. Drawing courtesy of The Virtis Company.

then downward through a spiral pathway, around the glass withdrawal tube (F), past the bearing flange at the bottom of the tube, then into the hollow center and up and out the tube. Rotation of the filter is provided by a magnetic stirrer external to the water bath in which the flask is immersed. The bearing upon which rotation occurs is provided by the flange at the end of the stationary glass tube (E). A seal to prevent loss of culture (including cells) is provided by the ring seal (G) at the top of the filter support. Culture vessel volumes can be varied from 200 to 1000 ml. The cover is provided with swage-lock fittings for the introduction of accessory tubes including those for medium inflow, gassing, sampling, and gas-exit. Hypodermic needle stock with Luer-Lok fittings is used for these various tubes. One large fitting is provided in the cover for the liquid level probe.

Filter Materials

The original description of this device[3] embodied the use of a membrane filter cemented to a stainless steel support structure. The presently available commercial device allows the use of several interchangeable filter media. We have used primarily the Balston filter cylinder[4] to which the dimensions of the filter holder were designed. However, since only the inside diameter of the Balston cylinders is closely controlled due to the method of manufacture, we have found it necessary to seal the outside junction of this filter and the rubber gasket at each end of the cylinder with Silastic.[5] This seal is particularly needed for flow rates above 100 ml per hour which often reveal the presence of leaks in the system (determined by the appearance of cells in the effluent liquid). There are also filters of the same geometry made from ceramic material, fused glass beads, sintered metal and wound wire, as well as a plastic holder for supporting a membrane filter sleeve.

The criteria which we apply in evaluating the use of any new filter material are twofold: (1) the material should be nontoxic (by test) for the particular cell line and medium to be used; (2) because of the physical principles involved, the surface of whatever filter medium is used must be smooth enough and of sufficiently small porosity at the surface that it has the nature of a membrane surface. It is not sufficient to have a filter material with a nominal porosity which depends upon the depth of the filter.

Ancillary Equipment

The pump for withdrawal of culture filtrate can be of a number of types, that is, rotary peristaltic, "finger" type, or a multiplace piston-type pump. The main requirements of this system will be satisfied by any pump which allows reasonable control of pumping rate over a range of 10 to 500 ml per hour depending upon the application and culture volume employed. We have found a multiplace piston pump[6] to be particularly suitable since it obviates the problems of wear and creeping of tubing. It also gives reproducible control both of rate and volume of pulse, the latter being independently controlled for each piston in the multiple array.

In our operation of the spin filter we have used a liquid level sensing probe with appropriate electronic circuitry to control the opening and closing of a solenoid which allows fresh medium to flow by gravity into the culture vessel. The probe is sufficiently sensitive that variations in culture volume are kept to a minimum. One essential feature of this device for use with cell culture media containing serum or other foam-producing materials is a glass sleeve around the probe with a hole at the bottom for admission of liquid and an opening at the top. The sleeve protects the sensor probe from contact with foam which is sufficient to give a false surface reading.

Depending upon the requirements of the cell line being cultured, it may be necessary to provide for flushing the culture with a gas mixture even though this

[4] Microfibre filter tubes, ¾ × 6 cm, Grade AA, Balston, Inc., Lexington, Massachusetts.

[5] Silastic 382, Medical Grade Elastomer, Dow-Corning, Midland, Michigan.

[6] Multiple metering pump, Model 1508, Harvard Apparatus Co., Inc., Dover, Massachusetts.

may not be required in ordinary spinner cultures. For example, with L1210 cells we have found it necessary to use a mixture of 5% CO_2 in air, bubbled through the culture medium with stainless steel needle stock. For a culture volume up to at least 0.4 liter, a sparger does not appear to be necessary.

In experiments such as short-term drug effect studies, repeated sampling several times a day necessitates some precaution to avoid contaminating the culture. For this purpose, we use a sampling tube permanently installed through one of the swage-lock fittings, such that samples of whatever volume desired can be withdrawn aseptically and delivered to a container.

Temperature control may be provided by a water bath, infrared lamp or by jacketing the vessel, depending upon facilities available. It is necessary to allow minimum bottom distance for proper coupling with the magnetic stirrer.

Since there is friction in the bearing and seal system upon which the filter rotates, a magnetic stirrer must be used with a strong enough magnetic field to ensure coupling. If the magnets decouple and rotation stops while pumping of culture fluid continues, cells will be drawn into the filter causing clogging.

General Operational Procedures

When using Silastic-sealed Balston filters, the filter assembly is cemented in place several days prior to use to allow for evaporation of the residual vapors. The glass outlet tube must be placed through the filter holder before the filter is put on. To prepare for sterilization, the level probe, complete sampling device, gas tube, and a lead for medium inflow are fitted through the appropriate ports in the cover. Each is plugged with cotton as needed for aseptic connections. The lead for medium inflow ends in an 18-gauge needle.

Culture medium is prepared separately and provision for connecting the stock bottle to the feed line is made through the use of a drying tube with a serum plug in the large opening. The flask and filter unit are sterilized by autoclaving at 110°C for 60 minutes. When cool, assembly is completed by connecting the media feed line, gas tube, and outflow line to their respective accessories, and attaching the inoculating flask, which is sterilized separately. With the Harvard Apparatus pump the actual pump head including the one-way valves and the piston plunger are sterilized and inserted in the pump housing at the time of assembly. With a new ring seal it is often necessary to run the apparatus without cells for a few days to assure a good bearing surface between the glass tube and the ring seal. For this purpose, the vessel is filled to a convenient volume with culture medium and the stirrer turned on, and the pump set at a low flow rate to provide lubrication. When satisfactory operation is achieved, the inoculation can be made. For this purpose, we use a separate flask from which a volume of medium containing the desired number of cells is introduced through the inlet tube to the vessel. If this results in a greater liquid volume than the final working volume of the culture, the excess is removed by pumping down to the level of the sensing probe.

Cell Propagation

The propagation of populations of L1210 cells has been effected with the use of Fischer's medium containing 10% horse serum. It has been found efficient to use a high level inoculum, e.g., 1–2 × 10^6 cells/ml, produced in a standard (batch) spinner culture to save time in the initial phases of reaching high population density. When cells at this level are inoculated into a 400-ml culture volume, an initial flow rate of approximately 30 ml/hour produces satisfactory further growth. As growth continues over a period of several days, the medium flow rate is increased gradually, gauged by cell density. When the population has reached 10^7 cells/ml, a flow rate of 50 to 75 ml/hour appears adequate in a 400 ml culture. It is possible by protracted incubation at somewhat increased flow rates to reach cell densities approaching 10^8 cells/ml. Although exhaustive studies have not been done, it would appear that the cell doubling time increases markedly at these high densities despite increases in flow rate (Fig. 2).

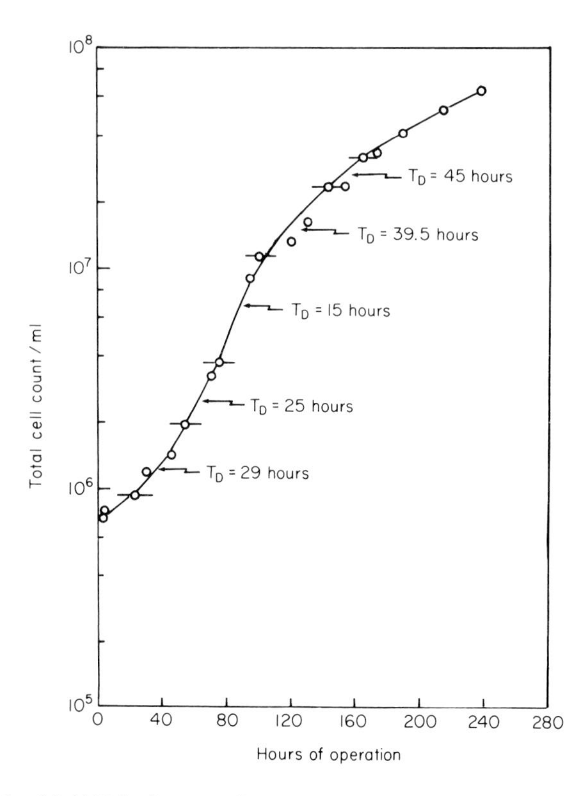

Fig. 2. Growth of L1210 leukemia cells in spin filter culture. Mean flow rates for successive time periods: 0–50 hours, 31 ml/hour; 50–170 hours, 53 ml/hour; 170–240 hours, 69 ml/hour. Fischer's medium, 10% horse serum, gassed with 5% CO_2 in air at 37°C. Culture volume, 400 ml. T_D, doubling time in hours.

Drug Perfusion

When drug is added to a growing culture in the spin filter device, it may be removed by any number of schedules. We have found it particularly pertinent in modeling *in vivo* studies to remove drug at an exponential rate, that is by a constant high rate of medium throughput.[7] The relationship between half-time (in hours) and flow rate is defined by the equation $T_{1/2} = 0.69 \times V/F$ where V equals the working volume in milliliters and F is the flow rate in ml/hours. For example, in a culture volume of 200 ml, a half-time of 60 minutes will be produced by a flow rate of 138 ml/hour. From this equation, further equations can be developed to express residual concentration at any time during perfusion and the integrated $C \times t$ of exposure.[3] Experimental data with tritiated drug have verified the predictions of this equation (Fig. 3).

For other purposes it may be desired to perfuse cells at a constant drug concentration. To do this, it is necessary to add a single dose of drug directly to the culture to bring the concentration up to the desired level and to add drug to the

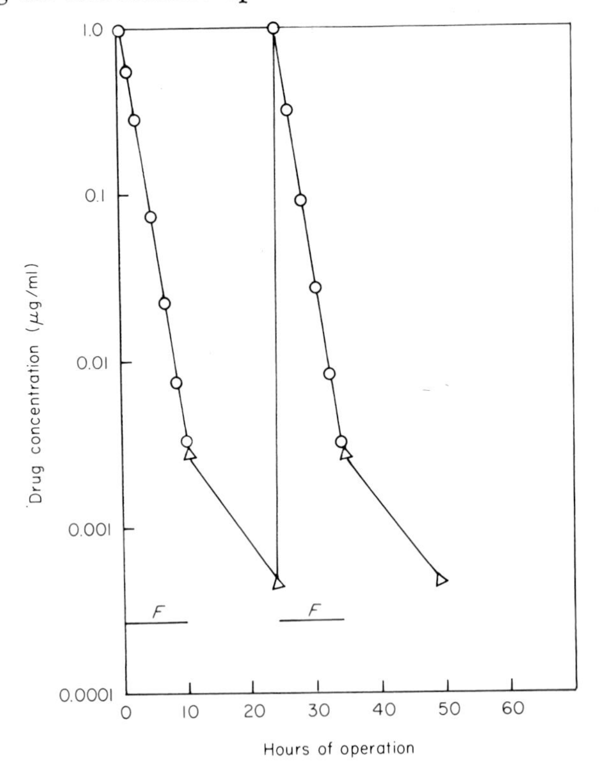

Fig. 3. Exponential removal of tritiated amethopterin in spin filter culture device. Residual concentration: △, calculated from flow rate, ○, observed by scintillation counting. Flow rate, in rapid flow periods (F), 255 ml/hour; in maintenance period 57 ml/hour. Total volume, 400 ml. Half-time: rapid, 1.15 hours; maintenance, 5.3 hours.

[7] P. S. Thayer, P. Himmelfarb, and De W. Roberts, *Cancer Res.* **30**, 1709 (1970).

perfusion medium at the same level. If drug is added only to the input medium, the level of drug in the culture vessel will approach the desired level asymptotically at a rate determined by the flow rate and culture volume.

Other Applications

The published work to date as indicated above has involved only a few of the possible applications of this device. It is expected that it might find application in virus multiplication studies, as a reactor for the formation of metabolic products, as a high cell density chemostat, and to explore the nutritional requirements for continued growth of cells.

CHAPTER 13

Automation of Continuous Suspension Cultures by Means of a Nephelostat

Carl Peraino and William J. Eisler, Jr.

The Nephelostat[1] is a photocell-controlled continuous culture device, capable of maintaining steady state growth in all types of suspension cultures. This instrument was originally developed for use with microorganisms,[2] but its flexibility of design permitted its ready adaptation to mammalian cell cultures.[3]

General Principles of Nephelostat Operation

Light from a tungsten exciter lamp passes through a cylindrical portion of the culture vessel (monitoring probe) which extends down into the Nephelostat detector assembly. The presence of cells in the medium produces light scattering, with the consequent excitation of a photocell offset at 30 degrees from the incident light beam. This excitation unbalances a Wheatstone bridge circuit

[1] Developed by W. J. Eisler, Jr. and R. B. Webb, Argonne National Laboratory. (This instrument is now being manufactured by Lab-Line Instruments Inc., Lab-Line Plaza, Melrose Park, Illinois 60160.)

[2] W. J. Eisler, Jr. and R. B. Webb, *Appl. Microbiol.* **16,** 1375 (1968).

[3] C. Peraino, S. Bachetti, and W. J. Eisler, Jr., *Science* **169,** 204 (1970).

which had previously been balanced with respect to the detector photocell and a reference photocell receiving light from the same exciter lamp. As the cell density increases, the bridge imbalance registers as a positive deflection on a meter on the Nephelostat control console, and this meter reading is also plotted on a strip-chart recorder. The latter provides a continuously recorded analog of cell density. Monitoring of the cell density in this manner occurs at intervals, the frequency of which can be selected by a plug-in timer to accommodate cells with a wide range of growth rates. If the culture is light sensitive, a built-in shutter can be activated which blocks the light from the exciter lamp between monitoring events. The meter mentioned above incorporates a relay which can be set at any point along the meter scale. When the meter reading (cell density) exceeds the relay setting, culture medium is pumped (for a predetermined time) into the growth vessel, with the simultaneous removal of an equivalent volume from the culture. Each time medium is added, the event is recorded on the strip-chart recorder as well as a digital counter. Continuous agitation of the growth vessel contents is provided by a vibrator and the culture is gassed through the bottom of the monitoring probe to ensure its constant equilibration with the main body of the growth vessel. Gassing is automatically interrupted briefly during monitoring to prevent interference with the light scattering measurements. Adjustable instrument settings include: (a) exciter lamp intensity; (b) signal amplification; (c) bridge balance adjustments; (d) monitoring frequency; and (e) pumping time.

The instrument currently available measures scattered light only. This mode of operation is advantageous in monitoring cultures with low cell density; a concentration of less than 10^3 mammalian cells per milliliter can be effectively monitored and controlled. The meter reading responds linearly with increases in cell density up to approximately 10^6 mammalian cells per milliliter. Above this density, the turbidity of the suspension is sufficiently high to prevent scattered light from reaching the detector photocell, and linearity of response is lost. In a recent modification, presently under evaluation in our laboratory, the detector photocell was placed on a movable, calibrated track so that its position is adjustable with respect to the beam from the exciter lamp. Thus, in cultures with high cell density it is possible to move the photocell into direct line with the exciter lamp beam. This, plus the incorporation of a reflectance-transmittance switch into the meter circuitry, permits the accurate monitoring of high cell densities by measuring the absorbance of transmitted light rather than the reflectance of scattered light.

The main features of the Nephelostat which render it useful for mammalian cell cultures are as follows.

1. Ease of operation. The operator need only refill the medium supply reservoir and replace full collection flasks at periodic intervals. All other operations are performed automatically.
2. Versatility. As indicated above, instrument control settings are broadly adjustable to accommodate the growth characteristics of a wide variety of cell types. In addition, the design of the culture vessel can be varied to

meet individual requirements. Theoretically, there is no upper limit on the size of the growth vessel. The only requirements are that: (a) a cylindrical detection probe, able to fit into the Nephelostat detector assembly, is attached in a manner that permits its ready equilibration with the contents of the main growth vessel; and (b) adequate provision is made for temperature control and agitation of the growth vessel contents.

3. Uniqueness of growth rate measurement and control. (a) Growth rate measurement—Since the volume of the culture remains constant, and the volume of each added aliquot of medium can be readily determined, the time required for the complete replacement of the culture vessel contents can be determined from the feeding frequency record. Multiplication of this time value by 0.693 gives the doubling time of the culture.[4] Thus, the growth rate of the cells at any preselected concentration can be determined while the cells are automatically maintained at that concentration. (b) Growth rate control—The Nephelostat is neither a fermentor nor an electronic chemostat.[4] In the chemostat, the growth rate is limited by the rate at which fresh medium flows into the growth vessel; this flow rate is controlled externally. In the Nephelostat, the supply of medium is internally controlled, i.e., by the growth rate of the cells, thus permitting the cells to adapt at their own rate to changing conditions without the danger of their being washed out of the culture. The Nephelostat can be easily converted to chemostat operation by simply bypassing the nutrient feed control and supplying the nutrient at a constant rate. This feature extends the versatility of the instrument because it permits a comparison between cells grown in the same medium under conditions of externally or internally controlled nutrient feeding.

Design of Mammalian Cell Culture System Incorporating a Nephelostat

Commercial Equipment Requirements

1. Mark II Nephelostat[1]
2. Two peristaltic pumps[5]
3. Medical-grade silicone rubber tubing[6]
4. Agitator[7]
5. Light-tight incubator capable of housing the culture vessel, detector assembly, and agitation equipment
6. Air—CO_2 supply

[4] H. E. Kubitschek, "Introduction to Research with Continuous Cultures." Biological Techniques Series, Prentice Hall, Englewood Cliffs, New Jersey, 1971.

[5] Holter micro-infusion roller pump, Model RL175, Extracorporeal Medical Specialties Inc., King of Prussia, Pennsylvania. (The Nylon tubing connectors were replaced by stainless steel copies fabricated at Argonne. These pumps have silicone rubber pumping chambers.)

[6] Silastic, ⅛ inch ID by ¼ inch OD, Dow Corning Corp., Midland, Michigan.

[7] Vibromixer, Model E1, Chemapec Inc., Hoboken, New Jersey. (The stainless steel mixing shaft was replaced with a longer one fabricated at Argonne.)

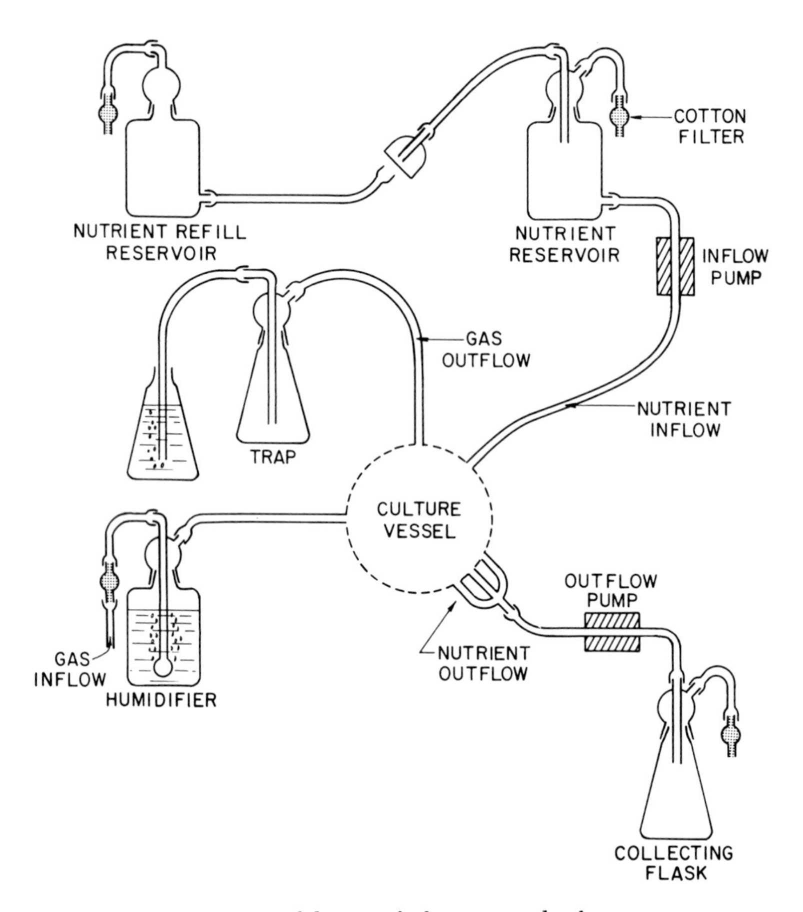

Fig. 1. General layout of glassware and tubing.

General Layout. A schematic arrangement of glassware and tubing is shown in Fig. 1. This assembly resembles that shown previously[3] with two exceptions. First, a trap is now included in the gas outflow line to prevent the sodium hydroxide solution, used as a contamination barrier, from inadvertently being aspirated into the culture vessel. Second, separate pumps are now used to control the inflow to, and outflow from the culture vessel. The inflow pump is controlled by the Nephelostat control console, and the outflow pump runs continuously at a slow rate, preventing the volume of the culture from exceeding the level of the lowest open outflow port on the culture vessel (see below). This was found to be the preferred method for maintaining the culture at constant volume. Since these pumps each have two channels, their pumping capacity has been doubled by connecting both channels in parallel, using silicone rubber tubing and stainless steel "Y" connectors fabricated at Argonne.

Culture Vessel. Figure 2 is a diagrammatic representation of the culture vessel now in use. The rotary stirrer and seal[3] have been replaced by a vertical vibrator

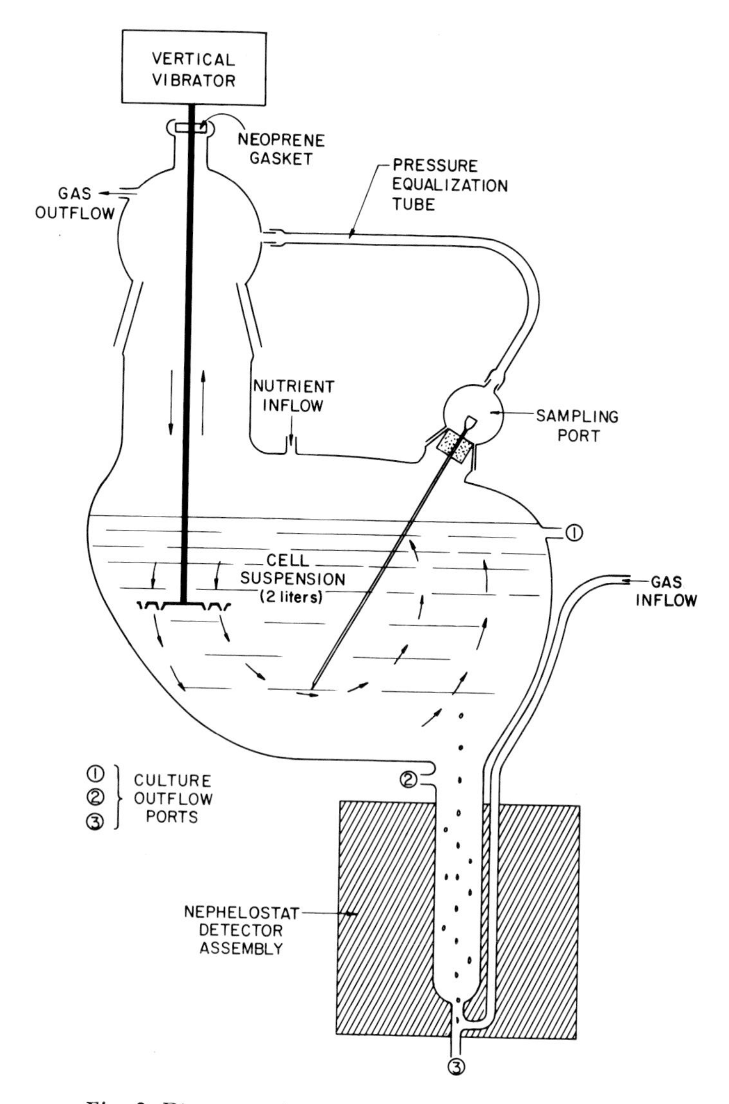

Fig. 2. Diagrammatic representation of culture vessel.

which enters the vessel through an autoclavable neoprene gasket. The turbulence produced in the cell suspension by this agitator is depicted by the circular pattern of arrows. The agitator paddle has been offset from the gas stream to prevent excessive foaming. The gas outflow port is now sufficiently far from the surface of the culture to prevent any foam from entering the outflow tube and causing back pressure. A pressure equalization tube now connects the sampling port cover with the main culture chamber. The vessel also contains three culture outflow ports which are connected to a glass manifold and then via a single tube to the outflow pump. During operation at maximum capacity, the tubes from ports 2 and 3 are clamped. Port 2 is opened to drain the main body of the culture vessel while still leaving the monitoring probe full. This permits long-term maintenance of the culture at low volume, which conserves medium. Port 3 is opened to drain the system entirely or for partial drainage to clear the system of debris (e.g., denatured serum) that might accumulate over a long period.

Setting up a Typical Experiment

Autoclaving. The interior of the culture vessel is siliconized to inhibit the attachment of cells to the glass. Tubing connections are made among the various parts of the system as shown in Fig. 1, using silicone rubber tubing throughout. (The nutrient refill reservoir is not connected to the nutrient reservoir.) The free ends of the gas outflow tubing and the tubing of the nutrient refill reservoir are covered with aluminum foil, the tubing between the cotton filter on the gas inflow line and the humidifier is closed with a pinch clamp, and the humidifier is three-quarters filled with distilled water. Aluminum foil is also placed over the glass bell of the connector on the nutrient reservoir, to which the tubing from the nutrient refill reservoir attaches during filling operations. The entire glassware assembly is then autoclaved for 4 hours to ensure the killing of all mold spores.

Mounting. The glassware assembly is placed in position, with the monitoring probe of the culture vessel extending into the Nephelostat detector and the pumping chambers mounted on the peristaltic pumps. The pinch clamp is removed from the tubing between the cotton filter and the humidifier and placed on the tubing between the glass bell connector and the nutrient reservoir. Other pinch clamps are placed on the tubing from the nutrient refill reservoir and outflow ports 2 and 3 on the culture vessel. The end of the gas outflow tube is unwrapped and placed in a solution of 1 *N* NaOH. The gas inflow tube is connected to the regulator from the Nephelostat which is, in turn, connected to the gas supply.

Filling. Medium is prepared (serum should be filtered to remove particles) and added to the nutrient refill reservoir. The glass bell fitting on the nutrient reservoir is then uncovered and flamed. The end of the tubing from the nutrient refill reservoir is held under the bell, uncovered, and pushed onto the tapered end of the glass connector inside the bell. The nutrient reservoir is allowed to fill by gravity, after which the pinch clamp on the nutrient reservoir is closed. The tubing from the nutrient reservoir is disconnected from the glass bell fitting, and the fitting is flamed thoroughly and covered with sterile aluminum foil. The nutrient reservoir is then placed in a refrigerator located beside the Nephelostat. The inflow pump is connected to a wall outlet and the culture vessel is pumped full of medium. The inflow pump is then connected to the control receptacle on the Nephelostat console. The vibrator and gas supply are turned on and adjusted to provide maximum agitation with minimum foaming. Cells are added through the sampling port by means of a sterile syringe.

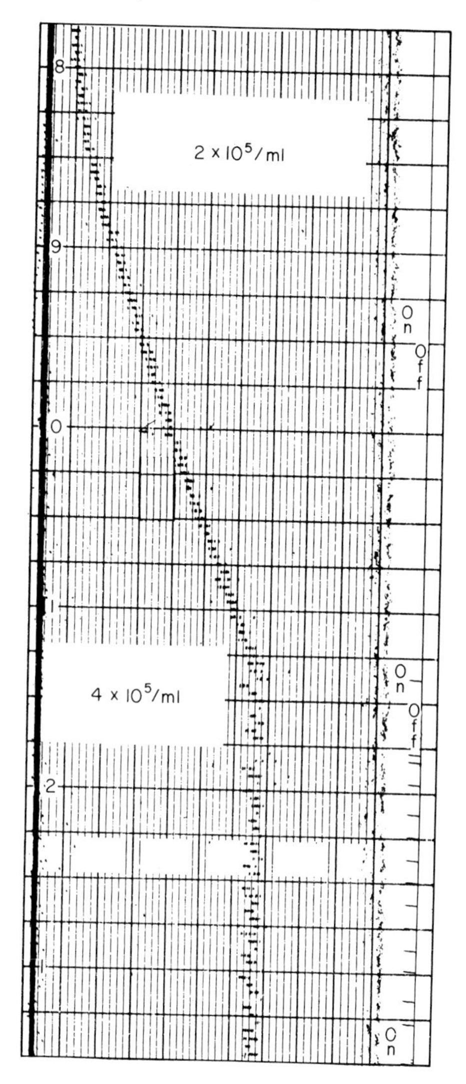

Fig. 3. Nephelostat recording showing monitoring and control of the growth of Chinese hamster cells.

Control and Monitoring. Specific directions for operating the Nephelostat controls are supplied with the instrument and will not be described here. As stated previously, the appropriate setting of these controls will automatically maintain the cell density at any desired level while simultaneously providing a record of growth rate. A sample of monitoring data recorded by the Nephelostat for a culture of Chinese hamster cells is given in Fig. 3. Prior to taking this record, the cells were allowed to grow in the Nephelostat to a concentration of 8×10^5 cells/ml, at which point they entered stationary phase. After 12 hours under these conditions they were diluted to a concentration of 2×10^5 cells/ml and, after a short lag, again entered log phase. The Nephelostat controls were set

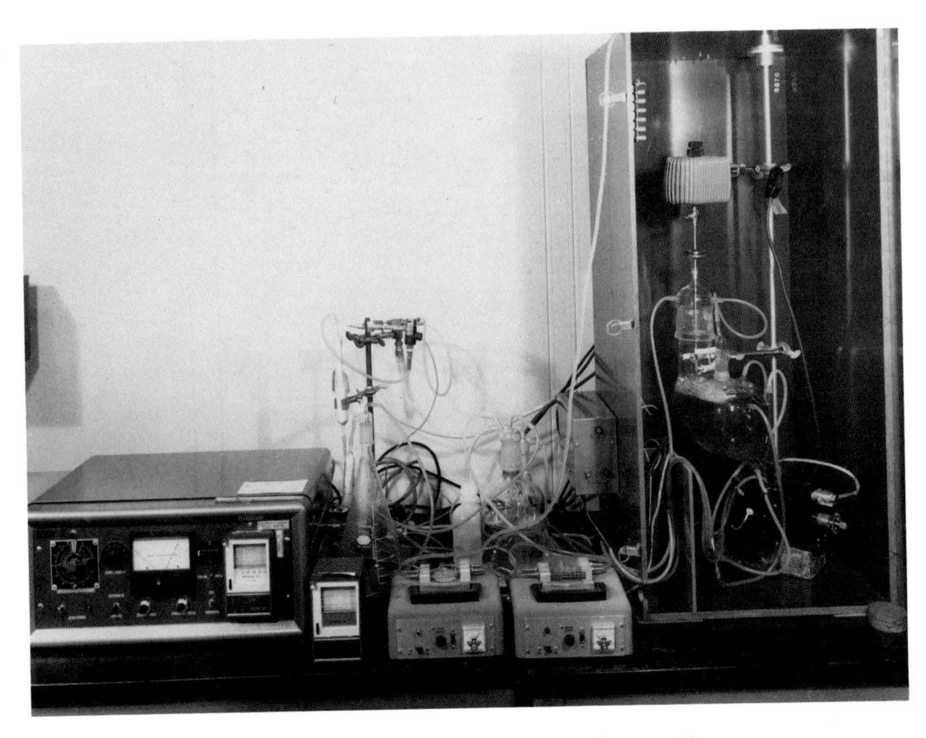

Fig. 4. View of the entire apparatus in operation.

to maintain the cell density at 4×10^5 cells/ml, and began adding medium when the cells reached that density. The record shows the lag period, the period of growth, and the feeding frequency (event marks on right side of chart) during maintenance of the cells at the preselected concentration.

Figure 4 is a photograph of the entire apparatus (nutrient reservoir is in refrigerator at right, not visible) maintaining a culture of Chinese hamster cells subline V79-350I.[8]

[8] S. Bacchetti and W. K. Sinclair, *Radiation Res.* **44**, 788 (1970).

CHAPTER 14

A Chemostat for Single Cell Cultures of Higher Plants

W. G. W. Kurz

A serious disadvantage in the use of plant cell suspension cultures for physiological or metabolic investigations is the strong tendency for the cells to grow as cell aggregates; so far no treatment or device has been found which will avoid this. A variety of culture vessels have been described for growing plant cells in liquid suspension culture.[1-5] These vessels employ the conventional methods of agitation and aeration using different types of stirrers or spargers. Because most plant cells form aggregates in suspension culture, only heterogeneous cell populations are obtained in these vessels. Such fermentors may be adequate for large-scale production of cell mass or where plant tissue cells are employed in the production of secondary metabolites, but they are unsatisfactory for physiological or metabolic investigations where single cells are desired.

The fermentor described here is based on a novel principle which prevents the cells from clumping by using pulses of compressed air for agitation and aeration.

Apparatus

The fermentor (Fig. 1) is made from a thin-walled Pyrex glass cylinder (55 × 750 mm), flat bottomed with rounded corners, providing a working capacity up to 1800 ml. A 25-mm long glass pipe (Ø 15 mm) at the center of the base serves as a port through which air for agitation and aeration enters the fermentor.

Compressed air, at 5 to 10 psi, is passed through a large fiberglass filter for sterilization, and pulsed via a Skinner magnetic valve Type R2DX41 (Skinner Electric Valve Division, New Britain, Connecticut) and a flexible tubing into the fermentor. A second fiberglass filter is placed between the magnetic valve and the fermentor to trap any metal powder abrased from the plunger of the

[1] G. Melchers and L. Bergmann, *Ber. Deut. Bot. Ges.* **71**, 459 (1959).
[2] C. Wang and E. J. Staba, *J. Pharm. Sci.* **52**, 1058 (1963).
[3] W. Tulecke, *Ann. N. Y. Acad. Sci.* **139**, 162 (1966).
[4] R. A. Miller, J. P. Shyluk, O. L. Gamborg, and J. W. Kirkpatrick, *Science* **159**, 540 (1968).
[5] I. A. Veliky and S. M. Martin, *Can. J. Microbiol.* **16**, 223 (1970).

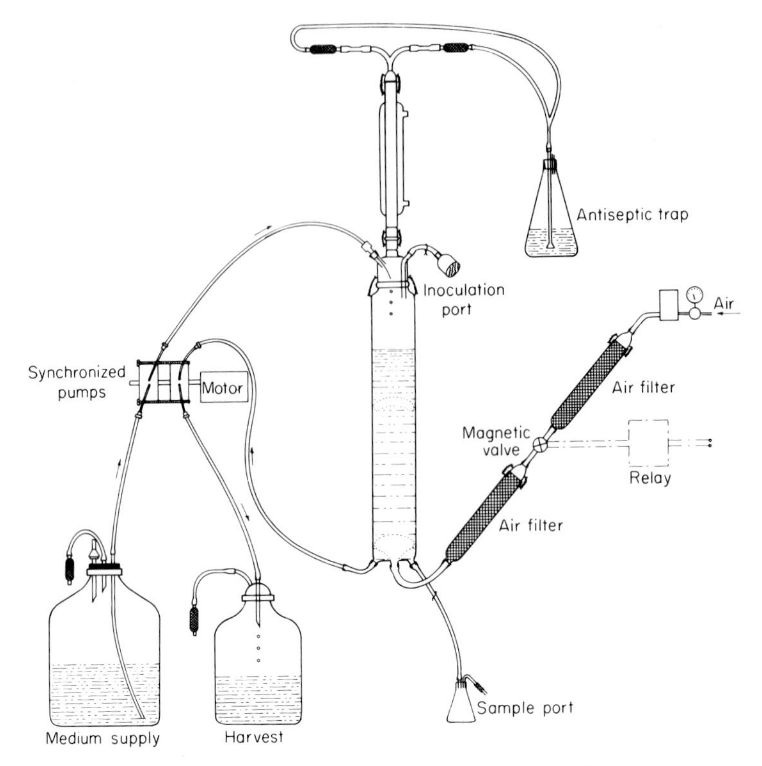

Fig. 1. Diagram of chemostat and auxiliary equipment.

valve during operation. All nonstainless steel parts of the valve are coated with Glyptal Cement (Canadian General Electric Co. Ltd.) to minimize corrosion. Operation of the Skinner magnetic valve is regulated by a control relay, for which a circuit diagram is shown in Fig. 2. This relay allows variable settings of the intervals between valve openings as well as the duration of openings.

At the base of the fermentor, opposite each other and at an angle of 120°,

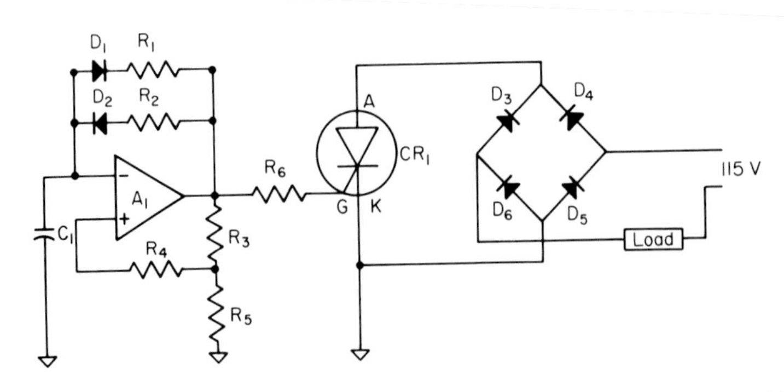

Fig. 2. Circuit diagram for control relay. A, anode; A_1, Burr-Brown 3267/12C; CR_1, 2N3228 (1 A, 1000 V SCR); C_1, 15 μF 15 V; D_1, D_2, silicone powder diodes 1 A, 100 V; D_3–D_6, silicone power diodes 1 A, 1000 V; G, gate; K, cathode; R_1, 500 K time off control; R_2, 250 K time on control; R_3, 100 K; R_4, 10 K; R_5, 1M; R_6, 1.8 K.

and vertically protruding, two smaller glass pipes are used as harvest and sampling outlets. The top of the fermentor has a standard taper joint (₸ 55/50). To it is fitted a top piece comprised of a condenser, where the effluent air is dehumidified, as well as inlet tubes for medium and inoculum. The inlet tube for medium is sealed by a rubber serum cap through which a hypodermic needle provides the medium to the vessel. After the effluent air has passed through the condenser and two fiberglass filters it is released via a sintered-glass plate into a vessel containing 1% $CuSO_4$ solution. This vessel serves as an antiseptic trap preventing possible contamination through the air exit line. All standard taper joints on the fermentor are secured by steel springs.

To ensure the constant flow of fresh medium into the fermentor as well as an equal flow of culture into the harvest bottle, two Masterflex Model 7014 roller pumps (Cole Parmer Co. Chicago, Illinois) are connected to a DC motor equipped with a rheostat for speed control.

Durrum 11304 latex rubber pumping tubes (Durrum Instrument Corp., Palo Alto, California) (1/16 ID × 1/16 inch wall) were used for all lines passing through the pumps.

The sterile culture medium B5[6] is kept in a 20-liter Pyrex glass bottle equipped with stainless steel inlet and outlet fittings. A similar bottle is used for collecting the harvested culture. Luer-Lok fittings (Becton Dickinson Co., Rutherford, New Jersey) are used to connect medium and harvest lines to the fermentor.

Operation

The apparatus is autoclaved in three sections: (a) the culture vessel, the top piece with the condenser and exit air filters, the magnetic valve minus the coil, the inlet air filters, and the sampling vessel; (b) the medium reservoir with the medium line; and (c) the harvest reservoir with the harvest line.

After sterilization the three sections are fitted together, the apparatus is connected to the air supply, the condenser to the cooling water, and the coil is fitted on the magnetic valve and connected with the relay. The speed of the pump motor is preset for delivery of the desired flow rate. An inoculum of actively dividing cells, growing in submersed culture on rotary shakers is used. Fresh medium containing 10% inoculum is poured into the fermentor through the inoculation port. The magnetic valve is activated and the relay is set to give pulses at 2- to 3-second intervals with a pulse duration of about one-tenth of a second.

As the compressed air enters the fermentor it expands into a large bubble having the same diameter as the fermentor. As the bubble slowly moves upward through the vessel the entire culture passes as a thin layer between the fermentor wall and the surface of the air bubble, thus being aerated and stirred effectively. The reduction of air pressure from 5 to 10 psi to atmospheric pressure and

[6] O. L. Gamborg, *Plant Physiol.* **45**, 372 (1970).

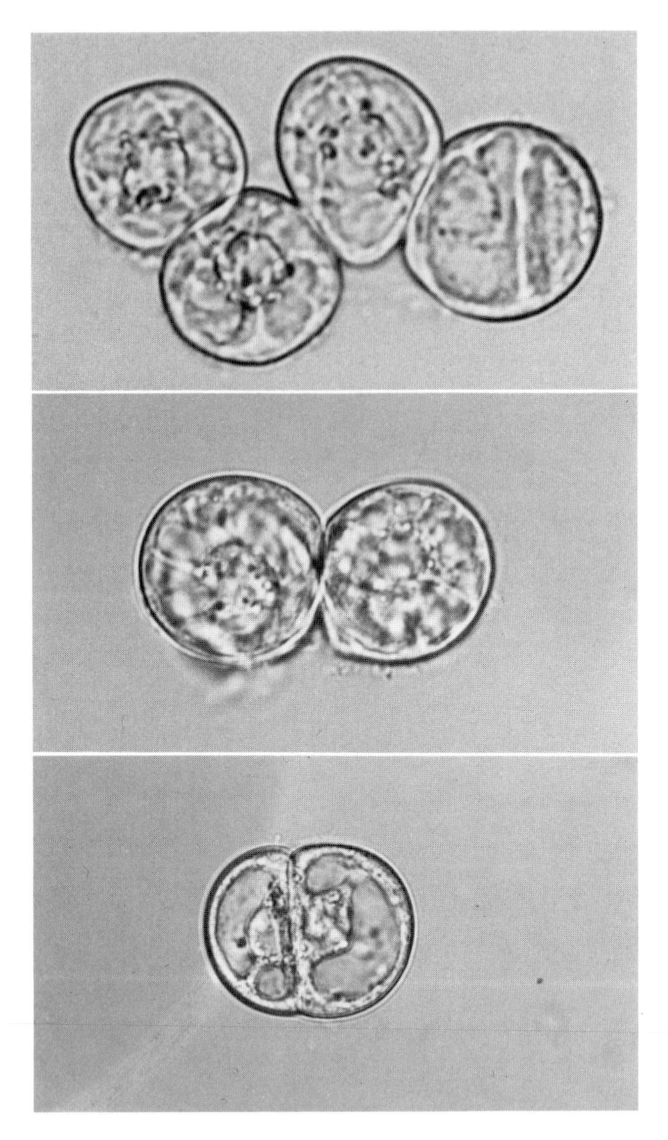

Fig. 3. Dividing soybean cells (*Glycine max* L.) grown in chemostat at a generation time of 30 hours.

expansion of the bubble causes a mild vibration in the culture, which probably is the main factor preventing the formation of cell aggregates.

The inoculum in the fresh medium is kept under batch conditions until the cells have adapted to the fermentor and start to divide again. Then the pump is started, and new medium is added and culture simultaneously withdrawn from the fermentor. The flow rate depends on the doubling or generation time (D_T) of the cells cultured. Generally it is between 25 and 35 hours, thus calling for a flow rate which would replace the volume of the culture according to the equation $D_T = R_T \times \log_e 2$, where R_T is the residence time of the cell in the fermentor.[7]

The fermentor design was tested on two cell lines, soybean root (*Glycine max* L.) and wheat (*Triticum monococcum* L.).[4–6] The fermentor was operated at culture volumes of 1500 and 1800 ml, maintained in a thermostated room at 27°C, and irradiated with 800 lux of light. The runs extended over a period of 2 to 6 weeks. In the case of the soybean cell culture the generation time was 30 hours. The cells were isodiametrical and the culture consisted mainly of single and double cells with a few quadruple cells being the largest cell aggregate found in the culture (Fig. 3). The amount of cell debris in the cultures was negligible. Because of the homogeneous character of the culture, cell growth could be measured on the basis of optical density; the relation between optical density and dry weight was linear. The dry weight of a soybean cell culture grown at a generation time of 30 hours equilibrated after 3 to 6 days at 1.3 mg/ml culture.

The main feature of this fermentor design is its ability to produce cultures consisting predominantly of single cells, in continuous cultivation for long periods of time. The unique stirring action may be particularly important in achieving this type of cell growth.

[7] W. G. W. Kurz, P. S. S. Dawson, and E. R. Blakley, *Can. J. Microbiol.* **15**, 27 (1969).

SECTION VII

Cell Propagation on Miscellaneous Culture Supports

Editors' Comments

Because of commercial availability, personal choice, and/or habit, cells *in vitro* are routinely carried in a variety of glass or plastic containers. One should be aware, though, that many types of vertebrate, invertebrate, and plant cells can be carried on culture supports or in containers made of other materials. The variety is demonstrated in this section where the reader will find information on cells which apparently thrived on such substrates as collagen-coated cellulose sponge, various other proteins, Sephadex beads, polyester film (see also Section VI), titanium metal, and filter paper. The tissue culture literature cites other materials which have been used for one reason or another on which or in which cells have been cultured. Current research in a number of laboratories also includes such promising possibilities as the lining with cells of various fabrics and devices for medical applications for, among other reasons, enhanced blood compatability. Elsewhere in this volume (Section IX) the reader will find examples of cells cultured on Millipore filters, Teflon, and decalcified eggshell membrane.

CHAPTER 1

Collagen-Coated Cellulose Sponge[1]

Joseph Leighton

In 1951 the author introduced a combined matrix of cellulose sponge and chicken plasma clot for organized growth in tissue culture.[2] In the same publication the use of two other three-dimensional sponges was described, natural sea sponge and a gelatin sponge (Gelfoam). The combined matrix of plasma clot and cellulose sponge was used in a number of studies on animal and clinical tumors, but several difficulties limited its use. Many cells lysed the plasma clot in a few days. When clot lysis appeared, viable histotypical outgrowth was disturbed. Sometimes the tissue retracted and all but the margin of the coalesced tissue became necrotic. On other occasions the cells became dislodged from the matrix as the fibrin clot dissolved.

Since most cells did not adhere to a cellulose substrate, the idea that a collagen-coated cellulose sponge might be a useful substitute was suggested by the work of Ehrmann and Gey.[3] These authors found that many types of cells adhere to and grow better on collagen-coated glass than they do on bare glass. Collagen-coated cellulose sponge combined the three-dimensional features of sponge with the surface properties of a collagen membrane. This matrix can be readily prepared in quantity weeks before the day on which tissue is to be set up in culture. The actual preparation on the day of cultivation takes a short time. Many types of tissue adhere very well to the collagen-coated trabeculae of the cellulose sponge and grow in the interstices in an organoid arrangement. We have used the matrix in the study of embryonic tissue, rodent tumors, and clinical cancer.[4-8]

[1] The development of methods described here was conducted under Research Grants CA-13219 from the National Cancer Institute and P-442 from the American Cancer Society. I also wish to acknowledge with gratitude the excellent technical assistance of Mrs. Martha Esper and Mrs. Ruth Alexander.

[2] J. Leighton, *J. Nat. Cancer Inst.* **12,** 545 (1951).

[3] R. L. Ehrmann and G. O. Gey, *J. Nat. Cancer Inst.* **16,** 1375 (1956).

[4] J. Leighton, *In* "The Proliferation and Spread of Neoplastic Cells" (E. Frei, ed.), p. 533. Ann. Symp. M. D. Anderson Hospital. Williams & Wilkins, Baltimore, Maryland, 1968.

[5] J. Leighton, *In* "Methods in Cancer Research IV" (H. Busch, ed.), p. 86. Academic Press, New York, 1968.

[6] J. Leighton, G. Justh, M. Esper, and R. L. Kronenthal, *Science* **155,** 1259 (1967).

[7] J. Leighton, R. Mark, and G. Justh, *Cancer Res.* **28,** 286 (1968).

Preparation of Collagen-Coated Cellulose Sponge

A collagen dispersion is available from the Research Division of Ethicon Inc., Somerville, New Jersey. It consists of a 1% suspension of collagen in fibrillary form dispersed in an equal volume of methanol and water base containing 2% cyanoacetic acid. A brief description of the method of preparation of this collagen dispersion has been published.[6]

1. Strips, measuring 1.0 × 0.5 cm on cross section, of dry, fine-pore cellulose sponge (DuPont) such as is used in photography, are compressed between a glass slide and cardboard, and cut transversely with a single-edged razor blade into slices. Place slices in water and discard those thicker than 1.0 mm. A large number of sponges of uniform thickness can be cut quickly using a commercial electric delicatessen slicer.
2. Wash the sponges thoroughly by boiling twice in glass-distilled water for a total of 1 hour. Immerse at room temperature in acetone, ethyl ether, and absolute alcohol for 30 minutes each. Return the sponge slices to water and repeat the boiling in distilled water twice as before.
3. Spread sponges out in Petri dishes to dry at room temperature.
4. Impregnate dry sponges with 1% Ethicon collagen dispersion. Invert the top of a Petri dish. Place several sponges in the center of the dish. Add about 5 ml of the dispersion. Allow the dispersion to soak into the sponge for a few seconds then, with a pair of curved forceps, gently work the dispersion into the sponge. Place the flat surface of the bottom of the Petri dish on the sponges. Gently press the sponges several times between the two glass surfaces.
5. Blot the wet sponges. Invert the top of a Petri dish. Insert a piece of No. 40 Whatman paper in the inverted cover. Place the sponges, wet with dispersion, on the dry filter paper. Cover the sponges with another piece of filter paper. Using the inverted bottom of the Petri dish, gently press out some of the dispersion. Remove the sponges from the filter paper and allow them to dry, standing on edge against the rim of a Petri dish, at room temperature.
6. Place about twelve dry sponges in a Rockefeller screw-cap tube. Add 50 ml of 50% methanol with ½% ammonium hydroxide. Place in the refrigerator overnight.
7. The next day change the methanol to fresh 50% methanol *without* ammonium hydroxide. Repeat with a fresh change of 50% methanol without ammonium hydroxide after 1 hour. For sterilization follow with two changes of 100% methanol, allowing the sponges to stand 1 hour in each exposure to 100% methanol.
8. Using sterile technique remove the sponges and place in a sterile Petri dish to dry at room temperature.
9. Store dry collagen-coated sponges in sterile container at room temperature.

---

[8] J. Leighton, G. Justh, and R. Mark, *In* "Recent Results in Cancer Research—Normal and Malignant Cell Growth" (R. J. M. Fry, M. L. Griem, and W. H. Kirsten, eds.), p. 147. Springer-Verlag, New York, 1969.

Procedure for Setting Up Cultures

1. Moisten sterile sponges with balanced salt solution or medium, at which time the sponges expand exceeding 10 × 5 mm. Trim to suitable surface dimensions, usually 10 × 5 mm.

2. Place explants of tissue, each about 1 mm in diameter, in a diagonal row on the surface of the sponge, four or five explants per sponge.

3. Transfer sponge with explants on top to a culture tube or medicine bottle. Add medium in sufficient volume to be flush with the upper surface of the sponge, just touching the explants.

4. Incubate at 36° or 37°C and replenish medium every other day, or as indicated by the fall in pH. After 24 hours culture vessels may be placed on a slow rocker in the incubator.

Histological Processing of Matrix Cultures

1. Fixation. Fixative (Bouin's or Stieve's), 4 hours to overnight; 80% alcohol two changes, at least 1 hour for each.

2. Trim sponges by making one cut with razor blade parallel to the diagonal row of explants and just beyond the row of explants. In this way when the sponge is embedded later on edge in paraffin with the diagonal cut down, the growth will be in an optimal orientation for histological sectioning.

3. Dehydration. Absolute alcohol, 2 hours; absolute alcohol, 1 hour; absolute alcohol, 1 hour.

4. Clearing. Toluene, 1 hour; toluene, 1 hour.

5. Infiltration. Paraffin, 64°C (Paraplast), ½ hour; paraffin, 64°C overnight (5:00 P.M. to 9:00 A.M.).

6. Embed sponge on edge with tissue toward microtome knife. Good histological sections require a perfectly sharp knife. The knife edge becomes dull quickly as cellulose is cut. It is well to have several sharp knives available.

7. Routine hematoxylin and eosin or special stains may be used.

Observation of Living and Fixed Cultures

The living cultures should be examined ideally with an inverted microscope to check for the appearance of cells at the margins of the sponge and on the adjacent glass. If the cultures are doing well, observation of the living culture is unsatisfactory. The sponge interstices become filled with living cells, making an opaque tissue. The quality of the growth in the matrix cannot be determined reliably without histological sections (Fig. 1).

Since the most elaborate growth patterns are found on the surface, at the gas-medium interphase, it may be economical to embed sponges flat, rather than on edge, to obtain a large sample of this outer crust in a few slides. It must be appreciated, however, that this view fails to provide for observation of the

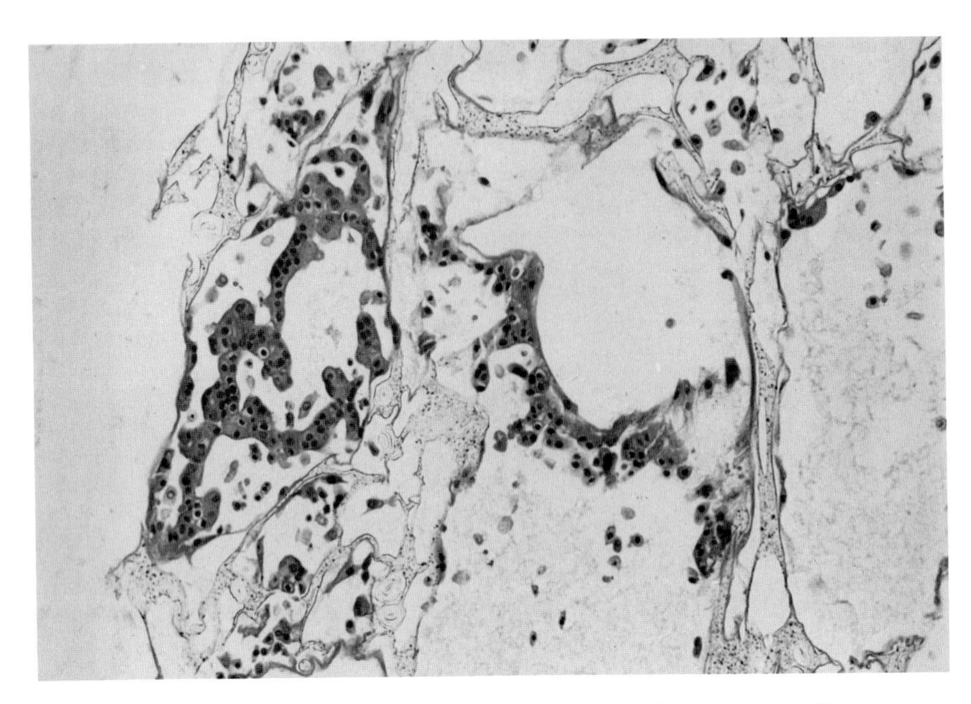

Fig. 1. Histological section of a 1-week-old culture in collagen-coated cellulose sponge of a fresh suspension of normal human amnion cells. Hematoxylin and eosin, X160. Multiple interstices of the sponge contain intersecting cords of cells as illustrated in the left part of the figure. The cellulose skeleton of the matrix appears as an angular branching network with a fine granularity. A collagen membrane coats the trabeculae of cellulose sponge but is not seen readily with this stain. The collagen membrane is best identified for light microscopic study with stains such as aniline blue.

changing quality of the growth with reference to the gradient of decreasing oxygen tension from the surface of the sponge to its depths.

Application of the Method

This procedure has been used successfully in the cultivation of fragments of chick embryonic organs and of solid tumors from patients and laboratory animals. Combinations of embryonic tissues and solid human tumors have been particularly successful in the preservation of the histotypical architecture of the tumor in *in vitro* growth[8] (Fig. 2). The most complex tissue formations are found at the medium-gas interphase, i.e., the upper surface of the sponge. Tissue growth in the deeper parts of the sponge appears as loosely arranged spindle-shaped cells. For some tissues, especially embryonic heart and liver, it consists of branching endothelial-lined channels as well as mesenchymal connective tissue.

When the inoculum consists of a suspension of cells, the procedure for preparation of the culture is modified, since a distinct diagonal line of inoculation is not used. Ascites tumors of rodent origin, or suspensions of cells derived

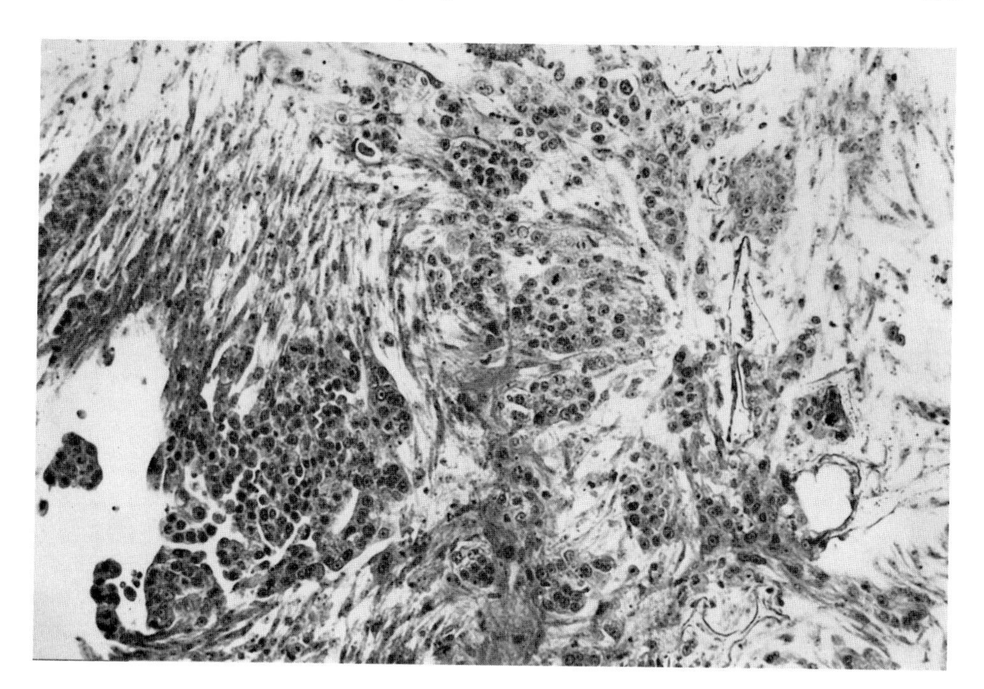

Fig. 2. Histological section of carcinoma of the ovary inoculated in combination with embryonic tissue and cultured in collagen-coated cellulose sponge. Hematoxylin and eosin, X160. The tumor inoculum consisted of cells obtained by centrifuging a peritoneal effusion from a patient with carcinomatosis. The embryonic tissue consisted of a fine mince of tissue from the mesonephric region of a 10-day-old chick embryo. After a week in culture, as seen in the figure, numerous carcinoma cells are seen in several arrangements, as cords, nests, and papillary structures. The groups of tumor cells are seen in a supporting connective tissue stroma derived from chick embryonic tissue.

from tissue culture cell lines may be inoculated on a moistened sponge as a dense cell suspension. A heavy inoculum consists of 0.1 ml of a 50% packed cell volume. As medium is added to the container to bring the level flush with the upper surface of the sponge, many of the freshly inoculated cells will be washed free of the sponge. With gentle handling, however, a substantial part of the inoculum remains in the sponge, adheres to the collagen, and in a few days develops extremely dense growth.

CHAPTER 2

Microcarrier Cultures of Animal Cells

A. L. van Wezel

Microcarrier culture[1] is a culture system in which tissue cells are grown as a monolayer on the surface of small solid particles and kept suspended in the culture medium by stirring. In this way the characteristics of both suspension culture and monolayer culture are brought together while retaining all advantages of both culture systems.

The microcarrier culture offers in the same way as the suspension culture better possibilities for measurement and control of the environmental conditions, such as, temperature, pH, CO_2, and O_2 tension, than the nonhomogeneous monolayer culture. Homogeneous culture systems also facilitate the mass cultivation of cells and reduce the risk of contamination by decreasing the number of manipulations. In contradiction to the suspension culture, microcarrier culture can be used for the homogeneous cultivation of primary cells and diploid cell strains. In addition, in this culture system the replacement of medium and the washing of the cells is facilitated as will be described below.

MICROCARRIERS

Materials. Certain properties are required of materials to be used as microcarriers. The cells should adhere easily and firmly to the microcarrier. As the microcarriers should be kept in suspension at low stirring speed, the density of the particles should be about the same as the density of the culture medium. They should be large enough to support a few hundred cells per particle. On the other hand, they should not be too large in order to maintain the homogeneous character of the culture. Particles with a diameter of 100 to 250 μm appear to be most suitable. Finally the material should not be toxic for the cells.

From all materials tested DEAE-Sephadex A50 appears to be most suitable to serve as microcarrier. The negatively charged tissue cells adhere very easily to the positively charged DEAE-Sephadex A50, and the density of DEAE-Sephadex beads, after swelling, is about the same as the density of the culture medium. A disadvantage is that a slight inhibition of cell growth is found at concentrations exceeding 1 mg DEAE-Sephadex A50 per milliliter culture medium. This can be obviated by coating the beads with a nitrocellulose product.

[1] A. L. van Wezel, *Nature* (*London*) **216**, 64 (1967).

Satisfactory results have been obtained also with similar products, such as DEAE-Sephadex A25 and QAE-Sephadex. However, the surface of these beads is not as smooth as that of DEAE-Sephadex A50 beads. Moreover, DEAE-Sephadex A50 beads are more transparent. This makes them more suitable for microscopic observations.

Plastic beads, such as specially treated polystyrene and Rilsan (nylon 11), have been tested too. Unfortunately the cells do not adhere firmly enough to these beads.

Workers at the Institut Pasteur[2] suggest the use of Spherosil beads (a special kind of porous glass beads) as microcarrier. They claim to have good results with these beads in microcarrier culture. A disadvantage may be that the beads are too heavy to be kept in suspension at low stirring speeds, while at higher speeds the cells may not easily attach to the beads.

Pretreatment of the DEAE-Sephadex A50; Sieving of the Sephadex. We use commercial DEAE-Sephadex A50 (Pharmacia, Uppsala, Sweden). The diameter of the dry beads varies from 40 to 120 μm and after swelling in a physiological solution from 60 to 280 μm.

As the cells appear to adhere preferably to the smaller beads, the variation in diameter of the beads results in an unequal distribution of the cells over the beads. It is therefore recommended to sieve the dry DEAE-Sephadex A50 beads beforehand, e.g., by an Alpine air sieve in fractions with a difference in diameter of not more than 20–25 μm. Most DEAE-Sephadex A50 is found back in the fractions with a diameter from 60 to 80 μm and from 80 to 105 μm. Both fractions may be used.

Washing Procedure. Ten grams DEAE-Sephadex A50 is suspended in 1 liter distilled water and transferred to a G 3 sintered-glass filter. The Sephadex is washed successively three times with each of the following fluids: 250 ml 0.5 N HCl, 250 ml distilled water, 250 ml 0.5 N NaOH, and 250 ml distilled water. Finally the beads are washed with phosphate-buffered saline (PBS: 8 g NaCl, 0.2 g KCl, 2.0 g Na_2HPO_4, 0.4 g KH_2PO_4 dissolved in 1 liter distilled water; pH 7.2–7.3) until equilibrium is reached. The DEAE-Sephadex A50 is suspended in about 500 ml PBS and autoclaved for 15 minutes at 120°C.

Coating Procedure. The Sephadex is coated in an apparatus of special design, developed for the dispersion of tissue cells.[3] Essentially the apparatus consists of a Jena glass cylinder with a Vibromixer at the top plate and a screen on the bottom plate (Fig. 1). In order to prevent sticking of DEAE-Sephadex A50 to the glass wall, the glass cylinder is siliconized before sterilization. The sterilized DEAE-Sephadex A50 is transferred aseptically to the coating apparatus. In order to remove all free water, the Sephadex is washed twice with approximately 500

[2] Institut Pasteur, Procedure for the cultivation of cells. Patent Appl. France Nr. EN 6923294 (1969).

[3] A. L. van Wezel, P. A. van Hemert, W. Parisius, N. B. Cucakovich, and H. G. Macmorine, *Appl. Microbiol.* **24**, 506 (1972).

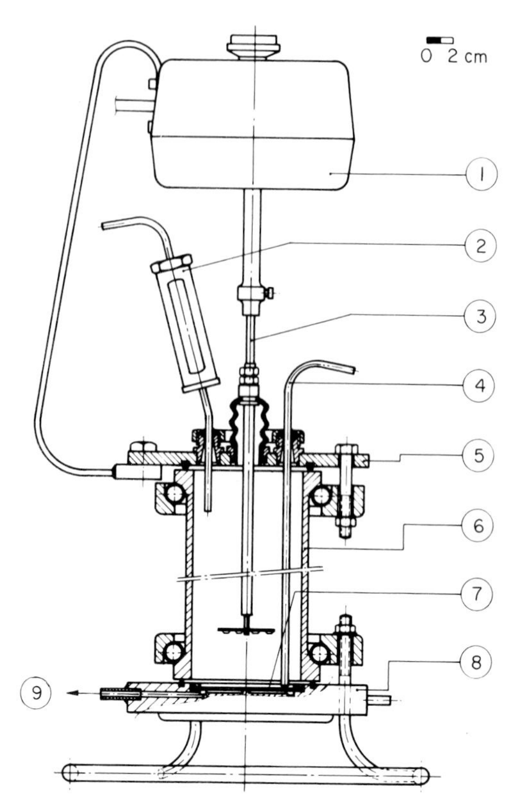

Fig. 1. Coating Apparatus. 1, Vibromixer (model E 1, Chemapec Inc.); 2, air filter; 3, Vibromixer shaft with impeller; 5, upper lid with six standardized stainless steel in- or outlet tubes (4); 6, standard Jena glass pipe 50 cm; 8, bottom lid with stainless steel screen of 100 μm (7) and outlet (9).

ml 100% methyl alcohol and resuspended in 500 ml methyl alcohol. An equal volume of an 0.1% solution of nitrocellulose (Celloidin, Edward Gurr Ltd, London) in methyl alcohol is added to this suspension under mixing. After 15 minutes the DEAE-Sephadex A50 is washed again with methyl alcohol to remove free celloidin, then once with 50% methyl alcohol in PBS and finally a few times with PBS for removing the alcohol. The DEAE-Sephadex A50 is resuspended in the proper tissue culture medium and kept at 4°C until use.

Procedure for Microcarrier Cultures

Cultivation Apparatus. For microcarrier cultures of animal cells the same fermentors can be used as for suspension cultures of bacteria and cell lines. Our experiments are performed in the "Bilthoven Unit."[4] Glass fermentors should be siliconized to prevent sticking of the DEAE-Sephadex A50 beads to the glass wall. This can be done by wetting the glass vessel before sterilization with a

[4] P. A. van Hemert, *Biotechnol. Bioeng.* **6**, 381 (1964).

2–5% solution of silicone oil (Silicone fluid MS 1107, Midland Silicones Ltd.) in ethyl acetate. After drying with air, the glass vessel is rinsed with hot distilled water.

Cells. Microcarrier cultures of cell lines and diploid cell strains are started with cells obtained from monolayer cultures. For the preparation of suspensions of primary cells the perfusion technique of Kammer[5] or the incubation method of van Wezel[3] is used. Both cell dispersion methods give high cell yields and much less cell debris than the conventional dispersion methods. In general, the cultures are inoculated with 100×10^3 cells/ml.

Media. Diploid cell strains and cell lines are cultivated in Eagle's Minimum Essential Medium (MEM) supplemented with 10% calf serum. Primary cells can be cultivated in lactalbumin hydrolysate medium supplemented with 10% bovine serum. However, much higher cell densities can be achieved in Eagle's MEM supplemented with 10% bovine serum and 0.2% lactalbumin hydrolysate.

Sephadex Concentration. In general, 1 mg DEAE-Sephadex A50 is added per milliliter culture medium. This corresponds with 8000 to 9000 beads per milliliter. As the mean diameter of the swollen beads is 0.2 mm, it gives a culture area of 8 to 10 cm^2/ml. This is more than five times the culture area of conventional monolayer cultures.

Culture Conditions. The cultures are started by bringing medium, Sephadex, and cells together in the fermentor. During the first 24 hours the cells adhere to the microcarriers and gradually a confluent monolayer is formed (Fig. 2). In general the culture conditions for microcarrier cultures are the same as for suspension cultures of cells from cell lines. For instance, optimal cell densities in microcarrier culture of primary monkey kidney cells are obtained when the cells are cultivated at a temperature of 37°C, pH 7.2–7.3, and an oxygen tension of 50 to 100% air saturation. The stirring speed is lower than in suspension culture and depends upon the type and the volume of the fermentor. In 3-liter cultures we are using a stirring speed of 60 rpm and in 40-liter cultures 75 rpm.

Cell Count and Microscopic Analysis. The cell concentration can be determined by counting nuclei on a Fuchs-Rosenthal hemocytometer. Beforehand the cells on the DEAE-Sephadex A50 beads are incubated at 37°C for 1 hour in 0.1% crystal violet in 0.1 *M* citric acid after removing the culture medium. In this way the nuclei are detached from the beads and stained with crystal violet. Microscopic observation of the cells on the beads can be done directly. However, a better impression can be obtained after staining with hematoxylin.

Medium Change. Changing medium can be easily done by syphoning off the spent medium. A 100 μm stainless steel screen on the outlet pipe prevents loss of Sephadex beads. When dealing with the multiplication of virus for the prepara-

[5] R. Kammer, *Appl. Microbiol.* **17**, 524 (1969).

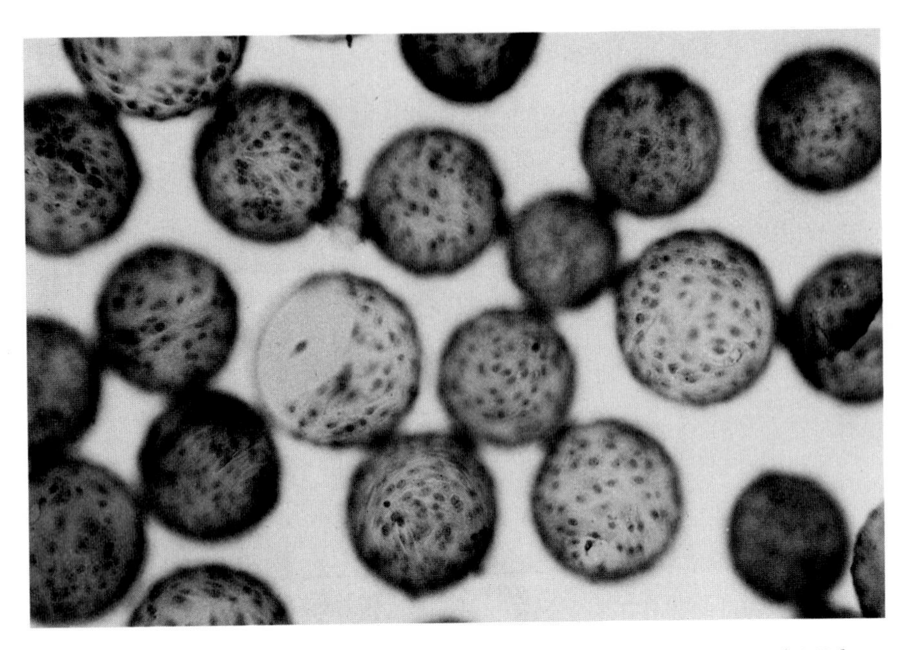

Fig. 2. Primary monkey kidney cells on microcarriers after a cultivation time of 140 hours.

tion of human vaccines, it is necessary to remove the culture medium and to wash the cells free from serum components. For this washing procedure the same system can be used. Another elegant system for medium change is the perfusion system. In this system medium is continuously added and withdrawn (Fig. 3). Good results are obtained also if the effluent in the perfusion system is continuously recycled.[6]

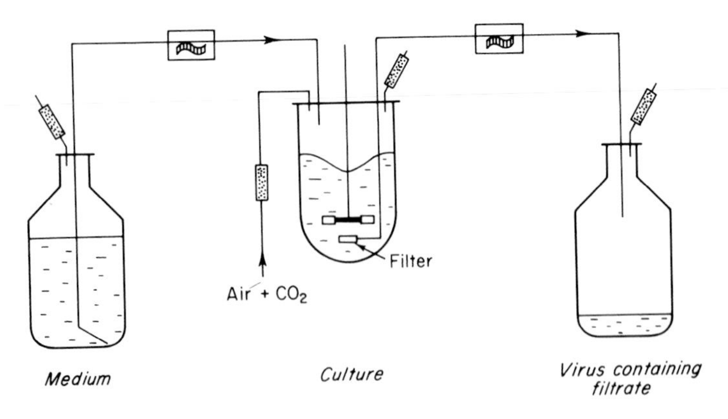

Fig. 3. Perfusion system for continuous refreshment of a cell or cell/virus culture on microcarriers. From P. A. van Hemert, D. G. Kilburn, and A. L. van Wezel, *Biotechnol. Bioeng.* **11**, 875 (1969).

[6] P. A. van Hemert, D. G. Kilburn, and A. L. van Wezel, *Biotechnol. Bioeng.* **11**, 875 (1969).

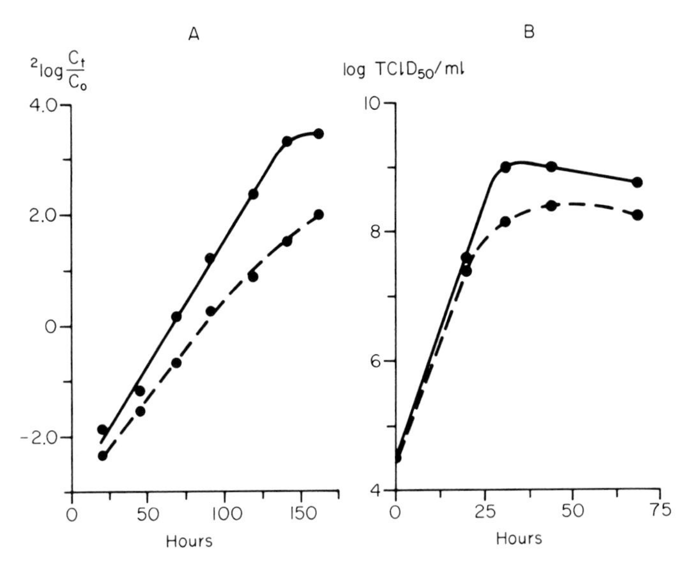

Fig. 4. (A) The effect of the O_2 tension on the growth of primary monkey kidney cells in microcarrier culture. ———, O_2 tension 75% air saturation; - - - - - - - -, O_2 tension 5% air saturation. The points correspond to the number of attached cells. Inoculum 100×10^3 cells/ml. C_t/C_0 = attached cells per milliliter at time t/cells inoculated. (B) The effect of the cell concentration on the multiplication of poliomyelitis virus type 1 Mahoney on primary monkey kidney cells in microcarrier culture. ———, 13×10^5 cells/ml; - - - - - $3{,}5 \times 10^5$ cells/ml. $TCID_{50}$ = 50% tissue culture infective dose.

Applications

The microcarrier culture offers an elegant cultivation technique for large-scale production of animal cells, especially of those cells which do not multiply in free suspension, such as, primary cells and diploid cell strains. As a result of the large culture area per milliliter and a better control of the environmental conditions, the cell yield is mostly higher than in conventional monolayer cultures, even without medium change[7] (Fig. 4A). When dealing with the preparation of virus vaccines in these cells, higher cell densities will result in a much higher virus concentration per milliliter culture (Fig. 4B). For the multiplication of viruses which do not destroy the cells directly, such as measles and rubella viruses, the simple batch method or the more sophisticated perfusion technique may be used for harvesting virus suspensions.

Summarizing, microcarrier culture offers the possibility to study the optimal conditions for cell growth and virus propagation and also for the large-scale production of cell and virus products.

[7] A. L. van Wezel, *Progr. Immunobiol.* Standardization **40**, 150 (1972).

CHAPTER 3

Polyester Sheeting

H. Firket

In several types of experiments, the use of thin glass support for cells in culture is not possible or presents difficulties: (a) glass coverslips are not very convenient when embedding *in situ* is required for electron microscopy. Whether carbon-coated or not, they do not separate readily from the resins; (b) when the cells have to be irradiated with heavy particles, glass absorbs too large a fraction of the incoming radiation; (c) in scanning electron microscopy of cells on glass, the support may give a secondary electron emission leading to considerable "noise" and decreased definition.

In these cases, glass (or the usual plastics) may be replaced with advantage by thin polyester sheets. Many cell types (cell strains, nerve cells, explant cultures) grow well on this material. It is not altered by the fixation and embedding procedures for electron microscopy, and it separates without any problem from the Epon or Araldite media, leaving the cells embedded.[1] It does not reduce considerably the intensity of alpha particles or neutron beams passing through it.[2] It has less secondary emission than glass in the scanning electron microscope.[3]

This material is also transparent enough to permit high magnification observation of living cells. It can be marked by knife for localization of interesting structures, and it can be cut to shape.

As received from the manufacturer, the polyester sheets may not be clean enough for tissue culture and have to be pretreated. It is available in sheets of various thicknesses (6–190 μm). Thin sheets have better transparency but may tear more easily.

Procedure

Polyethylene terephthalate[4] sheets of suitable thickness (75 μm is usual) are washed in lauryl sulfate detergent with a soft brush and rinsed thoroughly. They

[1] H. Firket, *Stain Technol.* **41**, 189 (1966).

[2] G. W. Barendsen and T. L. J. Beusker, *Radiation Res.* **13**, 832 (1960).

[3] A. Boyde, F. Grainger, and D. W. James, *Z. Zellforsch.* **94**, 46 (1969).

[4] Trade name Melinex "O" produced by Imperial Chemical Industries Ltd. (I.C.I.), Plastics Division, Welwyn Garden City, Herts, England. A similar material by the name of Hostaphen has been produced in Germany by Kalle A. G., Wiesbaden.

are soaked in alkaline permanganate (5% $KMnO_4$ in 2% KOH) for 1 hour at 60°C, and rinsed again very thoroughly with tap and distilled water. They are then cut to the required shapes with scissors. One side may be marked with a knife, drawing grid lines for later localization of chosen structures.

Sterilization may be achieved by autoclaving (which reduces the optical transparency) or by alcohol and careful drying.

The wetability of these sheets is about the same as that of very clean glass. They can be coated with collagen or chick plasma.

Barendsen and Beusker[2] have sealed them as bottoms to special dishes for irradiation.

Embedding of cells for electron microscopy is made easy as the flexible support bends inward on the Epon capsules during polymerization, preventing the formation of air bubbles. After the resin is hardened, these coverslips separate easily from it by a simple twist of the hand (no forceps!).[1]

The ease with which polyester sheets can be marked and cut to any shape could also be advantageous in a number of other types of cultures.

CHAPTER 4

Protein Polymers as a Substratum for the Modulation of Cell Proliferation in Vitro

A. Macieira-Coelho and S. Avrameas

Previous studies have suggested that the affinity of the cells with the substratum can regulate the expression of the loss of contact inhibition of division.[1,2] These findings have led us to develop a substratum on which cells could grow and where cell attachment could be modulated.[3]

A protein polymer made of bovine serum albumin (BSA) has been adapted to tissue culture using a method previously reported for the preparation of insoluble protein polymers. The growth of two mouse cell lines varies on the BSA polymer depending on the physiochemical properties of the surface.

[1] A. Macieira-Coelho, *Exp. Cell Res.* **47**, 193 (1967).
[2] A. Macieira-Coelho, *Intern. J. Cancer* **2**, 297 (1967).
[3] A. Macieira-Coelho and S. Avrameas, *Proc. Soc. Exp. Biol. Med.* **139**, 1374 (1972).

Preparation of the Substratum for Cell Growth

To 1 ml of a sterile 30% solution of BSA,[4] 1.2 ml of a 2.5% aqueous solution of glutaraldehyde is added at 4°C. One milliliter of this solution is quickly poured onto the surface of a 30-ml plastic bottle (Falcon) and the protein is allowed to polymerize for 3 hours at room temperature. Borate buffer (0.15 *M*, pH 8.4) is added and left for 24 hours. The borate buffer is then replaced with PBS–EDTA (0.1% EDTA in PBS without Ca and Mg) which is changed daily during 12 days. Before seeding the cells the surface is washed twice with tissue culture medium.

If one wants to attach a substance to this surface the BSA polymer is covered during the first 24 hours with borate buffer; then borate buffer containing the substance is added and replaced 24 hours later by PBS–EDTA. The latter is renewed during 12 days as described above. Different substances were used to treat the polymer at the following concentrations in borate buffer: 1 mg/ml polylysine or polyornithine, 0.1 mg/ml crude calf thymus histones, 1 mg/ml DEAE–dextran, 1 mg/ml polyglutamic acid, and 1 mg/ml heparin. Polyamino acids of different molecular weights give identical results.

Although the experiments described herein were all made on BSA polymers, protein polymers prepared with bovine and human serum and with human immunoglobulins were also found to support the growth of cells. The latter, however, were not studied so extensively as the BSA polymers.

Enzymes such as neuraminidase and hyaluronidase can also bind to the polymer using the procedure described above for the polyaminoacids.

Chemistry of Polymerization

The polymerization of BSA is due to cross-linkage originated by bonds between amino groups (mainly ε-amino groups of lysine) of proteins and active

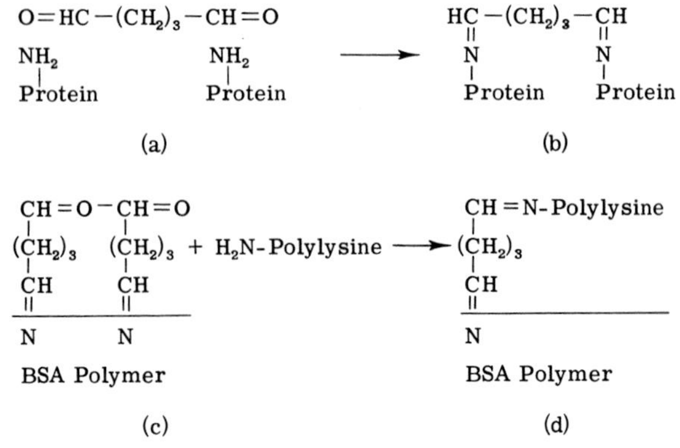

(a) $O{=}HC{-}(CH_2)_3{-}CH{=}O$; NH_2–Protein ; NH_2–Protein $\longrightarrow$ (b) $HC{-}(CH_2)_3{-}CH$ with each CH $=$N–Protein

(c) BSA Polymer–N$=$CH–$(CH_2)_3$–CH$=$O and BSA Polymer–N$=$CH–$(CH_2)_3$–CH$=$O + H_2N-Polylysine $\longrightarrow$ (d) BSA Polymer–N$=$CH–$(CH_2)_3$–CH$=$N-Polylysine

Fig. 1. Polymerization of BSA. For explanation see text.

[4] Available from Poviet Production N. V., Amsterdam, Holland.

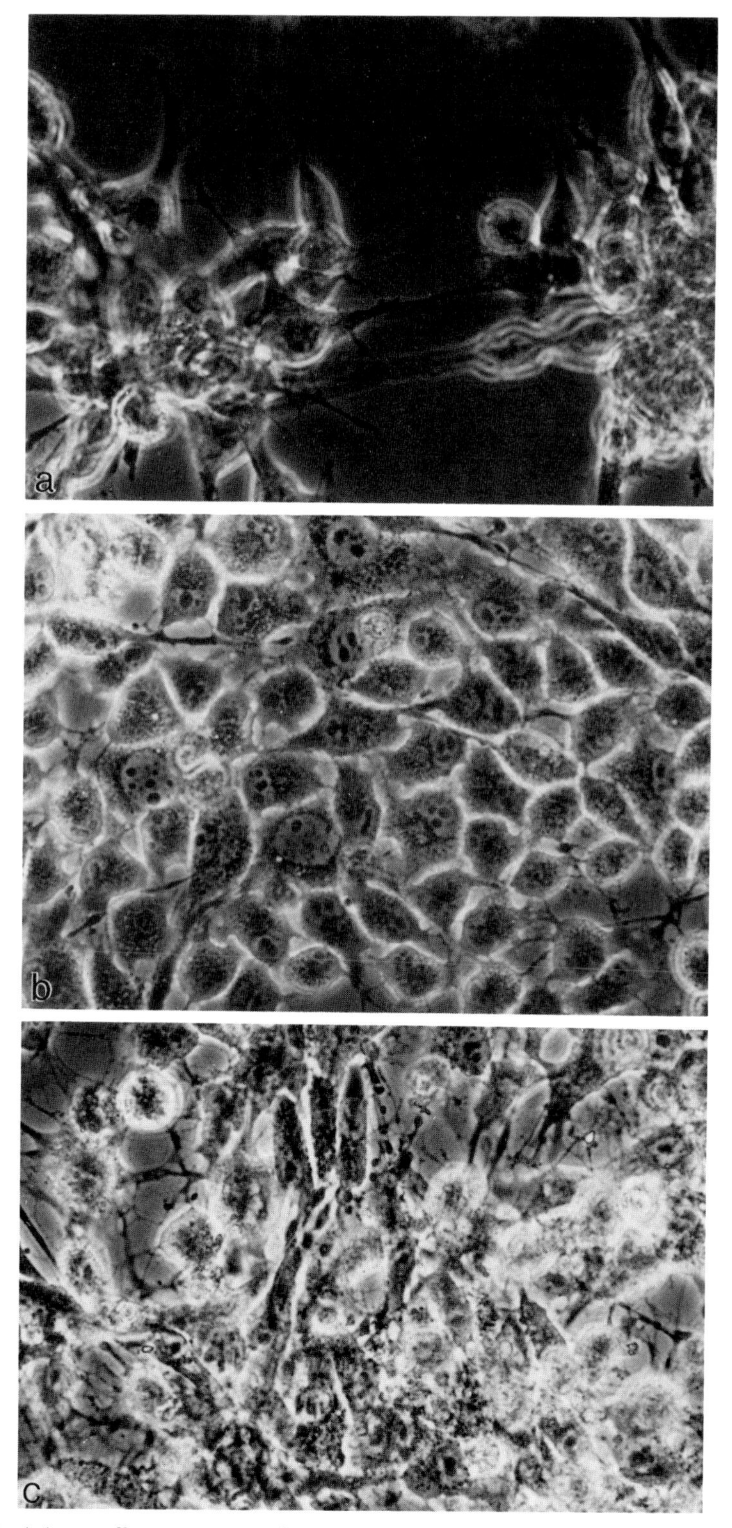

Fig. 2. (a) L cells growing in clusters on a negatively charged BSA polymer. (b) L cells attached to a plastic bottle. (c) L cells attached to a positively charged BSA polymer.

aldehyde groups of glutaraldehyde (Fig. 1a). It is assumed that these bonds are of the Schiff base type[5] (Fig. 1b). After polymerization of the protein, active aldehyde groups from glutaraldehyde remain free in the polymer (Fig. 1c). Substances possessing free amino groups (e.g., polylysine) will link by covalent bonds to these aldehyde groups (Fig. 1d).

Substances like heparin of DEAE–dextran which do not possess free amino groups, cannot bind covalently. These substances are presumably fixed to the BSA polymer by noncovalent bonds; heparin binds to positively charged groups (e.g., amino groups) and DEAE–dextran to negatively charged groups (e.g., carboxyl groups).

Cell Culture

Two different types of cells were cultivated on these surfaces, the L[6] and the L1210[7] mouse cells lines. The L cell line grows in monolayer on glass and the L1210 leukemic cells grow in suspension as nonagitated cultures. Both lines were maintained in Eagle's MEM supplemented with 10% calf serum and 50 μg/ml kanamycin.

Before the experiments, the cells were pooled and evenly distributed onto the different surfaces with an inoculum of 2×10^6 cells per flask.

The L Cell Line. The L cells when growing on the BSA polymer treated with negatively charged substances (polyglutamic acid or heparin) do not become confluent but form isolated aggregates or clusters more than one cell layer thick (Fig. 2a). Treatment with positively charged substances (histones, polylysine, polyornithine, or DEAE–dextran) results in a monolayer which is dis-

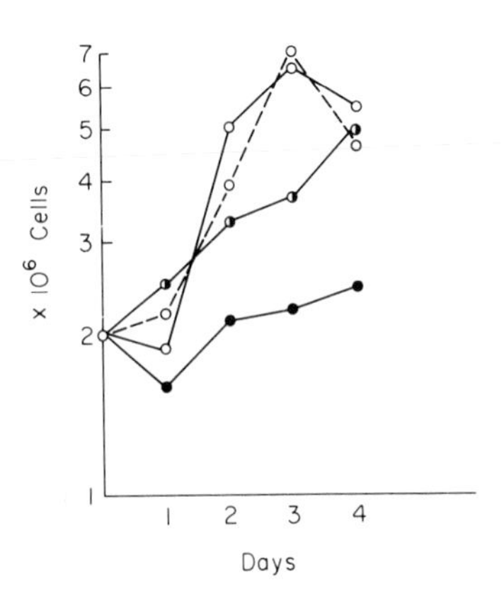

Fig. 3. Growth curves of L cells proliferating on the BSA polymer treated with: polylysine (○- - - - -○), DEAE–dextran (●———●), polyglutamic acid (◑———◑), and polyglutamic acid plus DEAE–dextran (○———○).

[5] S. Avrameas and T. Ternynck, *Immunochemistry* 6, 53 (1969).
[6] W. R. Earle, *J. Nat. Cancer Inst.* 4, 135 (1943).
[7] Kindly supplied by Dr. I. Gresser, CNRS, 94-Villejuif, France.

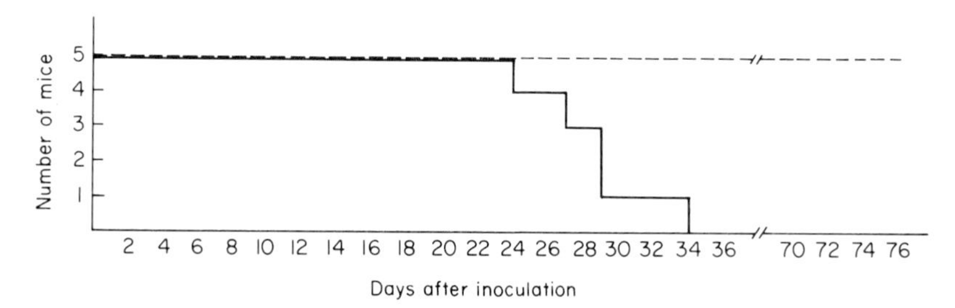

Fig. 4. Survival of DBA/2 mice inoculated with L1210 cells grown in suspension in glass bottles (———) and with L1210 cells grown as fibroblasts on BSA polymer treated with DEAE–dextran (- - - - - -). Each mouse was inoculated intravenously with 10^6 cells.

tinguishable from the cultures attached to a plastic surface (Fig. 2b) by the fibroblastic shape with disorderly arrangement of the former (Fig. 2c).

When a negatively charged surface (e.g., treated with polyglutamic acid) is washed during 24 hours after the second borate buffer treatment, with borate buffer containing 1 mg/ml DEAE–dextran, L cells form a monolayer on the negatively charged surface. Under these conditions, however, the maximum cell concentration per flask is lower than the cell concentration attained by cells growing in monolayer on a positively charged surface (Fig. 3).

The L1210 Leukemia Cells. When L1210 cells are subcultivated on a BSA polymer treated with positively charged substances, cells grow attached to the polymer. A few cells eventually lose the rounded shape and spread out assuming a fibroblastic form. The fibroblastic cells progressively invade the whole surface while the round cells disappear. The new cell type can be transferred by trypsinization. Fibroblastic L1210 cells differ from the parent L1210 cells growing in suspension in that they are not oncogenic for DBA/2 mice (Fig. 4).

CHAPTER 5

Titanium Disks

Jack Litwin

The use of metals as a surface substrate for cell growth *in vitro* has long been considered an attractive possibility because of the ease in manufacturing specialized structures of different geometries for large-scale mammalian cell cultivation. Such structures would not be subject to breakage as with glassware, so they could be used repeatedly for an indefinite period of time. Titanium is one of the few metals which accepts cells directly and combines good mechanical properties with resistance against corrosion. In addition, its surface properties are probably more uniform than those of the more complex structure of glass and does not show the same variability that exists with glass made by different manufacturers or from batch to batch. The metal seems to be acceptable for the growth of a variety of cells, for example, primary cells such as monkey kidney and human epithelial cells, and cells passed serially such as HeLa and human embryonic diploid fibroblast strains.

Cell Attachment

The properties of diploid fibroblast attachment and growth were studied on stationary titanium disks by Molin and Hedén[1] and were found to be in agreement with earlier studies on glass substrates by Easty *et al.*[2] using mouse ascites tumor cells. The attachment of cells to titanium proceeded much faster in a medium free of calf serum than when serum was present, but eventually the same number of cells attached in both cases (Fig. 1). When attachment

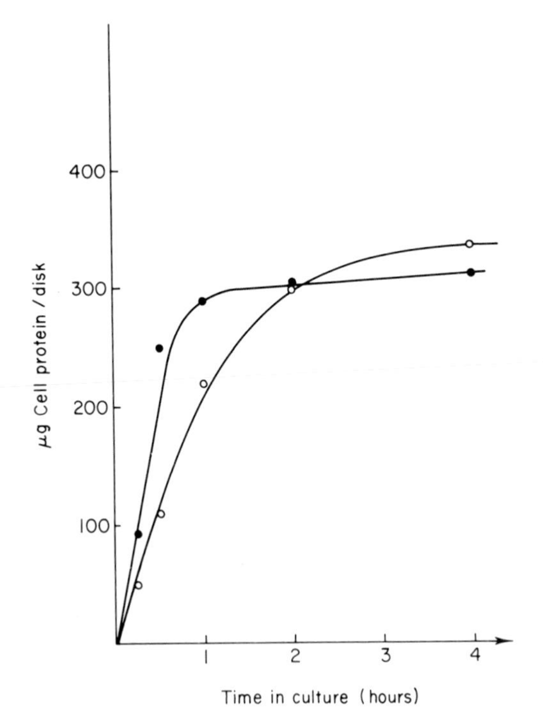

Fig. 1. Adsorption of human diploid lung cells to titanium surface in serum-free medium (●) and medium supplemented with 10% calf serum (○). The microgram cell protein expressed as bovine albumin equivalents is plotted against time in culture. (O. Molin and G. C. Heden.[1])

[1] O. Molin and C. G. Hedén, *Progr. Immunobiol. Standardization* 3, 106 (1969).
[2] G. C. Easty, D. M. Easty, and E. J. Ambrose, *Exp. Cell Res.* **19**, 539 (1961).

TABLE I

Maximal Cell Densities of Human Diploid Lung Cells Grown on Different Surfaces[a,b]

	Growth surface			
Days of incubation	Glass cells/cm^2 ($\times 10^4$)	Plastic cells/cm^2 ($\times 10^4$)	Titanium (static) cells/cm^2 ($\times 10^4$)	Titanium (rotating) cells/cm^2 ($\times 10^4$)
3	4.1	4.6	4.4	4.9
4	5.6	6.0	6.2	6.7

[a] The values represent an average of four or more experiments using cells in an active growth phase at low passage number.

[b] From O. Molin and C. G. Hedén.[1]

took place in a serum-free growth medium, the serum was added several hours after the cultures were seeded. The cells appeared to grow well after this treatment. This procedure was used when seeding the rotating disk machine (see below). A different result was obtained with cells on glass. If serum was added within a few hours after the cells were seeded, a large proportion of the attached cells rounded up and came off the glass. The cells remaining on the glass grew slower than expected. However, when the serum was added after overnight incubation, very few cells came off the glass and good subsequent growth was obtained.[3] These observations suggest that there may be some differences between the binding properties of cells to glass and metal substrates. However, the growth of diploid cells on titanium disks paralleled that obtained on glass surfaces and probably was slightly better, especially when the disks were rotated (Table I). Unfortunately rotating glass disks were not examined for comparison.

Molin and Hedén[1] observed that the titanium disks could be reused several times without intermittent washing. According to Weiss,[4] cells alter glass surfaces by leaving behind cellular substances which enhance subsequent cell attachment. Litwin[5] suggested that material from human diploid fibroblasts accumulated on the glass surface of a bottle repeatedly used without washing, which eventually slowed growth and reduced longevity. However, a more thorough investigation showed that Roux bottles used for over 1 year without washing had no influence on the growth rate or longevity of three independent strains of human diploid fibroblasts when compared with parallel cultures grown in clean Roux bottles with each passage. Therefore, there appears to be no limit to the length of time one can reuse a cell culture surface without washing.

Titanium Disk Apparatus

The application of titanium metal as a surface substrate in a large-scale mammalian cell cultivation device was first attempted by C. G. Hedén, and

[3] J. Litwin, unpublished results.

[4] L. Weiss, *Exp. Cell Res.* **25**, 504 (1961).

[5] J. Litwin, *12th European Symp. Poliomyelitis and Allied Diseases*, p. 271. Bucharest, 1969.

later was developed further by Molin and Hedén.[1] The device consisted of a 1-liter cylindrical glass bottle containing thirty-five titanium disks, 9 cm in diameter, spaced 3 mm apart along a shaft connected to a rotating drive system. During the initial inoculation of cells the vessel was oriented so that the disks were horizontal and the chamber was filled with the medium containing cells. After a sufficient period of adsorption, the vessel was reoriented so that the disks were covered only to their midpoint. They then could be rotated at different speeds. The growth of human embryonic diploid lung fibroblasts similar to those described by Hayflick and Moorhead[6] and Hayflick[7] was studied in this machine and on stationary titanium disks by Molin and Hedén.[1]

In the rotating titanium disk cultivator described by Molin and Hedén,[1] excellent growth was reported when the disks were rotated at 1 revolution per 2 minutes. Slower velocities could be used but faster velocities tended to strip cells from the surface.

Some of these experiments were repeated in a commercial model of this machine (Fig. 2) to study some of the conditions for cell attachment. It was found that when cells were allowed to attach to the metal in complete growth medium containing 10% calf serum and then the plates slowly rotated vertically in the chamber half filled with medium, most of the cells were stripped off the surface

Fig. 2. Biotec rotating titanium disk cell cultivator. The disks can be rotated in the vertical plane as shown in the figure or in the horizontal plane by placing the machine in the upright position on the projecting legs. (Courtesy of Biotec AB, Bromma, Sweden.)

[6] L. Hayflick and P. S. Moorhead, *Exp. Cell Res.* **25**, 585 (1961).

[7] L. Hayflick, *Exp. Cell Res.* **37**, 614 (1965).

by the surface tension when the disks were about 2 to 3 mm apart. The surface tension was sufficiently great to denature a considerable amount of the serum protein, causing a large flocculant precipitate to accumulate in the medium after 1 week continuous operation. When the disks were moved about 1 cm apart or if the chamber was completely filled with medium, this stripping effect did not occur. These observations suggest that the cells bind stronger to titanium in serum-free medium than when they attach in the presence of serum, perhaps sufficiently strong to resist the effects of surface tension. The smoothness of the metal surface may also be a factor affecting cell attachment and growth. Preliminary observations showed that slightly better attachment and growth in the rotating device was obtained when the metal surface was polished than with the rougher surface of the disks used earlier.

It is apparent that further studies on cell attachment to titanium surfaces in the presence or absence of serum are needed to yield more information on the cell-surface substrate binding mechanism. This information would be particularly useful for the type of culture system where the surfaces rotate in and out of an air-medium interface.

CHAPTER 6

Filter Paper Supports for Plant Cultures

René Heller

The use of ashless filter paper supports has been recommended for the culture of isolated plant tissues of the callus type.[1] These tissues are cultured according to Gautheret's technique,[2] which consists in culturing explants, 2–3 mm side measurement, weighing about 150 mg, in vertical tubes containing about 25 ml of nutrient medium generally solidified with agar, as these tissues cannot tolerate immersion in a liquid. Unfortunately agar is a source of impurities, particularly calcium, which forms part of the constituents and which the tissues are capable of extracting. It cannot be used for precise research in inorganic nutrition.

The technique described here attempts to maintain the simplicity of culture on agar, while guaranteeing a far greater purity. The ashless filter papers washed

[1] R. Heller, *C. R. Soc. Biol.* **142,** 947 (1948).

[2] R. J. Gautheret, "La Culture des Tissus Végétaux, Techniques et Applications." Masson et Cie, Paris, 1959.

with hydrochloric and hydrofluoric acids do not contain more than 0.0065% ash, i.e., 12 μg of ash per tube, 1000 times less than that provided by agar; moreover, it consists mainly of nonassimilable silica.

Nutrient Solutions

Nutrient solutions necessarily contain sucrose (3–5 g/liter) or dextrose (2–5 g/liter), auxin (β-indolylacetic acid 3×10^{-8} to 5×10^{-7} g/liter) or an equivalent growth substance (e.g., α-naphthaleneacetic acid), inorganic salts and, for certain tissues that are difficult to culture, certain vitamins (thiamine, pantothenic acid, biotin, etc.).

For the inorganic substances several formulas have been suggested. We advise the following,[3] seen in the tabulation below, which is suitable in most cases (expressed in mg/liter):

Major constituents		Minor constituents	
KCl	750	$FeCl_3 \cdot 6\ H_2O$	1
$NaNO_3$	600	$ZnSO_4 \cdot 7\ H_2O$	1
$MgSO_4 \cdot 7\ H_2O$	250	H_3BO_3	1
$NaH_2PO_4 \cdot H_2O$	125	$MnSO_4 \cdot 4\ H_2O$	0.1
$CaCl_2 \cdot 2\ H_2O$	75	$CuSO_4 \cdot 5\ H_2O$	0.03
		$AlCl_3$	0.03
		$NiCl_2 \cdot 6\ H_2O$	0.03
		KI	0.01

The pH should be adjusted to about 6.0 (with NaOH or HCl); buffer solution is not necessary.

Procedure

1. Prepare the nutrient solution and distribute about 25 ml per tube into 160 × 24 mm tubes.

2. Cut out round pieces of ashless filter paper of 55 mm in diameter (manipulate with forceps to avoid contaminating with impurities).

3. Prepare a stamper [Fig. 1(1)] with the following:

a. A glass tube of 21 mm in external diameter and a few centimeters in length, fixed onto a large cork, which serves as support to maintain it in a vertical position.

b. A tube of 23 mm internal diameter and a few centimeters in length with a flange at one end.

c. A glass push-rod that slides easily into the cylinder described above and which consists simply of a test tube of suitable diameter.

4. Place a round piece of filter paper on the end of the internal tube of the

[3] R. Heller, *Ann. Sci. Nat. Bot. Biol. Vég.* **14**, 1 (1953).

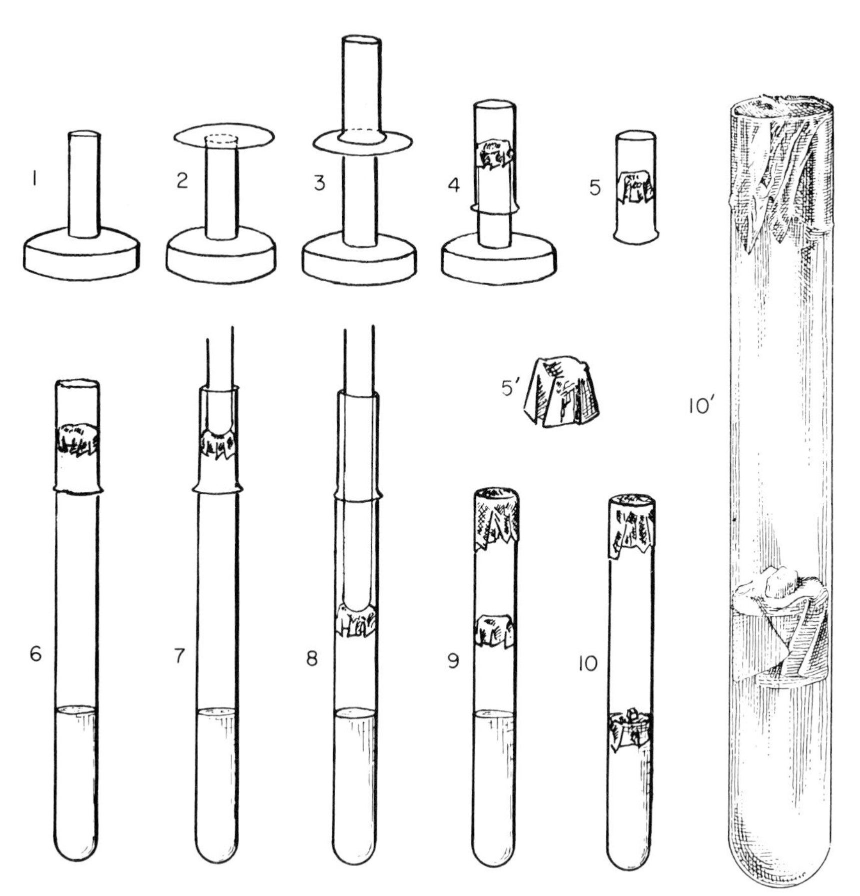

Fig. 1. Heller's technique for plant tissue cultures on filter paper supports, placing in position of the supports and transfer.

stopper [Fig. 1(2)], and apply the external tube on top of it [Fig. 1(3)], push the latter so as to surround the internal tube and give a slight twist, which causes the edges of the filter paper to fold downward [Fig. 1(4)]. The latter is then shaped in the form of a cap wedged between the two tubes. Remove the external tube which withdraws the paper cap with it [Fig. 1(5)] and apply over the opening of a culture tube [Fig. 1(6)]. By means of the push-rod, force down the filter paper support [Fig. 1(7)] until it is 5 to 6 cm above the free surface of liquid [Fig. 1(8)]; the pressure of the paper on the glass is sufficient to maintain it in position [Fig. 1(9)], so that during autoclaving this prevents it touching the solution and being spoiled.

5. Operate in the same way for the other tubes. Sterilize; after allowing to cool, aseptically transfer the fragments of callus (cubes, the sides measuring 2 to 3 mm) derived from cultures by placing them delicately on the paper support; at the same time, using the same forceps, plunge the support carrying the sample into the liquid in such a way that it completely bathes in the liquid and that the plane surface is 1 or 2 mm above the level of the liquid (Figs. 1 and 2). Stopper

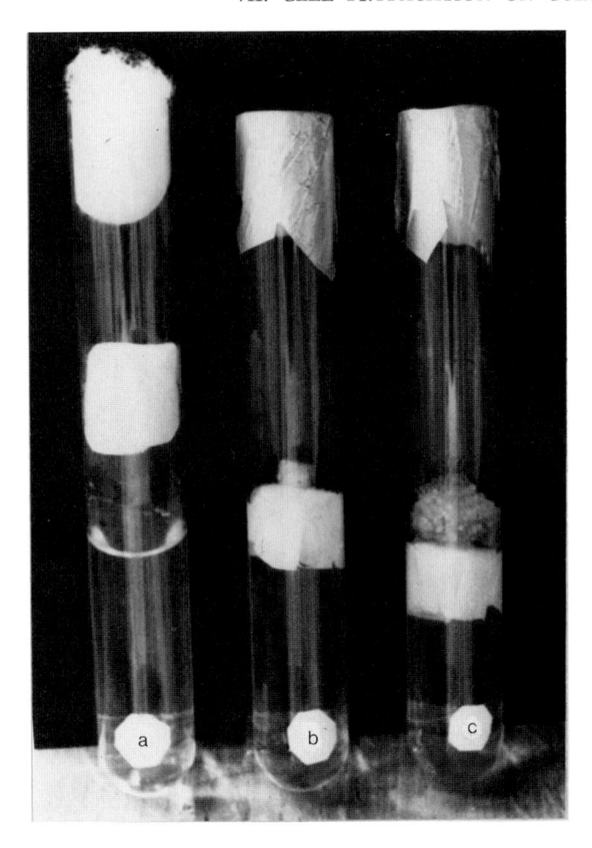

Fig. 2. Culture on liquid medium. (A) Position before autoclaving (ashless paper above the liquid); (B) just after transfer of tissue; (C) at the end of culture. From Heller.[3a]

the tubes with a plug of cotton wool and/or tinfoil stopper, or by any other means that maintains asepsis without totally preventing gaseous exchanges.

Notes

1. These paper supports can be made very quickly and placed in position; a hundred tubes can be equipped with them in one-quarter of an hour.
2. If the support is lowered too far into the liquid, a bubble of air remains enclosed under the support. This, however, is of no inconvenience.
3. After several weeks, the level of the liquid may become appreciably reduced; by shaking the tube several times the callus and its support are adjusted to a suitable level.
4. Growth in a liquid medium surmounted by a filter paper support is less abundant than on agar[2] (e.g., in 2 months increase in growth on agar is 2.700

[3a] R. Heller, *in* "International Conference on Plant Tissue Culture" (R. White and A. Grove, eds.). McCutchan Publ. Corp., Berkeley, 1965.

mg; in liquid, 1.300 mg), but it is sufficient for physiological experiments. The tissues are healthy and very similar to those cultured on agar.

Cultures in Renewed Medium

For certain research work it is of interest to renew the medium during the course of the experiment, e.g., to maintain a relatively constant composition in spite of the intake by the tissues, or on the contrary, so as to be able to deliberately modify its composition. The apparatus shown in Figs. 3 and 4, devised by M. Richez,[4] enables one to do so without displacing the callus.

It consists of a variation of the process described above in which the base of the tube is fitted with an attachment which enables it to be connected by a flexible tube to a recipient that can be either lowered (to empty the tubes) or raised (to fill them). The round pieces of filter paper are pierced with a small hole (2 mm diameter) to allow the passage of air during the emptying or filling process.

Three invaginations are made on the surface of the tubes just below the level of the support and form protrusions on the inside. These protrusions hold the paper support in place and prevent it from being displaced when the tube is emptied.

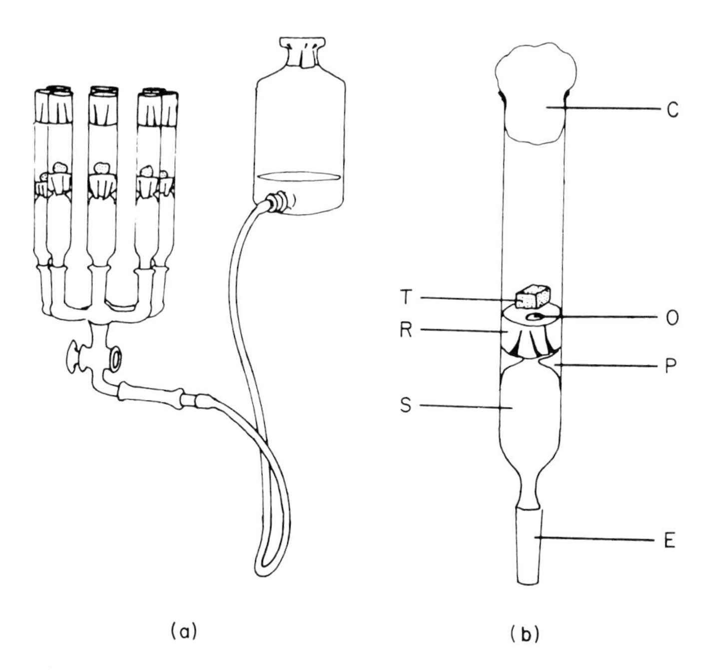

***Fig.** 3.* Richez's apparatus for periodically renewing the culture medium: (a) the complete apparatus; (b) detail of one tube C, cotton wool plug; T, explant; O, opening; R, filter paper; P, glass protrusions; E, ground-glass attachment; S, nutrient solution.

[4] M. Richez, *C. R. Acad. Sci.* (*Paris*) **260**, 655 (1965).

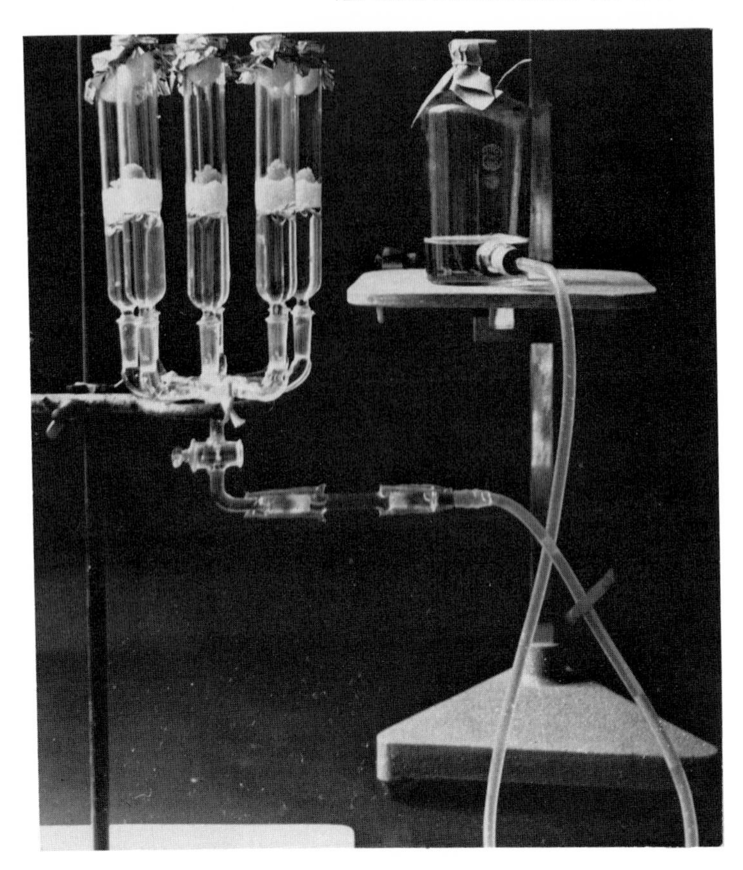

Fig. 4. Richez's method of renewing the medium in cultures on a liquid nutrient. From Heller.[3a]

The apparatus described above makes it possible not only to renew the medium, but also to periodically transfer back and forth the nutrient solution (by drawing off and replacing the solution by lowering and raising the reservoir bottle without any further addition). Growth is then definitely improved; it becomes practically as abundant as on agar medium. This improvement is probably due to the mixing of the atmosphere above the culture.[4]

Culture on ashless filter paper is suitable not only for tissue cultures, but also for the culture of embryos, seedlings and any other plant material not too large in size, and which it is required to culture aseptically on a nutrient medium of strictly calculated composition.

SECTION VIII

Evaluation of Culture Dynamics

Editors' Comments

Cell number is the most commonly used measurement for evaluating the progress and behavior of cells in culture. Basically the methods for measuring this parameter are simple and routine but can contain certain assumptions of which the investigator should be aware. These assumptions are alluded to in the individual chapters in this section. For example, all cells seen and counted in the hemocytometer are not necessarily viable cells, nor are the nuclei counted derived only from viable cells. Protein measurement contains a similar assumption. Thus, dye-exclusion tests for cell viability should be included in such measurements. If cells can be dispersed without severe damage, electronic enumeration of cells can be used but with the same limitations as those listed above. Moreover, this procedure can be used to measure changes in cell size, which may indicate either abnormal behavior of the cell under the culture conditions imposed or a change in ploidy (see also Section II, Chapter 19).

A recent innovation in labeling cell membranes and subsequent release of label as a measure of cell death are also described in this section.

A more accurate evaluation of the progress of cells in culture is achieved through a comparison of the "growth" curve and the "proliferative" fraction of the cell population. Cell doubling time as measured from a growth curve may not represent the true generation time of the cell since only a fraction of the population may proliferate. A truer evaluation of the culture dynamics is derived from the measurement of distribution of cells in the various phases of the cell cycle. The last two chapters of this section relate to this measurement. Further details for synchronization of cells are given in Section XIV, Chapter 5; for collection of mitotic cells, see Section IV, Chapter 4.

CHAPTER 1

Hemocytometer Counting

Marlene Absher

Enumeration of cells propagated *in vitro* may be conveniently determined using well-dispersed cell or nuclei suspensions in a standard hemocytometer chamber. The thick glass chambers are divided into sections of calibrated area and depth (Fig. 1). The total number of cells in a suspension then is easily calculated from the counts of cells in the hemocytometer chamber. Although the operation is easily performed, there are certain disadvantages to the method. If there are large numbers of samples to be counted on a routine basis, hemocytometer methods become time-consuming and laborious. In this case electronic methods of counting would be more convenient (see Section VIII, Chapter 2). There are several possible sources of error inherent in hemocytometer counting. According to Berkson *et al.*[1] and Sanford *et al.*[2] counting errors are of the order of 10%. Errors in sample preparation include inadequate dispersion of cells, loss of cells in the process of dispersion, inaccuracies in making dilutions, and incomplete mixing of cells prior to filling the hemocytometer chamber. Overfilling or underfilling the chamber, presence of air bubbles or dirt particles, and counting the cells before they settle will introduce counting errors. Another source of counting error arises from random distribution of cells in the chamber and according to Berkson *et al.*[1] the "error of field" is approximately 5%. The error can be reduced by counting more cells since the percentage field error varies inversely as a standard deviation.

Preparation and Counting of Cell Suspensions

Cell monolayers are gently dispersed with enzyme (see Section III) and resuspended in tissue culture medium (or other suitable salts solution). If the resulting cell suspension is too concentrated, a dilution should be made prior to filling the hemocytometer with the sample. For ease and accuracy of counting, chambers should be filled with cell suspensions containing approximately 30–45 cells/mm^2 on the ruled hemocytometer chamber.[2] Dilutions may be made in a test tube or in a blood dilution pipette. If dilutions are made in a test tube, a

[1] J. Berkson, T. B. Magath, and M. Hurn, *Amer. J. Physiol.* **128**, 309 (1939).

[2] K. K. Sanford, W. R. Earle, V. J. Evans, H. K. Waltz, and J. E. Shannon, *J. Nat. Cancer Inst.* **11**, 773 (1951).

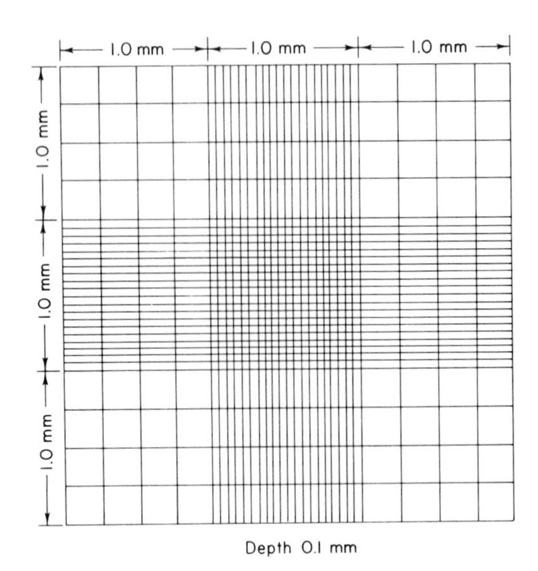

Fig. 1. Line drawing of ruling of Neubauer type hemocytometer chamber. Chamber units are 1 mm with a depth of 0.1 mm.

Pasteur pipette can be used for taking a sample to fill the hemocytometer chamber. The chamber may be filled directly from a blood dilution pipette. To minimize sample errors it is important that cell suspensions be well mixed when applied to the counting chamber. Care should be taken to add sufficient cell suspension to load the chamber but not to allow it to overflow. Allow the cells to settle a few minutes before counting. Routinely, using the Neubauer type chamber, cells in 10 of the 1-mm squares are counted (fill both sides of chamber, and count the four corner and the middle squares of each side). With a chamber depth of 0.1 mm the sample volume will then be 10×0.1 or 1 mm^3. As an example, if the number of cells counted in ten 1-mm squares is 300 then the sample contains 300 cells/mm^3 or 300,000 cells/ml. Multiplying by the dilution factor gives the amount of cells per milliliter in the undiluted sample, and multiplying this by the total volume gives the total number of cells in the original cell suspension.

Notes on Cell Counting

If determination of the viability of cells is desired, a vital stain may be incorporated into the diluting fluid allowing simultaneous enumeration of total cells and percentage of viable cells in the suspension (see Section VIII, Chapter 3).

Preparation and Counting of Cell Nuclei

Cells which are difficult to disperse or tend to aggregate in suspension may be enumerated by nuclei counts. A modification of the nuclei counting technique of Sanford *et al.*[2] will be described. Monolayer cultures are rinsed twice with phosphate-buffered saline, pH 7.2 (PBS), and drained well on absorbent paper. To a T-25 flask add 2.0 ml of 0.1 *M* citric acid solution warmed to 37°C (use 5.0 ml for a T-75 flask). Allow the citric acid solution to remain on the cells a sufficient time for release of the nuclei. The time may vary for different types of cells. Sanford *et al.*[2] suggest incubation in citric acid for 1 hour, but 30 minutes has been found to be adequate for human foreskin fibroblasts (personal observation). Frequent and vigorous shaking during the incubation aids in disrupting the cells and freeing the nuclei. The resulting nuclei suspensions are stable at 5°C for at least 7 days.[2] To count, fill the hemocytometer chamber with the nuclei suspension and proceed as outlined above for whole cell suspensions. To aid in visualizing the nuclei, 0.1% crystal violet may be incorporated in the citric acid solution.

Notes on Nuclei Preparations

Nuclei from aged, degenerating cells will disintegrate in the citric acid solution[2] and are, therefore, not countable. If desired, the nuclei suspensions may be used directly for determining total cell protein of the culture. The technique of Oyama and Eagle[3] for measuring cell growth in tissue culture by a modification of the method of Lowry *et al.*[4] for determining protein is used. We have found in culturing human foreskin fibroblasts that protein concentration per cell, as measured by the Lowry method,[4] varies with days in culture, media conditions, and passage level of the culture. Therefore, using cell protein concentration as a measure of cell growth, as reported by Oyama and Eagle,[3] is not applicable to all *in vitro* culture systems. Determination of total protein in the cell culture is easily accomplished by adding the Lowry reagents directly to the citric acid nuclei suspensions. If protein is to be determined in this manner, the crystal violet must be omitted from the citric acid solution as it will interfere with the colorimetric measurement. Blanks and protein standards must contain a proportional amount of citric acid as is present in the nuclei samples.

[3] V. I. Oyama and H. Eagle, *Proc. Soc. Exp. Biol. Med.* **91**, 305 (1956).

[4] O. H. Lowry, N. J. Rosebrough, A. L. Farr, and R. J. Randall, *J. Biol. Chem.* **193**, 265 (1951).

CHAPTER 2

Electronic Enumeration and Sizing of Cells

A. Primary Tissue Cells

W. F. Daly

A simple, quick, and reproducible method for determining cell yield from primary tissue has become a necessity for large-scale tissue culture production. Many methods of estimating mammalian cell populations have been reported[1,2] but the hemocytometer and the index-of-viability dyes, e.g., trypan blue[3,4] and eosine,[5,6] have been the most widely used means of determining viable cell concentrations. However, hemocytometer counts have proved to be time-consuming and subject to serious error,[7,8] and the dye exclusion tests are not always reliable indicators of cell cultivability.[9]

This laboratory utilizes an electronic counter which enumerates individual cells by a form of electronic "gating," for establishing the cell inoculum. The instrument has been evaluated for blood cell counting,[10,11] and its application in growth studies has been demonstrated.[12]

Culture Methods

Primary cell suspensions are prepared from freshly harvested chicken embryos, and rhesus and cercopithecus monkey kidneys. Following trypsinization[1,13,14] the cells are resuspended in appropriate growth media, thoroughly mixed, and samples are removed for enumeration. The remaining cell suspension is diluted, inoculated into culture bottles, and incubated at 36°C. Confluent

[1] J. S. Younger, *Proc. Soc. Exp. Biol. Med.* **85**, 202 (1954).

[2] V. I. Oyama and H. Eagle, *Proc. Soc. Exp. Biol. Med.* **91**, 305 (1956).

[3] A. M. Pappenheimer, *J. Exp. Med.* **25**, 633 (1917).

[4] W. F. McLimans, F. E. Giardinello, E. V. Davis, C. J. Kucera, and G. W. Rake, *J. Bacteriol.* **74**, 768 (1957).

[5] J. H. Hanks and J. H. Wallace, *Proc. Soc. Exp. Biol. Med.* **98**, 188 (1958).

[6] R. Schrek, *Arch. Pathol.* **35**, 857 (1943).

[7] J. Berkson, T. B. Magath, and M. Hurn, *Amer. J. Physiol.* **128**, 309 (1940).

[8] R. Biggs and R. L. Macmillan, *J. Clin. Pathol.* **1**, 269 (1948).

[9] J. R. Tennant, *Transplantation* **2**, 685 (1964).

[10] G. Brecher, M. Schneiderman, and G. Z. Williams, *Amer. J. Clin. Pathol.* **26**, 1439 (1956).

[11] C. F. T. Mattern, F. S. Brackett, and B. J. Olson, *J. Appl. Physiol.* **10**, 56 (1957).

[12] M. Harris, *Cancer Res.* **19**, 1020 (1959).

[13] D. Bodian, *Virology* **2**, 575 (1956).

[14] R. Dulbecco, *Proc. Nat. Acad. Sci. U. S.* **38**, 747 (1952).

monolayers developed in 16 to 18 hours for chicken embryo cultures and in 7 days for monkey kidney cultures.

ENUMERATION

Enumeration with the electronic counter is accomplished by a form of electronic "gating." As a particle is drawn through a small aperture (100 μm), an equal amount of electrolyte is displaced causing a voltage drop due to an increase in aperture impedance. The resultant pulses are amplified, recorded by a decade counter, and visualized on an oscilloscope.[10-12] The instrument was calibrated by using a relatively uniform particle (ragweed pollen) according to the method described by Coulter Electronics.

For accurate cell enumerations optimum instrument settings were predetermined by calculating the total volume of each cell type. For example, on a model A the settings for monkey kidney cells are threshold 20 with an aperture current setting of 3 and a gain of 2, whereas for chicken embryo cells the settings are 5, 2, and 5, respectively. Cells are suspended in phosphate-buffered saline[15] and four successive counts are taken on each sample, representing four aliquots of the same cell population, or a total of 2 ml.

The results of a study[16] showed that the electronic counter is suitable for routine tissue culture practices, since it is more consistent than the hemocytometer and possesses excellent reproducibility. It is evident that the degree of variation existing in hemocytometry could eventually be responsible for the establishment of erroneous planting rates causing either a decrease or increase in the number of total cells available, which would directly affect the number of cultures obtained from a primary suspension. The utility of the electronic counter for routine operation is further supported by a twofold saving in time and the elimination of many stress factors that may contribute to erroneous visual counts. The electronic counter has been in operation in this laboratory for the past 9 years as the standard enumerating method. During this period substantial numbers of cercopithecus and rhesus monkey kidneys and chicken embryos were trypsinized, planted, and in all instances gave rise to suitable monolayers for tissue culture production.

[15] R. Dulbecco and M. M. Vogt, *J. Exp. Med.* **99**, 167 (1954).
[16] W. F. Daly, *J. Pharm. Sci.* **55**, 426 (1966).

CHAPTER 2

Electronic Enumeration and Sizing of Cells

B. Tissue Culture Cells

M. Harris

Rapid and accurate determinations of cell number and cell size are essential for quantitative studies on growth and population dynamics. Electronic cell counters which fulfill these requirements have been developed by several manufacturers[1] and have come into wide use in recent years. Such instruments provide a means for routine growth measurements on any tissue culture cells that can be effectively dissociated with trypsin or other dispersing agents.

OPERATION OF ELECTRONIC CELL COUNTERS

The counters now in use are based on a difference in conductivity between cells and suspending medium. The sensing device is a tube containing a small cylindrical aperture (usually 100 μm in diameter for tissue culture cells) and which is immersed in a beaker containing cells to be enumerated. A small d-c voltage is applied across this gate. When a metered volume of suspension is drawn through the opening by vacuum, the passage of an individual cell causes a voltage drop, due to increased aperture impedance. The corresponding pulse can be visualized by an oscilloscope and is recorded by the instrument counting circuits. Further operational details and critical evaluation of electric sensing zone counters can be found in several published papers.[2–4]

PREPARATION OF CELL SUSPENSIONS

Electronic counting is limited to cell populations in culture that can be reasonably dissociated without conspicuous damage, formation of debris, or residual clumping.[5] Occasional cells are unsatisfactory in these respects, but most primary or established lines in culture can be suitably dispersed by one or another of the methods described earlier in this volume. Usually dissociation with 0.1% trypsin or 0.1% trypsin plus 0.1% disodium versenate (TV) gives optimal results. These solutions are prepared in calcium-magnesium free phosphate-

[1] Coulter Electronics, Inc., Hialeah, Florida. Particle Data, Inc., Elmhurst, Illinois.

[2] G. Brecher, M. Schneiderman, and G. Z. Williams, *Amer. J. Clin. Pathol.* **26**, 1439 (1956).

[3] R. B. Adams, W. H. Woelker, and E. C. Gregg, *Phys. Med. Biol.* **12**, 19 (1967).

[4] R. J. Harvey, *Methods Cell Physiol.* **3**, 1 (1968).

[5] M. Harris, *Cancer Res.* **19**, 1020 (1959).

buffered saline (PBS^-), centrifuged and filtered, and stored in frozen form until used. Cultures to be enumerated are first rinsed twice with PBS^- and drained briefly on absorbent tissue. A measured amount of TV is added and the culture incubated until the cell sheets slough off and dissociation is advanced. The cells in each flask are then suspended uniformly by gentle but thorough mixing with a pipette. This is a critical step in the procedure. The cell suspensions should be monitored visually with an inverted microscope, and pipetting continued until dispersion is complete. Serious artifacts result if counts are made on partially dissociated or clumped suspensions.

Counting Procedures

Trypsinized cell suspensions are usually too concentrated for optimal counting, and must be diluted to reduce errors from coincidence (if two or more cells pass through the aperture simultaneously they register as a single pulse). Diluting solutions are available commercially[6] but 0.85% NaCl is quite satisfactory for this purpose. Counts are conveniently performed in 30-ml beakers into which an appropriate volume of NaCl is added from a burette or pipetting device and mixed with a measured aliquot of trypsinized cells. The dilution is chosen to give an estimated machine count of 10,000 or less (on a metered volume of 0.5 ml), at which level coincidence is minor and can usually be ignored. For example, if a Petri dish culture contains approximately 1.5×10^6 cells and 2.0 ml TV are used for dissociation, an appropriate dilution would be 0.4 ml suspension plus 19.6 ml NaCl. At this concentration the machine count should be approximately 7500, based on a metered sample of 0.5 ml. Thus the cells per milliliter in the original trypsinized suspension would be $7500 \times 2 \times 50 = 750{,}000$, and the total cell number in the culture would be 1.5×10^6, as indicated.

It is important in making instrument counts to choose settings which register all pulses above background noise (i.e., all cells of the sample population). The aperture current and amplitude controls should be adjusted to give an oscilloscope pattern in which the impulses fill one-half to two-thirds of the screen height. Excessive aperture current may affect cell volume and should be avoided. The threshold switch should be set to give a minimum background without reducing the true cell count. Control counts on the NaCl diluting solution, as well as blank TV and PBS^- should be reproducibly low. All glassware should be scrupulously clean and care taken to avoid the presence of cotton fibers or other particulate debris that may occlude the aperture tube. Spurious counts will result if the aperture is partially or completely blocked. Blockage may be visible through the side telescope focused on the aperture, or may be signaled by a change in audible counting cadence. Other artifacts may be produced by growth of mold or other contaminants in the aperture tube. This can be prevented by terminal flushing with a mild disinfectant at the end of each day. A 1:1000 dilution of Roccal[7] is satisfactory for this purpose.

Electronic cell counting is readily incorporated into procedures for propa-

[6] "Isoton," Coulter Electronics Inc., Hialeah, Florida.

[7] Winthrop Laboratories, New York.

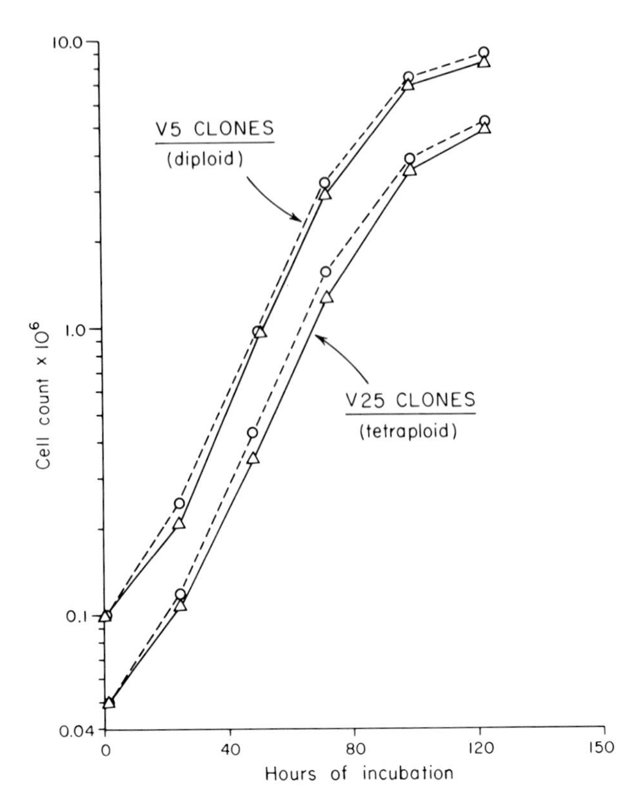

Fig. 1. Growth curves for clonal populations of Chinese hamster cells in monolayer cultures. Each point represents the average value for three determinations of cell number with a Coulter counter. From M. Harris, *Exp. Cell Res.* **66**, 329 (1972).

gating stock lines, and gives a more objective baseline than the traditional splitting of confluent populations. Machine counts are particularly useful for following changes in cell number during population growth cycles, or for determining the effects of specific agents on growth rates. Figure 1 illustrates the reproducibility of data obtained in this way with clones of diploid and tetraploid Chinese hamster cells. Each series was initiated with replicate cultures in 1-ounce prescription bottles. At successive time intervals groups of three cultures for each line were trypsinized and used to obtain an average cell count. Logarithmic growth rates derived from this data are essentially identical for the four populations concerned.

Cell Size Analyzers

In addition to enumeration, electronic counters can provide information on the distribution of particle sizes in a sample population.[8,9] The change in re-

[8] A. C. Peacock, G. Z. Williams, and H. F. Mengoli, *J. Nat. Cancer Inst.* **25**, 63 (1960).

[9] G. Brecher, E. F. Jakobek, M. A. Schneiderman, G. Z. Williams, and P. J. Schmidt, *Ann. N. Y. Acad. Sci.* **99**, 242 (1962).

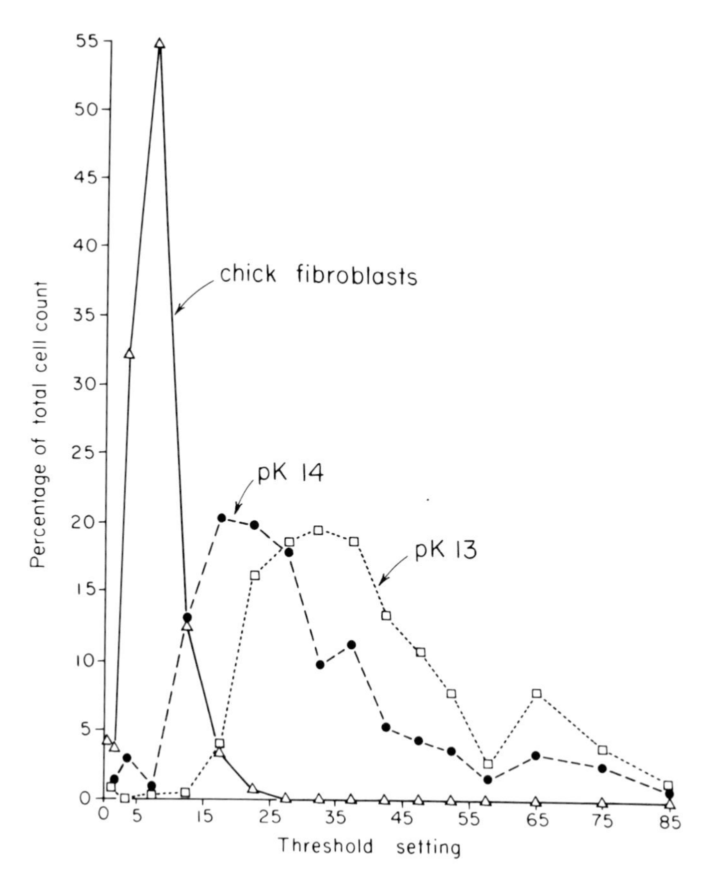

Fig. 2. Size distributions for three cell populations maintained separately as monolayer cultures. The cell lines respectively are chick fibroblasts from skeletal muscle, and diploid (pK 14) and tetraploid (pK 13) clones from an established line of pig kidney cells. From M. Harris.[5]

sistance at the instrument aperture and pulse height are proportional to the volume of particles passing through it. Thus a frequency distribution of particle sizes can be obtained by scanning at successive levels as the threshold setting is raised to exclude impulses below a minimum size. Early instruments[10] utilized this procedure, or employed a "window" between upper and lower threshold settings. The data obtained can be expressed in the form of a derivative curve, with the percentage of total population plotted against threshold settings (see Fig. 2).

A second generation of size analyzers is now available with improvements of instrumentation which permit simultaneous accumulation of size data in multiple channels,[10] thus eliminating the need for a single channel scan. In these instruments the total size range may be divided into 64 or 100 channels, and the overall

[10] Coulter Electronics, Hialeah, Florida. Early instruments (Model J) are now replaced by multichannel analyzers which include model P_{64} (64 channels) and the Channelyzer (100 channels).

size distribution may be visualized on an oscilloscope, expanded in resolution for particular areas, or recorded graphically on a plotter. These instruments provide maximum flexibility in the discrimination and analysis of particle sizes, although the range of useful applications for cell culture systems does not yet match the instrumental capabilities.

Perhaps the most important index for tissue culture populations is mean cell size. It is useful to be able to monitor this variable routinely during maintenance of stock populations, since a change in cell size may be indicative of unfavorable growth conditions, cross-contamination by a foreign cell type, or a heritable change in the precursor cells. The MCV/Hct accessory recently developed for the Coulter Electronic Cell Counter[1] is a very satisfactory device for this purpose, and can be operated quite simply as an adjunct to the normal counting procedure. This instrument is designed to give a digital readout for mean corpuscular volume in blood samples, but it can be readily adapted to measure the volume of larger tissue cells as well.

Since human erythrocytes usually have a mean cell volume of approximately 90 μ^3, the direct scale (0–200) provided on the MCV/Hct accessory falls below the range required for tissue cells in culture (volumes of 1000 μm^3 or more). By using larger particles of known size as standards, however, the instrument readings can be converted as a simple multiple to the desired volumes. Ragweed pollen is ideal for this purpose, since the particles are uniform spheres which correspond roughly in size to tetraploid Chinese hamster cells. The mean cell volume for ragweed pollen can be determined by measurements of diameter with an ocular micrometer. For example, the mean value from 100 measurements of one pollen sample was 19.4 μm. When converted to a volume basis by the formula $V = 4/3\ \pi r^3$ this gives a particle size of 3800 μ^3, which was used as a standard for further work. Stock suspensions for routine use can be prepared by mixing ragweed pollen in 1000 ml 0.85% NaCl containing 0.15 ml Tween 80 to minimize clumping. As preservatives, 3.0 ml neutral buffered formalin plus 0.1 ml tincture of Merthiolate (1:1000) are added as preservatives. The final concentration of pollen particles is adjusted to give machine counts of approximately 10,000 (i.e., 20,000 pollen particles per milliliter).

For size analysis with the MCV/Hct attachment, the controls for the model B Coulter counter are set at aperture current 2, amplification 16, and threshold 2. Variation between instruments and the position of other control switches may necessitate changes from these settings. The MCV/Hct accessory computes the mean particle size after counts are completed, and displays this result in the MCV dial. By using the calibration screw provided, the dial setting can be adjusted to read 95 for a standard suspension of ragweed pollen. The instrument size factor is then 40, since $95 \times 40 = 3800\ \mu^3$, the measured volume for ragweed. Under similar conditions, the volume of other particles can be obtained by multiplying the MCV readings by 40, assuming the response is linear for particles of different sizes.

A test of this assumption is provided by the data shown in Fig. 3. Ragweed pollen, mulberry pollen, and two lines of Chinese hamster cells were employed. In each case the particle volume was measured directly with a Coulter Model

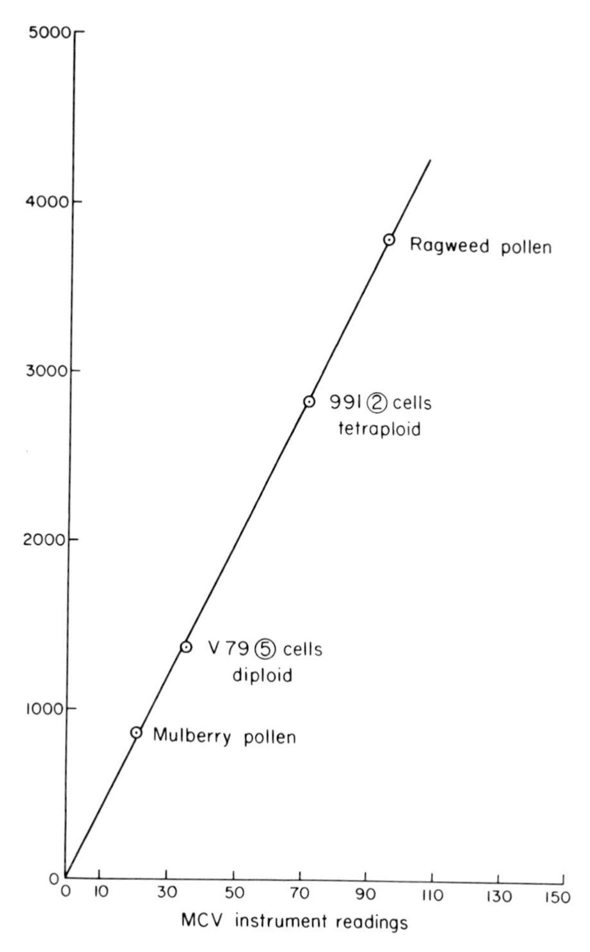

Fig. 3. Comparison of particle size by a Coulter Cell Size Analyzer, Model J, with similar determinations by an MCV/Hct accessory for the Model B Counter.

J cell size analyzer, and again with the MCV/Hct accessory as described above. When these independent determinations are plotted together the agreement is clear-cut, and the linearity of response extends to zero. The threshold setting for this work is critical and must be set just above the background (level of 2) for an optimal response. At higher threshold settings the mean volumes shift upward perceptibly, presumably because of a partial cutoff of smaller cells from the distribution pattern. If this artifact is avoided, however, the MCV readings provide a simple and direct measurement of relative cell size.

CHAPTER 3

Dye Exclusion Tests for Cell Viability

Hugh J. Phillips

The dye exclusion test for cell viability depends upon the fact that viable cells do not take up certain dyes whereas nonviable cells do. It has been shown that nonviable cells which take up the dye by this technique do not respire, glycolyse, or extend cellular processes when replanted in a tissue culture system. Further, they are readily digested by dilute solutions of trypsin.[1]

Numerous dyes have been used to differentially stain nonviable cells or tissues[2–5]; trypan blue and erythrosine B will be discussed. The former is probably used more extensively than the latter. Each dye, however, has staining characteristics which preclude its use under certain conditions. These conditions will be discussed briefly after the procedure for viability testing is described.

STOCK DYE SOLUTION

A 0.4% aqueous solution of either trypan blue or erythrosine B has a pH of about 6.5 and is hypotonic to animal cells. Therefore, stock solutions of the dye are made up as an isotonic salt solution buffered at pH 7.2–7.3. To about 95 ml distilled water in a 250-ml beaker are added: 0.4 g trypan blue or erythrosine B, 0.81 g sodium chloride, 0.06 g potassium phosphate monobasic, and 0.05 g methyl *p*-hydroxybenzoate as a preservative. The solution is heated to boiling to dissolve all materials, then cooled, and the pH is adjusted to 7.2–7.3 with 1 *N* sodium hydroxide (about eight drops). The final volume is adjusted with distilled water to 100 ml. This solution is stable indefinitely at room temperature.

TEST FOR VIABILITY

Cells in suspension can be stained directly for viability. However, cell suspensions have to be prepared when monolayer cultures are used. The procedure described is based on cells growing as a monolayer on glass in a 60-cm^2 surface area flask, but it can be adjusted for any size cell culture. Culture medium is siphoned or poured off and discarded. The cells and flask are rapidly washed

[1] H. J. Phillips and R. V. Andrews, *Exp. Cell Res.* **16**, 678 (1958).
[2] A. M. Pappenheimer, *J. Exp. Med.* **25**, 633 (1917).
[3] R. Schrek, *Arch. Pathol.* **37**, 319 (1944).
[4] H. J. Phillips and J. E. Terryberry, *Exp. Cell Res.* **13**, 341 (1957).
[5] J. H. Hanks and J. H. Wallace, *Proc. Soc. Exp. Biol. Med.* **98**, 188 (1958).

with 5 ml of trypsin solution[6] which is discarded. Five milliliters of trypsin solution are added to the flask and set aside until the cells detach from the flask (about 5 minutes). The cells are then dispersed about fifty times with a bulb and Pasteur pipette. One milliliter of the suspension is transferred to a 15-ml centrifuge tube and the cells collected at 1,000 *g* for 5 minutes. The supernatant fluid is siphoned off and discarded. Hanks' solution (1 ml) and 0.1 ml of stock dye solution are added to the cells. The mixture is dispersed ten times with a Pasteur pipette and the time noted. A drop of the suspension is placed on a hemocytometer and a viability count is made 4 minutes after dispersing cells. The number of stained cells and nonstained cells in a given area on the hemocytometer are counted. At least 100 cells should be counted. The percentage viable cells is equal to nonstained cells/(stained + nonstained cells) × 100.

Factors Affecting Accuracy of Test

The degree to which cells take up dye is pH dependent. Within a range of pH 6.6 to 7.6, trypan blue uptake occurs maximally at pH 7.5. For erythrosine B, the maximum uptake occurs at pH 7.0.[7]

Viable cells that have been subjected to trypsinization are changed so that they gradually take up dye when used at dye concentrations of 114 mg % (0.4 ml dye solution with 1.0 ml cell suspension). This does not occur if either dye is used at 36 mg % final concentration (0.1 ml dye with 1.0 ml cell suspension).

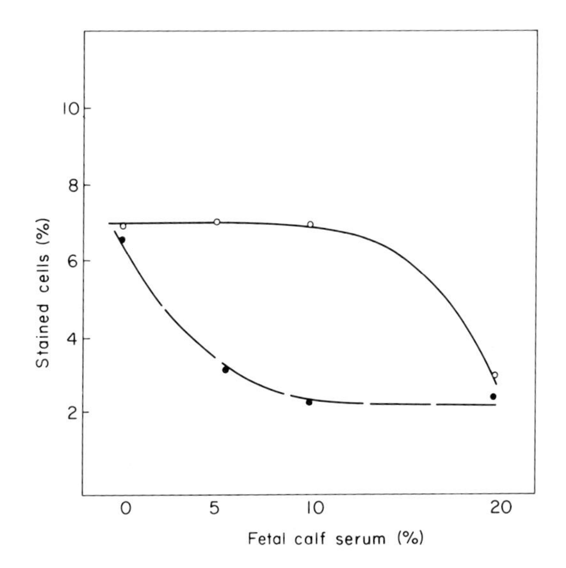

Fig. 1. The effect of fetal calf serum concentration on the staining of nonviable cells with trypan blue (—●—) or erythrosine B (—○—). All points represent a 1.0-ml sample from the same cell suspension of HeLa cells. The final dye concentration was 36 mg % in all cases.

[6] P. I. Marcus, S. J. Cieciura, and T. T. Puck, *J. Exp. Med.* **104**, 615 (1956).
[7] H. J. Phillips and L. A. Farr, unpublished data.

A major drawback for trypan blue is that the dye has a greater affinity for protein in solution (serum in medium) than it does for nonviable cells. This was first suggested by Pappenheimer[2] and recently verified by Phillips and Farr.[7] As can be seen in Fig. 1, the number of cells in suspension stained by trypan blue rapidly decreases as the serum concentration is increased. If the concentration of trypan blue is increased to exceed protein concentration, the cells cannot be seen. Therefore, trypan blue should be used only in the absence of soluble protein in the suspension. Apparently erythrosine B has a greater affinity for nonviable cells than soluble protein. If the serum concentration of the suspension is 20%, erythrosine B at a final concentration of 66.7 mg % (0.2 ml dye with 1.0 ml cell suspension) should be used, rather than the 36 mg % as shown in Fig. 1.

CHAPTER 4

Quantitative Methods for Measuring Cell Growth and Death

H. T. Holden, W. Lichter, and M. M. Sigel

Cell growth and death can be investigated by microscopic observations of cells in monolayers. More precise measurements are achieved by cell counts performed by means of a hemocytometer or by automatic electronic cell counters. Visual counting is usually accompanied by vital stains which distinguish viable cells from damaged or dead ones. However, microscopic evaluations are tedious and impractical when large numbers of cultures have to be counted. The automatic counters are expensive and not readily available in all laboratories and, furthermore, they do not discriminate between viable and nonviable cells.

For these reasons it is advantageous to employ chemical measurements whenever quantification of cell growth and death is required, especially when dealing with a large number of cultures. In our laboratory we have utilized two procedures; one measures increases in protein content of cultures and the other measures cytotoxic effects.

Protein Determination for the Study of Cell Growth and Death

The measurement of cell growth by determination of cell protein concentration is a rapid and reliable method which can be employed in large-scale testing.

This method, as described by Oyama and Eagle,[1] is a modification of the method of Lowry *et al.*[2] for measuring protein using a phenol reagent. The amount of cell growth can be determined with cultures grown in suspension and also with cultures adhering to a surface. The ease of determination of the initial protein concentration and of the subsequent concentrations during the time course of the experiment makes this method very attractive. With appropriate conversion factors the values may be changed to dry weight, protein nitrogen, or cell count. This method has found great utility for assessing cytocidal and cytostatic effects of extrinsic substances and is also useful for monitoring the dynamics of cell populations during the logarithmic phase of growth.

Procedure. Dissociated cells from stock monolayer cultures or cells from suspension cultures are diluted in growth medium to a desired concentration. The adjusted cell preparation, kept in uniform suspension by constant stirring, is delivered to culture vessels with a continuous pipetting device. Cultures are incubated at the temperature optimal for the cells in question or at the temperature dictated by the experimental objective. To serve as a baseline (initial value) for determining cell growth, the cellular protein per unit seed volume is determined. Tubes receiving only growth medium serve as cell-free controls.

The cellular protein after appropriate incubation periods is determined as follows: the medium is decanted from the cultures and the tubes washed twice with 5-ml aliquots of physiological saline. (When suspension cultures are employed, measured aliquots of cell suspensions are transferred to tubes and the cells sedimented by centrifugation prior to and after each wash.) To each tube is added 5 ml of Lowry's C solution.[3] As an aid in digesting the cells, the tubes are heated at 56°C for 5 minutes and then shaken for 5 minutes, following which 1 ml of distilled water is added to give a final volume of 6.0 ml. For each set of assays, protein standards are prepared consisting of 1 ml of various concentrations of bovine serum albumin (selected to cover the anticipated cellular protein range) plus 5 ml of Lowry's C solution. All tubes receive 0.5 ml of 1 *N* phenol reagent[4] jetted in by syringe to obtain rapid admixture. (The above volumes are employed for cultures grown in 16 × 125 mm tubes. When culture vessels of other sizes are employed, it may be advisable to proportionally alter the amounts of the respective reagents.) One hour after phenol addition, the mixtures are transferred to cuvettes and read in a spectrophotometer at 660 nm. Prior to reading, the instrument is adjusted to zero against reagent control (5 ml of Lowry's C + 1 ml of water and 0.5 ml of phenol reagent). The optical densities (O.D.) of bovine albumin standards are determined. To compensate for any residual growth medium protein in the assay cultures, the instrument is then

[1] V. I. Oyama and H. Eagle, *Proc. Soc. Exp. Biol. Med.* **91**, 305 (1956).

[2] O. H. Lowry, N. J. Rosebrough, A. L. Farr, and R. J. Randall, *J. Biol. Chem.* **193**, 265 (1951).

[3] Lowry's solution: solution A, 20 g Na_2CO_3, 4 g NaOH pellets water to give 1 liter (store in plastic bottle); solution B, equal parts of 1% cupric sulfate and 2.7% sodium potassium tartrate (unstable; prepare only as required); solution C, 100 ml solution A, 2 ml solution B, and 1 ml 1% sodium lauryl sulfate.

[4] Fisher Scientific Company, Atlanta, Georgia.

adjusted to zero against digests from the cell-free controls, following which the O.D. of the remaining preparations is determined.

From the O.D. values of the bovine albumin standard—representing known amounts of protein—is constructed a standard curve. Using this curve, the O.D. values of the experimental cultures are converted into micrograms of protein. The average cellular protein value for replicate cultures after a specified period of incubation under any given experimental conditions minus the initial cellular protein value represents the net protein synthesis under these conditions. If the respective O.D. values of replicate cultures differ by more than 0.1 the results are discounted and the test repeated.

As previously indicated, these techniques are very useful in assays of cytotoxic substances. To test 100 materials for their effect on cell growth, the following procedure is used. The test materials are delivered to 600 culture tubes (representing 100 materials × three dilutions each × two tubes per dilution) and Hanks' balanced salt solution (HBSS) is introduced into an additional 25 tubes. The adjusted cell suspension is added to all tubes containing test material and to twenty cell control tubes containing HBSS. (We favor volumes of 0.2 ml for test materials or HBSS and 1.8 ml of cell suspension for cultures in 16 × 125 mm tubes.) The remaining five tubes with HBSS receive cell-free medium and are used as medium controls. All cultures are incubated for the desired period and the effects of the test materials are reflected in the comparative net protein synthesis in cultures containing these materials and in the cell controls. Results are expressed in percentage as follows:

$$\frac{\text{Average value, test material} - \text{initial value}}{\text{Average value, cell controls} - \text{initial value}} \times 100$$

Release of ^{51}Cr as a Measure of Cell Death

This method, which is a modification of the procedure of Wigzell,[5] utilizes the release of ^{51}Cr to measure cell death. After the target cells are labeled, the isotope is slowly released (spontaneously) over a period of 24 to 48 hours; however, if the cell is injured or killed, the isotope is released very quickly. Quantitation of the amount of isotope released by the culture gives a measurement of the percentage of cells that were killed or died during a prescribed time interval. Excellent correlation has been shown with other techniques (e.g., dye exclusion) which measure cell damage. This technique has been successfully employed to measure cell death due to virus infection,[6] to detect cytotoxic lymphocytes, and to detect and quantitate cytotoxic antibodies—the latter described in detail here. Although this method can be used with various target cells, the description which follows is based on work with Rous sarcoma transformed hamster cells.

Procedure. MSR-5 cells, a line established in our laboratory from a tumor induced in LHC/LAK inbred hamsters by the Schmidt-Ruppin strain of Rous

[5] H. Wigzell, *Transplantation* 3, 423 (1965).

[6] H. M. Vickrey and S. S. Elberg, *J. Immunol.* **106**, 191 (1971).

sarcoma virus (RSV), are cultured in spinner flasks using the following growth medium: Eagle's Minimal Essential Medium (MEM) for suspension culture containing MEM nonessential amino acids and supplemented with 10% tryptose phosphate broth, 8% calf serum, and 2% heat-inactivated fetal calf serum. Methylcellulose (15 cps[4]) is added to a final concentration of 0.12%. Antibiotics (200 units of penicillin and 200 μg streptomycin per milliliter) are employed. The cells are centrifuged and reseeded in fresh medium at 24-hour intervals.

Inbred hamsters bearing transplanted RSV tumors serve as the source of the cytotoxic antibody; this antibody is directed toward the RSV-induced cell surface antigen. We use only sera free of noncomplement-mediated cytotoxicity as determined by incubation of the antibody with target cells in the absence of complement for 1 hour.

Fresh frozen rabbit serum serves as the source of complement. If the rabbit serum exerts a nonspecific toxic effect against the target cells, as is often the case, it must be diluted to a point where it causes less than 20% chromium release in the absence of antibody but retains a sufficient level of complement activity to effect specific lysis in the presence of antibody. In our laboratory the diluting factor is usually 1:5 to 1:10.

To label with isotope, 2×10^6 viable cells (as determined by dye exclusive technique) are washed twice in test buffer (gelatin-Veronal buffer[7] containing 25% heat-inactivated fetal calf serum) and resuspended in 0.1 ml of the same buffer to which is added 40 μCi of ^{51}Cr (as sodium chromate in sterile isotonic saline[8]). This mixture is incubated at 37°C for 30 minutes. After three 10-ml washes the cells are resuspended in 2.0 ml of cold buffer and kept on ice until use.

The assay is carried out in 12×75 mm plastic tubes and each test is done in duplicate or triplicate. Using 500 μl Hamilton syringes equipped with repeating dispensers reagents are dispensed in 10-μl volumes. Buffer is added to give a total volume of 40 μl. The procedure outlined in Table I is designed to determine the cytotoxicity of a serum sample. The reagents are listed in sequential order of addition. The assay tubes are capped and incubated at 37°C for 1 hour in a shaking water bath. To stop the reaction, 2.0 ml of cold buffer is added to each tube followed by mixing. The tubes are then centrifuged at 250 *g* for 10 minutes and 1 ml of the supernatant fluid is placed in another tube for determination of radioactivity. Counting is done in a Baird Atomic Spectrometer. To obtain a value for 100% release of the isotope, 2.0 ml of distilled water is added to a sample of labeled cells in buffer (100% lysis control tubes) and this mixture is frozen and thawed three times and treated as the other tubes.

To obtain the percentage of cells killed by the cytotoxic antibody (percentage ^{51}Cr release) the following formula is used:

$$\%\ ^{51}\text{Cr release} = (C_s - C_r)/(C_t - C_r) \times 100$$

where C_s is the number of counts in the sample tube, C_t is the number of counts

[7] E. A. Kabat and M. M. Mayer, "Experimental Immunochemistry," p. 149. Thomas, Springfield, Illinois, 1961.

[8] Amersham/Searle, Des Plaines, Illinois.

TABLE I
Experimental Design for Cytotoxicity[a]

	(1) Buffer	(2) Antibody	(3) Comple-ment	(4) Target cells	Actual results	
					CPM	% Release
Test	10[b]	10[b]	10[b]	10[b]	994	91
Antibody control	20	10	—	10	53	—
Complement control	20	—	10	10	195	0
Spontaneous release	30	—	—	10	52	—
100% Lysis control[c]	30	—	—	10	1075	100

[a] In order of addition of reagents.
[b] In microliters.
[c] Freeze–thaw three times.

in the 100% freeze–thaw control, and C_r is the number of counts in the complement control tube (^{51}Cr release due to rabbit complement alone).

In the results of a typical experiment presented in Table I, two important notes should be made. First, the release in the antibody control is no greater than the spontaneous release; therefore, the serum is not inherently cytotoxic. Second, the amount of isotope released by rabbit complement alone is less than 20% of that released by the 100% control. The initial test is performed with undiluted serum. If significant cytotoxicity is obtained (75–100%) the test is repeated with serial twofold dilutions, the titer being the highest dilution of serum causing 50% release of ^{51}Cr.

CHAPTER 5

Cell Cycle Analysis

A. Mammalian Cells

A. Macieira-Coelho

The growth curves of most cell lines show after an adequate inoculum an initial lag during the first 24 hours followed by an exponential phase and finally a plateau. Growth curves, however, do not give the exact information about the proliferative fraction of a cell population. A more accurate picture can be ob-

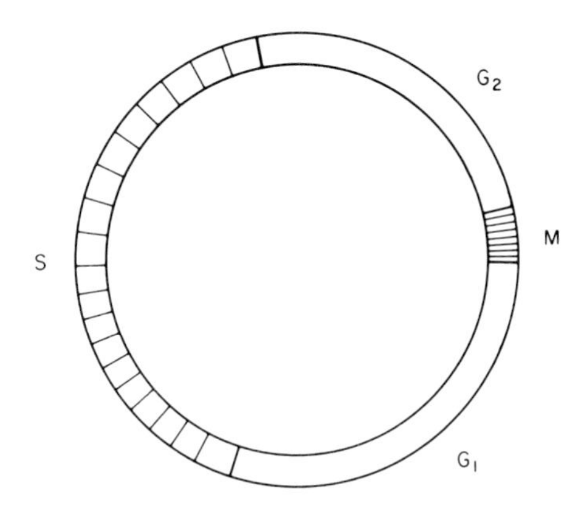

Fig. 1. Division cycle of mammalian cells.

tained if growth curves are compared with the percentage of cells labeled with radioactive thymidine ([^{3}H]TdR), measured by autoradiography.

One can distinguish four main periods in the division cycle of mammalian cells (Fig. 1). The first one, called G_1 (for gap) comprises the time between the end of telophase and the start of DNA synthesis; the second, the S period (for synthesis) corresponds to the time taken by a cell to double its DNA content; the third (G_2 period) comprises the time between the end of DNA synthesis and beginning of prophase; finally, the fourth period (M or mitosis) corresponds to the time needed to go through prophase, metaphase, anaphase, and to complete telophase. The sum of these different periods corresponds to the cell generation time (GT). The distinction of these four periods has been possible because of two markers: (1) the detection of the incorporation of radioactive precursors into DNA by autoradiography, and (2) the morphology assumed by cells during mitosis. It should be stressed that this subdivision of the cell cycle is certainly not definitive and that attempts have been made to find other markers and to further subdivide the cell cycle.[1-5] Moreover this subdivision does not include events such as the so-called nonscheduled DNA synthesis which occurs during G_2 or G_1.

Analysis of the Proliferative Fraction in Cell Monolayers

Human embryonic lung fibroblasts were subcultivated so that one group resulted from a 2:1 split (the cells of two dishes were subcultivated into one dish), another group from a 1:1 split, a third from a 1:2 split, and a fourth from a 1:4 split. [^{3}H]TdR (0.1 μCi/ml final concentration) was added to four cultures in each group at the time of subcultivation and to two cultures from each group on successive days thereafter. At each day after subcultivation the duplicate

[1] S. Kishimoto and I. Lieberman, *Exp. Cell Res.* **36**, 92 (1964).
[2] T. T. Puck and J. Steffen, *Biophys. J.* **3**, 379 (1963).
[3] E. W. Taylor, *J. Cell Biol.* **19**, 161 (1963).
[4] H. M. Temin, *J. Cell. Physiol.* **78**, 161 (1971).
[5] R. A. Tobey, D. F. Petersen, E. C. Anderson, and T. T. Puck, *Biophys. J.* **6**, 567 (1966).

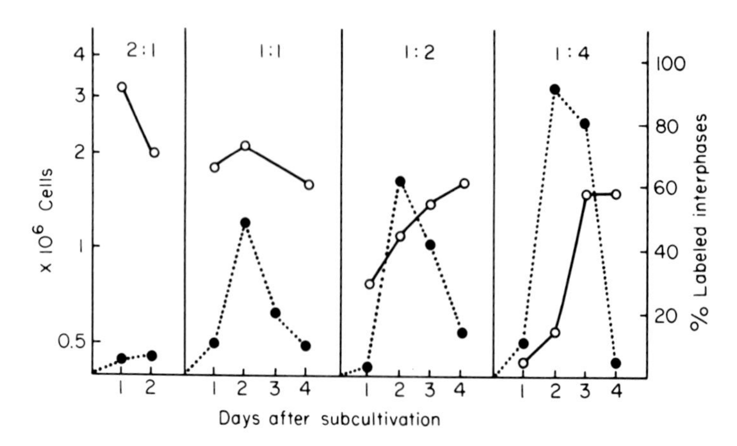

Fig. 2. Cell counts (—○—) plotted semilogarithmically and percentage labeled interphases plotted arithmetically (—●—) at different days after subcultivation of early passage human embryonic lung fibroblasts, with 2:1, 1:1, 1:2, and 1:4 splits.

cultures labeled for 24 hours were washed in phosphate buffer salt solution (PBS) and fixed in acetic acid:methanol (1:3) for at least 1 hour, then left for 5 minutes in cold 5% perchloric acid, washed with methanol, and dried. They are then processed for autoradiography.[6] Duplicate cultures labeled at subcultivation were also fixed at the end of the experiment in each group. All labeled cultures were processed for autoradiography. In the remaining cultures the nutrient medium was left unchanged and cells were counted each day after subcultivation up to the time when the cultures were stationary (Fig. 2). The cell counts obtained after subcultivation corresponded well to what would be expected from the different split ratios. The counts obtained at the time of confluency show that despite the different inocula the cultures had the same density at the time when DNA synthesis approached zero. It can also be seen that the maximum number of cells synthesizing DNA during a 24-hour period increased as the inoculum decreased. The analysis of cultures labeled continuously from the time of subcultivation to the time of confluency (Table I) reveals that the total amount of cells synthesizing DNA during that period also increased as the inoculum decreased.

The comparison between the data presented in Table I and Fig. 2 shows that

TABLE I
Percentage Labeled Interphases Found after a Continuous Labeling from Subcultivation to Resting Stage[a]

Split ratio	Percentage
2:1	5
1:1	21
1:2	75
1:4	97

[a] Same experiment as in Fig. 2.

[6] A. Macieira-Coelho, J. Pontén, and L. Philipson, *Exp. Cell Res.* **42**, 673 (1966).

in the 1:1 split 21% of the cells synthesized DNA during the time between subcultivation and confluency, while during the first day after subcultivation 10% of the cells synthesized DNA, during the second day 48%, during the third day 20%, and 10% during the fourth day. The same comparison made in the 1:2 split shows that a total of 75% of cells synthesized DNA between subcultivation and confluency, while 62% synthesized DNA during the second day, 42% during the third day, and 14% during the fourth day. In the 1:4 split 97% of the cells synthesized DNA between subcultivation and confluency, 11% during the first day, 91% during the second day, and 80% during the third day. This means that in each of the three groups there were some cells that either entered the S period at least twice or detached after going through the S period. The data also demonstrates that although the growth curve can show one population doubling (1:2 split), there are cells that do not divide and others that divide more than once.

The following experiments performed with human and bovine fibroblasts altered by oncogenic viruses illustrate how this pattern observed with normal cells can change and that the true picture is given only when cell counts are compared with the percentage of cells synthesizing DNA.

These experiments were performed as described above, i.e., the percentage labeled cells during each day after subcultivation (24 hours labeling) were determined by autoradiography and the cells grew without medium changes.[7,8] The following cells were utilized: normal human embryonic fibroblasts, the same cells transformed by Rous sarcoma virus (RSV), RSV-transformed bovine embryonic fibroblasts, SV40 transformed human adult skin fibroblasts,[7,8] and a cell line obtained from a human osteogenic sarcoma.[9] Results are illustrated in Fig. 3. In the normal cultures the percentage labeled interphases declined to very low levels when the growth curve reached a plateau. In the transformed cultures,

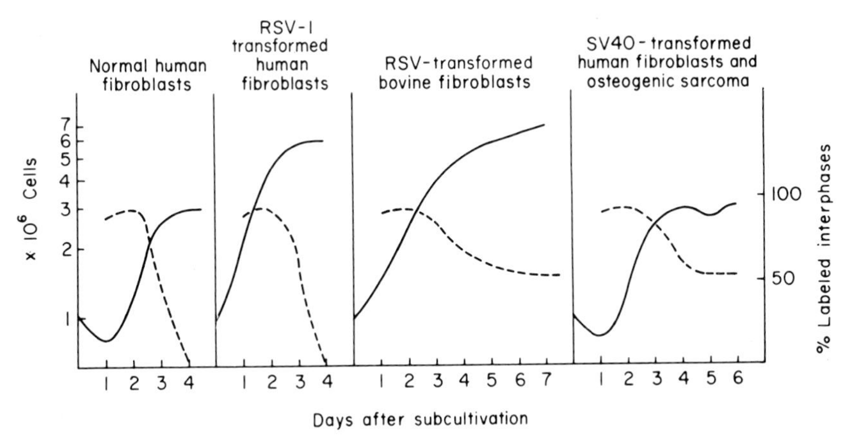

Fig. 3. Growth curves (———) plotted semilogarithmically and percentage labeled interphases plotted arithmetically (– – – – –) for each 24 hours after subcultivation of normal and transformed cells.

[7] A. Macieira-Coelho, *Exp. Cell Res.* **47**, 193 (1967).
[8] A. Macieira-Coelho, *Intern. J. Cancer* **2**, 297 (1967).
[9] J. Pontén and E. Saksela, *Intern. J. Cancer* **2**, 434 (1967).

TABLE II

Percentage of Cells Synthesizing DNA during a 24-Hour Period in Normal and Transformed Cells at the Terminal Cell Density of the Respective Normal Lines

Cells ($\times 10^4$)/cm^2	Normal line[a]	Percentage labeled interphases in Normal line	Percentage labeled interphases in Transformed line	Transformed line[a]
5.1	HEB	2	50	HEB-SR
1.1	74S	13	57	1S-SV40
3.3	B10	13	46	B10-EH
5.1	B11	11	53	B11-EH
1.1	2 S	10	50	2 T

[a] HEB, Human embryonic lung fibroblasts; HEB-SR, the same cells transformed by RSV; 74S, human adult skin fibroblasts; 1S-SV40, the same cells transformed by SV40 virus; B10 and B11, bovine embryonic lung fibroblasts; B10-EH and B11-EH, the same cells transformed by RSV; 2 S, normal human adult skin fibroblasts; 2 T, osteogenic sarcoma cells (2 S and 2 T came from the same patient).

however, DNA synthesis either slowed down but only at a cell density twice that of the controls (RSV-transformed human fibroblasts) or never slowed down during the whole experimental period. In the latter case the maximal cell density was either identical to the one found in the controls (SV40-transformed human fibroblasts and osteogenic sarcoma cells) or reached much higher values (RSV-transformed bovine fibroblasts). Furthermore, in the transformed bovine fibroblasts the growth curve instead of reaching a plateau, increased progressively although at a slower rate. In the SV40-transformed fibroblasts and in the sarcoma cells the growth curve at the plateau varies due to alternate loss and replacement of cells.

A quicker method can be used also to see if cell cycle inhibition is altered either spontaneously or after infection with oncogenic viruses. The percentage of cells synthesizing DNA during 24 hours is determined in control and transformed cultures at the cell density where the respective control growth curve reaches the plateau. Results obtained for the same cell lines utilized in Fig. 3 are presented in Table II. They show that the percentage of labeled interphases was higher in the transformed cultures and reveals an impaired growth control in the transformed population.

Analysis of the Cell Division Cycle

The duration of GT, S, G_2, and M can be measured directly. The duration of G_1 is obtained by subtracting the three other periods from GT.

As one could expect, the behavior of cells during the division cycle varies during the different growth stages mentioned above; initial lag, exponential phase and time between exponential phase, and the plateau of the growth curve. Thus different methods have to be used depending on the growth stage in which the division cycle is analyzed. The experiments described below were all done with human fibroblasts growing in a monolayer.

Generation Time. The GT can be measured on the biphasic curve representing the percentage labeled metaphases obtained after the pulse labeling of actively growing cells.[10] It should be used on a homogeneous population where the fraction of cells entering and leaving the division cycle is constant during the experimental time.

Cultures growing on coverslips are exposed to 1 μCi/ml tritium-labeled thymidine ([^{3}H]TdR) with a specific activity of about 2 mCi/m*M* for 15 minutes. After the labeling period the cells are washed twice with culture medium and fresh medium supplemented with nonlabeled thymidine (100 × the concentration of [^{3}H]TdR) is added. This is considered zero hour. Duplicate coverslips are removed every hour or each second hour, fixed, and processed for autoradiography. At least 100 metaphases should be counted to determine the percentage of labeled metaphases at each hour after labeling. The curve representing the percentage labeled metaphases at different hours after labeling is of the type illustrated in Fig. 4. The cells that were at the end of the S period at the time of labeling will, after going through G_2, appear as the first labeled mitoses. All cells that were in S at the time of labeling will progressively appear as labeled mitosis and after going through G_1 and through a second S and G_2 will start the second wave illustrated in Fig. 4. The GT will correspond to the time between addition of label and the lowest point of the descending limb of the labeled metaphases curve or to the time between the 50% points on the two ascending limbs of the biphasic curve. Both values are generally identical, i.e., 17 hours in Fig. 4.

The S Period. The length of the S period can be calculated from the proportion of the labeled interphases after a short period of labeling. In the experiment

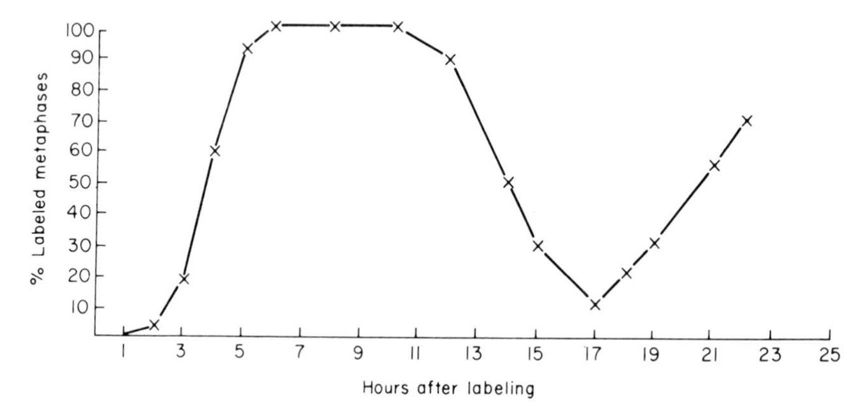

Fig. 4. Percentage labeled metaphases at different hours after a pulse labeling with [^{3}H]TdR, during the exponential growth of human embryonic lung fibroblasts in the eighteenth passage.

[10] H. Quastler and F. G. Sherman, *Exp. Cell Res.* **17,** 420 (1959).

illustrated in Fig. 4, 62% interphases were found labeled at zero hour. This means that the S period would correspond to 62% of the GT, i.e., 11 hours. The S period can be calculated also from the distance between the 50% points in the first wave of labeled metaphases (10 hours in Fig. 4). These two methods are very sensitive to changes of the synchronization within the culture. Another method which is more accurate and less dependent on synchronization can also be used to measure the S period.[11] The time interval between the first appearance of labeled metaphases and the point when the number of grains over metaphases reaches a plateau during continuous labeling corresponds to the length of the S period.

[^{3}H]TdR in a concentration of 0.01 μCi/ml is added to actively growing cultures. Duplicate samples are removed at hourly intervals thereafter and treated as described above. The number of grains found above 100 labeled metaphases for each hour after addition of [^{3}H]TdR is plotted as histograms. The cumulative percentage of such labeled metaphases plotted against grain counts on probability paper gives straight lines. The intersection of the straight lines with the 50% line on the paper represents the peak values of metaphase grain counts for each hour after addition of the [^{3}H]TdR. The interval between the first occurrence of detectable label and the time when the peak metaphase grain counts reaches a plateau (Fig. 5) corresponds to the duration of the S period, i.e., 6 hours in the experiment illustrated in Fig. 5.

The G_2 Period. Its length can be determined as the time interval between pulse labeling and the point when 50% of the metaphases are labeled on the ascending limb of the first wave of labeled metaphases, i.e., 4 hours in the experiment illustrated in Fig. 4. This measurement however, gives only the average length of the G_2 period in the cell population.

In situations where there is a delay in the G_2 period rather than a block, which is the case for human fibroblasts approaching resting phase, other procedures

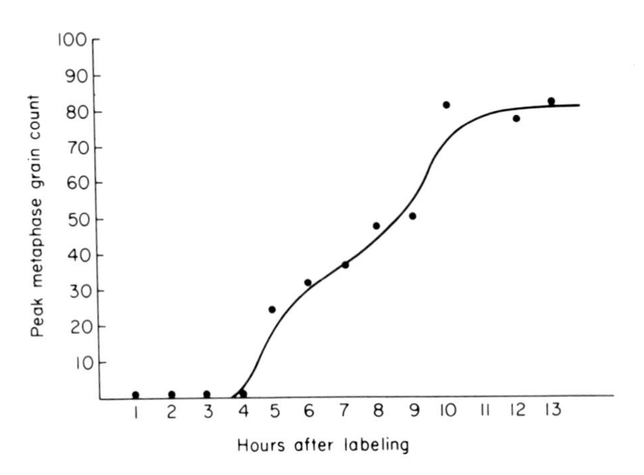

Fig. 5. Peak metaphase grain counts at each hour after addition of [^{3}H]TdR to fibroblastic cultures originated from a human adult skin biopsy.

[11] C. P. Stanners and J. E. Till, *Biochim. Biophys. Acta* 37, 406 (1960).

will have to be used to give an idea of the speed of movement in the cell cycle. Methods like microspectrophotometry, for instance, will be useless since they only indicate what happens at a given instant. The situation is similar to the information obtained from a pulse or a continuous labeling.

A more precise idea of the behavior of cells during the G_2 period can be obtained with a method illustrated in the following experiment. Human embryonic fibroblasts are subcultivated on coverslips. The first day after subcultivation, 0.01 μCi/ml [^{3}H]TdR are added to a group of identical cultures; then duplicate samples are fixed each second hour during a 10-hour period. This procedure is repeated every day until the growth curve reaches a plateau. The cells are processed for autoradiography and the percentage labeled metaphases during each continuous labeling is determined (Fig. 6). The surface of the shaded areas illustrated in Fig. 6 represents the G_2 compartment for each day after subcultivation. Measurement of these areas with a planimeter shows that the G_2 compartment is constant during the time the population increases (Fig. 7), i.e., 2nd, 3rd, and 4th day after subcultivation. When the growth curve reaches a plateau (5th and 6th day after subcultivation) the time spent in G_2 increases.

This delay in the G_2 period during cell crowding can also be observed by the following method. Starting at the time of subcultivation and each day thereafter 0.01 μCi/ml [^{3}H]TdR is added to four cultures. Twenty-four hours later, two cultures are fixed while the other two are carried for an additional 24 hours in 10 μg/ml cold TdR (100 $\times$ the concentration of [^{3}H]TdR) and then fixed. Cultures are processed for autoradiography. The ratio between the mean grain count in labeled interphases fixed after 24 hours in the presence of [^{3}H]TdR, and the mean grain count in identical cultures grown for an additional 24 hours in cold TdR is used to determine the proportion of cells that divides during the 24 hours following the removal of the labeled precursor. The mean number of grains

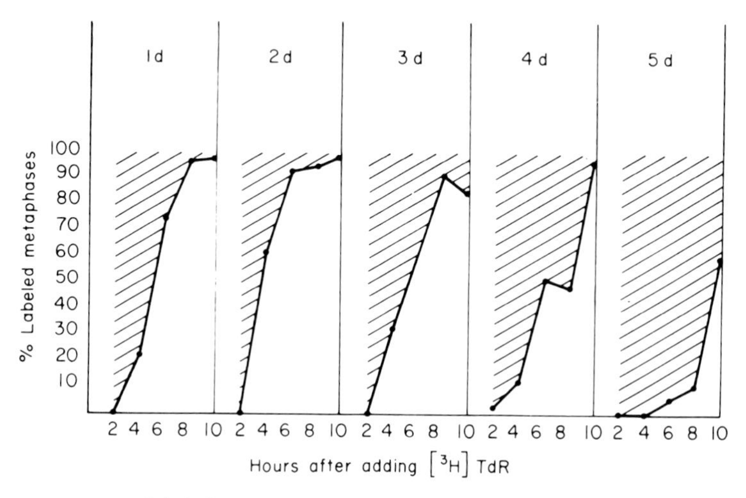

Fig. 6. Percentage labeled metaphases during 10-hour periods following the addition of [^{3}H]TdR on different days (d) after subcultivation of human embryonic lung fibroblasts in the sixth passage.

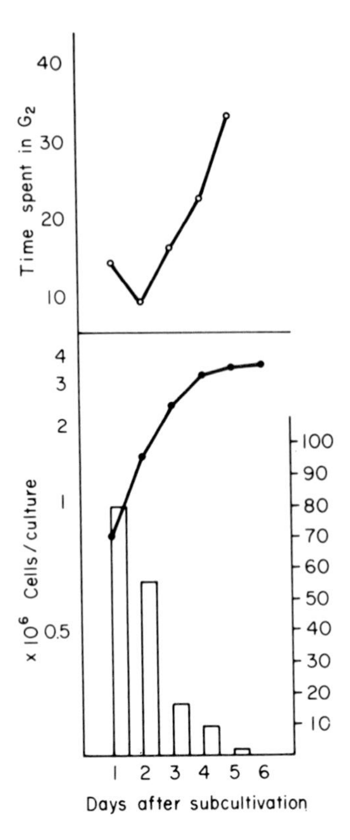

Fig. 7. Growth curve (—●—), percentage labeled interphases (bars), and surface (—○—) of the shaded areas illustrated in Fig. 6, in the same experiment.

should show a 50% reduction considering that all cells which synthesize DNA would have divided during the following 24 hours while kept in cold TdR. The experiment presented in Table III shows that cells divided during the first 2 days since the ratios are close to 2, and that the ratios in the cultures approaching confluency (day 3/day 4 and day 4/day 5) were close to 1. The distributions in

TABLE III

Ratios between the Mean[a] Grain Counts Found in Cultures Carried for 24 Hours in [^{3}H] TdR and in Sister Cultures Carried Further in Cold TdR, at Different Days after Subcultivation

	Days in the presence of [^{3}H]TdR		Days in the presence of cold TdR	
	1/2	2/3	3/4	4/5
Ratios	2.2	1.8	1.2	1.1
X^2 [b]	4.90	0.28	44.44	62.18
	(≤5.99)	(≤5.99)	(≤3.84)	(≤5.99)

[a] Mean of 50 cells.

[b] "Goodness of fit" between mean grain counts observed and those expected: had the number of mean grain counts halved, in the labeled cultures. Values within parentheses show the respective 95% confidence limits.

grain counts can be analyzed by the X^2 test: the actual grain count observed in cultures carried for an additional 24 hours in cold TdR as compared with the grain count expected, had the number of grains found after the labeling period halved during the following 24 hours. Results show that there is a good correlation between the observed and the expected distributions on day 1/day 2 and day 2/day 3, when the ratios between the mean grain counts were close to 2. As cultures approach confluency there is a decreased "goodness of fit" for the X^2 values (day 3/day 4 and day 4/day 5). Results again show that the G_2 period is prolonged when human embryonic fibroblastic cultures approach confluency.

In the methods described above for the measurement of the G_2 period cell kinetics are analyzed on various days after subcultivation, through exponential growth up to resting phase. Another procedure can also be used which follows the events after the stimulation of cultures in resting phase. This method gives an idea of the distribution of cells around the cycle at the moment the stimulus is applied. Cultures are combined, subcultivated, and counted each day. When a plateau is reached (two consecutive days with the same cell counts) the medium is changed in one-half of the cultures and 0.1 μCi/ml [^{3}H]TdR is added to all

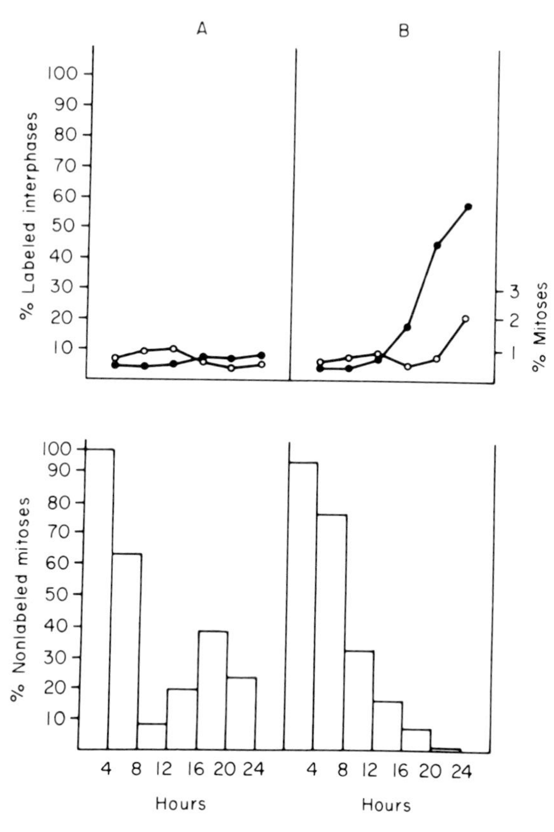

Fig. 8. Percentage labeled interphases (—●—), mitotic index (—○—), and percent nonlabeled mitoses (bars) during 4-hour periods following the addition of [^{3}H]TdR to cultures that had reached resting stage (A) and to identical cultures where medium was renewed (B) at the time of [^{3}H]TdR addition (0 hour).

cultures (0 hour). In each group (with and without medium change) four cultures receive 0.02 μg/ml Colcemid and are fixed 4 hours later; this procedure is repeated every 4 hours during 24 hours. The mitotic indices, percentage labeled interphases and percentage nonlabeled mitoses are determined (Fig. 8). Results show that only about 5% of the interphases are labeled during each 4-hour period in cultures kept in resting phase, and that nonlabeled mitoses appear with an irregular pattern during the whole experimental period. In the group stimulated with a medium change the percentage of labeled interphases starts increasing after the 12th hour and the mitotic index after the twentieth hour. The percentage of nonlabeled mitoses decreases progressively during the 24 hours following the medium change; this percentage is identical in both groups during the first 4 hours: however, during the two subsequent 4-hour periods, the percentage of nonlabeled mitoses is higher in the stimulated cultures, showing that the medium change accelerates the cells delayed in G_2.

Still another method can be utilized to see the distribution of cells around the cycle in resting phase cultures. Cells that reached stationary phase are combined and subcultivated and 0.01 μCi/ml [^{3}H]TdR is added to all cultures. Starting at the time of subcultivation and each 4th hour thereafter, Colcemid is added to four cultures which are fixed 4 hours later; this is done up to the 12th hour after subcultivation. Results show (Fig. 9) that most of the mitoses are nonlabeled up to 12 hours after subcultivation. Nonlabeled mitoses decrease progressively during the experimental period. These results as expected are identical to the ones obtained in the preceding experiment.

Mitoses. The best way to measure the length of mitosis is either by direct observation or by time-lapse cinematography. This can be done under phase contrast with a microscope placed in a warm room or with a microscope with a heated stage.

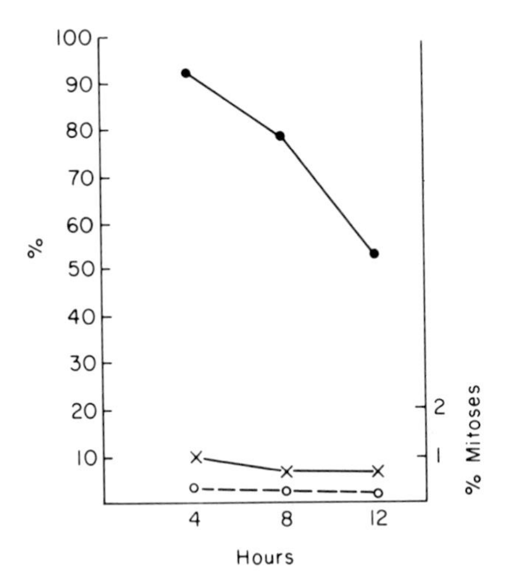

Fig. 9. Percentage nonlabeled mitoses (—●—), mitotic index (—×—), and percentage labeled interphases (– –○– –) during 4-hour periods following the subcultivation of resting stage cultures. [^{3}H]TdR was added at the time of subcultivation (0 hour).

CHAPTER 5

Cell Cycle Analysis

B. Control and Analysis of the Mitotic Cycle in Cultured Plant Root Meristems[1]

J. Van't Hof

One characteristic shared by higher plants and animals is complexity of morphology. The form of an individual reflects the results of cell reproduction and differentiation which, in turn, are the expressions of genetic factors. The study of the mitotic cycle is, therefore, important to developmental and cell biology alike. The conceptual and experimental guideline for most, if not all, work with eukaryotic cells is the model of Howard and Pelc[2] which separates the mitotic cycle into four parts. The first is mitosis (M) which begins with cells in prophase that have a 4C nuclear DNA content and is terminated after telophase with the formation of two daughter cells, each with a nuclear DNA content of 2C. The newly formed daughter cells are in the presynthetic G_1 period of interphase that ends when DNA synthesis is initiated. The S period is the time of DNA synthesis, when the chromosomes are replicated and when the nuclear DNA content is increased from 2 to 4C. Following S is the postsynthetic G_2 period from which the cells eventually divide thus initiating another cycle. Described in this article are experimental and analytical procedures to regulate and measure the progression of plant root meristematic cells through the mitotic cycle. Meristems characterized by three different cell distributions in the cycle are discussed below.

Meristems with Asynchronous Division

The first cell distribution is that found in meristems of growing roots of either intact seedlings or of primary root explants. In such tissue, division is asynchronous and the relative number of cells in a given mitotic cycle period is roughly proportional to its duration as shown in Fig. 1a. Thus, for example, if G_2 is of 2 hours duration and S is 4 hours, twice as many cells are in S as in G_2 at any given moment. A pulse (1 hour or less) of the radioactive DNA precursor tritiated thymidine [^{3}H]TdR labels cells currently in S and also those that enter

[1] Research carried out at Brookhaven National Laboratory under the auspices of the U. S. Atomic Energy Commission.

[2] A. Howard and S. R. Pelc, *Heredity Suppl.* 6, 261 (1953).

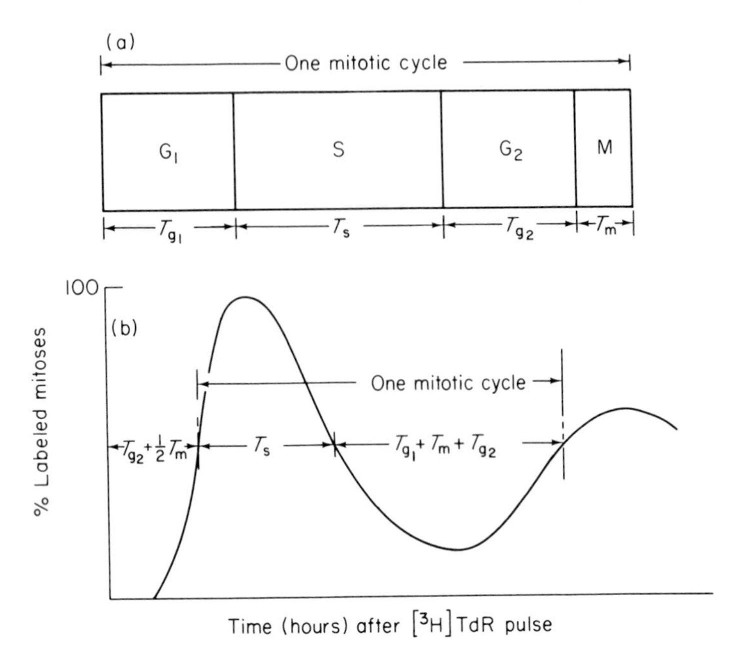

Fig. 1. (a) A highly diagrammatic representation of cell distribution in the mitotic cycle in root meristems with asynchronous cell division. T_{g_1}, T_s, T_{g_2}, and T_m are symbols for the duration of G_1, S, G_2, and M, respectively. (b) The rise, descent, and rise in the percentage labeled mitoses [(No. labeled mitoses)/(No. labeled + No. unlabeled mitoses) × 100] after a pulse of [^{3}H]TdR.

S while the tracer is present in the cells.[3] The labeled cells, which subsequently traverse G_2 and divide as a group, are easily distinguished by autoradiography during interphase and at mitosis. The sequential division of the marked cells produces a wavelike curve when the percentage labeled mitoses is expressed as a function of time after pulsing with [^{3}H]TdR (Fig. 1b). The manner in which the curve is generated is understood by imagining the block of cells in S in Fig. 1a to move slowly toward M with time. At first no labeled mitoses are observed because the cells are in G_2. However, after completion of G_2, labeled mitoses are seen and the number increases, maximizes, and decreases as the block of cells advance through M to G_1. The second rise in percentage labeled mitoses does not occur until the marked cells traverse G_1, S, and G_2. The use of the labeled mitoses curve to estimate the duration of the mitotic cycle and its component periods is demonstrated graphically in Fig. 1b and the reader is directed to footnotes 4 and 5 for a detailed discussion and analysis of such curves. The mitotic cycle duration of root meristem cells of unrelated species is longer for those with more nuclear

[3] J. Van't Hof and Y. H. Ying, *Nature* (*London*) **202**, 981 (1964).

[4] H. Quastler, *In* "Cell Proliferation" (L. F. Lamerton and R. J. M. Fry, eds.), p. 18. Davis, Philadelphia, Pennsylvania, 1963.

[5] J. Van't Hof, *In* "Methods in Cell Physiology" (D. M. Prescott, ed.), p. 95. Academic Press, New York, 1968.

DNA[6] and a compilation of data from twenty-seven species shows that the cycle and S period extend, respectively, 0.34 and 0.17 hour per pg of 2C DNA.[7]

Stationary Phase Meristems

The distribution of cells in the cycle is greatly altered if the primary root of *Pisum sativum* and that of 9 other species is excised and cultured in liquid or on solid agar media, e.g., White's[8] or Torrey's,[9] without exogenous carbohydrate for more than 48 hours. The meristematic cells arrest in G_1 and G_2 as shown diagrammatically in Fig. 2a because of a deficiency of proteins and energy for the $G_1 \rightarrow S$ and $G_2 \rightarrow M$ transitions[10,11] and remain stationary in these periods until supplied

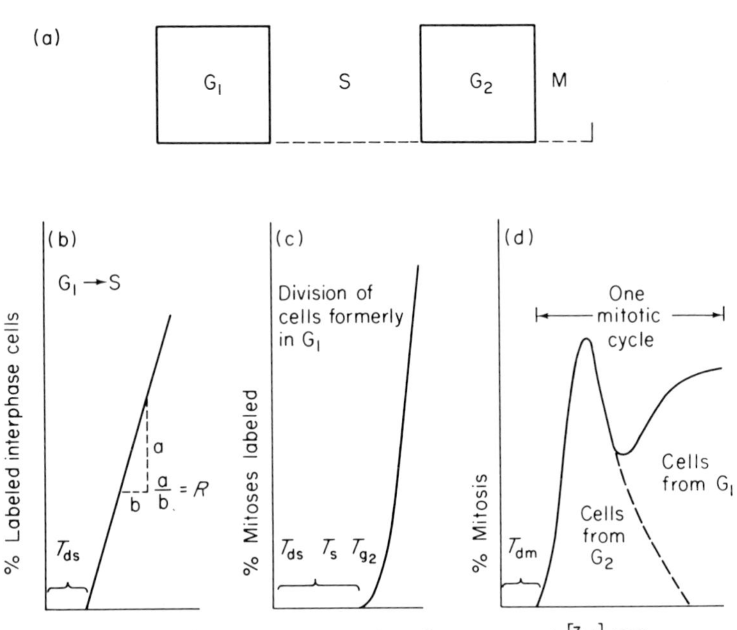

Fig. 2. (a) A diagram representing the location of arrested cells in carbohydrate-starved (stationary phase) root meristems. (b) The increase of [^{3}H]TdR-labeled interphase cells as they enter S after sucrose provision to stationary phase meristems: T_{ds}, delay in entry from $G_1 \rightarrow S$; R, rate of increase of labeled interphase cells. (c) The division of cells formerly arrested in G_1 measured in terms of an increase in percentage mitoses labeled [(No. labeled mitoses)/(No. labeled + No. unlabeled mitoses) × 100]. (d) The frequency of mitoses per 4000 cells expressed as percentage after sucrose provision to stationary phase meristems: T_{dm}, delay of cell entry into M from G_2; other symbols as in Fig. 1.

[6] J. Van't Hof, *Exp. Cell Res.* **39**, 48 (1965).
[7] J. Van't Hof, unpublished results.
[8] P. R. White, "A Handbook of Plant Tissue Culture." Cattell, Lancaster, Pennsylvania, 1943.
[9] J. G. Torrey, *Physiol. Plant.* **12**, 873 (1959).
[10] P. L. Webster and J. Van't Hof, *Exp. Cell Res.* **55**, 88 (1969).
[11] P. L. Webster and J. Van't Hof, *Amer. J. Bot.* **57**, 130 (1970).

with carbohydrate (2% sucrose, w/v). The ratio of the number of cells stopped in G_1 to those in G_2 is species specific[12] and such nondividing meristems are termed stationary phase. Upon addition of sucrose and with the inclusion of [^{3}H]TdR in the medium to mark cells synthesizing DNA, the stationary phase is terminated and cells resume progression in the cycle toward mitosis. To envision what occurs when progression in the cycle resumes, imagine the two blocks of cells shown in Fig. 2a to advance from G_1 toward S and from G_2 toward M. Typical data obtained and used to analyze resumption of cell progression in the cycle in starved meristems are shown in Figs. 2b, c, and d. First noted is that advancement into S from G_1 is not immediate when carbohydrate is supplied (Fig. 2b). A delay occurs, symbolized as T_{ds}, before DNA synthesis is initiated. The delay and the eventual entry into S are measured in terms of the percentage of labeled interphase cells as indicated in Fig. 2b. Entry is linear, at a rate R, expressed as percentage labeled interphase cells per hour. The absolute value of R varies directly with the number and inversely with the variance of distribution of cells that were once arrested in G_1. After initiation of DNA synthesis cells must traverse S and G_2 before they divide. The transit times or duration of S and G_2 are symbolized as T_s and T_{g_2} and their sum is determined by noting the time labeled mitoses are first observed. In Fig. 2c is a plot of the percentage mitoses labeled expressed as a function of time. The hour at which the extrapolated curve intercepts the abscissa registers when cells from G_1 begin to divide and is the sum of T_{ds}, T_s, and T_{g_2}. The minimum $T_s + T_{g_2}$ is estimated by the subtraction of T_{ds}.

Cells arrested in G_2 have completed DNA synthesis and consequently will not incorporate [^{3}H]TdR into their chromosomes. Cells in G_2 are also nearest M (Fig. 2a) and divide earlier than those in G_1. The interval of time between the observation of unlabeled mitoses and sucrose provision is T_{dm} (Fig. 2d). Immediately following this initial delay, the frequency of mitoses increases rapidly with time as cells formerly in G_2 divide (Fig. 2d). After most of the cells from G_2 divided, the frequency of mitoses descends to a low value at which time late cells from G_2 and early ones from G_1 are in division. Eventually, as more cells from G_1 divide the mitotic frequency again increases until finally only labeled mitoses are observed. The rapid rise to a maximum followed by a trough and then another rise in the frequency of mitoses with time (Fig. 2d) reflects the separation of the cells in G_2 and G_1 before sucrose is supplied.[13]

Both T_{ds} and T_{dm} increase with the duration of starvation[11] and each represents time to synthesize requisite factors for either the $G_1 \rightarrow S$ or $G_2 \rightarrow M$ transition.[10–12]

To determine the relative number of cells arrested in G_1 and G_2 cytokinetic and cytophotometric measurements are used. Cytokinetic determination requires that the length of the experiment be equal to or greater than one mitotic cycle duration so most of the dividing population is scored. The specific steps involved include: (1) establishment of stationary phase meristems; (2) reversal of the stationary phase by carbohydrate replenishment; (3) tracing cell progression

[12] J. Van't Hof and C. J. Kovacs, *In* "The Dynamics of Meristem Cell Populations" (M. W. Miller and C. C. Kuehnert, eds.), p. 15. Plenum, New York, 1972.

[13] J. Van't Hof, *Radiat. Res.* **41**, 538 (1970).

toward mitoses via continuous [^{3}H]TdR labeling; (4) scoring the meristems sampled serially with time for labeled and unlabeled mitoses; (5) expression of the data in terms of percentage mitoses as a function of time over one mitotic cycle; and finally (6) comparison of the respective areas under the percentage mitoses curve for labeled and unlabeled division figures (Fig. 2d). If more cells were stopped in G_2 by starvation, the area representing the unlabeled will exceed that of the labeled mitoses and vice versa. In our laboratory the ratio of cells arrested in G_1:G_2 is corroborated by cytophotometric measurements of nuclei of starved meristems.

Meristems with Cells Synchronized at the G_1/S Boundary

The third distribution is that of meristematic cells synchronized at or very near to G_1/S boundary (Fig. 3a).[14–16] Cells at G_1/S have the necessary factors

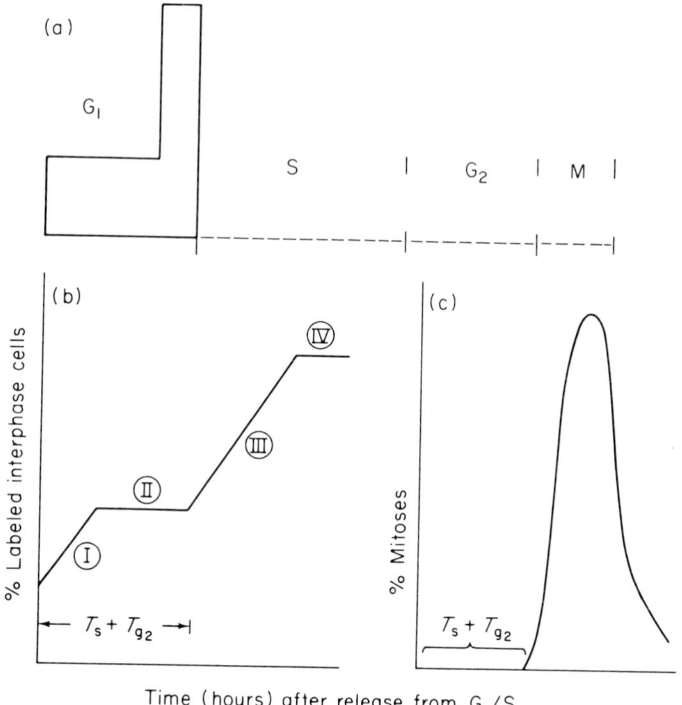

Fig. 3. (a) A diagram representing the cycle position of cells synchronized at the G_1/S boundary. (b) The increase of [^{3}H]TdR-labeled cells after release from the G_1/S boundary. See text for further information. (c) The division of cells previously synchronized at the G_1/S boundary expressed as percentage mitoses per 4000 cells. Symbols as defined in legend of Fig. 1.

[14] C. J. Kovacs and J. Van't Hof, *J. Cell Biol.* **47**, 536 (1970).
[15] J. Van't Hof and C. J. Kovacs, *Rad. Res.* **44**, 700 (1970).
[16] C. J. Kovacs and J. Van't Hof, *Rad. Res.* **48**, 95 (1971).

to initiate and complete DNA replication; only carbohydrate is lacking. Consequently, when released from G_1/S into S, they do not exhibit a delay in the initiation of DNA synthesis. The biochemical differences between cells of stationary phase meristems arrested in G_1 and those synchronized at G_1/S are useful in cytological and molecular studies on the nature of radiation-induced mitotic delay and on the regulatory mechanisms of the mitotic cycle.[15–17] To date, only cells of *Pisum* have been synchronized and this is accomplished by the following four-step culture protocol: (1) arrest cells in G_1 and G_2 by starvation; (2) incubate the tissue for 12 hours in medium with 2% sucrose, 10^{-6} *M* 5-fluorodeoxyuridine (FUdR), and 10^{-6} *M* uridine to retain cells in G_1 and allow those in G_2 to divide and proceed to G_1; (3) starve the tissue of carbohydrate for 12 hours to rid the cells of FUdR; and (4) provide carbohydrate to release the cells into S from G_1/S and label continuously with [^{3}H]TdR to trace their progression toward mitosis. At the time of release approximately one-half of the cells are at the G_1/S boundary and the remainder are within 6 hours of S as depicted in Fig. 3a. There are, of course, no cells in S, G_2, or M.

Scoring the autoradiographs for labeled interphase cells and expressing these data as a function of time after release into S produces a stepwise curve that traces progression toward mitosis (Fig. 3b). The curve has four phases.[15] Phase I has two parts, the intercept which corresponds to those cells immediately at the G_1/S boundary and a segment with a positive slope representing the entrance into S by cells near but not at the G_1/S boundary. Phase II is a plateau, established after all the cells have initiated DNA synthesis and which extends until they traverse S and G_2 and begin to divide. The division of each labeled mother cell produces two labeled daughter cells thus increasing the total number of such cells scored. Phase III is the segment that results from an increase in labeled cells from division and is initiated simultaneously with a sharp rise in the frequency of mitoses (Fig. 3c). After the synchronized cells have divided another plateau is observed (Phase IV) at a percentage approximately twice that of Phase II.

[17] C. J. Kovacs and J. Van't Hof, *Radiat. Res.* **49**, 530 (1972).

SECTION IX

Recent Techniques Facilitating Microscopic Observation of Cells

Editors' Comments

Over 300 years have elapsed since Robert Hooke used a primitive light microscope to ascertain the *cellular* structure of living tissue (Micrographia, 1665). A century and a half later the Scottish botanist Robert Brown first noted microscopically the existence of a nucleus in plant cells. Microscopy has no doubt been the single most important instrumentation made available to life scientists, and the technological advancements in microscopy in the last three decades have been of special value to tissue culturists. In this section a discussion is first presented on contrast methods in the microscopy of cells in which descriptions, specifications, and comparisons are made of dark phase, phase contrast, and Nomarski interference contrast methods. The remainder of the section is devoted to procedures which may be of help to those studying particular ultrastructural features of cells (see also Section XII, Chapter 4) or to those who follow the course of cells in culture cinematographically (see also Section VI, Chapter 3). The intent of this section, therefore, is simply to call attention to some of the basic principles and differences in light microscopy and to present several technical "tricks" in methodology involved in microscopy which hopefully are representative of a host of other specialized protocols found in the scientific literature.

CHAPTER 1

Contrast Methods in the Microscopy of Living Tissue

Robert F. Smith

In the microscopy of fixed and stained tissue, little or no difficulty is experienced in the study of structural detail. This naturally assumes that the best principles of microscopy are adhered to, i.e., *Köhler illumination,* perfect optical alignment and strict adherence to correct coverglass thickness when using high dry objectives without correction collars. However, in tissue culture we are interested in the vital functions of cells; therefore, the specimens are unstained and in media which all too often does not differ sufficiently in refractive index to produce an image of suitable contrast.[1–5] This very slight difference in refractive index is responsible for microscopic structures exerting a negligible influence on the light they transmit, and their effect is limited to a single character of light waves that cannot be detected by the eye.[6,7] What these structures do change is the phase or momentary vibrational state. It is for this reason that bright-field illumination will not reveal any differences in brightness between the structural elements and their surroundings.[8,9]

Today these obstacles need not be considered since we have at our disposal such illumination systems as *phase contrast,*[10] *Nomarski differential interference contrast,*[11] and *Jamin-Lebedeff interference.*[12] Each has its advantages and disadvantages. It is the purpose here to explain the merits of each method and then describe the most versatile combination that will not only produce optimum results, but convenience of operation.

[1] R. Barer, *J. Opt. Soc. Amer.* **47**, 545 (1957).
[2] G. C. Crossmon, *Stain Technol.* **24**, 61 (1949).
[3] J. H. Hall, "The Interference Microscope in Biological Research." Livingston, London, 1958.
[4] J. Meyer-Arendt, *Photogr. Forsch.* **5**, 121 (1952).
[5] J. M. Mitchison and M. M. Swann, *Quart. Microscop. Sci.* **49**, 381 (1953).
[6] R. F. Smith, *Photogr. Appl. Sci. Technol. Med.* **4**, 24 (1970).
[7] R. F. Smith, *J. Biol. Photogr. Ass.* **22**, 15 (1954).
[8] R. F. Smith, *J. Biol. Photogr. Ass.* **23**, 74 (1955).
[9] R. F. Smith, *Photogr. Appl. Sci. Technol. Med.* **7**, 21 (1972).
[10] C. P. Saylor, A. T. Brice, and F. Zernicke, *J. Opt. Soc. Amer.* **40**, 329 (1950).
[11] R. F. Smith, *Photogr. Appl. Sci. Technol. Med.* **6**, 19 (1971).
[12] R. F. Smith, *Photogr. Appl. Sci. Technol. Med.* **5**, 19 (1970).

Space does not permit a detailed description of the principles of each method, therefore we will deal with the practical aspects of the observation of living tissue.

Strict adherence to all proper microscopy procedures and *Köhler* illumination cannot be overemphasized, as optimum optical efficiency is dependent solely upon the degree of care exercised in adjusting the illumination system of the microscope.

Optical Considerations

Prior to the availability of the three methods mentioned for the study of unstained material, there was only dark-field and oblique illumination.[9] The dark-field method is still very useful in many areas of research, but is of little value in tissue culture microscopy. Oblique illumination is also a most valuable technique, but is far surpassed by the Nomarski interference contrast optics. Of all the methods mentioned we can only consider two in this section. Dark-field is not at all acceptable as will be shown, and the full potential of Jamin-Lebedeff interference cannot be demonstrated in the monochromatic renditions that we are restricted to in this volume. The differential optical staining potential of this method with living tissue must be seen to be appreciated.

It is now necessary to make a comparison between dark-field, phase-contrast, and Nomarski interference. Such a comparison is made in the photomicrographs of HeLa cells in Fig. 1. It is quite evident in Fig. 1A that the most refractile portions of the cells show a great deal of flare and brilliance. This is where dark field is at its best, i.e., for the detection of spirochetes, bacteria, or small particles. The shortcoming of this system in tissue culture is the lack of details that can be clearly observed due to excess flare. If this photomicrograph is compared with Fig. 1B (phase-contrast) it is immediately obvious that chromosome detail in the mitotic figure falls far short of being satisfactory. The phase-contrast image not only shows the chromosomes clearly, but also the surrounding cells. Now if this phase image is compared with Fig. 1C (Nomarski differential interference contrast) we get an entirely different impression of the cellular structures. The first and most obvious is that of a three-dimensional image. This is responsible for the appearance of greater contrast in the surrounding cells. It must also be brought out at this time that greater contrast is sometimes achieved with phase than with Nomarski interference. It will vary with the type of culture; therefore, a microscope equipped for both systems and not requiring removal or addition of optical components is desirable and shall be described in detail later in this chapter. Further comparison between the optical qualities of Fig. 1B and C will reveal the illusion of greater depth of field in Fig. 1C. This is a function of and is analogous to the high numerical aperture (N.A.) of the Nomarski system. The high N.A. limits the depth in which interference contrast is produced. Because the contrast is produced in such a thin section the image is free of out-of-focus details. The sectioning of the image creates the illusion of greater depth of field and is further enhanced by the strong shadow cast effect.

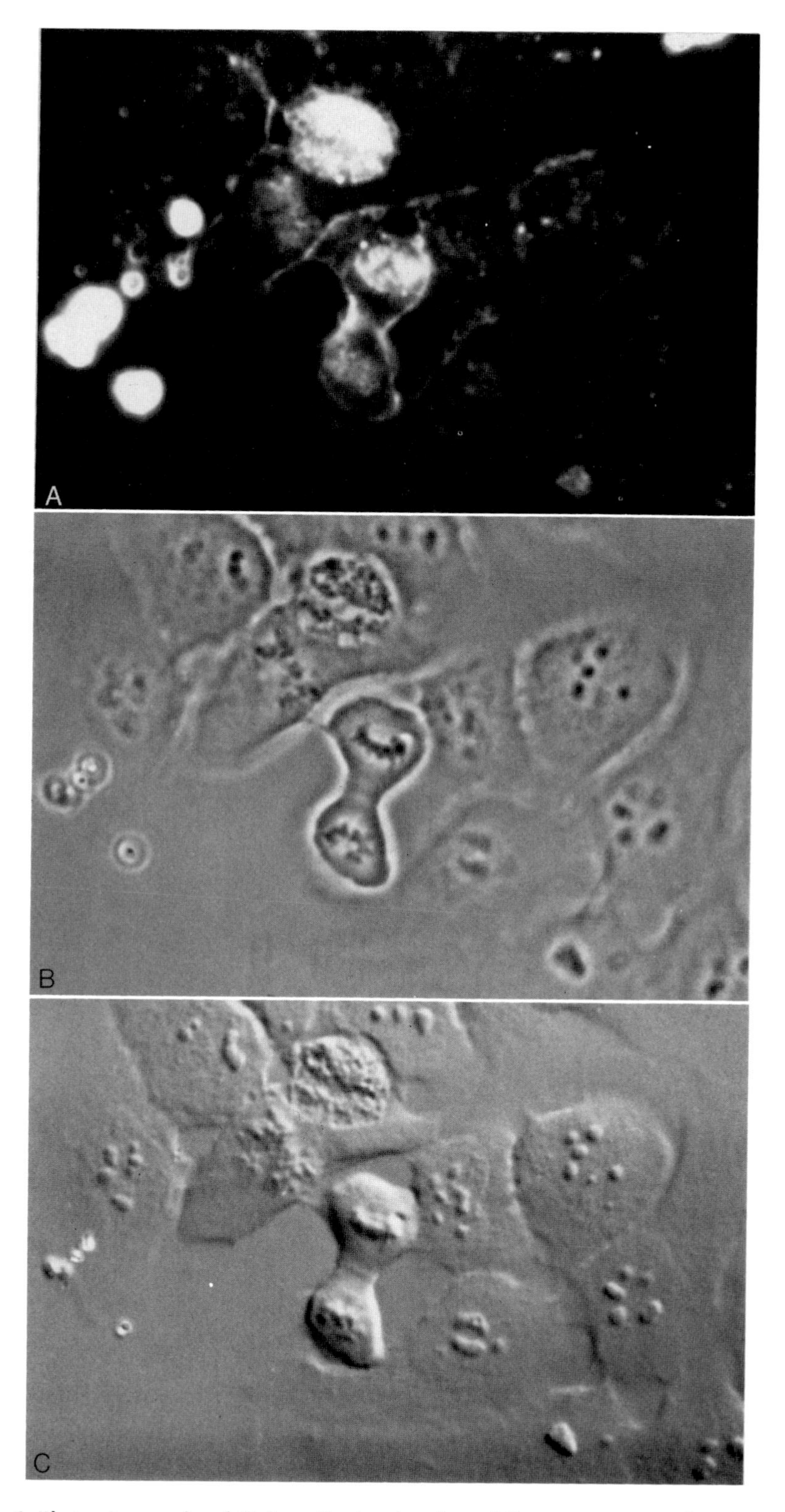

Fig. 1. Photomicrographs of HeLa cells showing three different contrast modes. (A) Dark-field, (B) phase-contrast, and (C) Nomarski differential interference contrast.

Microscope Considerations

If one wishes to observe the tissue cultures in their plastic growing flasks, an inverted microscope is necessary. The use of this instrument limits the mode of illumination to phase-contrast and bright-field. It cannot be overstressed that no one system is best for the observation of living tissue. The method selected will be governed by the type of preparation and whether differential color contrast and photomicrography are necessary. Coverslip preparations are the best choice if any flexibility in the type of microscope optics is to be realized. When using coverslip preparations a conventional microscope equipped optically for all methods of illumination may be used. The inverted microscope is very useful for checking the cultures in the flasks, and if possible should be a part of the complement of optical instruments. For detailed study the following instrument has been successfully used by the author. This instrument also takes into consideration the desirability of photographical recording and time study of tissue cultures. The instrument ideally suited for this purpose is the Zeiss Photomicroscope equipped with an aplanatic achromatic N.A. 1.40 combination bright-field, phase-contrast, and Nomarski interference condenser. It is fully automatic and, with suitable film and processing, photomicrographs may be made while observing the specimen, and without the need for special preparation or interruption of study. If photography is not a prime consideration then the large Universal stand with the same optics should be chosen. If at a later date photography becomes necessary, an automatic camera attachment is available. However, attachment cameras cannot be compared to the built-in system of the Photomicroscope from the standpoint of convenience and efficiency.

Oculars and Objectives

The Photomicroscope and large Universal stand contain a magnification changer (Optovar) that obviates the necessity of changing oculars. The factors of the Optovar are 1.25, 1.6, and 2. The author's choice with this device is the 8× Kpl. high eyepoint oculars. With this combination the ocular magnification is 10×, 12.8×, and 16×. Ocular magnification must be increased with caution as one can reach the point of empty magnification quite easily, depending on the N.A. of the objective. A safe rule to follow is never exceed 1000 times the N.A. of the objective being used. This in no way rules out the 10× or 12.5 wide-field oculars, provided there is no overamplification of the image.

The proper choice of objectives is of the greatest importance, and with a nosepiece providing five receptacles, matching phase, Nomarski and bright-field illumination is possible without the removal of optical components from the microscope. For Nomarski interference the Zeiss Planachromats are required. These are flat-field achromatic objectives and their use is dictated in preference to apochromats because of the polarizing components used in the Nomarski optical system. If apochromatic objectives were used the image quality of

TABLE I
Guide for Selecting Combination of Objectives

Designation	Initial magnification	N.A.	Focal length (mm)	Working distance (mm)	Coverglass thickness (mm)
Objectives for bright-field and Nomarski interference					
Planachromat 16×	16.1	0.35	10.4	2.7	—
Planachromat 40×	40.8	0.65	4.13	0.7	0.17 No. 1½
Planachromat 100× oil	100.6	1.25	1.66	0.09	0.17 No. 1½
Objectives for phase-contrast					
Neofluar 16×	16.1	0.40	10.8	0.9	0.17 No. 1½
Planapo. 40× with correction collar	40.4	0.95	4.25	0.09	0.11–0.23
Planapo. 100× oil	100.2	1.30	1.63	0.09	0.17 No. 1½

the Nomarski interference image would show a marked deterioration. Apochromatic objectives have anomalous birefringence due to the combinations of glass elements used to achieve their high order of correction. Obviously such a condition is not compatible with a Nomarski system where optimum optical quality is desired or where its full potential is necessary. The Planachromats are bright-field objectives and they serve a dual purpose when the examination of fixed and stained material is required. The 16× and 40× objectives are a good choice for both phase and Nomarski illumination. This will provide matching magnifications in both contrast modes as well as in bright-field. For the fifth position in the nosepiece either a 100× Planachromat for Nomarski interference and bright-field or a 100× Phase objective may be used as conditions dictate. Table I serves as a guide for selecting the most suitable combination of objectives.

Photomicrography

This is a most important facility in any research endeavor for documentation and publication purposes.

There is no simple explanation of all the aspects of this most complex but necessary adjunct to microscopy. However, the film and developer combinations shown in Table II have been found to be most suitable for 35 mm photomicrography of unstained tissue.[13]

The ASA ratings in Table II are for integrated exposure measurements. The Photomicroscope offers both spot and integrated measuring of the field, but the integrated measuring is to be preferred for the most consistent results. If spot metering is used slight variations in the ASA ratings will be necessary under some conditions.

Developer should be discarded after each roll of film is processed and agitation during development must be even and gentle. If a single roll of film is being processed in a Nikor tank, a dummy reel should be placed on top of the loaded

[13] R. F. Smith, *Photogr. Appl. Sci. Technol. Med.* **2**, 26 (1969).

TABLE II
Film and Developer Combinations

Film	Developer	Time (min); Temp. (°F)	ASA rating
Ilford PAN-F	DK-50	5; 70	100
Ilford PAN-F	D-19	6; 70	200
Kodak high contrast copy	DK-50	5; 70	25
Kodak high contrast copy	H & W control	16; 68	25
H & W control	H & W control	13; 72	80

reel to prevent overagitation when the tank is inverted during the agitation cycle.

To summarize, the following points must be remembered. In phase microscopy, phase details are made visible due to differences in refractive index or thickness in the specimen. When there are uniform phase details, only areas containing gradients of steep refractive index will appear as different intensities in the image. Phase details are also made visible by the Nomarski system, but appear as apparent relief or shadow cast images. The background image in Nomarski interference contrast is the area in the specimen, or in the case of cells the slide itself, where no object is present. With white light used to illuminate the object, the background can be made to appear colored, black and white, or gray. Therefore, color phenomena may be produced regardless of the presence of an object in the light path. This is an instrument characteristic and affords a multitude of optical staining possibilities.

CHAPTER 2

Rapid Embedding of Cell Culture Monolayers and Suspensions[1]

E. Robbins

The cell monolayer or suspension grown *in vitro* under controlled environmental conditions is a versatile tool for electron microscopic study since it is a simple matter to prepare and section such samples with presently available

[1] Supported by Grants GM12182, GM14582, and A14153 from the National Institutes of Health and Grants from the National Science Foundation and ACS.

techniques. A time-saving modification in routine procedure for embedding incorporates the potential for an improved electron microscopic image by avoiding exposure to 100% ethanol.

We have previously described the methods that we use for growing cells on carbon-coated slides,[2] as first suggested by Bloom,[3] and fixing these cells in glutaraldehyde[4] followed by a modified osmium tetroxide schedule.[5] Cells in suspension may be treated similarly. Following fixation it is possible to omit most of the usual embedding steps.

The preparatory schedule which we employ immediately after fixation is shown in Table I.

TABLE I
Postfixation Preparation for Final Embedding

Solution	Time
20% Ethanol–80% H_2O	30 Seconds
60% Ethanol–40% Epon 812	5 Minutes
30% Ethanol–70% Epon 812	5 Minutes
100% Epon 812	5 Minutes

The final embedding mixture succeeding the 100% Epon 812 consists of Epon 812, dodecylsuccinic anhydride, Nadic methyl anhydride, the amine catalyst, 2,4,6-tri(dimethylaminomethyl)phenol (DMP), and dibutyl phthalate, in a ratio of 50:25:25:4:1. In the case of monolayers, the cells are passed through three 5-minute changes in this mixture, following which gelatin or, preferably, BEEM (LKB Instruments, Rockville, Maryland) capsules filled with the embedding monomer are inverted onto the monolayer.[2] For suspensions it is desirable to pellet the cells in gelatin or BEEM capsules by centrifugation, in which case the monomer can be rapidly decanted. The pellet should by very thin, preferably two or three cells thick, so that exchange of embedding medium is not impeded. Polymerization is carried out at 55°C for 3 days. The critical innovation in the schedule is the use of ethanol-Epon solutions for dehydration. Water is quite soluble in these solutions; however, to ensure its complete removal from the cells before the addition of the pure Epon, slides or pellets must be drained thoroughly between each step.

Aside from the fact that the above embedding procedure is completed in about 40 minutes with monolayers (slightly longer for suspensions because of centrifugation), the method also eliminates the necessity of exposing cells to 100% alcohol as generally employed during routine embedding. We have the impression that this omission of absolute alcohol results in a less "extracted" electron microscopic image. The method is not intended for use on tissue blocks or thick solid pellets of cells. Its efficacy with thin specimens probably reflects the ready exchange between intra- and extracellular fluid in this type of preparation.

[2] E. Robbins and N. K. Gonatas, *J. Cell Biol.* **20**, 356 (1964).
[3] W. Bloom, *J. Biophys. Biochem. Cytol.* **7**, 191 (1960).
[4] D. D. Sabatini, K. Bensch, and R. J. Barnett, *J. Cell Biol.* **17**, 19 (1963).
[5] E. Robbins and N. K. Gonatas, *J. Cell Biol.* **21**, 429 (1964).

CHAPTER 3

Embedding in Situ[1]

B. R. Brinkley and Jeffrey P. Chang

Electron microscopic studies of tissue culture cells frequently require preselection of cells or cellular components prior to ultrathin sectioning. A variety of methods for embedding cultured cells *in situ* have been published.[1a-8]

In the present report we will describe two methods which have been developed in our laboratories and have been used routinely for several years. We feel that our methods have several advantages over other techniques in that: (1) cells may be grown under routine culture conditions; (2) cells or cellular components may be examined and photographed under an oil immersion objective lens followed by subsequent thin sectioning and examination with the electron microscope; and (3) instead of only a few cells, thousands of cells may be prepared at any one time.

T-Flask or Petri Dish Method

This procedure is a modification of the one published by Brinkley *et al.*,[7] and permits *in situ* embedding of cell monolayers grown in plastic containers such as Falcon plastic T-30 flasks or Falcon Petri dishes. Cell monolayers cultured in a routine manner are fixed by decanting the growth medium and adding two changes of 3% Millonig's phosphate-buffered glutaraldehyde[9] or other suitable fixatives. Fixation may be carried out at room temperature or in the cold by submerging the containers into an ice bath. It should be remembered that cold temperatures disrupt some cellular organelles such as microtubules and produce inferior fixation. After 1 hour, the cells are washed in two changes of phosphate buffer and postfixed in 1% osmium tetroxide for 30 minutes. After

[1] This study was supported in part by a Research Contract from the National Institutes of Health, NIH-NICHD-69-2139.

[1a] E. J. Borysko, *J. Biophys. Biochem. Cytol. Suppl.* **2,** 15 (1956).

[2] A. Howatson and J. P. Lameida, *J. Biophys. Biochem. Cytol.* **4,** 115 (1958).

[3] M. Mishivera and S. S. Rangan, *J. Biophys. Biochem. Cytol.* **1,** 411 (1960).

[4] E. Robbins and N. K. Gonatas, *J. Cell Biol.* **20,** 356 (1964).

[5] J. P. Persijn and J. P. Schereft, *Stain Technol.* **40,** 89 (1965).

[6] W. Bloom, *J. Biophys. Biochem. Cytol.* **1,** 191 (1966).

[7] B. R. Brinkley, P. Murphy, and L. C. Richardson, *J. Cell Biol.* **35,** 279 (1967).

[8] J. P. Chang, *J. Ultrastr. Res.* **37,** 370 (1971).

[9] G. Millonig, *J. Appl. Phys.* **32,** 1637 (1961).

fixation, embedding is accomplished by the following schedule; the original culture flask is used in all steps:

1. Wash thoroughly in three changes of distilled water.

2. Prestain by covering cell monolayers with filtered 2% aqueous uranyl acetate for 20 minutes.

3. Wash in three changes of distilled water.

4. Dehydrate either at room temperature or in the cold by the following schedule: (a) 35% ethanol for 5 minutes; (b) 50% ethanol for 5 minutes; (c) 75% ethanol for 5 minutes; (d) 90% ethanol for 5 minutes; (e) 90% hydroxypropyl methacrylate (HPMA, Polysciences, Warrington, Pa., 18976), three changes over a period of 15 minutes; (f) pure (97%) HPMA, two changes over a period of 15 minutes; (g) two parts HPMA:one part Luft's Epon 812[10] for 15 minutes; (h) one part HPMA:one part Luft's Epon 812 for 15 minutes; (i) one part HPMA:two parts Luft's Epon 812 for 30 minutes; (j) Pure Luft's Epon, three changes for 10 minutes each.

5. Drain off Epon until a thin layer about 2 to 3 mm thick is left covering the cells.

6. If T-flasks are used, burn holes in top of culture flask with a heated glass rod or wire (drilling is not recommended since shreds of the flask will fall into the Epon). Leave flask overnight in 37°C oven. Transfer to 60°C oven for at least 24 hours for final polymerization. If Petri dishes are used, simply remove the tops of the dishes and place directly into the embedding oven.

7. When polymerization is completed, the plastic container is cut or broken away leaving a thin plastic sheet containing Epon in which the cells are embedded and the bottom of the plastic culture flask (Fig. 1a). The bottom of the plastic container can be separated from the Epon in one of two ways. Grasp the Epon sheet between the thumb and forefinger of both hands and gently twist back and forth a few times. Because of difference in expansion coefficient, the Epon will usually separate from the bottom of the plastic container (Fig. 1b). Caution should be taken not to produce stress marks on the cell surface by forcibly detaching the two plastic surfaces. An alternate method involves the submersion of the Epon sheet into a beaker of liquid nitrogen followed quickly by plunging the sheet into tap water. This abrupt change in temperature usually produces a clean separation between the container and the Epon. The cell monolayer remains in the Epon layer and can be viewed by placing the Epon sheet, cell side up, under a phase microscope (Fig. 1c). If high resolution is needed, immersion oil can be placed directly onto the Epon without damage or contamination to the cells.

8. Cells to be sectioned for electron microscopy are photographed and then marked by scoring a circle around them with a microslide marker. The scored area is either cut out with a saw, bored out with a cork borer of slightly larger diameter, or punched out with a metal punch, producing a small disk. The disk is glued, cell side up, onto an Epon capsule with Eastman 910 adhesive. This cement hardens rapidly and within 1 minute the selected area can be trimmed

[10] J. H. Luft, *J. Biophys. Biochem. Cytol.* **9**, 409 (1961).

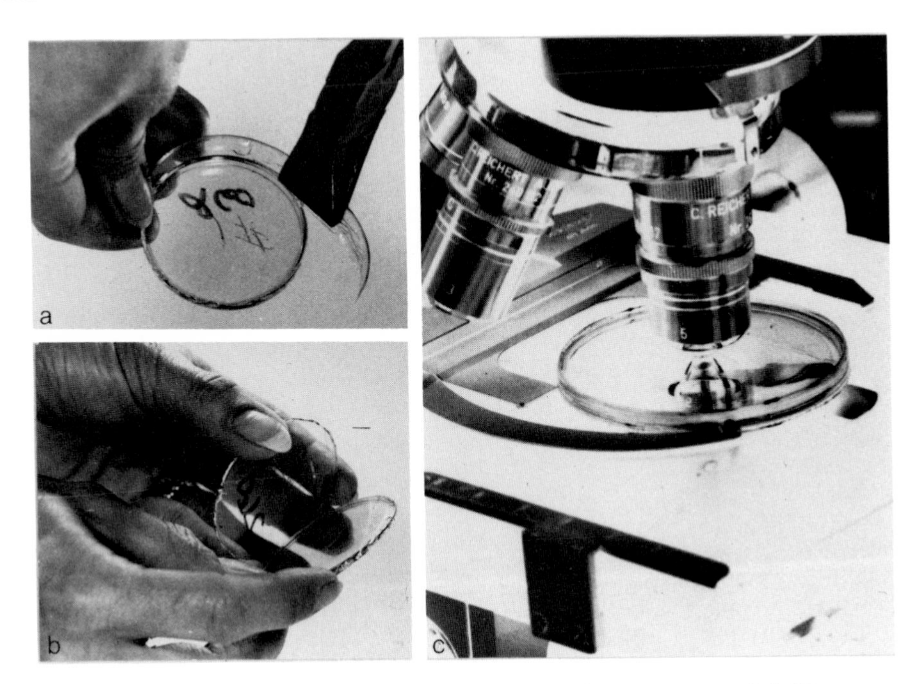

Fig. 1. Steps in separating polymerized Epon from plastic container. (a) Pliers are used to break away sides of plastic Petri dish from Epon. (b) Epon is separated from bottom of Petri dish as described in text. (c) Cells may be viewed with phase-contrast microscope.

for sectioning with either a glass or diamond knife. Since the cells to be sectioned are a single monolayer thick, care should be used in proper alignment of the block face with a diamond knife. Sections are picked up on unsupported copper grids and stained with uranyl acetate followed by lead citrate. For serial sections, we use LKB slotted grids coated with 1% collodion.

Teflon-Treated Coverglass Method

The Teflon-coverglass method is a new, simple, and reproducible technique for growing cells or attaching tissues on the coverglass for subsequent epoxy embedding.[8] Clean separation between the coverglass and embedding plastics can be assured by sudden cooling in liquid nitrogen. A specially designed mold (Figs. 2 and 3) is required for achieving the best result. The detailed procedures are outlined as below.

Preparation of Embedding Molds. Silicone rubber is used for making the embedding molds. First, aluminum squares are cut into sizes slightly smaller than the coverglasses. The thickness of the aluminum sheet will determine the thickness of the final Epon block. The squares are numbered and glued to the bottom of Falcon plastic Petri dishes with epoxy glue to make the master template (Fig. 2). Finally, the Petri dish is filled with silicone rubber and catalyst mixture prepared according to the directions supplied by the manufacturer. After

Fig. 2. Master template for preparing silicone rubber embedding molds.

polymerization, the silicone rubber mold (Fig. 3) can be taken out and is ready for use.

Preparation of Coverglasses. Chemically cleaned coverglasses are first soaked in absolute alcohol and then wiped clean with lint-free tissues. Small 11 × 22 mm coverglasses are preferred because many of them may be inserted into one regular T-30 and T-60 tissue culture flask, saving reagents and facilities. Any size coverglass can be used according to the purposes of the experiments.

The clean coverglasses are laid in rows on a piece of black paper and each upper left corner marked with a permanent ink dot. After shaking the pressurized can of Teflon, hold the can about 12 inches from the coverglasses and aim the spray parallel to the rows spraying three or four times on the side marked with the ink dot. Use different angles of light reflection to check the density of the Teflon droplets on the coverglasses. A matte surface indicates that sufficient

Fig. 3. Completed silicone rubber embedding molds.

numbers of droplets have been deposited onto the surface. The density can be determined with a light microscope by counting the exact number of droplets in a unit area. We have found that the density of the Teflon droplets is not at all critical and can be varied one- or twofold. Soon after spraying, put the coverglasses on ceramic racks and place into an oven at 250°C for about 30 minutes. The Teflon droplets will bond to the glass surface and become impervious to organic solvents, sonication, and any forms of sterilization. The treated coverglasses are then ready for growing monolayer cells after sterilization or for attachment of other materials. Be sure that the Teflon-treated surface of the coverglass is face up in the culture flask.

Fixation. For monolayer cells, the coverglasses are taken out of the T-flask and immersed immediately into proper fixatives in Columbia staining dishes. One staining dish can accommodate up to fourteen coverglasses placed back to back and diagonally.

Dehydration and Embedding. Specimens are dehydrated in the staining dish used for fixation. Dropwise replacement is preferred for the alcohol–propylene oxide dehydration sequence. Following the final change of propylene oxide, the staining dish is filled with propylene oxide–Epon mixture. After 2 hours or more at 37°–40°C, the mixture is replaced with two changes of fresh embedding Epon. At this point, the following steps are followed:

1. Fill each depression of the embedding mold (Fig. 3), with drops of Epon.
2. Remove the coverglass from the staining dish.
3. Drain off the excess Epon from the coverglass.
4. Wipe clean only the cell-free side of the coverglass.
5. Put the coverglass cell side down onto the depression of the mold as indicated by the ink dot at the right corner of the coverglass and be certain that no air bubbles are trapped beneath the coverglass.
6. Complete the polymerization at 60°C as usual.

Preparation for Sectioning. After polymerization, the Epon block can be separated from the coverglass by immersion in liquid nitrogen. A popping sound indicates a successful separation. Dry ice will occasionally function as well as liquid nitrogen. Excessive Epon which has overflowed onto the four sides of the coverglass may interfere with immediate separation. It is then necessary to file off the excess Epon to expose the edges of the coverglass before immersion in liquid nitrogen. Thus, clean separation of the specimen is always achieved.

After separation from the coverglass, the Epon block can be used immediately for light microscopy. The cell or cells can be examined, preselected, photographed, and marked with a microslide marker. A metal punch is used to remove the selected area which is glued onto a blank Epon block, molded with a BEEM capsule, in horizontal, vertical, or any other position desired. When the epoxy glue is hardened (about 2 to 3 hours at 60°C), the Epon block is ready for thin sectioning for electron microscopy.

The main advantages of this technique may be summarized as follows:

1. It is simple, reproducible, and separation between properly treated coverglass and embedding compound never fails.

2. All necessary materials can be obtained commercially: the silicone rubber mold, the Teflon spray (Osrow Magic Spray), and the Teflon-treated coverglasses are available from Electron Microscopy Sciences, Box 251, Fort Washington, Pennsylvania 19034; Polysciences, Inc., Paul Valley Industrial Park, Warrington, Pennsylvania 18976; silicone rubber, Dow Corning 3112RTV Ecapsulant, and Dow Corning RTV Catalyst S from Dow Chemical Company; ceramic racks for coverglasses and Columbia staining dishes from A. Thomas Company, Philadelphia, Pennsylvania; metal punch, No. 5 Junior Hand Punch from Robert Witney, Inc., Rockford, Illinois.

3. It allows preselection of cell components or cells for correlated light and electron microscopy and multiple sampling from a single culture flask or dish for time sequence or similar experiments.

4. It is applicable to processing of monolayer cells, frozen sections, fresh imprints, ascites cells, squash preparations, as well as blood and bone marrow smears.

CHAPTER 4

Vertical Sectioning

A. Cells on Millipore Filters[1]

H. Dalen

The search for a method of preparing cell monolayers for comparative light microscopic and electron microscopic observations led to a technique of *in situ* fixation and embedding of the cells without detachment from the growth surface.[2] In this way the cells preserved their lifelike shape and spatial arrangement. Since we were particularly interested in phenomena associated with cytoplasmic spreading and cell attachment, vertical sectioning through the cell monolayers permitted ultrastructural observations of the relationship between the cells and the growth surface. MF-Millipore filters were originally used as growth surfaces, but they were later replaced by transparent films made of the same material

[1] This work was supported by The Swedish Medical Research Council (Grant B71-12V-630-07).

[2] H. Dalen and T. J. Nevalainen, *Stain Technol.* 43, 217 (1968).

to facilitate observations of the living cultures by phase-contrast microscopy prior to fixation.

A similar vertical sectioning technique has also been applied to prepare suspended cells collected on Millipore filters for electron microscopy.[3] In this manner the original step of spinning down suspended cells into a pellet is avoided. It has been demonstrated that vacuum filtration is far less harmful for the preservation of the cellular fine structure than centrifugation.

The vertical sectioning technique described here will be demonstrated using Chang liver cells (purchased from Grand Island Biological Co., Grand Island, New York). The cells were grown in a modified Eagle's Basal Medium[4] using standard cell culture techniques.

Millipore Filters and Transparent Films

The MF-Millipore filters are composed of pure esters of cellulose and are suitable for cell attachment and growth. However, to obtain cell monolayers the selection of an appropriate pore size is important. In our studies of Chang liver cells we tested filters of different pore sizes (0.10, 0.22, 0.45, 1.2, and 5.0 μm), and found that the cells were unable to spread and form a monolayer unless the pore diameter was 0.45 μm or less. On the other hand, it is preferable to use as large pore size as possible for cell sampling by filtration.

Transparent films can be made by dissolving MF-Millipore filters in glacial acetic acid. To obtain a smooth film of constant thickness the filter is supported by a coverglass of same size and form and dipped briefly into the acid. After hardening for 1 day under dust-free conditions in a 45°C oven, the glass-supported film makes an excellent growth surface in an appropriate tissue culture chamber. If necessary, the edges of the film can be cemented (MF Cement, Millipore Filter Corp.) to the glass to prevent separation of the two materials.

Prior to use, the filters and films are washed in 70% ethanol, several baths of distilled water, and sterilized by autoclaving for 30 to 45 minutes at 121°C. This procedure is not necessary for filters used for cell sampling by filtration, unless the collected cells are to be reincubated in the medium for extended growth before fixation.

Cell Monolayers

The preferable method of obtaining cell monolayers is to cultivate the cells under standard conditions in simple tissue culture wells. These are constructed by cementing (MF Cement) a glass ring on top of the growth surface. After washing and autoclaving the wells are placed in sterile Petri dishes, and freshly trypsinized cells are inoculated. The cell concentration should be high enough to yield a nearly confluent monolayer after 1 to 2 days incubation.

[3] H. Dalen, *J. Microsc.* **91**, 213 (1970).
[4] R. S. Chang, *J. Exp. Med.* **113**, 405 (1961).

Cell Filtrates

The sampling of suspended cells on Millipore filters by vacuum filtration is demonstrated on mitotic Chang liver cells. A nearly 90% mitotic cell suspension was obtained by selective harvesting of mitoses according to the method introduced by Terasima and Tolmach.[5] The total mitotic yield could be increased four to five times, if the cell culture was synchronized by blocking DNA synthesis for 24 hours with excess thymidine (2.5 m*M*). When the mitotic wave reached its optimum 12 hours after release of the thymidine block, the cell harvesting was carried out. It is preferable not to filter a greater number of cells than necessary to form a monolayer on the filter. The vacuum filtration is carried out under ice-cold conditions with the vacuum gauge reading between 10 and 20 cm Hg. The filters carrying the cells are then immediately transferred to glass Petri dishes for fixation.

Preparation of Cells for Electron Microscopy

Principally the same procedure is followed to prepare both cell monolayers and cell filtrates for electron microscopy. If necessary, the first step is to wash the filters or films carrying the cells gently in growth medium to remove any cellular debris. The cells are then fixed under ice-cold conditions for 1 hour in 2% glutaraldehyde and for ½ hour in 1% OsO_4. Both fixatives are made up in 0.1 *M* cacodylate buffer at pH 7.2

After fixation in glutaraldehyde the filters or films are routinely cut into halves; one-half is used for light microscopy (Fig. 1), the other for electron microscopy (Fig. 3). The half selected for examination by light microscopy is not postfixed in osmium but instead stained *in situ* with hematoxylin. After dehydration, as described below, the filters or films carrying the cells are mounted in a drop of Epon 812[6] on a microscopic slide. A coverglass is carefully placed on top, avoiding trapping of air bubbles, and the resin is hardened overnight in a 60°C oven.

The dehydration of both filter halves is simultaneously carried out at 4°C in increasing concentrations of ethanol according to the following schedule: 70, 80 and 95%, each for 10 minutes, and finally twice for 10 minutes each in absolute ethanol. Because propylene oxide dissolves the MF-Millipore filter, toluene (twice for 5 minutes each) is used instead as a transitional solvent.

Prior to embedding, while still immerged in toluene, the filters are cut in small strips measuring about 2 × 15 mm. Excess toluene is drained off on a filter paper, and the strips are then placed in a drop of resin mixture on an object glass. The opaque porous filters now become transparent and can be examined under the phase-contrast microscope. However, this step can be omitted for the transparent films, since the living cultures can be examined prior to fixa-

[5] T. Terasima and L. J. Tolmach, *Exp. Cell Res.* **30**, 344 (1963).
[6] J. H. Luft, *J. Biophys. Biochem. Cytol.* **9**, 409 (1961).

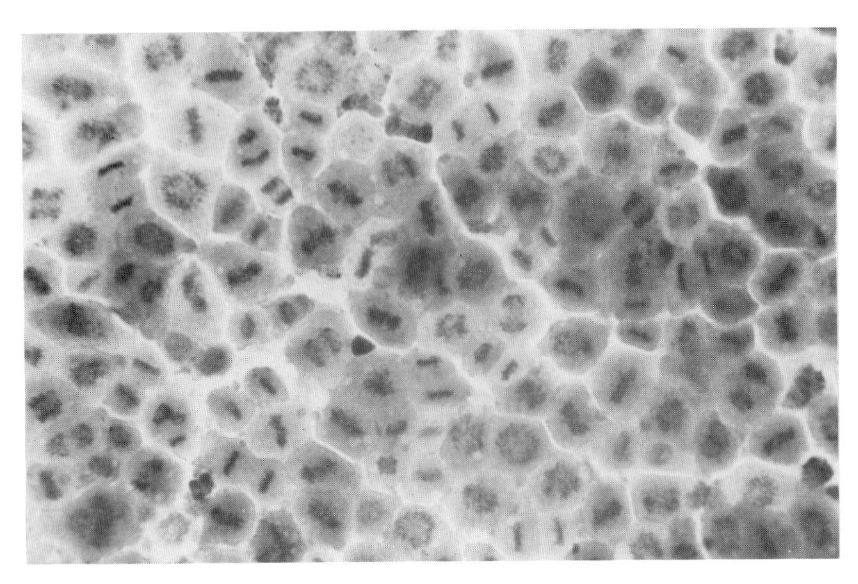

Fig. 1. A light micrograph of mitotic Chang liver cells collected by vacuum filtration on a Millipore filter. Fixed in glutaraldehyde and stained with hematoxylin. ×500.

tion. Since Epon monomer makes the filter strips very soft, selected strips are carefully transferred to BEEM flat-embedding molds. Polymerization is carried out overnight at 37°C, then 12 hours at 45°C followed by 1–2 days at 60°C.

An alternate procedure is not to cut the filters or films into strips but to flat-face embed them on an object glass. The amount of resin used should be enough to make an approximately 1-mm thick plate after polymerization. After about 12 hours in the 60°C oven, when the resin starts to solidify, the Epon plate can easily be separated from the glass by the aid of a scalpel. This step is performed while the plates are still warm, and afterward they are returned to the oven to complete polymerization.

The great advantage of the latter embedding procedure is that it facilitates both light microscopic and electron microscopic observations of the embedded monolayers. The cells are examined under the phase-contrast microscope, areas of interest are encircled with a diamond Object Marker, and Epon blocks of appropriate size are cut out and trimmed. In the cured blocks the porous filter has a golden color, which is easy to distinguish from the yellow embedding material. Regarding the transparent films, the cell monolayers have been separated from the growth surface during the embedding procedure. However, they are still attached to a thin electron-dense film, probably formed by precipitated proteins from the growth medium.

The trimmed blocks are sectioned in a plane vertical to the supporting filter or film. Thick sections (1 μm) are stained with toluidine blue[7] for orientation in the light microscope (Fig. 2). Ultrathin sections are cut on glass knives and

[7] B. F. Trump, E. A. Smuckler, and E. P. Benditt, *J. Ultrastr. Res.* **5**, 343 (1961)

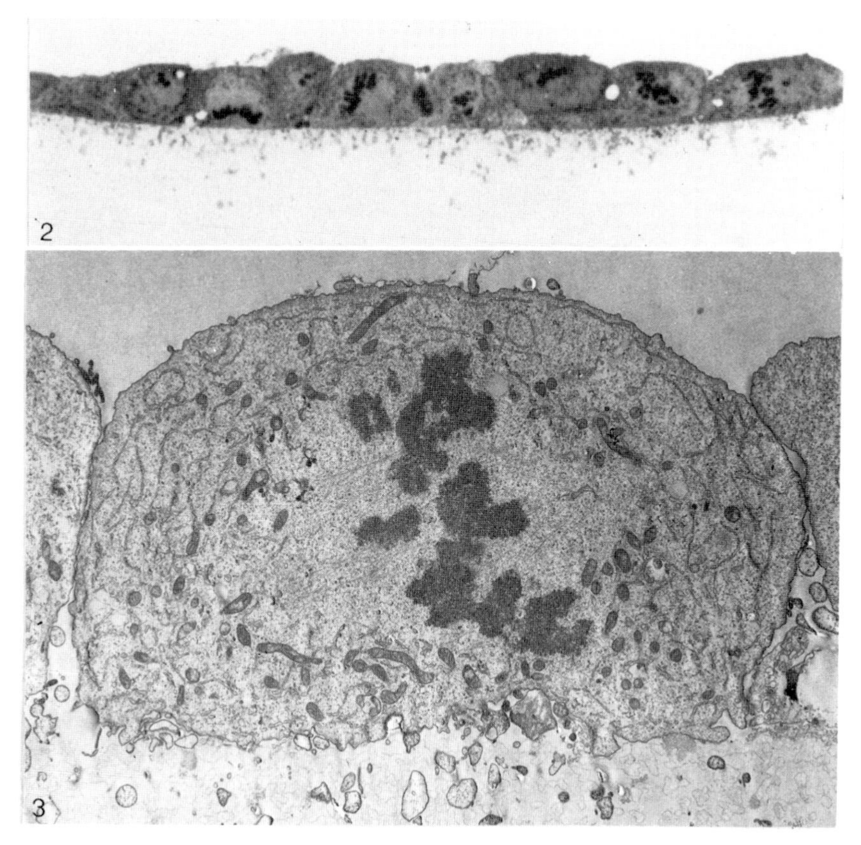

Fig. 2. A 1-μm vertical section through the Epon-embedded mitotic cell filtrate viewed in the light microscope, doubled-fixed in glutaraldehyde and OsO_4, and stained with toluidine blue. ×800.

Fig. 3. An electron micrograph of a metaphase cell resting on top of the porous Millipore filter. Doubled fixed in glutaraldehyde and OsO_4, and stained with uranyl acetate and lead citrate. ×5000.

picked up on single-hole grids coated with formvar film. The sections are double-stained with uranyl acetate[8] and lead citrate[9] for examination in the electron microscope (Fig. 3).

NOTES ON THE TECHNIQUE

The described vertical sectioning technique, with appropriate modifications, is applicable to other cell lines and filter types.[10] The MF-Millipore filters can

[8] M. L. Watson, *J. Biophys. Biochem. Cytol.* **4**, 475 (1958).
[9] E. S. Reynolds, *J. Cell Biol.* **17**, 208 (1963).
[10] R. Cornell, *Exp. Cell Res.* **56**, 156 (1969).

also successfully be dehydrated in isopropyl alcohol to reduce any swellings of the filters and be embedded in Araldite.[11] When necessary, the selected embedded cells can also be sectioned in a plane parallel to the growth surface.[12]

[11] B. Friedman, P. Blais, and P. Shaffer, *J. Cell Biol.* **39**, 208 (1968).
[12] B. R. Brinkley, P. Murphy, and L. C. Richardson, *J. Cell Biol.* **35**, 279 (1967).

CHAPTER 4

Vertical Sectioning

B. Cells in Plastic Flasks

*L. N. Keen, R. Reynolds, W. L. Whittle, and P. F. Kruse, Jr.**

Most electron micrographs of tissue culture cells have been oriented horizontal to the plane of the cultures. Some, however, have been oriented perpendicularly, i.e., from sections cut vertical to the cells and supporting surfaces.[1-3] In either case, preparation for electron microscopy has usually included a provision for separation of the embedded cells from culture supports prior to sectioning. Separations have been made by simply breaking the culture surface—glass,[4] mica,[5] or plastic[6]—away from the resin embedment, usually with precooling; often this procedure has been facilitated by prior interpositioning of films of palladium,[7] carbon,[8] and silicone.[9]

In recent years techniques have been described[10-14] for culture or collection of cells on Millipore or Nucleopore filters, with subsequent sectioning of the

[1] B. Goldberg and H. Green, *J. Cell Biol.* **22**, 227 (1964).
[2] A. Martinez-Palomo and C. Brailovsky, *Virology* **34**, 379 (1968).
[3] A. Martinez-Palomo, C. Brailovsky, and W. Bernhard, *Cancer Res.* **29**, 925 (1969).
[4] M. Kumegawa, M. Cattoni, and G. Rose, *J. Cell Biol.* **36**, 443 (1968).
[5] M. Kumegawa, M. Cattoni, and G. Rose, *Tex. Rep. Biol. Med.* **26**, 205 (1968).
[6] B. R. Brinkley, P. Murphy, and L. C. Richardson, *J. Cell Biol.* **35**, 279 (1967).
[7] H. W. Fisher and T. W. Cooper, *J. Cell Biol.* **34**, 569 (1967).
[8] E. Robbins and N. K. Gonatas, *J. Cell Biol.* **20**, 356 (1964).
[9] S. I. Rosen, *Stain Technol.* **37**, 195 (1962).
[10] R. M. McCombs, M. Benyesh-Melnick, and J. P. Brunschwig, *J. Cell Biol.* **36**, 231 (1968).
[11] H. Dalen and T. J. Nevalainen, *Stain Technol.* **43**, 217 (1968).
[12] B. Friedman, P. Blais, and P. Shaffer, *J. Cell Biol.* **39**, 208 (1968).
[13] R. Cornell, *Exp. Cell Res.* **56**, 156 (1969).
[14] H. Dalen, *J. Microsc.* **91**, 213 (1970).

embedded filter-cell composite for vertically oriented micrographs.[15] Also vertical sections are easily obtained of cells grown on decalcified eggshell membranes.[16]

We have found that vertical sections of cultures in Falcon plastic flasks can similarly be made, i.e., by sectioning directly through the plastic vessel-embedded cell "sandwich." This direct approach apparently preserves the shape and spatial arrangements of cells, and enables examination of cell to surface attachments as illustrated previously[17] and below. A similar procedure has been described by Eguchi and Okada[18] and improved by Douglas and Elser.[19]

Method

Cultures were set up in Falcon T-75 plastic flasks using 30 ml per flask of Medium 7a, a modification[20] of McCoy's Medium 5a, plus 10% whole calf sera. Cell populations of densities equivalent to at least several layers of cells were obtained by frequent medium changes of postconfluent Tb1Lu[21] cultures, or by perfusion methods[17] with WI-38VA13A[22] cultures.

Preparation for electron microscopy is made by decanting the medium and (a) fixing the tissue with 5% glutaraldehyde in Hanks' basic pH 7.4 salt solution for 20 minutes, (b) postfixing in 1% Veronal-buffered OsO_4, pH 7.4, for 30 minutes; (c) dehydrating in 15-minute changes of 70% ethanol containing 2% uranyl acetate followed with 80, 95, and twice with 100% ethanol; and (d) embedding with two 15-minute changes of Maraglas mixture E[23]–absolute alcohol mixtures followed with pure Maraglas mixture E. Propylene oxide is not used for infiltrating the cells or mixed with the Maraglas because of its tendency to dissolve the plastic culture flask. After refrigeration overnight at 4°C, the Maraglas is decanted and replaced with fresh Maraglas mixture. Holes are made in the top of the plastic flasks with heated glass rods[6] and the resin is polymerized at 60°C for 3 to 5 days.

Small blocks are cut out of the plastic flask-embedded tissue "sandwich," fastened to a Maraglas capsule with epoxy cement, and sectioned on a Porter-Blum MT-1 microtome with glass or diamond knives. Sectioning is done directly through the entire "sandwich" in a plane vertical to the embedded tissue sheet. The sections are picked up on copper grids and stained with uranyl acetate followed by lead citrate; they are then examined with an electron microscope.

[15] See Section IX, this volume, Chapter 4A.

[16] J. Leighton, S. Mansukhani, and L. W. Estes, *In Vitro* **6**, 251 (1971); see also Section IX, this volume, Chapter 4C.

[17] P. F. Kruse, Jr., L. N. Keen, and W. L. Whittle, *In Vitro* **6**, 75 (1970).

[18] G. Eguchi and T. S. Okada, *Develop. Growth Different.* **12**, 297 (1971).

[19] W. H. J. Douglas and J. E. Elser, *In Vitro* **8**, 26 (1972).

[20] P. F. Kruse, Jr., W. L. Whittle, and E. Miedema, *J. Cell Biol.* **42**, 113 (1969).

[21] Adult bat lung cells obtained from Dr. A. Kniazeff, University of California, San Diego, La Jolla, California 92037.

[22] SV40 virus-transformed WI-38 human embryonic diploid lung fibroblasts; obtained from Dr. V. Cristofalo, The Wistar Institute, Philadelphia, Pennsylvania 19104.

[23] A. M. Glavert, *In* "Techniques for Electron Microscopy" (D. H. Kay, ed.), p. 197. Davis, Philadelphia, Pennsylvania, 1965.

Applications

Figure 1 illustrates micrographs of vertical sections from cultures of the two cell types. The plastic surface contains many spherical inclusions; those close to the surface are approximately 6000 Å diameter. The inclusions are thought to be lubricant material (inert) admixed during manufacture of the plastic material. A thin shell of plastic which covers the inclusions at the uppermost surface of the flasks is plainly visible.[19] An advantage of using plastic flasks for these studies of vertical sections is that no deterioration of the surface is apparent in the electron beam, while appreciable distortion and sublimation has been encountered on occasion with Millipore filter culture supports.[12]

In Fig. 1A the large intercellular spaces and extensive cytoplasmic processes typical of Tb1Lu bat cells is shown. Close apposition of cell processes to the plastic surface is evident. Figure 1B shows the random and close association of

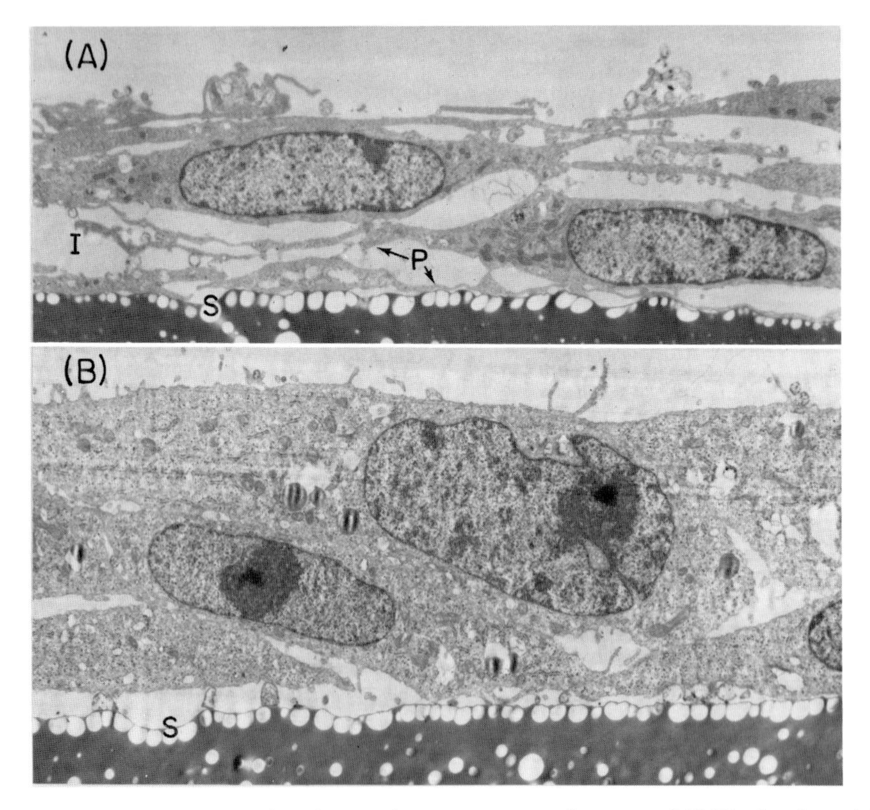

Fig. 1. Electron micrographs of vertical sections prepared *in situ* of Tb1Lu bat lung (A), and WI-38VA13A (SV40-transformed WI-38 human lung) (B) tissue cultures in plastic flasks. Sectioning was done directly through the plastic surface-embedded tissue "sandwich." See text for explanation of spherical inclusions in plastic surfaces, S. (A) Large intercellular spaces, I, of Tb1Lu cells and close apposition of extensive cell processes, P, to plastic surface. (B) Random and close association of multiple-layered WI-38VA13A cells and frequent spaces between cells and plastic surface. ×3200.

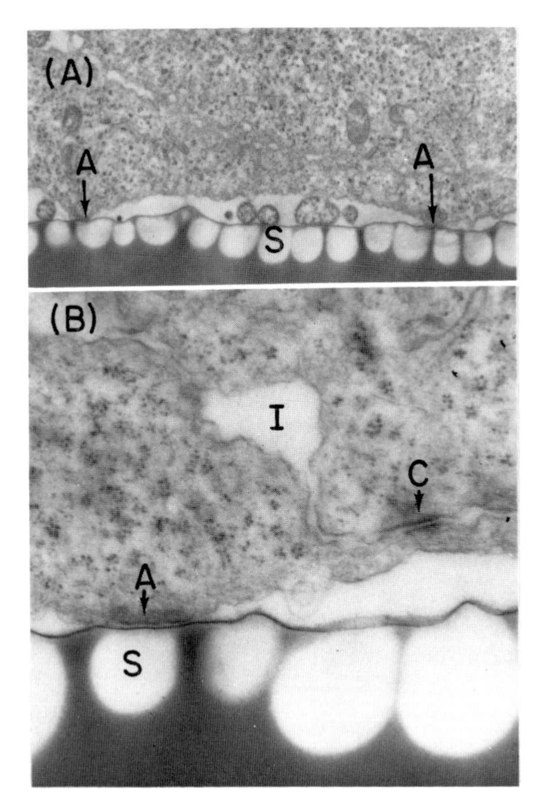

Fig. 2. Electron micrographs of vertically sectioned (cf. Fig. 1 and text) WI-38VA13A tissue cultures illustrating cell-to-surface attachments and a cell-to-cell contact. (A) Two points of attachment, A, of a cell to the underlying plastic surface, S. ×8000. (B) Higher magnification of cell attachment, A, to plastic surface, S, and illustrating a cell-to-cell contact, C. I, intercellular space. ×27,500.

WI-38VA13A cells in multiple layers. As with most tissue cultures so far examined, the bottom most cells are separated from the plastic flask surface except at relatively infrequent points of contact as shown in more detail in Fig. 2. Here two attachment points of a cell are shown (Fig. 2A); a portion of another cell is visible which overlies the one attached. Figure 2B shows a cell-to-surface attachment at higher magnification; it is similar in appearance to the "half-desmosome" observed between epithelial cells and basement membranes.[24] A region of cell-to-cell contact is also apparent in Fig. 2B, similar in appearance to the attachment sites illustrated by Ross and Greenlee.[25]

[24] A. S. Zelickson, "Ultrastructure of Normal and Abnormal Skin," p. 61. Lee and Febiger, Philadelphia, Pennsylvania, 1967.

[25] R. Ross and T. K. Greenlee, Jr., *Science* **153**, 997 (1966).

CHAPTER 4

Vertical Sectioning

C. Cells Grown on Decalcified Eggshell Membrane with Special Reference to Electron Microscopic Study[1]

Joseph Leighton, Nabil Abaza, and Sunder Mansukhani

A number of surface substrates have been used for the cultivation of monolayers of cells to be examined with electron microscopy. Each of them, whether commercial filter membranes or transparent membranes, has particular useful applications and particular limitations. We have found that decalcified eggshell membrane provides a suitable surface for the cultivation of monolayers of a tissue culture cell line.[2] The advantages of this substrate are its simplicity and economy, its insolubility in the solvents used in histological or electron microscopic preparation, and its ultrastructure. Its major limitations are its opacity and the complex protein nature of its fibrillary structure.

Method of Preparing Membranes

1. Cut holes 2 or 3 cm in diameter in the blunt ends of three or four fresh raw eggs. Pour out the fluid contents.
2. Rinse the remaining calcareous cups inside and out with water.
3. Submerge the cups completely in 5% acetic acid for a week or more. This can be accomplished easily by placing one or more glass balls in each empty eggshell.
4. During the week of decalcification, rinse the sacs of shell membrane daily in tap water. Each day gently remove the proteinaceous residue of decalcified shell mechanically as far as possible. Return the membranes to fresh 5% acetic acid for further decalcification. Repeat the procedure daily until the membranes are soft, and are completely free of brittle eggshell remnants.
5. When the membranes are completely free of shell remnants, cut the sacs into two or three pieces, rinse briefly in distilled water, and place in a liter of distilled water overnight in the refrigerator.
6. On the following day cut the membranes into smaller pieces with straight dissecting scissors and blunt forceps. In our laboratory we use pieces about 5 × 8 mm. Pool the individual pieces in a few bottles containing distilled water, and

[1] The development of methods described here was conducted under Research Grants CA-13219 from the National Cancer Institute and P-442 from the American Cancer Society.

[2] J. Leighton, S. Mansukhani, and L. W. Estes, *In Vitro* 6, 251 (1971).

sterilize by autoclaving at 15 pounds for 20 minutes. The membranes may be stored indefinitely in sterile distilled water in the refrigerator. When cultures are to be prepared, pieces are removed as needed.

Preparation of Cultures with Cell Line Growing on Both Sides of the Membrane

1. Prepare a fresh suspension of cells from a continuous line that normally adheres to a glass or a plastic surface *in vitro*.
2. Add the cells suspended in medium to a series of tubes or bottles each containing one or more pieces of shell membrane.
3. Feed the cultures in 1 or 2 days, when many cells have become adherent to the shell membrane and are proliferating. Transfer one or more membranes to new containers and add fresh medium.
4. The principles of subsequent maintenance of these cultures are the usual ones for tissue culture. Replenish the culture medium according to the indices in use in the individual laboratory. Although adhesion and growth take place on both surfaces of the membrane, the surface closer to the gas phase has a denser population of cells.

Preparation of Cultures with Cell Line on One Side of the Membrane Cemented to a Coverslip

Growth on both surfaces of the shell membrane is not always desirable and can be restricted to the upper surface by cementing the pieces of membrane to glass surfaces.

Plasma Clot. The method for preparation of a plasma clot is given below.

1. Dip membranes from distilled water into balanced salt solution. Transfer from balanced salt solution to a small pool of heparinized chicken plasma.
2. Apply a thin film of heparinized chicken plasma to a glass (or plastic) coverslip.
3. Apply two drops of chick embryo extract to the coverslip. Promptly place the shell membrane on the coverslip. Remove excess of clotting mixture and permit clot to form.

Coagulated Egg Albumin. The protocol used for preparing coagulated egg albumin is given in the following steps.

1. Wipe a large area over the air space of the shell of a raw fresh egg with 1% iodine solution in 80% alcohol.
2. Open the treated shell aseptically, and remove shell fragments and shell membrane with sterile forceps.
3. Transfer 1 or 2 ml of egg albumin to a sterile tube.
4. Place the required number of pieces of membrane in a Petri dish. Allow excess water to drain off on glass.

5. Place one drop of egg albumin on a coverslip. Spread the albumin to cover an area slightly greater than that of the piece of membrane.

6. Place membrane immediately on the coated area of the coverslip and press gently with a spatula.

7. Coagulate the albumin by placing the coverslip in an appropriate culture container, and autoclaving (10 pounds for 10 minutes) or immersing in boiling water for 15 minutes.

Inoculation of "Cemented" Membranes. Place coverslip with adherent shell membrane into a tube, bottle, or Petri dish. Inoculate membrane with a suspension of cells in medium. During subsequent incubation cells will settle, adhere, and grow on the membrane as well as on the adjacent surface of the coverslip.

Processing of Growth on Membranes for Histology

1. Remove the coverslip from culture vessel after the period of cultivation, and dip into 10% formalin for 30 to 60 seconds.

2. Shave the membrane from the coverslip surface with a razor blade. Fix membrane overnight using 10% formalin in 95% ethyl alcohol or other appropriate fixative.

3. Transfer membrane to 80% alcohol, and trim membrane to the required dimensions for eventual embedding in paraffin.

4. The same sequence of dehydration, clearing, paraffin infiltration, and embedding may be used as for other tissues in routine histology. Since the specimen is thin, however, the time intervals required for each step may be cut in half.

5. Embed on edge, cut 6 μm sections, and stain as required.

Processing for Electron Microscopy

1. According to the number of specimens to be processed, prepare an adequate volume of Sorensen phosphate buffer.[3] If preferred, other buffers may be used such as cacodylate[3] or *s*-collidine.[4] In such case the chosen buffer replaces phosphate buffer in the following procedure.

2. To prepare 200 ml of phosphate buffer, pour 28 ml of 0.2 M NaH_2PO_4, and 72 ml of 0.2 M Na_2HPO_4 into a flask. Add 98.5 ml of distilled water and mix thoroughly. With constant stirring add 1.5 ml of 1% $CaCl_2$.

3. Prepare 3.5% glutaraldehyde solution in phosphate buffer, pH 7.2–7.4, and store overnight at room temperature.

4. Prepare 2% solution of osmium tetroxide in distilled water. Combine equal volumes of osmium tetroxide solution and phosphate buffer to make 1% phos-

[3] A. M. Hayat, "Principles and Techniques of Electron Microscopy: Biological Applications," Vol. 1, p. 342. Van Nostrand Reinhold, New York, 1970.

[4] H. S. Bennett and J. H. Luft, *J. Biophys. Biochem. Cytol.* **6**, 113 (1959).

phate-buffered osmium tetroxide. Store in refrigerator. Remember to handle osmium tetroxide carefully and under the hood. CAUTION: Osmium vapors are dangerous and should be used carefully.

5. With a fresh razor blade, quickly shave off the shell membrane with its overlying monolayer from the glass surface, and place in a vial with an excess of 3.5% phosphate-buffered glutaraldehyde at room temperature. Cap the vials and fix at room temperature for 30 to 45 minutes with occasional swirling. Note that the time of fixation and subsequent dehydration of monolayers is less than is required for routine processing of tissues. Procedures given by other authors call for even shorter periods for tissue cultured cells. We have found that our schedule gives consistently good results.

6. Wash the membranes for 1 hour, with four changes at 15-minute intervals, in phosphate buffer at 4°C with occasional swirling.

7. Postfix in 1% phosphate-buffered osmium tetroxide for 15 to 30 minutes at 4°C. Make sure that handling of osmium tetroxide is done carefully and under the hood. Cap vials and swirl occasionally.

8. Pour off osmium into a waste jar in the hood, and dehydrate specimen at 4°C. Start dehydration with 50% ethyl alcohol, followed by 70%, and 95%, for 10 minutes each.

9. If schedule necessitates a hold in the procedure, this is best done in glutaraldehyde (see step 5), or in 70% ethyl alcohol.

10. Bring egg membrane and the overlying monolayer to room temperature in 95% ethyl alcohol.

11. Continue dehydration with four changes of 100% ethyl alcohol for 10 minutes each. During the last alcohol change, pour off contents including the membrane culture into a Petri dish with excess of alcohol. Cut membrane into strips or squares as required while still completely immersed. This cutting is required only if BEEM capsule embedding is planned. In case of flat embedding this step is unnecessary, and further trimming is done after curing the EM resin.

12. Pour off alcohol and add propylene oxide to the egg membrane in vials. Change reagent twice, each time for 5 minutes. Work quickly to avoid drying of the membrane and monolayer.

13. While membrane is in alcohol, prepare the complete resin mixture including the accelerator. Epon-Araldite mixture or Epon alone may be used.

14. To prepare Epon-Araldite mixture combine 10 ml of Araldite 502, 12.5 ml Epon 812, and 30 ml dodeceylsuccinic anhydride (DDSA), and mix thoroughly and continuously for 5 minutes. With a disposable syringe add 2.5% of the accelerator 2,4,6-tri(dimethylaminomethyl)phenol (DMP-30). Mix for 2 to 3 minutes until the color of the mixture turns into a deep golden yellow. This resin provides for adequate hardness of the final blocks. If Epon resin is preferred, weigh 54.9 g of Epon 812. Add to it 29.1 g of DDSA, and 30.75 g of Nadic methyl anhydride (NMA). Mix thoroughly in a glass beaker, then add 0.9 g of the accelerator benzyldimethylamine (BDM), and mix quickly and thoroughly. While preparing the resin mixture, make sure to wipe dry all stock bottles before closing to avoid hardening of the resins on the screw caps. Also immerse glassware used for resin in propylene oxide or acetone.

15. Add part of the resin to propylene oxide in the ratio of 1:1. Keep the remaining resin in the freezer for subsequent embedding.

16. Pour off last change of propylene oxide (see step 12) and quickly deliver approximately 3 ml of resin propylene oxide mixture to each vial. Take extra care not to allow the tissue to dry out. Leave vials uncapped overnight under the hood. Most of the propylene evaporates.

17. Next morning remove resin (see step 15) from freezer to thaw. Transfer membrane into clean vials with fresh Epon-Araldite mixture. Use a pointed applicator stick to transfer membrane or membrane fragments. Leave for 2 to 3 hours in fresh Epon-Araldite mixture for better infiltration.

18. For flat embedding, pour enough Epon-Araldite mixture into commercially available aluminum dishes to have a layer 3 to 4 mm thick. Place membrane on surface of resin so that it lies flat. Place aluminum dishes in 60°C oven. After 1 or 2 hours check specimen for orientation. At that time the viscosity of the resin is lowest, and orientation is facilitated.

19. For capsule embedding, fill BEEM capsules with fresh Epon-Araldite mixture. Avoid dripping the resin onto the outside of the capsules, otherwise the embedded specimen is difficult to separate from the plastic capsule after curing. With a pointed applicator stick, remove strips or fragments of membrane, and drop them into the filled capsules. After 1 or 2 hours, check and orient the position of the strips as required. Frequent checking and orientation are desirable if accurate vertical or horizontal sections of the monolayer are required.

20. Leave aluminum dishes or capsules in 60°C oven for 48 hours for resin to harden.

21. Remove from the oven and leave at room temperature for 24 hours to cool before cutting because cured resin will be slightly soft after removal from the oven.

22. Use a jigsaw to cut required zones of embedded membrane if flat embedding was done. Use flat-peaked chucks, or fasten sawed pieces into dummy capsules by epoxy resin. If BEEM capsules were used, make two parallel slits into the plastic of the BEEM capsule with a fresh razor blade, and remove the cured resin.

23. For 1 μm sections trim the block face into a trapezoid. Use the ultramicrotome and glass knives to produce these sections. Stain with 0.3 to 1% toluidine blue or other stains such as methylene blue–azure II[5] or *p*-phenylenediamine[6] and examine with light microscope.

24. Cut ultrathin sections 600 to 700 Å. Use a diamond knife because of the toughness of the membrane.

25. Pick up the sections on a copper grid. Stain with 3% uranyl acetate[7] and/or lead citrate.[8] Examine the grids with the electron microscope.

[5] K. C. Richardson, L. Jarrett, and E. H. Finke, *Stain Technol.* **35**, 313 (1960).

[6] J. F. Estable-Puig, W. C. Bauer, and J. M. Blumberg, *J. Neuropathol. Exp. Neurol.* **24**, 531 (1965).

[7] M. L. Watson, *J. Biophys. Biochem. Cytol.* **4**, 475 (1968).

[8] E. S. Reynolds, *J. Cell Biol.* **17**, 208 (1963).

COMMENTS

There is a growing interest in the study of monolayer growth on solid substrates. The practice of scraping the monolayer before or after fixation introduces obvious artifacts. The monolayer cannot be examined *in situ*. It also fails to provide the investigator with the proper orientation to examine any intimate relationship between the cell surface and the supporting substrate.

Since the introduction of electron microscopy it has been necessary to modify the routine technique of processing cultured cells together with their supporting substrate for subsequent EM examination. Evaporated carbon film[9] and cured EM resin[10] are two of several solid substrates that may be applied to the glass surface of slides or coverslips before cell growth takes place. These techniques, however, are tedious and time-consuming because of the difficulty of separating the cured EM resins from the glass surface. Several commercially available filter membranes,[11] gelatin,[12] and agar[13] have also been applied as substrates to glass

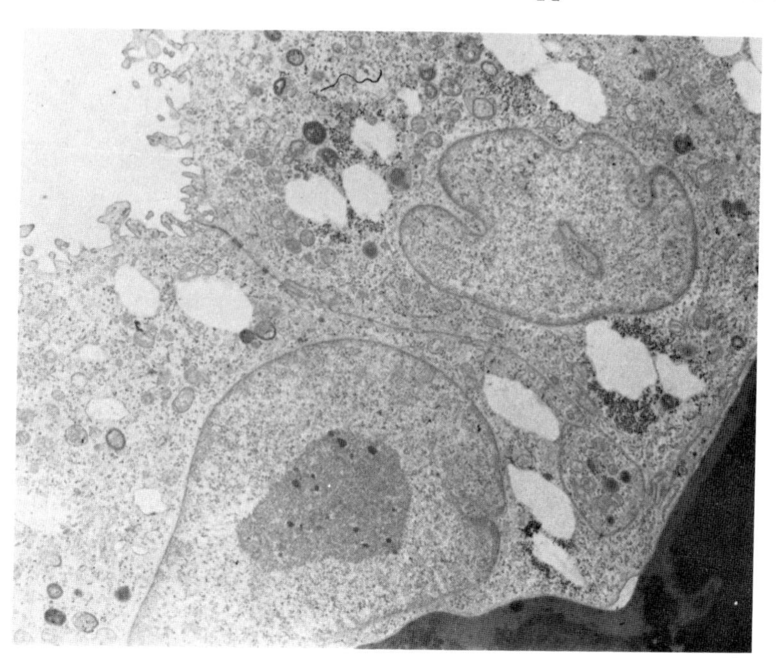

Fig. 1. Cell line MDCK cultivated for 1 week on a substrate of decalcified eggshell membrane, which was previously fastened to a glass slide with coagulated egg albumin. Growth appeared as a monolayer of low columnar epithelium. The culture was fixed in osmium tetroxide in phosphate buffer. The dark osmophilic area in the lower right corner is the combination of shell membrane and coagulated egg albumin. Magnification ×5200.

[9] E. Robbins and N. K. Gonatas, *J. Cell Biol.* **20**, 356 (1964).
[10] W. Smith, E. W. Gray, and J. M. K. Mackey, *J. Microsc.* **89** (pt. 3), 359 (1968).
[11] H. Dalen and T. J. Nevalainen, *Stain Technol.* **43**, 217 (1968).
[12] R. S. Spiers and M. X. Turner, *Exp. Cell Res.* **44**, 661 (1966).
[13] J. K. Koehler, *Stain Technol.* **36**, 94 (1961).

surfaces. These substrates are easy to separate from glass, but some of them dissolve or separate from the monolayer during processing for either light or electron microscopy. The embedded monolayer, in this case, is left without attached substrate and the *in situ* examination is not fulfilled.

Decalcified eggshell membrane provides a good substrate for monolayer growth (Fig. 1). Its advantages include its low cost, easy separation from the glass surface, and feasibility of processing for both light and electron microscopy. Strips of fixed membrane with attached monolayer may be embedded flat, or oriented into a BEEM capsule. Therefore, the monolayer and its supporting substrate may be sectioned vertically, obliquely, or horizontally to the plane of growth. The opacity of the substrate, however, may be circumvented by multiple sectioning of either paraffin or plastic embedded strips of the shell membrane, and staining for examination with light microscopy. Our conclusion is, therefore, that the membrane is a suitable substrate for any cell type that can adhere and grow on it.

CHAPTER 5

Hydraulically Operated Microscope Stage[1]

Frederick H. Kasten

The microscopist occasionally finds it necessary to carry out extra manipulations, in addition to the usual ones required for routine microscope operations. For example, physiologists interested in doing bioelectric recording from single cells must be prepared to manipulate one or more microelectrodes and adjust the position of the cells simultaneously. Similar needs are encountered when carrying out microsurgical operations under the microscope. In addition to focusing the microscope, the operator must be prepared to move the mechanical stage. The stage controls are generally separate knobs, which are difficult or impossible to handle simultaneously with one hand, limiting motion to one direction at a

[1] This research was supported in part by USPHS Research Grants NS-09524 from the National Institute of Neurologic Diseases and Stroke and CA-12067 from the National Cancer Institute. It has also benefited from USPHS Training Grant 5-T01-DE-00241-04 from the National Institute of Dental Research.

time. Any path not on a line with one of the stage slides must be attained by sequential motion of the two stage slides, with the field of view following a staircase pattern.

The apparatus to be described was developed in response to the need to obtain smooth motion of the object field during movie filming of living cells in culture. It was necessary to obtain smooth mechanical scanning in any direction at all levels of magnification, leaving one hand free to operate the camera controls or microscope focus. A hydraulically operated microscope stage was developed which was capable of one-hand control of both direction and rate of translation, making it possible to follow any arbitrary path at controlled rates of movement. The mechanical dimensions of the stage for this model were designed to be used with a Zeiss Photomicroscope and to hold a 3 × 2 inch Rose multi-

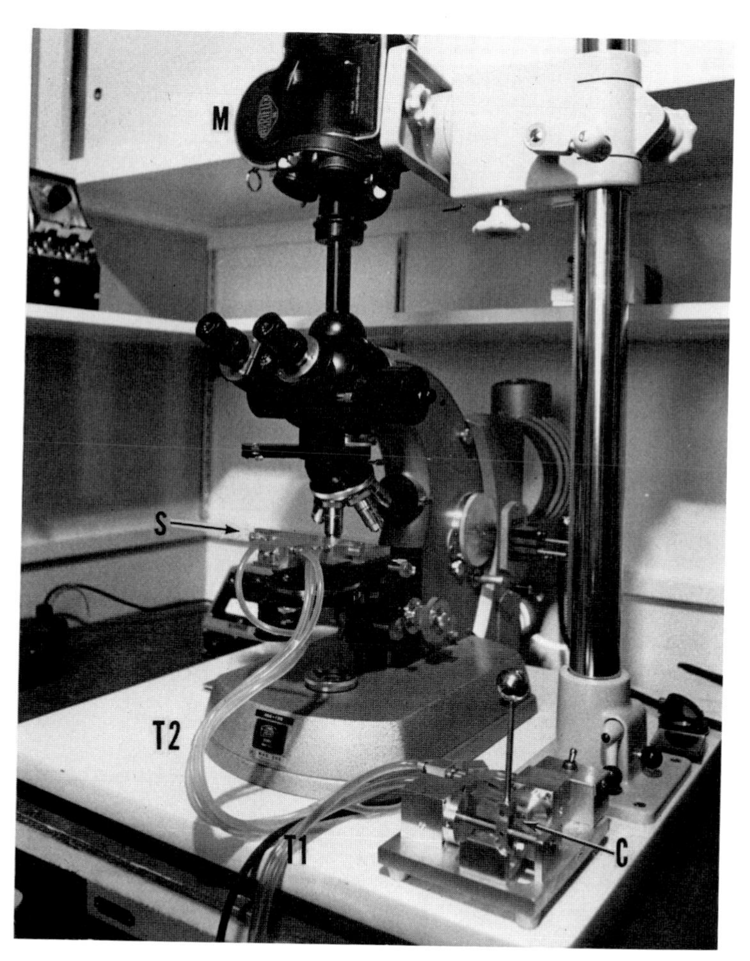

Fig. 1. General view of the control unit (C), tubing (T1) from motor driven oil pump unit, tubing (T2) to stage unit, stage unit (S), and 16-mm Arriflex movie camera (M) in position for filming. Stage unit is attached to Zeiss Photomicroscope.

purpose chamber,[2] but the stage could easily be fabricated and adapted to work with other microscopes and chambers. A standard 3 × 1 inch slide will fit the unit without modification. The description which follows is taken from the original publication where the stage was first described.[3]

The original equipment consisted of a three-piece system. A motor driven oil pump unit, with oil and relief valve, is remotely located and connected by flexible tubing to a control unit near the microscope. The third unit is the stage assembly. Figure 1 is a general view of the control unit (C) and stage (S). A close-up view of the control unit (Fig. 2) shows a single joystick lever (L) movable around two axes. These intersect at right angles and operate two valves, which govern the direction and flow rate of oil supplied to the stage assembly. A switch

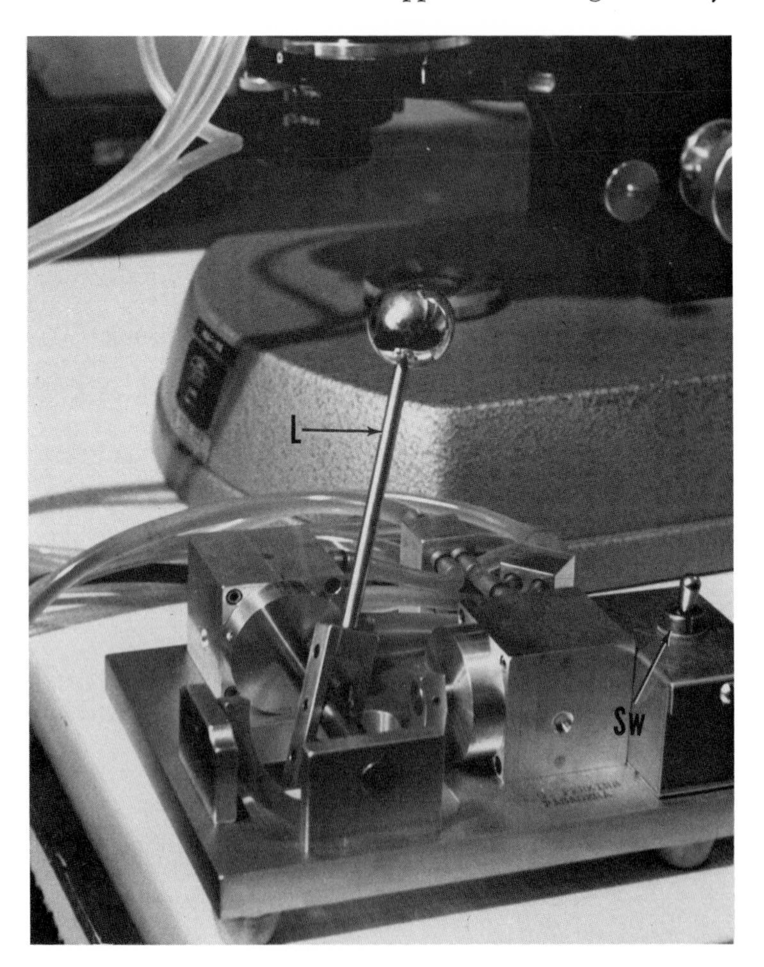

Fig. 2. Close-up of the control unit shows the joystick lever (L), which can be moved in any direction to govern direction and rate of travel of stage and switch (Sw), which turns on the motor.

[2] G. Rose, *Tex. Rep. Biol. Med.* **12**, 1074 (1954).
[3] F. H. Kasten and T. G. Perkins, *Exp. Cell Res.* **63**, 473 (1970).

Fig. 3. Close-up of the stage assembly shows the flexible tubing connected to the cylinders. The assembly is designed in this case to accommodate and move a Rose chamber (R) for tissue culture work.

(Sw) turns on the pump and puts the system in readiness, according to the position of the joystick. The stage assembly (Fig. 3) has two double-acting cylinders bored into a one-piece aluminum block. The "fore-and-aft" cylinder moves a plate dovetailed into the bottom surface of the block. This plate has dowel pins and screw which clamp the plate, and hence the entire unit, to the fixed stage of the microscope. The same holes are used which mount the original mechanical stage, so no modification of the microscope is required.

The fore-and-aft cylinder thus moves the entire assembly with respect to the optical axis. The right or left cylinder moves a slide dovetailed to the front face of the block, and specimen clips are attached to this slide.

Four flexible tubes connect the cylinders to the control valve assembly, and are so arranged that the field of view moves in the same direction the control lever is pushed, i.e., if the joystick lever is moved to the right, the apparent motion of the object as seen in the eyepiece is also to the right. Since the lever controls both cylinders, the stage moves in whichever direction the lever is moved, and at a rate which increases the further it is moved from its center position. The speed range of both cylinders is continuously variable from 1 to 100 μm/second, and the travel is 2.5 × 5 cm.

This first model performs very well in routine mechanical scanning with all objectives, from ×10 to ×100. It has been employed to obtain film records from salamander cultures of ciliated lung cells,[4] isolated neurons of chick embryo spinal ganglia,[5] and beating heart cells of newborn rats.[6,7] In combination with the Arriflex movie camera, cultures have been scanned mechanically with this stage at film speeds of 16 frames/second to as high as 60 frames/second.

Three faults have been detected in the instrument which it is expected a revised design will correct. The cylinder assembly is centered behind the optical axis, and is high enough to interfere with some of the objectives when several objectives are mounted simultaneously on the lens turret. One objective at a time may be used, as a simple solution. However, the problem can be avoided altogether by placing the unit in front instead of behind, but this puts it between the microscopist and his specimen slide, which is not very convenient. It would seem better to mount the actuating assembly off to one side. In this model, the cylinders and supply tubes are at zero pressure when at rest. When a valve is opened to supply oil from the pump, the pressure must rise high enough in the cylinder to overcome static friction before the piston will move. Since the connecting tubes are elastic, they expand slightly, increasing their volume.

These two considerations lead to a time delay before any motion occurs, and when motion does start there is a small "broad-jump" effect as the tubes adjust to the lower moving friction resistance. Both these effects become increasingly prominent as the valve opening is made smaller and as higher magnification objectives are used. A third objection is that the valves have too wide a deadband around their center positions, and must be moved several degrees before any oil flows. This, in conjunction with the response delay, makes control behavior objectionable. Design changes to correct these three conditions are now in process, and a quite valuable attachment should be the result.

It is expected that the hydraulic stage will be useful to cell biologists doing filming, bioelectric recording, microsurgery, and other specialized operations. The apparatus, after modification as indicated above, may likewise prove worthwhile for routine observation and scanning of tissue cultures and stained slides.

[4] F. H. Kasten, T. Okigaki, and J. A. DiPaolo, *Proc. Amer. Ass. Cancer Res.* **10**, 45 (1969).
[5] F. H. Kasten, Z. Lodin, and J. Booher, *J. Cell Biol.* **43**, 166A (1969).
[6] F. H. Kasten, *In Vitro* **4**, 150 (1969).
[7] F. H. Kasten, *Acta Histochem. Suppl.* **9**, 775 (1971).

CHAPTER 6

Culture Dish for High Resolution Microscopy[1]

A. D. Deitch, A. F. Miranda, and G. C. Godman

Observations of living cells grown in monolayer culture can be performed at high magnification with the phase-contrast microscope using cells grown in a variety of laminate slides, perfusion chambers, and similar vessels.[1a-16] One of the simplest vessels which has been described, the plastic Cooper dish,[17] as modified

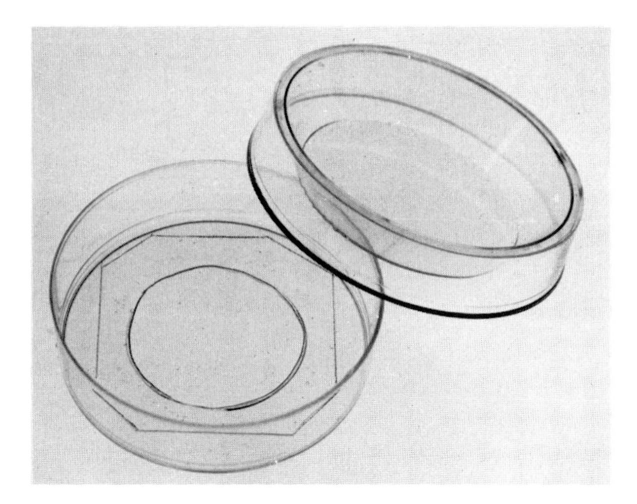

Fig. 1. The modified Cooper dish. In the bottom of the plastic dish (left), a 28 mm hole has been cut and a 40 mm^2 No. 1½ coverglass (with corners broken off to fit the dish) has been cemented to the inside of the dish over the hole.

[1] This work was supported by Grant Nos. CA-13835 and AI-05708 from the National Institutes of Health.

[1a] C. M. Pomerat, *In* "Methods in Medical Research" (M. B. Visscher, ed.), Vol. 4, pp. 275–280. Yearbook, Chicago, Illinois, 1951.

[2] G. B. Mackaness, *J. Pathol. Bacteriol.* **64**, 429 (1952).

[3] G. S. Christiansen, B. Danes, L. Allen, and P. J. Leinfelder, *Exp. Cell Res.* **5**, 10 (1953).

[4] R. Buchsbaum and J. A. Kuntz, *Ann. N. Y. Acad. Sci.* **58**, 1303 (1954).

[5] G. G. Rose, *Tex. Rep. Biol. Med.* **12**, 1074 (1954).

[6] K. M. Richter and N. W. Woodward, Jr., *Exp. Cell Res.* **9**, 585 (1955).

[7] E. O. Powell, *J. Roy. Microsc. Soc.* **75**, 235 (1956).

[8] J. Paul, *Quart. J. Microsc. Sci.* **98**, 279 (1957).

[9] B. L. Toy and W. A. Bardawil, *Exp. Cell Res.* **14**, 97 (1958).

[10] J. A. Sharp, *Exp. Cell Res.* **17**, 519 (1959).

[11] J. A. Sykes and E. B. Moore, *Proc. Soc. Exp. Biol. Med.* **100**, 125 (1959).

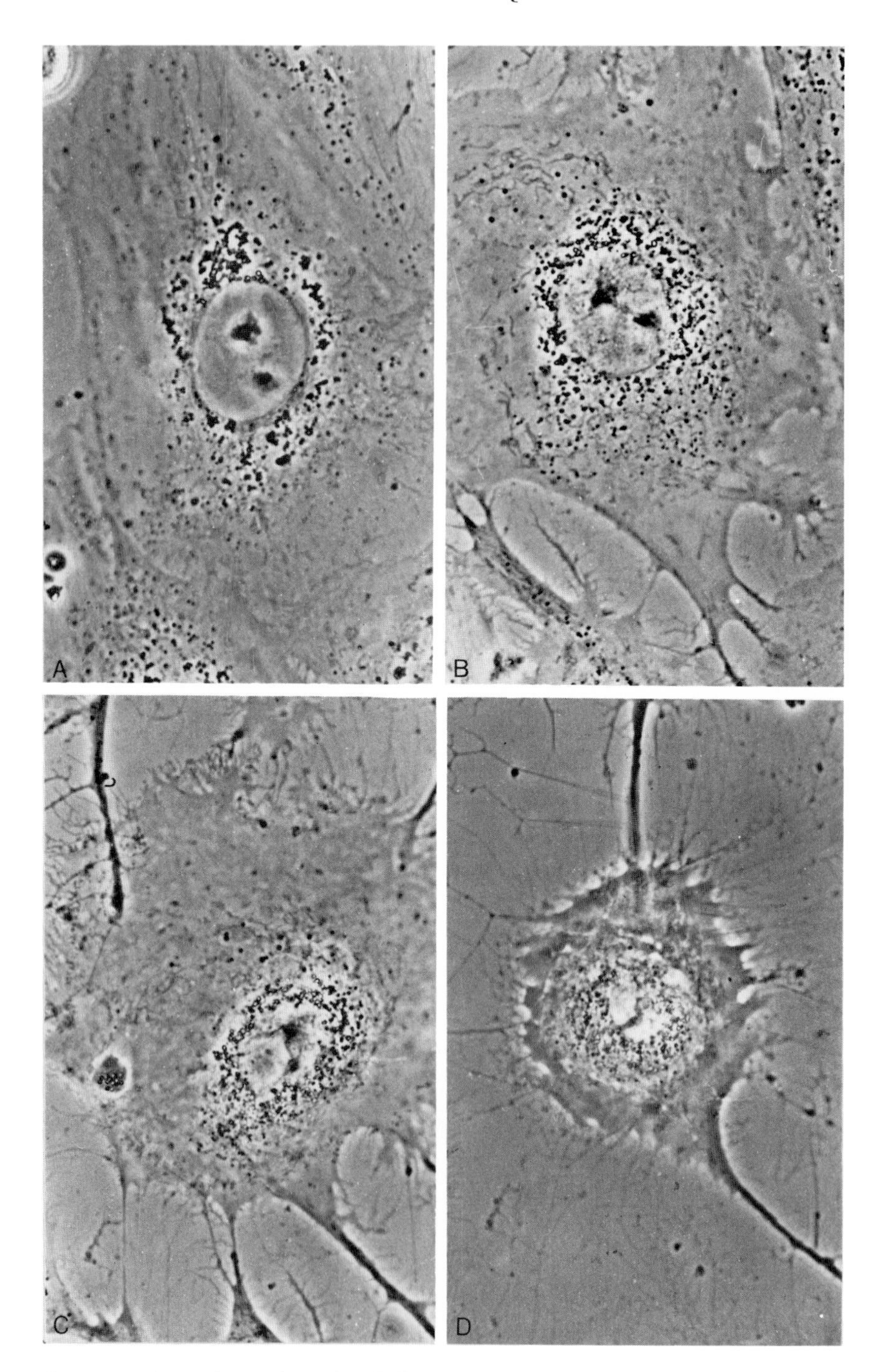

Fig. 2. Primary culture of monkey kidney cells infected with ECHO 9 (Coxsackie A23) virus. Successive photographs of the same cell taken at intervals after infection, ×780, Wild M 40 inverted phase-contrast microscope. (A) At 2 hours, prior to the onset of cytopathic changes, the highly flattened cell is in contact with its neighbors. The nucleus is slightly

by Deitch and Godman,[13] offers ease of assembly and ready maintenance of the cell monolayer while allowing long-term microscopy with oil immersion objectives on an inverted phase contrast microscope. The Cooper dish as commercially obtained[18] is a polystyrene Petri dish, 60 mm in diameter, the top half of which has a depressed 40 mm central zone (Fig. 1). Microscopic observation is performed through the 2.5 mm thick central area of the assembled dish. The dish will contain 3–4 ml of medium and it has an adequate air space for gas exchange at the sides. The thickness of the plastic bottom of the Cooper dish (0.35 mm) however, seriously limits the usefulness of the unmodified dish for critical microscopic observations. Most objectives used in transmitted light microscopy are designed to perform through a coverglass of 0.17 to 0.18 mm thickness (No. 1½) and marked impairment of the image can be noticed when 40× objectives are used through glass or plastic of greater thickness. Since most oil immersion objectives are designed for working distances of 0.12 to 0.30 mm, they cannot be used at all with the dish in its original form. These disadvantages are overcome by making a hole in the center of the bottom half of the plastic dish and cementing a No. 1½ coverglass on the inner surface of the dish over the hole using an epoxide resin[19] which has proved to be nontoxic to primary cultures (mouse and rat muscle explants) and various cell lines (HeLa, Vero, L, etc.). The dishes may be modified rapidly and after preparation and cleaning, they are easily sterilized by ultraviolet light or other cold sterilization procedures. After the cells and medium are added, the dishes should be sealed with plasticine to prevent loss of CO_2 if bi-

[12] E. V. Orsi, *Exp. Cell Res.* **20**, 139 (1960).
[13] A. D. Deitch and G. C. Godman, *Stain Technol.* **41**, 19 (1966).
[14] J. Molé-Bajer and A. Bajer, *La Cellule* **67**, 257 (1968).
[15] R. O. Poyton and D. Branton, *Exp. Cell Res.* **60**, 109 (1970).
[16] J. A. Dvorak and W. F. Stotler, *Exp. Cell Res.* **68**, 144 (1971).
[17] W. G. Cooper, *Proc. Soc. Exp. Biol. Med.* **106**, 801 (1961).
[18] Falcon plastics division, B-D Laboratories, Los Angeles, California 90045.
[19] P. W. Brandt, J. P. Reuben, L. Gerardier, and H. Grundfest, *J. Cell Biol.* **25**, 233 (1965).

thicker than the cytoplasm and contains two prominent phase-dark nucleoli and homogeneous chromatin. In the juxtanuclear region, numerous small phase-dense granules (representing chiefly lysosomes and pinocytotic vesicles) all lie essentially in the same focal plane. Filamentous mitochondria are clearly visible. (B) At 5¾ hours, early in the cytopathic process, the cell has begun to withdraw from neighboring cells. Broad and fine filamentous processes of the central and neighboring cell are visible; most of these are in contact with adjacent cells or the coverglass substrate. As cell contraction occurs, the juxtanuclear zone thickens. Long filamentous mitochondria are prominent in the thinner subcortical cytoplasm. The nucleus has become more flattened and the chromatin more granular. (C) At 6¼ hours, the cytoplasmic processes are attenuated and beset with fine fibrillar extensions. The small zones of rarefaction seen in the cortical and subcortical cytoplasm (phase-lucent areas) represent vacuoles enlarging by fusion. The small granules and vacuoles of the juxtanuclear cytoplasm are heaped around the shrunken, crenated nucleus. Filamentous mitochondria are visible around the perinuclear mass of granules. (D) At 7½ hours, the cell has almost completely retracted from neighboring cells, leaving behind an arboreal network of fibrillar processes. The cytoplasmic periphery is markedly scalloped because of the migration of vacuoles to the periphery and then bursting. The residual cytoplasm is more dense, the juxtanuclear granular mass more compact, and the nucleus shriveled and convoluted by deep incisures.

carbonate buffered culture fluids are used. If HEPES[20]-buffered culture fluids are employed, it is not essential to seal the dishes for pH control. However, it is still desirable to seal them to prevent evaporation of the medium. Culture fluid changes or addition of drugs or viruses can be made by heating a sterile disposable hypodermic needle briefly and piercing the side of the dish. The needle is then withdrawn and discarded since it may have become plugged with plastic and a second needle of the same gauge introduced into the hole and a plastic syringe added.

Although the original report[13] suggested drilling the holes with a hole saw on a drill press, it has since been found more efficient to use a heated tool to cut out a circle in the Cooper dish. This can be done as follows. The bottom half of the dish is inverted and centered on a flat metal surface which is slightly smaller than the dish. While the dish is held firmly in place, a thin-walled metal tube (such as a culture tube closure of stainless steel) of appropriate diameter (20–28 mm), heated briefly in a gas flame, is pressed rapidly against the bottom of the plastic dish. A scalpel is used to ream the edges of the resulting hole if necessary.

The epoxide used to cement the coverglass[19] can be made in advance and stored for several weeks at $-20°C$. Coverglasses of appropriate size (30 or 40 mm^2, the latter having the corners scored and removed in order to fit the dish) of a brand nontoxic to cells in tissue culture[21,22] are cleaned by a procedure appropriate for tissue culture glassware (see Section XIV, this volume), rinsed well, and dried. The container of epoxide is warmed to room temperature before being opened. A small amount of resin, sufficient to seal the coverglass to the dish, is placed around the hole, a coverslip is centered over the hole, and the resin is polymerized at 60°C for 48 hours. Subsequently the dishes are soaked overnight in a tissue culture detergent and rewashed. The completed dishes may be sterilized by irradiation with ultraviolet light such as that emanating from a sterilamp used in culture hood.

An example of a sequence of photographs taken over a period of 7½ hours is shown in Fig. 2. It has been possible to follow the same field of cells for at least 3 days.

[20] C. Shipman, Jr., *Proc. Soc. Exp. Biol. Med.* **130**, 305 (1969); cf. Section XIV, this volume.

[21] Gold Seal, Clay Adams, Inc., New York, 10010.

[22] E. R. Peterson, A. D. Deitch, and M. R. Murray, *Lab. Invest.* **8**, 1507 (1959).

SECTION X

Cell Hybridization

Editors' Comments

The development of techniques within the past decade for the production of hybrid cells containing the genetic complement of two or more genetically different parent cells has opened new vistas to our understanding of somatic cell genetics. The procedures for producing hybrid cells can be divided into three parts: (a) fusion of cells of different genetic complement; (b) isolation of the resultant hybrids; and (c) characterization of these cells. This section emphasizes the first of the three parts; isolation techniques, e.g., cloning (see Section V), and characterization (see Section XIV, Chapters 14–16) are described elsewhere in this volume. Selective media for isolation of the hybrid cells are described in this section.

CHAPTER 1

Somatic Hybridization in Studies of Heredity of Cell Malignancy

Georges Barski

The first experiments which led to the discovery of somatic cell hybridization[1,2] were undertaken to check the possible transfer of genetic or epigenetic information between high cancer (N1), and low cancer (N2) cell lines[3] of C3H mouse origin, intimately associated in mixed cultures. The identification of the hybrid cells which appeared in these cultures was possible by direct karyological analysis, which revealed a nearly complete integration of the two chromosomal sets of the N1 and N2 cells, including distinct marker chromosomes from both parental cells, into the somatic hybrid. During continuous cultivation of the mixed cultures, the N1 cells diminished, probably due to their poor adherence to the glass walls, and the low malignant N2 cells accumulated progressively along with the hybrids, which showed an "intermediary" morphology and adherence capacity.

The mixed cell cultures, when inoculated into syngeneic, histocompatible C3H mice produced tumors, in most cases composed exclusively of hybrid cells. In addition, hybrid clones could be isolated directly from mixed cultures, without passing in animals, by picking up from sparsely seeded cultures hybrid colonies recognizable by the large size of cells and "rough" edges characteristic of the "noninhibited" type of growth. A similar natural selective advantage, in favor of the hybrid cells emerging in mixed cultures, was also observed in some homologous mouse cell hybrids,[4] and in Chinese hamster × Armenian hamster somatic hybrids.[5] This can, however, hardly be considered as a general rule and a manifestation of "hybrid vigor." In most cases, the recognition, accumulation, and isolation of hybrid cells requires a recourse to more or less artificial conditions concerning the choice of cell partners for mating, of selective media, and of different *in vitro* culture systems or animal inoculation procedures.

[1] G. Barski, S. Sorieul, and F. Cornefert, *C. R. Acad. Sci.* (*Paris*) **251**, 1825 (1960).

[2] G. Barski, S. Sorieul, and F. Cornefert, *J. Nat. Cancer Inst.* **26**, 1269 (1961).

[3] K. K. Sanford, R. M. Merwin, G. L. Hobbs, J. M. Young, and W. R. Earle, *J. Nat. Cancer Inst.* **23**, 1035 (1959).

[4] R. L. Davidson and B. Ephrussi, *Nature* (*London*) **205**, 1170 (1965).

[5] G. Yerganian and M. B. Nell, *Proc. Nat. Acad. Sci. U. S.* **55**, 1066 (1966).

In a general way, the following methodological points have to be considered in relation to the production, isolation, and identification of the somatic hybrids.

Appropriate Choice of Cell Candidates for Hybridization

The procedure obviously calls for crossing cells which have distinct chromosomal, morphological, and functional markers, reasonably stable and identifiable by available methods. The karyological differences have to be clear enough to distinguish the new hybrid cells from 2S hyperploid cells resulting from endoreduplication, which frequently occurs *in vitro* cultures. This demand is relatively easily satisfied when cells of different species are crossed. The recent development of differential chromosome staining methods[6-8] considerably enlarged the choice of appropriate distinct cell partners for hybridization. Also, differences in growth potential, in cell morphology, and in a pronounced capacity to grow either in suspension or in monolayers, as well as differences in nutritional requirements, should determine the choice of cell candidates for hybridization. The presence of distinct antigenic or isoenzyme markers, detectable by immunological or biochemical methods,[9,10] may similarly be important.

Procedures Favoring and Intensifying the Cell Fusion Event

It was noticed that maintenance for a few hours at lowered temperatures of the order of 4°C can by itself contribute to a more intense coalescence and subsequent fusion of cells in mixed suspensions. However, the most commonly practiced method for promoting cytoplasmic cell fusion is the use of ultraviolet-inactivated HVJ or Sendai myxoviruses following the procedure devised by Okada[11] and developed by Harris and Watkins.[12] According to this technique, concentrated, ultraviolet-inactivated Sendai virus preparation is first adsorbed at +4°C on a mixed cell suspension which is subsequently incubated at 37°C and seeded on the surface of culture vessels. A difficulty in using this method may arise from the fact that the fusion process once started is not easy to regulate and leads frequently to a very extensive cell fusion resulting in a generalized formation of huge polykaryocytes containing dozens of nuclei. However, only polykaryocytes containing few nuclei are viable and may lead to the production of nuclear hybrids. According to a variant of this technique,[13] cells of the two types intended for hybridization are mixed in very unequal numbers (3 to 4 log difference) which may by itself favor productive hybridization. The mixture is

[6] T. Caspersson, L. Zech, and C. Johansson, *Exp. Cell Res.* **60**, 315 (1970b).

[7] F. E. Arrighi and T. C. Hsu, *Cytogenetics* **10**, 81 (1971).

[8] H. J. Evans, K. E. Buckton, and A. T. Sumner, *Chromosoma* **35**, 310 (1971).

[9] R. A. Spencer, T. S. Hauschka, D. B. Amos, and B. Ephrussi, *J. Nat. Cancer Inst.* 33, 893 (1964).

[10] F. H. Ruddle, T. R. Chen, T. B. Shows, and S. Silagi, *Exp. Cell Res.* **60**, 139 (1970).

[11] Y. Okada, *Exp. Cell Res.* **26**, 98 (1962).

[12] H. Harris and J. F. Watkins, *Nature* (*London*) **205**, 640 (1965).

[13] R. Davidson, *Exp. Cell Res.* **55**, 424 (1969).

seeded first and allowed to attach on the surface of the culture vessel for 24 hours at 37°C. After removal of the medium the cells are treated with inactivated Sendai virus suspension at 0°C for 10 minutes, rinsed twice with serum-free medium, incubated for a few minutes with the same medium and finally, after its elimination, transferred to 37°C and fed with complete medium or transferred directly to a selective medium (see later). Under these conditions, a hybridization efficiency amounting to 1:25 of seeded cells can be obtained. In a more general way it is important to realize that the hemagglutination titer of the Sendai virus suspension used is only grossly related to its fusing activity. This activity is, on the other hand, less or more pronounced depending on the type of target cells. This makes the whole procedure still quite empirical.

Selective Conditions Promoting Preferential Growth and Isolation of Hybrid Cells in Mixed Cultures

Several artificial systems were devised in order to select and to accumulate hybrid cells in mixed cell populations. The best known is the one devised by Littlefield.[14] It uses cell mutants resistant to 8-azaguanine and lacking hypoxanthine guanine phosphoribosyltransferase and other cell mutants resistant to 5-bromodeoxyuridine and missing thymidine kinase. Hybrids obtained by crossing these two types of mutants are selected by cultivation in a medium containing hypoxanthine, aminopterin, and thymidine (HAT medium) in which the parental mutants are unable to synthesize DNA and to grow, but in which the hybrid cells, carrying, by mutual compensation, the genes for production of both enzymes, proliferate normally. However, it has to be mentioned that in practice the selection of hybrids and, later, the elimination of enzymatically deficient mutants may be difficult in dense monolayers and, to be successful, requires handling of rather sparsely seeded cultures.

Half-selective systems were devised[15] in which only one parent cell lacks one of the two enzymes necessary for growth in the HAT medium. The other partner can be a metabolically normal cell provided that it multiplies slowly or has another growth handicap in the given experimental conditions. This system was successfully applied for obtaining hybrids, e.g., between nonproliferating normal human lymphocytes and permanent mouse cell strains[16] or between intensely proliferating malignant mouse cells and normal Akodon Urichi fibroblasts.[17] Similarly, Harris *et al.*[18] used a half-selective system in which 8-azaguanine-resistant A9 cells growing easily *in vitro* were fused with different ascites tumor cells which grew poorly *in vitro*.

Other selective systems have been proposed and successfully used. The system

[14] J. W. Littlefield, *Science* **145**, 709 (1964).

[15] R. Davidson and B. Ephrussi, *Nature (London)* **205**, 1170 (1965).

[16] M. Nabholz, V. Miggiano, and W. Bodmer, *Nature (London)* **223**, 358 (1969).

[17] J. Belehradek, Jr. and G. Barski, unpublished data, 1972.

[18] H. Harris, O. J. Miller, G. Klein, P. Worst, and T. Tachibama, *Nature (London)* **223**, 363 (1969).

employed by Kao and Puck,[19] involved auxotrophic mutants of Chinese hamster cells, deficient for some amino acids like glycine and proline. These cells grow in a complete medium supplemented with the macromolecular fraction of fetal calf serum but do not grow when the specific auxotrophic requirements are omitted. Hybrid cells obtained following fusion of these auxotrophs, e.g., with normal human fibroblasts, form colonies in the deficient medium and can be easily isolated. Another type of selective system using Chinese hamster cells resistant to amethopterin (1.0 μg/ml) fused with isologous cells resistant to actinomycin D (0.01 μg/ml) was described by Sobel *et al.*[20] Hybrid cells grew in a selective medium containing both amethopterin and actinomycin D which suppressed the growth of both parental cells. In some cases, when resistance to 8-azaguanine could additionally be induced in cells already resistant to actinomycin D, hybrids obtained by crossing these cells with enzymatically normal, actinomycin D-sensitive cells, were easily isolated in media containing the two drugs.[21]

Selection of Tumorigenic Somatic Hybrids and Chromosomal Shifts

If malignant and nonmalignant cells are crossed and if the malignant cell partner is eliminated either spontaneously or with the help of one of the above mentioned selective systems, the inoculation of the mixed culture into a histocompatible host (inbred animal in the case of isologous, and F1 hybrid animal in the case of homologous somatic hybrids) results in a further selection in favor of malignant hybrids. It is clear, however, that this selection eliminates hybrid cells that lack tumor-producing capacity or in which this capacity is expressed to a lesser degree. This is an important point since, as was demonstrated in, among others, some of our homologous mouse somatic hybrids,[22] the tumor-producing capacity expressed in the hybrid clones, is by no means an "all or none" phenomenon. On the contrary, it usually appears as a graded cell property which possibly depends on the equilibrated participation in the hybrids of different genetic (chromosomal) and/or extrachromosomal determinants of somatic cell heredity. This is an alternative to the concept of "one hit" chromosomal deletion as responsible for cell malignancy. In this connection, it becomes important not only to recognize the very fact of heritance of malignant capacity of the somatic hybrids but also to determine the degree of this capacity by inoculation in appropriate hosts of graded numbers of hybrid cells, preferably developed from hybrid clones, and quantitative evaluation of (a) the proportion of takes, (b) the average latency period, (c) the rate of increase of tumor size, and (d) the rate and delay of survival of inoculated animals.[23]

A large body of evidence has been accumulated demonstrating a general tendency in the hybrid cells for chromosomal deletion from the complete karyotype of the "ideal" primary hybrid resulting from fusion of the two parental

[19] F. T. Kao and T. T. Puck, *Genetics* **55**, 513 (1967).

[20] J. S. Sobel, A. M. Albrecht, H. Riehm, and J. L. Biedler, *Cancer Res.* 31, 297 (1971).

[21] M. Berebbi and G. Barski, *C. R. Acad. Sci.* (*Paris*) **272**, 351 (1971).

[22] J. Belehradek, Jr. and G. Barski, *Intern. J. Cancer* **8**, 1 (1971).

[23] G. Barski and F. Cornefert, *J. Nat. Cancer Inst.* **28**, 801 (1962).

cells. The loss occurs following *in vitro*, and, usually more intensely, after *in vivo* passages.[10,18,22] The karyological instability of the hybrids, especially in early passages, is undoubtedly related to frequent mitotic abnormalities. This can be directly demonstrated by comparative counts of abnormal mitoses on fixed and stained preparations of monolayer cultures, untreated with colchicine, of parental and hybrid cells. In some homologous mouse hybrids the abnormal mitoses were found to be at least ten times more frequent that in the parental cell cultures.[22]

According to Harris, Klein, and their group,[18] the chromosomal deletion in the somatic hybrids, obtained by crossing malignant and nonmalignant cells, is essential for the expression of tumorigenic properties in the hybrids. Furthermore, this loss, to be effective, has to involve specific chromosomes coming from the nonmalignant cell and capable of compensating for an original genetic defect of the malignant cell partner.

In some cases, the preferential loss in the hybrids of chromosomes originating from one of the two cell partners can be expected according to data already accumulated. This is the case in some interspecies hybrids like human × mouse, where, in several combinations, the human chromosomes showed the tendency to disappear progressively.[24,25] Similarly, in Chinese hamster × mouse somatic hybrids, the mouse chromosomes were preferentially maintained and even augmented as a result, obviously, of a fusion between two interphase or one dividing mouse cell with one of the other cell partner.[26]

It is evident that in interspecies hybridization experiments the identification of individual chromosomes, as originating from corresponding parental cells, or being interchromosomal hybrids resulting from interspecies translocation, is greatly facilitated, and may be much more so by the use of recently developed techniques of differential chromosome staining.[6-8] An interesting method aimed at a predetermination of preferential loss of chromosomes originating from one partner cell intended for crossing was recently proposed.[27] It consists of X or gamma irradiation (600 rad) of one of the two partner cell populations before fusion. It may result in a directional, though random, elimination in the hybrids of damaged chromosomes from the irradiated parental cell.

Expression of Malignancy in Somatic Hybrids in Absence of a Histocompatible Host

A crucial point in the experimentation dealing with the mechanism of expression of malignant properties in the somatic hybrids is, naturally, a relevant assessment of their tumor-producing capacity. This is, as outlined, relatively simple in mouse isologous or homologous systems but becomes a problem in other systems when an appropriate histocompatible host is lacking.

Indirect methods of recognition of tumorigenic cell properties, according to

[24] M. C. Weiss and H. Green, *Proc. Nat. Acad. Sci. U. S.* **58**, 1104 (1967).
[25] Y. Matsuya and H. Green, *Science* **163**, 697 (1969).
[26] G. Barski, M. G. Blanchard, and J. K. Youn, unpublished data, 1972.
[27] G. Pontecorvo, *Nature (London)* **230**, 367 (1971).

cell morphology, less or more contact-inhibited type of growth, following assays of colony growth in soft agar[28] or agglutinability by phytohemagglutinins like concanavalin A,[29] applied either to parental cells or to the hybrids, have only indicative value and may be misleading.

Direct methods, depending on the production of tumors by inoculation into immunologically deficient hosts, have the advantage of a direct demonstration of tumorigenicity according to the standard histopathological criteria of neoplastic growth.

Two such methods are currently used in our laboratory and have been applied in studies of several intraspecies and interspecies somatic hybrids. These were assays of inoculation either into the cheek pouch of weanling Syrian hamsters or onto the chorioallantoic membrane (CAM) of 10-day-old embryonated chick eggs, according to the artificial air chamber technique.[30] In both cases suspensions of trypsinized cells, containing a standard number of 3×10^6 in 0.1 ml of tissue culture medium, were used as inocula. The hamsters received on the day of inoculation an intramuscular injection of 2.5 mg of cortisone acetate, which was repeated twice a week during an observation period of 60 days, and their cheek pouches were inspected under light ether anesthesia every 3–4 days. Outgrowing nodules showing on histological sections, an evidence of invasive growth, were considered as positive. If desired, the excised nodules could be easily recultured after trypsinization for study of the karyotype and phenotypical properties of the cells composing the tumor. Not infrequently these cells appeared as a selection of variants more tumorigenic than the initial hybrid cell population. Usually no host cells were present in these cultures.

The inoculated embryonated eggs were maintained at 37°C in an humidified incubator for 6 to 7 days, then the CAM's were excised, spread in Petri dishes, and inspected under a stereomicroscope. Nodules measuring at least 2 mm in diameter were considered as positive if the malignant and invasive character of growth was confirmed after examination of the sections according to the usual histological techniques and criteria. Here again, the excised nodules could be easily subcultured *in vitro* for further examination.

In preliminary studies, using several cell lines developed from inbred C3H or C57BL mice and having a known tumorigenic potential, it was demonstrated that there exists a quite satisfactory correlation between the production of tumors in the syngeneic animal and in the two heterologous, immunologically deficient systems we used, though the syngeneic assay system appeared as being, generally, by one or two logs more sensitive.

On the basis of these data the heterologous system of checking of malignant properties was used in the study of Chinese hamster malignant × nonmalignant cell hybrids,[21] as well as in investigations on expression of malignant properties in interspecies hybrids obtained by crossing malignant mouse cells either with nontumorigenic Chinese hamster cell variants or with normal Akodon Urichi

[28] I. Macpherson and L. Montagnier, *Virology* **23**, 291 (1964).

[29] M. Inbar and L. Sachs, *Proc. Nat. Acad. Sci. U. S.* **63**, 1418 (1969).

[30] R. E. Billingham, *In* "Transplantation of Tissues and Cells" (R. P. Billingham and W. S. Silvers, eds.), p. 61. Wistar Inst. Press, Philadelphia, Pennsylvania, 1961.

fibroblasts.[17] In both cases a good concordance was found between the results obtained in the two heterologous checking systems used. In particular, it could be concluded that determinants of malignancy are present and can be expressed in the interspecies hybrids and that in certain combinations of genetic (or epigenetic) ingredients these determinants remain unexpressed or repressed.

In this way a technical possibility is open for an analytical study of interdependence between the different karyotypes of interspecies hybrids—in which the participation of chromosomes from the two parental cells is relatively easy to recognize—and the phenotypical characteristics of these hybrids concerning in particular, their tumorigenic potential defined by standard methods.

CHAPTER 2

Production and Characterization of Proliferating Somatic Cell Hybrids

Richard E. Giles and Frank H. Ruddle

Since the initial observations of mammalian cell fusion *in vitro*[1–5] and the subsequent development of Sendai virus as an experimental tool to facilitate cell fusion at a high frequency[4–6] as well as the development of biochemical selective systems[7] useful for the isolation of fusion products,[8] the phenomena of cell fusion has been utilized in an ever increasing variety of *in vitro* systems as a parasexual process in somatic cell genetics. The demonstration of interspecific cell fusion[3–5,9] and the production of hybrid cells (mononuclear cells containing genomic elements from two or more genetically different parents) capable of indefinite propagation has created new opportunities for the genetic analysis of higher organisms. Cell hybrids have been used to obtain information in such areas as linkage

[1] G. Barski, S. Sorieul, and F. Cornefert, *C. R. Acad. Sci.* (*Paris*) **251**, 1825 (1960).
[2] Y. Okada, *Biken J.* **1**, 103 (1958).
[3] B. Ephrussi and M. C. Weiss, *Proc. Nat. Acad. Sci. U. S.* **53**, 1040 (1965).
[4] H. Harris and J. S. Watkins, *Nature* (*London*) **205**, 640 (1965).
[5] G. Yerganian and M. B. Nell, *Proc. Nat. Acad. Sci. U. S.* **55**, 1066 (1966).
[6] Y. Okada, *Curr. Topics Microbiol. Immunol.* **48**, 102 (1969).
[7] W. Szybalski, E. N. Szybalska, and G. Ragnie, *Nat. Cancer Inst. Monogr.* **7**, 75 (1962).
[8] J. W. Littlefield, *Science* **145**, 709 (1964).
[9] H. Harris, "Cell Fusion." Harvard University Press, Cambridge, Massachusetts, 1970.

analysis[10,11] cytodifferentiation,[12] mitotic regulation,[13] and viral replication.[14,15] The basic procedures involved in producing cell hybrids for research purposes may be conveniently divided into three major areas: (a) techniques for the fusion of parental cell populations (spontaneous or virus-mediated); (b) selective systems and procedures for the elimination of nonhybrid cell types; and (c) characterization of hybrid cells on the basis of such properties as morphology, antigens, isozymes, and chromosomal constitution. The protocol for cell fusion is presented in terms of general steps common to most fusion protocols coupled with a discussion of the procedure. Selective systems useful in the isolation of cell hybrids are described and a number of procedures useful in the characterization of hybrid cells are discussed. Watkins[16] has also recently reviewed methodologies in these areas.

Both spontaneous and Sendai-mediated cell fusion can give rise to hybrid cells. Littlefield[8] hybridized two subclones of strain L by growing them together in suspension culture for 4 days and then plating 5×10^6 cells from the mixture in Petri dishes. Davidson and Ephrussi[17] mixed A9 cells and cells from secondary cultures of skin from newborn CBA mice carrying the T-6 translocation and noted the formation of viable hybrids after cocultivation in monolayer. Interspecific cell hybrids resulting from spontaneous cell fusion were first reported by Ephrussi and Weiss,[3] who mixed LM(TK$^-$) cells with euploid rat embryo cells. Thus, the possibility of obtaining hybrid cells by spontaneous cell fusion is well established, although it is conceivable that such "spontaneous" fusion may be due to an undetected virus.

The fusion of mammalian cells in suspension by Sendai virus has been extensively studied by Okada and his collaborators[6,18-24] utilizing Ehrlich ascites tumor cells (ETC) passaged in mice. From the results of these experiments, Okada[6] has formulated a four-step model to describe Sendai-mediated cell fusion: (a) adsorption and agglutination of the cells by Sendai at 4°C. Cells are frequently treated with virus at 4°C to synchronize the process of adsorption, as subsequent steps in viral replication usually do not proceed until the temperature is raised; (b) Damage to the cell membrane at the site of adsorbed Sendai particles at 37°C.

[10] F. H. Ruddle, *Symp. Intern. Soc. Cell Biol.* **9**, 233 (1970).
[11] F. H. Ruddle, *Advan. Hum. Genet.* **3**, 173 (1972).
[12] R. L. Davidson, *In Vitro* **6**, 411 (1971).
[13] P. N. Rao and R. T. Johnson, *Nature* (*London*) **225**, 159 (1970).
[14] H. Koprowski, *Fed. Proc.* **30**, 914 (1971).
[15] H. Green, R. Wang, C. Basilico, R. Pollock, T. Kusano, and J. Salas, *Fed. Proc.* **30**, 930 (1971).
[16] J. F. Watkins, *In* "Methods in Virology" (K. Maramorosch and H. Koprowski, eds.), Vol. 5, p. 1. Academic Press, New York, 1971.
[17] R. L. Davidson and B. Ephrussi, *Nature* (*London*) **205**, 1170 (1965).
[18] Y. Okada, F. Murayama, and K. Yamada, *Virology* **27**, 115 (1966).
[19] Y. Okada, *Biken J.* **4**, 209 (1961).
[20] Y. Okada and F. Murayama, *Exp. Cell Res.* **52**, 34 (1968).
[21] Y. Okada and J. Tadokoro, *Exp. Cell Res.* **26**, 108 (1962).
[22] Y. Okada, *Exp. Cell Res.* **26**, 119 (1962).
[23] Y. Okada and J. Tadokoro, *Exp. Cell Res.* **32**, 417 (1963).
[24] Y. Okada, K. Yamada, and J. Tadokoro, *Virology* **22**, 397 (1964).

Hosaka and Koshi[25] studied Sendai fusion of ETC using the electron microscope and noted that the absence of intimate contact between cell membranes of adjacent cells in step (a) was followed by the formation of intimate cell membrane associations, particularly near virus particles, after 1 minute at 37°C; (c) Repair of the damaged sites, which is energy requiring and which may result in the connection of the cell membranes of adjacent cells with opposed sites of damage. Hosaka and Koshi[25] noted the formation of cytoplasmic bridges between cells as early as 1 minute after transfer to 37°C, many of which appeared to enlarge upon further incubation; and (d) Spherical giant cell formation, which also requires energy. For a diagrammatic representation of steps (a)–(c) see Fig. 1.

Okada fused the ETC cells by mixing an aliquot of cell suspension, generally containing 10^7 cells in 1 ml, with an equal volume of Sendai virus, at various concentrations of hemagglutinating units (HAU)/ml, and allowed agglutination to take place at 4°C for 10 to 60 minutes. The agglutinated cells were then incubated at 37°C with gentle shaking for 60 minutes. The extent of cell fusion was evaluated by calculating the fusion index (FI). Using this system, Okada and his associates[6] described several parameters of the cell fusion reaction, such as: (a) an optimum pH range of 7.3–7.6; (b) a requirement for energy; (c) a requirement for calcium ions[26]; and (d) temperature dependence. Croce *et al.*[27] have studied the effect of pH on Sendai-induced cell fusion and the recovery of hybrid cells from mouse × human hybridization experiments. These investigators

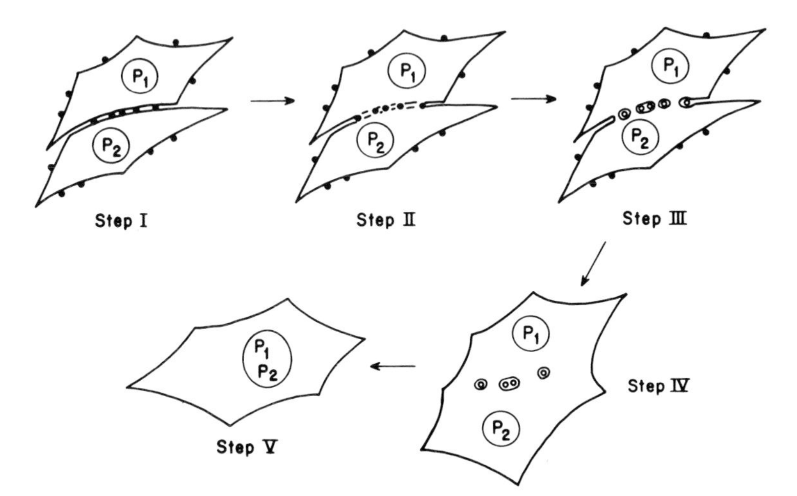

Fig. 1. Sendai-mediated cell fusion. Step I. Adsorption of virus and agglutination of cells at 4°C. Step II. Damage to the membrane at the site of adsorbed Sendai particles at 37°C. Step III. Repair of damaged membrane sites and the formation of cytoplasmic bridges between adjacent cells. Step IV. Enlargement of the cytoplasmic bridges and heterokaryon formation. Step V. Hybrid cell formation by the reconstitution of parental genomes in a single nucleus.

[25] Y. Hosaka and Y. Koshi, *Virology* **34**, 419 (1968).
[26] Y. Okada and F. Murayama, *Exp. Cell Res.* **44**, 527 (1966).
[27] C. M. Croce, H. Koprowski, and E. Eagle, *Proc. Nat. Acad. Sci. U. S.* **69**, 1953 (1972).

found an optimum pH range of 7.6 to 8.0 for both processes, although the optimum pH for growth of both parental and hybrid cells was 7.3.

Three protocols for cell fusion are described below. The first is for spontaneous fusion by cocultivation in monolayer, while the second and third protocols describe Sendai-mediated cell fusion in suspension and monolayer systems, respectively. Finally, comments on useful variations of the fusion protocols are included. It has been reported[28,29] that lysolecithin may be used as a means of inducing cell fusion. The original reports should be consulted for cell fusion methods employing this agent. Aseptic technique is used throughout.

Spontaneous Cell Fusion in Monolayer (Protocol I)

1. The parental cell lines are harvested by routine procedures, the cell number determined, and aliquots of each parent mixed together to give a total cell concentration of 5 to 10 × 10^5 cells/milliliter in growth medium at any desired ratio, e.g., 1:1–1:10,000.

2. Falcon[30] flasks No. 3012 or Falcon tissue culture dishes No. 3002 are inoculated with 5 ml of the cell suspension in growth medium and incubated for 24 to 48 hours at 37°C.

3. The growth medium is replaced with selective medium after the 24- to 48-hour incubation period. Colonies should appear after 1 to 3 weeks of further incubation.

Sendai-Mediated Cell Fusion in Suspension (Protocol II)

1. The parental cell lines are harvested by routine procedures, the cell number determined, and aliquots of each parent mixed together to give a total cell concentration of approximately 5 × 10^6 cells in 1 ml of growth medium at any desired parental cell ratio.

2. One milliliter of the cell suspension is chilled at 4°C for 5 minutes in a 25 ml screw-cap Erlenmeyer flask. One milliliter of inactivated Sendai virus[31] (1000 HAU/ml) is also chilled.

3. One milliliter of the Sendai virus (1000 HAU/ml) is added to the cell suspension in the Erlenmeyer flask. The virus is allowed to absorb for 30 minutes at 4°C with intermittent agitation. Clumps of agglutinated cells should be evident after this incubation.

4. The cell suspension is incubated on a shaker at 60–100 rpm for 30 to 60 minutes at 37°C.

5. The cell suspension is diluted to 5 to 10 × 10^4 cells/ml with growth medium and 5-ml aliquots plated in Falcon flasks No. 3012 or dishes No. 3002. Selective medium may be used at this stage or after 24 to 48 hours.

[28] A. R. Poole, J. I. Howell, and J. A. Lucy, *Nature (London)* **227**, 810 (1970).
[29] C. M. Croce, W. Sawicki, and D. Kritchevsky, *Exp. Cell Res.* **67**, 427 (1971).
[30] Falcon Plastics, Division of BioQuest, 5500 West 83 Street, Los Angeles, California 90045.
[31] R. J. Klebe, T. Chen, and F. H. Ruddle, *J. Cell Biol.* **45**, 74 (1970).

Sendai-Mediated Cell Fusion in Monolayer (Protocol III)

1. The parental cell lines are harvested by routine procedures, the cell number determined, and aliquots of each parent mixed together to give a total cell concentration of approximately 5 to 10 × 10^5 cells/milliliter in growth medium at any desired ratio.
2. Small Falcon flasks, No. 3012, are seeded with 5 ml of the cell suspension and incubated for 12 to 24 hours to allow the cells to attach and flatten.
3. The growth medium is removed and the cell monolayer is washed once with growth medium.
4. The cells and virus are chilled at 4°C for 5 minutes.
5. One milliliter of inactivated Sendai virus[31] (1000 HAU/ml) is added to the Falcon flask and the flask is incubated at 4°C for 30 minutes for virus adsorption. The flask is tilted from side to side intermittently to distribute the virus suspension over the monolayer.
6. Two milliliters of growth medium at 37°C are added and the flask incubated at 37°C for 60 minutes.
7. All fluid is removed and the cell monolayer is washed twice with growth medium.
8. The cells are harvested as before, diluted with growth medium to 5 to 10 × 10^4 cells/milliliter, and planted in small Falcon flasks, No. 3012, or 60-mm tissue culture dishes, No. 3002, at 5 ml per dish. Selective medium may be used instead of growth medium at this time or applied after 24 to 48 hours incubation.

Procedural Notes

Fusion Potential of Parental Cells. Several different cell types were tested by Okada and Tadokoro[23] to determine the efficiency of Sendai-mediated cell fusion of cells of markedly different characteristics. Mouse, human, and rabbit leukocytes, a variety of mouse ascites tumors, and several tissue culture cell strains and lines were used. Normal leukocytes from man, mouse, and the rabbit did not appear to fuse, though fusion was observed when cells of a leukemic patient were employed. The continuous tissue culture lines fused readily and gave a higher FI at a lower virus concentration than the mouse ascites tumor lines, which also showed substantial cell fusion. All the cell types agglutinated at 4°C in the presence of Sendai virus independent of subsequent fusion after incubation at 37°C. Cells from secondary cultures of chick embryos, mouse embryos, bovine embryo kidney, and monkey kidney showed a very low capacity to fuse in the presence of Sendai virus. Despite the failure of normal human leukocytes to fuse with themselves under the conditions used by Okada and Tadokoro,[23] peripheral blood leukocytes have been shown to fuse with mouse cells in a system described by Miggiano *et al.*[32]

[32] V. Miggiano, M. Nabholz, and W. Bodmer, *Wistar Inst. Monogr.* **9**, 61 (1969).

Selection of Parental Cultures. Klebe *et al.*[31] have suggested using stationary phase cells to initiate cell fusion experiments to avoid possible chromosome pulverization, which can result from the fusion of cells at different positions in the cell cycle.[33,34] Stadler and Adelberg[35] found an increase in the fusion of the mouse leukemic cell line L-5178Y when cells were exposed to Colcemid and mitotic cells accumulated in the population prior to treatment with Sendai virus. Treatment of the L5198Y cell with trypsin was also noted to increase the frequency of cell fusion.[35]

Harvesting Procedures. Cell monolayers may be harvested with trypsin[36] (see also Section III) or Viokase.[37] Using Viokase solution[37] it is convenient to harvest in the following manner: (a) The growth medium is removed and the monolayer washed with Viokase solution at one-half the volume of growth medium; (b) A second aliquot of Viokase solution which is equal in size to the volume used to wash the monolayer is allowed to remain in contact with the monolayer for 30 seconds. It is then removed and discarded; (c) The cells are incubated for 5 to 10 minutes at 37°C. It is very important to adjust carefully the time of incubation for each cell line, since excessive incubation will result in extensive cell damage and a loss of viability; (d) The flask is rapped sharply against the heel of the palm or against the edge of a table to dislodge the cells; and (e) The cells are resuspended in balanced salt solution or growth medium by pipetting, and the cell number determined[36] (see also Section VIII this volume).

Parental Cell Density and Cell Ratio. For cell fusion in monolayer by Sendai virus or by spontaneous cell fusion, it is necessary to have the cell mixture at a density sufficiently high to ensure intimate cell contact. Consequently, it is better to have too many rather than too few cells in the initial culture. It may be useful to apply the following method to establish the number of cells of each parent to be used in a monolayer fusion protocol. The density of each parental line at confluency is determined and the mixed culture initiated with each parent present at one-half its confluent density. This procedure adjusts for variation in cell size, spreading, etc. If it is desired to use greatly unequal parental cell ratios, then the predominant parent may be plated at a density just below its confluent density. In the absence of appropriate biochemical markers, it may be useful to employ a combined monolayer–suspension system. Utilizing this approach, the monolayer parent should be planted at one-third its confluent density to minimize homokaryon formation, and the other parent introduced as a cell suspension at or above its confluent density.

When Klebe *et al.*[31] crossed RAG and LM(TK^-) mouse cells at a constant

[33] N. Takagi, T. Aya, H. Kato, and A. A. Sandberg, *J. Nat. Cancer Inst.* **43**, 335 (1969).

[34] R. T. Johnson and P. N. Rao, *Nature* (*London*) **226**, 717 (1970).

[35] J. K. Stadler and E. Adelberg, *Proc. Nat. Acad. Sci. U. S.* **69**, 1929 (1972).

[36] D. J. Merchant, R. H. Kahn, and W. H. Murphy, Jr., "Handbook of Cell and Organ Culture." Burgess, Minneapolis, Minnesota, 1964.

[37] Ten × Viokase (GIBCO cat. 517) is diluted with Ca- and Mg-free 1 × PBS (GIBCO cat. 419). Disodium ethylenediaminetetraacetic acid (EDTA) may be added at 0.2 g/1000 ml.

cell density and virus concentration, they found the greatest yield of hybrid clones occurred at a parental cell ratio of 1:1, a result which might be expected for cells of equal size based on the binomial distribution. Other investigators[38-40] have observed enhanced hybrid recovery at unequal parental cell ratios (1:10 through 1:10,000) for both spontaneous and Sendai-mediated cell fusion. Davidson and Ephrussi[39] observed a mating rate of greater than 1 hybrid/100 T-6 cells, when they spontaneously fused T-6 diploid CBA mouse cells with A9 cells at a ratio of one T-6 cell/10,000 A9 cells. The reasons for increased hybrid formation at unequal parental cell ratios, under certain conditions, are not known, but may be partially explained on the basis of preferential adhesion or sorting out and/or the preferential fusion of like cells. Mukherjee *et al.*[41] have demonstrated the preferential fusion of like cells by comparing the frequency of heterokaryons to homokaryons using Sendai-induced fusion of: (a) human fibroblasts × A9; (b) human fibroblasts × RAG; and (c) RAG × chick fibroblasts. A deficiency of heterokaryons was noted for each cross.

It has been shown by Ricciuti and Ruddle[42] that hybrid cells may be formed by the fusion of three parental cells. In view of this observation, it is likely that the number of parental genomes initially present in hybrid cells may be partially dependent on the density and ratio of parental cells. The number and ratio of parental genomes present in hybrids may control some aspects of the hybrid cell phenotype. An example of this situation has been described by Fougère *et al.*,[43] who have suggested that pigment synthesis in mouse × hamster melanoma cell hybrids is partially dependent on the dosage of melanoma genes. Consequently, it may be desirable to manipulate the parental cell densities and ratio to favor the formation of hybrids composed of parental genomes at different multiplicities.

Cell Fusion. The fusion reaction may be carried out in growth medium containing 10% serum,[31] although Davidson[44] has suggested that the presence of serum during the 37°C incubation period reduces the final yield of hybrid cells at low virus concentrations. If a particular cell line proves difficult to fuse, serum should be eliminated from the fusion mixture. For cells sensitive to damage by Sendai virus, the presence of serum may be desirable as a protective agent, or unadsorbed virus may be removed prior to the 37°C incubation.[44]

A number of variations are possible in the virus adsorption step in protocols II and III or in a monolayer–suspension system. One parent may be preferentially exposed to virus in an attempt to enhance heterokaryon formation when one of the parents shows a strong tendency to agglutinate with itself or has a low level of fusion. For example, if it desired to cross a cell line which shows a strong

[38] R. L. Davidson, B. Ephrussi, and K. Yamamoto, *Proc. Nat. Acad. Sci. U. S.* **56**, 1437 (1966).

[39] R. Davidson and B. Ephrussi, *Exp. Cell Res.* **61**, 222 (1970).

[40] H. G. Coon and M. C. Weiss, *Proc. Nat. Acad. Sci. U. S.* **62**, 852 (1969).

[41] A. B. Mukherjee, V. G. Dev, and O. J. Miller, *Exp. Cell Res.* **68**, 130 (1971).

[42] F. Ricciuti and F. H. Ruddle, *Science* **172**, 470 (1971).

[43] C. Fougère, F. Ruiz, and B. Ephrussi, *Proc. Nat. Acad. Sci. U. S.* **69**, 330 (1972).

[44] R. Davidson, *Exp. Cell Res.* **55**, 424 (1969).

tendency to self-agglutinate with a cell line of low fusing potential, the former line could be plated at one-third its confluent density, incubated 24 hours, and then treated with Sendai virus at 4°C. After removal of excess virus, the second parent could be added at a high cell density. Presumably, this treatment should favor hybrid formation by enhancing the agglutination of unlike cells at 4°C. Such a variation should also be useful for the fusion of two cell lines which both show a strong tendency to form homokaryons. There appears to be a substantial amount of flexibility in the amount of time which can be utilized for the adsorption of virus at 4°C (10–60 minutes); consequently, modifications of the protocol at this step are probably unlikely to have severe deleterious effects.

Postfusion Incubation. During the period of incubation in selective medium it is advisable to change the medium every 24 hours during the first few days to avoid nonspecific cell death as a result of toxicity of cell breakdown products. This rule is also applicable to feeding schedules of cell populations undergoing any selection characterized by substantial cell death or utilizing a labile selective agent. Temperature and pH deviations from optimum conditions must be avoided during this period due to the marked sensitivity of cells at a low population density to variations in these two factors. A 50% medium change may be preferable to a complete medium change.

When biochemically marked parental cells are used to selectively isolate hybrid cells, it is of prime importance to determine the frequency of revertants in parental cell populations carrying biochemical markers, under the conditions used to select hybrid cells. This can be accomplished by plating each parent separately in selective medium. If the combined frequency of revertants in the parental cell population is at the same level as the frequency of hybrid cell formation, it will be difficult to isolate hybrids efficiently and have confidence in the retention of specific genetic markers in hybrid cells. Some lines carrying specific biochemical markers such as A9 and LM(TK^-) have failed to show revertants in the hands of a number of different investigators. Nevertheless, such lines should be tested for the presence of revertants for two reasons: (a) Cell lines derived from a common source may show a substantial divergence in phenotype when propagated in different laboratories under different conditions over a period of time; and (b) substantial growth of individual parental lines in selective medium is a good indication of an error in medium preparation.

When a series of hybrid clones is analyzed it is often important to examine clones of independent origin.[10] If only one clone is picked from a single culture flask or dish (protocol I, step 2; protocol II, step 5; and protocol III, step 8), this series of clones will be of independent origin because each clone arose from a different fusion event.

Selective Systems

Systems used to select hybrid cell lines from mixed populations of fused cells and parental lines may be separated into two general categories: (a) biochemical methods; and (b) methods based on growth characteristics and morphology. Combinations of the two methods are frequently used.

Production of HGPRT⁻ Cells. Szybalski *et al.*[7] developed a system for the isolation of cell lines lacking hypoxanthine-guanine phosphoribosyltransferase (HGPRT, E.C. 2.4.2.8), which catalyzes the transfer of ribose-5-phosphate to either hypoxanthine or guanine (Fig. 2). These investigators utilized purine analogs [8-azaguanine (8AG), 8-azaguanosine (8AGR), and 8-azahypoxanthine (8AH)] to select drug-resistant cell lines. Conversion of the purine analogs to analog intermediates of the purine salvage pathway by HGPRT results in the death of the cell. Drug-resistant cells are able to proliferate in the presence of the purine analogs and may be isolated by growing the sensitive cell population at increasing drug concentrations. In the case of 8AG treatment, Szybalski *et al.*[7] noted a two-step survival curve when they plotted the percentage survival against drug concentration. A cell line resistant to low levels of 8AG (6 μg/ml) did not show a marked decrease in HGPRT activity but appeared to be resistant to the drug as a result of reduced uptake 8AG as measured by [^{14}C]8AG incorporation. Littlefield[8] has reported the isolation of a HGPRT⁻L cell line (A3-1) by selection with 8AG at a level of 3 μg/ml. Since there appears to be a possibility of selecting at least two different types of mutants resistant to 8AG over the concentration range of 1 to 10 μg/ml, it is necessary to show reduced enzyme activity by a direct biochemical assay if deficiency mutants at the HGPRT locus are being sought. It is also advisable to use 8AG at a relatively high concentration in the final selection treatment (e.g., 40 μg/ml). Another approach is to use a combination of purine analogs sequentially to select HGPRT-deficient cells, for example, 8AH at 2 to 500 μg/ml,[45] or 6-thioguanine at 5 to 10 μg/ml.[46] HGPRT⁻ cell lines may also be isolated from patients exhibiting the Lesch-Nyhan syndrome, an X-linked recessive genetic disorder with severe clinical manifestations characterized by a deficiency in HGPRT activity in all cells. The HGPRT deficiency is a stable property of such cells in tissue culture.

HAT Medium. If *de novo* purine synthesis is inhibited with aminopterin (4×10^{-7} *M*)[8] or amethopterin (50 μg/ml[45] or 10^{-5} *M*[46]), which inhibit the enzyme dihydrofolate reductase (E.C. 1.5.1.3), the cells become dependent on exogenous sources of purines and pyrimidines. Cells possessing normal HGPRT activity are able to use hypoxanthine as an exogenous source of purines in the presence of aminopterin. Glycine must be supplied either in the medium or as an additive since inhibition of dihydrofolate reductase also blocks the formation of glycine from serine. Cells possessing normal thymidine kinase activity can use thymidine as an exogenous source of pyrimidines. Littlefield[8] demonstrated the feasibility of making use of the HAT system[7] (hypoxanthine, aminopterin, thymidine) for the isolation of hybrid cells by the selective biochemical elimination of parental cells

[45] J. S. Sobel, A. M. Albrecht, H. Biehm, and J. L. Biedler, *Cancer Res.* **31**, 297 (1971).
[46] K. Sato, R. S. Slesinski, and J. W. Littlefield, *Proc. Nat. Acad. Sci. U. S.* **69**, 1244 (1972).

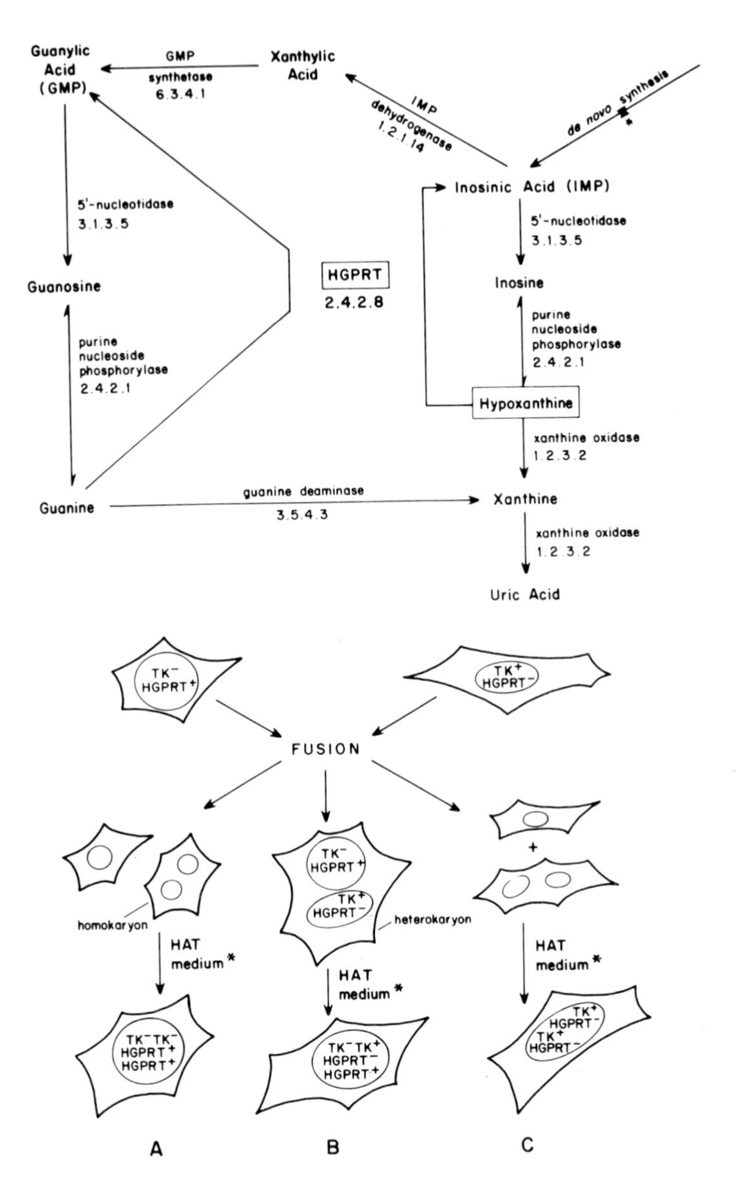

Fig. 2. Purine interconversions and HAT selection. *In the presence of aminopterin (A) or amethopterin the *de novo* synthesis of purines and pyrimidines is blocked, making the cell dependent on exogenous sources of purines (hypoxanthine, H) and pyrimidines (thymidine, T) and HGPRT activity. (A) Homokaryons and parental cells die because they cannot use exogenous thymidine to form pyrimidines. (B) Hybrids survive because they can use both hypoxanthine and thymidine. (C) Homokaryons and parental cells die because they cannot use exogenous hypoxanthine to form purines.

under conditions which permit the growth of the hybrids but not the parents. The concentrations of some of the components of the HAT medium described by Szybalski *et al.*[7] have been altered by other investigators for use in the selection of hybrid cells from mixed populations (see Table I).

TABLE I
Components of HAT[a,b] Medium

Additive	Molecular weight	Molarity	Amount (μg/ml)
Hypoxanthine	136.1	3.67×10^{-5}	5.0[a]
Hypoxanthine		1.00×10^{-4}	13.61[c]
Hypoxanthine		1.00×10^{-4}	13.61[d]
Aminopterin	440.4	4.54×10^{-8}	0.02[a,e]
Aminopterin		4.00×10^{-7}	0.176[c]
Aminopterin		1.00×10^{-5}	4.40[d]
Thymidine	242.2	2.06×10^{-5}	5.0[a]
Thymidine		1.60×10^{-5}	3.88[c]
Thymidine		4.0×10^{-5}	9.69[d]
Glycine	75.1	3×10^{-6}	0.225[c]
Glycine		1×10^{-5}	0.751[d]

[a] HAT designation originated by W. Szybalski, E. H. Szybalska, and G. Ragnie.[7]
[b] It is recommended that the concentrations of HAT components described below in either footnote c or d be used for the isolation of hybrid cells.
[c] J. W. Littlefield.[8]
[d] S. Marin and J. W. Littlefield, *J. Virol.* **2**, 69 (1968).
[e] In the system described by Szybalski *et al.*,[a] this was twenty times the 50% inhibitory level.

Hybrid Selection Using HAT Medium. In order to achieve selection against both parental cell types, Littlefield[8] employed one parent lacking HGPRT activity and another parent lacking thymidine kinase (TK) activity. The absence of either HGPRT activity or TK activity is a lethal cell defect in the presence of aminopterin. Parental cells and homokaryons die in selective medium because of their conditionally lethal biochemical deficiencies. Hybrid cells survive in selective medium because each parent contributes either an HGPRT$^+$ or TK$^+$ phenotype to the hybrid cell, enabling the hybrid to utilize the exogenous purine and pyrimidine sources provided in the selective medium (Fig. 2).

Davidson and Ephrussi[39] noted that the transfer of cells grown on HAT to HAT-free medium resulted in a cessation of growth and the onset of cell degeneration. This cell death could be prevented if cells were transferred to medium supplemented with hypoxanthine and thymidine as an intermediate step. Davidson and Ephrussi[39] pointed out that it might be feasible to select against one parent in a hybridization experiment by pretreating this parent with HAT medium and planting a mixture of both parents in *normal* growth medium. Presumably, the pretreated parent and its homokaryons will die while the hybrids and unpretreated parent will survive.

Hybrid Selection Using Alanosine–Adenine Medium. Adenosine-5′-phosphate (AMP) is synthesized endogenously by the conversion of inosine-5′-phosphate (IMP) to AMP. This reaction is inhibited by the antibiotic alanosine.[47,48] Kusano

[47] G. R. Gale and G. B. Schmidt, *Biochem. Pharmacol.* **17**, 363 (1968).
[48] G. R. Gale, W. E. Ostrander, and L. M. Atkins, *Biochem. Pharmacol.* **17**, 1828 (1968).

et al.[49] have used alanosine to make cells conditionally dependent on adenine phosphoribosyltransferase (APRT, E.C. 2.4.2.7) activity. APRT activity enables the cell to convert exogenous adenine (A) to AMP. Thus, when *de novo* synthesis of adenine is blocked by alanosine, the cell is dependent on APRT activity and exogenous adenine. Alanosine–adenine medium (AA, each component at 5×10^{-5} *M*) was used by Kusano *et al.*[49] to select hybrid cells produced by the fusion of a mouse cell lacking APRT activity with a normal human diploid cell (Fig. 3). The $APRT^-$ mouse cells are killed by AA medium. The human diploid cells grow slowly in AA medium. Hybrid cells survive and grow rapidly because they have APRT activity derived from the human parent and possess the rapid growth rate of wild-type mouse cells in AA medium. The $APRT^-$ mouse line was produced by a two-step selection procedure: (a) 3T6 mouse cells were treated with 2,6-diaminopurine (DAP) over a 4-month period at concentrations ranging from 2.5 to 100 μg/ml; and (b) 3T6 cells resistant to 100 μg/ml DAP were treated with 2-fluoroadenine (2FA) over a 1-month period at concentrations ranging from 0.1 to 25 μg/ml. The 2FA selection was necessary because 3T6 cells from step (a) were not killed in AA medium. A subclone from the step (b) cell populations, sensitive to AA medium, was designated 3T6-DF8, and used in the hybridation described by Kusano *et al.*[49] In considering the utility of the AA selective system, it should be noted that Kusano *et al.*[49] observed decreased effectiveness of drug selection against 3T6-DF8 at cells densities greater than $5 \times 10^3/\text{cm}^2$.

Hybrid Selection Using Ouabain Resistance. The recent development of ouabain-resistant cell lines[49a,b] provides a drug selection system which promises to be very useful in the selection of hybrid cells, particularly mouse × human hybrid cells. Ouabain resistance appears to behave as a codominant trait when hybrid cells produced from a sensitive and a resistant parent are exposed to levels of ouabain lethal for the sensitive parent.[49c,d] Hybrid resistance to ouabain appears to be intermediate with respect to parental sensitivities and may be skewed toward either the sensitive or resistant parent.[49c] When a doubly mutant parental cell line is available which is resistant to both ouabain and 8AG(HGPRT-deficient), it should be possible to use drug selection to eliminate *both* parents in a hybridization between the doubly marked parental cell and any other unmarked parental cell showing normal sensitivity to ouabain by using HAT medium containing ouabain at a level sufficient to kill the sensitive parent. Thompson and Baker[49b] have noted that human cells are killed at very low concentrations of ouabain (10^{-7} *M*) while the lethal dose for mouse cells is substantially higher (10^{-3} *M*). Consequently, for the isolation of hybrid cells from mouse × human cell fusions, it may not be necessary to select for ouabain resistance in the mouse cells. There are several advantages in effecting fusions

[49] J. Kusano, C. Long, and H. Green, *Proc. Nat. Acad. Sci. U. S.* **68**, 82 (1971).

[49a] E. Mayhew, *J. Cell. Physiol.* **79**, 441 (1972).

[49b] L. H. Thompson and R. M. Baker, *In* "Methods in Cell Physiology" (D. M. Prescott, ed.), Vol. 6. Academic Press, New York, in press.

[49c] R. M. Baker, personal communication.

[49d] R. Mankovitz, as cited in footnote 49b.

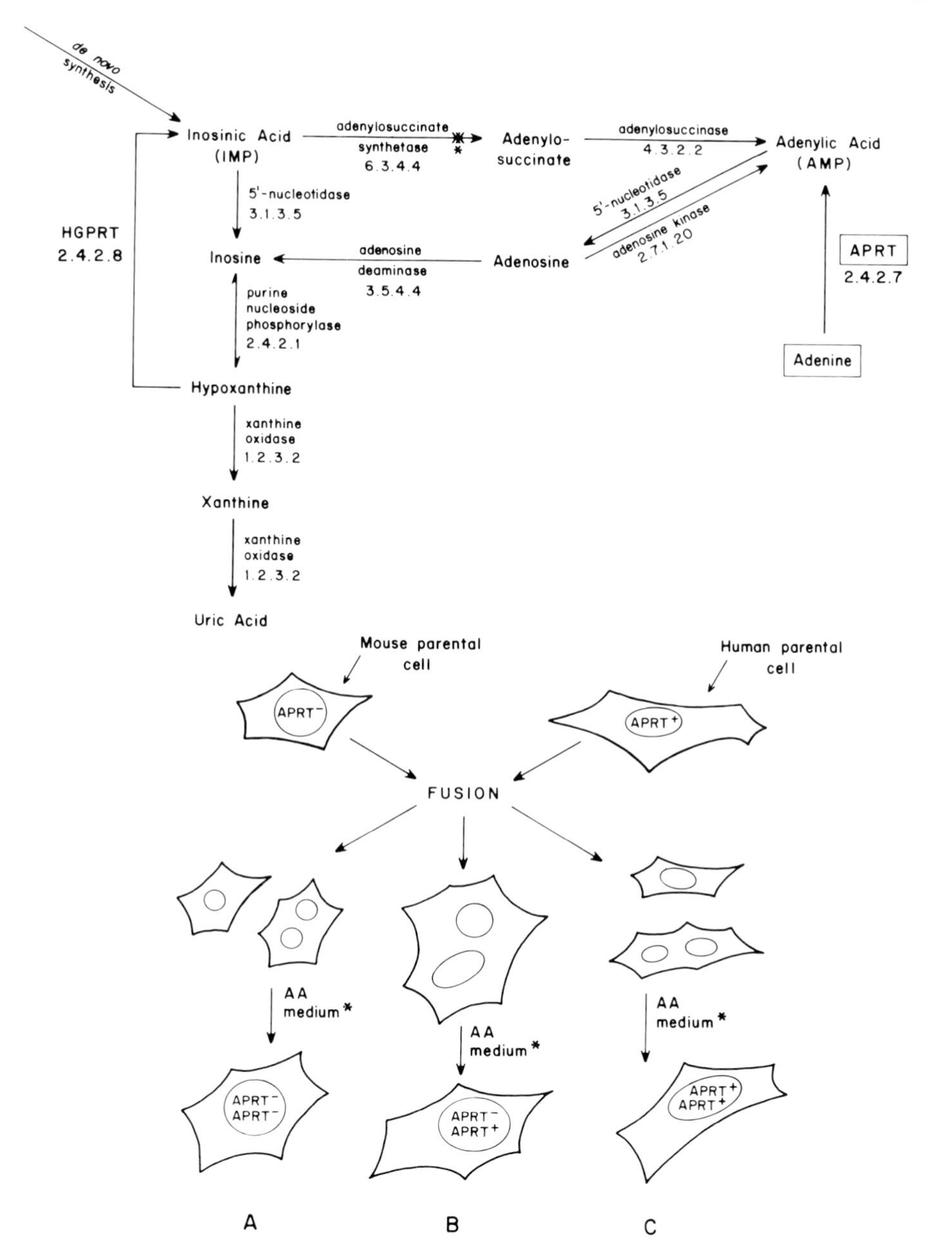

Fig. 3. Purine interconversions and AA selection. *In the presence of alanosine (A) the *de novo* synthesis of AMP is blocked, making the cell dependent on exogenous adenine (A) and APRT activity. (A) Homokaryons and parental cells die because they cannot use exogenous adenine to form AMP. (B) Hybrids survive because they can use exogenous adenine. Hybrids also grow rapidly in AA medium. (C) Parental human diploid cells and homokaryons grow slowly in AA medium and are rapidly overgrown by the hybrid cells.

between a drug-resistant cell line and an unselected parental cell, such as (a) differentiated or recently explanted cells may lose important phenotypic markers during drug selection; (b) selection and characterization of drug-resistant mutants are expensive and time-consuming processes; and (c) an increasing number of limited life span human fibroblast lines with various genetic markers are becoming available which do not have drug resistance markers and may not be capable of surviving *in vitro* for a sufficient number of cell generations to produce useful drug resistant cell lines.

Hybrid Selection Using Auxotrophic Mutants. The development of procedures for producing auxotrophic mutants in mammalian cells by Kao and Puck[50] has provided several biochemically marked cell lines. Auxotrophic mutants promise to be of great use in somatic cell genetics[51-54] by virtue of the ability to preferentially select hybrids and eliminate parental cells on the basis of nutritional requirements. The system which Kao and Puck[50] developed was based on the initial observation made by Stahl *et al.*[55] that bacteriophage DNA (T2) could be sensitized to "visible light" by growing phage in the presence of 5-bromodeoxyuridine (BUdR). (Visible light produced from 40 watt General Electric "Cool White" fluorescent lamps.[50]) Puck and Kao[56] were able to demonstrate that mammalian cells could be sensitized to visible light by growth in the presence of BUdR. They also demonstrated the feasibility of using the technique to specifically select auxotrophic cell lines by preferential killing of prototrophic cells under nutritionally restrictive conditions. This system[56] consists of the following basic steps listed below.

1. A cell population containing both prototrophs and auxotrophs, grown on "complete" or "enriched" growth medium, is transferred to "minimal" medium for a period of time sufficient to deplete the auxotrophs of the metabolites absent from the minimal medium.
2. BUdR is added to the minimal medium after the preliminary incubation and is incorporated into the DNA of prototrophic cells, which continue to grow despite the absence of nutrilites. Auxotrophs fail to incorporate lethal amounts of BUdR because of their inability to grow or metabolize at a high rate in minimal medium.
3. BUdR is removed after one to two cell generations.
4. The cells are irradiated with visible light[50] for 30 minutes to kill prototrophic cells, which have incorporated BUdR, and are then transferred to complete medium.
5. Colonies which appear after 1 to 2 weeks of incubation are isolated by

[50] F. T. Kao and T. T. Puck, *Proc. Nat. Acad. Sci. U. S.* **60**, 1275 (1968).
[51] F. T. Kao, R. T. Johnson, and T. T. Puck, *Nature (London)* **164**, 312 (1969).
[52] F. T. Kao, L. Chasin, and T. T. Puck, *Proc. Nat. Acad. Sci. U. S.* **64**, 1284 (1969).
[53] F. T. Kao and T. T. Puck, *Nature (London)* **228**, 329 (1970).
[54] T. T. Puck, P. Wuthier, C. Jones, and F. T. Kao, *Proc. Nat. Acad. Sci. U. S.* **68**, 3102 (1971).
[55] F. W. Stahl, J. M. Crasemann, L. Okun, E. Fox, and C. Laird, *Virology* **13**, 98 (1961).
[56] T. T. Puck and F. T. Kao, *Proc. Nat. Acad. Sci. U. S.* **58**, 1227 (1967).

standard cloning methods and tested for their specific nutritional requirements. In order to isolate auxotrophs by this method, it was found necessary to treat the wild-type cell population with either *N*-methyl-*N*-nitro-*N*-nitrosoguanidine (MNNG), 0.5 μg/ml for 16 hours, or ethylmethane sulfonate (EMS), 200 μg/ml for 16 hours.[56] The majority of mutants isolated by Puck and Kao[56] were auxotrophic for glycine, while others were found to require glycine + thymidine + hypoxanthine, or inositol.

Hybrids were produced between various Chinese hamster ovary (CHO) lines auxotrophic for the same or different metabolites[51,52] and between CHO auxotrophs and normal human diploid cells.[53,54] Selective elimination of parental cells and homokaryons was achieved by selection in minimal medium. Some hybridization experiments between different glycine auxotrophs did not yield hybrids, presumably due to the failure of these mutants to complement.[52] Noncomplementation between auxotrophs mutant at different loci could occur if one or both of the mutant loci behaved as a dominant. In the case of the CHO auxotroph × human diploid fibroblast, only the Chinese hamster parent could be eliminated by a specific nutritional deficiency. The human parent was eliminated by a failure to form colonies when plated in a medium supplemented with the macromolecular fraction of fetal calf serum.[53]

Miscellaneous Biochemical Selective Systems. Human cells carrying an X-linked deficiency for glucose-6-phosphate dehydrogenase (E.C. 1.1.1.49) may be useful as a selectable parent in the production of cell hybrids according to Siniscalco.[57] The enzyme galactose-1-phosphate uridyltransferase (E.C. 2.7.7.12) deficient in human cells from patients with galactosemia, may also provide a selective lever to eliminate parental cells carrying this lesion by virtue of the inability of such cells to utilize galactose.[58]

Hybrid Selection Using Growth Characteristics. Aspects of cell phenotype, such as growth rate, contact inhibition, adhesion to glass or plastic, and cell morphology have been used to isolate hybrids in many instances when selective elimination of both parents is not possible on the basis of drug toxicity. Growth rate and/or contact inhibition (CI) are often useful characteristics to consider in fusions involving a rapidly growing biochemically marked cell line and normal human diploid fibroblasts. Leukocytes from peripheral blood or from permanent cell lines established from peripheral blood cultures do not adhere to either glass or plastic. These cells are an ideal parent for hybridization experiments and have been shown to be capable of fusion with other cells via Sendai virus.[32,59] Cells taken directly from peripheral blood do not proliferate *in vitro,* while cells from established leukocyte lines may be readily removed from a suspension–monolayer fusion system by frequent medium

[57] M. Siniscalco, *Symp. Intern. Soc. Cell Biol.* **9**, 205 (1970).

[58] H. L. Nadler, C. M. Chocko, and M. Rachmeler, *Proc. Nat. Acad. Sci. U. S.* **67**, 976 (1970).

[59] M. Nabholz, V. Miggiano, and W. Bodmer, *Nature (London)* **223**, 358 (1969).

changes at 24-hour intervals for 5 to 10 days. Less satisfactory than either of the above methods is gross cellular morphology.[31,60,61] Often, hybrids may show a morphology intermediate between parental cells,[31,60] enabling the investigator to select colonies on the basis of the appearance of the cells. In other cases, hybrids may closely resemble one parent initially, but subsequently segregate variants which resemble the other parent.[61]

Selective Fixation and Back Selection. In terms of producing cell hybrids, an adequate selective system is of primary importance for obtaining hybrid cells at a high frequency with or without the use of Sendai virus. Selective systems play an additional role in the subsequent analysis of isolated hybrid cell populations. Utilizing the HAT system, it is possible to "fix" the human locus controlling HGPRT by growing the hybrid cells on HAT medium. Back selection with respect to the HGPRT locus may be accomplished in hybrid cells originally isolated and maintained in HAT medium by applying selective methods used to isolate $HGPRT^-$ cells. Selective fixation and back selection are also possible in the AA selective system. Fixation of APRT is achieved by growth in AA medium while back selection is carried out by treating the hybrid cells with DAP or 2-fluoroadenine in standard medium.[49] Back selection should be feasible in the BUdR visible light auxotrophic system by labeling the prototrophic hybrid cells with BUdR in minimal medium and exposing the hybrid cell population to visible light. BUdR in standard medium has been used to back select for the loss of TK activity from mouse × human hybrids originally isolated in HAT medium.[62-64] It is often important to conduct such reverse selection in order to confirm suspected gene-gene or gene-chromosome linkages.

Other Selective Methods. Two selective methods have recently been reported which were used to analyze hybrid cells rather than to obtain hybrid cells initially. Treatment of human × mouse hybrids with polio virus for the purpose of selecting polio-resistant hybrids was undertaken by Wang *et al.*[65] and Kusano *et al.*,[66] who were able to isolate such resistant hybrids and make a comparison between the ability of resistant and sensitive hybrid cells to adsorb virus and support viral replication. In addition to viral selection, another type of selection system applied after hybrid isolation has been developed by Puck *et al.*,[54] who treated human × hamster cell hybrids with rabbit anti-human cell antiserum. It is possible that viral and *in vitro* immunological selection may be adaptable to the initial isolation of hybrid cells and might prove useful in isolating and analyzing a greater spectrum of hybrid phenotypes.

[60] H. G. Coon and M. C. Weiss, *Wistar Inst. Symp. Monogr.* **9**, 83 (1969).
[61] F. Wiener, A. Cochron, G. Klein, and H. Harris, *J. Nat. Cancer Inst.* **48**, 465 (1972).
[62] B. R. Migeon and C. S. Miller, *Science* **162**, 1005 (1968).
[63] Y. Matsuya, H. Green, and C. Basilico, *Nature (London)* **220**, 1199 (1968).
[64] C. M. Boone, T. R. Chen, and F. H. Ruddle, *Proc. Nat. Acad. Sci. U. S.* **69**, 510 (1972).
[65] R. Wang, R. Pollack, T. Kusano, and H. Green, *J. Virol.* **5**, 677 (1970).
[66] T. Kusano, R. Wang, R. Pollack, and H. Green, *J. Virol.* **5**, 682 (1970).

Characterization of Presumptive Hybrid Cells

After isolation, hybrid cells are best characterized on the basis of such properties as chromosomal constitution, isozymes, and antigens.

Chromosomes. One of the primary features used to characterize hybrids is the chromosome complement. The number and morphology of the chromosomes present in the hybrid cells may be compared to the chromosomal patterns of the parental cells as a method of determining whether or not the presumptive hybrids are true hybrids. This method of verifying the hybrid nature of presumed hybrids is most useful when each parental cell line possesses one or more distinctive marker chromosomes or when the parental karyotypes exhibit many prominent differences in chromosome number and morphology. Ricciuti and Ruddle[42] demonstrated that a hybrid cell could arise from the fusion of three different parental cells by fusing a mixture of RAG, A9, and WI38 cells with Sendai virus. These investigators isolated a clone which showed chromosomes specifically characteristic of each parent. The triple hybrid nature of this clone was also supported by the enzyme phenotype for several enzymes assayed in a starch gel electrophoresis system. A number of situations make the analysis of the karyotype of presumed hybrids difficult. (a) Intraspecific hybrids between diploid cells cannot be distinguished from parental tetraploid cells unless one of the parents contains a marker chromosome. (b) Interspecific hybrids often show a rapid and substantial loss of chromosomes.[3,5,59] (c) Some parent cells may have similar karyotypes. (d) Structural rearrangements may complicate the identification of chromosomes in hybrid cells. (e) Hybrid cell populations may exhibit marked heterogeneity in the chromosomal complement. Despite these drawbacks, a careful characterization of the chromosomal pattern of presumed hybrids at the earliest possible time after isolation constitutes a key step in analysis.

Chromosome preparations suitable for staining procedures that demonstrate constitutive heterochromatin[67-71] or chromosome banding[72-79] (see also Section

[67] F. E. Arrighi and T. C. Hsu, *Cytogenetics* **10**, 81 (1971).
[68] T. R. Chen and F. H. Ruddle, *Chromosoma* **34**, 51 (1971).
[69] R. Gagné, R. Tanguay, and C. Laberge, *Nature (London), New Biol.* **232**, 29 (1971).
[70] J. J. Yunis, L. Roldan, W. G. Yasmineh, and J. C. Lee, *Nature (London)* **231**, 532 (1971).
[71] M. Bobrow, K. Madan, and P. L. Pearson, *Nature (London), New Biol.* **238**, 122 (1972).
[72] T. Caspersson and L. Zech, *In* "Perspectives in Cytogenetics, The Next Decade," (S. W. Wright, B. F. Crandal, and L. Boyer, eds.), Chapter 12, pp. 163–185. Thomas, Springfield, Illinois, 1972.
[73] T. Caspersson, L. Zech, C. Johansson, and E. J. Modest, *Chromosoma* **30**, 215 (1970).
[74] O. J. Miller, D. A. Miller, R. E. Kouri, P. W. Allderdice, V. G. Dev, M. S. Grewal, and J. J. Hutton, *Proc. Nat. Acad. Sci. U. S.* **68**, 1530 (1971).
[75] P. Aula and E. Saksela, *Exp. Cell Res.* **71**, 161 (1972).
[76] A. T. Sumner, H. J. Evans, and R. A. Buckland, *Nature (London), New Biol.* **232**, 31 (1971).
[77] M. E. Drets and M. W. Shaw, *Proc. Nat. Acad. Sci. U. S.* **68**, 2073 (1971).

XIV) may be made using procedures based on the air-drying technique of Rothfels and Siminovitch.[80]

The chromosome banding procedures promise to be of great use in analyzing the chromosome constitution of hybrid cells because of the ability of these techniques to demonstrate subtle differences in morphologically similar chromosomes. Chen and Ruddle[68] have discussed the use of constitutive heterochromatin in the analysis of chromosomes in mouse × human hybrids. These investigators noted that mouse biarmed chromosomes have prominent *symmetrical* staining of heterochromatin in the region of the centromere while many human biarmed chromosomes show asymetric or very faint staining. Blocks of heterochromatin were also found in regions some distance from the centromere in certain chromosomes. Chen and Ruddle[68] were able to identify specifically human Al, C9, E16, and Y chromosomes in mouse × human hybrids on the basis of constitutive heterochromatin staining, demonstrating the usefulness of this technique for chromosome analysis of hybrid cells. Quinacrine and Giemsa stains have been used to produce "banding patterns" in chromosomes which have made possible a detailed karyotypical analysis of human[72,73,75–79] and mouse[74] chromosomes. Protocols for staining include: (a) Giemsa staining for bright-field examination; (b) aceto-orcein staining for phase-contrast examination; (c) ASG (acetic-saline-Giemsa) technique developed by Sumner *et al.*[76]; (d) trypsin banding described by Tolby[79]; and (e) quinacrine mustard (QM) staining based on the methods of Caspersson.[72,73] A sample karyotype using the ASG technique is shown in Fig. 4 and Fig. 5 shows one using the quinacrine mustard stain technique.

Isozymes. Isozymes are defined as multiple molecular forms of an enzyme demonstrating similar or identical catalytic properties.[81] Three major categories of isozymes are tabulated by Harris.[81] (a) Isozymes resulting from multiple gene loci coding distinctly different polypeptide chains; (b) isozymes arising as a result of multiple alleles at a single locus; and (c) isozymes which are formed as a result of secondary modifications, which includes covalent modifications, conformational isomers, and the formation of monomers, dimers, trimers, etc., of a basic subunit which may be composed of more than one polypeptide chain. The separation of isozymes on the basis of surface charge (and to a lesser extent on molecular weight) may be achieved by electrophoresis in starch gel, acrylamide gel, agarose, cellulose acetate or Cellogel under conditions of pH, ionic strength, and ionic composition appropriate for a specific enzyme.[11] (See also Section XIV.)

Evolutionary divergence has led to alterations in the structure of many homologous enzyme proteins which can be resolved as interspecific differences

---

[78] W. Schnedl, *Nature* (*London*), *New Biol.* **233**, 93 (1971).

[79] B. Tolby, personal communication.

[80] K. H. Rothfels and L. Siminovitch, *Stain Technol.* 33, 73 (1958).

[81] H. Harris, "The Principles of Human Biochemical Genetics," pp. 53–57. American Elsevier, New York, 1970.

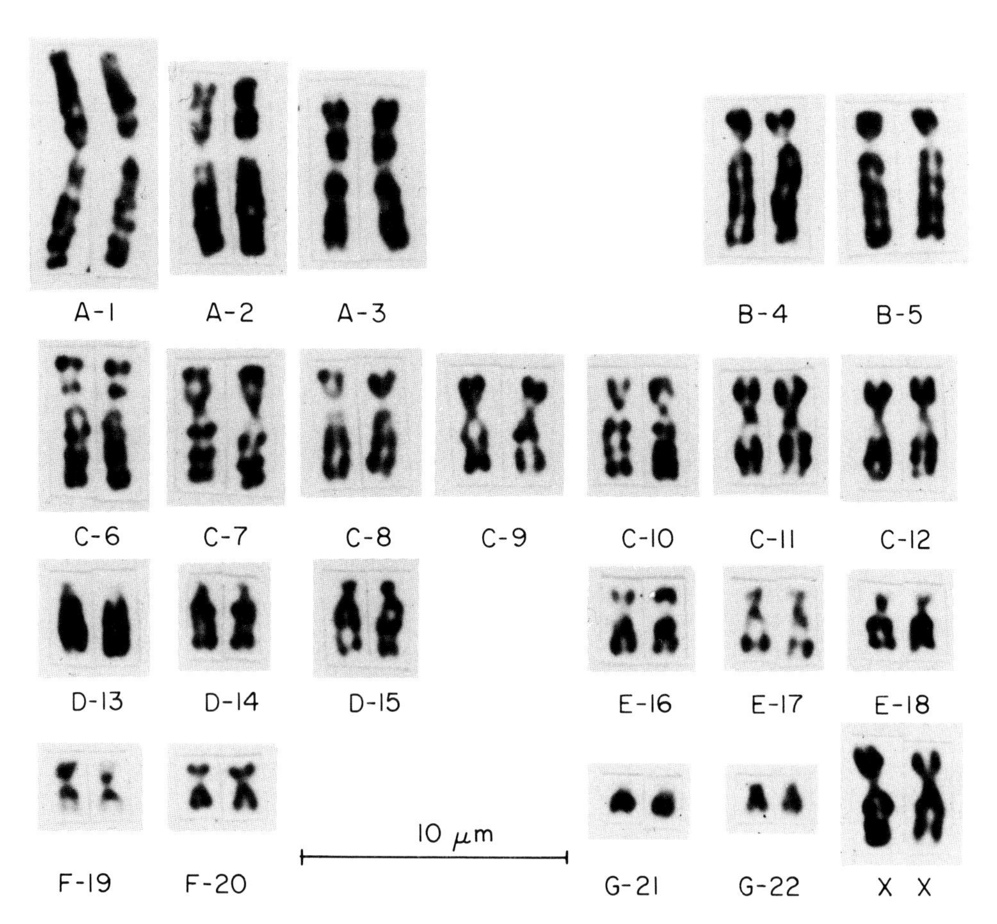

Fig. 4. Giemsa banding (ASG) of WI38, a human diploid cell. (J. A. Tischfield, Ph.D. Thesis, Yale University, 1972.)

in isozyme electrophoretic mobilities. Complex isozyme patterns are often observed when the isozyme phenotype for a particular enzyme is determined for hybrid cells. The complexity of these hybrid cell patterns is due to the expression of some or all of the isozymes from both parents. Further complexity may arise in the case of oligomeric enzymes as a result of the formation of unique isozymes composed of subunits derived from both parents. The presence of such "hybrid" enzyme forms constitutes primary proof of the hybrid nature of the presumptive hybrid cell clone, since such forms are not observed as a result of cocultivation of cells or the extraction of cell mixtures, but only occur when the parental genomes are functioning in the same cell.[11]

A number of additional characteristics make isozymes extremely valuable as markers for characterizing hybrid cells: (a) a high frequency of genetic polymorphism at loci controlling various enzymes[11] has resulted in the accumulation of information on the genetic control and linkage of certain isozymes

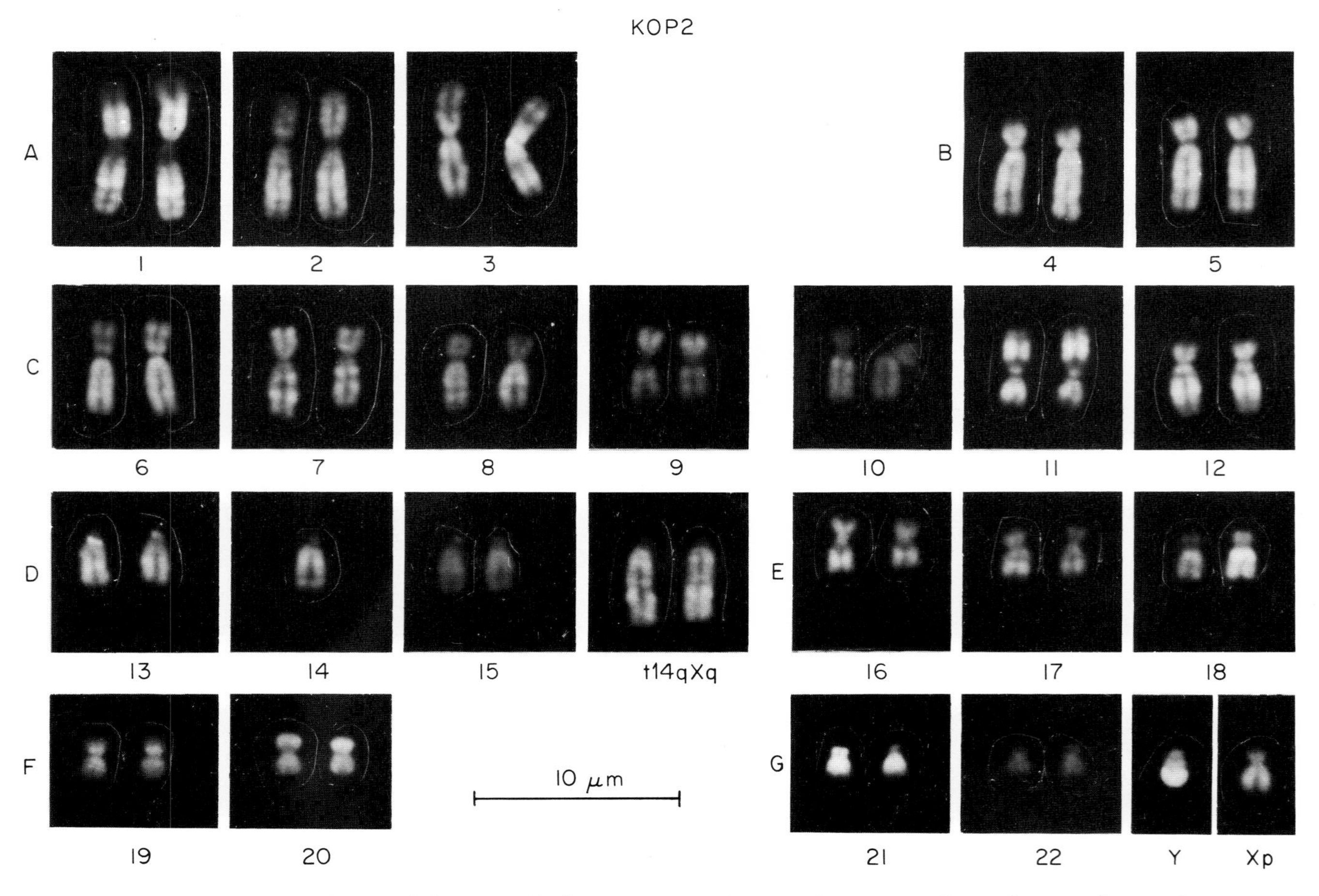

Fig. 5. Quinacrine mustard stained human cell chromosomes with a translocation involving the X and D14 chromosomes. (F. Ricciuti, Ph.D. Thesis, Yale University, 1972.)

through studies on isozyme phenotypes in the mouse[82] and man.[83,84] This characteristic provides the potential of correlating results concerning genetic linkages determined *in vitro* by the techniques of somatic cell genetics with data obtained from Mendelian sexual segregation of isozyme phenotypes in intact organisms. An example of this potential is provided by the demonstration by family studies in man that HGPRT and glucose-6-phosphate dehydrogenase (G6PD) are X-linked and the subsequent confirmation of this linkage by interspecific somatic cell hybrids[85]; (b) the existence of isozyme variants within a species often permits the resolution of interspecific differences in the electrophoretic mobilities of homologous isozymes normally showing no difference; (c) some isozymes are expressed constitutively, making them excellent markers for hybrid characterization, while other isozymes are expressed facultatively and are useful in studies of eukaryotic regulation[11,85]; (d) a substantial number of clinical conditions are associated with specific enzyme deficiencies.[11,86] (e) the expression of certain enzymes, such as HGPRT, can be made conditionally essential or lethal by drug selection; and (f) there is a large, well-characterized repertoire of isozymes which can be separated electrophoretically (see Table II).[10,11,81,87]

Starch gel electrophoresis has been the method most commonly used to separate isozymes. Buffer systems and assay reactions have been established for a number of enzymes useful in characterizing hybrid cells.[87–91] Three slices can be obtained from a single gel slab, enabling the investigator to determine the isozyme pattern for three different enzymes from one gel, providing the buffer systems are compatible. Disadvantages of the starch gel system are: (a) time-consuming gel preparations; (b) quality control of starch lots; (c) large sample size required; and (d) long running time. Recently, cellulose acetate and Cellogel (cellulose acetate coated with a thin protein gel)[90,92] have become popular because of the ease of preparation of the strips, small sample size, and short running time. The repertoire of enzymes characterized on Cellogel is small in comparison to the number of enzymes which have been run in starch gel systems, but the conditions of electrophoresis developed in the starch systems can probably serve as a basis for developing systems for the separation of isozymes using Cellogel. Examples of isozyme separations are provided in Figs. 6 and 7 for lactic dehydrogenase (LDH) and glucose-6-phosphate dehydrogenase

[82] T. H. Roderick, F. H. Ruddle, V. M. Chapman, and T. B. Shows, *Biochem. Genet.* **5**, 457 (1971).

[83] H. Harris, *Can. J. Genet. Cytol.* 13, 381 (1971).

[84] D. A. Hopkinson and H. Harris, *Annu. Rev. Genet.* **5**, 5 (1971).

[85] F. H. Ruddle, *Fed. Proc.* **30**, 921 (1971).

[86] B. R. Migeon, *Ann. N. Y. Acad. Sci.* **171**, 396 (1970).

[87] G. S. Omenn and P. T. W. Cohen, *In Vitro* **7**, 132 (1971).

[88] C. R. Shaw and R. Prasad, *Biochem. Genet.* **4**, 297 (1970).

[89] F. H. Ruddle and E. A. Nichols, *In Vitro* **7**, 120 (1971).

[90] P. Meera Khan, *Arch. Biochem. Biophys.* **145**, 470 (1971).

[91] G. J. Brewer, "Introduction to Isozyme Techniques." Academic Press, New York, 1970.

[92] Chemetron Technical Bulletin, "Cellogel Electrophoresis," Chemetron via G. Modena, 24, 20129, Milan, Italy.

TABLE II
Isozyme Repertoire

Enzyme		EC No.	M vs. Hu[a]	CH vs. Hu[b]	References[d]
Adenosine deaminase	ADA	3.5.4.4	+	+	1-3
Adenine phosphoribosyltransferase	APRT	2.4.2.7	+		4–6
Alcohol dehydrogenase	ADH	1.1.1.1	+		7–10
Catalase	CAT	1.11.1.6	+	+	11–14
Fructose-1,6-diphosphate aldolase A, B, C	ALD	4.1.2.13	−A, +B, +C		10, 15–18
Glucose-6-phosphate dehydrogenase	G6PDH	1.1.1.49	+	+	2, 10, 19–24
Glucosephosphate isomerase	PGI-GPI	5.3.1.9	+	+	2, 14, 25
β-Glucuronidase	GRD	3.2.1.31	+		26–28
α-Galactosidase	α-GAL	3.2.1.22	+	+	29
β-Galactosidase	β-GAL	3.2.1.23			30, 31
Glutamic oxaloacetic transaminase M, S[c]	GOT	2.6.1.1	−M, +S	−M, +S	32–34
Glutamic pyruvic transaminase	GPT	2.6.1.2	+		35
α-Glycerol-3-phosphate dehydrogenase	GAPD	1.1.1.8			10, 14
Hypoxanthine-guanine phosphoribosyltransferase	HGPRT	2.4.2.8	+		10, 24, 36–40
Lactate dehydrogenase A, B	LDH	1.1.1.27	+A, −B	+A	2, 10, 14, 19, 24, 41–44
Indophenol oxidase A, B	IPO		+A, +B		24, 45
Isocitrate dehydrogenase M, S[c]	IDH	1.1.1.42	+S	+S	2, 10, 24, 46
Malate dehydrogenase M, S	MDH/MOD	1.1.1.37	+S	−S	2, 24, 47–49
Malic enzyme M, S	MDH/MOR	1.1.1.40	−M, +S	−M, +S	2, 24, 50
Mannosephosphate isomerase	MPI	5.3.1.8	+	+	51
Peptidase A, B, C, D	PEP-A, -B, -C, -D	3.4.3.−	+, +, +, +		2, 10, 52–56
6-Phosphogluconic dehydrogenase	6PGD	1.1.1.43	—	+	2, 10, 24, 57, 58
Phosphoglucomutase 1,	PGLuM/PGM	2.7.5.1	+1	+1	2, 10, 57, 59–64
Phosphoglycerate kinase	PGK	2.7.2.3	+	+	14, 24, 65–67
Purine nucleoside phosphorylase	NP	2.4.2.1	+		68, 69
Sorbitol dehydrogenase	SDH	1.1.1.14			10
Thymidine kinase	TK	2.7.1.21			10, 70–72
Tyrosine aminotransferase	TAT	2.6.1.5			73, 74
Xanthine oxidase	XOD	1.2.3.2	+		10, 75–77

[a] Mouse versus human. [b] Chinese hamster versus human. [c] Mitochondrial, supernatant.

[d] References: 1. H. Spencer and D. A. Hopkins, *Ann. Hum. Genet.* **32,** 9 (1968); 2. F. H. Ruddle and E. A. Nichols, *In Vitro* **7,** 120 (1971); 3. J. C. Detter, G. Stamatoyamopoulos, E. R. Giblett, and A. G. Motulsky, *J. Med. Genet.* **7,** 356 (1970); 4. W. N. Kelley, R. I. Levy, F. M. Rosen-

bloom, J. F. Henderson, and J. E. Seegmiller, *J. Clin. Invest.* **47,** 22 (1968); 5. W. Nyhan, J. A. James, A. J. Teberg, L. Sweetman, and L. G. Nelson, *J. Pediat.* **74,** 20 (1969); 6. B. Bakay and W. L. Nyhan, *Biochem. Genet.* **5,** 81 (1971); 7. M. Smith, D. A. Hopkinson, and H. Harris, *Ann. Hum. Genet.* **34,** 251 (1971); 8. D. A. Hopkinson and H. Harris, *Amer. Rev. Genet.* **5,** 5 (1971); 9. R. F. Murray and A. G. Motulsky, *Science* **171,** 71 (1971); 10. C. R. Shaw and R. Prasad, *Biochem. Genet.* **4,** 297 (1970); 11. W. E. Nance, J. E. Empson, T. W. Bennet, and L. Larson, *Science* **160,** 1230 (1968); 12. J. G. Scandalios, *Ann. N. Y. Acad. Sci.* **151,** 274 (1968); 13. S. Matsubara, H. Suter, and H. Aebi, *Humangenetik* **4,** 29 (1967); 14. G. S. Omenn and P. T. W. Cohen, *In Vitro* **7,** 132 (1971); 15. E. E. Penholt, T. V. Rajkumar, and W. J. Rutter, *Proc. Nat. Acad. Sci. U. S.* **56,** 1275 (1966); 16. W. J. Rutter, T. Rajkumar, E. Penholt, M. Kochman, and R. Valentine, *Ann. N. Y. Acad. Sci.* **151,** 102 (1968); 17. H. G. Lebherz and W. J. Rutter, *Biochemistry* **8,** 109 (1969); 18. F. Farron, H. H. T. Hsu, and W. E. Fox, *Cancer Res.* **32,** 302 (1972); 19. C. M. Boone and F. H. Ruddle, *Biochem. Genet.* **3,** 119 (1969); 20. H. N. Kirkman, *Advan. Hum. Genet.* **2,** 1 (1971); 21. M. C. Rattazzi, L. F. Gernini, G. Fiorelli, and P. Mannucci, *Nature (London)* **213,** 79 (1967); 22. W. D. Peterson, Jr., C. S. Stulberg, N. K. Swanborg, and A. R. Robinson, *Proc. Soc. Exp. Biol. Med.* **128,** 772 (1968); 23. A. Yoshida, S. Watanabe, and S. M. Gartler, *Biochem. Genet.* **5,** 533 (1971); 24. P. Meera Khan, *Arch. Biochem. Biophys.* **145,** 470 (1971); 25. R. J. DeLorenzo and F. H. Ruddle, *Biochem. Genet.* **3,** 151 (1969); 26. M. C. Weiss and B. Ephrussi, *Genetics* **54,** 1111 (1961); 27. D. Robinson, R. G. Price, and N. Dance, *Biochem. J.* **102,** 525 (1967); 28. R. E. Ganscho and B. G. Bunker, *Biochem. Genet.* **4,** 127 (1970); 29. E. Beutler and W. Kuhle, *Amer. J. Hum. Genet.* **24,** 237 (1972); 30. M. W. Ho and J. S. O'Brien, *Science* **165,** 611 (1969); 31. S. Okada and J. S. O'Brien, *Science* **160,** 1002 (1968); 32. R. G. Davidson, J. A. Cortner, M. C. Rattazzi, F. H. Ruddle, and H. A. Lubs, *Science* **169,** 391 (1970); 33. R. J. DeLorenzo and F. H. Ruddle, *Biochem. Genet.* **4,** 259 (1970); 34. S. H. Chen and E. R. Giblett, *Amer. J. Hum. Genet.* **23,** 419 (1971); 35. S. H. Chen and E. R. Giblett, *Science* **173,** 148 (1971); 36. V. M. DerKaloustian, R. Byrne, W. J. Younge, and B. Childs, *Biochem. Genet.* **3,** 299 (1969); 37. B. Bakay and W. L. Nyhan, *Proc. Nat. Acad. Sci. U. S.* **69,** 2523 (1972); 38. S. Shin, P. Meera Khan, and P. R. Cook, *Biochem. Genet.* **5,** 91 (1971); 39. W. N. Kelley and J. C. Meade, *J. Biol. Chem.* **246,** 2953 (1971); 40. W. J. Arnold and W. N. Kelley, *J. Biol. Chem.* **246,** 7398 (1971); 41. F. H. Ruddle, *Fed. Proc.* **30,** 921 (1971); 42. T. B. Shows and F. H. Ruddle, *Proc. Nat. Acad. Sci. U. S.* **61,** 574 (1968); 43. F. H. Ruddle, V. M. Chapman, T. R. Chen, and R. J. Klebe, *Nature (London)* **227,** 251 (1970); 44. A. S. Santachiara, M. Nabholz, V. Miggiano, A. J. Darlington, and W. Bodmer, *Nature (London)* **227,** 248 (1970); 45. G. J. Brewer, "Introduction to Isozyme Technique." Academic Press, New York, 1970; 46. N. S. Henderson, *J. Exp. Zool.* **158,** 63 (1965); 47. N. S. Henderson, *Arch. Biochem. Biophys.* **117,** 28 (1966); 48. T. B. Shows and F. H. Ruddle, *Science* **160,** 1356 (1968); 49. T. B. Shows, V. M. Chapman, and F. H. Ruddle, *Biochem. Genet.* **4,** 707 (1970); 50. B. R. Migeon, *Biochem. Genet.* **1,** 305 (1968); 51. E. A. Nichols, V. M. Chapman, and F. H. Ruddle, *Biochem. Genet.* **8** (No. 1) (1973); 52. W. H. P. Lewis and H. Harris, *Nature (London)* **215,** 351 (1967); 53. W. H. P. Lewis and H. Harris, *Ann. Hum. Genet.* **32,** 317 (1969); 54. W. H. P. Lewis and G. M. Truslove, *Biochem. Genet.* **3,** 493 (1969); 55. S. Rapley, W. H. P. Lewis, and H. Harris, *Ann. Hum. Genet.* **34,** 307 (1971); 56. P. J. L. Cook, S. Povey, and E. B. Robson, *Ann. Hum. Genet.* **36,** 89 (1972); 57. A. Westerveld and P. Meera Khan, *Nature (London)* **236,** 30 (1972); 58. R. G. Davidson, *Ann. Hum. Genet.* **30,** 355 (1967); 59. N. Spencer, D. A. Hopkinson, and H. Harris, *Nature (London)* **204,** 742 (1964); 60. J. M. Parrington, G. Cruckshank, D. A. Hopkinson, E. B. Robson, and H. Harris, *Ann. Hum. Genet.* **32,** 27 (1968); 61. R. A. Fisher and H. Harris, *Ann. Hum. Genet.* **36,** 69 (1972); 62. T. B. Shows, F. H. Ruddle, and T. H. Roderick, *Biochem. Genet.* **3,** 25 (1969); 63. P. J. McAlpine, D. A. Hopkinson, and H. Harris, *Ann. Hum. Genet.* **34,** 169 (1970); 64. D. M. Dawson and S. Jaeger, *Biochem. Genet.* **4,** 1 (1970); 65. S. H. Chen, L. A. Malcolm, A. Yoshida, and E. R. Giblett, *Amer. J. Hum. Genet.* **23,** 87 (1971); 66. E. Beutler, *Biochem. Genet.* **3,** 189 (1969); 67. P. Meera Khan, A. Westerveld, K. H. Grzeschik, B. F. Deys, D. M. Garson, and M. Siniscalco, *Amer. J. Hum. Genet.* **23,** 614 (1971); 68. E. H. Edwards, B. J. McGee, and H. Harris, *Ann. Hum. Genet.* **34,** 347 (1971); 69. Y. H. Edwards, D. A. Hopkinson, and H. Harris, *Ann. Hum. Genet.* **34,** 395 (1971); 70. H. Green, *Wistar Inst. Symp. Monogr.* **9,** 51 (1969); 71. B. R. Migeon and C. S. Miller, *Science* **162,** 1005 (1968); 72. B. R. Migeon, S. W. Smith, and C. L. Leddy, *Biochem. Genet.* **3,** 583 (1969); 73. R. L. Blake, *Biochem. Genet.* **4,** 215 (1970); 74. E. B. Thompson, G. M. Tomkins, and J. F. Curran, *Proc. Nat. Acad. Sci. U. S.* **56,** 296 (1966); 75. T. T. T. Yen and E. Glassman, *Genetics* **52,** 977 (1965); 76. E. Glassman, T. Shinoda, E. J. Duke, and J. Collins, *Ann. N. Y. Acad. Sci.* **151,** 263 (1968); 77. M. L. Sachler, *J. Histochem. Cytochem.* **14,** 326 (1965).

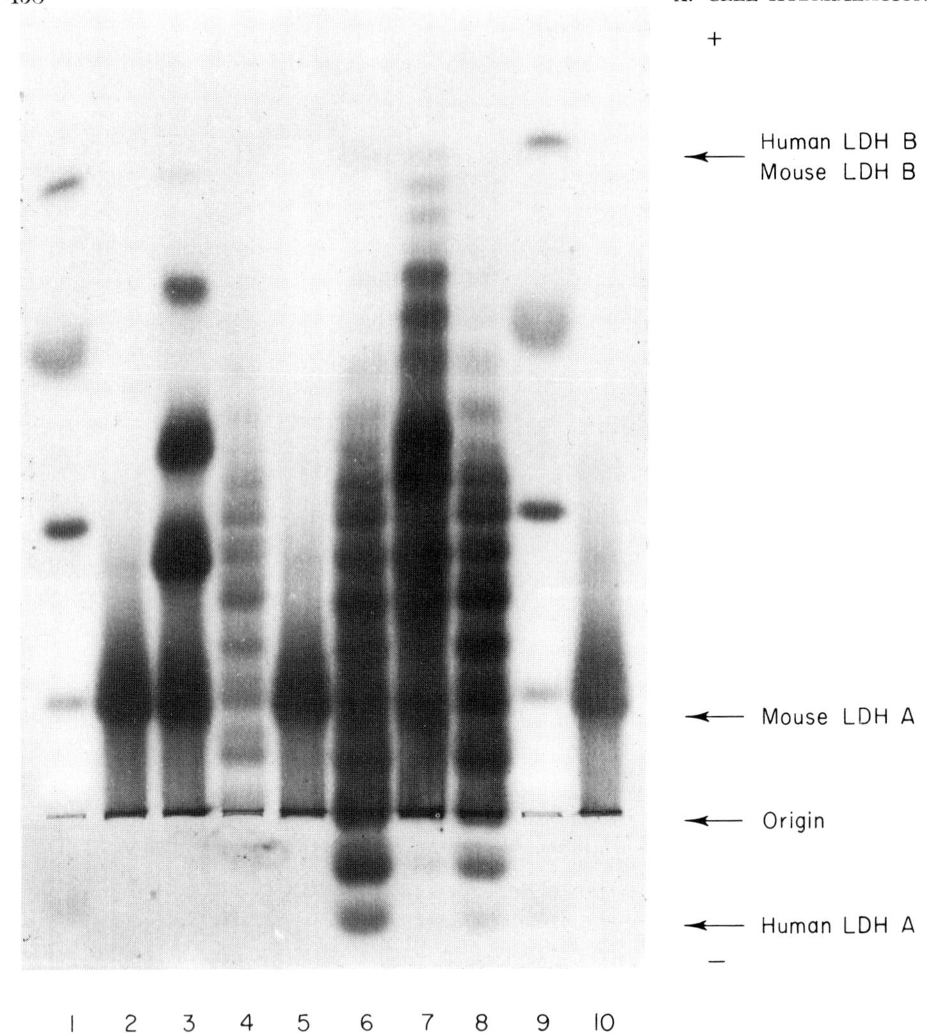

Fig. 6. Lactic dehydrogenase isozymes (starch gel electrophoresis). Channel (1) human KB; (2) mouse RAG; (3) mouse kidney; (4) mouse × human hybrid expressing human A; (5) mouse RAG; (6) mouse × human hybrid expressing human A; (7) mouse × human hybrid expressing human B; (8) mouse × human hybrid expressing human A; (9) human KB; and (10) mouse RAG.

(G6PD), respectively. Brewer[91] has described the use of a variety of electrophoretic methods for the separation of isozymes (see also Section XIV).

Surface Antigens. Cell surface antigens have been used to characterize short-term heterokaryons[93,94] and hybrid cells.[59,95–97] Several classes of surface

[93] J. F. Watkins and D. M. Grace, *J. Cell Sci.* **2**, 193 (1967).

[94] L. D. Frye and M. Edidin, *J. Cell Sci.* **7**, 319 (1970).

[95] G. Klein, U. Gars, and H. Harris, *Exp. Cell Res.* **62**, 149 (1970).

[96] G. Grundner, E. M. Fenyö, G. Klein, E. Klein, U. Bregular, and H. Harris, *Exp. Cell Res.* **68**, 315 (1971).

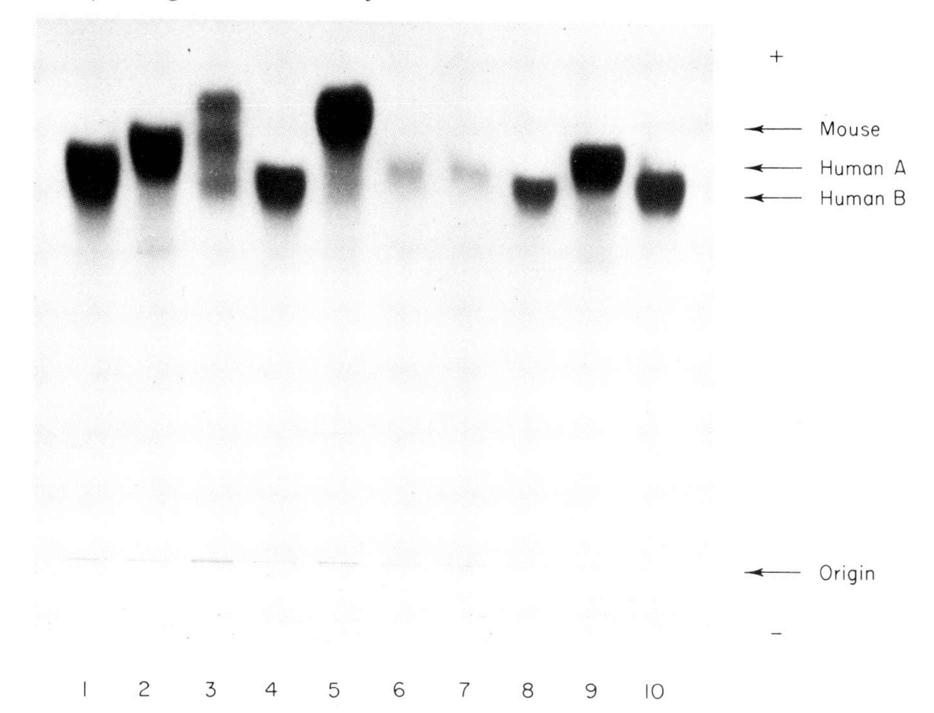

Fig. 7. Glucose-6-phosphate dehydrogenase (starch gel electrophoresis). Channel (1) human WI38; (2) human KB; (3) mouse × human hybrid (Hal-5); (4) human leukocyte; (5) mouse RAG; (6) human leukocyte; (7) human leukocyte; (8) human leukocyte; (9) human KB; and (10) human WI38.

antigens have been studied in hybrid cells such as species-specific antigens,[59,93,94] histocompatibility antigens[94–97] (*H-2* locus in the mouse and *HLA* locus in man; see also Section XIV this volume), and virus-specific surface antigens.[95–97] Three major types of assay systems were employed to detect these surface antigens: (a) cytotoxicity,[59,95] (b) mixed agglutination or mixed hemadsorption,[93,96,97] and (c) immunofluorescence.[94–97]

The expression of species-specific and histocompatibility antigens appears to be remarkably stable *in vitro*.[98–100] Some mouse ascites tumors may show changes in *H-2* expression during immunoselection.[101] Klein *et al.*[95] found that fusion of Ehrlich ascites tumor cells which have little detectable *H-2* activity with three different L cell derivatives expressing the *H-2*^{k} complex resulted in varying degrees of suppression of *H-2*^{k} expression in several hybrid clones. A

---

[97] E. M. Fenyö, G. Grundner, G. Klein, E. Klein, and H. Harris, *Exp. Cell Res.* **68**, 313 (1971).

[98] D. Franks, *Biol. Rev.* **43**, 17 (1968).

[99] D. Klein, D. J. Merchant, J. Klein, and D. C. Shreffler, *J. Nat. Cancer Inst.* **44**, 1149 (1970).

[100] V. M. Papermaster, B. W. Papermaster, and G. E. Moore, *Fed. Proc.* **28**, 379 (1969).

[101] T. S. Hauschka, L. Weiss, B. A. Holdridge, T. L. Cudney, M. Zumpft, and J. A. Planinsek, *J. Nat. Cancer Inst.* **47**, 343 (1971).

virus-specific antigen and an L cell-specific antigen were also suppressed.[96] During subsequent propagation these antigenic specificities reappeared independently in certain malignant segregants arising from the hybrid clones.[96] Fenyö *et al.*[97] crossed A9 cells with two different Moloney lymphoma lines, YAC and YACIR, and found that several parental antigenic specificities ($H\text{-}2^k$, $H\text{-}2^d$, L cell antigen, Moloney specific surface antigen) were expressed in the hybrids.

Detailed protocols have been described for the assay of species-specific surface antigens by mixed agglutination[36] and by mixed hemadsorption.[16] Protocols for the assay of other antigenic specificities discussed above may be obtained by consulting the cited references. The potential to assay individual cells or small clones for a phenotypic character by means of a nonlethal reaction and the potential to conduct *in vitro* and *in vivo* immunoselection promises to make the study of surface antigens in hybrid cells important in the genetic analysis of somatic cells.

CHAPTER 3

Plant Protoplast Culture[1]

R. A. Miller, K. N. Kao, and O. L. Gamborg

Although protoplasts were isolated from plant cells and studied before the turn of the century, experimentation was limited by the lack of an isolation technique which could supply large quantities of protoplasts. This deficiency was overcome with the isolation of enzymes capable of degrading plant cell walls. Since that time experimentation with protoplasts has advanced rapidly to the stage where these units have been induced to synthesize new walls, undergo cell division, establish new colonies, and subsequently differentiate into intact plants.

Materials

Plant Cell Sources. Protoplasts have been isolated from both intact plants and cell cultures of many plant species.[1a-3] Leaves are an especially suitable cell source

[1] NRCC No. 13002.

[1a] E. C. Cocking, *Nature* (*London*) **187**, 927 (1960).

[2] R. U. Schenk and A. C. Hildebrandt, *Crop Sci.* **9**, 629 (1969).

[3] A. W. Ruesink, *In* "Methods in Enzymology" (S. P. Colowick and N. O. Kaplan, eds.), Vol. 23A, p. 197. Academic Press, New York, 1971.

since mesophyll cells have relatively thin walls which are readily degraded. However, it may be necessary to remove the epidermis since enzymes have difficulty in penetrating the cuticle. Pretreatment with α-glucuronidase or pectinglucosidase can partially substitute for epidermis stripping.[4] Other plant parts such as root tips and embryos may also be used but generally have cells with more resistant walls.

Rapidly growing cell cultures are a most suitable source of protoplasts. Old cells, or cells from slowly growing cultures often have walls which are highly resistant to enzyme degradation.

Enzymes. Enzymes from various sources have been used to produce plant protoplasts. Most commonly these have been crude preparations of cellulases, hemicellulases, and pectinases produced by *Tricoderma, Myrothecium,* or *Aspergillus* and marketed under various trade names. Although crude, these preparations can yield viable plant protoplasts.[5,6] More specific enzyme preparations may be produced by growing the molds on a particular plant material. Also, highly purified enzymes will produce protoplasts from some plant sources.[7]

After testing a number of crude commercial enzyme preparations, we selected the following on the basis of highest activity and lowest toxicity: (a) Onozuka Cellulase SS,[8] (available in a range of specific activities) (b) Rhozyme HP150[9] (a hemicellulase), and (c) pectinase.[10] We desalt these by gel filtration before use.[11] Since the enzymes are highly stable, they can be readily freeze-dried and stored for long periods without appreciable loss of activity. Before use they are sterilized by filtration through a 0.5-μm membrane filter. Other sources of enzymes are now becoming available and may prove superior to those used in the past.

Protoplast Production

Enzyme Incubation. Routinely we use cellulase at 1 to 2% along with pectinase and/or hemicellulase at 0.5 to 1%. The incubation medium must contain an osmotic stabilizer which may be ionic or nonionic. Sorbitol and mannitol alone or in combination have been the most widely used nonionic stabilizers, while calcium chloride has been used as the ionic stabilizer.

The medium should be slightly hypertonic since it appears that plasmolysis hastens the wall degradation, perhaps by allowing enzyme attack from the inner surface of the wall. However, many cells will deplasmolize before protoplast

[4] L. Schilde-Rentschler, *Z. Naturforsch.* **27,** 208 (1972).
[5] K. N. Kao, W. A. Keller, and R. A. Miller, *Exp. Cell Res.* **62,** 338 (1970).
[6] T. Nagata and T. Takebe, *Planta* (*Berlin*) **99,** 12 (1971).
[7] W. A. Keller, B. Harvey, O. L. Gamborg, R. A. Miller, and D. E. Eveleigh, *Nature* (*London*) **226,** 280 (1970).
[8] Kinki Yakut Mfg. Co., Ltd. 8–21 Shingikan-Cho, Nishinomiya, Japan.
[9] Rohm and Haas Co. of Canada, West Hill, Ontario.
[10] Sigma Chemical Co., St. Louis, Missouri.
[11] K. N. Kao, O. L. Gamborg, W. A. Keller, and R. A. Miller, *Nature N. Biol.* **232,** 124 (1971).

production is complete. When the cell walls have been weakened by partial digestion, a hypotonic shock can rupture the wall and allow protoplast release. However, this procedure leads to substantial protoplast fusion producing multinucleated homokaryons.

The pH of the incubation mixture should be between 5 and 6 which is about optimum for enzyme activity and protoplast stability. Elevated temperatures hasten wall degradation, but most plant cells deteriorate rapidly at temperatures above 40°C. If the protoplasts are to be used in culture experiments the production step should not be carried out above 30°C.

Gentle agitation aids protoplast release but since these are extremely fragile structures they are highly vulnerable to mechanical damage, although this feature varies with the cell source. Agitation also helps to provide adequate aeration of the cells during the incubation period—seemingly a critical time for the cells. Droplets or thin sheets of the incubation mixture facilitate aeration due to the high surface-to-volume ratio.

The shorter the incubation time necessary to produce protoplasts the greater the chance for obtaining healthy protoplasts. However, we have successfully cultured protoplasts after 18 hours enzyme incubation. Protoplasts appear to be more stable than cells in our incubation mixtures. Generally all cells not converted to protoplasts are dead after 18 hours incubation with the enzyme mixture.

Enzyme Removal. Several methods have been used to remove the enzyme following protoplast production. Dense sucrose solutions can effect protoplast flotation. Protoplasts can then be collected from the surface liquid.[12] This procedure provides a separation of protoplasts and undigested cells if these differ in density. (Cells generally are slightly more dense than protoplasts.)

When lower concentrations of sugar or a nonionic stabilizer are used, protoplasts can be centrifuged at low speeds. The supernatant can be aspirated, the protoplasts resuspended in fresh medium, and the centrifugation repeated. Protoplasts having dense inclusions are easily destroyed at other than very low centrifugal force. Undigested cells are included with the protoplasts unless a sucrose gradient is used.

Since protoplasts must be completely freed of enzymes and contaminants, adequate repetition of the above procedures is time-consuming. Satisfactory washing of protoplasts can be done by gravity filtration on an 8 μm membrane filter (no vacuum). Addition of small aliquots of wash medium with intermittent draining can yield clean protoplasts of good quality. Complete draining of the liquid causes breakage of the protoplasts. Undigested cells can be removed by prefiltration through 30- or 40-μm nylon or stainless steel screen but this procedure damages or destroys many protoplasts.

Protoplast Culture

Generally the same or a similar medium used for cell culture is suitable for protoplast culture after the addition of an osmotic stabilizer. The optimum os-

[12] D. W. Gregory and E. C. Cocking, *J. Cell Biol.* **24**, 143 (1968).

molality (Osm) varies with the species being cultured and may vary from 0.3 to 1.0 Osm. In some cases there may be an advantage in using two or more osmotic compounds rather than a single one. Sorbitol, mannitol, and sucrose have been widely used.

Both supplemented and defined medium have been used for successful protoplast culture. The supplements may be coconut milk, casein hydrolysate, nucleosides, and amino acids in similar concentrations to those found to be beneficial to cell cultures. The requirements for medium enrichment will depend upon the species of protoplasts. The vitamin and hormone requirement is essentially the same as for cell cultures. Although auxin may cause instability of protoplasts at certain concentrations, 2,4-D can be successfully employed at 0.1 to 2.0 mg per milliliter. Other hormones such as NAA and kinetin have been used on leaf protoplasts.[12]

Both liquid and solid medium may be used for culture. In either case thin sheets or droplets must be used for many plant species in order to provide adequate aeration. We have found that 5 mg/ml DEAE-cellulose enhances plating efficiency of protoplasts embedded in agar.

A minimum concentration of protoplasts must be maintained; again this varies with the protoplast source. As low as 10^3 tobacco leaf protoplasts per milliliter yielded colonies in solid medium.[6] It is important that fresh medium be made available to many protoplast species as soon as small colonies have been established. In liquid culture this can be done by simply adding a small quantity of fresh medium (at the time the osmolality can be lowered). In agar medium colonies have to be transferred to fresh plates. We have found that flooding agar plates with liquid medium, or overlaying with more agar medium has not been successful to date.

Plants from Protoplasts

Plant development from somatic cells can occur by embryogenesis or be induced by exogenously supplied hormones. The cells derived from protoplasts possess the same properties for differentiation and morphogenesis as the original cells. Nagata and Takebe[6] obtained tobacco plants from leaf protoplasts after plating and culturing in a completely defined medium containing naphthaleneacetic acid and benzyladenine.

We prepared protoplasts from carrot cell suspension cultures.[13] The cultures consisted of cell clusters, free cells, and proembryoids. Subculturing into fresh medium twice each week ensured the best cell material for protoplast production. Filtering the cells and using a fraction of smaller aggregates facilitated protoplast production, which required 10–12 hours. The conditions for incubation and removal of enzymes followed the same procedure as described for protoplasts in general. Unchanged cell aggregates can be removed by filtering the suspension through a 44-μm stainless steel sieve or through nylon cloth. Droplets of 20 to 50 μl of protoplast suspension (about 5×10^5 protoplasts per milliliter) were placed

[13] H. J. Grambow, K. N. Kao, R. A. Miller, and O. L. Gamborg, *Planta* (*Berlin*) **103**, 348 (**1972**).

in 30 × 15 mm plastic Petri dishes sealed with Parafilm and incubated at room temperature (23°–27°C). Light does not appear to be a critical factor, although higher intensity light in excess of 1000 lux may damage protoplasts.

Wall formation takes place almost immediately and the first cell divisions can be observed after 1 to 2 days. After 1 week fresh 2,4-D-free B5 medium[14] was added carefully in amounts of 5 to 10% by volume at 3 to 4 day intervals.

In the carrot protoplast cultures two types of cell clusters were formed. One of these was a callus and the other consisted of very dense and tightly packed cells. After 3 to 4 weeks the latter types developed into recognizable proembryoids, which were transferred to agar plates made up with hormone-free B5 medium. The plantlets formed in the succeeding 3–5 weeks were placed on B5 agar slants, and later transplanted into vermiculite. The plants were grown to the flowering stage and morphologically resembled the normal carrot.

Interspecific Gene Transfer

Protoplast Fusion. Controlled fusion of protoplasts from different plant species would provide the basis for somatic cell hybridization. This could serve as a means of gene transfer between species which cannot be crossed sexually. Since, in some cases, single plant cells can be cultured into complete plants, the ultimate products of protoplast fusion would be new plant genera and species.

To date only one method has been reported to mediate protoplast fusion.[15] This procedure uses a treatment with $NaNO_3$ to cause the plasma membranes to change into a state conducive to fusion. Other salts did not elicit the same response. In our laboratory we have tested a very large number of compounds with respect to their effect on the fusion of free protoplasts. None of the chemicals tested have proved satisfactory in the sense that membranes are induced to fuse without deleterious effects on the membrane and general health of the protoplasts. Thus, chemicals which weaken membranes and allow fusion to take place may preclude the subsequent successful culture of the heterokaryon.

Obviously, protoplast contact must be attained before fusion can take place. However, excessive force that exerts pressure on the protoplasts does not contribute to fusion. Furthermore, high forces may damage the protoplasts.

An increase in temperature causes some increase in protoplast fusion. At temperatures below that which is detrimental to protoplast health, the benefit to fusion is small.

A fusion technique which employs a specific attack at the contact point of protoplasts is not available now. Possible techniques involving an immunological approach or a laser beam are being investigated at present.

Transformation. Cell transformation is another potential technique for interspecific gene transfer. Isolated DNA from bacteria has been integrated into the

[14] O. L. Gamborg, R. A. Miller, and K. Ojima, *Exp. Cell Res.* **50,** 151 (1968).

[15] J. B. Power, S. E. Cummins, and E. C. Cocking, *Nature* (*London*) **225,** 1016 (1970).

genome of plants species.[16] The experiments were performed with intact plants. The use of protoplasts may have several advantages compared with whole plants. Each protoplast which is transformed has a potential to divide and form a mass of cells that may differentiate and form plants.

Protoplasts adsorb viruses and metabolites[17] and we have demonstrated that they also absorb DNA.[18] Protoplasts from *Ammi visnaga*, soybean, and carrot cell cultures took up *E. coli* DNA-^{14}C. The uptake amounted to 0.5 to 2% in 4 hours of incubation at 28°C, and up to 50% of the absorbed DNA was TCA-insoluble. The presence of poly-L-ornithine (mol. wt. 60–100 × 10^3)[19] enhanced substantially the rate of absorption.

Poly-L-ornithine and DNA are toxic to protoplasts if used in concentrations above 2.5 μg/ml. Soybean and *Ammi* protoplasts formed cell walls and divided in the presence of 0.5 μg/ml of poly-L-ornithine and 1 μg/ml of *Ammi* DNA. Each protoplast source may have separate tolerance levels.

[16] L. Ledoux, R. Huart, and M. Jacobs, *In* "Uptake of Informative Molecules by Living Cells" (L. Ledoux, ed.), p. 159. North Holland, Amsterdam, 1971.

[17] S. Aoki and J. Takebe, *Virology* **39**, 439 (1969).

[18] K. Ohyama, O. L. Gamborg, and R. A. Miller, *Can. J. Bot.* **50**, 2077 (1972).

[19] New England Nuclear Corp., Powal, Quebec, Canada.

SECTION XI

Virus Propagation and Assay

Editors' Comments

Besides their use in basic cell research, the major application of animal cells in culture is as a substrate for virus multiplication, principally for vaccine production. Thus, the virologists propagate viruses in cell cultures by intent, whereas other tissue culturists probably do so from time to time inadvertently. In either case, the experimentalist needs to have at his disposal representative protocols for the propagation and/or assay of wanted or unwanted viruses. The purpose of this section is to present several methods illustrative of both of these situations. The use of cell cultures in medical diagnostic virology is discussed in Section XIII.

CHAPTER 1

Microculture Procedures

A. Vesicular Stomatitis Virus[1]

Leonard J. Rosenthal and Isaac L. Shechmeister

The standard method for the determination of viral concentration by plaque count requires large quantities of costly reagents and considerable time. Consequently, micromethods for tissue culture are being developed that prove to be accurate, more economical, and less time-consuming than the macromethods. Microtitrations are based upon observation of the extent of cytopathic effect produced by a virus. Such procedures have evolved, in part, from the work of Takatsy[2] who introduced calibrated spiral loops and pipettes to perform serial dilutions in the wells of specially designed plates. Many of the components of Takatsy's system were modified by Sever,[3] and also by Rosenbaum *et al.*[4] The latter workers showed a close agreement between virus titers determined by the micromethod and those derived by the conventional tissue culture tube technique. More recently Fuccillo *et al.*[5] performed multiple virus titrations on minicultures of different mammalian cell lines. Their results demonstrate excellent reproducibility and correlation with the standard tube methods for each of the many viruses tested.

The following description is an adaptation of Rosenbaum's procedure for vesicular stomatitis virus (VSV) and compares the sensitivity of the micromethod under different experimental conditions with the conventional plaque assay. In addition, other viruses, including infectious bovine rhinotracheitis (IBR) and bovine virus diarrhea (BVD), are currently being routinely assayed by these microculture procedures in our laboratory.

Materials

Plastic disposo-trays and micropipettes are available from commercial sources such as Linbro Chemical Company, Inc., New Haven, Connecticut and Cooke

[1] This work has been supported by Grant AI00155 from the National Institute of Allergy and Infectious Diseases and Training Grant CA05018 of the National Institutes of Health.

[2] G. Takatsy, *Acta Microbiol.* 3, 191 (1955).

[3] J. L. Sever, *J. Immunol.* 88, 320 (1962).

[4] M. J. Rosenbaum, I. A. Phillips, E. J. Sullivan, E. A. Edwards, and L. F. Miller, *Proc. Soc. Exp. Biol. Med.* 113, 224 (1963).

[5] D. A. Fuccillo, L. W. Catalano, Jr., F. L. Moder, D. A. Dubus, and J. L. Sever, *Appl. Microbiol.* 17, 619 (1969).

Engineering Company, Alexandria, Virginia. Several types of disposo-trays can be employed for tissue culture procedures. The plates most commonly used in these procedures either contain forty-eight flat-bottom wells (6 × 8 rows), each with a working capacity of 0.5 ml or ninety-six round-bottom wells (8 × 12 rows), each with a working capacity of 0.25 ml. The commercial disposo-trays are purchased already processed to permit cell attachment and growth and are packaged ready for tissue culture.

The micropipettes are made of semitransparent polypropylene and can be autoclaved. The pipettes in combination with the calibrated dropping tip deliver accurately either 0.025 ml/drop or 0.05 ml/drop. In addition, modified, considerably less expensive pipettes which also provide uniform delivery 37 drops/ml can be assembled from a bevel-free 17-gauge needle, an observation tube, and a rubber bulb.

Cell Culture Procedures

Preparation of primary cultures of chick embryo fibroblasts (CEF) are prepared according to the method of Dulbecco and Vogt[6] as modified by McClain and Hackett.[7] Similar procedures for the preparation of primary cultures are described by Rubin in Section II (Chick Embryo Cells). Cell suspensions (2–3 × 10^6 cells/ml) are made in lactalbumin hydrolysate in Earle's salt solution (LE) supplemented with 5% lamb serum (LE + S). All media contain penicillin (100 units/ml), streptomycin (0.5 mg/ml), and neomycin (0.1 mg/ml).

Preparation and Titration of Preformed Monolayers in Disposo-Trays

CEF monolayers are prepared in flat-bottom disposo-trays by delivering three drops (0.075 ml) of LE + S in each well. This is followed by the addition of 1.5 to 2.0 × 10^5 cells per well contained in three drops of LE + S. The trays are rotated gently to ensure mixing. Each well is either overlaid with 0.075 ml sterile mineral oil or the entire tray may be covered with commercially available covers of styrene, acetate, or formed clear polystyrene. The plates can then be stacked and incubated at 37°C for 24 to 36 hours.

Prior to titration, growth medium is discarded by either aspirating off the fluids with a Pasteur pipette or carefully pouring off spent medium by inverting the plate. The monolayers are then washed twice with 0.025 ml of Dulbecco's buffered saline. When ready for titration, two drops of LE are delivered to each well which is to be infected with virus, while wells designated as controls receive three drops of LE. One drop of virus dilutions, previously prepared by tenfold increment dilutions in test tubes with LE as diluent, is then delivered to six replicate wells. The plates can be either covered or overlaid with mineral oil and reincubated for 48 hours at 37°C. After this time the monolayers are observed for

[6] R. Dulbecco and M. Vogt, *J. Exp. Med.* **99**, 167 (1954).
[7] M. E. McClain and A. J. Hackett, *J. Immunol.* **80**, 356 (1958).

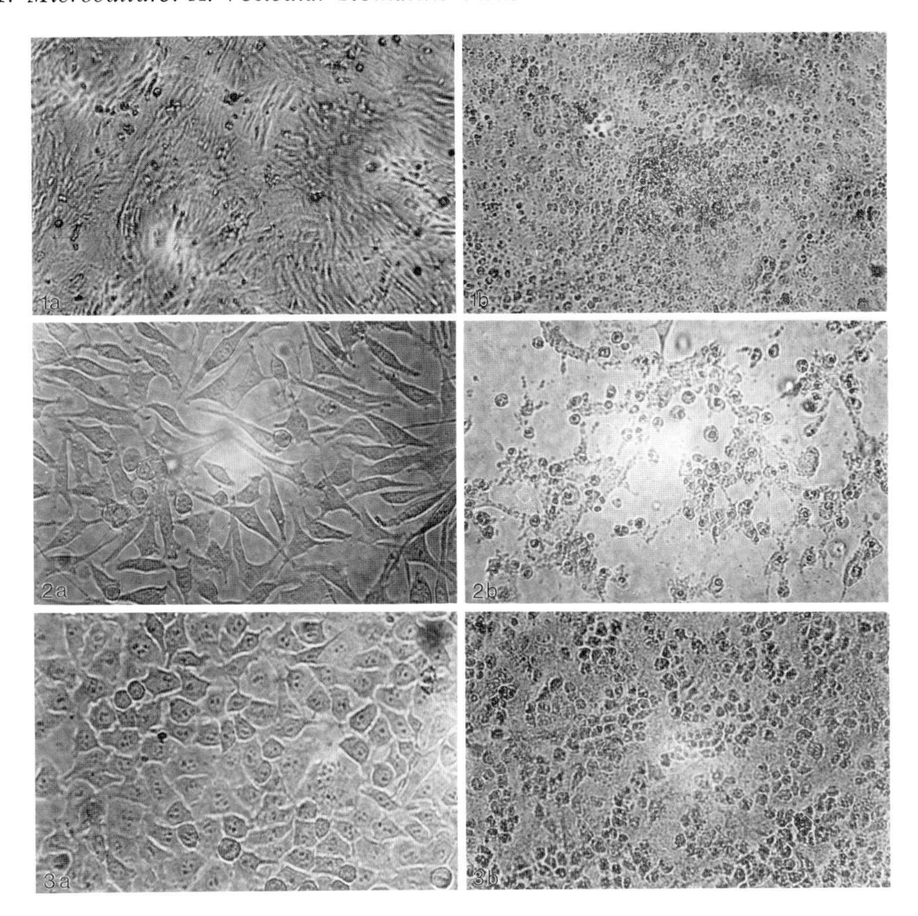

Fig. 1. Appearance of normal and VSV infected tissue cultures in flat bottom microplate wells. (The authors express their gratitude to Dr. Robert Truitt for these photographs.) la,b. Uninfected CEF monolayer in microplate well; cytopathic changes at 48 hours in a similar monolayer infected with W^+ strain of VSV. 180×. 2a,b. Uninfected L monolayer in microplate well; cytopathic changes at 48 hours in a similar monolayer infected with W^+ strain of VSV. 180×. 3a,b. Uninfected HeLa monolayer in microplate well; cytopathic changes at 48 hours in a similar monolayer infected with W^+ strain of VSV. 180×.

cellular degeneration and "rounding up" of infected cells, i.e., CPE (Fig. 1). Cytopathic effect is scored on the conventional basis of 0 through 4+. Zero indicates a normal confluent monolayer; 1+ represents a visual estimation that 25% of the cells show CPE; 2+, 3+, and 4+ represent, respectively, conditions where 50, 75, and 100% of the cells are involved.

Infectivity titer is expressed as the highest dilution of virus which infects the monolayers in 50% of the wells. The titer, designated at ID_{50}, is computed according to the procedure described by Reed and Muench.[8] A graded cytopathic response 1+ to 4+ is observed during the first 12 to 36 hours following incubation of infected microplates. An all-or-none response is present at the end of the

[8] L. J. Reed and H. Muench, *Amer. J. Hyg.* **27**, 493 (1938).

assay period of 48 hours, since monolayers infected with VSV always show 4+ CPE while uninfected monolayers show no evidence of cellular degeneration.

A similar procedure is followed to prepare CEF monolayers in round bottom microplates. Since these plates contain more wells, they allow for a greater number of assays as well as occupy less incubation space. Two drops of LE + S are added to each well, followed by the addition of one drop of CEF cell suspension ($2–3 \times 10^6$ cells/ml). The wells are either overlaid with 0.05 ml mineral oil or the entire plate is fitted with a commercial cover and incubated at 37°C for 24 to 36 hours.

For production of other monolayers in round-bottom microplates, 7000–8000 L cells or 2000–4000 HeLa cells suspended in 0.075 ml of Eagle's Minimum Essential Medium (MEM) with 10% fetal bovine serum are delivered to each well. Porcine kidney (PK) monolayers are also prepared by suspending 4000–6000 PK cells in 0.075 ml of LE + S. The wells are then overlaid with mineral oil or covered and then incubated at 37°C until monolayers are confluent at 72 hours.

Suspended Cell Cultures

Similar concentrations of CEF or cell line suspensions described for monolayer production are used for experiments with suspended cells. In this case, 0.025 ml of each virus dilution prepared in test tubes is added to each well, followed by addition of cell suspensions in the appropriate growth medium. Plates are covered and incubated at 37°C. The main advantage of this technique is that the cells and virus are added at the same time, thus eliminating the time necessary for production of monolayers as well as washing and titration procedures. Following incubation, wells are scored for the presence of CPE and ID_{50} titers are then determined.

Plaque Assay

Infectivity titers are determined according to the plaque assay method,[9,10] Each of the cell monolayers in 3-ounce prescription bottles, washed twice with 5 ml of Dulbecco's buffered saline, is inoculated with 0.1 ml of the virus dilution.

Following 1 hour of adsorption at room temperature, CEF monolayers are overlaid with 8 ml of nutrient agar medium containing neutral red. L, HeLa, and PK monolayers are each initially overlaid with 8 ml of nutrient agar medium without neutral red; a second overlay incorporating neutral red is added after 48 hours to these monolayers. Plaques appear as clear, unstained areas against a reddish background after 72-hour incubation. Titers are expressed as plaque-forming units (PFU) per milliliter of undiluted viral suspension. The plaque titers reported represent the average plaque count of five bottles used per dilution.

[9] G. D. Hsiung and J. L. Melnick, *Virology* **1**, 533 (1955).
[10] R. Dulbecco, *Proc. Nat. Acad. Sci. U. S.* **38**, 747 (1952).

Applications of Microtiter Procedure

The above procedures describing detailed methods for carrying out microculture assays were applied to determine residual infectivity of VSV following neutralization and heat inactivation tests. For the neutralization study, anti-VSV stock serum was employed to assay varying concentrations of VSV. The serum was heat-inactivated at 56°C for 30 minutes prior to use to destroy labile substances that have antiviral activity or affect neutralization. Serial tenfold dilutions of virus and serum prepared separately in test tubes were made with LE as diluent. An aliquot (1.0 ml) of each serum dilution was then mixed with decreasing virus dilutions (1.0 ml) in test tubes and the mixtures incubated at room temperature for 30 minutes. The concentration of the remaining unneutralized virus was determined by comparative titrations using the microtiter and plaque methods previously described. Monolayers in microplate wells were inoculated with 0.025 ml amounts of the same serum-virus mixtures used to inoculate 3-ounce bottles. After 48-hour incubation, ID_{50} titers were calculated and were based on the degree to which the serum inhibited CPE. The inhibition of CPE was compared to the degree by which the serum reduced plaque counts.

For the thermoinactivation study, virus samples were diluted tenfold in minimum essential medium (MEM) in French square bottles, pretempered to 54°C for 1 hour in a water bath. The inactivation time course was begun at the time the virus stock was added to the MEM. Heated samples were taken at various intervals and immediately cooled in an ice bath. Serial tenfold dilutions were prepared separately in test tubes with LE as diluent and aliquots were delivered to CEF monolayers in microplate wells and in 3-oz prescription bottles. Following incubation CPE was observed by microscopic observation, and infectivity titers were calculated. This was then compared to PFU titers obtained from plaque counts.

Discussion

VSV titrations were compared using microtiter suspended cell and preformed cell monolayer methods with the routine plaque assay. The results shown in Table I indicate that the microtiter procedure using preformed monolayers produced higher infectivity titers than cell suspensions or plaque formation. A two-way analysis variance showed that these differences are statistically significant ($p = 0.05$). No statistically significant difference in sensitivity was revealed between the cell suspensions and the plaque assay method. However, the microtiter procedure facilitated a more rapid assay of VSV with no loss in accuracy. The presence of lamb serum throughout the incubation of microtiter suspended cell plates did not affect virus titers, since neutralizing antibodies to VSV were not observed.

The reproducibility of the microtiter system has been investigated, in work

TABLE I
Comparative Titrations of VSV by Microtiter and Plaque Methods Using Suspended Cells and Preformed Monolayers[a]

Expt. No.	Cell line	Microtiter ID_{50}-average of two microplates: Suspended cells	Microtiter ID_{50}-average of two microplates: Preformed monolayers	Plaque assay PFU/ml
1	CEF	5.5×10^8	8.7×10^8	1.2×10^8
	L	2.2×10^8	1.7×10^9	3.6×10^8
2	CEF	1.0×10^9	1.4×10^9	2.5×10^8
	HeLa	8.6×10^7	—[b]	1.8×10^8
	PK	1.0×10^8	—	1.6×10^8
3	CEF	9.8×10^8	3.0×10^9	3.5×10^8
	PK	—	8.5×10^7	5.2×10^8
4	CEF	—	1.3×10^9	3.7×10^8
	HeLa	1.4×10^8	3.4×10^8	6.2×10^8
5	CEF	6.3×10^8	8.6×10^8	2.0×10^8
	L	1.5×10^8	4.1×10^8	3.0×10^7
	PK	8.2×10^6	1.5×10^7	7.0×10^7
	HeLa	1.7×10^7	3.5×10^7	4.8×10^7

[a] From L. J. Rosenthal and I. L. Shechmeister (see footnote 12).
[b] Not done.

presented elsewhere.[3-5,11,12] In our laboratory, using ID_{50} titers obtained from a series of five microplates with cell suspensions and preformed monolayers, the standard error of the mean was computed to be 0.48 $\log_{10}$ and 1.10 $\log_{10}$, respectively. In the data presented in Table I standard errors are not determined since the ID_{50} reported represented the average ID_{50} from two microplates.

In other experiments presented in Fig. 2, using a sample of VSV stored at 4°C, titrations of residual infectivity were carried out periodically for various times up to 100 days by microtiter and plaque methods on preformed CEF monolayers. The ID_{50} titers were calculated from two replicate microplates using six wells per dilution, while plaque titers were determined from the average count of four bottles used per dilution. Regardless of the storage time, the ID_{50} titers proved statistically higher ($p = 0.05$) than the plaque counts. Both methods also showed a close agreement in the rate of inactivation of VSV stored at 4°C.

Comparative virus neutralization tests were also carried out using both microtiter and plaque methods on CEF monolayers. Similarly, the microtiter procedure was statistically ($p = 0.10$) more sensitive in detecting unneutralized virus than the plaque procedure. The data are presented in Fig. 3 and the individual experiments are designated with the capital letter A, B, and C. In these experiments, the ID_{50} titers paralleled plaque neutralization results and revealed similar viral neutralization characteristics.

An evaluation of thermoinactivation experiments demonstrated again that the microtiter procedure with preformed cell monolayers was statistically ($p = 0.10$)

[11] L. J. Rosenthal, Master's thesis, Southern Illinois University, Carbondale, Illinois, 1967.
[12] L. J. Rosenthal and I. L. Shechmeister, *Appl. Microbiol.* **21**, 400 (1971).

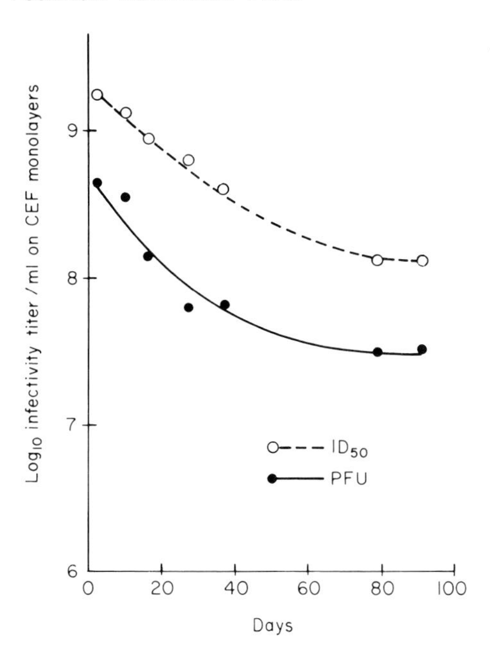

Fig. 2. Survival curve of VSV stored at 4°C as measured by microtiter and plaque assay methods on CEF monolayers. (From L. J. Rosenthal and I. L. Shechmeister.[12])

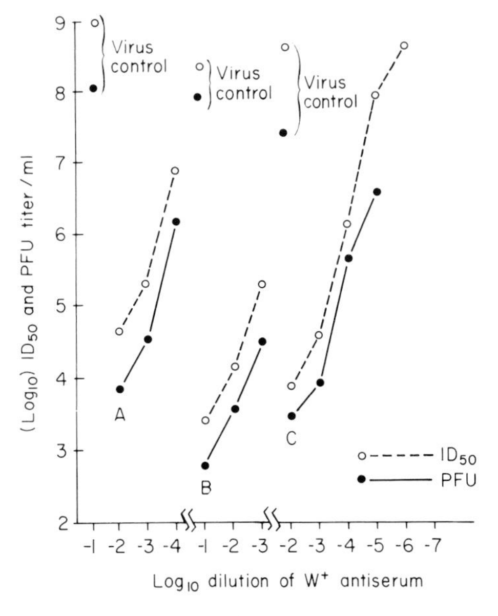

Fig. 3. Comparative neutralization tests of VSV measured by microtiter and plaque assay methods on CEF monolayers. (From L. J. Rosenthal and I. L. Shechmeister.[12])

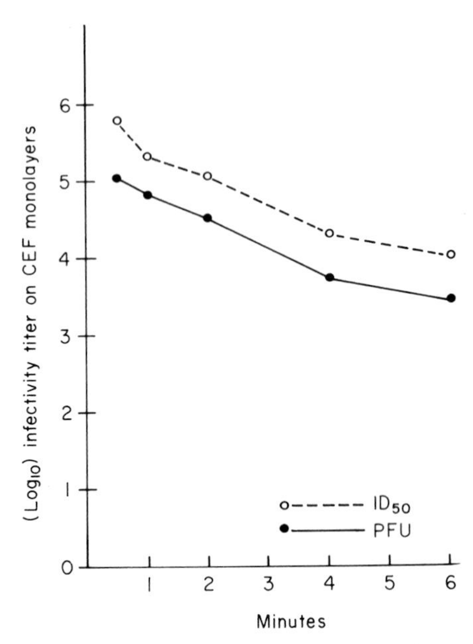

Fig. 4. Inactivation curve of VSV at 54°C as measured by microtiter and plaque assay methods on CEF monolayers. (From L. J. Rosenthal and I. L. Shechmeister.[12])

a more sensitive measure of viral infectivity than plaque formation. The results of a typical inactivation experiment with VSV are presented in Fig. 4. In all samples taken during inactivation, the ID_{50} titers are higher than plaque counts. The data also showed good agreement in the rate of inactivation of VSV as determined by either the microplate or plaque method. Thus, the microtiter procedure is applicable for inactivation studies of VSV in addition to being simpler, less costly, and as accurate as the plaque assay.

We conclude that the infectivity titer obtained from cell cultures in disposo-trays either with cell suspensions or preformed monolayers is a useful parameter in titration, neutralization, and inactivation experiments with VSV. The micromethod not only compares favorably with the plaque method for accuracy, but also has the advantage of economy and less effort. The disposo-trays are currently being used for insect tissue culturing and would appear applicable in cell fusion and virus rescue studies.

CHAPTER 1

Microculture Procedures

B. Simian Virus 40

James A. Robb

Simian virus 40 (SV40) is an icosahedral virus composed of 72 capsomeres. The virion contains about six structural proteins and a supercoiled double-stranded DNA of about 3×10^6 daltons. This virus is oncogenic for all mammalian cells so far tested, including human cells, and has also transformed cells from the gecko lizard. Although SV40 can infect almost any animal cell, it replicates efficiently only in primary African green monkey kidney (AGMK) cells and cell lines derived from AGMK cells such as TC7, CV-1, BSC-1, AH-1, VERO, and MK-134. These cells are termed productive cells for SV40 infection. Some cell lines such as mouse Balb/3T3 cells do not permit the replication of SV40 DNA or produce progeny virions, although these cells efficiently synthesize the SV40-specific tumor (T) antigen. This type of cell is termed an abortive or nonproductive cell for SV40 infection. Other cells such as human WI-38 cells and primary Chinese hamster cells permit SV40 DNA synthesis and progeny virus production in about 1 to 5% of the cells under conditions where more than 90% of the cells are synthesizing T antigen. This type of cell is termed semiproductive for SV40 infection. Productive, semiproductive, and abortive cells can be transformed by SV40 with appropriate techniques.[1-3]

There are two SV40-specific intranuclear antigens that can be used to monitor the early and late phases (prior to and after viral DNA synthesis) of the SV40 infectious cycle in Microtest plates. SV40-specific T antigen is synthesized during productive, semiproductive, and abortive infections and in SV40-induced tumor cells. This antigen is not present in mature virions, and its synthesis is not dependent upon prior synthesis of SV40 DNA (T antigen synthesis is an "early" function). The SV40 virion (V) antigen is the viral capsid. Synthesis of SV40 DNA is required prior to the initiation of V antigen (V antigen synthesis is a "late" function). The immunofluorescent micromethods for assaying T and V antigen in Microtest plates are described below. Although the following microtechniques have been developed for SV40, they should be applicable, with appropriate modifications, to many animal viruses.

[1] P. H. Black, *Annu. Rev. Microbiol.* **22**, 391 (1968).

[2] J. S. Butel, S. S. Tevethia, and J. L. Melnick, *Advan. Cancer Res.* **15**, 1 (1972).

[3] G. J. Todaro, *In* "Fundamental Techniques in Virology" (K. Habel and N. Salzman, eds.), p. 220. Academic Press, New York, 1969.

Infection of Cells in Microtest Plates

Cells in Microtest plates (Falcon Plastics, No. 3034, 2 × 3 inches, sixty wells per plate) can be infected in suspension or as monolayers.[4-6] The inoculation, feeding, and handling of cells in Microtest plates have been described in Section V (Microtest Plates: Replica Plating). To infect cells in suspension, the virus stock is diluted with a cell suspension of 1 to 5 × 10^5 cells/ml depending upon the desired percentage of confluency in the well. Each well receives 0.002 ml virus–cell suspension using 2–5% serum in the medium. The type of serum depends upon the cell being used (e.g., fetal bovine serum, Flow Laboratories, Rockville, Maryland, for monkey and human cells; calf serum, Colorado Serum Co., Denver, Colorado, for mouse cells). At least 2% serum is required for adequate cell attachment. Serum levels above 5% inhibit SV40 infection in most cell lines. When human and monkey cells are used for first cycle infections (described below), the plates can be incubated with only 0.002 ml per well until fixation. When mouse cells are used, an additional 0.010 ml per well medium is added at 4 to 6 hours after inoculation of the virus–cell suspension. The 4- to 6-hour delay is necessary for cell attachment and virus adsorption. If 0.010 ml per well is used initially, the efficiency of adsorption is greatly reduced (by about 90%) due to the large volume of medium in which adsorption must take place. This type of infection in suspension results in a "chronic" infection; the virus is left in contact with the cells during the length of the experiment.

Another method can be used when short-term virus–cell contact is desired. The plates are inoculated with cells alone and are incubated for 8 to 24 hours to allow complete cell attachment and monolayer formation. After the cells have attached, the medium is aspirated, and 0.010 ml per well phosphate-buffered saline (PBS) at pH 7.4 or serum-free medium (SFM) is added (see Section V). After aspirating again, 0.002 ml per well of the virus dilution in SFM is added, and the plates are incubated for 2 hours at the desired temperature. Following this viral adsorption period, 0.010 ml per well PBS or SFM is added, and the wells are aspirated. Then either 0.002 ml per well (for human and monkey cells) or 0.010 ml per well (for mouse cells) medium with 2 to 5% serum is added, and the plates are incubated until fixation. If the experiment requires that all non-penetrated virus be neutralized, 5% rabbit SV40 neutralizing serum (Grand Island Biological Co.) can be included in the final incubation medium.

Immunofluorescent Detection of SV40-Specific Intranuclear T and V Antigens

Commercially available antisera are used for detecting SV40-specific T and V antigens. Hamster anti-T ascitic fluid and bovine anti-SV40 neutralizing

[4] J. A. Robb and R. G. Martin, *Virology* **41**, 751 (1970).
[5] J. A. Robb and R. G. Martin, *J. Virol.* **9**, 956 (1972).
[6] J. A. Robb, H. S. Smith, and C. D. Scher, *J. Virol.* **9**, 969 (1972).

serum (anti-V antibody) are available from Flow Laboratories. Fluorescein-conjugated rabbit or goat anti-hamster-globulin globulin and anti-bovine-globulin globulin are available from Sylvana Co., Millburn, New Jersey. The anti-globulin globulin must be used, since the anti-globulin serum is not adequate.

In general, the following dilutions are used, but should be checked for each serum lot number and for each type of SV40-cell infection. The hamster anti-T ascitic fluid and both fluorescein-conjugated anti-globulin globulins are brought to a final dilution of 1:19 (v/v) in PBS, pH 7.4. The bovine SV40 neutralizing serum (anti-V) is brought to a 1:4 (v/v) final dilution in PBS. Twenty-milliliter volumes are prepared, and 1- to 3-ml aliquots of the diluted ready-to-use solutions are stored at −20°C. The ready-to-use aliquots should not be thawed and refrozen more than once for best results. Increased sensitivity of the anti-T antibody can be achieved by adsorbing the ascitic fluid on noninfected cell monolayers at a 1:2 dilution with PBS for 30 minutes at 37°C, with agitation every 5–10 minutes. The adsorbed fluid is centrifuged at 1500 *g* for 5 minutes to remove cellular debris. This adsorption is very important when T-antigen is being assayed in acutely infected abortive cells that synthesize relatively little T antigen, such as Balb/3T3 cells.

The fixation of cells in Microtest plates is accomplished with absolute methanol by the following procedure. After the appropriate incubation time, the plate is rinsed once with PBS from a plastic squirt bottle. The PBS should not be shaken from the wells; the excess should be poured off. Mouse cells must be rinsed with PBS containing 1% fetal bovine serum to prevent the extrusion of nuclei during the methanol fixation. Absolute methanol is then applied from a plastic squirt bottle with the plate held in the vertical position, so the methanol–PBS mixture can rapidly run off. This fast methanol–PBS run-off step is necessary for optimal preservation of T and V antigen. After the plate is shaken hard to remove the residual methanol–PBS solution, it is flooded with absolute methanol so that all wells are filled. The plate is then allowed to stand at least 30 seconds (up to 10 minutes) before the methanol is shaken off. After the methanol is removed, the plates are immediately dried with compressed air before immunofluorescent staining or storage. The plates can be stored at −55° to −70°C for at least 1 year at this stage. Do not store the plates at room temperature or 4°C.

Indirect immunofluorescent staining of the infected cells in the Microtest wells is the most economical method, although direct staining with fluorescein-conjugated anti-T or anti-V antibody can be done with the same techniques. Enhanced sensitivity can be achieved by using fluorescein-conjugated anti-T or anti-V antibody for the first step in the indirect method, so that two layers of fluorescein-conjugated antibody are applied instead of one. The following procedure is used for indirect staining. After the plates have been fixed and air-dried for at least 5 minutes or allowed to come to room temperature after freezing, the wells are rinsed once with PBS and thoroughly shaken to remove as much PBS as possible. This PBS rinse prevents air bubble formation when the antiserum is applied. Then 0.005 ml per well T or V antiserum is applied using a Hamilton repeating dispenser (see Section V). The plates are then incubated at 37°–41°C for 2 hours followed by one PBS rinse from a squirt bottle. After thoroughly shaking off the

PBS, the fluorescein-conjugated anti-hamster (for T antigen) or anti-bovine (for V antigen) solutions are added at 0.005 ml per well, and the plates are again incubated at 37°–41°C for 2 hours. The 2-hour incubation time for the antisera is minimal for optimal staining. Either of the antiserum applications can also be incubated overnight at 4°C in order to fit into the daily work schedule. The plates are then rinsed once with PBS, and 0.005% Evans' blue (Chroma-Gesellschaft, Schmid and Co., Stuttgart, Germany, supplied by Roboz Surgical Instrument Co., Washington, D. C.) in PBS with 1% fetal bovine serum is applied with a squirt bottle as a counterstain. The Evans' blue solution is left in the wells for 30 to 120 seconds depending upon the cell line and antigen being studied. If the Evans' blue is left on too long, the green fluorescence of the infected nuclei will be masked by the red fluorescence of the Evans' blue. Destaining of the Evans' blue can be accomplished by applying 0.005 ml per well bovine SV40 neutralizing serum (anti-V solution) at a 1:4 dilution with PBS for 30 minutes at 37°–41°C. The cells then can be restained appropriately with Evans' blue. After the Evans' blue is rinsed off with PBS (one rinse is sufficient), the wells are refilled with PBS for microscopic observation or storage. The plates can be read immediately or stored at 4°C for several months. If the PBS evaporates during storage, the wells can be refilled with PBS prior to microscopic examination. All of the above techniques can be used for cells in Petri dishes with the only modification being the use of 0.5 ml antisera per dish. The Petri dish technique eliminates the use of conventional coverslip techniques.

An excellent system for observing the fluorescent nuclei of infected cells involves the use of a Leitz fluorescence microscope equipped with a Leitz dry dark-field condenser and a Leitz or Zeiss FITC filter with matching barrier filter. A dry dark-field condenser is necessary to eliminate the mess produced by the immersion oil that is required when an oil dark-field condenser is used. The use of bright-field fluorescence microscopy has not been successful using Microtest plates.

First Cycle Infection and End-Point Dilution Titration (Multiple Cycle Infection)

The ability of the input virus to initiate T and/or V antigen synthesis is determined in the following manner (first cycle infection). Cells are infected in suspension or as monolayers as described above. The observation of first cycle T and V antigens by the above immunofluorescent technique is maximal at about 40 to 48 hours after infection at 39°–41°C; 68–76 hours at 37°C; and 90–102 hours at 33°C. If longer incubation periods are required, the appearance of T and V antigen produced by progeny virus infection (second cycle infection) can be prevented by adding rabbit SV40 neutralizing serum directly to the incubation medium in the wells at 20 to 24 hours after infection. In this way, a final concentration of 5 to 9% is achieved (e.g., 0.001 ml neutralizing serum is added per 0.010 ml incubation medium per well or 0.001 ml 15% neutralizing serum in PBS is added per 0.002 ml incubation medium per well).

The above method permits the determination of the relative number of T-positive nuclei (T-forming units, TFU) or V-positive nuclei (V-forming units, VFU) per well bottom. This method is useful for comparing different virus preparations on a relative basis when equal numbers of input virions are present (equal multiplicity of infection, m.o.i.). The method is modified slightly when the absolute number of TFU per milliliter or VFU per milliliter of a given virus stock is required.[4] Serial tenfold dilutions of the virus stock are made and the wells inoculated at one plate per dilution per virus stock. The plates are fixed and stained for first cycle T and V antigen as described above. The wells are examined with a 10× objective so that every cell in any given well can be observed. The wells that do not contain any antigen-positive nuclei are counted, and the TFU per milliliter or VFU per milliliter titer is calculated from the Poisson distribution. Titers derived from plates that have most or very few of the wells infected are subject to large statistical error. The Poisson distribution is

$$P_r = (s^r e^{-s})/r$$

where s is the average number of infectious particles per sample, r is the actual number in a given sample, and P_r is the probability of having r particles in a given sample. Thus, by counting the number of wells that do not have any T or V antigen-positive nuclei (an indication that virus infection has not occurred, i.e., $r = 0$) one can determine the titer (s times the dilution factor) of the virus sample from $P_0 = e^{-s}$. Table I allows the calculation of titers simply by knowing the number of negative wells (P_0) in the sixty-well plate. To use this table, the

TABLE I

TFU, VFU, and IU Titer Determination Table Based upon the Number of Negative Wells (P_0) in a Sixty-Well Microtest Plate[a]

P_0	T	P_0	T	P_0	T
1	2045	21	525	41	190
2	1700	22	500	42	179
3	1500	23	480	43	168
4	1360	24	460	44	155
5	1240	25	438	45	144
6	1150	26	418	46	133
7	1075	27	400	47	122
8	1005	28	381	48	112
9	950	29	363	49	101
10	895	30	347	50	91
11	850	31	330	51	81
12	805	32	315	52	72
13	752	33	299	53	62
14	730	34	284	54	53
15	695	35	269	55	44
16	660	36	255	56	34
17	630	37	242	57	25
18	600	38	228	58	17
19	575	39	215	59	8
20	550	40	203	60	

[a] The number of TFU/ml, VFU/ml, or IU/ml is determined by multiplying the number in column T corresponding to the number of negative wells, P_0, by the reciprocal of the virus dilution. See text for an example.

number in column T that corresponds to the number of negative wells, P_0, is multiplied by the reciprocal of the dilution used in that particular plate. For example, if a 10^{-6} dilution plate has thirty-three negative wells ($P_0 = 33$), multiply 299 (column T) by 10^6 ($1/10^{-6}$ = reciprocal of dilution) to get 3.0×10^8 IU/ml, VFU/ml, or TFU/ml. This table has been corrected for the 500-fold dilution that occurs when the plates are inoculated at 0.002 ml per well.

The propagative titer or infectious units per milliliter (IU/ml, the number of virions that are able to propagate per milliliter of viral stock) can be determined by the following end-point dilution titration micromethod. Plates are made in the identical manner as for the first cycle TFU and VFU titers above, except that 0.010 ml medium per well is added at 20 to 24 hours after infection, and the plates are incubated for 8 to 9 days at 39°–41°C, 10–12 days at 37°C, and 13–15 days at 33°C. Productive monkey cells must be used for the infectious unit titer. At the appropriate times, the cells are fixed and stained for V antigen. The wells that have less than five V antigen-positive nuclei are counted as negative, and the IU/ml titer is determined as above for the TFU/ml and VFU/ml titers. The IU/ml titer of a viral stock, as determined by this end-point dilution titration micromethod, is the same as the plaque forming titer (PFU/ml) determined by standard plaque assay on the same cells. All three titers (TFU/ml, VFU/ml, and IU/ml) should be similar in virus stocks that contain few defective particles.[7]

Isolation of Temperature-Sensitive Mutants

A convenient method for isolating temperature-sensitive mutants blocked prior to the assembly of viral capsids (V antigen-negative mutants) combines the virus replica plating technique described in Section V and the above cell infection and microimmunofluorescent techniques. A mutagenized virus stock[4] is end-point dilution titered at 33°C and inoculated as a virus–cell suspension at one IU/well in 0.002 ml per well using monkey cells. At 20 to 24 hours after infection at 33°C, 0.010 ml medium per well is added. These "master" plates are then incubated at 33°C for 4 to 5 weeks and frozen at −70°C. Two plates should be fixed and stained for V antigen at about 2 weeks, so that the number of negative wells can be determined. There should be 30–70% negative wells, so that the optimal number of wells containing a single cloned virus (about 30 to 37%) will be present. If P_0 in the Poisson equation varies from 30 to 70%, P_1 will vary between 30 and 37%. Each master plate is replica plated into two recipient plates. One recipient plate is incubated at 41°C for 8 days and the other recipient plate is incubated at 33°C for 15 days. Following incubation the plates are stained for V antigen. Wells that have no V-positive nuclei at 41°C and many V-positive nuclei at 33°C contain potential temperature-sensitive mutants. The master plates that were restored at −70°C after replica plating are thawed, and the virus stocks are removed from the appropriate wells with disposable 1-ml tuberculin syringes. These cloned potential mutant stocks are then grown to

[7] S. Uchida, K. Yoshiike, S. Watanabe, and A. Furuno, *Virology* **34**, 1 (1968).

sufficient concentration for further testing. Any mutants thus isolated should be named according to the proposed nomenclature for SV40 mutants.[8]

Characterization of Temperature-Sensitive Mutants Using Microtest Plates

The micromethods described below have been used to characterize the *ts*101* temperature-sensitive mutant during productive and abortive infections.[5,6] The temporal location of the mutational block at restrictive temperature is determined as follows. Equal m.o.i. is used for the wild-type parental virus and all mutants (the IU/ml titer is determined at 33°C in monkey cells). Duplicate plates are inoculated using a different viral stock in each of the six rows per plate. One plate is incubated at 41°C and the other at 33°C. The plates are fixed and stained for first cycle T and V antigen using five wells per row for both T and V antigen. The internal control is very good because T and V antigen are assayed in the same plate for all the virus stocks. The number of antigen-positive nuclei per well bottom is counted and averaged for the five wells per antigen per virus. The 33°/41°C TFU and VFU ratios for the wild-type are used as a baseline for the mutant ratios. The wild-type ratio should be about 1 to 5 while the mutant ratios can vary from 10 to 1000. The mutants can therefore be rapidly screened for the temporal location of the mutational block and for their relative leakiness of T and V antigen synthesis as compared to the wild-type virus.

The following microcomplementation test can be used for mutants that are blocked prior to V antigen synthesis. Duplicate plates containing wild-type and mutant viruses in single infections (virus stock 1:1 with the cell suspension) and mixed infections (viral stocks 1:1 with each other) are incubated at 41°C. One plate is made with the single infections only and incubated at 33°C. After 40 to 48 hours at 41°C, the duplicate plates are fixed. One is stained for T antigen and the other is stained for V antigen. If one row is used for each single and mixed infection, the ten wells can be averaged per infection per antigen for a very accurate number of TFU and VFU per well bottom. Complementation between two mutants occurs if the number of VFU per well bottom in the mixed infection is more than three times the number of the sum in the two corresponding single infections. Negative complementation is determined by observing whether the number of wild-type VFU per well bottom in the single infection is decreased by any mutant in a mutant–wild-type mixed infection. The plate stained for T antigen is used to determine whether complementation between T-negative mutants occurs. The 41°C T antigen plate is also a control for T-positive, V-negative mutants. The 33°C single infection plate is stained for T and V antigen at 96 hours using five wells per row for both T and V antigen. This plate is used as a control to ensure that all mutants behave as mutants in the single infections.

The Microtest plates permit complicated reciprocal temperature-shift experiments to be performed with relative ease during the characterization of tempera-

[8] J. A. Robb, P. Tegtmeyer, R. G. Martin, and S. Kit, *J. Virol.* 9, 562 (1972).

ture-sensitive mutants. Replicate plates are made containing wild-type virus and one to five different mutant viruses using one row for each viral stock. Equal numbers of plates are incubated at 41° and 33°C. After varying periods of time, the plates are either stained for T and V antigen (five wells per row per antigen) or shifted to the reciprocal temperature and allowed to incubate for varying periods of time at the new temperature before T and V staining. This general type of experiment will determine how long the mutant genome survives at 41°C (as determined by the down-shift) and how much time is required to overcome the mutational block at 33°C (as determined by the up-shift).

CHAPTER 1

Microculture Procedures

C. Antibody Neutralization Surveys[1]

S. S. Kalter

The neutralization test is considered to be the most specific and reliable laboratory procedure for the identification of viruses and for the precise determination of viral antibody development. Many modifications of the neutralization procedure are available, but the basic principle is the same: specific neutralization of infection by antibody. Accordingly, when an active virus and serum containing homologous antibody are brought together and this mixture is placed in contact with susceptible host cells under specific conditions, no lesions result. If the virus and the antibody are not related, neutralization does not occur, and the host cells become infected. Since the neutralization test is dependent upon infection, living cells (animals, embryonated eggs, or cell cultures) are required. Details concerning the basic techniques of the neutralization test are well documented in a number of sources.[2-4] Other investigators have recommended differ-

[1] Funded in part by U.S.P.H.S. Grant RR00361 and Contract NIH 71-2348 and WHO Grant Z2/181/27. This laboratory serves as the WHO Regional Reference Center for Simian Viruses.

[2] E. H. Lennette and N. J. Schmidt (eds.), "Diagnostic Procedures for Viral and Rickettsial Infections," 4th ed. Amer. Public Health Assoc., Inc., New York, 1969.

[3] J. Casals, *Methods Virol.* 3, 113 (1967).

[4] S. S. Kalter, "Procedures for Routine Laboratory Diagnosis of Virus and Rickettsial Diseases." Burgess, Minneapolis, Minnesota, 1963.

ent procedures for inoculating cells simultaneously with their neutralization tests.[5,6] We have found the method described herein to eliminate any tendency to disrupt the cells or scarify the well surface of the plates, thereby making readings difficult. This technique avoids the use of dilution loops and permits one-step inoculation of microplates of established monolayers. Furthermore, it permits the use of primary cell or serial cell cultures which are allowed to grow within each well.

This section is concerned with the application of microprocedures in surveying for neutralizing antibodies. These procedures have also been well described and are adaptations of the serological microtechniques developed by Takatsy *et al.*[7] and modified by Sever.[8] Microneutralization protocols and necessary equipment for the satisfactory performance of these tests have been previously reported.[9-14]

Plates

Sterile, nontoxic, disposable microculture plates are commercially available from a number of sources: for example, Falcon Plastics Division, BBC, Oxnard, California and Linbro Chemical Co., New Haven, Connecticut. These plates are of rigid construction, sterile, and consist of ninety-six flat-bottomed wells per unit. Sterile, nontoxic lids and pressure-sensitive tapes are also available.

Cell Cultures

Primary or serial cell cultures derived from any suitable source may be used satisfactorily. Trypsinization of tissue for primary cell culture or removal of cell monolayers from glass or plastic bottles is accomplished by any of the existing accepted procedures. For primary cultures, growth medium consists of Hanks' buffered saline supplemented with 0.65% lactalbumin hydrolysate and 2% inactivated newborn calf serum. For cell maintenance, Eagle's Minimal Essential Medium with 0.5% inactivated newborn calf serum is used after the cells achieve confluency, usually 4–6 days after seeding. Antibiotics which should be added to all media include: 1 unit bacitracin, 100 units penicillin, 100 μg neomycin per

[5] M. J. Rosenbaum, E. A. Edwards, and E. J. Sullivan, *Health Lab. Sci.* **7**, 42 (1970).

[6] L. W. Catalano, D. A. Fuccillo, and J. L. Sever, *Appl. Microbiol.* **18**, 1094 (1969).

[7] G. Y. Takatsy, J. Furesz, and E. Farkas, *Acta Physiol.* **5**, 241 (1954).

[8] J. L. Sever, *J. Immunol.* **88**, 320 (1962).

[9] M. J. Rosenbaum, I. H. Phillips, E. J. Sullivan, E. A. Edwards, and L. F. Miller, *Proc. Soc. Exp. Biol. Med.* **113**, 224 (1963).

[10] M. Kende and M. L. Robbins, *Appl. Microbiol.* **13**, 1026 (1965).

[11] D. A. Fuccillo, L. W. Catalano, Jr., F. L. Moder, D. A. Debus, and J. L. Sever, *Appl. Microbiol.* **17**, 619 (1969).

[12] L. W. Catalano, D. A. Fuccillo, and J. L. Sever, *Appl. Microbiol.* **18**, 1094 (1969).

[13] M. J. Rosenbaum, E. A. Edwards, and E. J. Sullivan, *Health Lab. Sci.* **7**, 42 (1970).

[14] R. J. Helmke, R. L. Heberling, and S. S. Kalter, *Appl. Microbiol.* **20**, 986 (1970).

milliliter. Media is removed by inverting plates or by vacuum aspiration of individual wells. Other cell systems might require different media, for example Kinney *et al.*[15] recommend the following growth media for Vero cells (African green monkey kidney): Eagle's Minimal Essential Medium in Earle's balanced salt solution containing 0.1% bicarbonate and 10% fetal calf serum with 100 μg streptomycin and 100 μg neomycin per milliliter. For maintenance, a reduction of the fetal calf serum to 5% and an increase of bicarbonate to 0.2% is recommended.

Each well is seeded with approximately 2×10^4 cells. Cell counts are based on the erythrosine B dye exclusion method. Cells and media may be dispensed with 1.0-ml Cornwall automatic syringes. Since each well accommodates volumes of 0.2 to 0.3 ml, it is unnecessary to add mineral oil overlays. The plates with lids are held in a humid, 5% CO_2 incubator or sealed with pressure tape and maintained in an ordinary incubator at 37°C. Individual plates and lids may be sealed with masking tape wrapped around the edges. Multiple units with lids may be stacked and covered with Saran Wrap (Dow Chemical Co.).

Virus

Pools of virus for surveys may be made in any suitable medium. It is preferable to employ culture conditions which will not interfere with the test by virtue of nonspecific inhibitors or because of the presence of anticellular antibody in the serum to be examined. The desired virus should be antigenically broad enough to include all the strain variants but specific enough not to cross with other members of the virus group. For such purposes, a prototypical viral strain that has been well characterized and obtained from a reputable source should be used. These pools should be of sufficient volume to supply all virus for the particular survey. Titers of each virus pool are ascertained in order to determine the $TCID_{50}$ per 0.1 ml. In a test, 100 $TCID_{50}$ virus per 0.1 ml is generally used, although other amounts may give satisfactory answers. Controls and interpretation are required for final evaluation of results regardless of the infecting dose.

Sera

For survey work, a single dilution of serum, generally 1:10 (initial) in phosphate-buffered saline at pH 7.0 to 7.2 is satisfactory. More concentrated serum levels are often cytotoxic in microprocedures. All sera are usually inactivated at 56°C for 1 hour after dilution. When processing large numbers of sera it is preferable to prepare the serum dilutions in advance, place aliquots in sterile tubes, and store them at -20°C. One tube of each serum dilution is removed, thawed, and used for testing. We generally store sera in units of 0.1 ml; this volume has been found to be sufficient for routine microtest surveys. Larger volumes of 0.2 or 0.3 ml may be similarly stored.

[15] M. T. Kinney, K. L. Albright, and R. P. Sanderson, *Appl. Microbiol.* **20**, 371 (1970).

TEST

To 0.1 ml diluted serum, add 0.1 ml virus which has been diluted to 100 $TCID_{50}$ per 0.1 ml. Agitate the mixture for a few moments in a Vortex mixer, and allow it to stand approximately 1 hour at room temperature. Then add 0.4 ml of maintenance medium to the reaction tube. Addition of this medium serves as a transfer vehicle and also provides a change of medium for the cells. Two-tenths of each serum–virus mixture is then inoculated into two wells with an ordinary 1.0-ml serological pipette. Control titrations include tenfold dilutions of each virus diluted 1:6 in maintenance medium and inoculated into four wells per dilution. Microtests are read on an inverted microscope at low power (40–60×) 48 hours after cytopathology is observed in the virus (100 $TCID_{50}$) control wells. Titrations are kept 7–14 days to determine the actual virus dose used in the test.

NOTES

A number of variations on the above procedure may be introduced for any given virus–serum situation. Generalizations are difficult, since many virus-cell reactions vary from virus strain to virus strain and cell system to cell system. Prior to initiation of any large scale survey, a pilot test is strongly recommended to determine appropriate cell systems and suitable virus–serum dilutions. Macrotests, run concurrently with the pilot test, may be necessary in order to obtain comparative information.

CHAPTER 2

Quantal and Enumerative Titration of Virus in Cell Cultures

C. H. Cunningham

Procedures for biological assay of viruses require standardized quantitative methods to be reliably reproducible. The value of any method is ultimately dependent upon its ability to yield statistically accurate data. The degree of statistical probability desired may differ, however, in different laboratories.

Nevertheless, there are certain principles involved in standard methods. The technical details of statistical measurement and probability are beyond the scope of this portion of the book and the reader is referred to other sources[1-4] for this information.

General Considerations

Quantitative studies of viruses[5] are interpreted from titration of the virus and the quantal or enumerative type of dose-response employed. Titration is a quantitative assay by serial dilution, differing by a constant dilution factor, of the relationship between the amount of virus-containing material, and the frequency of response of the indicator system. The response must be characteristic of infection of the indicator system with the specific virus and one that is readily interpretable without bias.

Factors that must be considered in the statistical concept of quantitative methods are (1) the overall dilution error of the series of dilutions, (2) the dilution factor required to give the accuracy desired, and (3) the number of replicate measurements required for significance. For uniformity of comparison of results, the titer should be expresed in doses per milliliter.

The cumulative error in a dilution series is represented by $E \times \sqrt{n}$, where E is the error of an individual dilution and n is the number of dilution steps. Measurement of the material to be diluted and also of the diluent itself is subject to the same error. If a pipette having an accuracy of $\pm 2\%$ were used, the minimum error of any dilution would be $2\sqrt{2}$ or about 2.8%. If nine dilution steps were used, the minimum error of the final dilution would be $2.8\sqrt{9}$ or about 8.4%.

Any dilution factor can be employed. Tenfold (10^{-1}) dilutions, one volume of virus suspension + nine volumes of diluent, are used especially when the titer of the virus is unknown and may be within a large range of values. "Half-log" ($10^{-0.5}$) dilutions can also be used. The factor is $\sqrt{10}$ (3.16), one volume of virus suspension +2.16 volumes of diluent, so that two successive dilutions equal one tenfold dilution ($10^{-0.5} \times 10^{-0.5} = 10^{-1}$). Within the critical dose range, it is desirable to have the dilution factor small and on a geometric progression, one volume of virus suspension plus one volume of diluent, such as ½, ¼, ⅛, etc. Individual pipettes should be used for making the dilution series to avoid any carry-over from tube to tube.

Five tubes of cell cultures per dilution are recommended for quantal assay and four plates for enumerative assay. For uniformity of comparison of results,

[1] W. R. Bryan, *Ann. N. Y. Acad. Sci.* **69**, 698 (1957).

[2] P. D. Cooper, *Advan. Virus Res.* **8**, 319 (1961).

[3] R. M. Daugherty, *In* "Techniques in Experimental Virology" (R. J. C. Harris, ed.), p. 169. Academic Press, New York, 1964.

[4] A. Kleczkowski, *In* "Methods in Virology" (K. Maramorosch and H. Koprowski, eds.), Vol. 4, p. 615. Academic Press, New York, 1968.

[5] C. H. Cunningham, "A Laboratory Guide in Virology," 7th ed. Burgess, Minneapolis, Minnesota, 1973.

the titer of the virus should be expressed per milliliter. If the inoculum were 0.1 ml, multiply by 10 (add 1.0 log), if 0.2 ml, multiply by 5 (add 0.7 log), etc.

Quantal Assay

The quantal type of assay is the ratio of positive responses to the total number of possible responses. It is employed with the 50% "end-point" method of titration in which groups of tubes of cell cultures are inoculated with certain dilutions of virus. For each member of a group the response is scored from an assigned numerical value. The percentage or proportion of positive responses per group is used for calculation of the statistical dose of virus contained in the original sample from which dilutions were made for titration. This is expressed as the median tissue culture dose (TCD_{50}).

The cytopathic effect (CPE) characteristic of that produced by the virus under study can be assigned numerical value as follows:

0 = no CPE
1 = ¼ of the cell culture affected
2 = ½ of the cell culture affected
3 = ¾ of the cell culture affected
4 = 4/4 or all the cell culture affected

The results are expressed as a fraction in which the denominator represents the number of possible responses and the numerator the number of observed responses. If five tubes of cell culture were used per dilution, the maximum possible score would be 20. If the sum of the positive scores for each of these five tubes per dilution were 15, for example, the results for that dilution would be recorded as 15/20.

The 50% end-point by the Reed and Muench[6] method is determined by interpolation from the cumulative frequencies of positive and negative responses to occur in a dilution in which there would be 50% positive responses and 50% negative responses. The range of test dilutions should extend from the highest with all positive responses through the highest with all negative responses. For explanation, assume that the data in Table I are the results obtained from titration of a virus using the above numerical values for scoring.

In this example, the dilution in which there would be expected to be 50% positive responses lies between the 10^{-7} and the 10^{-8} dilutions. The proportionate distance between these two dilutions is inferred by linear interpolation according to the following:

$$\frac{(\%\text{ pos. next above }50\%) - 50\%}{(\%\text{ pos. next above }50\%) - (\%\text{ pos. next below }50\%)} = \text{proportionate distance}$$

$$\frac{66.7 - 50.0}{66.7 - 14.3} = 0.3\ (0.32)\ \text{proportionate distance}$$

[6] L. J. Reed and H. A. Muench, *Amer. J. Hyg.* **27**, 493 (1938).

TABLE I

Procedure for Interpolation of 50% End-Point

		Dilution of virus			
		10^{-6}	10^{-7}	10^{-8}	10^{-9}
	Positive rate	20/20	12/20	4/20	0/20
	Positive-number	20	12	4	0
	Negative-number	0	8	16	20
Accumulation totals	Positive (←)	36	16	4	0
	Negative (→)	0	8	24	44
	Positive rate	36/36	16/24	4/28	0/44
	% Positive	100	66.7	14.3	0
			50%		

The following formula is used to calculate the 50% end-point:

$$\begin{array}{lcr} \text{Log lower dilution (dilution in which pos. next above 50\%)} & = & -7.0 \\ \text{Proportionate distance (0.3)} \times \text{log dilution factor (10)} & = & \underline{-0.3} \\ (0.3 \times -1.0) & & \\ \text{Sum (50\% end-point)} & & -7.3 \\ \text{Log TCD}_{50} = -7.3 & & \end{array}$$

$$\text{TCD}_{50} = 10^{-7.3} = \frac{1}{2.0 \times 10^{7}}$$

This is the end-point dilution, the reciprocal of which is the titer in number of infective doses per unit of inoculum. If the inoculum were 0.2 ml per tube of cell culture the final expression would be:

$$\frac{2.0 \times 10^{7}}{0.2} = 1 \times 10^{8} \text{ TCD}_{50} \text{ per milliliter}$$

When the 50% end-point method is used for dilutions other than tenfold, caution must be exercised to use the correct log of the dilution factor. Assume that twofold dilutions were made and the accumulation totals were 60% positive in the $\frac{1}{16}$ dilution and no positives in the $\frac{1}{32}$ dilution. The proportionate distance is 0.2. The 50% end-point is calculated as follows:

$$\begin{array}{lcr} \text{Log lower dilution (dilution in which pos. next above 50\%)} & = & -1.20 \\ \text{Proportionate distance (0.2)} \times \text{log dilution factor (2)} & = & \underline{-0.06} \\ (0.2 \times -0.3010) & & \\ \text{Sum (50\% endpoint)} & & -1.26 \\ \text{Log TCD}_{50} = -1.26 & & \end{array}$$

$$\text{TCD}_{50} = 10^{-1.26} = \frac{1}{1.8 \times 10^{1}}$$

If the inoculum were 0.2 ml per tube of cell culture the final expression would be:

$$\frac{1.8 \times 10^1}{0.2} = 9.0 \times 10^1 \text{ TCD}_{50} \text{ per milliliter}$$

With the Kärber[7] method for determining the 50% end-point, the sum (S) of the actual proportion of positive responses to each inoculum is substituted in the formula below. It is not necessary to use accumulation totals of positive response ratios as with the Reed and Muench[6] method, although these can be used. The range of dilutions of the inoculum should extend from the highest with all positive responses through the highest with all negative responses;

$$\text{Log TCD}_{50} = m - df[(S) - 0.5]$$

where m is log dilution containing highest concentration of virus; df is log of dilution factor; S is sum of actual proportion of position responses to each inoculum; and 0.5 is a constant that is used in all cases.

Using the data in Table I, the following is obtained:

$$\text{Log TCD}_{50} = -6 - (1.0)[(0.2 + 0.6 + 1.0) - 0.5] = -7.3$$

Enumerative Assay

The enumerative type of assay is based on the presence of focal lesions such as plaques in cell cultures with agar overlay in Petri dishes, bottles, etc. These are "real" units and not "statistical" units as calculated by the 50% end-point method. The most important criterion for assay by the focal lesion or plaque method is that, at least over the range of statistical practicability, the count must be proportional to the concentration of the virus. The virus suspension should be sufficiently diluted so that discrete, uniformly distributed plaques are formed without overlap on the cell culture.[2] The lesions are expressed as plaque-forming units (PFU).

The values of the focal lesion counts are directly proportional to the amount of infectious virus in the inoculum. The titer as the number (N) of infectious units of virus per milliliter is calculated from the average number of focal lesions $(\bar{X})$, the volume of inoculum (V) in milliliter, and the dilution (D) according to $N = \bar{X}/VD$. In a series of dilutions differing by a constant factor such as 2, the dilutions will differ by a constant amount ($\log \frac{1}{2} = 10^{-0.3010}$; $\log \frac{1}{4} = 10^{-0.6021}$; $\log \frac{1}{8} = 10^{-0.9031}$; . . . .). If 0.5 ml of a $\frac{1}{4} \times 10^{-4}$ dilution of virus produced on the average 53.75 lesions, the titer as N is $53.75/(\frac{1}{2} \times \frac{1}{4} \times 10^{-4}) = 4.3 \times 10^{-6}$. The titer as $\log_{10} N$ is $1.7304/(-0.3010) + (-0.6021) + (-4.0000) = 1.7304 + 0.3010 + 0.6021 + 4.000 = 6.6335 = 4.3 \times 10^6$.

Data such as those above (TCD_{50}, PFU) can be used to determine the factors necessary for preparing dilutions of stock virus to contain a desired number of infective units for various purposes.

The above examples of quantal and enumerative assays are based upon the

[7] G. Kärber, *Arch. Exp. Pathol. Pharmakol.* **162**, 480 (1931).

Poisson distribution and the single-particle concept of active virus. Providing there are no errors in dilution, there is a relationship between the 50% end-point units and focal lesion units. From the single-particle concept, half the hosts will receive no viral particles and the other half, which are infected, will have received more than one particle; a calculable proportion average of 1.38 particles. The overall average including those hosts not infected is 0.69 particle per host. If the systems for assay by both the end-point dilution and focal lesion methods have equal sensitivity, 0.69 focal lesion units equal one TCD_{50}. On this basis, titers expressed as 50% end-points will be 1.44× those expressed as focal lesions per unit of inoculum. The end-point unit equivalent to one focal lesion unit is the TCD_{63}, i.e., 63% positive and 37% negative responses ($e^{-1} = 0.367879 = 37\%$).

The ideal efficiency ratio 1:1 with respect to the virus-host systems is seldom exactly attained. For practical purposes, the focal lesion count (PFU) or the 50% end-point may be used for estimating absolute units of virus because of the greater simplicity of methods and many tables and references are designed for this.

The average number of "infectious particles" (d) present per inoculum may be estimated from p, the proportion of positive responses, according to $d = -\log_e(1 - p) = -2.303 \log_{10}(1 - p)$. If $p = 0.5$, then $d = -2.303 \log_{10}(1 - 0.5) = -2.303(-0.3010) = 0.6932$.

CHAPTER 3

Diagnosis of Virus-Infected Cultures with Fluorescein-Labeled Antisera

G. Shramek and L. Falk, Jr.

Studies of the interaction of viruses with tissue culture cells have been greatly facilitated by fluorescent antibody (FA) techniques as first described by Coons *et al.*[1] in studies of mumps virus antigens in tissues of experimentally infected animals. Direct and indirect staining are the basic types of FA tests and in both methods a specific antiserum is used that has been conjugated with a dye, usually fluorescein isothiocyanate (FITC) which fluoresces when excited by ultraviolet

[1] A. H. Coons, J. C. Snyder, F. S. Cheever, and E. S. Murray, *J. Exp. Med.* **91**, 31 (1950).

light. In the direct test FITC-conjugated serum containing antiviral antibodies is the only serum used, whereas in the indirect test the primary serum, containing antiviral antibodies, is used first followed by a FITC-conjugated antiserum to the species from which the primary serum was obtained.

Application of FA methods in virological studies have centered around visualization of viral antigens in infected tissue cultures and the isolation and identification of viruses from clinical specimens. Examination of fixed tissue culture cells by FA methods at various time intervals after infection gives some indication in which cell regions viral antigens have been synthesized. Antigens can develop only in the cytoplasm or the nucleus, or with certain viruses they can be found in both. Staining of living cells can reveal viral-induced cell membrane antigens.

Preparation and Inoculation of Cell Cultures

Cell cultures are selected which best support the replication of the virus being studied. Using standard procedures, primary cell cultures, diploid cell strains, or continuous cell lines are grown as monolayer cultures either in 16 × 125 mm culture tubes, on coverslips in Leighton tubes, or on tissue culture chamber slides (Lab-Tek, Division of Miles Lab, Westmont, Illinois). Cultures are inoculated when large islands of cells or a 70% monolayer has formed. The medium is removed and the inoculum is adsorbed onto the cells for 1 hour at 37°C. The cell monolayer is then washed with Hanks' balanced salt solution (HBSS) to remove unadsorbed virus and the appropriate maintenance medium is added. High titered virus preparations should be diluted to contain 10^2 to 10^3 $TCID_{50}$ to avoid a too rapid degeneration of the cell monolayer. Studies involving the isolation and the identification of viral agents from clinical specimens, however, require no dilution, since only small amounts of virus are usually present.

Cell cultures which grow in suspension are prepared and infected as follows: The cells are centrifuged and the cell pellets are washed with HBSS. Cell pellets are resuspended in 1.0 ml of an appropriate dilution of virus and incubated at 37°C for 1 hour. Unadsorbed virus is removed after centrifugation and the cells are resuspended to a desired cell density in growth or maintenance medium.

Preparation of Cell Cultures for FA Staining

The optimum time to fix cell cultures for staining will depend on the replication rate of the virus and the type of experiment being performed. Generally, cultures infected with rapidly replicating viruses such as herpes are fixed at hourly intervals, whereas daily intervals are used for others such as myxoviruses. Cell cultures inoculated with clinical specimens can best be examined at days 1, 3, and 5 postinfection. The medium from cells grown as monolayers in culture tubes is decanted and the cell monolayer is washed one or two times with phosphate-buffered saline (PBS pH 7.2). The cells are trypsinized, centrifuged and washed once with PBS. The cell pellet is resuspended in a small amount of PBS (0.5 ml) and several smears are made on microscope slides (Corning-thickness

0.97–1.06 mm) with a Pasteur pipette. These are air-dried either at room temperature or at 37°C and fixed in cold acetone for 10 minutes. Infected suspension cultures are prepared and fixed as described above. Cell cultures seeded onto coverslips or on tissue culture chamber slides are washed in PBS after removing the medium, air-dried, and directly fixed in acetone.

For the demonstration of viral-induced membrane antigens, viable cells are used in the staining procedure and infected suspension or trypsinized monolayer cultures are suitable for these tests. However, with some virus-infected cells trypsinization may destroy the membrane antigens. Membrane antigens can be regenerated by incubating for 6 hours in tissue culture medium. Cell suspensions ($0.5–1.0 \times 10^6$ cells/ml) are washed twice in PBS before being stained immediately.

Indirect FA Staining Procedures

The indirect FA staining procedure is carried out in two steps: (1) fixed cells are flooded with specific antiserum at a predetermined dilution and incubated for 30 minutes at 37°C or overnight at 4°C for antigens which stain weakly. The serum is removed with two 10-minute washes in PBS. (2) The cultures are then flooded with the FITC-conjugated antiserum prepared against the animal species of the specific serum, also at a predetermined dilution, and incubated for another 30 minutes at 37°C. The smears are washed as above and mounted with a coverslip using either buffered glycerol (nine parts glycerol and one part PBS) or a semipermanent mounting medium[2] containing a polyvinyl alcohol (Elvanol, DuPont) with a pH of at least 7.4. Viable cells are stained as follows: washed cell pellets are resuspended in 0.05 to 0.1 ml of specific antiserum and incubated for 30 minutes at 37°C. Serum is removed by washing the cells twice in PBS and after the last wash, cell pellets are resuspended in 0.05 to 0.1 ml of FITC-conjugated antiserum. After incubation and washing, cell pellets are resuspended in a drop of buffered glycerol and the cells are mounted on microscope slides with coverslips. Cultures mounted with buffered glycerol must be examined before the smears dry out, however; cultures mounted with a semipermanent medium can be held for several days at 4°C without loss of fluorescence. The following controls should be included in each test for each virus under study: (1) a positive control (infected cells + specific antiserum + FITC-conjugated antibody), (2) negative controls (noninfected cells + specific antiserum + FITC-conjugated antibody and infected cells + serum with no specific antiviral antibodies + FITC-conjugated antiserum) and (3) a conjugate control (infected and noninfected cells stained with FITC-conjugated antibody only). The optimum dilution of the specific antiserum and the FITC-conjugated antiserum is determined by making serial twofold dilutions in PBS and reacting these with positive and negative control cultures. The dilution of specific serum and FITC-conjugated antiserum which gives the best fluorescence with the least amount of nonspecific fluorescence is used.

[2] J. Rodriquez and F. Deinhardt, *Virology* **12**, 316 (1960).

INTERPRETATION OF RESULTS

A test is positive when specific fluorescence is observed in a cell culture stained with the specific antiserum and the FITC-conjugated antibody but not in the negative or conjugate controls. If staining is observed in these controls the specificity of the test cannot be interpreted. Nonspecific staining can be a result of reactions between the cellular components and the globulins of the antiserum or more often when there is an excess amount of FITC in the conjugate. These nonspecific reactions normally can be removed by adsorption of the specific antiserum and the conjugate with acetone-dried liver powders.[3]

Weak or no fluorescence in a known positive test can be due to prolonged storage of the fixed smears or a low pH of the buffer and the mounting medium. The antigens in fixed cells are usually stable for a few days at 4°C or for several weeks at −20°C, but deterioration occurs with prolonged storage.

The excitation of the fluorescein dye is pH dependent and no fluorescence or a very dull reaction occurs when the pH of the buffer and mounting medium is below 7.0. The intensity of the reaction, however, can be increased by raising the pH. Generally the pH of the mounting medium should not be below 7.4.

[3] R. C. Nairn, "Fluorescent Protein Tracing" p. 136. Williams & Wilkins, Baltimore, Maryland, 1969.

CHAPTER 4

Rapid Method for Detection of Budding Virus by Electron Microscopy[1]

B. Kramarsky

Mouse mammary tumor virus (MTV) production by cell cultures of mammary tumors cannot be detected by common virological procedures such as the observation of cytopathic effect or hemadsorption. In order to detect and quan-

[1] The work reported in this article was developed with the support of U. S. Public Health Service Grant CA-08740, Contract PH 43-68-1000 from the National Cancer Institute, General Research Support Grant FR-5582 from N.I.H., and Grant-in-Aid Contract M-43 from the State of New Jersey.

titate MTV production in these cells, negatively stained whole cell mounts were prepared for electron microscopic examination. The procedure, which is an adaptation of the technique of Dales,[2] was reported at the 19th Annual Meeting of the Tissue Culture Association and was subsequently published in *Cancer Research*.[3] Because of the ease and rapidity of the technique for the visualization of budding viruses, we have extended our work to the study of a number of viruses which are human pathogens or are related to such pathogens.

Cells infected with the following viruses have been studied: influenza, mumps, measles, sindbis, West Nile fever, rubella and vesicular stomatitis. It was shown that direct electron microscopic observation of specific antibody binding by the infected cells was an effective technique for the identification of budding virus. Electron micrographs of whole mount preparations of tissue culture cells infected with five different budding viruses with and without treatment with specific antisera are shown in Fig. 1.

Preparation for Electron Microscopy

The cells are prepared for electron microscopy as follows:

1. Cells from monolayers are brought into suspension by treatment with a trypsin-Versene mixture.
2. The cells are sedimented by low-speed centrifugation and are resuspended in 3 ml of tissue culture medium.
3. After standing at room temperature for 60 minutes, the cell suspension is sedimented by low-speed centrifugation.
4. The cells are resuspended in 3 ml of a hypotonic buffer solution containing one part isotonic phosphate-buffered saline (PBS), pH 7.2 and four parts distilled water. The suspension is chilled in an ice bath and allowed to stand for 10 minutes in the cold.
5. One milliliter of a 2% aqueous solution of OsO_4 is added to the suspension, and the cells are fixed in suspension for 20 to 30 minutes in the ice bath.
6. The cells are again sedimented, washed two times with the hypotonic buffer solution, and then resuspended in 2 ml of this solution.
7. One drop of the cell suspension is placed on a Formvar-coated 300 or 400 mesh copper grid and allowed to stand for 1 to 2 minutes. Excess fluid is removed with filter paper and the cells are stained with 2% sodium phosphotungstate.

In cases where the cell membranes appear to be somewhat labile, such as in cells infected with mumps, measles, or rubella virus, a slight modification in the procedure improves cell preservation. In place of steps 4 and 5, the cells are resuspended in a solution containing equal parts of the hypotonic buffer and the aqueous solution of OsO_4. The suspension is chilled in an ice bath and cells are fixed in suspension for 20 to 30 minutes in the cold.

[2] S. J. Dales, *J. Cell Biol.* **13**, 303 (1962).
[3] B. Kramarsky, E. Y. Lasfargues, and D. H. Moore, *Cancer Res.* **30**, 1102 (1970).

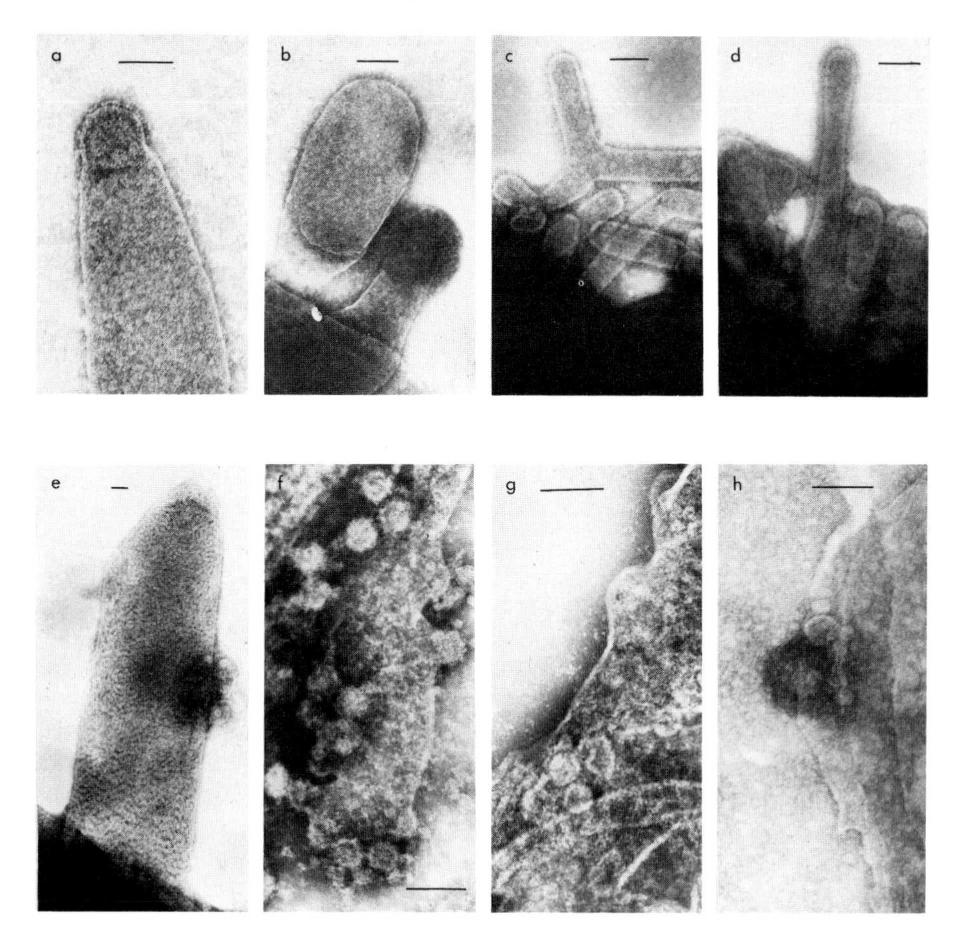

Fig. 1. Electron micrographs of whole cell mount preparations of virus-producing tissue culture cells. The bar represents 0.1 μm. (a) MTV budding from a microvillus of a mouse mammary tumor cell. (b) Same preparation as (a). Treated with rabbit anti-MTV serum and goat anti-rabbit globulin. (c) Influenza type A-Japan 305-57 virus budding from a chick embryo fibroblast. (d) Same preparation as (c). Treated with rabbit antiserum to influenza A-Japan 305-57 virus and goat anti-rabbit globulin. (e). Mumps virus budding from a chick embryo fibroblast. (f). Sindbis virus budding from a chick embryo fibroblast. (g) Rubella virus budding from a BHK 21 cell. (h). Same preparation as (g). Treated with rabbit anti-rubella virus serum and goat anti-rabbit globulin.

Preparation for Immunoelectron Microscopy

1. The cells are prepared as in steps 1 through 3 above.
2. The cells are then resuspended in two drops of serum-free medium. Two drops of heat inactivated antiserum, specific to the virus tested, are added to the suspended cells.
3. After incubation for 1 hour with occasional shaking at room temperature, the cells are sedimented, washed two times in 10 ml of isotonic PBS, and resuspended in two drops of PBS.

4. Two drops of an antiserum prepared against the gamma-globulin of the antiviral serum used in step 2 are added and the cells are incubated and washed as in step 3.

5. The cells are then prepared for electron microscopy according to steps 4 through 7 above.

It should be borne in mind that the antiviral serum used must bind specifically to viral envelope antigens rather than to the internal "soluble" antigens.

Electron Microscopy

Preparations of negatively stained whole cell mounts are first scanned in the electron microscope at low magnification in order to select the cells to be studied. Cells partly overlying a grid wire are usually chosen since cells lying away from the wire tend to break away when struck by the focused electron beam. The margin of the selected cells is then scanned for virus.

Quantitative Evaluation of Virus Production

Whole cell mount preparations are examined for the presence of virus. They are classified as either positive or negative. A single virus bud observed on a cell is sufficient to establish the cell as positive, but a cell is not considered negative unless more than one-third of its circumference is scanned and found to be free of virus. At least 100 cells are scored in each sample. Virus production is defined as the ratio of cells with budding virus to the total number of cells. Figure 2 illustrates an application of this technique.

Aliquots of cells from an RIII mammary tumor cell line, harvested at increasing intervals were taken for evaluation of MTV production by means of whole-cell electron microscopy and immunofluorescence. While the methods gave very

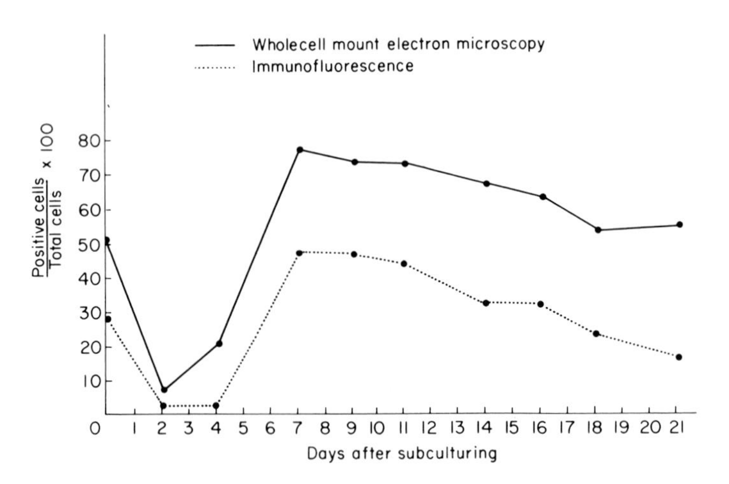

Fig. 2. Virus production of the mouse mammary tumor cell line RIIIMT assayed by whole cell mount electron microscopy and by immunofluorescence at increasing intervals after subculture.

similar results, whole cell mount electron microscopy preparations had consistently higher ratios, indicating that this method was more sensitive than immunofluorescence.

General Comments

Whole cell mount electron microscopy is an effective supplement to other techniques for the detection of budding viruses. It is especially valuable for the confirmation of infection with tumor viruses where the morphology of the budding virus serves to define the type of infection (leukemia virus, mammary tumor virus). Specific antibody binding to the budding virion further provides a means of unambiguous identification of the infecting virus.

CHAPTER 5

Monitoring for Presence of Oncogenic Virus in Tissue Culture

A. J. Hackett

Several methods exist which enable detection of oncogenic virus infection. When applied routinely to mammalian cell cultures used as substrates, these methods can also reveal the presence of adventitious and/or latent viruses as well as mycoplasma contaminants. The manipulations necessary to cultivate cells, together with the use of mammalian sera as nutrient supplements, invite the introduction of unwanted agents.[1-4] A virus may persist at equilibrium with the host cells,[5] or it may be latent, but inducible.[6-8] Particularly in the case of prolonged cultivation during experimentation, continued monitoring for these viruses is essential to determine whether the cells maintain the characteristics peculiar

[1] E. Stanbridge, *Bacteriol. Rev.* **35**, 206 (1971).
[2] J. Fogh, N. Holmgren, and P. P. Ludovici, *In Vitro* **7**, 26 (1971).
[3] C. Hallauer, G. Kronauer, and G. Siegel, *Arch. Ges. Virusforsch.* **35**, 80 (1971).
[4] G. McGarrity and L. L. Coriell, *In Vitro* **6**, 257 (1971).
[5] W. A. Nelson-Rees, L. E. Hooser, and A. J. Hackett, *J. Nat. Cancer Inst.* **49**, 713 (1972).
[6] Z. Steplewski and H. Kaprowski, *Exp. Cell Res.* **57**, 433 (1969).
[7] D. R. Lowry, W. P. Rowe, N. Teich, and J. W. Hartley, *Science* **174**, 155 (1971).
[8] B. Hampar, J. G. Derge, L. M. Martos, and J. L. Walker, *Proc. Nat. Acad. Sci. U. S.* **69**, 78 (1972).

to them when the experiments were initiated or whether newer infections have occurred or other viruses become detectable.

Electron Microscopy

Electron microscopy has been utilized to survey cells for the presence of virus particles for many years and various techniques of sample preparation have been described.[9-10] The search for human oncogenic viruses in primary tumor tissue, as well as in cultured cells, remains one area where electron microscopic surveys are extensively used. Although the results have been largely negative in surveys of human cells, both feline and murine tissues have revealed particles. The techniques successful in these systems are described.[11,12]

A few precautions can assist in the distinction between fact and artifact in this type of survey. The techniques of sample preparation for tissue cultured cells used in this laboratory have been described.[13] In the case of cell cultures to be monitored by electron microscopy and where trypsin is used, the cells should be fixed 3 days after seeding and while they are replicating, so that recovery from trypsinization is complete. A sample of 200 individual cells is surveyed before a culture is considered negative. It is essential that the microscopist be thoroughly familiar with the prototypes of the oncogenic viruses, their structure, morphogenesis, and the changes produced in the infected host cell. The structural characteristics of the RNA oncogenic viruses have been compared and illustrated[11,14] and the A, B, and C type viruses of mouse, cat, monkey, and human cells compared.[12] Haguenau[15] has illustrated and discussed the most commonly encountered cellular organelles and structural formations which resemble virus particles and are most frequently mistaken for them. The most reliable criterion for the identification of resident viruses is the observation of particles in various stages of maturation.

Isotope Labeling Techniques

Biophysical properties and nucleic acid content can be utilized to indicate the presence of contaminants and/or viral agents in cultures. The monitoring of cell cultures for the detection of isotope labeled particles by isopycnic centrifugation (DIPIC) has been routinely performed in this laboratory. The RNA-

[9] D. Pease, "Histological Techniques for Electron Microscopy," 2nd Ed. Academic Press, New York, 1964.

[10] R. W. Horne, *In* "Methods in Virology" (K. Maramorosch and H. Koprowski, eds.), Vol. 3, p. 521. Academic Press, New York, 1967.

[11] A. J. Dalton *et al., Proc. Intern. Conf. on Leukemia-Lymphoma,* 1968. Lea & Febiger, Philadelphia, Pennsylvania.

[12] A. J. Dalton, *J. Nat. Cancer Inst.* **49,** 323 (1972).

[13] J. S. Manning and A. J. Hackett, *J. Nat. Cancer Inst.* **48,** 417 (1972).

[14] A. J. Dalton and F. Haguenau, *In* "Ultrastructure of Animal Viruses and Bacteriophages: An Atlas" (A. J. Dalton and F. Haguenau, eds.), in press. Academic Press, New York, 1973.

[15] F. Haguenau, *In* "Ultrastructure of Animal Viruses and Bacteriophages: An Atlas" (A. J. Dalton and Haguenau, eds.), in press. Academic Press, New York, 1973.

containing tumor viruses characteristically band, under the conditions used, at density 1.16 g/cm,[3] forming a peak of radioactivity in that region of the gradient.[16] Confirmation that the peak region contains the virion can be obtained by electron microscopy, by immunological methods when the reagents are available, and by infectivity assay if a method is known.

DNA-containing tumor viruses can be detected by this technique as well. In human lymphoblastoid cells, a virus ([^{3}H]thymidine-labeled) related to the Epstein-Barr virus (EBV) is detectable by the DIPIC technique as a broad peak which encompasses the 1.22–1.14 g/cm^3 density region of the gradient.[17] In this connection, a recent report indicates that lymphoblastoid cultures derived from patients with Hodgkin's disease characteristically release RNA-containing particles which form a band in the same density region as the DNA ([^{3}H]thymidine-labeled) EBV-like particles.[18] As yet, the nature of these particles is unknown but there is some evidence suggesting an infectious RNA-containing virus.[19] In this laboratory, a variety of other human tumor-derived lymphoblast cultures display this phenomenon, but no evidence that the RNA-containing material is viral in nature has been obtained.[20]

The DIPIC technique may also be used as a sensitive indicator for mycoplasma contaminants of tissue cultures,[21] and it is particularly useful when the mycoplasma strain is difficult to detect by bioassay. Mycoplasmas band below the density of the oncornaviruses at 1.22 g/cm.[3]

Procedures for detection of oncornaviruses by isotope labeling in tissue cultures are the following:

1. Labeling. Replace the growth medium from subconfluent cultures with half volume growth medium containing 20 μCi/ml of [5-^{3}H]uridine, specific activity 25 Ci/mmole (10 ml/250-ml Falcon flask). Continue incubation for 24 hours at 37°C. Harvest supernatant fluid only and clarify at 10,000 rpm for 10 minutes in a preparative Sorval centrifuge. Freeze the supernatant at −70°C until ready to process.

2. Concentration and purification. Centrifugation to a sucrose density interface removes some of the cellular components contaminating the preparation. A discontinuous sucrose gradient is prepared as follows. Pipette a 0.3 ml "cushion" of 52.5% sucrose solution[22] on the bottom of a size ½ × 2 inch cellulose nitrate tube (Beckman Instruments) for use in the SW 50.1 or 50L rotor. Slowly layer another 0.5 ml 15% solution of sucrose over the cushion. Carefully layer 4.2 ml of thawed sample over the last sucrose layer. Centrifuge at 50,000 rpm for 30 minutes at 4°C–brake on. Remove the tube from the rotor and carefully attach it to the harvesting unit (Fig. 1). Puncture the bottom of the tube using a 23-gauge, 1-inch needle. Harvest the region of the tube at and just above the interface in about four drops.

[16] G. J. Todaro, V. Zeve, and S. A. Aaronson, *Nature* (*London*) **226**, 1077 (1970).
[17] A. J. Hackett, unpublished results, 1972.
[18] M. Eisinger, S. M. Fox, E. De Harven, J. L. Biedler, and F. K. Sanders, *Nature* (*London*) **233**, 104 (1971).
[19] F. K. Sanders, personal communication.
[20] A. J. Hackett, unpublished observation, 1972.
[21] G. J. Todaro, S. A. Aaronson, and E. Rands, *Exp. Cell Res.* **65**, 257 (1970).

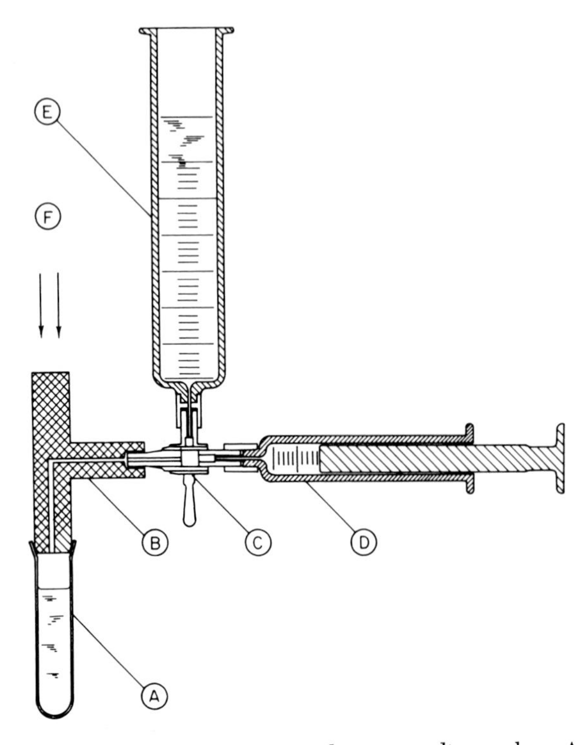

Fig. 1. Device for removing samples from density gradient tubes. A, Density gradient tube; B, transparent plastic block machined to fit density gradient tubes and stopcock; C, stopcock with Luer-lok syringe connections; D, displacement syringe (1, 2, or 5 ml); E, mineral oil reservoir (large syringe without plunger); F, light beam. The gradient tube is attached to the device while the stopcock is open to the reservoir, forcing out the air above the gradient. A hole is punctured in the bottom of the tube with a needle and the stopcock turned to connect the tube with the displacement syringe, which previously had been filled with oil from the reservoir. Samples can be collected from the hole in the bottom of the tube as measured volumes of oil are introduced at the top from the displacement syringe. The viscosity of the oil prevents rapid flow. Droplets can be collected without the use of the displacement syringe, by opening the stopcock to the reservoir. (F. L. Schaffer and L. H. Fromhagen, *Virology* **25,** 662, 1965; reprinted with permission of authors.)

3. Isopycnic zonal banding. Dilute the density interface solution to 1.4 ml in TEN buffer.[22] Layer over a continuous density gradient consisting of 2 ml of 52.5% and 2 ml of 15% sucrose. Gradients can be made manually by successive layering of five steps of decreasing density and allowing to stand overnight at 4°C; however, there are gradient makers available from several laboratory supply houses. Alternatively, a device can be readily constructed by a machinist.[23] Refractive indices or weight measurements are required, however, for accurate confirmation of densities through the gradient.

[22] TEN Buffer: 0.1 *M* NaCl, 0.001 *M* EDTA, and 0.01 *M* Tris-HCl, pH to 7.4. Sucrose (w/w) is made with 0.1% bovine serum albumin (Armour Fraction V) in TEN buffer. Filter through Millipore, 0.45 μm.

[23] R. J. Britten and R. B. Roberts, *Science* **131,** 32 (1960).

A virus with known characteristics, under the conditions of centrifugation used, can be a convenient means to mark a particular region in the gradient. For routine monitoring, we use ^{32}P-labeled murine leukemia virus to mark the 1.16 g/cm^3 zone of the gradient.

Centrifuge the gradients at 50,000 rpm for 3 hours in the SW 50.1 or 50L rotor and harvest by bottom puncture in three-drop fractions directly onto the middle of individual strips of chromatographic paper, ¾ × 3 inch, which have been appropriately numbered in pencil (Fig. 2). Alternate samples can be collected in tubes for infectivity assays and density measurements.

4. Acid-precipitable radioactivity. Allow the papers to air-dry at room temperature. Soak them in a beaker in 0.5 *M* cold perchloric acid (PCA) for 20 minutes at 4°C. Decant PCA and again cover with cold 0.5 *M* PCA. Swirl beaker by hand for 10 minutes to separate individual strips and allow to stand at 4°C for 10 minutes. Decant PCA and transfer papers to a clean beaker with cold 95% ethanol, to cover, and swirl. Refrigerate at 4°C for 15 minutes. Decant and add fresh 95% ethanol. Wait 5 minutes, pour off alcohol and spread papers out to dry. The dry papers are rolled and stuffed into scintillation vials, covered with scintillation fluid,[24] and counted.

Fig. 2. Stainless steel tray with edges bent to hold strips of Whatman No. 3 chromatographic paper, ¾ × 3 inches.

[24] Scintillation Fluid for use in Beckman L 250: 16.79 g Permablend and 1.00 gallon toluene.

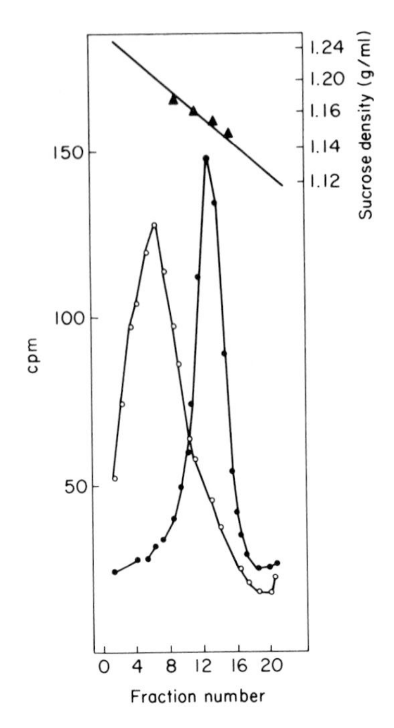

Fig. 3. Banding pattern typical for mycoplasma in a sucrose gradient following isopycnic centrifugation. ○——○, ^{3}H-labeled thymidine mycoplasma; ●——●, ^{32}P-labeled marker virus (MLV), ▲——▲, density.

The counts per sample are plotted on arithmetic graph paper. Figure 3 shows typical results obtained when a culture of mouse cells infected with and producing leukemia virus is also contaminated with mycoplasma. The [^{3}H]uridine has been incorporated not only by the newly formed C type, RNA leukemia viruses which budded from the cell membrane and were collected from the supernatant, but also by the more dense, newly forming mycoplasmas.

Marker Virus

Moloney leukemia virus-infected mouse cells are fluid changed to Eagle's Medium (phosphate-free)[25] and 10% dialyzed calf serum containing 20 μCi/ml of ^{32}P (carrier-free $H_3\,^{32}PO_4$). The medium is harvested after 24 hours incubation at 37°C, and processed as described above. After isopycnic zonal (IZ) centrifugation the gradient is harvested in 0.5-ml fractions into small tubes. Samples of 10 μl are counted and the peak fraction identified. This fraction is diluted 1:3 in TEN buffer, divided into several aliquots and stored frozen. For use as marker, a volume calculated to give 200 counts per minute is added to the IZ gradient with the unknown sample. Figure 4 shows typical data when murine leukemia, ^{3}H-labeled, is banded with the ^{32}P-labeled marker virus.

[25] Grand Island Biological Company, Berkeley, California.

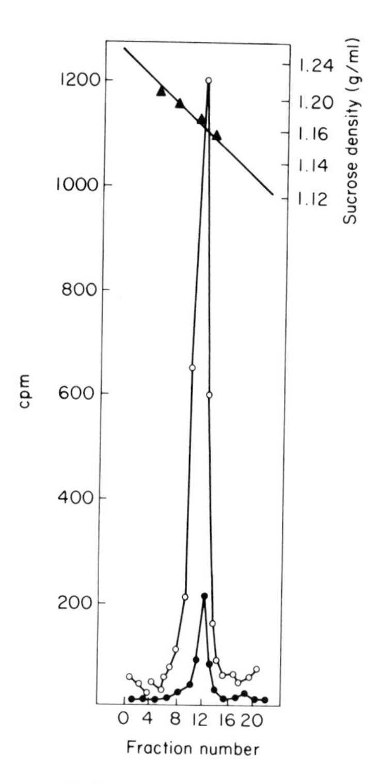

Fig. 4. Banding pattern typical for unknown C-type virus and marker (MLV) virus. ○——○, Unknown virus, [^{3}H]uridine label; ●——● marker virus, [^{32}P]MLV; ▲——▲, density.

The Reverse Transcriptase as an Indicator of Oncornavirus Replication

The RNA-directed DNA polymerase (RDP) characteristic for the oncornaviruses[26] is a useful indicator of the presence of these agents. It has been utilized as an assay for the murine leukemia viruses,[27] and as a means to monitor mouse cells for the production of virus.[28,29] The enzyme assay is particularly useful when no suitable biological assay for infectivity is available[30] or is low in titer.[29] The DNA product of the enzyme has been utilized to detect RNA complementary to a murine leukemia virus in human cells.[31,32]

[26] H. M. Temin and S. Mizutani, *Nature (London)* **226**, 1211 (1970); D. Baltimore, *ibid.*, 1209.

[27] G. J. Kelloff, M. Hatanaka, and R. V. Gilden, *Virology* **48**, 266 (1972).

[28] S. A. Aaronson, G. J. Todaro, and E. M. Scolnick, *Science* **174**, 157 (1971).

[29] S. A. Aaronson, *Proc. Nat. Acad. Sci. U. S.* **68**, 3069 (1971).

[30] P. J. Fischinger, S. Nomura, P. T. Peebles, D. K. Haapala, and R. H. Bassin, *Science* **176**, 1033 (1972).

[31] D. Kufe, R. Hehlmann, and S. Spiegelman, *Science* **175**, 182 (1972).

[32] S. Spiegelman and J. Schlom, *In* "Molecular Studies in Viral Neoplasia," 25th Annual Symposium on Fundamental Cancer Research. The University of Texas at Houston, M. D. Anderson Hospital and Tumor Institute, 1972.

The major problem in utilization of this enzyme as an indicator of oncogenic virus replication is one of specificity. Contaminating DNA-directed RNA polymerase must be carefully excluded.[33] Verification can be obtained quite readily in systems where large quantities of purified virus are available; specificity of the enzyme has been elegantly demonstrated by immunoprecipitation techniques.[34] Hybridization of product DNA with purified viral RNA is probably the most stringent test for specificity.[31] Spiegelman *et al.*[31,32] applied this to establish homology of the product made by enzyme isolated from human milk, with the RNA from murine mammary tumor virus. Similarly the product made by enzyme isolated from other human tumor tissues was homologous with the RNA of the Rauscher leukemia virus.[32] Human tumor cells grown in tissue culture have been monitored for the enzyme[35] and the results have thus far been negative.

However, the method has been useful for monitoring cells other than human, and is included here. Caution is advised in drawing conclusions without verification by other means.

Assay for RNA-Dependent DNA Polymerase[36]

1. Clarify the tissue culture supernatant by centrifugation at 5000 rpm for 15 minutes at 4°C.

TABLE I
Solution A for the Endogenous Template Method for the RNA-Dependent DNA-Polymerase Assay

Solution A	μl	Final molarity
1.0 *M* Tris-HCl, pH 7.8	3	0.03
3.0 *M* KCl	2	0.06
0.6 *M* $Mg(C_2H_3O_2)_2$	1	6×10^{-3}
0.05 *M* dATP	1	5×10^{-4}
0.05 *M* dCTP	1	5×10^{-4}
Triton 1.4% (v/v)[a]	1	0.014% (v/v)
H_2O	1	
DTT (3 mg/ml)[b]	5	2×10^{-3}
TTP-^{3}H (5000 cpm/pmole)[b]	10	1×10^{-5}
	25	

[a] Triton X: Rohm and Haas, Independence Mall West, Philadelphia, Pennsylvania 19105.
[b] The above components can be made up as a stock solution. DTT and [^{3}H]TTP can be added just before use.

[33] M. S. Robert, R. G. Smith, R. C. Gallo, P. S. Sarin, and J. W. Abrell, *Science* **176**, 798 (1972).

[34] J. Ross, E. M. Scolnick, G. J. Todaro, and S. A. Aaronson, *Nature N. Biol.* **231**, 163 (1971).

[35] R. McAllister and S. Rasheed, Children's Hospital and University of Southern California Medical School, Los Angeles, California.

[36] I am indebted to Dr. E. M. Scolnick, National Cancer Institute, Bethesda, Md., and Dr. S. D. Huff, Radiobiology Laboratory, University of California, Davis, Ca., for the protocols for analysis for the RNA-dependent DNA polymerase.

TABLE II

Solution B for the Polymer-Dependent Template Method for the RNA-Dependent DNA-Polymerase Assay

Solution B	μl	Final molarity in 100 μl RM
1.0 *M* Tris-HCl, pH 7.8	3.00	0.03
0.1 *M* $MnCl_2$	0.25	2.5×10^{-4}
3.0 *M* KCl	2.00	0.06
Triton 10%	1.00	1%
H_2O	4.00	
DTT (3 mg/ml)[a]	5.00	2.0×10^{-3}
[^{3}H]TTP (5000 cpm/pmole)[a]	10.00	1×10^{-5}

[a] The above components can be made up as a stock solution. DTT and [^{3}H]TTP can be added just before use.

2. Centrifuge the clarified supernatant through 10% glycerol in TEN buffer (2 ml glycerol is overlaid with 8 ml supernatant and centrifuged at 100,000 *g* for 60 minutes at 4°C).

3. Resuspend the pellets as 500× concentrate of the initial volume in buffer (0.02 *M* Tris-HCl, pH 7.8, 0.10 *M* NaCl, and 0.001 *M* dithiothreitol (DTT).

4. Assay the concentrate for polymerase using 2, 10, and 20 μl per 100 μl of reaction mixture (RM) with either the endogenous (Table I) or polymer-dependent (Table II) method. The components included in the RM for each method is calculated to give the specified concentration in a total volume of 100 μl (i.e., when 25 μl of solution A is mixed with the other components listed in Table II and brought to 100 μl).

A stock of solution A can be made omitting DTT and [^{3}H]TTP. To run a 10-component test as illustrated in Table III, use 0.10 ml of stock solution A and add 0.10 ml [^{3}H]TTP and 0.05 ml DTT. DTT should be made up fresh each week and stored frozen, protected from light.

5. Preincubate mixture 5 minutes at 24°C prior to initiation of reaction by

TABLE III

Example of a Positive Polymerase Assay Using Endogenous Template

	Solution A (μl)	0.05 *M* dGTP	Pelleted virus	H_2O (μl)	Cpm[a]
1	25	—	—	75	200
2	25	—	—	75	200
3	25	1	—	74	200
4	25	1	—	74	200
5	25	—	2	73	300
6	25	—	10	65	500
7	25	—	20	55	800
8	25	1	2	72	3,000
9	25	1	10	64	10,000
10	25	1	20	54	20,000

[a] An example of a positive result. The data show that DNA synthesis requires four triphosphates and is dependent on the virus.

TABLE IV
Example of Polymerase Assay Using Polymer-Dependent Template[a]

	Solution B (μl)	$rA \cdot dT_{12-18}$[b]	Pelleted virus	H_2O (μl)	Cpm[c]
1	25	—	—	75	200
2	25	—	—	75	200
3	25	—	2	73	220
4	25	—	10	65	230
5	25	—	20	55	500
6	25	1	2	72	500,000
7	25	1	10	64	>500,000
8	25	1	20	54	>500,000

[a] More sensitive (100–1000 ×) than the endogenous template method.
[b] 0.01–.02 OD_{260} units/ml $rA.dT_{12-18}$. Source: Research Products Division, Collaborative Research, Inc., 1365 Main Street, Waltham, Massachusetts 02154.
[c] An example of counts per minute (CPM) obtained from a positive assay.

the addition of dGTP for the endogenous method and $rA \cdot dT_{12-18}$[37] for the polymer-dependent method. The example shown in Table IV was taken at one time interval, but there is an advantage in taking samples at several time intervals. Demonstration of increasing activity from 0 time, 15, and 30 minutes following initiation of the reaction gives added reliability to the data.

For counting radioactivity, the sample is deposited on strips of Whatman No. 3 filter paper and air-dried.

6. When the virus is well characterized—such as avian and murine RNA-tumor viruses—the specificity of the DNA product can be obtained by use of specific antisera to the viral enzyme[31] or by hybridization of the product to RNA obtained from purified virus.[33]

In cases where a DNA product is to be used as an indicator of the presence of a tumor virus which cannot be bioassayed or is produced in very low titer, more involved experiments to identify and confirm specificity are required.[32]

[37] Research Products Division, Collaborative Research, Inc., 1365 Main Street, Waltham, Massachusetts 02154.

CHAPTER 6

Tobacco Mosaic Virus in Plant Tissue Culture

A. C. Hildebrandt

Tobacco mosaic virus (TMV) (Fig. 1) and the mosaic disease induced by it have been studied extensively. Isolated cells and tissues growing *in vitro* have been increasingly used to study this virus and other viruses for their host cell interactions.[1-4]

General methods for the isolation and growth of plant cells, tissues and organs have been described elsewhere in this volume. The various manipulations involved in working with the tissue cultures are carried out under aseptic conditions.

TMV-Infected Cultures

Several methods have been used to obtain TMV-infected tissue cultures. Callus cells infected with TMV may be secured by explanting stem tips or stem sections from plants infected with the virus (Fig. 2) to agar medium. Otherwise, the callus cells may be artificially inoculated *in vitro*.[3,5-7] Similarly, root tips from plants infected with TMV can be isolated to agar or liquid media. Tomato root tips infected with TMV were isolated aseptically directly from infected plants by White,[1] but Melchers and Bergmann[8] also artificially infected tomato roots *in vitro*. In either case, the callus and root tip cultures are maintained for unlimited periods by periodic subculture to fresh liquid or agar media. In the case of liquid cultures, it is necessary to provide aeration by shaking or rolling the culture vessels, or by otherwise passing sterile air through the culture medium.

TMV Propagation

Agar cultures of tobacco callus may also be inoculated with a sterile, filtered preparation of TMV. The tissue pieces are lightly dusted with a sterile car-

[1] P. R. White, *Phytopathology* **24**, 1003 (1934).
[2] G. Morel, *Ann. Epiphyt.* **14**, 123 (1948).
[3] A. C. Hildebrandt, *Proc. Nat. Acad. Sci. U. S.* **44**, 354 (1958).
[4] L. Hirth, Thesis, Faculty of Sciences, University of Paris, p. 115 (1960).
[5] L. Bergmann, *Nature (London)* **184**, 648 (1959).
[6] L. Hirth and G. Segretain, *Ann. Inst. Pasteur (Paris)* **91**, 523 (1958).
[7] H. H. Murakishi, *Virology* **27**, 236 (1965).
[8] G. Melchers and L. Bergmann, *Verh. 4th Intern. Pflanzenchutzkongr., Hamburg* (1957).

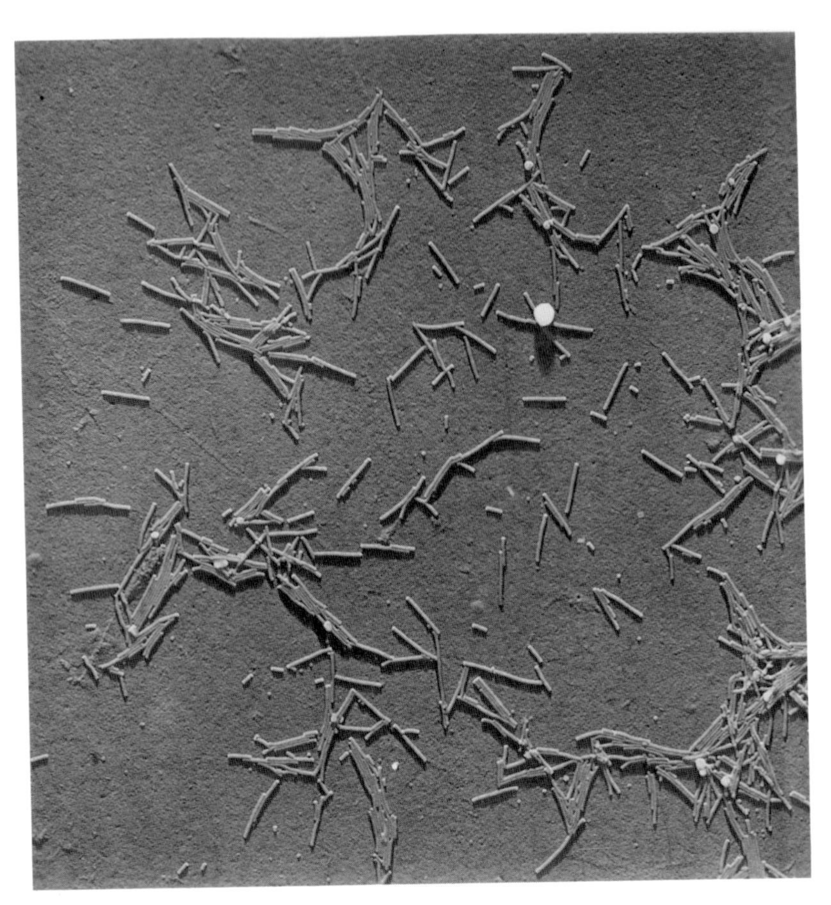

Fig. 1. Electron micrograph of a very mild strain of TMV from tissue culture showing typical rods of TMV. ×40,000.

borundum powder and rubbed lightly with a sterile spatula previously dipped in the TMV solution. Alternately, a drop of the sterile filtered TMV preparation may be placed on the tissue piece and the area rubbed with the sterile spatula.[3,9] The success in artificially inoculating tissues was influenced by the type and number of injuries produced at the time of inoculation.[10] Sterile TMV for inoculations may also be obtained directly in the exudate forced from TMV-infected tobacco stems or petioles exposed to air pressure. Tobacco callus cells growing in liquid suspension culture may be infected without added abrasives by adding the sterile filtered TMV inoculum directly to the liquid medium–tobacco cell suspension on a shaker.[3,9] The shaking process evidently facilitates the infection of the cells. The susceptibility of the cells was influenced by the age of the cells.[11]

[9] A. C. Hildebrandt and A. J. Riker, *Fed. Proc.* **7**, 980 (1958).
[10] B. Kassanis, T. W. Tinsley, and F. Quak, *Ann. Appl. Biol.* **46**, 11 (1958).
[11] J. H. Wu, A. C. Hildebrandt, and A. J. Riker, *Phytopathology* **50**, 587 (1960).

Fig. 2. Isolated segment of stem tissue on top of stem section (top center) removed under aseptic conditions to agar medium in 6-ounce prescription bottle (top center) after several weeks produced callus (bottom center) which was then transferred to liquid medium (left center) or to agar medium. Healthy cell cultures were thus established from stem segment of healthy plant (left) or TMV-infected cultures from TMV-infected plants (right). (From M. Singh and A. C. Hildebrandt, unpublished photograph.)

Single cell clones of tobacco tissue also were shown to vary in the susceptibility to TMV infection.[3,12,13]

Environmental Effects on TMV

The virus activity in the infected cells is influenced by a variety of environmental conditions. TMV-infected tobacco cell cultures showed increased TMV infectivities on culture media with increased levels of nitrate, even at concentrations inhibiting cell growth.[3] Decreased TMV activity was noted in extracts of tissue cultured cells grown on culture media with progressively increased levels of phosphate. The level of potassium in the culture media seemed to have little effect on the host cell virus infectivity. The incubation temperature (Fig. 3) and pH of the medium (Fig. 4) also influenced the amount of TMV activity.

The concentrations in the culture medium of growth substances, nucleic acids, purines, pyrimidines, and some analogs have also been shown to influence the TMV infectivity in tobacco cells in culture.[3,9,11,14,15]

[12] A. Sampath, A. Lutz, and L. Hirth, *Bull. Soc. Fr. Physiol. Veg.* **12**, 273 (1966).
[13] L. Hirth and A. Durr, *Colloq. Intern. Cent. Nat. Rech. Sci.* **193**, 481 (1971).
[14] R. J. Kutsky, *Science* **155**, 19 (1952).
[15] R. H. Kurtzman, A. C. Hildebrandt, R. H. Burris, and A. J. Riker, *Virology* **10**, 432 (1960).

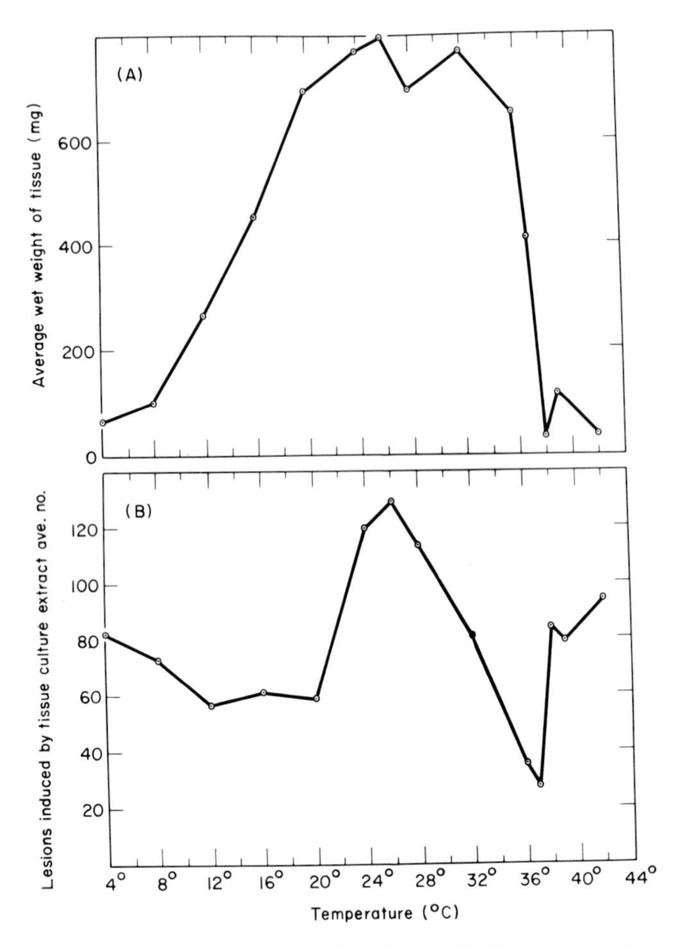

Fig. 3. (A) Average wet weight after 6 weeks' incubation at temperatures indicated of tobacco cell cultures infected with tobacco mosaic virus. (B) Average number of local lesions per half leaf produced on hybrid tobacco assay plant by tissue homogenate from above cultures incubated at the respective indicated temperatures. (From A. C. Hildebrandt.[3])

TMV Assays

The virus presence in cultured cells and tissues may be determined in several ways.[4,16] Physical and chemical methods include electron microscopical virus particle counts, phase-contrast microscopical observations (Fig. 5), fluorescent antibody staining, colorimetric, optical density, and spectrophotometric determinations. Grafts of tissue pieces may be made to whole plants. Also, serological methods may be used. In many cases, however, the TMV activity and infectivity in cultured cells has been determined with local lesion assay tobacco plants. The infected tissues are ground in a mortar and pestle and diluted with distilled water or with 0.05 *M* phosphate mono- and dibasic sodium phosphate buffer in proper

[16] J. Hansen and A. C. Hildebrandt, *Virology* **28**, 15 (1966).

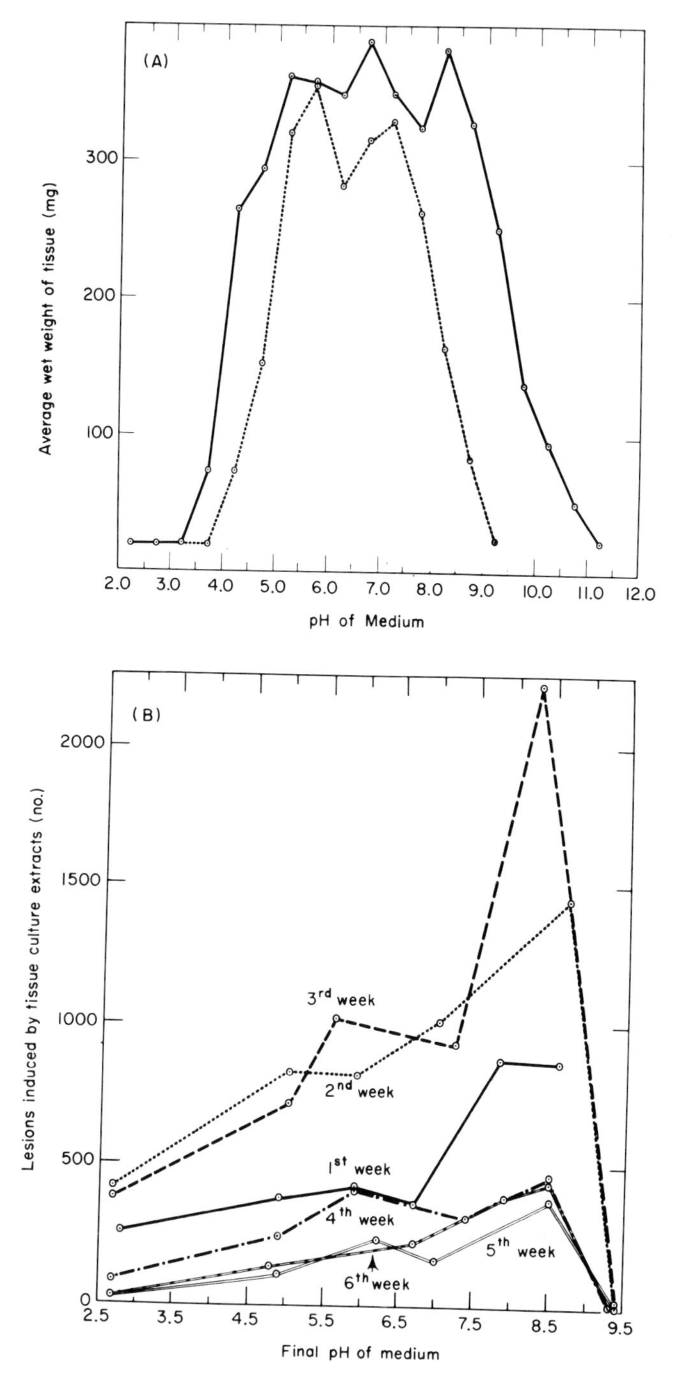

Fig. 4. (A) Average wet weight of TMV-infected tobacco cell cultures after 6 weeks' incubation on basal medium with respective acidities as indicated. (B) Tobacco mosaic virus infectivity of homogenates of tobacco cell cultures incubated on media with acidities as indicated and assayed at six weekly intervals. (From A. C. Hildebrandt.[3])

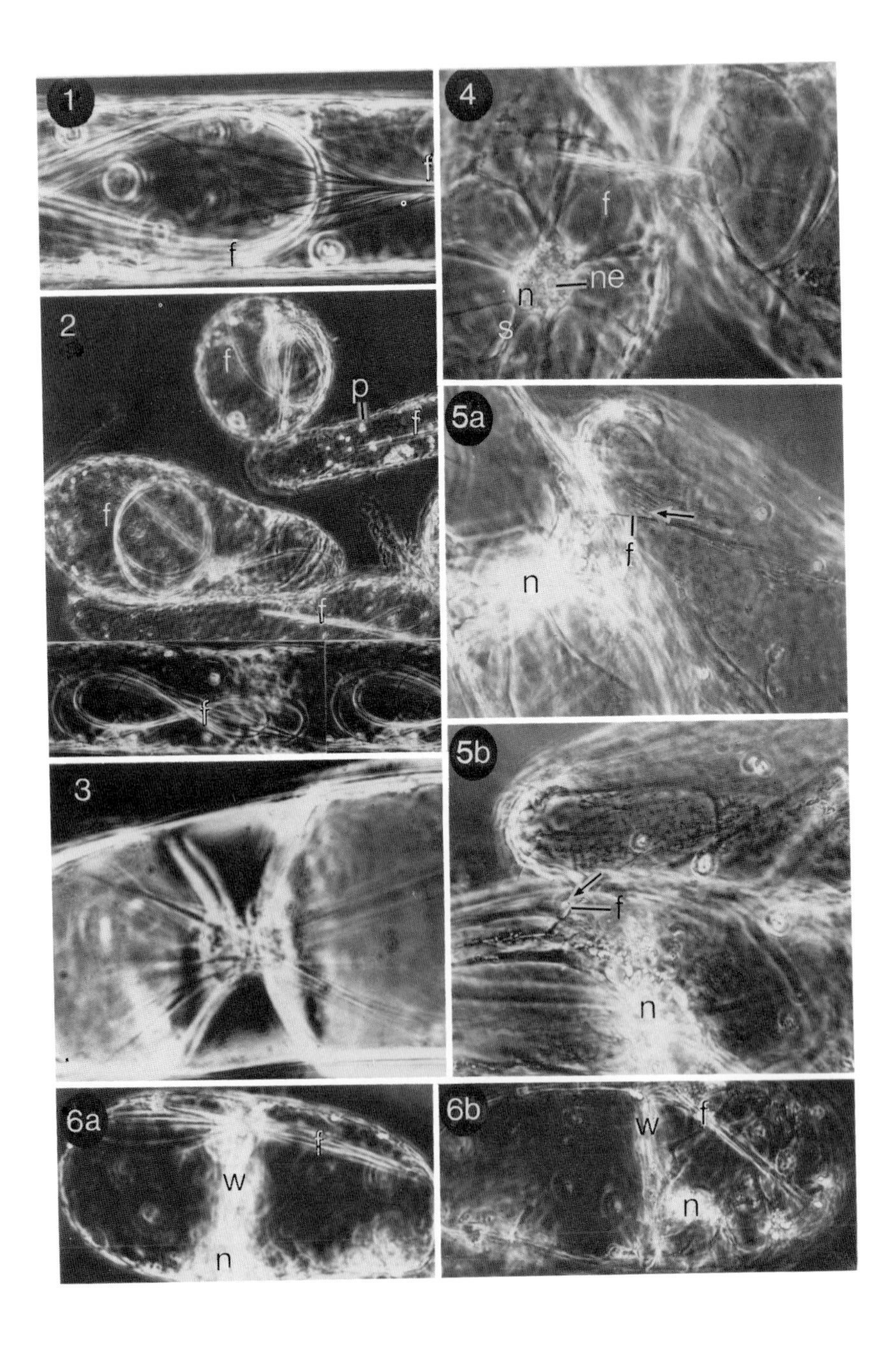
1
f
f
2
f
p
f
f
f
f
3
4
f
n
ne
s
5a
f
n
5b
f
n
6a
f
w
n
6b
w
f
n

concentrations. The whole or diluted extract is then rubbed on the upper leaf surface of a local lesion host tobacco plant such as, *Nicotiana tabacum* × *Nicotiana glutinosa*, *Nicotiana glutinosa* or *Nicotiana tabacum* var. *xanthi*. The half leaf method (Fig. 6) is preferably used with a Latin square plan to compensate for variations in plant susceptibility, inoculation procedures, and the position effect of the leaves on the plants. It is of utmost importance to have uniform, young, vigorous, healthy assay tobacco plants and to use a standard inoculation practice. The virus samples may be applied to the leaves with a sterile glass spatula previously dipped in the virus homogenate. Otherwise some workers use the finger directly for inoculation or apply the inoculum with a uniform 25-cm square of cheesecloth as a swab dipped in the preparation. Practice is needed to make the procedure uniform and the results reproducible. The center five to six leaves on five or six plants, respectively, are usually best and, when inoculated according to the 5 × 5 or 6 × 6 Latin square design, may minimize the inherent variability in the plants and inoculation procedure. Plants that are old, diseased, or wilted are to be avoided. The leaves are dusted lightly with carborundum powder. Some workers prefer to give the plants a 12-hour dark period before inoculation. One-half of each test leaf is inoculated with the homogenate of the cultured tissues and the opposite half leaf with a TMV standard solution (purified TMV at 0.2 μg/ml) (Fig. 6). The inoculations are replicated at least five times on half leaves of five plants with the sample of each homogenate applied at a different leaf position on each plant. Lesions often develop in 3 to 5 days depending on temperature and light conditions. With both the tissue culture cell homogenate and the standard TMV solution it will probably be necessary to run preliminary test assays to get the dilution adjusted so that too many or too few lesions are avoided. The dilutions should be such that the local lesions are not merged, but can be counted separately. The relative infectivities of the homogenates of the tissues may be then calculated by dividing the number of local lesions produced by the homogenate by the number of lesions produced by the TMV standard.

Fig. 5. TMV virus inclusions in living, unstained tobacco callus cells as seen in microcultures with phase-contrast microscopy. (1) Cell with long fibrous inclusions (f) forming a loop. Several individual thin fibers composing the fibrous inclusion body are seen diverged into the cytoplasm. Magnification ×350. (2) Long fibrous inclusions (f) in callus cells are seen forming various figures −6, −8, or a ring. The shining small bodies in the cells are the plastids (p). Magnification ×240. (3) Nucleus (n) of a cell with several fibrous needles (f) passing through it. Magnification ×400. (4) A needle (f) has passed through the walls of adjoining cells. The cell on the left shows a centrally placed nucleus (n) with nucleolus (ne) and cytoplasmic strands (s) converging on it. Magnification ×300. (5a) A needle (f) has passed through the walls of adjoining cells. The needle pushed by the streaming cytoplasm is seen in the cell on the right (arrow). (5b) Picture taken 10 hours later shows that more than three-fourths of the needle has passed into the second cell. Magnification ×300. (6a) A long fibrous inclusion (f) intercepted by a newly formed wall (w) in a dividing cell. (6b) The active cytoplasmic streaming in the daughter cell on the right withdrew the inclusion into its cytoplasm. The large part of the needle forming a loop bent downward and backward is out of focus. Magnification ×300. (From M. Singh and A. C. Hildebrandt, *Virology* **30**, 134, 1966.)

Fig. 6. (A) Lower surface of a leaf on hybrid tobacco (*Nicotiana tabacum* x *N. glutinosa*) used for the local lesion TMV assay 4 days following inoculation with virus. The right-hand side was inoculated with an homogenate of known dilution of the virus-infected tissue culture that had incubated on a test medium. The left-hand side of the leaf was inoculated with a standardized TMV preparation containing a known amount of virus. The number of necrotic leaf spots or local lesions was proportional to the concentration of virus in the two samples. (B) Best center five leaves inoculated with TMV as above. Each, respectively, of five different tissue culture homogenates rubbed on the right side of the leaf in each case according to a Latin square plan and standard TMV rubbed on left side of leaves. (From A. C. Hildebrandt, unpublished photograph.)

Fig. 7. Lesions induced by homogenates of tobacco cell cultures infected with three strains of TMV and applied on the hybrid tobacco assay plant leaf. Left half: Johnson's mild strain. Right half: Johnson's severe strain (large lesions), and very mild strain screened out earlier from tobacco tissue culture (small lesions). All inoculations made at the same time, and the picture was taken 12 days after inoculation. (From J. H. Wu *et al.*, *Phytopathology* **50**, 1960.)

Different severe and mild strains of TMV have been successfully artificially inoculated to and screened out by tobacco cell cultures *in vitro* and assayed on local lesion host plants[4,11] (Fig. 7).

TMV assays have also been made using individual cells mechanically separated from TMV-infected callus cultures of *Nicotiana tabacum* Havana 38, of White Burley, and of a hybrid between Havana 38 and *N. glutinosa*.[16] A microassay of the homogenized single cells in 0.2 μl of buffer on local lesion host plants showed that 30–40% of the cells contained TMV.

Certain strains of tobacco callus tissue infected with TMV that have been repeatedly subcultured for over 10 years have maintained the TMV.[3,4] Other strains of tobacco cell cultures may lose the virus after a few months and several subcultures. Cooper *et al.*[17] showed that callus inoculated and maintained in liquid suspension shake cultures remained infected longer than agar-grown cultures, but that a certain number of cultures lost the TMV, whereas others remained infected through 7 monthly subcultures. Thus, the original mode of in-

[17] L. Cooper, A. C. Hildebrandt, and A. J. Riker, *Plant Physiol. Suppl.* **37**, 53 (1962).

fection, and subsequent methods of callus maintenance appear to have a decisive influence on the distribution of the virus in the callus, and upon the possibility of maintaining infected tissue cultures for extended periods.

Recently it has been possible to artificially infect tobacco mesophyll cell protoplasts with TMV.[18] The mesophyll cell protoplasts were prepared by cellulase enzymatic treatment of the intact cells to remove the cell walls. The purified virus suspension was added to the protoplast suspension in the presence of poly-L-ornithine. Rapid TMV multiplication resulted in the infected protoplasts during a 24-hour period. This method may provide an improved means of obtaining a greater efficiency of infection and multiplication and to study host cell–virus relations further.

Plant cell culture methods thus provide a fine tool to study TMV activities at the cellular level. Cultures should be assayed at regular intervals to verify the virus content. Liquid suspension cultures may be more useful than agar cultures to maintain long-term cultures. The origin of the tissue, cell, or protoplast and the method of virus inoculation appear equally important.

[18] I. Takebe and Y. Otsuki, *Proc. Nat. Acad. Sci. U. S.* **64**, 843 (1969).

SECTION XII

Production of Hormones and Intercellular Substances

Editors' Comments

As is true for a number of other scientific disciplines in biology, the development of tissue culture methodology enables endocrinologists and others to study hormones and expressions of specific cell function at the cellular level. The protocols in this section include those developed primarily in the study of mechanisms involved in expressions of cell function. It will be obvious that others are mostly the result of interests in the potential uses of tissue culture to produce products of pharmaceutical and medicinal value. On both counts the future role of tissue culture, as it relates to the welfare of man, is very bright.

CHAPTER 1

Growth Hormone and Prolactin from Rat Pituitary Tumor Cells[1]

P. S. Dannies and A. H. Tashjian, Jr.

Clonal strains of rat pituitary tumor cells which secrete growth hormone and prolactin into the medium have been serially propagated in culture for 6 years. Production of these two hormones in intact animals is controlled by a variety of factors; these pituitary cell strains respond to many of the same factors as normal pituitary cells. These strains therefore provide a homogeneous cell population that is a useful model for determining the mechanism of action of factors which influence production of growth hormone and prolactin.

ESTABLISHMENT OF PITUITARY CELL LINES

Cultures were established in 1965 from a multihormone-producing rat pituitary tumor, MtT/W5,[2] and were adapted to tissue culture by alternate culture and animal passage.[3] Dispersed tumor cells were allowed to attach and grow in plastic culture dishes for 3 to 5 days. The surviving attached cells were then harvested and reinjected into a rat of the Wistar-Furth strain. Cells from the new tumor, derived from the cultured cells, were recultured and the process repeated several times. Cells will attach and grow more easily when plated from such culture-derived tumors.

Several epithelial cells lines were cloned,[3] have been maintained in culture for over 6 years, and have maintained hormone production. Extensive studies have been formed with the GH_3 strain which produces both growth hormone and prolactin. At least twelve serial clones of GH_3 have been isolated, each derived from a single cell of the preceding clone, and each producing both growth hormone and prolactin. This suggests that both hormones are produced by a

[1] Preparation of this review and the results of original experiments described were supported in part by a research grant from the USPHS (AM 11011). P. S. Dannies is a Postdoctoral Research Fellow of The Arthritis Foundation.

[2] H. Takemoto, K. Yokoro, J. Furth, and A. I. Cohen, *Cancer Res.* **22,** 917 (1962).

[3] A. H. Tashjian, Jr., Y. Yasumura, L. Levine, G. H. Sato, and M. L. Parker, *Endocrinology* **82,** 342 (1968).

single cell, but it is possible that cells differentiate after dividing into types which produce only one hormone. In the GH_3 strain the capacity to produce growth hormone actually increased over the first 3 years (15–30 μg compared to 4–6 μg growth hormone/mg cell protein/24 hours). However, in a matter of months at the start of 1972, the serially propagated cells suddenly produced little or no detectable basal levels of growth hormone (0.06 μg or less hormone/mg cell protein/24 hours). This strain, now designated GH_4, which still produces prolactin (6–12 μg hormone/mg cell protein/24 hours) can produce growth hormone when stimulated by factors which induce increased synthesis of growth hormone.[4] The hormone-producing GH_3 cells have been preserved in liquid nitrogen and maintain their original properties when thawed and reestablished in culture.

As well as the cells which produce both hormones, or prolactin alone, there is a spindle-shaped subclone of the GH_1 strain,[3] called GH_12C_1, which produces no detectable levels of prolactin.

These strains are aneuploid with a modal chromosome number ranging from 69 to 76.[5] They are grown on plastic or glass in Ham's F 10 Medium[6] supplemented with 15% horse serum and 2.5% fetal calf serum at 37°C in a humidified atmosphere of 5% CO_2 and 95% air. GH_3 cells can be grown as a monolayer or in suspension cultures.[7]

Determination of Amounts of Growth Hormone and Prolactin

Growth hormone and prolactin can be measured in the culture medium or in cell homogenates by the technique of microcomplement fixation.[8-10] The lower limits of detection of the hormones by this method are 0.025–0.1 μg/ml medium. The amount measured in the medium is not necessarily equivalent to the amount synthesized, since the amount that appears in the medium may be the result of both secretion and degradation. The quantity of a hormone measured in the medium is, therefore, referred to as production of that hormone and is expressed as the amount of hormone which accumulates in the medium during a defined period of time divided by the cell protein at the time of collection.[11]

Both growth hormone and prolactin can be measured by complement fixation inside the cell. There is about as much growth hormone stored in the cells as appears in the medium after 15 to 30 minutes of incubation while the amount of

[4] A. H. Tashjian, Jr., P. M. Hinkle, and P. S. Dannies, *Proc. 4th Intern. Congr. Endocrinol., Excerpta Med.* Amsterdam, 1973.

[5] C. Sonnenschein, U. I. Richardson, and A. H. Tashjian, Jr., *Exp. Cell Res.* **61**, 121 (1970).

[6] R. G. Ham, *Exp. Cell Res.* **29**, 515 (1963).

[7] F. C. Bancroft and A. H. Tashjian, Jr., *Exp. Cell Res.* **64**, 125 (1971).

[8] L. Levine and H. Van Vunakis, *In* "Methods in Enzymology" (C. H. W. Hirs, ed.), Vol. XI, p. 928. Academic Press, New York, 1967.

[9] A. H. Tashjian, Jr., L. Levine, and A. E. Wilhelmi, *Ann. N. Y. Acad. Sci.* **148**, 352 (1968).

[10] A. H. Tashjian, Jr., F. C. Bancroft, and L. Levine, *J. Cell Biol.* **47**, 61 (1970).

[11] F. C. Bancroft, L. Levine, and A. H. Tashjian, Jr., *J. Cell Biol.* **43**, 432 (1969).

intracellular prolactin is equal to that appearing in the medium after 1 to 2 hours of incubation.[10,11]

Synthesis of a hormone in the cells can be detected by incubating the cells with radioactive amino acids and isolating the newly synthesized, labeled hormone from the cell homogenate by immunoprecipitation. The precipitate can then be analyzed by sodium dodecyl sulfate acrylamide gel electrophoresis, which dissociates the hormone from the antibody and spreads out any nonspecific background.[12] The labeled hormone is a discrete band in the acrylamide gel. Labeled hormones that have been secreted into the medium may be detected by the same method.

Factors Which Influence Production of the Hormones

Hydrocortisone. The effect of hydrocortisone (cortisol) on growth hormone will be discussed in Section XII, Chapter 2, but we wish to emphasize here that hydrocortisone affects not only the production of growth hormone but also prolactin. Growth hormone production by GH_3 cells is increased four- to eightfold, while that of prolactin is decreased to one-quarter that of control cells (Fig. 1). This effect is seen with a hydrocortisone concentration as low as 5×10^{-8} *M*. There is a lag period of 24 to 36 hours and a maximum effect is seen

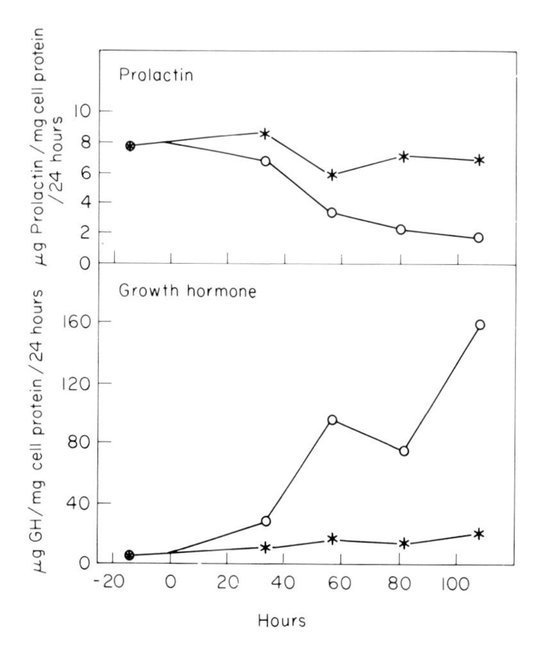

Fig. 1. Effects of hydrocortisone on prolactin and growth hormone production by GH_3 cells. Hydrocortisone was added to duplicate plates at zero time to give a final concentration of 3×10^{-6} *M*. (○) Hydrocortisone-treated; (*) control. (Reproduced from Tashjian *et al.*[10])

[12] D. Cashman and H. C. Pitot, *Anal. Biochem.* **44,** 559 (1971).

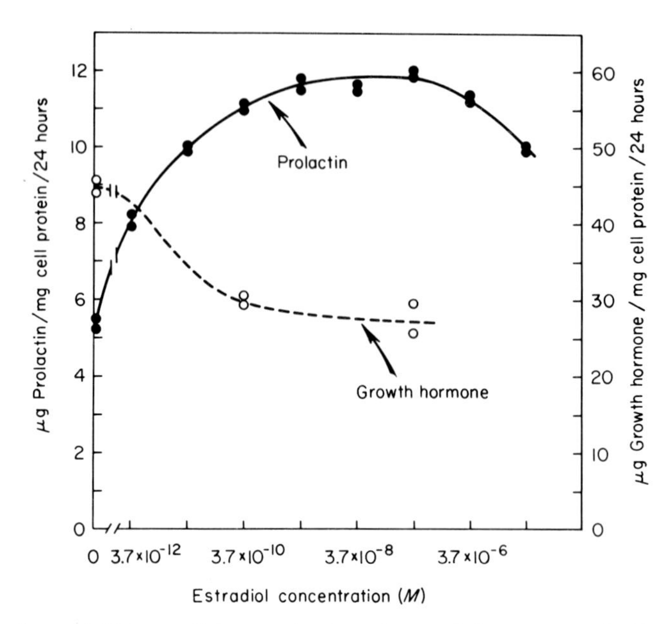

Fig. 2. Effects of 17β-estradiol on prolactin and growth hormone production by GH_3 cells. The data shown are for the collection period 3–7 days after the addition of estrogen. (Reproduced from Tashjian *et al.*[4])

at 70 to 100 hours.[10,11] It has been shown that hydrocortisone stimulates growth hormone production in primary cultures of monkey pituitary glands,[13] indicating this response is not unique to the GH_3 cells.

17β-Estradiol. The data in Fig. 2 show that estradiol, at concentrations as low as 10^{-11} *M*, increases prolactin production twofold; at the same time, growth hormone production is decreased by about 30%. The stimulation of prolactin production, measured by complement fixation, is detected after about 48 hours and reaches a maximum at about 4 days. Whether estrogenic steroids affect prolactin secretion in the intact animal by a direct action on the pituitary as well as through the hypothalamus has not been resolved; however, the discovery of estrogen receptors in the pituitary[14,15] as well as the report of a direct effect of estrogen on explanted pituitary fragments[16] suggests that the effects of estrogen on prolactin production in the intact animal are due to actions on both the hypothalamus and the pituitary. The direct effect of estrogen on prolactin production in GH_3 cells supports this view.

Ergocryptine. Drugs which influence hormone secretion in the intact animal are found to influence the production of these hormones in cell culture. The synthetic ergot alkaloid 2-Br-α-ergocryptine, an analog of ergocornine, is known

[13] P. O. Kohler, W. E. Bridson, and P. L. Rayford, *Biochem. Biophys. Res. Commun.* **33,** 834 (1968).

[14] I. Kahwanago, W. L. Heinrichs, and W. L. Herrmann, *Endocrinology* **86,** 1319 (1970).

[15] A. C. Notides, *Endocrinology* **87,** 987 (1970).

[16] K. H. Lu, Y. Koch, and J. Meites, *Endocrinology* **89,** 229 (1971).

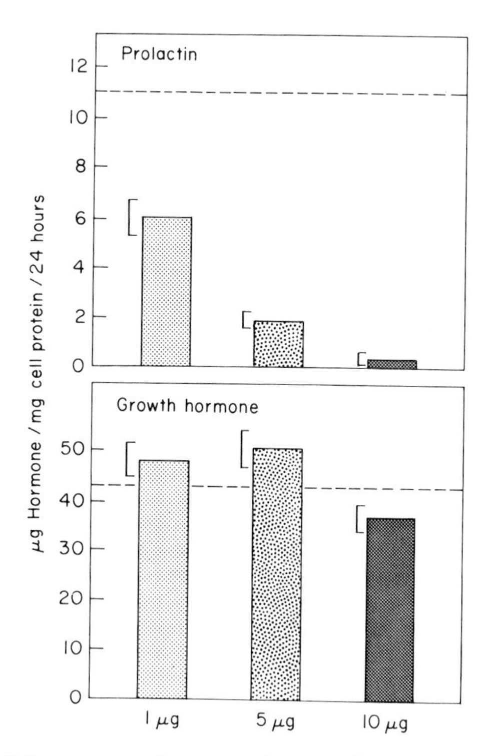

Fig. 3. Effects of 2-Br-α-ergocryptine on prolactin and growth hormone production by GH_3 cells. The data shown are for the collection period 3–6 days after the addition of the drug. The dashed horizontal lines give levels of prolactin and growth hormone in control cultures. The doses given are per milliliters of culture medium. (Reproduced from Tashjian *et al.*[22])

to inhibit prolactin secretion in animals.[17-19] This alkaloid specifically and strongly inhibits prolactin production without affecting that of growth hormone, as shown in Fig. 3. In the responses caused by many of the factors which affect the production of growth hormone and prolactin, the production of one hormone is related in a reciprocal manner to the production of the other. However, with ergocryptine, production of prolactin may be changed without influencing growth hormone, indicating a change of production of one hormone is not necessarily linked to the other.

Cyclic AMP. Cyclic AMP has been found to affect the release of prolactin and growth hormone from the pituitary, as well as increasing synthesis of pituitary hormones.[20] The effect of dibutyryl cyclic AMP on GH_3 cells is illustrated in Fig. 4. The cyclic nucleotide stimulates production of both hormones three- to fourfold, but only at relatively high extracellular concentrations. Addition of theophylline, an inhibitor of phosphodiesterase, does not enhance the

[17] M. D. Shelesnyak, *Amer. J. Physiol.* **179**, 301 (1954).
[18] P. V. Malven and W. R. Hoge, *Endocrinology* **88**, 445 (1971).
[19] W. Wuttke, E. Cassell, and J. Meites, *Endocrinology* **88**, 737 (1971).
[20] F. Labrie, G. Béraud, M. Gauthier, and A. Lemay, *J. Biol. Chem.* **246**, 1902 (1971).

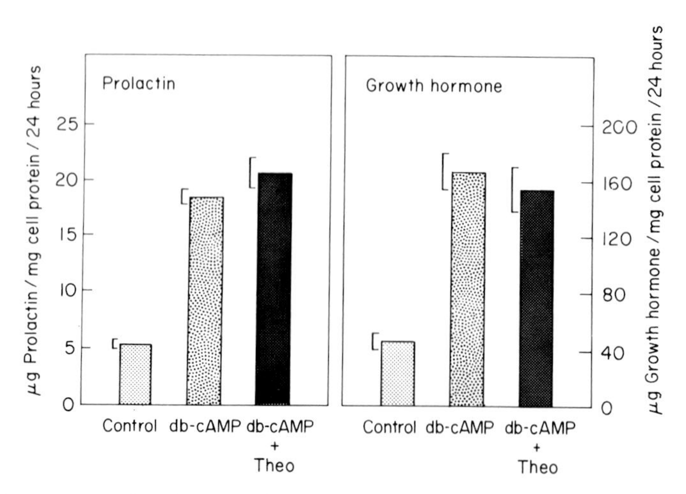

Fig. 4. Stimulation of prolactin and growth hormone by dibutyryl cyclic AMP in GH_3 cells. Dibutyryl cyclic AMP (db-cAMP) was used at a concentration of 10^{-3} *M* and theophylline at 10^{-4} *M*. Theophylline alone had no effect. (Reproduced from Tashjian *et al.*[22])

effect of cyclic AMP. Whether cyclic AMP is in fact a mediator of any of the factors which affect production of the hormones in GH_3 cells remains to be determined. Since cyclic AMP increases production of both hormones, any proposed mechanism involving cyclic AMP as a mediator will have to be able to explain how production of only one hormone may be increased.

Thyrotropin Releasing Hormone. GH_3 cells have proved useful not only because they respond to factors known to influence hormone secretion in the intact animal; they have also been useful in discovering the prolactin-releasing activity of thyrotropin releasing hormone (TRH). The hypothalamic tripeptide, TRH, causes the rapid release of pituitary thyrotropin both in intact animals and from isolated pituitary explants, but consistent effects of TRH on the synthesis or release of other pituitary hormones had not been described until Tashjian and co-workers reported a direct effect of TRH on prolactin production by GH_3 cells.[21] The effect of TRH on prolactin and growth hormone production is shown in Fig. 5. TRH increases prolactin production at least twofold and inhibits growth hormone production with no effect on cell growth. The stimulation of prolactin production is detected by complement fixation from 4 to 6 hours after the addition of the tripeptide and reaches a maximum within 24 to 48 hours. An effect is observed at 0.28 n*M* TRH, and the effect increases up to a concentration of 28 n*M* TRH. Unlike the other factors described here, the effect of TRH does not begin to disappear as soon as TRH is removed from the cells. Two experiments are shown in Fig. 6 that illustrate this long-term effect; after exposure to 10 ng/ml (28 n*M*) TRH for 6 hours prolactin production remains elevated for 3 days or more. TRH also causes morphological changes in the GH_3 cells which can be observed by simple phase-contrast microscopy or by the scanning

[21] A. H. Tashjian, Jr., N. J. Barowsky, and D. K. Jensen, *Biochem. Biophys. Res. Commun.* **43,** 516 (1971).

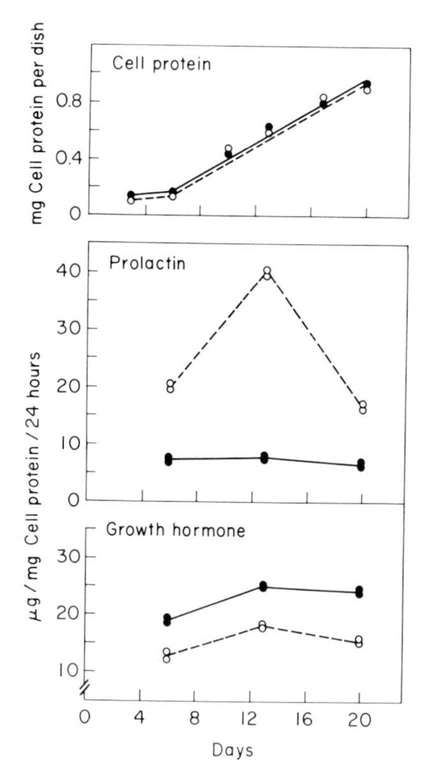

Fig. 5. Effects of TRH on cell protein, and on prolactin and growth hormone by cultures of GH_3 cells. TRH at a concentration of 10 ng/ml (28 n*M*) was added at zero time. (○) Treated with TRH; (●) control. (Reproduced from Tashjian *et al.*[21])

electron microscope.[22] These morphological changes become apparent at the same time prolactin secretion is increased.

Since the discovery of the effect of TRH on GH_3 cells, it has been shown by several laboratories that in species such as human, monkey, and cow, TRH causes a release of prolactin in the intact animal.[23,24] It is clear this response to TRH is not restricted to tumor cells in culture but occurs in the intact organism as well. However, the physiological role of TRH in the control of prolactin secretion and its relation to other factors which control prolactin release remain to be elucidated.

Cations. Calcium ion plays an important role in a variety of cell secretory processes including, possibly, the pituitary gland. When the Ca^{2+} of the culture medium (normally 0.8 m*M*) is reduced selectively by the addition of increasing

[22] A. H. Tashjian, Jr. and R. F. Hoyt, Jr., *Proc. Symp. Molecular Genet. Develop. Biol. Soc. Gen. Physiol., Woods Hole, Massachusetts, Sept. 4–7,* 1971, p. 353. Prentice-Hall, Englewood Cliffs, New Jersey, 1973.

[23] C. Y. Bowers, H. G. Friesen, P. Hwang, H. J. Guyda, and K. Folkers, *Biochem. Biophys. Res. Commun.* **45,** 1033 (1971).

[24] L. S. Jacobs, P. J. Snyder, J. F. Wilbur, R. D. Utiger, and W. H. Daughaday, *J. Clin. Endocrinol.* 33, 996 (1971).

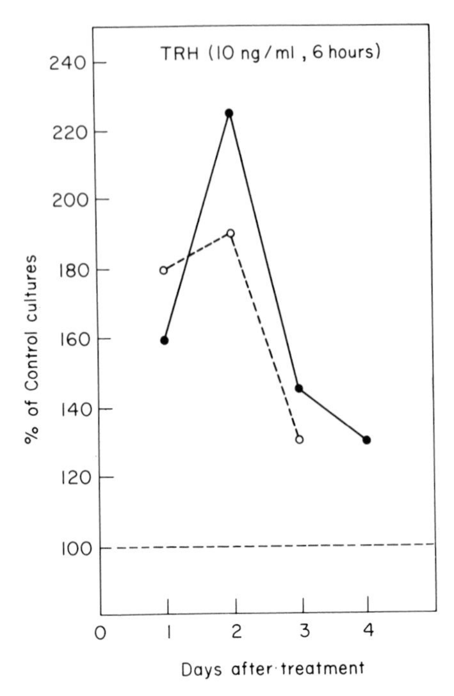

Fig. 6. Effects of removal of TRH on the production of prolactin by GH_3 cells after a 6-hour exposure to the tripeptide (28 n*M*). Results of two separate experiments are shown; each point is the mean of duplicate plates. Results are expressed as the percentage of prolactin production of the TRH-treated cultures as compared to the controls. (Reproduced from Tashjian *et al.*[4])

quantities of sodium ethyleneglycol-bis(β-aminoethyl ether)-*N,N'*-tetraacetic acid (NaEGTA), there is a parallel decrease in the rate of production of both growth hormone and prolactin. The time course of the effect of reduced Ca^{2+} on hormone production is shown in Fig. 7. Hormone production remained unchanged for 2 hours; subsequently the rates of production of both hormones decreased progressively from 4 to 24 hours. Total protein synthesis, measured by the incorporation of labeled amino acids into material precipitated by trichloroacetic acid, remained unchanged. The effects of exposure of the cells to low Ca^{2+} were reversed merely by restoration of the Ca^{2+} to 0.8 m*M*.

Increasing the Ca^{2+} to 6.0 m*M* stimulated the production of both hormones while high Mg^{2+} (6 m*M*) depressed the rate of growth hormone and prolactin production. Intracellular levels of both hormones always paralleled the concentrations of hormone in the medium suggesting that the observed effects are mainly on the synthesis of the two hormones.

Use of the Tumor Cells for Further Research

Clonal strains of rat pituitary tumor cells, which have been propagated for up to 6 years in culture, produce growth hormone, prolactin or both hormones.

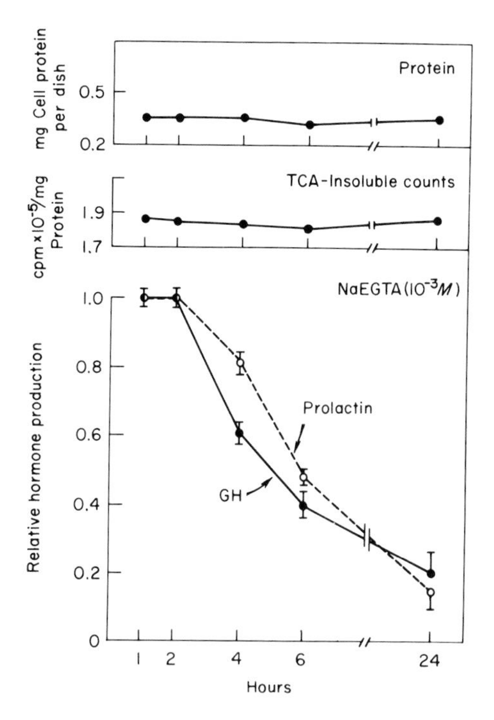

Fig. 7. Time course of the effect of 1.0×10^{-3} M EGTA on hormone production. The results are expressed as relative hormone production (NaEGTA-treated divided by control). At the top are shown the total cell protein values and labeled amino acid incorporation data. (Reproduced from A. H. Tashjian, Jr. and K. M. Gautvik, *4th Int. Ferment. Symp., March 19–25, Kyoto, Japan, 1972,* in press.)

The production of these hormones by cells in culture can be altered by many factors which have been found to influence the pituitary gland in the intact animal. However, unlike the pituitary, these cell strains represent clonal populations of cells in an environment that can be highly controlled, and, as such, provides an important tool for future studies on the binding sites of the various factors that influence hormone production, the fate of the factors after binding, and the primary molecular events that occur in the cells as a result. The fact that the cells are tumor cells means all events may not be identical in the intact animal; for example, in the pituitary, one cell does not produce both prolactin and growth hormone. However, although tumor cells can lose functions or express them inappropriately, it seems unlikely to us that they can invent either completely new functions or new modes of action. Control experiments with normal pituitary cells or explants will always be necessary to confirm that mechanisms deduced from tumor cells correspond to normal mechanisms.

CHAPTER 2

Enhancement of Growth Hormone Production by Cortisol

P. O. Kohler and W. E. Bridson

While attempting to reverse the gradual decline of hormone synthesis by normal monkey anterior pituitary cells in explant culture, we found that physiological concentrations of cortisol stimulated growth hormone production.[1] This effect appeared to be limited to steroids with glucocorticoid activity. Studies with inhibitors suggested that both RNA and protein synthesis, but not DNA synthesis, were required for the response. The enhancement of growth hormone production by cortisol also occurred in primary cultures of human pituitary adenomas and normal human pituitary glands.[2] However, these cultures contained a mixed population of cells secreting a variety of different protein hormones.[3] The availability of cloned lines of growth hormone-producing rat pituitary tumor cells[4,5] provided improved model systems for examining the metabolic events in a homogeneous cell population during the regulation of protein hormone synthesis by cortisol.

The methods we have used to examine the important correlation between the induction of growth hormone by glucocorticoids and the effect of cortisol on protein and nucleic acid synthesis are described.

METHODS

Cell Culture. The GH_1 line[5,6] of rat pituitary tumor cells was used for these studies. The cells were maintained in Ham's F10 Medium[7] containing 3.2% fetal bovine serum and 13.5% horse serum with 50 units penicillin and 50 μg streptomycin per milliliter. Replicate cultures were made by dispersing the GH_1 cells with 0.05% trypsin, 0.02% sodium versenate (GIBCO), and plating 2×10^6 cells

[1] P. O. Kohler, W. E. Bridson, and P. L. Rayford, *Biochem. Biophys. Res. Commun.* **33**, 834 (1968).

[2] W. E. Bridson and P. O. Kohler, *J. Clin. Endocrinol. Metab.* **30**, 538 (1970).

[3] P. O. Kohler, W. E. Bridson, P. L. Rayford, and S. E. Kohler, *Metabolism* **18**, 782 (1969).

[4] Y. Yasumura, A. H. Tashjian, and G. H. Sato, *Science* **154**, 1186 (1966).

[5] A. H. Tashjian, Y. Yasumura, L. Levine, G. H. Sato, and M. L. Parker, *Endocrinology* **82**, 342 (1968).

[6] GH_1 cells may be obtained from the American Type Culture Collection, 12301 Parklawn Drive, Rockville, Maryland 20852.

[7] R. G. Ham, *Exp. Cell Res.* **29**, 515 (1963).

in each 30-ml plastic flask (Falcon). (Plastic dishes would be preferable in most instances.) The total 5 ml volume of medium was changed every 24 hours for each flask. Experiments were initiated 3 days after plating. The cultures were kept in a humidified 95% air–5% CO_2 gas mixture at 37°C with the flask caps loosened so that the gas phase over the culture could equilibrate with that in the incubator.

All steroids used were first dissolved in ethanol which was then added to the whole medium. The final concentration of ethanol in the medium was less than 0.5%. Cortisol (Hydrocortisone, Δ^4-pregnene-11β,17α,21-triol-3,20-dione) was obtained from Sigma Chemical Co.

Experimental Design. Replicate cultures of GH_1 cells were incubated with control medium and medium containing concentrations of cortisol from 5×10^{-9} to 5×10^{-6} *M*. Growth hormone levels in the medium were measured by a radioimmunoassay method previously described.[8,9]

To examine the rates of protein and nucleic acid synthesis, replicate control and cortisol-treated cultures were taken at intervals of 2, 12, 24, 48, and 72 hours after the addition of cortisol at 5×10^{-6} *M*. At each time period the medium was completely changed and replaced with 2 ml of medium containing labeled precursors for protein and nucleic acid synthesis. The incubation time for the RNA precursor ([5-^{3}H]uridine, Schwarz/Mann) and protein precursor (^{14}C-reconstituted protein hydrolysate, Schwarz/Mann) was 1 hour. By using concentrations of 5 μCi/ml [^{3}H]uridine and 0.2 μCi/ml of the amino acid mixture, rates of synthesis of both protein and RNA could be measured simultaneously by the procedures listed below. The incubation time for the DNA precursor ([^{3}H]methyl thymidine, Schwarz/Mann) at a concentration of 4 μCi/ml was 6 hours.

At the end of the incubation period, the cells were washed with plain medium and frozen at −70°C. At the time of analysis the tops of the flasks were removed and the cells scraped out of the flask with a rubber policeman. The cells from each flask were homogenized in 2 ml of chilled saline with ten strokes of a Dounce homogenizer (Kontes Glass Co.).

A 10-μl aliquot of the homogenate from each culture was placed on a Whatman 3MM paper disk for measurement of labeled amino acid incorporation into protein after the method of Mans and Novelli.[10] Duplicate 20-μl aliquots were also saved for colorimetric protein determinations using the method of Lowry.[11] The remainder of the homogenate was precipitated in a final trichloroacetic acid (TCA) concentration of 10% for 1 hour in ice. The tubes were centrifuged at

[8] L. A. Frohman and L. L. Bernardis, *Endocrinology* **82**, 1125 (1968).

[9] Materials for radioimmunoassay of rat growth hormone may be obtained by writing: Hormone Distribution Officer, Office of the Director, Bldg. 31-9A47, National Institute of Arthritis, Metabolism, and Digestive Diseases, National Institutes of Health, Bethesda, Maryland 20014.

[10] R. J. Mans and G. D. Novelli, *Arch. Biochem. Biophys.* **94**, 48 (1961).

[11] O. H. Lowry, N. J. Rosebrough, A. L. Farr, and R. J. Randall, *J. Biol. Chem.* **193**, 265 (1951).

1200 *g* for 20 minutes and the supernatant was discarded. The precipitate was washed with 2 ml of 10% TCA at 0°C, resuspended with a vortex mixer and recentrifuged. The precipitate was then sequentially washed with 2 ml of cold 95% ethanol, 2 ml of boiling 95% ethanol, and 2 ml of ether at room temperature. After evaporation to dryness the precipitate was resuspended in 2 ml of 5% TCA and heated for 30 minutes in a 90°C water bath in parafilm-covered tubes. After this hydrolysis, the tubes were again centrifuged at 1200 *g* and the supernatant saved for analysis. For determination of labeled precursor incorporation, aliquots of the supernatant were dissolved in a 0.5 ml volume of NCS (Nuclear-Chicago) and counted in a Liquiflour–toluene solution in a liquid scintillation counter. Quenching was found to be uniform and, therefore, no corrections were necessary for these studies. RNA content of the supernatant was determined by a colorimetric method employing an orcinol reagent,[12] and DNA was measured by the diphenylamine method of Burton.[13]

The rate of incorporation for each precursor was determined by calculating total counts of specific precursor into the total mass of specific protein or nucleic acid in each flask. Therefore, total counts per minute (CPM) of ^{14}C-labeled amino acid were divided by total protein determined in each culture; total CPM of [^{3}H]uridine by total RNA, and total [^{3}H]thymidine by total DNA. Usually, each point in each experiment consisted of three replicate cultures. The mean of the specific activities (CPM/unit mass) could, therefore, be determined for statistical purposes. The significance of differences between treated and untreated groups was examined by a two-way analysis of variance.

Results

The effects of concentrations of cortisol between 5×10^{-9} *M* and 5×10^{-6} *M* in the medium on growth hormone production by the GH_1 cells are shown in Fig. 1. Cortisol in the medium from 5×10^{-8} *M* to 5×10^{-6} *M* caused marked increases in growth hormone production by the GH_1 cells. Although significant increases in growth hormone in the medium were usually detectable only after 48 hours, the intracellular content of growth hormone was increased by 24 hours.[14] Removal of cortisol from the medium resulted in a decrease of growth hormone production per cell to pretreatment levels. Cortisol treatment for 72 to 96 hours increased growth hormone production per cell maximally to six to eight times the level in untreated cultures. The increase in growth hormone production between replicate cortisol-treated and control cultures was usually only about fourfold. The reason for this difference was the finding that cells in the cultures receiving cortisol multiplied at a slower rate.[14] We therefore studied the effect of cortisol on protein and nucleic acid synthesis to correlate the induction of growth hormone synthesis with other metabolic events in the treated cells.

[12] G. Ceriotti, *J. Biol. Chem.* **214,** 59 (1955).

[13] K. Burton, *Biochem. J.* **62,** 315 (1956).

[14] P. O. Kohler, L. A. Frohman, W. E. Bridson, T. Vanha-Perttula, and J. M. Hammond, *Science* **166,** 633 (1969).

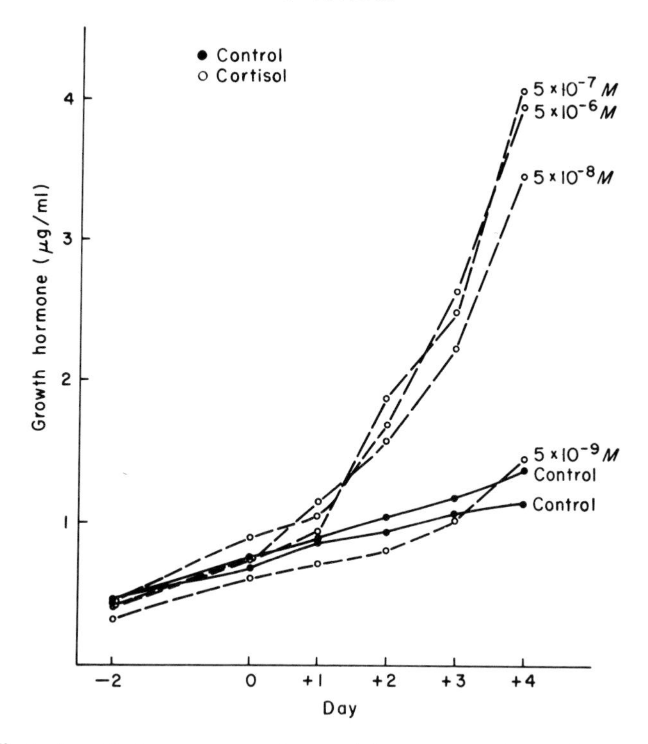

Fig. 1. Effect of cortisol on content of growth hormone in the medium. Cortisol in concentrations indicated was added to replicate cultures of GH_1 cells on day 0, and medium was changed every 24 hours. (From Kohler *et al.*[14])

Within 12 hours after the addition of cortisol, cells showed a significant decrease in the rate of DNA ($p < 0.01$) and protein synthesis ($p < 0.05$) in comparison to control cultures. This effect was even more pronounced by 24 to 48 hours (Fig. 2). Although protein synthesis remained suppressed for the remainder of the 72-hour treatment period, DNA synthesis appeared to show some escape from the suppression at 72 hours. In contrast, total RNA synthesis was significantly increased by 24 hours ($p < 0.05$) and remained increased over control levels at the 48- and 72-hour sampling periods. These studies indicate that cortisol has a differential effect, stimulating RNA synthesis during the induction of growth hormone, while decreasing DNA and total protein synthesis. These studies also indicate the selectivity of the cortisol effect in that production of the protein hormone, growth hormone, is increased while total protein synthesis is decreased. However, protein synthesis is required for the effect since inhibitors of protein synthesis such as cycloheximide at 0.1 μg/ml will block the stimulation of growth hormone production by cortisol.[14]

Comments

These studies illustrate some of the types of experiments which can be performed with responsive hormone-producing cells. The finding that physiological

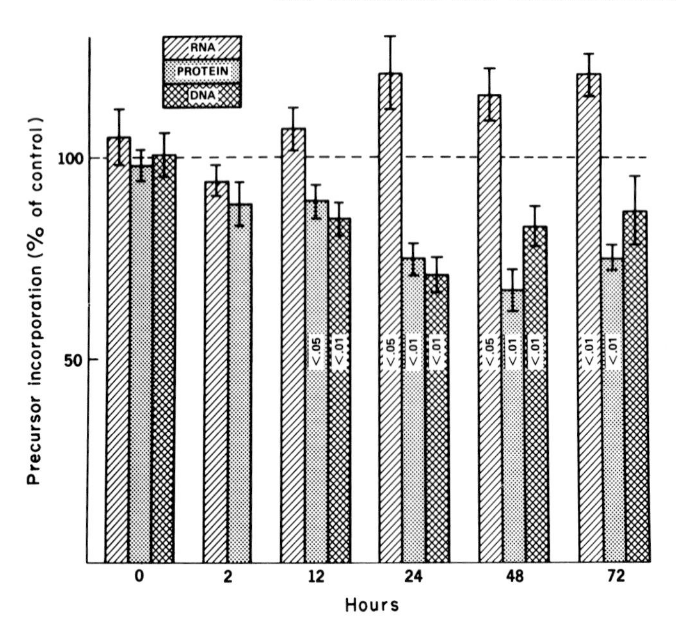

Fig. 2. Summary of effect of cortisol ($5 \times 10^{-6}\,M$) on RNA, protein, and DNA synthesis by GH_1 cells. Each bar represents specific precursor incorporation: ^{3}H-labeled uridine into RNA, mixed ^{14}C-labeled amino acids into protein, or [^{3}H]thymidine into DNA. Each specific activity was expressed as a percentage of control untreated cultures. The brackets represent $\pm$SEM. When values are significantly different from control rates, the p values are shown within the bars. Depression of both protein and DNA synthesis was present at 12 hours. Significant stimulation of RNA synthesis was detectable by 24 hours.

concentrations of glucocorticoids, such as corticosterone and cortisol, enhance growth hormone production in cultures of pituitary cells from the rat, monkey, and man suggests that glucocorticoids have a direct effect on growth hormone production which is perhaps obscured *in vivo* by factors influencing the hypothalamic-pituitary axis.

The effect of cortisol on growth hormone synthesis is specific to the extent that synthesis of all hormones is not increased by glucocorticoids. Tashjian and co-workers[15] have found that cells which synthesize both growth hormone and prolactin respond to cortisol with an increase in growth hormone production alone, whereas prolactin synthesis appears to be suppressed. Similarly, we have noted that monkey anterior pituitary cells in mixed culture show a marked increase in growth hormone production after cortisol, whereas luteinizing hormone synthesis remains unchanged.

The use of a cloned line of cells for these studies has enabled us to study protein and nucleic acid synthesis in a homogeneous cell population. These studies would be impossible in mixed cell cultures where cells other than the target cell of interest might be primarily or secondarily affected by glucocorticoid administration. The effect of cortisol on DNA and protein synthesis was

[15] A. H. Tashjian, Jr., F. C. Bancroft, and L. Levine, *J. Cell Biol.* **47**, 61 (1970); cf. Chapter 1, this Section.

anticipated from the action of glucocorticoids on other types of cells in culture.[16] The stimulation of RNA synthesis, including rapidly labeled nuclear RNA,[17] by cortisol is of interest since glucocorticoids decrease RNA synthesis in fibroblasts.[16] Glucocorticoids stimulate RNA synthesis in some target tissues *in vivo* such as liver, and in thymus cells *in vitro*.[18] However, these effects are usually more rapid, occurring within a few minutes rather than hours. The effect of cortisol on the rat pituitary cells might include an increased synthesis of specific messenger RNA, or might represent a delayed secondary effect on RNA synthesis. Further studies regarding the nature of the stimulated RNA will be necessary to answer these questions.

[16] W. B. Pratt and L. Aronow, *J. Biol. Chem.* **241**, 5244 (1966).
[17] W. E. Bridson, L. A. Frohman, and P. O. Kohler, in preparation.
[18] A. Munck, *Perspect. Biol. Med.* **14**, 265 (1971).

CHAPTER 3

Human Chorionic Gonadotropin

R. A. Pattillo

Human chorionic gonadotropin (HCG), the placental glycoprotein hormone, produced by the cytotrophoblast,[1] is comparable in biological action to pituitary luteinizing hormone (LH) or interstitial cell stimulating hormone (ICSH).[2] Minor immunological differences between LH and HCG[3] do not appear to be of significance for characteristic biological activity of the two hormones. Physiochemical and structural analysis[4-7] of highly purified preparations of the pregnancy HCG, which assayed approximately 12,000 IU/mg have revealed two nonidentical subunits (alpha and beta) with a combined molecular weight of 67,000 to 68,000; carbohydrate moieties consisted of hexoses: galactose and mannose, 8.6%; hexosamines: *N*-acetylglucosamine and *N*-acetylgalactosamine, 8.87%; fucose 0.67% and sialic acid 7.2%. The 57–66% polypeptide compo-

[1] R. A. Pattillo and G. O. Gey, *Cancer Res.* **28**, 1231 (1968).
[2] M. T. Scurry, J. Bruton, and K. G. Barry, *Arch. Intern. Med.* **128**, 561 (1971).
[3] D. A. Goss and John Lewis, Jr., *Endocrinology* **74**, 83 (1964).
[4] N. Swaminathan and Om P. Bahl, *Biochem. Biophys. Res. Commun.* **40**, 422 (1970).
[5] R. Got and R. Bourrillon, *Biochim. Biophys. Acta* **42**, 505 (1960).
[6] Om P. Bahl, *J. Biol. Chem.* **244**, 567 (1969).
[7] F. J. Morgan and R. E. Canfield, *Endocrinology* **81**, 1045 (1971).

nent consisted of amino acids: lysine, histidine, arginine, aspartic acid, threonine, serine, glutamic acid, proline, glycine, alanine, half-cysteine, valine, methionine, isoleucine, leucine, tyrosine, phenylalanine, and tryptophan. The two subunits are believed to be attached noncovalently, since they are dissociable by urea and recombine to give near full biological activity. The carbohydrate and protein moieties are important for biological activity while the carbohydrate component is not involved in antigenic structure.[8] Although the biological activity is markedly reduced with neuraminidase, there was no decrease in affinity of the desialylated derivative for gonadotropin binding sites in the rat testis *in vitro*.[9] These findings indicated that the sialic acid residues of gonadotropin hormones are not essential for biological activity at the target cell level.

At present, the chorionic hormone is the only human hormone for which a bulk hormone producing source has thus far been established *in vitro*. Pattillo and Gey,[1] in 1966, explanted trophoblastic tissue from the malignant placental tumor, choriocarcinoma, which had been serially transplanted to the hamster's cheek pouch by Hertz[10] from a patient's fatal cerebral metastasis. This ultimately pure culture of cytotrophoblast was the first human hormone synthesizing cell system to be established in continuous cultivation (CCL No. 98, American Type Culture Collection Repository—Rockville, Maryland). Bridson and associates reported cloned lines of similar trophoblastic tumors in 1969.[11] Four additional lines and clones were subsequently established by Pattillo and co-workers[12] directly from patient biopsy specimen without animal transplantation. Short-term (3–4 months) *in vitro* propagation and gonadotropin hormone production from the trophoblast of the normal placenta was reported by Pattillo *et al.*[13]; however, hormone production could be maintained for several weeks only.

Culture Methods

Explantation of trophoblastic tumor tissue is accomplished by selecting viable peripheral tissue margins from the typical tumor bolus and sharply incising 1–2 mm fragments with crossed scalpels bearing No. 11 blades. Using a curved Pasteur pipette, the tissue fragments are either implanted directly on fibroblast monolayers previously irradiated[14] with 4000 rads to inhibit mitosis, while permitting continued metabolism as feeder layer conditioning of the *in vitro* milieu. Alternatively, implantation on a reconstituted tropocollagen[15] matrix adjacent

[8] F. K. Mori, *Endocrinology* **86**, 97 (1970).
[9] M. L. Dufau, K. J. Catt, and T. Tsuruhara, *Biochem. Biophys. Res. Commun.* **44**, 1022 (1971).
[10] R. Hertz, *Proc. Soc. Exp. Biol. Med.* **102**, 77 (1959).
[11] W. E. Bridson, G. T. Ross, and P. O. Kohler, *J. Clin. Endocrinol. Metab.* **33**, 145 (1971).
[12] R. A. Pattillo, A. Ruckert, R. Hussa, R. Bernstein, and E. Delfs, *In Vitro* **6**, 398 (1971).
[13] R. A. Pattillo, G. O. Gey, E. Delfs, and R. F. Mattingly, *Amer. J. Obstet. Gynecol.* **100**, 582 (1968).
[14] T. T. Puck, P. I. Marcus, and S. J. Cieciura, *J. Exp. Med.* **103**, 273 (1956).
[15] R. A. Pattillo and G. O. Gey, *Cancer Res.* **28**, 1231 (1968).

to a parabiotic fibroblast layer is additionally enhancing. Finally, when the trophoblast is accompanied by sufficient stromal tissue, explanation directly on unconditioned polyanionically charged Falcon plastic flasks (Falcon Plastics, Bioquest Division, P. O. Box 243, Cockeysville, Maryland 21030) may be sufficient. About 25 to 50 tissue fragments are explanted per flask, accompanied by 2–3 ml nutrient medium.

Culture Media

A medium containing glucose concentrations of 180 to 300 mg% supplemented by 10–20% human serum is utilized. Waymouth's Medium (50%) 752/1 diluted with 30% Gey's balanced salt solution and supplemented by 20% human umbilical cord serum has been most satisfactory. A phenol red indicator permits fluid exchanges dictated by metabolic activity. Antibiotics (penicillin, 100 units, streptomycin, 50 mg) may be used if the tissue source is not entirely aseptic.

Culture Development

Mitotic activity is not observed prior to approximately 4 weeks when small foci of gray-white replicating trophoblast colonies may be noted. Mechanical, microsurgical subculture of such colonies should be carried several days after identification. This can be accomplished with the naked eye or with minimum ocular magnification. The trophoblast colonies are mechanically subcultured in juxtaposition to fibroblast feeder layers or on irradiated fibroblast layers to provide medium conditioning until sufficient numbers of cells and colonies are pooled. Progressively, the trophoblast colonies are increased while fibroblast colonies are decreased, until a pure culture is obtained.

Enzymatic dispersion for subculture should not be attempted until sufficient number of trophoblast colonies are available to provide close to 10^6 dispersed cells. Until that point, medium conditioning is provided by progressively decreasing numbers of feeder cells.

Assay for gonadotropin hormone production is carried out by various methods including bioassay,[16] immunoassay,[17] or radioimmunoassay. Steroid production[18] may be monitored by column chromatography, fluorometry, or gas liquid chromatography.

HCG, extracted from chemically defined growth medium approximates 1000 IU per 10^8 cells per 24 hours. Progesterone and estrogen production varies, depending upon precursors included in the growth medium.[19] Increased effi-

[16] E. Delfs, *Ann. N. Y. Acad. Sci.* **80**, 125 (1959).

[17] H. Lau, E. Ferraz, M. Butler, and G. S. Jones, *Johns Hopkins Med. J.* **127**, 247 (1970).

[18] R. A. Pattillo, R. O. Hussa, W. Y. Huang, E. Delfs, R. Mattingly, *J. Clin. Endocrinol. Metab.* **34**, 59 (1972).

[19] R. A. Pattillo, W. Y. Huang, M. Knoth, H. G. Friesen, G. O. Gey, E. Delfs, L. Hause, J. Garancis, J. Amatruda, J. Bertino, and R. Mattingly, *Ann. N. Y. Acad. Sci.* **172**, 288 (1971).

ciency of hormone production has recently been achieved by Knazek *et al.*[20] in a method which employs an *in vitro* artificial capillary network. This method permits perfusion of nutrients and collection of perfusate for extraction of its hormone-rich content.

[20] R. A. Knazek, P. M. Gullino, P. O. Kohler, and R. L. Dedrick, *Science* **178**, 65 (1972); cf. Section VI, Chapter 7, this volume.

CHAPTER 4

Parathyroid Hormone and Calcitonin[1]

K. M. Gautvik and A. H. Tashjian, Jr.

Parathyroid Hormone

The parathyroid glands synthesize and secrete a polypeptide hormone which regulates the concentration of calcium in plasma. The hormone which has been extracted from bovine and porcine parathyroid glands consists of 84 amino acid residues although a prohormone of larger size has recently been described.[2,3] Parathyroid glands have been cultivated *in vitro* primarily in order to obtain information about hormone biosynthesis and secretion. Organ or cell culture of parathyroid tissue is, in certain aspects, less complex than that of other endocrine glands. There is probably only one major secretory cell type, and only one type of hormone secreted. Furthermore, the function of the parathyroid glands is controlled principally by the concentration of circulating calcium ions by a direct negative feedback mechanism.[4-8] In addition, there is no evidence

[1] Preparation of this review and the results of original experiments described were supported in part by research grants from the USPHS (AM 10206 and AM 11011). K. M. Gautvik is a USPHS International Research Fellow (F05 TW 1766) on leave of absence from the Department of Physiology, University of Oslo, Oslo, Norway.

[2] B. Kemper, J. F. Habener, J. T. Potts, Jr., and A. Rich, *Proc. Nat. Acad. Sci. U. S.* **69**, 643 (1972).

[3] D V. Cohn, R. R. MacGregor, L. L. H. Chu, and J. W. Hamilton, *In* "Calcium, Parathyroid Hormone and the Calcitonins" (R. V. Talmage and P. L. Munson, eds.), pp. 173–183. Excerpta Medica, Amsterdam, 1972.

[4] H. M. Patt and A. B. Luckhardt, *Endocrinology* **31**, 384 (1942).

[5] L. G. Raisz, *Nature* (*London*) **197**, 1115 (1963).

[6] L. G. Raisz, W. Y. W. Au, and P. H. Stern, *In* "The Parathyroid Gland: Ultrastructure,

that the function of the parathyroid glands is under important nervous control. *In vitro* systems have, therefore, been used in several laboratories to study various aspects of parathyroid function since Gaillard[9] showed that these glands could be maintained in organ culture, and would release into the medium a bone resorption-stimulating factor, presumably parathyroid hormone (PTH).

Methods of Culture

Organ Culture. Normal glands from several species[6,10,11] have been used as well as hyperplastic and adenomatous glands from humans.[8,9,11] Small fragments of tissue are placed on fine metal grids or on Millipore filters, transferred to a culture chamber or Petri dish, and then incubated in various media in a humidified atmosphere of 95% O_2 and 5% CO_2 at 37°C. The composition of the medium varies according to the purpose of the study. Often commercially available culture media are used supplemented with 10 to 50% serum. Antibiotics, usually a mixture of penicillin and streptomycin, are then often added to the medium.

Disperse Cell Systems. So far no permanent cell line has been established from the parathyroid gland. Mixed cell cultures of human parathyroid adenomas have been cultivated for periods of several months.[12] Before the adenomas were plated in culture dishes, they were minced or treated with Viokase for 10 minutes at 37°C.[12] The cells were cultivated in Ham's F 10 Medium supplemented with 15% horse serum and 2.5% fetal calf serum. Martin *et al.*[13] have also reported the production of parathyroid hormone of human origin in monolayer cultures. In order to improve the growth of parathyroid cells in culture, Deftos *et al.*[14] transformed human parathyroid adenoma cells by infection with simian virus 40. The transformed cells were reported to outgrow the uninfected cells and could be serially passed over several generations. The transformed cells retained the ability to produce parathyroid hormone for more than 9 months, but they eventually died.

Secretion and Function" (P. J. Gaillard, R. V. Talmage, and A. M. Budy, eds.), pp. 37–52. University of Chicago Press, Chicago, Illinois, 1965.

[7] J. T. Potts, Jr., R. M. Buckle, L. M. Sherwood, C. F. Ramberg, Jr., C. P. Mayer, D. S. Kronfeld, L. J. Deftos, A. D. Care, and G. D. Aurbach, *In* "Parathyroid Hormone and Thyrocalcitonin (Calcitonin)" (R. V. Talmage and L. F. Bélanger, eds.), pp. 407–415. Excerpta Medica, Amsterdam, 1968.

[8] L. M. Sherwood, I. Herrmann, and C. A. Bassett, *Arch. Intern. Med.* **124**, 426 (1969).

[9] P. J. Gaillard, *Develop. Biol.* **1**, 152 (1959).

[10] W. Y. W. Au and L. G. Raisz, *Pharmacologist* **100**, 201 (1968).

[11] L. M. Sherwood, W. B. Lundberg, J. H. Targovnik, J. S. Rodman, and A. Seyfer, *Amer. J. Med.* **50**, 658 (1971).

[12] A. H. Tashjian, Jr., *Biotechnol. Bioeng.* **21**, 109 (1969).

[13] T. J. Martin, P. B. Greenberg, and R. A. Melick, *J. Clin. Endocrinol. Metab.* **34**, 437 (1972).

[14] L. J. Deftos, A. S. Rabson, R. M. Buckle, G. D. Aurbach, and J. T. Potts, Jr., *Science* **159**, 435 (1968).

Studies of Parathyroid Tissue Cultivated *in Vitro*

Production of Hormone. During the first 2–3 days of organ culture a rapid release of preformed hormone takes place. This is due, at least partly, to cell damage caused by the handling of tissue and transfer to *in vitro* conditions. Figure 1 shows the release of human parathyroid hormone from hyperplastic parathyroid tissue cultivated *in vitro*.[15] The explants secreted parathyroid hormone for more than 2 weeks. Because of the high rate of spontaneous release during the first several days in culture, studies of control during this period must be viewed with considerable caution. Sherwood *et al.*[8] have reported that bovine glands and human adenomas remained histologically intact during a period of more than 2 weeks in organ culture. *De novo* synthesis of parathyroid hormone has been documented in explants[2,3,8,10,16] and in dispersed cell cultures of hyperplastic and adenomatous tissue[12,13] as well as with transformed parathyroid cells.[14]

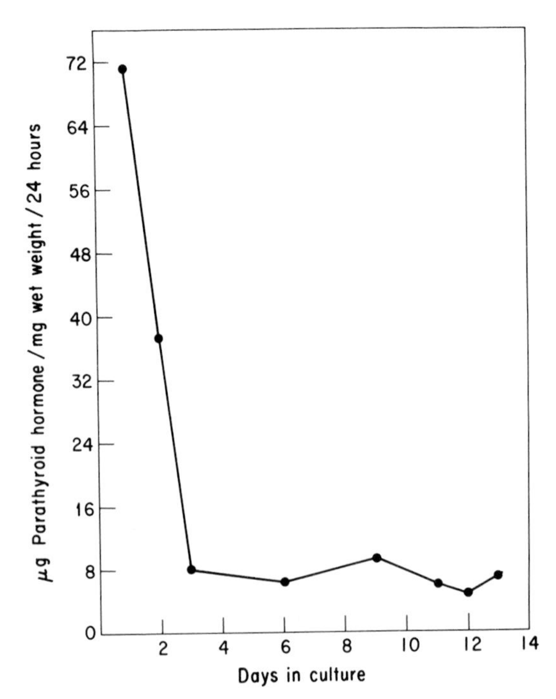

Fig. 1. Release of parathyroid hormone from human hyperplastic parathyroid tissue *in vitro*. The concentration of hormone in medium was measured by radioimmunoassay using bovine parathyroid hormone as standard. Medium (Ham's F 10 supplemented with 15% horse serum and 2.5% fetal calf serum) was changed and collected every 24 hours for assay.

[15] N. Matsuo, K. M. Gautvik, and A. H. Tashjian, Jr., in preparation.

[16] L. M. Sherwood, M. Abe, J. S. Rodman, W. B. Lundberg, Jr., and J. H. Targovnik, *In* "Calcium, Parathyroid Hormone and the Calcitonins" (R. V. Talmage and P. L. Munson, eds.), pp. 183–196. Excerpta Medica, Amsterdam, 1972.

Control of Secretion. The early *in vivo* studies of Patt and Luckhardt[4] and others (for review see footnote 7) of the regulation of parathyroid hormone secretion by plasma calcium were confirmed in organ culture by Raisz.[5] There appears to be a linear inverse relationship between the rate of parathyroid hormone secretion and calcium concentration over the range of calcium of 4 to 12 mg/100 ml.[7] Several studies using organ culture methods[8,11,17] have shown that not only calcium, but also magnesium, is of importance for the secretion of parathyroid hormone. Targovnik *et al.*[17] reported that calcium and magnesium were equivalent in blocking the release of parathyroid hormone from bovine glands in organ culture and that the total concentration of these cations was inversely related to parathyroid hormone release in a nonlinear fashion (Figs. 2 and 3). Recent studies[18] from the same laboratory have shown that high concentrations of dibutyryl cyclic adenosine monophosphate and theophylline can stimulate parathyroid hormone secretion, and these authors suggested that calcium may act through adenylate cyclase and intracellular cyclic AMP.

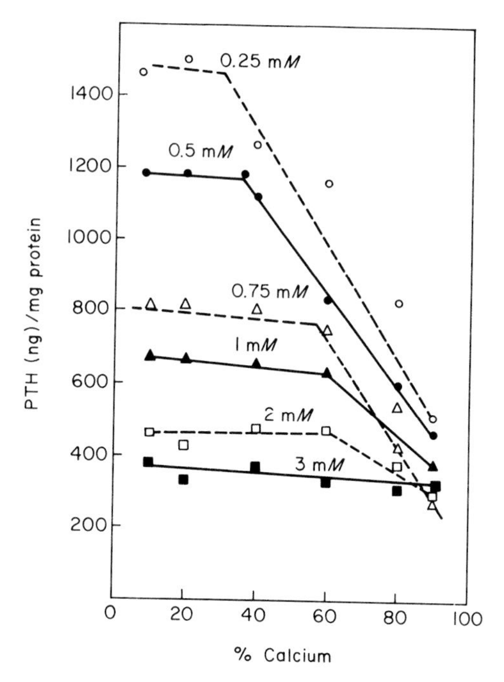

Fig. 2. Daily release of parathyroid (PTH) hormone from organ culture at six divalent cation concentrations and at varying ratios of calcium to magnesium. Hormone release was independent of cation ratio, except at very low magnesium concentrations (usually <0.3 m*M*), where secretion fell sharply. The fall in parathyroid hormone release at low magnesium concentrations was significant at $p < 0.01$ under the following conditions: 0.25 m*M* total (Mg = 0.15 m*M*), 0.5 m*M* total (Mg = 0.2 m*M*), 0.75 m*M* total (Mg = 0.15 m*M*), 1 m*M* total (Mg = 0.1 m*M*), 2 m*M* total (Mg = 0.2 m*M*). (Reproduced from Targovnik *et al.*[17])

[17] J. H. Targovnik, J. S. Rodman, and L. M. Sherwood, *Endocrinology* **88**, 1477 (1971).
[18] M. Abe and L. M. Sherwood, *Biochem. Biophys. Res. Commun.* **48**, 396 (1972).

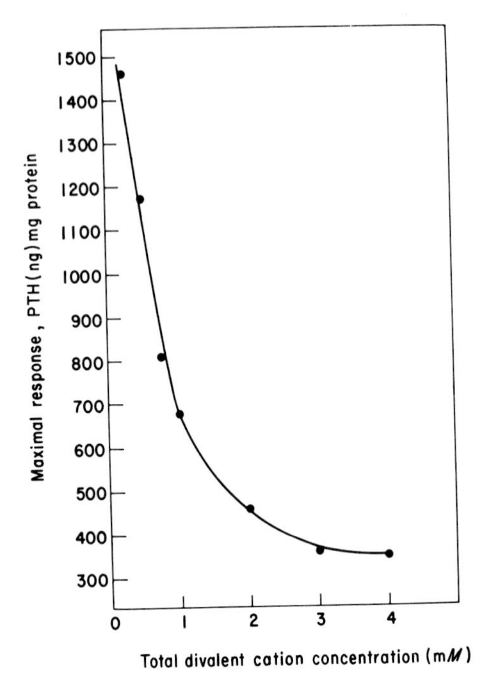

Fig. 3. Maximal release of parathyroid hormone plotted against divalent cation concentration using the data from Fig. 2. (Reproduced from Targovnik *et al.*[17])

Morphological studies performed *in vivo*[19] as well as *in vitro*[20,21] have shown that the ultrastructure of the chief cells changes in response to alterations in the ambient calcium concentration, and furthermore, the changes in ultrastructure were interpreted as correlating with the changes in parathyroid hormone secretion.[20]

Biosynthesis and Metabolism. In studies of the effects of various calcium concentrations on protein synthesis and hormone secretion in rat parathyroid gland explants, Raisz observed that high calcium concentrations (2.25 m*M*) reduced amino acid uptake, general protein synthesis, and inhibited protein secretion.[20] More recently experiments on bovine parathyroid slices[3] have shown that newly synthesized glandular parathyroid hormone varies inversely with ambient calcium concentration. In addition, Cohn *et al.*[3] have found that the gland synthesized a prohormone which was biologically active and larger in size than the extracted form of the hormone. By radioactive amino acid pulse-chase experiments, Kemper *et al.*[2] showed that bovine glands synthesize a poly-

[19] S. I. Roth, W. Y. W. Au, A. S. Kunin, S. M. Krane, and L. G. Raisz, *Amer. J. Pathol.* **53,** 631 (1969).

[20] L. G. Raisz, *Arch. Intern. Med.* **124,** 389, 1969.

[21] S. B. Oldham, J. A. Fisher, C. C. Capen, G. W. Sizemore, and C. D. Arnaud, *Amer. J. Med.* **50,** 657 (1971).

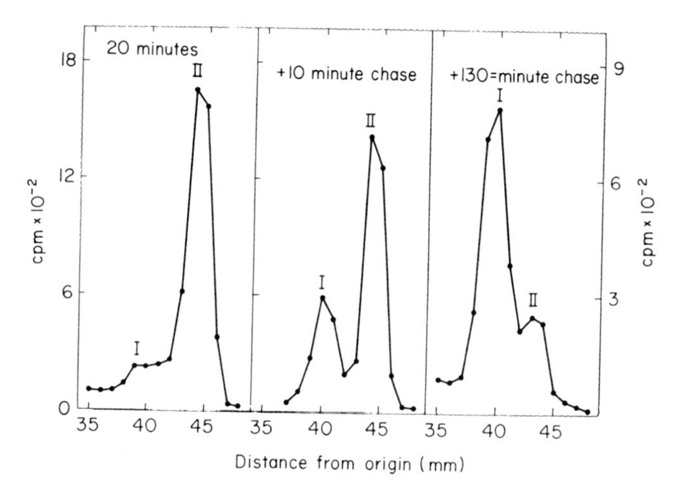

Fig. 4. Conversion of radioactivity in protein from peak II (presumed prohormone) to peak I (1-84 hormone) in the absence of additional incorporation of radioactive amino acids. Amino acids labeled with ^{14}C were incorporated into proteins of bovine parathyroid slices distributed in three separate vessels. After 20 minutes of incubation, the radioactive medium was removed, and fresh medium containing unlabeled amino acids was added and incubation continued for an additional 10 or 130 minutes. Incorporation was stopped and protein extracted and analyzed on urea-polyacrylamide gels at pH 4.4 after 20 minutes of incubation (left panel), after a 10-minute chase (middle panel), and a 130-minute chase (right panel). (Reproduced from Kemper *et al.*[2])

peptide (the prohormone) which is 15–20 amino acid residues larger than the stored, extracted hormone.[21a] Figure 4 shows the transfer of radioactivity in the prohormone to the 1–84 form of the hormone during a chase experiment with unlabeled amino acids. Thus, the results of recent *in vitro* experiments lead to the conclusion that parathyroid hormone is synthesized as a prohormone which is rapidly converted in the gland to an 84 amino acid polypeptide which is the stored form of the hormone. In response to a fall in plasma calcium levels the hormone is subsequently released from the gland.

Studies with Abnormal Parathyroid Glands. Parathyroid adenomas and hyperplastic glands from patients with chronic renal disease or after renal homotransplantation have been studied in organ culture.[8,9,11] Most of the adenomas studied behaved autonomously (no response to changes in medium calcium levels) while hormone secretion from glands from patients with secondary hyperparathyroidism generally responded normally to variations in calcium concentrations.[11] Further experience and studies with pathological human parathyroid glands should lead eventually to the establishment of functional cell lines.

[21a] Recent unpublished experiments have led Kemper and co-workers to the conclusion that proparathyroid hormone may contain only six additional amino acid residues rather than 15–20 (personal communication).

Calcitonin

Calcitonin is a polypeptide, consisting of 32 amino acid residues, which is synthesized in the parafollicular or C-cells of the mammalian thyroid gland and in the ultimobranchial body of many nonmammalian vertebrates.[22] Calcitonin was discovered by virtue of its potent hypocalcemic action, especially in young, actively growing mammals. There are striking differences in the sequences of amino acid residues in the calcitonins whose primary structures have been determined, including porcine, ovine, bovine, salmon, and human calcitonin isolated from a tumor.

Methods of Culture and Production of Calcitonin *in Vitro*

To date there are very few published articles on the culture of calcitonin-producing tissues. Dispersed cell cultures have been initiated from human medullary carcinoma tissue (a calcitonin-producing tumor) by Tashjian and Melvin[23] and by Grimley *et al.*[24] In neither case were permanent cell lines established. In the first report,[23] the authors emphasize that the tumor cells adhere poorly to glass or plastic surfaces and grow in association with fibroblasts. This has been a persisting observation extended to about twenty primary cultures. Feeder layer experiments with heterologous fibroblasts have not yet proved successful in establishing a permanent cell line. Grimley and colleagues[24] infected and transformed cells of a human calcitonin-producing tumor by simian virus 40 in order to enhance growth *in vitro*, and they reported production of calcitonin and prostaglandins (PGE_2 and $PGF_{2\alpha}$) by the tumor cells. Calcitonin secretion was detected for 7 months during serial culture of the transformed cells, although the measurements were made with an immunoassay for porcine calcitonin which others have shown cross-reacts poorly, if at all, with the human peptide. Careful light and electron microscopic examination of the cultures did not reveal amyloid formation, a frequent characteristic of the calcitonin-producing thyroid tumor *in vivo*.

Secretion, Biosynthesis, and Physiological Role of Calcitonin

In mammals, the secretion of calcitonin can be stimulated by elevation of the level of calcium perfusing the thyroid gland.[22,25] It has also been shown that certain gastrointestinal polypeptides, such as gastrin, pentagastrin, glucagon, pancreozymin, and secretin, have effects on calcitonin secretion. Calcitonin lowers

[22] P. F. Hirsch and P. L. Munson, *Physiol. Rev.* **49**, 548 (1969).

[23] A. H. Tashjian, Jr. and K. E. W. Melvin, *N. Engl. J. Med.* **279**, 279 (1968).

[24] P. M. Grimley, L. J. Deftos, J. R. Weeks, and A. S. Rabson, *J. Nat. Cancer Inst.* **42**, 663 (1969).

[25] R. V. Talmage and P. L. Munson (eds.), "Calcium, Parathyroid Hormone and the Calcitonins." Excerpta Medica, Amsterdam, 1972.

plasma calcium levels rapidly by an acute inhibition of both spontaneous and parathyroid hormone-induced bone mineral resorption.[22,25] Many have, therefore, concluded that the physiological role of calcitonin is to act in concert with, but in direction opposite to, parathyroid hormone in the moment to moment control of plasma calcium levels.[26] Although this concept may be, in part, correct, it seems too limited, for it has been difficult to measure the expected changes in calcitonin levels in plasma in response to physiological changes in calcium concentrations. Furthermore, apparent removal of the source of calcitonin has not usually resulted in acute and dramatic abnormalities of plasma calcium homeostasis. Finally, in nonmammals, calcitonin does not seem to affect plasma calcium levels or to respond readily to changes in plasma calcium concentrations.[22,25,26]

In vitro systems have also been used to study the control of secretion and the biosynthesis of calcitonins in different species. Organ culture or tissue-slice experiments using porcine,[27,28] rat,[29] and sheep[30] thyroid glands, and chick ultimobranchial, parathyroid, and thyroid tissues[31] have, in general, confirmed the *in vivo* findings that high ambient calcium levels stimulate calcitonin secretion and probably also, at least for short periods, synthesis as well. The preliminary results with avian tissues[31] are provocative in view of the differences observed from mammalian thyroid glands. For example, unlike fetal rat thyroid, which secretes calcitonin in direct proportion to the calcium concentration of the medium, no differences in calcitonin secretion by chick ultimobranchial glands were observed when medium calcium was varied from 0.50 to 2.5 m*M*. Chick thyroid secreted no hypocalcemic activity, but chick parathyroid glands did, and the secretion was stimulated by low, not high, medium calcium.

Studies of Calcitonin in Medullary Carcinoma of the Thyroid

Despite ignorance of the significance of calcitonin in normal human physiology, studies of the peptide in one disease state, medullary carcinoma of the thyroid gland, have proved informative. This thyroid neoplasm, which comprises about 5 to 10% of all thyroid cancers, occurs both sporadically, and as an inherited disorder. The tumor has been found regularly to secrete large amounts of calcitonin, and measurements of the peptide in plasma have proved extremely useful in the early diagnosis of the inherited form of the tumor and also in evaluation of prognosis and therapy of patients with the disease.[32]

[26] D. H. Copp, *Ann. Rev. Physiol.* **32**, 61 (1970).

[27] I. C. Radde, D. K. Parkinson, E. R. Witterman, and B. Höffken, *In* "Calcitonin 1969" (S. Taylor, ed.), p. 376. Heinemann Medical Books, London, 1970.

[28] N. H. Bell, *J. Clin. Invest.* **49**, 1368 (1970).

[29] J. D. Feinblatt and L. G. Raisz, *Endocrinology* **88**, 797 (1971).

[30] N. Sorgente and M. L'Heureux, *Proc. Soc. Exp. Biol. Med.* **136**, 80 (1971).

[31] J. D. Feinblatt and L. G. Raisz, *In* "Calcium, Parathyroid Hormone and the Calcitonins" (R. V. Talmage and P. L. Munson, eds.), p. 51. Excerpta Medica, Amsterdam, 1972.

[32] K. E. W. Melvin, H. H. Miller, and A. H. Tashjian, Jr., *N. Engl. J. Med.* **285**, 1115 (1971).

In future studies, it will be of importance to attempt to establish functional calcitonin-secreting cells from various species in dispersed cell culture and, if possible, to clone the cells. Since all of the studies thus far have used heterogeneous populations of cells derived from tissues which are known to secrete several different types of polypeptides, a considerable improvement in methodology and clarity of interpretation will follow from studies of cloned populations.

Some Limitations in the Use of Cell and Organ Cultures

Culture systems permit the administration of physiological signals to endocrine tissues or cells under carefully defined conditions and allow the study of their effects on hormone secretion and synthesis. Such systems are, however, devoid of normal nervous and vascular influences, and the signals must reach the responsive cells solely by diffusion. In addition, the functional polarization of secretory cells is usually not achieved in *in vitro* systems. The concomitant release from organ cultures of substances other than the specific hormone, such as, proteolytic enzymes, may seriously influence the structure of the hormone secreted into the culture medium. Findings obtained in such culture systems must, therefore, be interpreted with caution, and the results should, if possible, be confirmed in *in vivo* experiments in order to strengthen the possible physiological relevance of the outcome.

CHAPTER 5

Measurement of Collagen Synthesis

B. Goldberg

Both fibroblastic and nonfibroblastic mammalian cell types synthesize collagen in culture.[1] The collagen is secreted into the medium in a soluble form, but it eventually aggregates as extracellular fibers entrapped in the cell layer. Measurements of total collagen content of cultures should therefore include both medium and cell layer.

Measurements of collagen in the presence of other proteins have usually

[1] H. Green and B. Goldberg, *Symp. Intern. Soc. Cell Biol.* **7**, 123 (1968).

relied on the unique presence of hydroxyproline in collagen.[2] Hydroxyproline in protein hydrolysates may be measured directly by procedures employing a relatively specific chemical conversion of the molecule, or it may be chromatographically separated from the other amino acids before measurement. Colorimetric assays for hydroxyproline are available, but we have found a radioisotopic method to be ideal for tissue culture systems. Since proline is the precursor of the hydroxyproline residues of collagen, cultures are labeled with [^{14}C]- or [^{3}H]-L-proline; the hydroxyproline and proline residues in the protein hydrolysates are isolated by ion-exchange chromatography and their isotope content determined. From these data, the ratio of collagen synthesis to cell protein synthesis can be calculated. The detailed procedure follows.

Procedure

Cells are grown in plastic Petri dishes in Dulbecco's modification of Eagle's Medium with 10% calf or fetal calf serum. The medium is supplemented with sodium ascorbate (75 μg/ml) at least 24 hours before addition of the isotope. Following a medium change, 2–5 μCi of [^{14}C]-L-proline (>180 mCi/m*M*) or [2,3-^{3}H]-L-proline (>40 Ci/m*M*) are added to a plate containing 10^6–10^7 cells. After an incubation period of 12 to 24 hours, the plate is chilled, 10 mg each of unlabeled L-hydroxyproline and L-proline are added, and the media is made 5% in trichloroacetic acid. The plate is scraped with a Teflon policeman, and the suspension of cells and medium transferred to a dialysis sac with the aid of 1 to 2 ml of 0.1 *M* NaOH. After exhaustive dialysis against dilute phosphate-buffered saline and water, the retentate is hydrolyzed in 6 *N* HCl at 120°C for 12 hours. After addition of activated charcoal, the hydrolysate is filtered, 1 mg of carrier hydroxyproline added, and the filtrate evaporated to dryness under a nitrogen stream. The evaporate is taken up in 0.5 ml of 0.2 *N* sodium citrate buffer, pH 3.25, and eluted by that buffer from a cation-exchange column (0.9 × 25 cm, Bio-Rad Aminex Q-150S) according to the method of Moore *et al.*[3] Elution is at room temperature at a rate of 25 ml/hour, and 1.8 ml of each 2.0-ml tube fraction is put into Brays cocktail for scintillation counting. The early eluting hydroxyproline is separated from the proline peak by at least eight tube fractions and the proline elution is generally complete by 70 ml. For very low levels of collagen synthesis, the location of the hydroxyproline peak may be verified by a colorimetric assay on 0.05-ml aliquots of tube fractions.[4]

Calculations.[5] Radioactivity in proline represents both collagen and noncollagen protein synthesis, but the hydroxyproline counts are assumed to represent only collagen synthesis. Since collagen generally contains 1.25 proline (Pro) residues for each hydroxyproline (Hypro) residue, 1.25 × Hypro counts are sub-

[2] Elastin also contains hydroxyproline, but in smaller amounts.

[3] S. Moore, D. H. Spackman, and W. H. Stein, *Anal. Chem.* **30**, 1185 (1958).

[4] J. F. Woessner, Jr., *Arch. Biochem. Biophys.* **93**, 440 (1961).

[5] H. Green and B. Goldberg, *Nature* (*London*) **204**, 347 (1964).

tracted from the total Pro counts to give noncollagen incorporated Pro counts (Pro_{NC}). The ratio Hypro cpm/Pro_{NC} cpm can serve as a measure of collagen synthesis, but this value is usually multiplied by 4.1/12.2 (the ratio of the Pro residue content of cell protein to Hypro residue content of collagen) to give a final ratio of collagen synthesis to noncollagen protein synthesis.

Alternative Methods

Radioactive Pro and Hypro may also be isolated from the protein hydrolysate by thin-layer and paper chromatography.

The radioactive Hypro content of the protein hydrolysate may be determined without prior chromatographical separation of the amino acid. In this procedure the Hypro residues of the hydrolysate are oxidized to pyrrole and then extracted into toluene for scintillation counting.[6,7]

Collagen may be purified from other proteins by a variety of extraction procedures before quantitative assays for Hypro are applied. One method is to extract media and cell layers with 5 to 6% trichloroacetic acid at 90°–95°C.[8,9] The extracted collagen and precipitated noncollagen protein are then analyzed separately.

In another procedure radioactive protein is digested with purified, protease-free bacterial collagenase under optimal conditions. Radioactivity remaining in the supernatant after addition of 5% trichloroacetic–0.25% tannic acid is a measure of collagen synthesis, while radioactivity in the pellet represents noncollagen protein synthesis.[10]

[6] K. Juva and D. J. Prockop, *Anal. Biochem.* **15**, 77 (1966).
[7] K. J. Kivirikko, O. Laitinen, and D. J. Prockop, *Anal. Biochem.* **19**, 249 (1967).
[8] S. M. Fitch, M. L. R. Harkness, and R. D. Harkness, *Nature* (*London*) **176**, 163 (1955).
[9] G. Manner, *Exp. Cell Res.* **65**, 49 (1971).
[10] B. Peterkofsky and R. Diegelmann, *Biochemistry* **10**, 988 (1971).

CHAPTER 6

Evaluation of Milk Protein Synthesis

B. L. Larson and Corine R. Andersen

Synthesis of milk proteins *in vitro* has been demonstrated in cultures and in reconstituted cell-free systems derived from mammary tissue. Topper and Turkington and colleagues[1-4] have utilized the organ culture technique extensively with explants of mammary tissue from mid-pregnant mice and have shown milk protein synthesis on a medium containing prolactin and insulin after induction with insulin and hydrocortisone. Larson and colleagues[5-8] have utilized chiefly the dispersed cell technique with lactating bovine mammary tissue and have shown that collagenase-dispersed cells on a complete medium continue to synthesize specific milk proteins for varying lengths of time in culture.

Evaluation of net milk protein synthesis in these cultures has necessitated the development of intricate methods.[9,10] Some are only semiquantitative and most are applicable to systems of reconstituted cell-free components.[10] Milk contains a variety of proteins. The system best defined is that of bovine milk, from which the nomenclature system has been derived,[11] and may not be applicable to all other species. For example, a protein corresponding to α-lactalbumin is found in the milks of all species that synthesize lactose, but only the milks of a few contain β-lactoglobulin. The specific proteins in the casein complex

[1] R. W. Turkington, W. G. Juergens, and Y. J. Topper, *Biochim. Biophys. Acta* **111**, 576 (1965).

[2] W. G. Juergens, F. E. Stockdale, Y. J. Topper, and J. J. Elias, *Proc. Nat. Acad. Sci. U. S.* **54**, 629 (1965).

[3] D. H. Lockwood, R. W. Turkington, and Y. J. Topper, *Biochim. Biophys. Acta* **130**, 493 (1966).

[4] R. W. Turkington, D. H. Lockwood, and Y. J. Topper, *Biochim. Biophys. Acta* **148**, 475 (1967).

[5] K. E. Ebner, E. C. Hageman, and B. L. Larson, *Exp. Cell Res.* **25**, 555 (1961).

[6] T. D. D. Groves and B. L. Larson, *Biochim. Biophys. Acta* **104**, 462 (1965).

[7] D. J. Schingoethe, E. C. Hageman, and B. L. Larson, *Biochim. Biophys. Acta* **148**, 469 (1967).

[8] C. R. Andersen and B. L. Larson, *Exp. Cell Res.* **61**, 24 (1970).

[9] E. Rivera, *In* "Methods in Mammalian Physiology" (J. C. Daniel, Jr., ed.), p. 442. Freeman, San Francisco, California, 1971.

[10] B. L. Larson and G. N. Jorgensen, *In* "Lactation: A Comprehensive Treatise" (B. L. Larson and V. R. Smith, eds.). Academic Press, New York, in press.

[11] D. Rose, J. R. Brunner, E. B. Kalan, B. L. Larson, P. Melnychyn, H. E. Swaisgood, and D. F. Waugh, *J. Dairy Sci.* **53**, 1 (1970).

may also differ and they are tenacious in complexing with other proteins. This problem is compounded since homogenates contain cellular proteins, including phosphoproteins, which are similar to some of the caseins and are carried along when casein is precipitated from a mixed system. Reliable methods are dependent on the separation and/or identification of specific milk proteins. The ones considered here are representative of those which appear to give reasonable results.

Culture Conditions

A common dispersed cell culture experiment[5-8] involves about 2×10^7 cells disengaged from lactating mammary tissue placed in each T-30 flask containing 6 ml of nutrient medium (Eagle's Minimal Essential Medium with 5% dialyzed bovine serum and antibiotics) and, if tracers are used, 0.5 μCi of a ^{14}C- or ^{3}H-labeled essential amino acid for a pulse period of 6 to 18 hours. A common organ culture technique[1-4] utilizes several small explants of mammary tissue in each culture dish (totaling about 5 mg), 2 ml of nutrient medium (199), antibiotics, various hormone additions, and 15 μCi/ml of [^{32}P]inorganic phosphate or 5 μCi of ^{14}C-labeled amino acids for a pulse period of 4 hours.

Immunological Evaluation of Whey Protein Synthesis

The two major proteins of bovine milk whey, β-lactoglobulin and α-lactalbumin, have been evaluated immunologically at the levels found in dispersed cell cultures of bovine lactating mammary tissue by the Oudin single diffusion procedure.[12-14] The medium and cells are homogenized, concentrated by lyophilization if necessary, and clarified by centrifugation. A portion of the supernatant is layered over a mixture of equal parts of specific antisera to β-lactoglobulin (or α-lactalbumin) and 1% agar containing 0.9% NaCl and 0.2 mg/ml Merthiolate in 10-cm tubes having an inside diameter of 2 mm. The length of the turbid zone which forms divided by the square root of time is a linear function of the log of the protein concentration at constant temperature in the approximate 8 to 100 μg/ml range. A standard curve is prepared using known amounts of β-lactoglobulin (or α-lactalbumin). Suitable specific rabbit antisera for each protein must be prepared (available from Antibodies, Inc., Davis, California) and standardized. Other immunological methods used include the double-diffusion procedure[15] and the fluorescent antibody technique[14] which are useful in a qualitative sense.

[12] B. L. Larson and J. M. Twarog, *J. Dairy Sci.* **44**, 1843 (1961).

[13] B. L. Larson and E. C. Hageman, *J. Dairy Sci.* **46**, 14 (1963).

[14] J. M. Twarog and B. L. Larson, *Exp. Cell Res.* **28**, 350 (1962).

[15] D. C. Beitz, H. W. Mohrenweiser, J. W. Thomas, and W. A. Wood, *Arch. Biochem. Biophys.* **132**, 210 (1969).

Chemical Fractionation of Whey Proteins

The net synthesis of β-lactoglobulin and α-lactalbumin have also been determined in dispersed bovine cell cultures by utilizing labeled amino acid tracers in the culture medium.[6,7] These milk proteins, labeled with different amino acids and varying specific activities depending on the amount of radioactive amino acid used, are readily prepared and usable for other purposes.[6] If a quantitative determination of net synthesis is desired, the amount of β-lactoglobulin (or α-lactalbumin) must be determined in the milk used as the carrier (usually by the Oudin procedure described above) and the percentage that the protein contains of the same amino acid used as the tracer must be known.

The medium and cells scraped out of four T-30 flasks are pooled and homogenized with an equal volume of raw bovine skim milk. The pH is lowered to 4.6 with 1 *N* HCl and the solution centrifuged (precipitate may be saved for the analysis of β-casein). The supernatant is raised to 43% saturation with ammonium sulfate and centrifuged. The clear supernatant is adjusted to pH 2.0 with concentrated HCl and centrifuged. The resulting supernatant containing the β-lactoglobulin is adjusted to pH 6.0 with concentrated ammonia and dialyzed against a large excess of 0.1% solution of cold amino acid, the same as that used as the tracer, and then exhaustively against water. The solution is lyophilized and the incorporated isotope in the β-lactoglobulin determined by scintillation counting using 5-mg amounts dissolved in hyamine hydroxide. The pH 2.0 precipitate containing α-lactalbumin and several other proteins is further purified according to the scaled-down procedures of Aschaffenburg and Drewry,[16] and the isotope counts in the α-lactalbumin isolated are evaluated.

A somewhat similar procedure has been described also for the isolation of these two proteins from organ cultures of mouse mammary tissue.[3,4] However, since mouse milk does not contain β-lactoglobulin, the identity of the protein(s) identified as such is in question.

Electrophoresis and Other Methods for Whey Proteins

Whey proteins containing tracer amino acids may also be analyzed by electrophoresis on a variety of mediums including cellulose polyacetate and paper as well as on starch and polyacrylamide gels. These are more qualitative in nature but are useful for identifying the individual whey proteins in other species[3,4] and for the separation of genetic polymorphs of the milk proteins of the same species, particularly those of the bovine which have been well defined.[17] α-Lactalbumin has also been isolated from homogenates of guinea pig mammary tissue by Sephadex column fractionation[18] and determined enzymatically in

[16] R. Aschaffenburg and J. Drewry, *Biochem. J.* **65**, 273 (1957).
[17] R. Aschaffenburg, *J. Dairy Sci.* **48**, 128 (1965).
[18] E. Fairhurst, D. McIlreavy, and P. N. Campbell, *Biochem. J.* **123**, 865 (1971).

organ cultures of mouse and rat mammary tissue functioning as one of the component proteins of the lactose synthetase complex.[19]

Determination of Casein Synthesis (Crude)

The crude casein precipitated from the cellular homogenate with added milk carried in a tracer experiment may serve as a useful qualitative index of cellular differentiation and milk protein synthesis.[1]

Chemical Isolation of Casein Proteins

Procedures of Aschaffenburg[20] for bovine milk have been scaled-down to evaluate β-casein synthesis in both organ and dispersed cell cultures of bovine mammary tissue.[7,8] The crude casein precipitated at pH 4.6 (see previous section on β-lactoglobulin analysis) is washed with water, redissolved to pH 7.5 with 1 *N* NaOH, and reprecipitated at pH 4.6. The precipitate is dissolved in urea and 1 *N* NaOH to give a final 3.3 *M* urea concentration at pH 7.5. The pH is reduced to 4.6 and the precipitate removed. The crude β-casein remaining in the supernatant is subjected to several more reprecipitation steps,[20] dialyzed against a cold amino acid solution and water, lyophilized, and 5-mg samples counted similar to the procedures used with the β-lactoglobulin analysis.

Beitz *et al.*[15] similarly have scaled-down procedures originally developed with bovine milk for α_s-casein, β-casein, and κ-casein in determining the synthesis of these specific casein proteins in homogenates of reconstituted cell-free components of bovine mammary tissue.

Gel Electrophoresis Methods for Casein Proteins

Turkington *et al.*[1] and Juergens *et al.*[2] have determined the synthesis of casein in organ cultures of mouse mammary tissue by precipitation of the casein from the cellular homogenate with rennin and subsequent gel electrophoresis.

At the time of assay, explants from each dish are weighed and homogenized in 7 ml of solution at pH 6.7 containing the following: 0.15 *M* KCl, 0.004 *M* sodium phosphate, 0.01 *M* imidazole, and 3 mg of mouse casein precipitated at pH 4.5 from mouse milk. Nucleic acids may be hydrolyzed by incubation with deoxyribonuclease and ribonuclease. The homogenate is centrifuged at 100,000 *g* for 60 minutes, the supernatant fluid is made 0.01 *M* with respect to $CaCl_2$, and 50 μg of crystalline rennin (Nutritional Biochem.) are added followed by incubation at 37°C for 30 minutes with frequent stirring. After centrifuging at 800 *g* for 10 minutes at room temperature, the precipitate is washed twice with homogenizing solution (without casein) and made 0.01 *M* with respect to $CaCl_2$.

[19] R. W. Turkington and M. Riddle, *Endocrinology* **84,** 1213 (1969); ibid *Cancer Res.* **30,** 127 (1970).
[20] R. Aschaffenburg, *J. Dairy Res.* **30,** 259 (1963).

The isolated crude casein is dissolved in 0.12 ml of 7 *M* urea containing 2% unboiled starch and applied to the gel whose final composition is 11.4% starch and 7 *M* urea in Tris–citrate buffer, pH 8.6. Electrophoresis is carried out by the vertical starch-gel technique. After staining for the carrier protein pattern with amido black, the gels are radioautographed and cut sections counted in a liquid scintillation counter. The peaks of radioactivity were not identified as specific casein proteins (which have not been characterized for the mouse casein system) but rather as zones which increased under the hormonal treatment of the organ cultures.

Using the same method for the isolation of casein with rennin,[1,2] Turkington and Riddle[19] have utilized polyacrylamide gel electrophoresis with the casein isolated from organ cultures of explants from normal mouse and rat mammary gland and mammary carcinomas. Electrophoresis was carried out using a lower gel of 7.5% acrylamide buffered at pH 8.83 with Tris-HCl and containing 8 *M* urea. The upper gel was 5% acrylamide and 8 *M* urea. After electrophoresis at 25°C, the gels were fixed in 7.5% acetic acid, stained with amido black, sectioned, and counted in a liquid scintillation counter with Bray's solution. Comparison of the profiles of the electropherogram radioactivity of the hormonally treated cultures and the rat carcinoma with those of normal tissue indicated an increased synthesis level. Presumably many of the peaks correspond to specific protein components of casein.

CHAPTER 7

Interferon Yield Versus Cell Genotype

J. M. Moehring and W. R. Stinebring

Interferon may be the body's first line of defense against invading viruses. Interest in the therapeutic use of this specific native protein has been growing, and it is at least theoretically feasible that human interferon can be produced in cell culture and processed for clinical use. It was, therefore, of interest to us to determine what cell lines were the most prolific producers of interferon. We have studied 21 cell lines (9 presumed diploid and 12 established) for their ability to produce interferon in response to a virus challenge. In addition, the different abilities of these cell lines to become resistant to virus in response to the application of a standard preparation of human interferon has been evaluated.

Cell Cultures

All cell cultures were maintained routinely in Eagle's Basal Medium (BME) supplemented with the seven nonessential amino acids, 10% fetal calf serum, 100 units of penicillin per milliliter, and 100 μg of streptomycin per milliliter. Cultures were incubated at 36°C in an atmosphere of 5% CO_2 in air and were passaged every 3 or 4 days. Primary cell cultures were prepared by trypsinization of tissue samples and seeding of dispersed cells in 30-ml plastic tissue culture flasks.[1] These presumed diploid cell lines were always transferred by two for one splitting, so that their passage number was approximately equal to their generations in culture.[2]

Production of Interferon in Cell Cultures

For each cell line, equal numbers of cells were seeded in 10 ml of growth medium in 8-ounce (capacity 240 ml) prescription bottles. Inocula varied from 10^5 to 2.5×10^5 cells/ml, governed by the rate of growth of the cell line. When the cell sheets were almost confluent, from 48 to 72 hours, a count was made of the cells in one or two replicate cultures. Medium was removed from the other cultures by aspiration and Newcastle disease virus (NDV; vaccine strain) was added at a concentration of 4×10^{-5} hemagglutination units/cell in 2 ml of serum-free BME. The virus was allowed to adsorb to cells for 1 hour at 36°C, with shaking of the bottles at 15-minute intervals. The supernatant medium was then removed, and cell sheets were washed with 10 ml of Hanks' balanced salt solution (HBSS). A 10-ml amount of test medium was then added, and cultures were incubated for 24 hours at 36°C. Three test media were employed in these studies: (1) BME containing 5% fetal calf serum; (2) Neuman and Tytell (NT) medium,[3] a complex serum-free medium; and (3) "2XE" medium, unsupplemented BME in which twelve amino acids, eight vitamins, and glutamine are incorporated at twice their usual concentration.[4]

Cytopathic effect of the virus on the cells was scored, and culture fluid was harvested at 24 hours. Samples were titrated to pH 2 with 1 *N* HCl and were stored at 4°C for 4 to 8 days. Samples were then neutralized with 2 *N* NaOH before assay. A random selection of eight samples for sterility testing was made on two occasions. A 0.1-ml amount of each sample was inoculated into two 9-day-old embryonated hen's eggs, and the eggs were incubated and candled over a 5-day period. All embryos were alive after this period, and complete inactivation of live virus was assumed. Some samples were also centrifuged at 100,000 g after the pH 2 treatment to remove any residual virus.

[1] Available from BioQuest, Cockeysville, Maryland 21030.

[2] L. Hayflick and P. S. Moorhead, *Exp. Cell Res.* **25,** 585 (1961); cf. Section IV, this volume.

[3] R. E. Neuman and A. A. Tytell, *Proc. Soc. Exp. Biol. Med.* **104,** 252 (1960).

[4] D. J. Merchant and K. B. Hellman, *Proc. Soc. Exp. Biol. Med.* **110,** 194 (1960).

INTERFERON ASSAY

Interferon content of all samples was assayed by the plaque reduction method on monolayers of human foreskin (HF) fibroblasts in 60-mm plastic Petri dishes.[1] Samples were serially diluted in BME containing 2% fetal calf serum and were left in contact with full 48-hour monolayers of HF cells for 18 to 22 hours. HF-5 cells of passage number 20 to 38 were used in all interferon production assays. After draining and one 4-ml wash with HBSS, plates were inoculated with 40 to 80 plaque-forming units of vesicular stomatitis virus in 0.5 ml of serum-free medium. Virus was adsorbed for 1 hour at 36°C in the CO_2 incubator, with shaking at 15-minute intervals. A 4-ml amount of an overlay medium consisting of Eagle's Minimum Essential Medium (MEM) containing 5% inactivated fetal calf serum and 0.85% Noble agar[5] was then added, and plates were incubated for 48 hours. A second agar overlay containing 0.1% 2-(*p*-iodophenyl)-3-(*p*-nitrophenyl)-5-phenyltetrazolium chloride[6,7] was added, and the plates were incubated for 8 hours or longer, until the living cells reduced the tetrazolium salt to a purple-red formazan and plaques could be counted. One unit of interferon was that dilution which reduced the plaques by one-half, and the reciprocal of that dilution was the titer of a given preparation.

An interferon standard was included in all assays, as a check on cell condition and reproducibility. A standard interferon preparation, HIF-3, was produced in human foreskin cells by the same procedure utilized in these studies. It was exposed to pH 2 for 5 days before being neutralized, centrifuged at 100,000 *g* for 1 hour, and divided into small samples for routine use. Its titer on HF-5 cells averaged 500 in eleven assays.

PRODUCTION OF INTERFERON IN SERUM-CONTAINING MEDIUM

Table I shows the yields of interferon from 21 cell lines when challenged with NDV. Yields are given as units of interferon produced per cell to eliminate the differences in titer which were due to different numbers of cells in the challenged cultures. Although cultures were used when grown to almost confluent layers, the varying sizes and growth habit of the cell lines caused considerable differences in cell number per 8-ounce bottle; e.g., a "full" bottle of HF-5 had 3.2×10^6 cells, a number characteristic of the diploid lines used, HeLa showed 5.6×10^6 cells and typified most of the established epithelioid lines, and RPMI 2650 had 10.5×10^6 cells (it was a notably smaller cell than most).

Although it has been observed that some continuous cell lines, especially those of human epidermoid origin, are poor producers of interferon, it has been difficult to draw any firm generalizations from the available data regarding comparative

[5] Available from Difco, Detroit, Michigan 48201.

[6] Available from Aldrich Chemical Co., Milwaukee, Wisconsin 53210.

[7] E. C. Herrmann, J. Gabliks, C. Engle, and P. L. Perlman, *Proc. Soc. Exp. Biol. Med.* **103**, 625 (1960).

TABLE I
Interferon Yields from Various Human Cell Lines

Type	Designation[a]	Tissue of origin	Units of interferon produced per cell
Established aneuploid cell lines	Chang liver[b]	Normal liver	12[c]
	Chang (HT line)[b]	Subline of Chang liver	11
	Detroit 562[b]	Foreskin (Down's syndrome)	4
	HeLa[d]	Carcinoma, cervix	5
	HCAAT[b]	Subline of HeLa	135
	HEp-2[b]	Carcinoma, larynx	51
	Intestine 407[b]	Embryonic intestine	1
	J-111[b]	Monocytic leukemia	29
	KB[e]	Carcinoma, oral	65
	KB-R2[e]	Subline of KB	69
	WISH[b]	Amnion	57
Established quasidiploid cell line	RPMI 2650[b]	Tumor	0
Presumed diploid cell lines	Detroit 532, passage 12[f]	Foreskin (Down's syndrome)	217
	HEL, passage 28[b]	Embryonic lung	330
	HF-5, passage 8[g]	Foreskin (normal)	176
	HF-6, passage 28[g]	Foreskin (normal)	273
	NS-1, passage 4[g]	Synovial membrane (normal)	425
	NS-3, passage 4[g]	Synovial membrane (normal)	417
	VB, passage 22[g]	Synovial membrane (rheumatoid arthritis)	119
	SS, passage 4[g]	Synovial membrane (chronic synovitis)	200
	SJ, passage 4[g]	Kidney fibroblasts (lupus)	115

[a] Passage numbers given only for finite, presumed diploid cell lines.
[b] Laboratory of D. J. Merchant.
[c] Values expressed $\times 10^{-5}$.
[d] Laboratory of W. I. Schaeffer, University of Vermont.
[e] Laboratory of T. J. Moehring, University of Vermont.
[f] Laboratory of C. S. Stulberg, The Child Research Center, Detroit, Michigan.
[g] Produced from tissue.

production in established and diploid lines.[8,9] When interferon is produced by cell-virus interaction, the specific interactions must be evaluated separately for each individual case. Certain cell lines which produce little or no interferon in response to challenge with one virus may produce considerable interferon in the presence of another virus. By using in our study only one virus which is known to be capable of simulating high levels of interferon in some human cell lines and by carefully regulating the numbers of virus added per cell and the length of exposure, we have attempted to achieve a valid comparison of the production

[8] M. Ho, *In* "Interferons" (N. B. Finter, ed.), p. 36. North Holland, Amsterdam, 1966.
[9] J. Vilcek, "Interferon," p. 74. Springer-Verlag, New York, 1969.

capabilities of the various lines. Our study, which included cells of varied histories and origins, clearly indicated that, on a unit per cell basis, established human cell lines produce much less interferon than diploid human lines. The aneuploid lines produced low titers of interferon; only one of them (HCAAT) even approached the production of any of the presumed diploid lines. The diploid lines varied considerably among themselves with respect to interferon production, two normal synovial membrane lines (NS-1 and NS-3), and a human embryonic lung line (HEL) being exceptionally good producers. The average units per cell figured for all the established aneuploid lines was 3.7×10^{-4}. The average for the diploid lines was 25×10^{-4}.

The established quasidiploid tumor cell RPMI 2650 was unique in that it was the only cell in this study which produced no detectable interferon in any of the media used when challenged with the vaccine strain of NDV. Some further experiments were done with this line to determine whether it could produce interferon in response to a different challenge. No detectable interferon was produced when cultures were exposed to Sindbis virus or to polyinosinic:polycytidylic acid by the method used by Youngner and Hallum.[10] When cultures were exposed to the Roakin strain of NDV, a small amount of antiviral activity could be detected in supernatant medium, with a titer of less than 6.

Production of Interferon in Serum-Free Medium

For some experimental procedures and for ease of purification and concentration, it would be desirable to produce interferon in a medium free of any serum. We tested the interferon production of several aneuploid and diploid human cell lines in two serum-free media as well as medium with 5% fetal calf serum. Our findings are shown in Table II. All of the cell lines which produced interferon in 5% serum medium also produced in serum-free medium. The diploid lines in most cases had considerably reduced titers with no serum, but there were two notable exceptions: the HEL line gave an excellent yield in 2XE medium and the VB line had increased titers in both NT and 2XE medium. The aneuploid cells were affected less by the omission of serum. In general, they produced nearly the same amounts of interferon per cell in NT and 2XE medium as they did in BME with 5% fetal calf serum.

Sensitivity of Cell Lines to Exogenous Interferon

It has been noted that there seems to be a relationship between the ability of cells to produce interferon and their sensitivity to the action of externally applied interferon.[9] It has also been noted that some continuous cell lines have a low sensitivity or are even totally insensitive to interferon action.[8] We exposed eight diploid lines, ten aneuploid lines, and the quasidiploid tumor cell RPMI 2650 to serial dilutions of our standard human interferon preparation HIF-3. The results

[10] J. S. Younger and J. V. Hallum, *Virology* **37**, 473 (1969).

TABLE II

Interferon Yields in Serum-Containing and Serum-Free Media

Cell line	Units of interferon produced per cell[a]		
	$MEM_{95}FC_5$[b]	NT[c]	2XE[d]
Diploid lines			
Detroit 532, passage 12	217	90	56
HEL, passage 28	330	174	399
HF-5, passage 8	176	38	35
NS-1, passage 4	425	127	202
VB, passage 22	119	259	259
Aneuploid lines			
KB	65	71	76
HEp-2	51	70	51
HCAAT	135	102	145
J-111	29	15	11
WISH	57	—	57

[a] Values expressed $\times 10^{-5}$.

[b] Eagle Minimum Essential Medium supplemented with 5% fetal calf serum.

[c] Neuman and Tytell serum-free medium. (From Neuman and Tytell.[3])

[d] Eagle basal medium containing twice the original concentration of amino acids, vitamins, and glutamine and no serum added. (From Merchant and Hellman.[4])

TABLE III

Sensitivity of Various Cell Lines to Standard Interferon HIF-3

Cell line	Titer of HIF-3 on this cell line
Diploid lines	
Detroit 532, passage 14	820
HEL, passage 24	100
HF-5, passage 8	368
HF-5, passage 26	500
HF-6, passage 26	640
NS-1, passage 6	310
NS-3, passage 6	520
RS-1,[a] passage 5	540
Quasidiploid line	
RPMI 2650	56
Aneuploid lines	
Chang liver	48
Chang (HT)	28
Detroit 562	48
HeLa	48
HCAAT	32
HEp-2	48
Intestine 407	22
J-111	17
KB	43
WISH	82

[a] Synovial membrane (rheumatoid arthritis).

of these experiments are shown in Table III. Although none of the cell lines tested was totally insensitive to the action of the interferon, it was evident that the established cell lines were much less sensitive than were the diploids. In general, it requires the application of approximately ten times as many units of interferon to an established cell culture as are needed on a diploid culture to confer equal antiviral resistance. The quasidiploid tumor cell RPMI 2650 clearly falls in with the aneuploid lines with respect to its sensitivity to interferon. It is, however, very distinctly sensitive to the action of interferon despite the fact that it is a virtual nonproducer.

It has been observed[11] that human cell lines of embryonic origin are less sensitive to the action of interferon than are cell lines of neonatal origin. We included only one cell line of embryonic origin in our studies, HEL, a human embryonic lung line, and it is indeed considerably less sensitive to the action of HIF-3 than were the other presumed diploid lines of neonatal and adult origin. Siewers *et al.*[11] also observed that, when considering human diploid cell lines only, high sensitivity to interferon action did not correlate with ability to produce high titers of interferon in culture. This is also true of the diploid lines in our study: the most sensitive lines were not necessarily the best producers of interferon.

[11] C. M. F. Siewers, C. E. John, and D. N. Medearis, Jr., *Proc. Soc. Exp. Biol. Med.* **133**, 1178 (1970).

CHAPTER 8

Immunoglobulins

Yuji Matsuoka

A large number of cell lines of human lymphoid origin are now available in long-term suspension cultures; these have been established since 1964 from diseased[1] or healthy[2] individuals at Roswell Park Memorial Institute[3] (over 500 lines) and elsewhere.[4,5] Many of them were found to produce various kinds of

[1] S. Iwakata and J. T. Grace, Jr., *N. Y. State J. Med.* **64,** 2279 (1964).
[2] G. E. Moore, R. E. Gerner, and A. Franklin, *J. Amer. Med. Ass.* **199,** 519 (1967).
[3] A unit of the New York State Department of Health.
[4] M. A. Epstein and Y. M. Barr, *Lancet* **i,** 252 (1964).
[5] R. S. V. Pulvertaft, *Lancet* **i,** 238 (1964).

immunoglobulins or their subunits in spite of numerous generations in culture, and the cellular and biochemical mechanisms of immunoglobulin production have been extensively investigated using these fascinating materials.[6] In this chapter, the recent findings concerning immunoglobulin production by human established cell lines are briefly summarized mainly on the basis of our findings and experiences at Roswell Park.

Methods for Detecting Immunoglobulin Production

Various techniques detecting immunoglobulin can be used for the established cell lines, and most of them are based on specific reactions of antibodies directed against the immunoglobulins or the component subunits. The specificity of antibody reagents, therefore, is most important. Cross-reactions with other components in serum, cells, or with supplements of culture media such as fetal calf serum should be carefully examined and absorbed out completely by proper means. In most cases, addition of cross-reacting antigens in excess may be applicable for this purpose, but in special cases such as paired-labeled fluorescent antibody technique, it is desirable to remove the cross-reacting antibodies from the reagents by means of insoluble antigen adsorbents as described by Takahashi *et al.*[7,8]

Immunoglobulins in cells or on the surface of cells can be detected by staining the fixed or viable cells with anti-immunoglobulin antibodies labeled with fluorochromes or with radioisotopes. The coexistence of two different components of immunoglobulin in individual cells can be analyzed by the paired-label fluorescent technique using two antibody reagents such as fluorescein-labeled and tetramethylrhodamine-labeled antibodies.[7-9] Cell smears treated with radioisotope-labeled anti-immunoglobulin antibody give both qualitative and quantitative information on the cellular content of immunoglobulin; the total content is determined in smears by direct counting of radioactivity and the content in individual cells by counting silver grains after autoradiography.[10,11]

Immunoglobulins or their subunits secreted into spent culture media can be characterized by immunodiffusion[12] or immunoelectrophoresis.[13] Their concentration in media may be estimated by radial diffusion technique.[14] Since, in general, the concentration of immunoglobulins in media is low, most of the speci-

[6] Y. Yagi, *Symp. Intern. Soc. Cell Biol.* **9**, 121 (1970).
[7] M. Takahashi, Y. Yagi, and D. Pressman, *J. Immunol.* **100**, 1169 (1968).
[8] M. Takahashi, Y. Yagi, and D. Pressman, *J. Immunol.* **102**, 1268 (1969).
[9] J. J. Cebra and G. Goldstein, *J. Immunol.* **95**, 230 (1965).
[10] N. Tanigaki, Y. Yagi, G. E. Moore, and D. Pressman, *J. Immunol.* **97**, 634 (1966).
[11] M. Takahashi, Y. Yagi, G. E. Moore, and D. Pressman, *J. Immunol.* **103**, 834 (1969).
[12] "Methods in Immunology and Immunochemistry" (C. A. Williams and M. W. Chase, eds.), Vol. III, p. 103. Academic Press, New York, 1971.
[13] "Methods in Immunology and Immunochemistry" (C. A. Williams and M. W. Chase, eds.), Vol. III, p. 234. Academic Press, New York, 1971.
[14] "Methods in Immunology and Immunochemistry" (C. A. Williams and M. W. Chase, eds.), Vol. III, p. 213. Academic Press, New York, 1971.

mens should be concentrated by pervaporation or ammonium sulfate treatment before being subjected to analyses. For this reason, the more sensitive radioimmunoassay[15,16] is preferred; here active *de novo* synthesis is demonstrable by incorporation of ^{14}C- or ^{3}H-labeled amino acids into immunoglobulins.

Populations of Immunoglobulin-Producing Cells in Cultures

Most of the human lymphoid cell lines so far established were found to produce immunoglobulins or their component subunits. By using paired-labeled fluorescent antibody technique, Takahashi *et al.*[17] have investigated many lymphocytoid cell lines with respect to the component chains of immunoglobulin in individual cells. As shown in Table I, the cell lines can be divided into the following four groups, (A) cell lines with mixed populations of cells each producing a single immunoglobulin, (B) cell lines with a single population of cells producing a single immunoglobulin, (C) cell lines with a single population of cells producing only a light chain or only a heavy chain, and (D) cell lines with cells containing two different heavy chains.

Many cell lines in the early period after establishment (within less than 6 months) consisted of two or three distinct populations of cells, each producing a single but different immunoglobulin; after being maintained for long periods (more than 1 year), they became more homogeneous and produced a single immunoglobulin. For example, RPMI 8195, 8235, and 8265 were found to produce immunoglobulins of more than one class in early 1966,[10] but the results in 1968 indicated that immunoglobulin produced by each cell line is of a single heavy chain class and single light chain type.[17,18] Thus, it is evident that some selections are occurring during long periods of culture.

As can be seen in Table I, only 30–40% of the cells stained with fluorescent anti-immunoglobulin reagents. This does not, however, necessarily mean that the unstained cells did not produce immunoglobulins, since the fluorescent antibody technique had limitations of sensitivity and the cells used were not synchronized. In fact, practically all the cells of NK-9 line in the early S phase of synchronized culture were found to produce immunoglobulin by the more sensitive ^{125}I-labeled antibody technique,[11] whereas only 26% of NK-9 cells in conventional culture were stained by fluorescent antibody reagent (Table I).

Two abnormal features have been observed in significant number of individual cells in certain cell lines. The first is an extreme imbalance of the heavy and light chain synthesis (group C). The cells producing only light chain appear to be similar to those which are producing Bence-Jones protein in myeloma patients. RPMI 8226 cells, derived from a myeloma patient, were found to produce only free λ-type light chains which cannot be distinguished from the urinary Bence-Jones protein of the donor by either immunological or chemical anal-

[15] Y. Yagi, P. Maier, and D. Pressman, *J. Immunol.* **89**, 736 (1962).
[16] S. E. Salmon, G. Mackey, and H. H. Fudenberg, *J. Immunol.* **103**, 129 (1969).
[17] M. Takahashi, Y. Yagi, G. E. Moore, and D. Pressman, *J. Immunol.* **102**, 1274 (1969).
[18] Y. Matsuoka, Y. Yagi, G. E. Moore, and D. Pressman, *J. Immunol.* **103**, 1176 (1969).

TABLE I

Immunoglobulin-Related Products in Individual Cells of 27 Human Lymphoid Cell Lines as Determined by Staining with Fluorescent Anti-Immunoglobulin Antibody Reagents

Cell line	Diagnosis of donor	Determinant of heavy and light chains detected in individual cells[b]		
Group A				
RPMI 8767	Normal	$\underline{\gamma},\kappa$ (13%)	$\underline{\mu},\lambda$ (8%)	
RPMI 3287	Normal	$\underline{\gamma},\kappa$ (20%)	$\underline{\mu},\lambda$ (10%)	
RPMI 4467	Normal	$\underline{\alpha},\lambda$ (32%)	$\underline{\gamma},\kappa$ (12%)	
RPMI 6467	Normal	$\underline{\mu},\kappa$ (26%)	$\underline{\gamma},\kappa$ (10%)	$\underline{\alpha},\lambda$ (1%)
RPMI 1358	Malignant melanoma	$\underline{\gamma},\kappa$ (12%)	$\underline{\mu},\lambda$ (2%)	
Group B				
Jijoye	Burkitt lymphoma	$\underline{\mu},\kappa$ (7%)	$\underline{\mu}$,0 (5%)	
B35 M	Burkitt lymphoma	$\underline{\mu},\kappa$ (8%)	$\underline{\mu}$,0 (5%)	
EB-3	Burkitt lymphoma	$\underline{\mu},\lambda$ (14%)	$\underline{\mu}$,0 (8%)	
Raji	Burkitt lymphoma	$\underline{\mu},\kappa$ (6%)	$\underline{\mu}$,0 (4%)	
NK-9[a]	Burkitt lymphoma	$\underline{\mu},\lambda$ (26%)		
NK-10a	Burkitt lymphoma	$\underline{\mu},\lambda$ (12%)		
RPMI 4666[a]	Myelogeneous leukemia	$\underline{\alpha},\kappa$ (35%)	0,κ (4%)	
RPMI 8195[a]	Myelogeneous leukemia	$\underline{\alpha},\kappa$ (21%)	0,κ (4%)	
RPMI 8235	Myelogeneous leukemia	$\underline{\gamma},\kappa$ (15%)		
RPMI 8265[a]	Myelogeneous leukemia	$\underline{\gamma},\kappa$ (10%)	0,κ (3%)	
RPMI 5866	Lymphosarcoma	$\underline{\gamma},\lambda$ (15%)	0,λ (7%)	
Group C				
RPMI 8226	Multiple myeloma	0,λ (32%)		
Gor	Burkitt lymphoma	0,λ (19%)		
AL-1	Burkitt lymphoma	$\underline{\mu}$,0 (24%)		
Ogun	Burkitt lymphoma	$\underline{\mu}$,0 (28%)		
SL-1	Burkitt lymphoma	$\underline{\mu}$,0 (13%)		
Kudi	Burkitt lymphoma	$\underline{\mu}$,0 (2%)		
Group D				
LK1D	Lymphocytic leukemia	$\underline{\alpha},\underline{\gamma},\kappa$ (16%)		
RPMI 3157	Normal	$\underline{\alpha},\lambda$ (26%)	$\underline{\alpha},\underline{\gamma},\kappa$ (2%)	
RPMI 7437	Myelogeneous leukemia	$\underline{\gamma},\underline{\mu},\kappa$ (6%)	$\underline{\mu},\kappa$ (2%)	$\underline{\mu},\lambda$ (1%)
RPMI 8337	Multiple myeloma	$\underline{\gamma},\underline{\mu},\kappa$ (16%)	$\underline{\mu},\lambda$ (6%)	
NK-15a	Burkitt lymphoma	$\underline{\mu},\kappa$ (13%)	$\underline{\mu},\lambda$ (11%)	$\underline{\gamma},\underline{\mu},\kappa$ (5%)

[a] Free light chains were found in the spent media along with the whole immunoglobulin.

[b] Values in parentheses are percentages of cells belonging to each group. The rest of the cells was not stained with fluorescent antibody reagents. Heavy chain determinants are underlined. (From Takahashi *et al.*[17])

ysis.[19,20] Cells stained only with anti-heavy chain reagents are all derived from Burkitt lymphoma and stained with only anti-μ. Since the content of this anti-μ

[19] Y. Matsuoka, Y. Yagi, G. E. Moore, and D. Pressman, *J. Immunol.* **102,** 1136 (1969).

[20] Y. Matsuoka, Y. Yagi, G. E. Moore, and D. Pressman, *J. Immunol.* **103,** 962 (1969).

reactive substance in these cells was very low and no active secretion into culture media has been detected, the nature of this substance has not been characterized well. It is not certain yet whether this is the whole μ chain or only a part of heavy chain similar to those in "heavy chain disease." It is a possibility, however, that the anti-μ reactive substance in these cell lines may represent an incomplete form of antibodylike receptors which are usually detected on the cell surface of normal antibody-forming cell precursors. Klein *et al.*[21] have reported that 7 S IgM and free κ-chains were detected on the surface of cells of a Burkitt lymphoma-derived cell line, and the intracellular content of them was quite small.

The presence of two heavy chains within a single cell is the second abnormal feature (group D).[17,22] A significant number of individual cells in three cell lines were found to contain γ and μ heavy chains and, in two cell lines, γ and α heavy chains were found. In contrast, no cells were found in which two types of light chains existed. Even the cells containing two heavy chains had light chains of a single type. In this connection, the recent proposal by Wang *et al.*[23] seems to be of great significance. On the basis of their findings obtained from the analysis of paraproteins, they proposed that the cells within the same clone are potentially capable of producing more than one class of immunoglobulins which share the same structures of the light chains and the variable segments of heavy chains. In the normal situation, the cells produce only a single class of immunoglobulin at once, but they are changeable to another differentiation stage in which they produce another class of immunoglobulin. Then as a transition stage, it would be possible that the cells produce two classes of immunoglobulins at once. In the normal antibody response, Nossal *et al.*[24] detected both IgG and IgM antibodies within single cells in significant numbers. The detailed features of immunoglobulins of two different classes, detected within the same cells in culture, have not yet been characterized. It may be speculated that these cell lines represent a transition stage of differentiation, which is stably fixed in long-term culture by probable loss of control of the switching mechanism(s). More detailed characterization of the cells and immunoglobulins in these cell lines should provide much information about cell differentiation connected with immunoglobulin production.

Nature of Immunoglobulins Produced by Established Lymphoid Cell Lines

No peculiar features were found in the immunoglobulins produced *in vitro* by established cell lines so far tested. Homogeneity seems to be a common feature, and it is often comparable to that of myeloma proteins. In many cases, a

[21] E. Klein, G. Klein, J. S. Nadkarni, J. J. Nadkarni, H. Wigzell, and P. Clifford, *Cancer Res.* **28,** 1300 (1968).

[22] M. Takahashi, N. Tanigaki, Y. Yagi, G. E. Moore, and D. Pressman, *J. Immunol.* **100,** 1176 (1968).

[23] A. C. Wang, S. K. Wilson, J. E. Hopper, H. H. Fudenberg, and A. Nisonoff, *Proc. Nat. Acad. Sci. U. S.* **66,** 337 (1970).

[24] G. J. V. Nossal, A. Szenberg, G. L. Ada, and C. M. Austin, *J. Exp. Med.* **119,** 485 (1964).

limited range of electrophoretic mobility was demonstrated.[18,25–28] In addition to the antigenic determinants common to other immunoglobulins of the same heavy chain class and light chain type, individually specific antigenic determinants (idiotype determinants) were found with immunoglobulins of several cell lines.[18,19,28] Even among the immunoglobulins which are produced by cell lines derived from a single donor and belong to the same IgG-1 of κ type, differences in idiotypic determinants were observed.[18] The presence of such idiotypic determinants indicates that these cell lines represent the progeny of one or a few related clones of stem cells in the starting material.

Antibody activity has not been demonstrated so far in these immunoglobulins. Although there is a preliminary indication that the antibodylike activity against a known antigen could be induced in some cell lines,[29,30] it has not yet been confirmed that the activity is actually due to the immunoglobulins produced by these cell lines. However, there remains a good possibility that any antigenic substances which are able to react specifically with some immunoglobulins produced by established cell lines may be detected by an intense search. If some antibody activity were to be found, the established cell lines would be a fascinating tool for study of homogeneous antibodies.

Immunoglobulin Production in Connection with the Proliferation of Cells

Taking advantage of the tissue culture technique, the kinetics of immunoglobulin production was analyzed in connection with cell proliferation. First, the relation between cell proliferation and immunoglobulin production was analyzed by following the cell counts and the amounts of immunoglobulin accumulated in the media using several cell lines.[27] Figure 1 and Table II show a representative result obtained with RPMI 8235. It is evident that immunoglobulin is produced most actively during the phase of active proliferation and the accumulation of immunoglobulin in the media is mostly due to active secretion by live cells and is not due to the release from dead cells. The synthesis and secretion of immunoglobulin by these cells seem to be closely associated with their proliferation, since the cells that are alive but are not actively proliferating do not appear to produce immunoglobulin. Even when the cells were maintained in a nonproliferating state in a high density culture by daily renewal of fresh media, they produced little immunoglobulin in spite of the presence of sufficient nutrients.[6]

The maximal rate of immunoglobulin production differs greatly depending upon the cell lines. In several cell lines which seemed to be active producers, in-

[25] I. Finegold, J. L. Fahey, and H. Granger, *J. Immunol.* **99**, 837 (1967).

[26] J. D. Wakefield, G. J. Thorbecke, L. J. Old, and E. A. Boyse, *J. Immunol.* **99**, 308 (1967).

[27] Y. Matsuoka, M. Takahashi, Y. Yagi, G. E. Moore, and D. Pressman, *J. Immunol.* **101**, 1111 (1968).

[28] Y. Matsuoka, Y. Yagi, G. E. Moore, and D. Pressman, *J. Immunol.* **104**, 1 (1970).

[29] H. Kamei and G. E. Moore, *Nature (London)* **215**, 860 (1967).

[30] H. Kamei and G. E. Moore, *J. Immunol.* **101**, 587 (1968).

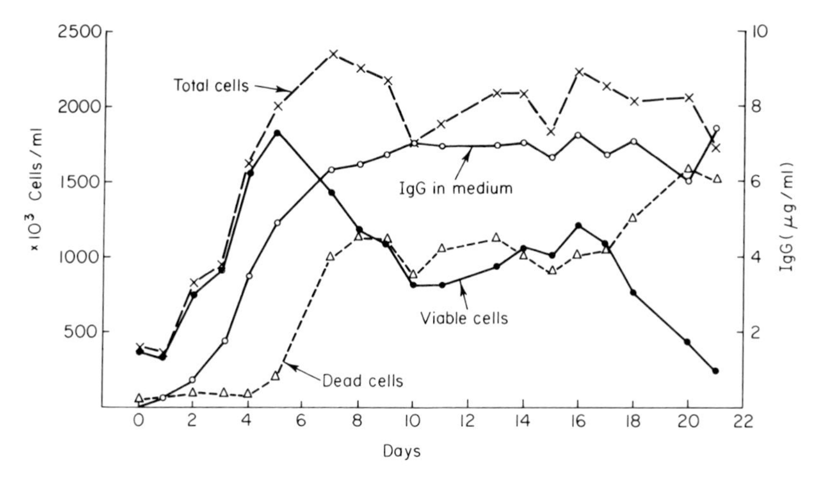

Fig. 1. Accumulation of IgG in the culture media of RPMI 8235. Cells in good condition were harvested and washed once with culture media (RPMI 1640 containing 10% fetal calf serum), then resuspended in totally fresh media. No fresh media were added thereafter throughout the experiment. A 10-ml portion of the spent media was taken at each time. Samples taken in earlier periods were concentrated about 40- to 50-fold; samples in later periods were concentrated about 20-fold by pervaporation. The concentration of IgG in the samples was determined by the radial diffusion method using anti-γ antibody plate. The IgG concentration in the unconcentrated media which is shown in the figure was calculated from the results obtained with the concentrate of the media considering the degree of concentration. (From Matsuoka *et al.*[27])

cluding RPMI 8235, the maximal rates were in the range of 2 to 3 μg per 10^6 cells per day, while in most other cell lines the rates were much less. There is one exceptional cell line (RPMI 8226) which was derived from a myeloma patient and is producing only λ-type light chain as described above. This cell line produced as much as 18 μg of free λ type light chain per 10^6 cells per day or 6000

TABLE II
Rate of IgG Production by RPMI 8235 Cells

Day after transfer	Viable cell concentration (cells/ml × 10^{-3})	IgG: Concentration in the media (μg/ml)	IgG: IgG produced by 10^6 cells per day[a] (μg)
0	360	0	
1	330	<0.2	<0.5
2	740	0.9	2.1
3	890	2.2	1.8
4	1560	4.4	2.5
5	1830	6.3	1.2
7	1410	8.0	0.5
8	1170	8.2	0.1

[a] The value for each period was calculated by dividing the increment with the length of the period (day) and the concentration of viable cells at the start of the period in units of 10^6. (From Matsuoka *et al.*[27])

λ chain molecules per cell per second. Such a rate is exceptionally high for established cell lines and seems to be in line with the rate found for myeloma cells or antibody-producing cells *in vivo*.[31] Nilsson[32] also reported that one IgE myeloma cell line (266B1) produced IgE protein at the rate of 4 to 9 μg per 10^6 cells per day. This high rate, suggests that the activity of stem cells at the initiation of culture may be stably inherited by the progeny cells for long periods *in vitro*.

The close association of immunoglobulin production with cell proliferation appears to be analogous to antibody formation *in vivo*. Although the terminal nondividing mature plasma cells seem to contain as much or even more antibody globulin than the dividing immature plasmablasts, the synthesis of RNA and proteins as assessed by the incorporation of [^{3}H]uridine (RNA precursor) or [^{3}H]leucine (protein precursor),[33] appears to be much less active in such mature cells than in immature cells. Thus, dividing immature plasmablasts appear to be more active producers of antibody globulin.

The availability of synchronized cells is an advantage of tissue culture technique, and immunoglobulin production was analyzed in reference to the generation cycle of cells using synchronized cultures of two cell lines; NK-9 producing IgM of λ type and RPMI 4666 producing IgA of κ type.[11] Cells were synchronized in the very early S phase by double exposure to excess thymidine. They were harvested at intervals after the release from the second thymidine block, and the amount of cellular immunoglobulin determined by treating cell smears with ^{125}I-labeled anti-immunoglobulin antibody. Under the condition used, the amounts of [^{125}I]antibody fixed on the smears and the number of grains found on the autoradiograph were shown to be proportional to the amount of immunoglobulin present. The results obtained with NK-9 are shown in Fig. 2, and quite similar results were obtained with RPMI 4666. It is clear that there is a marked fluctuation of immunoglobulin level in the cell during the course of its cell cycle. Cells in late G_1 and early S phases seem to produce immunoglobulin most actively. Another finding to be noted is that there is a large variation in the cellular immunoglobulin level among individual cells, even in such a synchronized culture. Although practically all the cells in this line are capable of producing immunoglobulin, the number of silver grains found at the peak level of cellular immunoglobulin in the early S phase (0.5 hour) ranged from 5 to 59 with an average of 18.9 grains per cell. Buell and Fahey[34] also reported that immunoglobulin synthesis is most active in the late G and S phases.

As is evident from the above, long-term cultures of established cell lines have many advantages, but they also have some limitations. Many of them produced homogeneous immunoglobulins comparable to myeloma proteins; however, except for a few cell lines such as RPMI 8226 the amounts that can be obtained from them are quite small and are not sufficient for more detailed analyses of chemical and immunological features unless the culture systems on large scale such as those of the Roswell Park Memorial Institute are available. Cell lines

[31] G. J. V. Nossal and O. Makela, *Ann. Rev. Microbiol.* **16**, 53 (1962).

[32] K. Nilsson, *Intern. J. Cancer* **7**, 380 (1971).

[33] J. Schooley, *J. Immunol.* **86**, 331 (1961).

[34] D. N. Buell and J. L. Fahey, *Science* **164**, 1524 (1969).

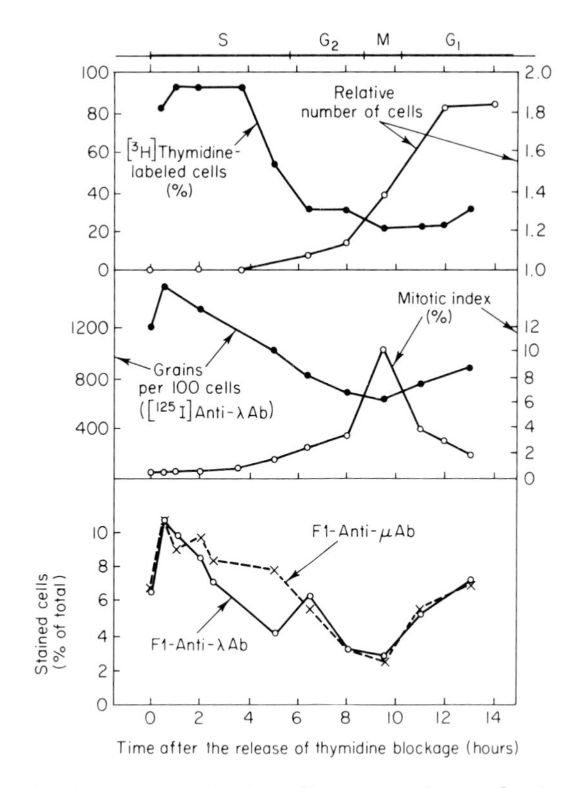

Fig. 2. Immunoglobulin content of NK-9 cells in a synchronized culture. The concentration of cells at the start of the synchronized culture was 1.12×10^6 cells/ml, and the viability was 97.9%. The percentage of cells labeled with [^{3}H]thymidine and of cells in the mitotic period was determined by counting more than 500 cells. Silver grains were counted for 100 cells on radioautograph of cell smears treated with ^{125}I-labeled anti-λ antibody. The percentage of cells stained by fluorescein-labeled anti-λ and anti-μ antibodies was determined by examining more than 500 cells. (From Takahashi *et al.*[11])

comprise more or less homogeneous populations of cells and show stable characteristics for long periods under various culture conditions. These are certainly advantageous features that cannot be attained in experiments *in vivo* or in short-term cultures of mixed cell populations, such as lymphoid tissue fragments. At the same time, however, these are disadvantageous features for some studies such as antibody induction, since the induction of antibody response to various antigens seems to require the interaction of cells of different types, i.e., thymus-derived lymphocytes and bone marrow-derived lymphocytes and/or macrophages,[35,36] and the cells seem to be compelled to differentiate during the course of antibody response. At present, no ways to change arbitrarily the differentiation stage of cells in culture are available. No matter what the limitations, future investigations on these cell lines should help to elucidate the cellular and biochemical events of immunoglobulin production in lymphoid cells.

[35] M. Fishman, *J. Exp. Med.* **114,** 856 (1961).
[36] J. F. A. P. Miller, A. Basten, J. Sprent, and C. Cheers, *Cell. Immunol.* **2,** 469 (1971).

SECTION XIII

Diagnosis and Understanding of Disease

Editors' Comments

This section includes representative examples of the use of tissue culture in the diagnosis and understanding of disease. It is apparent that some of the techniques overlap those described in earlier sections. Therefore, less emphasis is placed on establishing the cultures, for example, from biopsy tissue (discussed in Sections I and II) and on some of the more routine procedures used in diagnostic work such as karyoligical analysis (see Section XIV, Chapter 15) and virology assays (see Section XI). Specific tests for diagnosis of certain "inborn errors" are included.

From the practical standpoint, experimental models are desirable in elucidating human disease mechanisms. Thus, included in this section are two systems that result in the transformation of cells to the apparent oncogenic state. Only the details of the additions of the oncogenic factors and the subsequent characterization of the transformed cells as being "malignant" are cited since the establishment (see Sections I and II) and the clonal isolation (see Section V) are described elsewhere in the book.

CHAPTER 1

Some Cell Culture Procedures in Diagnostic Medical Virology

Margaret A. J. Moffat

During the last 20 years cell culture has become an invaluable tool in diagnostic medical virology. Cultured cells had been used previously for the detection of virus but it was not until type 2 poliovirus was grown in cultures of nonneural tissue from human embryos[1] that the technique began to receive due recognition. An easily observed destructive effect or cytopathogenic effect (CPE) was produced in the cells by the virus and this effect could be prevented by type-specific immune sera providing a convenient neutralization test.[2] Many viruses other than poliovirus could be detected and neutralized using this technique and a large number of previously unknown viruses was isolated. There have been continuous developments in cell culture and with each new technique comes the possibility of detecting the more elusive viruses. As a result of this, there are now many different methods of cell culture used in diagnostic virology and it is impossible to describe all of these here. I will attempt to give representative examples and references with the accent more on different methods than on preparation of cultures; such technical details will be found elsewhere in this book. The main uses of cell culture in diagnostic medical virology are to demonstrate the presence of virus and to detect humoral antibodies in patients' sera. The production of antigens will not be discussed.

Monolayer Cultures

General Method of Preparation. Culture of cells in monolayers is the most common method used in diagnostic virology laboratories. Cell suspensions are prepared from tissue or from stock cultures by using one of the dispersing agents. The suspension is then diluted in growth medium to contain from 50,000 to 200,000 cells per milliliter and dispensed into suitable containers; these may be large bottles for stock cultures, test tubes for roller culture, Petri

[1] J. F. Enders, T. H. Weller, and F. C. Robbins, *Science* **109**, 85 (1949).
[2] F. C. Robbins, J. F. Enders, and T. H. Weller, *Proc. Soc. Exp. Biol. Med.* **75**, 378 (1950).

dishes, or coverslips in Leighton tubes. Growth medium usually consists of a synthetic medium such as 199 or Eagle's Basal Medium with added 10% calf serum. The serum may be varied in type or amount according to the virus or cells being cultured. After about 3 days the cultures are changed to maintenance medium which may contain 2% serum or be serum-free. Antibiotics are usually included. The optimum temperature for growth is 37°C, but after inoculation a lower temperature may be more satisfactory. Incubation may be stationary or in a roller drum. Containers which are not stoppered, e.g., Petri dishes, must be incubated in an atmosphere containing 5% CO_2 in air.

Types of monolayer cultures may be divided into (a) primary or secondary, (b) continuous, and (c) semicontinuous.

Primary or Secondary Cultures. These cells retain their diploid chromosome number, a factor which makes them particularly susceptible to certain viruses. Their disadvantage lies in the fact that they must be prepared from fresh tissue which may vary in susceptibility. Kidney of various species is particularly suitable and monkey kidney is used routinely in most diagnostic laboratories. Simian viruses may be present in monkey kidney and apart from herpesvirus simiae, which may be a hazard to personnel, SV5 and SV40 may interfere with isolation of other viral agents. Human amnion and human thyroid also provide satisfactory cell cultures although amnion cultures require more care in preparation. Monkey kidney cultures are satisfactory for the isolation of the myxoviruses and the enteroviruses except coxsackie A. Human amnion cells support the multiplication of measles and enteroviruses including some types of coxsackie A.

Continuous Cell Cultures. The discovery of a cell line, e.g., HeLa cells, which could be subcultured indefinitely was of considerable importance. These cells were isolated from an epidermoid carcinoma of the cervix[3] and were found to support the growth of a large number of viruses. Since then, other cell lines, HEp-2 (human epithelium), have been isolated and it has also been possible to transform normal cells, e.g., BHK 21 (baby hamster kidney) and RK 13 (rabbit kidney). The Bristol strain of HeLa cells is particularly useful for the isolation of respiratory syncytial virus, and RK 13 cells produce a recognizable CPE when infected with rubella virus. Adenoviruses and herpes simplex produce a characteristic CPE in HeLa and HEp-2 cells.

Although the advantages of continuous cell cultures seemed so considerable that they might replace primary and secondary cultures, this has not been the case. During transformation of the cells the chromosome number becomes abnormal and the susceptibility to certain viruses is lowered.

Semicontinuous Cell Cultures. The ideal cell system for a diagnostic virology laboratory would be a continuous line which retained its diploid chromosome number. An attempt to attain this ideal system was made by subculturing

[3] G. O. Gey, W. D. Coffmann, and M. T. Kubicek, *Cancer Res.* **12**, 264 (1952).

cells from human embryonic tissue.[4] The cells, of the fibroblast type, remained diploid and could be subcultured up to fifty times. Such cultures can be preserved by freezing and thus theoretically provide a constant supply of cells.

These semicontinuous cells may be as susceptible to certain viruses as primary or secondary cells but this is not always the case. They provide a particularly satisfactory host system for cytomegalovirus, varicella zoster, herpes simplex, and certain rhinoviruses.

Routine Isolation of Viruses

It would be impossible in a routine laboratory to inoculate each specimen into every type of cell culture. A routine procedure is therefore employed and the specimen is inoculated into three types of cells which will support the growth of as many viruses as possible. For this purpose the cells are usually grown in stoppered test tubes which may or may not be "rolled." Rotation only seems to be essential for the isolation of certain rhinoviruses which also require a lower temperature (33°C) and a lower pH (6.8). A typical routine method for virus isolation may include the inoculation of: (a) secondary monkey kidney or human amnion cells, and (b) HeLa or Hep-2 cells. If the presence of rhinoviruses, cytomegaloviruses, or varicella zoster is suspected, human embryo fibroblasts should also be inoculated; rubella virus can be isolated in RK 13 cells while respiratory syncytial virus requires the more susceptible Bristol strain of HeLa cells.

Two tubes of each type of culture are inoculated with a small amount of the specimen (0.2 ml) and the tubes are then incubated. They are examined microscopically using a low-power objective every 2 days for 10 days or longer depending on the virus. If there is no evidence of virus after this time the cells may be inoculated into fresh cultures (blind passage).

Detection of Virus in Cell Cultures

Degenerative Changes (CPE) in the Cells. Different virus groups produce characteristic changes in the monolayers which can be seen microscopically, e.g., adenoviruses produce a clumping effect while enteroviruses produce rounding and shrinking of the cells. The CPE and the type of cells in which it is seen help to diagnose the virus. Identification is carried out by neutralization test in which the infected fluid is treated with known antisera; these virus–serum mixtures are inoculated into cell cultures and after incubation the cells are examined to determine which antiserum has neutralized the effect of the virus. In the case of large groups of viruses such as the echoviruses it is usual to combine several antisera in pools.

The CPE of a virus may be made more visible by the staining of cells, e.g.,

[4] L. Hayflick and P. S. Moorehead, *Exp. Cell Res.* **25**, 585 (1961); cf. Sections II and IV, this volume.

for identification of respiratory syncytial virus a 1:8000 dilution of neutral red is added to the inoculated cultures.

Hemadsorption Test. This technique has been found useful for the detection of certain viruses, such as parainfluenza, which do not always give a definite CPE in tissue culture.[5] The test is performed by adding 0.2 ml of 0.4% guinea pig or human "O" erythrocytes to roller tube cultures which have been incubated for 5 to 7 days following inoculation; uninoculated controls must be included. The tubes are incubated at 4°C for 20 minutes and then examined microscopically. A positive test is indicated by the adherence of red cells to the cultures. A hemadsorption inhibition test is then carried out by inoculating infected fluid into more cultures and incubating them until the hemadsorption test is positive. The medium is then replaced with fresh medium containing specific antisera and the tubes are incubated at 37°C for 1 hour. The fluid is removed and a hemadsorption test is carried out. Inhibition of hemadsorption by a particular antiserum indicates the identity of the virus.

Fluorescent Antibody Technique. Virus may be detected in cell cultures by this method *before* CPE occurs, thus providing a quick specific diagnostic test. The method has been successfully used for the diagnosis of respiratory syncytial virus infections.[6] The tube cultures are inoculated and incubated in the usual way. For fluorescent antibody examination the cells are scraped off the tubes and transferred to slides. After fixing they are then stained by the indirect method with rabbit antiserum, e.g., against respiratory syncytial virus, followed by fluorescein-conjugated anti-rabbit globulin. The virus can easily be detected within the cells when the preparations are examined in the ultraviolet microscope.

Interference Test. This test is not used routinely but may be useful to detect the presence of viruses which cannot be demonstrated by conventional methods. For example some rhinoviruses and rubella virus were first isolated by this technique.[7,8] African green monkey kidney cell cultures were originally used for rubella. Throat washings or other suitable specimens are inoculated into each of two tube cultures and after 1 hour at room temperature the inoculum is removed and replaced by nutrient medium. Following incubation for 7 days, one of each pair of cultures is tested by challenge with 1000 $TCID_{50}$ of echovirus type 11; control cultures are also inoculated and all cultures are incubated. When the control cultures show an advanced CPE, fluids are collected from control and test cultures and assayed for hemagglutination. Reduction of the hemagglutination titer in the test culture indicates the presence of a virus in the initial inoculum.

[5] R. M. Chanock, R. H. Parrott, K. Cook, B. E. Andrewes, J. A. Bell, T. Reichelderfer, A. Z. Kapikian, F. M. Mastrota, and R. J. Huebner, *N. Engl. J. Med.* **258**, 207 (1958).

[6] P. S. Gardner and J. McQuillin, *Brit. Med. J.* **3**, 340 (1968).

[7] G. Hitchcock and D. A. J. Tyrrell, *Lancet* **i**, 237 (1960).

[8] E. L. Buescher, P. D. Parkman, M. S. Artenstein, and S. B. Halstead, *Fed. Proc.* **21**, 466 (1962).

Mixed Culture Technique (Cocultivation)

This method has proved of value when a viral agent is difficult to isolate from the original tissue, e.g., isolation of measles from a brain biopsy.[9]

Primary cultures are prepared from such tissues by standard methods and allowed to grow into monolayers. They are then trypsinized and suspended in growth medium. One volume of this suspension is then combined with two volumes of HeLa cells and this mixture is cultured (see monolayer cultures). It is possible that a CPE will develop in the HeLa cells although no CPE was seen in the primary culture of the tissue.

Organ Culture

Cultures of chopped tissues have been used for many years to support the growth of viruses but the techniques were not readily applicable to diagnostic virology. However, a simple method of cultivating fragments of ciliated epithelium was described[10] which resulted in the isolation of new rhinoviruses and a new member of the coronavirus group.[11]

For the isolation of respiratory viruses, tracheas of 14- to 22-week-old human embryos are chopped into fragments; these are planted with the ciliated surface uppermost on the scratched surface of a plastic Petri dish and medium (e.g., Medium 199) is added. After approximately 2 days the cultures are inoculated by dropping a small amount of the specimen onto the fragments and incubating.

Virus may be detected in the following ways: (a) stopping of ciliary activity which may take 7–10 days, (b) fragments may be rubbed gently in distilled water and the suspension examined by negative contrast in the electron microscope, (c) the medium may be collected, sedimented in the ultracentrifuge, and examined in the electron microscope, or (d) the tissue may be examined for pathological changes.

A recent report describes the use of human embryo liver cultures for experiments with Australia antigen.[12] The antigen was detected by complement fixation test and by means of electron microscopy after freezing and thawing of the cells.

Detection of Viral Antibodies

Immunofluorescence Studies

1. *Monolayer cultures.* Cells may be cultured on coverslips, in Leighton tubes, or in Petri dishes. A suspension of cells is prepared in growth medium and

[9] L. Horta-Barbosa, D. A. Fuccillo, and J. L. Sever, *Nature* (*London*) **221**, 974 (1969).
[10] B. Hoorn, *Acta Oto-Laryngol. Suppl.* **188**, 138 (1964).
[11] D. A. J. Tyrrell and M. L. Bynoe, *Brit. Med. J.* **1**, 1467 (1965).
[12] A. J. Zuckerman, P. M. Baines, and J. D. Almeida, *Nature* (*London*) **236**, 78 (1972).

added in a concentration which will produce a monolayer. If Petri dishes are used it is helpful to make the coverslips adhere to the glass with a small spot of wax or silicone grease. Petri dishes may need to be incubated in an atmosphere of 5% CO_2 in air but the use of HEPES buffer will overcome this. The method has been found particularly useful in the diagnosis of rubella to detect rubella-specific IgM antibody.[13] Coverslip cultures of BHK 21 cells are prepared and inoculated with rubella virus. After incubation the cultures are fixed in acetone and can be preserved at $-20°C$ in the presence of silica gel. Uninfected cultures are also necessary. To detect rubella-specific IgM in human serum, dilutions of the serum are overlaid on the rubella-infected coverslip cultures. After incubation for 30 minutes followed by a thorough washing the coverslips are treated with fluorescein-labeled anti-human IgM. Specific fluorescence indicates the presence of rubella-specific IgM antibody due to recent infection.

2. *Suspended cell cultures.* Indirect immunofluorescence is used routinely to measure antibodies to EB virus.[14] Coverslip cultures cannot be prepared in the usual way since EB cells are cultured in suspension and do not attach to the surface of a glass vessel. Cell smears are therefore prepared. After centrifugation of the cultures the supernatant fluid is removed and the cells are resuspended in the remaining fluid. A small drop of this suspension is placed on a coverslip, evenly distributed over the surface, and allowed to dry thoroughly at 37°C. The coverslips are then fixed. Using the indirect method the coverslip preparations are exposed to diluted test serum (1:8) for 1 hour at 37°C. After washing, fluorescein isothiocyanate-labeled serum prepared to the appropriate globulin is added and left for a further incubation period; the coverslips are then washed and mounted. Specific fluorescence in the cells indicates EB antibody and the test can be made quantitative by using dilutions of serum.

Plaque Inhibition Test. The original demonstration[15] that poliovirus could produce plaques in monolayer cultures of monkey kidney cells has led to the demonstration of plaque production by most viruses. Cell monolayers are prepared in flat bottles or Petri dishes. The growth medium is removed and the cells are inoculated with virus. After a suitable time for adsorption (1 hour) the virus is removed and the monolayers are overlaid with growth medium containing agar (1.5%); neutral red may be incorporated in the medium. An accurate measurement of the number of plaque-forming units may be calculated after incubation. The neutralizing capacity of serum may be measured by mixing the serum with a known amount of virus and comparing the number of plaques produced by the mixture with the number produced by virus without serum. This method is sensitive and provides a more accurate neutralization test for enteroviruses than tube cultures[16] but it is more cumbersome and is therefore not used widely in routine laboratories. The plaque test has also been used to differentiate between wild and vaccine strains of poliovirus.

[13] M. Haire and D. S. M. Hadden, *J. Med. Microbiol.* **5**, 237 (1972).
[14] G. Henle and W. Henle, *J. Bacteriol.* **91**, 1248 (1966).
[15] R. Dulbecco and M. Vogt, *J. Exp. Med.* **99**, 167 (1954).
[16] G. D. Hsiung and J. L. Melnick, *Virology* **1**, 533 (1955).

Metabolic Inhibition Test. This test depends on the ability to use the metabolism of cell cultures as an indication of virus growth.[17] Acid is produced by actively growing cultures but if the metabolism is inhibited by the presence of virus the pH will be different; such differences can be measured if an indicator is incorporated in the medium. Neutralization tests can therefore be carried out using this principle, i.e., if virus is neutralized by antiserum and inoculated into cell cultures the cells will continue to metabolize whereas if there is no neutralization metabolism will be reduced. This test can be carried out in tubes, macroplates or microplates. Rigid disposable plastic plates have proved satisfactory. Serum dilutions in 0.025-ml amounts can be prepared in the plates and an equal quantity of virus added. After incubation of the serum-virus mixtures for 1 hour a suitable dilution of cells in 0.1 ml medium is added. The plates are then sealed with transparent tape and incubated. If phenol red is incorporated in the medium, neutralization will be indicated by a yellow color; the virus control will be pink. Time of incubation will vary for different viruses and modifications such as the addition of extra glucose and/or magnesium chloride have been made. The test has been used for coxsackie B viruses and rhinoviruses and has proved valuable to distinguish between wild and vaccine strains of poliovirus. It is economical on reagents and space.

[17] F. L. Shand, *J. Med. Lab. Technol.* **18**, 75 (1961).

CHAPTER 2

Diagnostic Use of Cell Cultures Initiated from Amniocentesis

Carlo Valenti

The prenatal detection of genetic disorders, by the analysis of cultivated fetal cells obtained by amniocentesis during the second trimester, has recently added a new useful and very promising dimension to genetic counseling, once based on empirical statistical data.[1-3] The indications to genetic diagnostic

[1] A. Dorfman (ed.), "Antenatal Diagnosis." The University of Chicago Press, Chicago, Illinois, 1972.

[2] A. Milunsky, J. W. Littlefield, J. N. Kaufer, E. H. Kolodny, V. E. Shih, and L. Atkins, *N. Engl. J. Med.* **283**, 1370, 1441, 1498 (1970).

[3] C. Valenti, *Ric. Clini. Lab.* **1**, 443 (1971).

amniocentesis can be grouped as follows: (a) cytogenetic conditions; (b) X-linked disorders; and (c) inborn errors of metabolism.

Cytogenetic Conditions

This group includes high risk couples such as balanced carriers of chromosome translocations. Typical is the case of a mother with a balanced D/G translocation [45,XX,D-,G-,t(DqGq)+] who has a 15% risk of giving birth to a child with the translocation or hereditary type of Down's syndrome.[4] Such risk is considerably lower when the father is the carrier. Amniocentesis has also been carried out in mothers over 35 years of age and in those who have already had a child with the trisomic or regular type of Down's syndrome. For these individuals the risk is probably less than 2%, only slightly higher than the risk of complications from amniocentesis. The indication here is not based on genetic grounds, but on the status of anxiety of the parents, which may indeed be significant.

X-Linked Disorders

In pregnancies at risk for X-linked disorders, such as hemophilia and the progressive type of muscular dystrophy, the fetal sex can be determined prenatally by scoring the uncultured amniotic fluid cells for the sex chromatin and Y fluorescent bodies. According to Valenti *et al.*[5] however, even when the two techniques are combined the degree of accuracy is not to be considered sufficient, and confirmation is deemed necessary by the karyotypical analysis of cultivated elements.

Inborn Errors of Metabolism

More than forty hereditary metabolic conditions can be diagnosed at the cellular level, most of them on amniotic fluid cells.[2] Although it has been suggested that a diagnostic impression can be obtained from the analysis of cell-free amniotic fluid and of its uncultured cellular constituents, the final diagnosis should only be based on the study of a suitable number of cells cultivated *in vitro*.[6]

Technique

The method to be described here has been used in this laboratory since 1969.[7] The most suitable time for transabdominal amniocentesis is between the 14th

[4] P. A. Jacobs, J. Aitken, A. Frackiewica, P. Law, M. S. Newton, and P. G. Smith, *Ann. Human Genet.* **34**, 119 (1970).

[5] C. Valenti, C. C. Lin, A. Baum, M. Massobrio, and A. Carbonara, *Amer. J. Obstet. Gynecol.* **112**, 890 (1972).

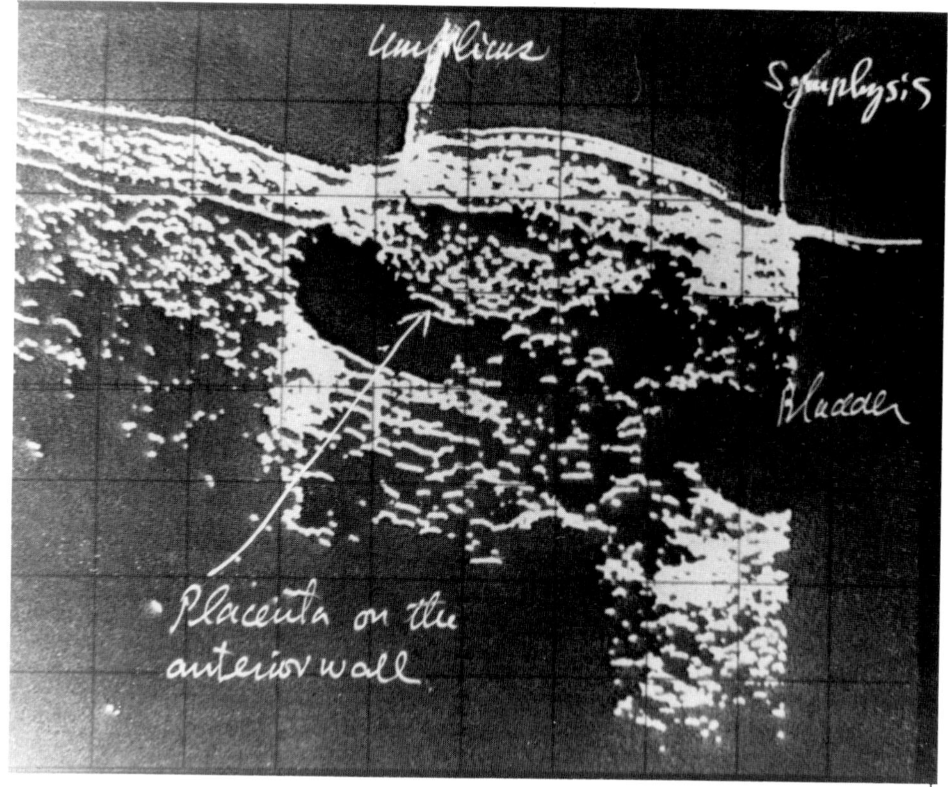

Fig. 1. Ultrasonic placentogram, longitudinal scan, at 18 weeks of gestation. The placenta is implanted on the anterior uterine wall. A placenta-free area is clearly seen at the fundus of the uterus. The needle was inserted through this area and a clear sample of amniotic fluid was obtained.

and the 16th week of amenorrhea. The placenta is located by ultrasounds (Fig. 1), a method already useful as early as the ninth week.[8] Avoidance of the placenta by the amniocentesis needle is important in lessening the risk of abortion or hemorrhage. From a diagnostic standpoint, it is also important to avoid contamination of the amniotic fluid by maternal blood elements, which may lead to misdiagnosis both in the cytogenetic and biochemical studies. Under strictly aseptic conditions and with local anesthesia, a 20-gauge 3½ inch spinal needle with stylet is inserted through the abdominal wall, preferably on the midline and in the lower third of the pubic-umbilical space, but always away from the placental site. After the stylet is removed, the amniotic fluid is slowly withdrawn into a plastic syringe. The volume varies according to the stage of pregnancy; at 16 weeks, approximately 40 ml will give good results in cell culture and no untoward effects. The first 2 or 3 ml are discarded. The amniotic fluid is transferred into a screw-capped

[6] H. L. Nadler and A. Gerbie, *Obstet. Gynecol.* 38, 789 (1971).
[7] C. Valenti and T. Kehaty, *J. Lab. Clin. Med.* 73, 355 (1969).
[8] M. Kobayashi, L. M. Hellman, and E. Cromb, "Atlas of Ultrasonography in Obstetrics and Gynecology," p. 53. Appleton-Century-Crofts, New York, 1972.

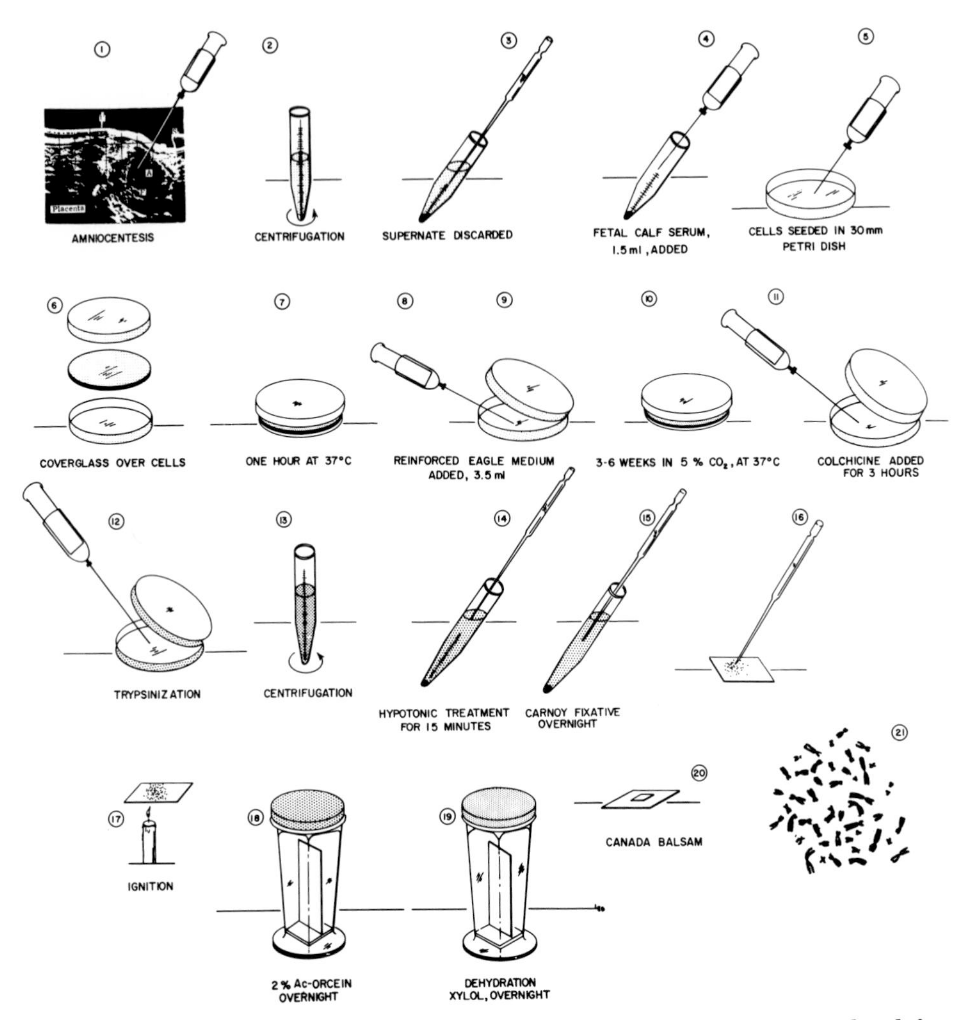

Fig. 2. Harvest and culture of fetal cells obtained by amniocentesis. (Reproduced from C. Valenti.[3])

plastic tube and processed as soon as possible. The success in cell cultures seems to be somewhat reduced by an interval of time longer than 24 hours, particularly when the sample is shipped. If necessary, the fluid should be stored at 5°C. Amniotic fluid is centrifuged at 42 *g* for 7 to 10 minutes (Fig. 2). The supernatant is decanted and centrifuged again. The two cell pellets are set up in culture separately. The cells are resuspended in fetal calf serum. Due to the low viability of amniotic fluid cells, which ranges from 8 to 30% between the 16th and 18th week of gestation,[5] a rather heavy suspension is advisable. Usually from the centrifugation of a 40-ml sample of amniotic fluid three cultures are set up; one culture is established from the second centrifugation. For each culture, the cells are suspended in 1.5 ml of fetal calf serum and placed in the center of a 60 × 15 mm plastic Petri dish. No cell viability count is deemed necessary as routine. A No. 1, 43 × 50 mm coverglass, trimmed to fit the Petri dish is placed over the cells and the dish is stored for 1 hour at 37°C. Eagle's[9] Medium (3.5 ml) with a concentration of amino acids and vitamins greater than usual is added. This is also known as Dulbecco's modified Eagle's Medium (Grand Island Biological Co.). The dish is then placed in an atmosphere of 8% CO_2. The culture medium is changed on the third or fourth day and approximately once a week thereafter.

In the beginning, separate colonies of epithelioid and fibroblastic cells are often recognized. These elements may show some different enzymatic activities *in vitro*. For the diagnosis of hereditary metabolic disorders it could be useful to grow these elements separately by cloning.[10] After 2 to 3 weeks a monolayer of clear, fusiform fibroblastlike cells will be noted (Fig. 3). The culture is then split by inverting and transferring the coverglass to another Petri dish. Enough elements are adherent to the coverglass to assure survival. Both cultures are refed. The elements on the coverglass can be maintained as a cell line, with a weekly 1:2 subcultivation ratio, for future references. If the cells are to be subjected to chromosome analysis, those in the original container are exposed to Colcemid (0.006%) for 4 hours and dislodged by trypsin (0.25%, for 5 minutes at 37°C) as well as by meticulous scraping by a rubber policeman under an inverted microscope. The elements are then suspended in warm hypotonic saline solution (0.54%

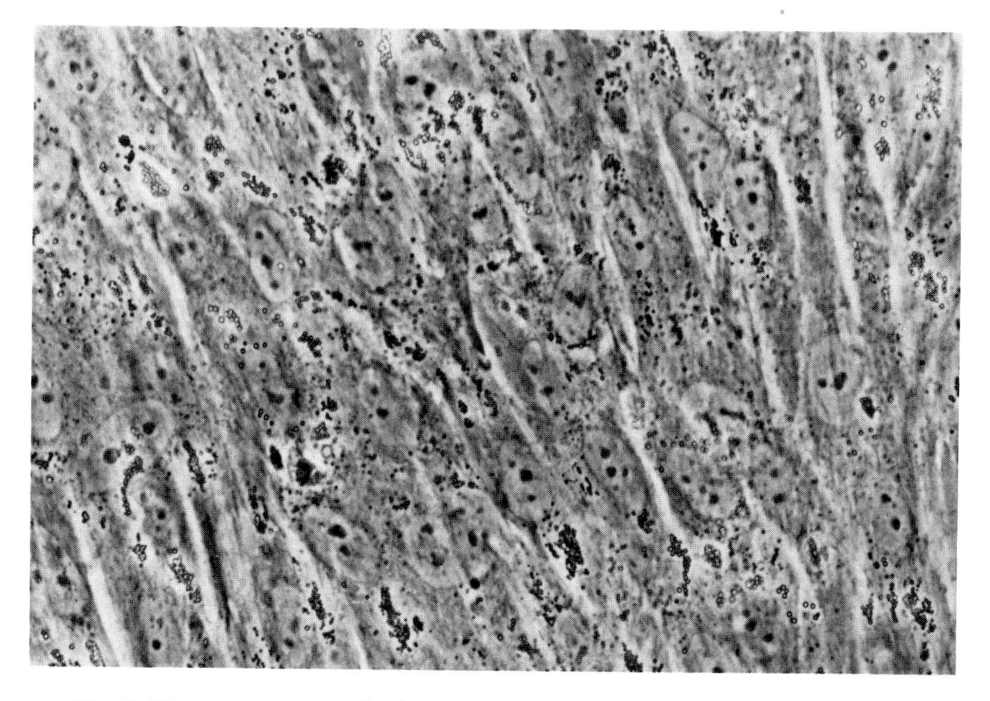

Fig. 3. Phase-contrast microphotograph of a monolayer of amniotic fluid cells, after 18 days in culture. ×490.

[9] H. Eagle, *Science* **130**, 432 (1959).

[10] M. M. Koback and C. O. Leonard, *In* "Antenatal Diagnosis" (A. Dorfman, ed.), p. 81. University of Chicago Press, Chicago, Illinois, 1972.

of KCl) for 10 minutes, spread on slides, air-dried, and stained with 2% acetic orcein, with the same technique adopted for chromosomal preparations from peripheral blood cultures.[11] If the cells are to be subjected to enzymatic analysis, the culture usually has to be continued for a longer period of time, since a larger number of cells is needed. The harvest is carried out in the same fashion.

Results

The low viability and small number of amniotic fluid cells available, the uncertainty as to their tissue origin, and culture contamination are the reasons for the relatively low success rate in cultivating amniotic fluid cells *in vitro*. Out of a total of 122 amniocenteses performed in this laboratory from the 13th to the 24th week of gestation, 94 samples yielded cultures suitable for either chromosomal or biochemical analysis. In twenty-five pregnancies the amniocentesis was repeated and a diagnosis reached in 94 of 101 pregnancies studied. Presumably a higher success rate would have been achieved if some of the patients had not refused a second and even third amniocentesis. No maternal nor fetal complications were noted following repeat taps of the amniotic sac.

Occasional failures are due to inadequacy of the chromosome preparations after satisfactory cell cultures. In four instances normal female karyotypes were obtained from amniotic fluid studies, followed by the birth at term of normal male fetuses. Most probably maternal cells, unrecognizable *in vitro*, had actually been cultivated. To avoid this serious mishap, it is suggested to regard with suspicion results obtained too early (less than 10 days after amniocentesis) and to repeat the analysis on later harvests.

The time interval between amniocentesis and cell harvest has ranged from 8 to 36 days (average of 19.7 days) for chromosome analysis and from 15 to 40 days (average 28 days) for enzyme analysis.

Conclusion

Amniocentesis for prenatal genetic diagnosis has proved to be a safe, reliable, and useful method. Improvement of both success rate in cell culture and diagnostic reliability may derive not only from modifications of the tissue culture technique but also from a different way of obtaining fetal cells. An endoamnioscopic technique has been successfully experimented in this laboratory, by which skin specimens[12] and fetal blood have been obtained under direct vision. Cultures of skin biopsy specimens have yielded cell preparations suitable for cytogenetic and biochemical analysis within less than 2 weeks, whereas lymphocyte culture has provided karyotypes within 72 hours. Blood sampling, if the technique will prove safe, will open the field of prenatal genetics to a number of hereditary conditions, such as sickle cell anemia, until now undiagnosable by amniocentesis.

[11] C. Valenti and S. K. Vethamany. *Amer. J. Obstet. Gynecol.* **99**, 434 (1967); cf. Section XIV, this volume.

[12] C. Valenti, *Amer. J. Obstet. Gynecol.* **114**, 561 (1972).

CHAPTER 3

Expressions of Citrullinemia in Skin Fibroblast Cultures

T. A. Tedesco and W. J. Mellman

Citrullinemia is a rare disease in man believed to be inherited as an autosomal recessive condition. The defect is due to a deficiency of the urea cycle enzyme argininosuccinic acid synthetase (ASA synthetase)[1] which catalyzes the condensation of citrulline with aspartic acid to form argininosuccinic acid. This enzymatic activity has been demonstrated in human diploid fibroblast cell cultures in three ways: (a) by direct enzyme assay, (b) by measurement of the incorporation of isotopically labeled citrulline into the acid-insoluble fraction (cell protein), and (c) by measurement of cell multiplication when citrulline is subsituated for arginine as an essential cell nutrient. Since cells with ASA synthetase deficiency as well as argininosuccinase (ASAase) deficiency, the next enzyme in the urea cycle which converts argininosuccinic acid to arginine, grow poorly when citrulline is substituted for arginine in the medium, these mutant cell lines have been of interest as auxotrophic cells in genetic experiments. Two other urea cycle enzymes, ornithine transcarbamylase and arginase are either absent or not at detectable levels in human fibroblast cell cultures. Citrulline, but not ornithine, can satisfy the arginine requirement of human skin fibroblast cells in culture. This was first suggested by Eagle in his studies on the essential amino acid requirements of heteroploid human cell cultures.[2]

Most of the experience with citrullinemic cells has been obtained with a single cell line (ATCC line CCL76).[3] Enzyme studies on a cell line derived from another citrullinemia patient have recently been reported.[4]

The enzymes of the urea cycle have been found to be adaptive to dietary protein in both rats[5] and primates.[6] ASA synthetase and ASAase activities of certain human and mouse heteroploid cell lines can be increased by incubating them in growth-limiting concentrations of arginine.[7] This effect has not been observed with human diploid fibroblast cell lines derived from skin biopsy.[3]

[1] V. E. Shih and M. L. Efron, *In* "The Metabolic Basis of Inherited Disease" (J. B. Stanbury, J. B. Wyngaarden, and D. S. Frederickson, eds.), 3rd ed., p. 379. McGraw-Hill, New York, 1972.

[2] H. Eagle, *Science* **130**, 432 (1959).

[3] T. A. Tedesco and W. J. Mellman, *Proc. Nat. Acad. Sci. U. S.* **57**, 829 (1967).

[4] A. Scott-Emuakpor, J. V. Higgins, and A. F. Kohrman, *Pediat. Res.* **6**, 626 (1972).

[5] R. T. Schimke, *J. Biol. Chem.* **237**, 459 (1962).

[6] C. T. Nuzum and P. J. Snodgrass, *Science* **172**, 1042 (1971).

[7] R. T. Schimke, *J. Biol. Chem.* **239**, 136 (1964).

Enzyme Assays: ASA Synthetase

Preparation of Cell Lysates. Trypsinized cells are washed in a balanced salt solution and collected by centrifugation. The cell button is suspended in 10 mM Tris-HCl buffer, pH 8.5, at a concentration of 20 to 30 $\times$ 10^6/ml. The cells are then lysed by freezing and thawing until a homogeneous preparation is obtained (five to six times). Both sonicates and acetone powders have been used with results comparable to those obtained by freezing and thawing.[7] The supernatant obtained can then be used in either colorimetric or radioisotope enzyme assays. The sensitivity of the radioisotope assay permits the use of fewer cells.

Colorimetric Assay. In principle, the urea formed from citrulline and aspartic acid substrates in the presence of excess amounts of ASAase and arginase is quantitatively estimated colorimetrically.[8] Under these conditions ASA synthetase is the rate-limiting step. Therefore the amount of urea produced per unit of time is a measure of enzyme activity. Pyruvate kinase and phosphoenolpyruvate are added to provide an ATP regenerating system.

The complete reaction mixture in a total volume of 0.4 ml contains: Tris-HCl, pH 8.5, 80 mM; citrulline, 10 mM; aspartic acid, 5 mM; $MgCl_2$, 2.4 mM; ATP, 2.4 mM; phosphoenolpyruvate, 10 mM; KCl, 20 mM; pyruvate kinase, 0.4 units/ml; ASAase, 400 units/ml; arginase, 100,000 units/ml (1 unit of enzyme converts 1 micromole of substrate/minute); and up to 0.2 ml of lysate.

Two blanks should be included in the assay: (a) one in which the lysate has been heat inactivated by boiling for 5 minutes before adding it to the reaction mixture, and (b) one in which citrulline has been deleted from the reaction mixture.

After incubation the reactions are stopped by placing the tubes in a boiling water bath for 5 minutes. The tubes are then centrifuged and aliquots of the supernatant are used for the quantitative estimation of urea by the colorimetric method of Archibald as modified by Ratner.[8]

Radioisotope Assay (Based on Schimke[7]). The reaction system is the same as that described for the colorimetric assay except that ureido[^{14}C]citrulline is used as the substrate and the final reaction volume is reduced to 0.2 ml. After incubation, heating, and centrifugation, [^{14}C]urea produced during the incubation is chromatographically separated from the [^{14}C]citrulline substrate and quantitatively estimated in a liquid scintillation spectrometer. Aliquots (20 μl) of the supernatant solution are chromatographed with carrier urea (10 μg) added to each spot applied to the chromatogram.

Chromatography is performed on 8 $\times$ 8 inch silica gel thin-layer sheets.[9] Twelve equal size channels are scribed on each sheet so that there are gel-free spaces between each sample. Chromatograms are developed in isopropanol-

[8] S. Ratner, *In* "Methods in Enzymology" (S. P. Colowick and N. O. Kaplan, eds.), Vol. IV, p. 356. Academic Press, New York, 1955.

[9] Available as Chromagram, Rochester, New York (Eastman type K-301-R-2).

ammonia-water (20:1:2). The end channels at each side of the thin-layer sheet are spotted with both citrulline and urea standards. Citrulline remains on the origin (R_f 0–0.06) and urea migrates with an R_f of 0.46–0.62. After developing the chromatogram with the various supernatant solutions, the individual strips are cut from the sheet. The position of [^{14}C]urea in each strip can be localized with a strip scanner but quantitation of the radioactivity in each spot is best accomplished by liquid scintillation counting of this region. Alternatively, urea can be localized in each of the standard channels from each side of the chromatogram. Provided the solvent front migrated without deformity, the location of these standards can be used as a guide for cutting out the urea spots in each of the other strips on the same chromatogram.

In a citrullinemic cell line (CCL76), ASA synthetase activity was comparable to that of control cells when assayed by these procedures (i.e., in the presence of 10 mM citrulline). The K_m for citrulline of that mutant cell line was markedly increased (K_m for control cells was 4×10^{-4} M and for the citrullinemic cells $10\text{–}100 \times 10^{-3}$). The one other reported citrullinemic cell line appears to be different in enzymatic behavior.[4] Citrullinemic mutations are likely to be heterogeneous and may be expressed by various alterations of kinetic parameters such as maximal velocities, K_m for either citrulline or aspartic acid, pH optimum, and heat stability.

Measurement of Citrulline Incorporation into Cell Protein

Schimke[7] has shown with HeLa cells, using ureido[^{14}C]citrulline, that 94–98% of the ^{14}C incorporated into cell protein is recovered as [^{14}C]arginine. Hence, for practical purposes, radioactivity recovered in the acid-insoluble fraction as cell protein after incubation with labeled citrulline can be assumed to represent substrate converted to arginine by the actions of ASA synthetase and ASAase.

In our experience measurements of citrulline uptake by cells and incorporation into cell protein can be done with the most sensitivity when the intracellular pool of arginine is depleted. This is accomplished by preincubating the cells in medium free of arginine, or by incubating with isotopic citrulline in arginine-free medium for a sufficient period of time that the intracellular arginine pool is depleted during the first part of the incubation. The latter process will result in a longer lag before maximal rates of isotope incorporation are attained. Prolonged periods of incubation may obscure differences between mutant and control cells. Cells incubated continuously with isotopically labeled precursors of protein reach a radioactive steady state due to the normal turnover of cell proteins. Control cells that convert citrulline to arginine more rapidly than mutant cells may reach maximal protein labeling before the mutant cells. Evidence should be obtained that the rate of isotope incorporation by the procedure used is proportional to time of incubation.

Incubation Procedure. Mutant and control skin fibroblast cultures are grown in milk dilution bottles to confluent density in standard growth medium (Eagle's

MEM with 10% fetal calf serum). The cells are incubated overnight in arginine-free, serum-free medium, which is then replaced with 10 ml of arginine-free, serum-free medium containing 0.02 m*M* ureido[^{14}C]citrulline (1.15 mCi/m*M*). After 6 hours at 37°C the radioactive medium is decanted and the cell sheet is briefly rinsed three times with balanced salt solution. Cells are removed by trypsinization, an aliquot of the cell suspension is removed for DNA determination by a modification of the method of Burton,[10] and the remaining cells are pelleted by centrifugation, frozen and thawed in 1 ml of H_2O, and precipitated with an equal volume of cold 10% TCA. The TCA precipitate is washed twice with 5% TCA, solubilized in 0.5 ml of 1.5 *N* NH_4OH (or one of the commercially available solubilizers used in scintillation counting), and counted in a scintillation spectrometer. Radioactivity of the TCA precipitate is expressed per unit of DNA.

By this procedure a citrullinemic cell line (CCL 76) was found to incorporate 2% of the isotope found in the TCA-precipitable fraction of control cell line.[3] This procedure can be applied to smaller numbers of cells and has been described with cell preparations grown on coverslips that were TCA precipitated, fixed with acetone, and counted directly on the coverslips.[11] Jacoby *et al.* enhanced the sensitivity of the method by increasing the ^{14}C specific activity of the citrulline. They were able to discriminate between skin fibroblast cells from individuals who were wild type, heterozygous, and homozygous for ASAase deficiency.

Growth Studies

Arginine (or citrulline) is essential for the multiplication of human skin fibroblasts. The rate of increase in numbers of control cells is the same whether the medium contains 0.2 m*M* citrulline or 0.2 m*M* arginine. Citrullinemic cells (CCL 76) do not multiply when citrulline is substituted for arginine. There is no difference between replicate cultures of citrullinemic cells incubated in arginine-free medium with or without supplemental citrulline.

Since serum is essential for the multiplication of human skin fibroblast cells, serum may need to be dialyzed if it is to be used in genetic experiments with citrullinemic cells where the ability to utilize citrulline is being used selectively. Fetal calf serum contains approximately 20–40 μM arginine.

The growth of cells on various experimental medium can be estimated by either cell count, protein, or DNA determination.

[10] T. A. Tedesco and W. J. Mellman, *Exp. Cell Res.* **45**, 230 (1966).

[11] L. B. Jacoby, J. W. Littlefield, A. Milunsky, V. E. Shih, and R. S. Wilroy, Jr., *Amer. J. Hum. Genet.* **24**, 321 (1972).

CHAPTER 4

The Hurler Cell in Culture[1]

B. Shannon Danes

The Hurler syndrome is a genetic disorder of mucopolysaccharide metabolism in which mucopolysaccharides are stored in various tissues of the body.[2] Progressive storage leads to the development of the clinical phenotype (dwarfism, grotesque skeletal deformity, restriction of joint movements, deafness, hepatomegaly, cardiac abnormalities, and mental retardation) during the first year of life.[3] The mode of inheritance is autosomal recessive.

From analyses of tissue specimens it was known[4] that the connective tissue cell, the fibroblast, synthesizes mucopolysaccharides. Thus, fibroblasts cultured from both homozygotes and heterozygotes for the Hurler syndrome may be used to study this inherited disorder of mucopolysaccharide metabolism at the cellular level.

Methodology

Split-thickness biopsies are taken without anesthesia from the extensor surface of the arm (Fig. 1). The establishment of the cell lines from skin biopsies by standard culture methods[5] requires about 6 weeks so that the cells studied have been grown as monolayer cultures from 1 to 2 months (two to six subcultures by trypsinization) prior to studies. Culture medium used in all cultures is reinforced Eagle's Medium containing 10% by volume of newborn calf serum. Prior to each experiment, cell lines are grown in large round bottles on a roller apparatus so that cell samples for both morphological and chemical studies come from one parent culture.

Histochemical Studies. Approximately 20,000 cells are inoculated into each 2-ounce glass flask which contains a coverslip. One week later the coverslip is removed, washed in warm balanced salt solution for 1 minute, and cut into three parts. The first part is immediately immersed in methanol for 5 minutes and

[1] This research was made possible by a grant from The National Foundation-March of Dimes.

[2] V. A. McKusick, *Amer. J. Med.* **47**, 730 (1969).

[3] V. A. McKusick, "Heritable Disorders of Connective Tissue," 3rd ed., p. 325. Mosby, St. Louis, Missouri, 1966.

[4] K. Meyer, M. M. Grumbach, A. Linker, and P. Hoffman, *Proc. Soc. Exp. Biol. Med.* **97**, 275 (1958).

[5] B. S. Danes and A. G. Bearn, *J. Exp. Med.* **129**, 775 (1969).

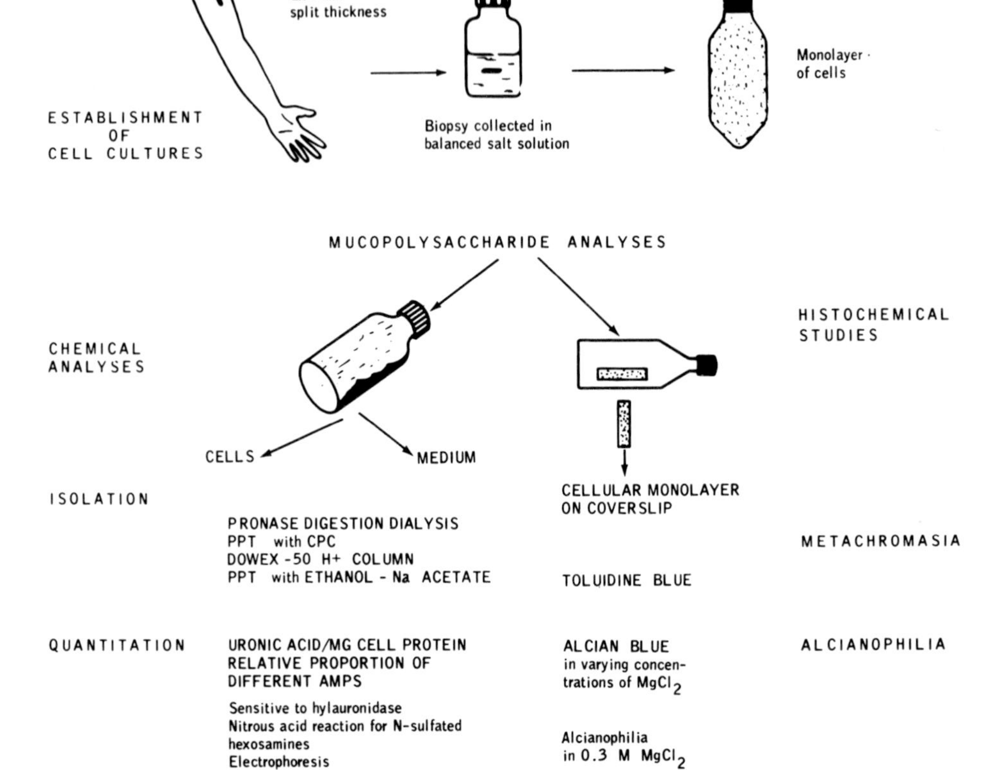

stained with the metachromatic dye, toluidine blue O.[6] The other two parts are stained for the presence of specific mucopolysaccharides in a simple one-step procedure, using Alcian blue 8 GX solutions containing 0.1 or 0.3 M $MgCl_2$.[7]

Chemical Studies. Roller bottle cultures of a cell line are established with approximately 10^8 cells and grown for about 2 weeks (with one medium change) prior to harvesting for chemical analyses.

Isolation of mucopolysaccharides. After decanting the medium the monolayer is trypsinized (0.25%) for 10 minutes and the sheet of cells shaken into suspension. The mucopolysaccharides of the cell suspension and used medium are then isolated by digestion with Pronase, dialysis, and precipitation with cetylpyridinium chloride (CPC).[8] Total polysaccharide is estimated as uronic acid by the carbazole method.[9]

[6] B. S. Danes and A. G. Bearn, *J. Exp. Med.* **123,** 1 (1966).
[7] B. S. Danes, J. E. Scott, and A. G. Bearn, *J. Exp. Med.* **132,** 765 (1970).
[8] B. S. Danes and A. G. Bearn, *J. Exp. Med.* **124,** 1181 (1966).
[9] T. Bitter and H. M. Muir, *Anal. Biochem.* **4,** 330 (1962).

Qualitative studies. The CPC precipitate is redissolved and reprecipitated with $CuSO_4$.[10] The precipitate (containing dermatan sulfate and hyaluronic acid) and the supernatant (containing heparan sulfate and chondroitin sulfates) are individually neutralized, dialyzed, passed over Dowex-50-X8 H^+ column, and concentrated.[10] Portions of both fractions are subjected to testicular hyaluronidase digestion and the undigested material reprecipitated with ethanol–sodium acetate. Hyaluronic acid and chondroitin sulfates are determined as the material susceptible to testicular hyaluronidase. Heparan sulfate is detected directly by the nitrous acid reaction for *N*-sulfated hexosamines.[11] Dermatan sulfate is calculated as the difference between the total and the other three mucopolysaccharides measured.

Electrophoresis. The pattern of the CPC-precipitated mucopolysaccharides is observed on cellulose acetate strips in a Beckman microzone cell (Model R-101) using two buffer systems to identify individual mucopolysaccharides.[12] In the calcium lactate buffer, hyaluronic acid and heparan sulfate can be separated whereas dermatan sulfate and chondroitin sulfates are electrophoretically indistinguishable. In barium acetate buffer, heparan sulfate and chondroitin sulfates migrated separately and hyaluronic acid moved together with dermatan sulfate.

Characteristics of the Hurler Cell in Culture

The Hurler cell in culture is distinguished from the normal cell, but not from cells from other mucopolysaccharidoses by (1) staining characteristics (metachromasia,[6] alcianophilia in 0.3 *M* $MgCl_2$[7]), (Fig. 2) (2) total mucopolysaccharide content, (3) increase in the relative proportion of dermatan sulfate,[13] and (4) decrease in α-L-iduronidase activity[14] (Table I).

Metachromasia reflects the intracellular presence of negatively charged substances, which include many cellular constituents including mucopolysaccharides. A histochemical procedure specific for the mucopolysaccharides [staining with alcian blue in various concentrations of magnesium chloride (alcianophilia)][7] combined with the chemical analyses of mucopolysaccharides is required to determine increased intracellular mucopolysaccharide content (Table I). However, in cultured fibroblasts which show increased mucopolysaccharide content (micrograms uronic acid/milligram cell protein), three patterns have been reported[13]: (1) increase in dermatan sulfate (the mucopolysaccharidoses, infantile amaurotic idiocy) (2) increase in hyaluronic acid (Marfan syndrome), and (3) a normal qualitative distribution of polysaccharides (Fabry's disease, Gaucher's disease, Krabbe's disease, cystic fibrosis, and Morquio syndrome). Thus, increased total mucopolysaccharide content (based on histochemical staining and uronic acid content) is not specific for a primary mucopolysaccharide disorder but may re-

[10] J. A. Cifonelli, J. Ludowieg, and A. Dorfman, *J. Biol. Chem.* **233**, 541 (1958).

[11] D. Lagunoff and G. Warren, *Arch. Biochem. Biophys.* **99**, 396 (1962).

[12] E. Wessler, *Anal. Biochem.* **26**, 439 (1968).

[13] R. Matalon and A. Dorfman, *Lancet* **ii**, 838 (1969).

[14] R. Matalon, J. A. Cifonelli, and A. Dorfman, *Biochem. Biophys. Res. Commun.* **42**, 340 (1971).

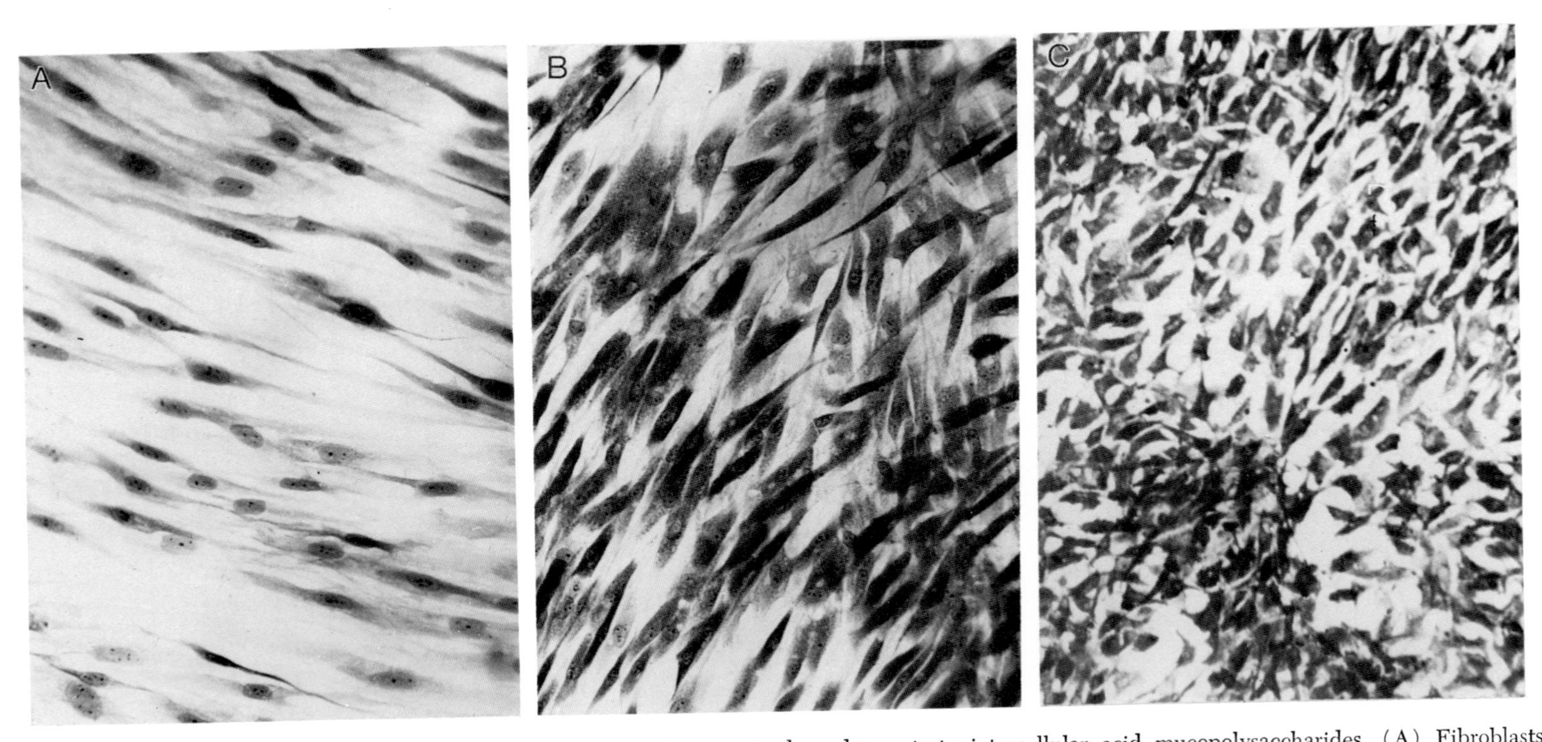

Fig. 2. Monolayers of skin fibroblasts grown in tissue culture stained to demonstrate intracellular acid mucopolysaccharides. (A) Fibroblasts from a normal individual stained light blue with toluidine blue O. ×1000. (B) Fibroblasts from a patient with the Hurler syndrome stained metachromatically (pink) with toluidine blue O. ×1000. (C) Fibroblasts from a patient with the Hurler syndrome showed alcianophilia (blue-green) when stained with alcian blue in 0.3 *M* $MgCl_2$. ×1000.

TABLE I
Mucopolysaccharides of the Human Skin Fibroblast Cultures

Characteristics	Skin fibroblast cultures		
		Hurler	
	Normal	Homozygote	Heterozygote
Histochemical[a]			
Metachromasia[b]	−	+	+
Alcianophilia in 0.3 *M* $MgCl_2$[c]	−	+	+
Chemical analyses			
Total cellular content[d]			
(μg uronic acid/mg cell protein)	1.8 ± 0.5	11.2 ± 2.1	9.1 ± 3.4
Relative proportion of mucopolysaccharides			
Hyaluronic acid	68%	28%	32%
Chondroitin sulfates	12%	6%	8%
Dermatan sulfate	16%	64%	58%
Heparan sulfate	4%	2%	4%
Incorporation of $^{35}SO_4$[d]		Increased	Increased
Hurler factor[e]	Present	Decreased	Decreased
α-L-Iduronidase activity[f]	Present	Decreased	Decreased

[a] Cytoplasmic staining with toluidine blue O is referred to as metachromasia (pink) (+) or orthochromasia (blue) (−); cytoplasmic staining with alcian blue solution containing 0.3 *M* $MgCl_2$ is alcianophilia (+) and no staining is (−).
[b] B. S. Danes and A. G. Bearn.[6]
[c] B. S. Danes, J. E. Scott, and A. G. Bearn.[7]
[d] B. S. Danes and A. G. Bearn.[8]
[e] J. C. Fratantoni, C. W. Hall, and E. F. Neufeld, *Proc. Nat. Acad. Sci. U. S.* **64**, 360 (1969).
[f] R. Matalon, J. A. Cifonelli, and A. Dorfman.[14]

flect other abnormal genotypes (Table II). Thus, the qualitative as well as the quantitative content of mucopolysaccharides (Table I) must be known before the Hurler cell in culture can be identified on the basis of cellular mucopolysaccharides.

Clinical Application

Heterozygote Detection. Identification of the homozygote for the Hurler syndrome is made on the basis of clinical phenotype, mucopolysacchariduria, and skeletal survey.[3] Analyses of the cell cultures confirms the clinical diagnosis (Table I). The heterozygotic state, although not demonstrable *in vivo*, can be clearly detected *in vitro* (Table I). The cultured fibroblast from an individual carrying one abnormal gene for the Hurler syndrome shows the characteristics of the Hurler cell; the disorder becomes dominant in culture.[6,8]

Prenatal Diagnosis. On the basis of mucopolysaccharide analyses (histochemical and chemical) of the cultured amniotic cells, prenatal diagnoses have been

TABLE II

Intracellular Mucopolysaccharides (Histochemical and Chemical) of Skin Fibroblast Cultures from Patients with a Diverse Number of Genetic Disorders[a]

Mucopolysaccharide characteristics			Disorders studied	
Histochemical staining[b]		Total cellular uronic acid		
Metachromasia	Alcianophilia			
−	−	Normal	Achondroplasia	Ehlers-Danlos syndrome
			Amyloidosis	Friedreich's ataxia
			Amyotonia	Gardner syndrome
			Cartilage hypoplasia	Incontinenti pigmenti
			Cerebral giantism	Metatrophic dwarfism
			Cholestasis	Muscular dystrophy
			Chediak-Higashi syndrone	Neurofibromatosis
			Congenital blindness	Osteogenesis imperfecta
			Cruzon's disease	Progeria
			Cutis laxa	Sarcoidosis
			Diastrophic dwarfism	Werner syndrome
			Dystonia	Wilson's disease
+	−	Normal	Cystic fibrosis	Gm_1 gangliosidosis
			Fibrodysplasia ossificans	Pelizaeus-Merzbacher disease
			Glycogen storage disease	Myotonic muscular dystrophy
+	−[c]	Increased	Marfan syndrome	
−	+	Increased	Larsen syndrome	
+	+	Increased	Fabry's disease	Lipomucopolysaccharidosis
			Gaucher syndrome	Multiple epiphyseal dysplasia
			Infantile amaurotic idiocy	Thanatophoric dwarfism

[a] Mucopolysaccharidoses not included.

[b] Cytoplasmic staining with toluidine blue O is referred to as metachromasia (pink) (+) or orthochromasia (blue) (−); cytoplasmic staining with alcian blue solution containing 0.3 *M* $MgCl_2$ is alcianophilia (+) and no staining (−).

[c] In the Marfan syndrome cultured cells stain with alcian blue in 0.1 *M* $MgCl_2$ as hyaluronic acid stains at this $MgCl_2$ concentration but not at 0.3 *M*.

made.[15] However, the specific test for *in utero* detection will most likely be the assay for α-L-iduronidase activity of the cultured amniotic fluid cell which, as yet, has not be established.

Treatment. Cell culture studies have shown that mucopolysaccharide storage can be reduced in the cultured Hurler cell by substituting human for calf serum in the culture medium.[16] These observations suggested that cellular storage might be impeded in the affected patient by plasma infusion. Clinical evidence initially

[15] J. C. Fratantoni, E. F. Neufeld, W. Uhlendorf, and C. B. Jacobson, *N. Engl. J. Med.* **280,** 686 (1969).

[16] M. C. Hors-Cayla, P. Maroteaux, and J. De Grouchy, *Ann. Genet.* **11,** 265 (1968).

suggested that mobilization of stored mucopolysaccharides might be enhanced in the Hurler patient by normal plasma infusions.[17] Subsequent studies have not continued this observation.[18-20]

The cultured Hurler fibroblast can be used as an *in vitro* model for such studies. The establishment of permanent lymphoid Hurler lines[21] having the same abnormal biochemical markers provides an alternate cell system for this research on the Hurler syndrome.

[17] N. Di Ferrante, B. L. Nichols, P. V. Donnelly, G. Neri, R. Hrgovcic, and R. K. Berglund, *Proc. Nat. Acad. Sci. U. S.* **68**, 303 (1971).

[18] B. S. Danes, M. Degnan, L. Salk, and F. J. Flynn, *Lancet* **2**, 883 (1972), **2**, 1044 (1972).

[19] A. S. Dekaban, K. R. Holden, and G. Constantopoulos, *Pediatrics* **50**, 688 (1972).

[20] R. E. Erickson, R. Sandman, W. v. B. Robertson, and C. J. Epstein, *Pediatrics* **50**, 693 (1972).

[21] B. S. Danes, T. H. Hütteroth, H. Cleve, and A. G. Bearn, *J. Exp. Med.* **136**, 644 (1972).

CHAPTER 5

Characteristics of Fibroblasts Cultured from Patients with Galactosemia or Galactokinase Deficiency[1]

Robert S. Krooth and Elizabeth K. Sell

Galactosemia[2] is a rare autosomal recessive disease characterized by milk intolerance. The acute symptoms following milk ingestion include vomiting, convulsions, jaundice, hepatosplenomegaly, and sometimes death. The chronic symptoms of the disease are cataracts and mental retardation. Virtually all the disagreeable features of galactosemia can be prevented—or greatly attenuated—by placing a newborn patient on a diet free of milk and related natural products. Figure 1 summarizes the three principal reactions involved in the conversion of galactose to glucose. In addition, a fourth reaction, which may also participate to a minor extent in this conversion, is shown. The tissues of patients with galacto-

[1] This research was supported by Program Project Grant GM 18153-02 from the National Institutes of Health, United States Public Health Service. E. K. Sell was supported by fellowship funds from Training Grant 5-TO1-GM-71-14 from the National Institutes of Health, United States Public Health Service.

[2] S. Segal, *In* "The Metabolic Basis of Inherited Disease" (J. B. Wyngaarden and D. S. Fredrickson, eds.), p. 174. McGraw-Hill, New York, 1972.

$$\text{Galactose} + \text{ATP} \xrightarrow{\text{galactokinase}} \text{Galactose-1-PO}_4 + \text{ADP}$$

$$\text{Galactose-1-PO}_4 + \text{UDPglucose} \underset{}{\overset{\text{galactose-1-phosphate: uridyltransferase}}{\rightleftharpoons}} \text{Glucose-1-PO}_4 + \text{UDPgalactose}$$

$$\text{UDPGalactose} \overset{\text{UDPgalactose-4-epimerase}}{\rightleftharpoons} \text{UDPGlucose}$$

$$\left[\text{Galactose-1-PO}_4 + \text{UTP} \overset{\text{UDPglucose (?) pyrophosphorylase}}{\rightleftharpoons} \text{UDPGalactose} + \text{pyrophosphate}\right]$$

Fig. 1. Summary of the reactions involved in the conversion of galactose to glucose.

semia and their cultured fibroblasts lack activity for the second enzyme of the sequence: α-D-galactose-1-phosphate: UDPglucose uridylyltransferase. Heterozygous persons have intermediate levels of activity in their cells and are clinically normal.

Another autosomal Mendelian mutation known in man affects the first enzyme of the sequence. Persons who are mutant homozygous at this locus have deficient or absent levels of galactokinase and tend to develop cataracts early in life.[3] Heterozygotes again have intermediate levels of activity of the affected enzyme and are clinically normal. The enzyme deficiency can be demonstrated both in the erythrocytes[2] and cultured fibroblasts of persons carrying the mutant gene.[4,5]

The abnormal phenotype of cultured fibroblasts from patients who are heterozygous or mutant homozygous at the galactosemia locus can be observed in three ways: by the nutritional properties of the cells, by their metabolic activity, and by the deficiency of transferase activity in cell extracts. The relevant nutritional response of the cells is determined by measuring cell growth in media in which, respectively, glucose and galactose are the sole hexoses present.[6,7] The relevant metabolic activity can be studied by separately incubating live cells with [1-^{14}C]-galactose or [1-^{14}C]glucose and measuring the release of $^{14}CO_2$.[6,8] Several methods for assaying the transferase activity of cultured fibroblasts have now been published.[4,9–14]

[3] R. Gitzelman, *Pediat. Res.* **1**, 14 (1967).
[4] J. S. Mayes and R. Guthrie, *Biochem. Genet.* **2**, 219 (1968).
[5] E. K. Sell, unpublished observations, 1970.
[6] R. S. Krooth and A. N. Weinberg, *J. Exp. Med.* **113**, 1155 (1961).
[7] R. S. Krooth, *In Vitro* **2**, 82 (1966).
[8] J. C. Petricciani, M. C. Binder, C. R. Merril, and M. R. Geier, *Science* **175**, 1368 (1972).
[9] T. A. Tedesco and W. J. Mellman, *J. Clin. Invest.* **48**, 2390 (1969).
[10] T. A. Tedesco and W. J. Mellman, *In* "Galactosemia" (D. Y. Y. Hsia, ed.), p. 66. Thomas, Springfield, Illinois, 1969.

Two interesting genetic phenomena involving galactosemic fibroblasts have recently been described. Nadler *et al.*[15] have reported that a somatic cell hybrid formed by the fusion of cells from different galactosemic patients contained significant (and kinetically abnormal) transferase activity. The parental lines had no detectable activity. Merril *et al.*[14] described the acquisition of transferase activity by galactosemic fibroblasts after exposure of the cells to a bacteriophage (lambda) whose genome included the galactose operon of *E. coli.*

Growth Experiments

Growth experiments are performed by making replicate cultures of cells of each of the strains to be tested. Both normal and galactosemic strains are incubated concurrently in identical media. The media contain dialyzed sera[6] rather than whole sera. The absence of glucose in the sera and media is verified chemically.[6] Once the basic hexose-free medium[6] has been compounded, it should be used to prepare four experimental media containing, respectively: (A) 100 mg% galactose[16]; (B) 100 mg% glucose; (C) 5 mg% glucose; (D) no added hexose ("hexose-free medium").

Twenty-four hours following subculture in complete medium,[17] the cells are washed twice (while still attached) in medium D and are then overlaid with experimental media. Five replicate cultures of each strain are removed for determination[18] of the initial cell protein as are five cell-free flasks. The flasks removed for this determination should be washed and then momentarily overlaid with medium D (in the same manner as the flasks which will be subsequently incubated). The cell-free flasks[7] are used to estimate the protein contributed to the glass by the medium. The mean protein values for these flasks is subtracted from the values obtained for flasks containing cells to obtain an estimate of *cell* protein. The cell protein of the individual replicate flasks (both those containing cells and the cell-free ones) should agree within 15%.

Once the experimental period has begun, three flasks of cells growing in each media and of each strain are removed every 72 hours for determination of cell protein. Three cell-free flasks containing media D are also removed from the experiment at each time point for determination of the protein blank, as described

---

[11] J. D. Russell and R. DeMars, *Biochem. Genet.* **1**, 11 (1967).

[12] J. D. Russell, *Biochem. Genet.* **1**, 301 (1968).

[13] J. D. Russell, *In* "Galactosemia" (D. Y. Y. Hsia, ed.), p. 204. Thomas, Springfield, Illinois, 1969.

[14] C. R. Merril, M. R. Geier, and J. C. Petricciani, *Nature* (*London*) **233**, 398 (1971).

[15] H. L. Nadler, C. M. Chacko, and M. Rachmeler, *Proc. Nat. Acad. Sci. U. S.* **67**, 976 (1970).

[16] Commercial preparations of galactose sometimes must be recrystallized once or twice in aqueous 70% ethanol to free them of detectable glucose (cf., H. M. Kalckar, K. Kurohshi, and E. Jordan, *Proc. Nat. Acad. Sci. U. S.* **45**, 1776, 1959).

[17] R. S. Krooth, *Cold Spring Harbor Symp. Quant. Biol.* **29**, 189 (1964).

[18] V. G. Oyama and H. Eagle, *Proc. Soc. Exp. Biol. Med.* **91**, 305 (1956); cf. Section VIII, this volume.

above. The experiment should normally run for 18 to 21 days. Once the cells in media C have become nearly confluent, flasks may be removed from the experiment for the determination of cell protein every 4 days rather than every 3. The cultures should be fed with fresh medium every 48 hours.

Under appropriate circumstances, the normal cells will grow in media A, B, and C, while in hexose-free medium (D) the cells will cease growth within 72 hours. Thereafter an ever increasing proportion of the cells in medium D will lyse.[7] The galactosemic cells behave like the normal ones, except that the galactosemic cells show the same arrest of growth and eventual degeneration in galactose-containing medium (A) as they do in hexose-free medium (D).

As noted above, galactosemic cells will not grow in a medium in which galactose is the sole hexose present, under circumstances in which normal and heterozygous cells will.[6,7] However, the selective system for this locus works consistently only when the cells are at a high population density. At low population density, a significant fraction of normal or heterozygous cells frequently die in galactose, and the discrimination between the cellular phenotypes on the basis of growth is obscured. The minimum population density at which the selective system works varies from experiment to experiment, but a confluent monolayer of galactosemic cells will invariably die in galactose, whereas a confluent monolayer of normal or heterozygous cells will invariably survive. The dependence of the selective system for this locus on cell population density may be related to the fact that the specific transferase activity of the cell protein progressively increases with cell population density.[13]

The competence of a culture to grow in galactose can be predicted from its ability to grow in reduced quantities of glucose. Repeated experiments have shown that only in circumstances where the cells are capable of growth in a medium containing 5 mg% glucose will they grow in galactose. Normal cells will usually grow in galactose or 5 mg% glucose when the initial population density is greater than 0.015 mg of cell protein per cm^2 available for growth. However, if the pH of medium is kept at 6.6[19] or if the cells are pregrown in reduced quantities of glucose,[7] a monolayer can be formed by normal cells in galactose from a much lower population density.

Of course, a system which requires a high cell population density to assess genotype is of little use for the purpose of selecting from the population a rare galactosemic cell that may have changed genetically due to mutation or some other cause. However, even with this limitation, such a system has a certain advantage over enzyme assays or histochemical methods for determining phenotype. The culture, if heterozygous or normal, is not lost to the investigator by the procedure of determining its phenotype. Hence, the system can be used to screen clones (but not to isolate them) on which other procedures can subsequently be performed.

Another curious feature of the selective system for the galactosemia locus is that, at least in our hands, it seems to work consistently only when the medium contains dialyzed human sera.[20] It is often not possible to demonstrate the hexose

[19] C. L. Baugh, R. W. Lecher, and A. A. Tytell, *J. Cell. Physiol.* **70**, 225 (1967).

[20] R. S. Krooth and E. K. Sell, *J. Cell. Physiol.* **76**, 311 (1970).

dependence of human diploid cell strains in medium containing dialyzed fetal calf sera. Not only do galactosemic cells sometimes grow in galactose, but cells of all genotypes frequently grow in hexose-free medium. When first placed in hexose-free medium the cells usually cease growth for a few days, and a minority of them may even die, but then, mysteriously, the culture recovers and growth resumes. Other laboratories have not had this problem and it may be that the mycoplasma which frequently contaminate commercially obtained fetal calf sera[21] are responsible for the phenomenon. We have not, however, thus far successfully cultured mycoplasma from media or cells employed in experiments where we noticed survival and growth in hexose-free medium. In any event, the selective system for the galactosemic locus, using dialyzed human sera, enables one to determine cellular phenotype when the cells being tested are at a high population density.

Metabolic Experiments

The incubation of suspensions of freshly trypsinized cells for 90 minutes in a buffered salt solution containing either [1-^{14}C]glucose or [1-^{14}C]galactose discriminates sharply between galactosemic and normal cells.[6] The galactosemic cells have negligible ability, compared to normal cells, to release $^{14}CO_2$ from [1-^{14}C]galactose. The ratio of counts in CO_2 derived, respectively, from galactose and glucose averages[22] about 0.3 in the case of normal cells and virtually zero in the case of galactosemic cells. Over very long periods of incubation (200–350 hours) galactosemic fibroblasts can convert quite significant quantities of [1-^{14}C]-galactose to $^{14}CO_2$, although at a rate which is only a few percent of the normal rate.[8] Interestingly enough, the ability of galactosemic fibroblasts to effect this conversion is markedly diminished in the presence of unlabeled glucose, which therefore further sharpens the discrimination between the two cellular phenotypes.[8] Moreover, even over these long periods of incubation, a cell strain from a patient with galactokinase deficiency could not release measurable amounts of $^{14}CO_2$ from [1-^{14}C]galactose.[8] These results suggest to us that UDPglucose pyrophosphorylase may have sufficient affinity for galactose-1-phosphate as a cosubstrate (Fig. 1) to effect a small but detectable synthesis of UDPgalactose. If this were the mechanism, the reaction would be inhibited if the fibroblasts contained an increased intracellular pool of the preferred substrate (glucose-1-phosphate), as might be expected when the incubation medium contains glucose. Also, fibroblasts lacking galactokinase would not have access to this alternative pathway.

Enzyme Assays

Galactokinase. To assay the galactokinase activity of cultured cells, we presently use a modification[4] of the method originally described by Sherman.[23] The

[21] E. Stanbridge, *Bacteriol. Rev.* **35**, 206 (1971).
[22] The buffer system employed contains cold glucose; see footnote 6.
[23] J. R. Sherman, *Anal. Biochem.* **5**, 548 (1963).

reaction mixture includes [1-^{14}C]galactose and ATP. The amount of product ([1-^{14}C]galactose-1-phosphate) formed is determined by spotting an aliquot of the reaction mixture on diethylaminoethylcellulose (DE-81) paper (Reeve Angel & Co., Inc., 9 Bridewell Place, Clifton, New Jersey 07014). Only negatively charged molecules bind to DE-81. Hence, after washing the paper several times at 55°C with distilled water, [1-^{14}C]galactose-1-phosphate, but not [1-^{14}C]galactose, is bound to the paper.

[^{14}C]Galactose preparations contaminated with glucose, or with other compounds which are phosphorylated when incubated in the presence of ATP and cell extract, will of course give spurious results. Moreover, [^{14}C]hexoses tend to decompose at an appreciable rate and, as a result, are liable to become contaminated with labeled nonhexose carbohydrates during storage. It is possible to purify [^{14}C]galactose preparations on a Whatman DE-52 column according to the method of Beutler *et al.*[24]

α-D-Galactose-1-phosphate:UDPglucose Uridylyltransferase. The reported assays of this enzyme activity in cultured fibroblasts are based either on spectrophotometry[4,9,10] or on the use of a radioactive cosubstrate and the isolation of labeled product.[13,14] Tedesco and Mellman[9] have reported that the *ratio* of transferase to galactokinase activity is particularly helpful in assessing cell genotype at the transferase locus.

The assay we presently employ is the one used by Merril *et al.*[14] and is a modification of the method of Russell and DeMars.[11] Cell extracts are incubated with uniformly labeled [^{14}C]galactose-1-phosphate and UDPgalactose. The newly synthesized radioactive UDPgalactose is separated from the [^{14}C]galactose-1-phosphate in two steps on PEI-cellulose thin-layer plates.

Higher levels of enzyme activity are obtained if the cell extract is preincubated for 15 minutes in the presence of dithiothreitol.[10] The preincubation mixture contains 10 μl of crude extract (3×10^6 cells/ml), 5 μl of glycine buffer (1 *M*, pH 8.7), and 5 μl of dithiothreitol (0.39 *M*). Following the preincubation, 10 μl UDPglucose (4×10^{-4} *M*) and 5 μl [^{14}C]-D-galactose-1-phosphate (214 mCi/m*M*, 100 μCi/ml) are added. The blank contains water in place of UDPglucose. This mixture is then incubated for an additional 15 minutes after which a 5-μl sample is removed and applied to the PEI-cellulose thin-layer plates. Then unlabeled UDPgalactose (5 μl of 10 μm/ml), as a marker, is applied to the origin at the same site as the sample. The radioactive UDPgalactose is separated from the [^{14}C]galactose-1-phosphate in two steps on PEI-cellulose thin-layer plates. In the first step, the solvent, 1 *N* formic acid, is allowed to migrate 4 cm from the origin. In the second step, the solvent (a solution of sodium formate which is made by titrating 2 *N* sodium formate with 1 *N* formic acid to pH 3.6) is allowed to migrate 13 cm beyond the previous solvent front. The position of the radioactive UDPgalactose is determined by the ultraviolet fluorescence of the unlabeled UDPgalactose. The UDPgalactose spots can be cut out and counted in TLA (Beckman) scintillation fluid.

[24] E. Beutler, N. V. Paniker, and F. Trinidad, *Biochem. Med.* **5**, 325 (1971).

CHAPTER 6

Evaluation of Amino Acid Metabolism in Maple Syrup Urine Disease

Joseph Dancis, Joel Hutzler, and Rody P. Cox

Clinical Manifestations

Maple syrup urine disease (MSUD) is an autosomal recessive disease in which there is an inability to degrade the three branched-chain amino acids (BCAA)—leucine, isoleucine, and valine. In the classical form, anorexia and vomiting appear a few days after birth followed soon by progressive neurological signs including rigidity, opisthotonus, convulsions, and coma.[1] If the dietary intake of BCAA is not carefully controlled, permanent neurological damage and death rapidly ensue.[2] Variants of the disease have been described in which there is some residual enzyme activity so that symptoms are often delayed in onset and are not as severe.[3]

Biochemistry

The enzymatic defect involves the decarboxylation of the three branched-chain keto acids (BCKA) (Fig. 1).[4,5] At present, it is uncertain as to whether each keto acid (KA) has an individual decarboxylase.[6] However, it is clear that single gene defects reduce the decarboxylation of all three KA.

Decarboxylation of the three BCKA is easily demonstrable in many tissues. Clinically, the skin fibroblast grown in tissue culture and the leukocyte are most accessible and have been most useful. The former has been of particular advantage for several reasons:

1. The investigator obtains a uniform population of cells which have been grown under controlled conditions, and the accumulation of inhibitory products may be avoided.

[1] J. H. Menkes, P. L. Hurst, and J. M. Craig, *Pediatrics* **14**, 462 (1954).

[2] S. E. Snyderman, P. M. Norton, E. Roitman, and L. E. Holt, Jr., *Pediatrics* **34**, 454 (1964).

[3] J. Dancis, J. Hutzler, S. E. Snyderman, and R. P. Cox, *J. Pediat.* **81**, 312 (1972).

[4] J. Dancis, M. Levitz, and R. G. Westall, *Pediatrics* **25**, 72 (1960).

[5] J. Dancis, J. Hutzler, and M. Levitz, *Pediatrics* **32**, 234 (1963).

[6] J. A. Bowden and J. L. Connelly, *J. Biol. Chem.* **243**, 3526 (1968).

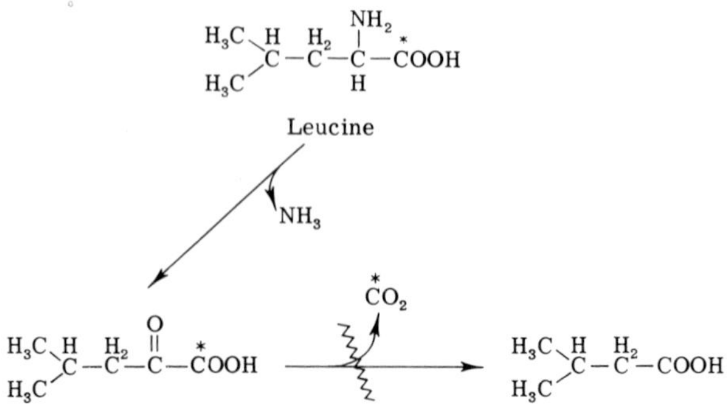

$$(H_3C)_2CH-CH_2-CH(NH_2)-\overset{*}{C}OOH$$

Leucine

$\downarrow$ NH_3

$$(H_3C)_2CH-CH_2-C(=O)-\overset{*}{C}OOH \xrightarrow{\overset{*}{C}O_2} (H_3C)_2CH-CH_2-COOH$$

Keto acid MSUD Block Isovaleric acid

Fig. 1. The pathway and metabolic block is analogous with isoleucine and valine. Asterisk indicates radioactive carbon in substrate when performing assay.

2. Cells may be stored, frozen, for indefinite periods. This has permitted the accumulation of rare specimens with review at an appropriate time.
3. The enzyme activity reflects fairly accurately the total BCKA decarboxylase capacity of the individual, so that clinically useful information is obtained.

Tissue Culture Methods

Routine methods are used, requiring no special description. We have generally used Waymouth's Medium with 10% fetal calf serum and antibiotics. However, we have recently observed that cysteine concentrations that approximate that in Waymouth's may, under some conditions, reduce BCKA decarboxylation, so that other media may be preferable.

Harvesting Cells

Cells are permitted to grow to heavy confluence before harvesting. Add 1 ml 0.1 *M* sodium phosphate buffer per 10 ml medium. Decant and discard about two-thirds of the medium and scrape cells into the remaining medium with a rubber policeman. Transfer approximately equal volumes of cells into incubation vials (see below). A cell population of 2×10^6 has been used for assays in most of our studies, but a reasonably accurate determination can be generally performed with as little as one-twentieth that number of cells.

Assay

Principle. One of the BCAA, labeled with radioactivity in the carboxyl carbon, is incubated with a suspension of fibroblasts. The CO_2 is trapped in a center well and assayed for radioactivity (Fig. 1). The BCAA is used as substrate, instead of the BCKA, because it provides a lower and more consistent blank. The first degradative step, transamination, does not appear to be rate limiting in the normal subject, and is certainly not rate limiting in MSUD. Suspensions of cells, instead of homogenates or cell extracts, are used because much of the enzyme activity is lost when the cells are broken.

The amount of radioactive CO_2 that is evolved is compared to the incorporation of the radioactive amino acid into protein. The ratio relates the decarboxylase activity to the amount of metabolically active tissue (Fig. 2). This has been more satisfactory than comparing the decarboxylase activity to the total amount of cellular protein though satisfactory results can be obtained with this method with careful control of culture conditions, and, particularly, within any single set of experiments.

Incubation Vials. Small flat-bottomed glass vials with a center well are used (Kimble Products, No. 60975-L, 19 × 48 mm, 2 dram, Opticlear vials; stoppers from Kontes Glass Co., No. K882310, and plastic No. K882320, or glass, No.

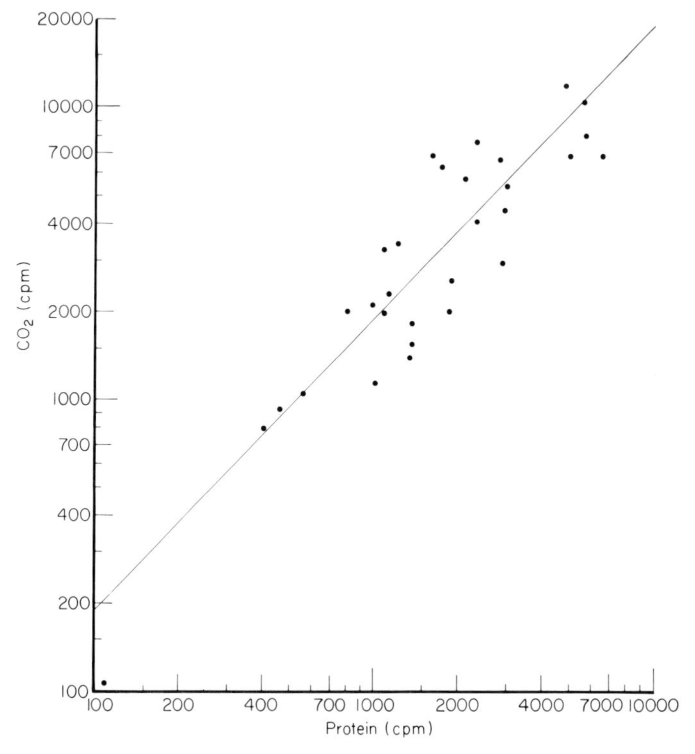

Fig. 2. Relation of rate of decarboxylation of [1-^{14}C]valine ($^{14}CO_2$—cpm) to rate of incorporation into protein of fibroblasts (cpm in trichloroacetic acid precipitate) in control subjects, plotted logarithmically. The line is constructed from the mean of the observed ratios. (From Dancis *et al.*[3])

K882330, center wells). In previous studies we have used 21 × 70 mm homeopathic vials (VWR Scientific) and No. 0 Silicone rubber stoppers from which glass wells were supported by several turns of No. 22 stainless steel wire.

Reagents

In the assay, less than 1% of the substrate is metabolized. It is therefore essential that the substrate be extremely pure and particularly free of volatile contaminants. The BCAA have been more satisfactory than the BCKA in this respect. The BCAA are checked for radioactive contaminants by paper chromatography in two solvent systems: *n*-butanol-1:acetic acid:water (12:3:5 v/v) and phenol:water (4:1 w/v). If there are radioactive contaminants, the same systems can be used for preparative chromatography. The radioactivity is diluted to 10 μCi/ml and carrier amino acid is added to prepare a stock solution of 0.5 μCi/μmole in NaCl, ca. 0.08 *M*, and pH is adjusted with HCl or NaOH to approximately 3. [1-^{14}C]-L-Leucine and [1-^{14}C]-L-valine have been obtained from New England Nuclear Corp., Boston, Massachusetts, and [1-^{14}C]-L-isoleucine from Calatomic, Los Angeles, California.

Volatile radioactive contaminants are removed by bubbling N_2 through the amino acid solution for 1 hour. The solution is then agitated in an Erlenmeyer flask fitted with a stopper and a center well containing 1 *N* KOH for 1 hour at room temperature in the dark. The radioactivity in the center well is determined and the process is repeated until the blank radioactivity is acceptably low. Occasionally it has been helpful in removing contaminants to add 0.1 ml 2,4-dinitrophenylhydrazine, 0.2% in 2 *N* HCl, and extract in 1 hour with peroxide-free and alcohol-free ether. The solution is stored at −20°C in a flask with a center well containing KOH. Immediately prior to use, it may be necessary to repeat the treatment with N_2 and agitation as described above because of decomposition during storage.

1. *Thiamine·HCl solution (100 mg/ml).* Prepare a solution of 1 g thiamine·HCl in 5 ml H_2O and correct pH by adding 5 ml 2% $NaHCO_3$ with stirring at 0°–5°C. The pH at completion of reaction should be 4.5–5.0. Store at −20°C.

2. *Krebs-Ringer-phosphate buffer.*[7]

3. *Scintillation/cocktail* (for CO_2 and digested protein). Methanol, 640 ml; toluene, 320 ml; phenethylamine, 20 ml (scintillation grade, Fisher Scientific Co.); glycerol, 20 ml; methylbenzethonium chloride (Hyamine 10×), 6 g (Sigma Chemical Co.); PPO (2,5-diphenyloxazole), 6 g; dimethyl-POPOP {1,4-bis[2-(4-methyl-5-phenyloxazolyl)]benzene}, 120 mg.

Incubation Procedure. The suspended cells are distributed into incubation vials as described above (see Harvesting Cells) and centrifuged at 600 *g* for 10 minutes. The supernatant fluid is removed with a Pasteur pipette and replaced with 0.3 ml freshly prepared incubation medium containing 0.025 ml

[7] P. P. Cohen, *In* "Manometric Techniques" (W. W. Umbreit, R. H. Burris, and J. F. Stauffer, eds.), pp. 147–150. Burgess, Minneapolis, Minnesota, 1957.

[1-^{14}C]-L-amino acid (0.5 μmoles, 0.25 μCi), 0.005 ml thiamine·HCl (0.5 mg), and 0.275 ml Krebs-Ringer-phosphate buffer with a final pH in the medium of 7.1 to 7.3. The vials are oxygenated and stoppered. The center wells contain 0.1 ml 1 *N* KOH. Extreme care must be exercised in handling the KOH because any misplaced alkali may trap radioactive CO_2. The cells are suspended by gentle swirling, and incubation is conducted with agitation in a Dubnoff metabolic incubator at 35°C for 90 minutes. The evolution of CO_2 is proportional to time for at least 120 minutes.

The reaction may be terminated by injecting 0.3 ml of 5% trichloroacetic acid (TCA) into the medium through the stopper very carefully and continuing the incubation for 30 minutes to complete the collection of dissolved CO_2. The procedure may be simplified by adding the TCA after the removal of the center well with the loss of about 15% of the radioactive CO_2 and a small change in ratio. With shorter incubation times, the loss becomes relatively greater.

The contents of the wells are transferred quantitatively to scintillation vials containing 12 ml of scintillation cocktail. The TCA precipitated cells are transferred to 12-ml centrifuge tubes using about 10 ml water, and permitted to stand for 20 minutes in a 60°C bath. The suspension is centrifuged at 800 *g* for 5 minutes and the supernatant discarded. The wash is repeated twice more and the cell button dissolved in 0.2 ml 4 *N* KOH by incubating at 60°C for 2 hours. The digest is transferred to the scintillation cocktail to determine radioactivity. If it is decided to do a protein assay by the Lowry method,[8] digest the cells in 4 *N* LiOH, dilute to 0.8 ml with water, transfer 0.5 ml to scintillant, and retain the rest for the protein assay.

Results

The amount of radioactive CO_2 liberated per milligram normal fibroblasts may be quite variable (2000–20,000 cpm), but is usually consistent within any series of experiments. The ratio of $^{14}CO_2$ to radioactivity incorporated into protein varies from 0.8 to 3.0 with a mean of about 1.5 for leucine and 1.9 for isoleucine and valine. The ratio in classical MSUD is 0–2% of the normal mean, and in the variant forms from 2–15%, depending on the severity of the defect.[3]

[8] O. H. Lowry, N. J. Rosebrough, A. L. Farr, and R. J. Randall, *J. Biol. Chem.* **193**, 265 (1951).

CHAPTER 7

Malignant Transformation of Cells in Culture Using Oncogenic Chemicals

Charles Heidelberger, Catherine A. Reznikoff, and David F. Krahn

Two systems have been used successfully for quantitative studies of chemical oncogenesis in tissue culture: the hamster embryo cell system first reported by Berwald and Sachs,[1,2] and the C3H mouse prostate system developed by Chen and Heidelberger.[3-5] A third system has recently been established in this laboratory, which involves the use of 3T3-like cell lines derived from C3H mouse embryo cells.[6,7,7a] Although considerable work with other systems has been reported, it appears at the time of this writing that they have not yet been developed to the point of yielding reproducible and quantitative information. Much attention is currently being devoted to develop systems for the malignant transformation of cultured liver cells, but this is also not sufficiently advanced to merit a description here.

In the hamster embryo cell system the cells are diploid at the time of treatment and transformation can be scored within 10 days. However, the nontransformed cells die after a few generations and cannot be cloned with high efficiency. The mouse prostate and 3T3 systems involve aneuploid cell lines that can be cloned with high efficiency.[8] Therefore, it is possible to compare for any properties a clone of nontransformed cells and a clone of transformed cells derived therefrom.[9] Scoring for malignant transformation may take 4–8 weeks, but requires less experience and eyestrain than the hamster cell systems when clones with low saturation densities are used. Thus, it is apparent that each system has its advantages and disadvantages. In all these systems, the transformed clones give rise to fibrosarcomas on inoculation into the proper host.[4-7,10]

[1] Y. Berwald and L. Sachs, *Nature (London)* **200**, 1182 (1963).
[2] Y. Berwald and L. Sachs, *J. Nat. Cancer Inst.* **35**, 641 (1965).
[3] T. T. Chen and C. Heidelberger, *J. Nat. Cancer Inst.* **42**, 903 (1969).
[4] T. T. Chen and C. Heidelberger, *J. Nat. Cancer Inst.* **42**, 915 (1969).
[5] T. T. Chen and C. Heidelberger, *Intern. J. Cancer* **4**, 166 (1969).
[6] C. A. Reznikoff, D. Brankow, and C. Heidelberger, in preparation.
[7] C. A. Reznikoff, J. Bertram, R. J. Langenbach, D. Brankow, and C. Heidelberger, in preparation.
[7a] A description of a rather similar 3T3 system has just been published: J. A. DiPaolo, K. Takano, and N. C. Popescu, *Cancer Res.* **32**, 2686 (1972).
[8] S. Mondal and C. Heidelberger, *Proc. Nat. Acad. Sci. U. S.* **65**, 219 (1970).
[9] S. Mondal, P. T. Iype, L. M. Griesbach, and C. Heidelberger, *Cancer Res.* **30**, 1593 (1970).
[10] J. A. DiPaolo, R. L. Nelson, and P. J. Donovan, *Science* **165**, 917 (1969).

Hamster Embryo Cell System

Cell Cultures. Primary cultures of hamster embryonic cells are prepared from 12- to 14-day-old fetuses of Syrian hamsters. Pregnant hamsters are killed by cervical dislocation and the uterus is aseptically removed and placed in a Petri dish. The uterus is washed thoroughly in phosphate-buffered saline (PBS) without Ca^{2+} and Mg^{2+} and the embryos are removed and subsequently washed in PBS to remove red blood cells. After washing, the embryos are decapitated and eviscerated (in more mature embryos the limbs are also removed) and the remaining body trunk is finely minced with a small scissors. The resulting tissue preparation is placed in a 200-ml trypsinization flask and washed with PBS for 8 minutes with magnetic stirring. This operation and all trypsinizations are carried out in a 37°C water bath. The stirring is terminated and a few minutes are allowed for the tissue particles to settle before the supernatant fluid is removed and discarded. This step serves to remove red blood cells.

About 200 ml of a 0.1% solution of trypsin in PBS is added to the flask and the tissue is stirred for 8 minutes. Tissue fragments are again allowed to settle, and the supernatant fluid is removed and discarded. This initial trypsinization is discarded since it contains tissue and cells damaged in the mincing procedure. Second and, if necessary, third trypsinizations are performed in a manner identical to the first except that the supernatant fluid remaining after the settling of tissue fragments is drawn off and placed in 40-ml centrifuge tubes containing 5 ml of heat-inactivated fetal calf serum, which inhibits the activity of the trypsin. These tubes are then centrifuged for 4 minutes at a speed sufficient to obtain a cell pellet (700 rpm). The supernatant fluid is removed and the cells are resuspended in Dulbecco's modified medium without phenol red, supplemented with 10% heat-inactivated fetal calf serum and antibiotics (100 units/ml penicillin, 50 μg/ml streptomycin). Phenol red provides a useful means to monitor the pH of the cell culture medium. However, commercial Dulbecco medium contains a concentration of the indicator (15 mg/liter) that is toxic to hamster cells. On those occasions when the pH is monitored, 5 mg/liter of phenol red is used. For the remainder of this description the mixture of medium, serum, and antibiotics will be referred to as complete medium.

The cells are counted with a Coulter counter or hemocytometer and are plated at 1×10^7 cells per 100-mm dish in 10 ml of complete medium. All cultures are incubated in a humidified atmosphere of 10% CO_2 at 37°C.

Rat primary cells are prepared by a procedure identical to that for the hamster cells and are utilized in the transformation assay as described in the next section.

Both hamster and rat secondary cultures are prepared from 2- to 3-day-old primary cultures. After removal of medium and a PBS wash, cells of the primary culture are removed from the dish by treatment for 3 to 4 minutes with 0.1% trypsin, the resulting cell suspension is centrifuged to give a pellet, and the supernatant fluid is discarded. The cells are suspended in complete medium, counted, and plated at 5×10^6 cells/100-mm dish in a 10-ml volume. Twenty-four hours later the number of attached cells per dish is about 4×10^6 and by the 3rd day the maximal cell density of $10–12 \times 10^6$ cells per 100-mm dish is attained.

Transformation Assay. On the first day of the experiment 2- to 3-day-old rat secondary cells are subjected to 4500 rads of X irradiation and plated at 60,000 cells per 60-mm dish in 2 ml of complete medium. The irradiated cells, which are referred to as "feeders," do not divide, but apparently perform a metabolic function necessary for the growth of small numbers of hamster secondary cells, and are also thought to aid in the activation of chemical oncogens to their proximal form.

Twenty-four to 40 hours after plating of the rat feeder cells, 2- to 3-day-old hamster secondary cells are plated in these dishes at 500–1000 cells per dish in 2 ml of complete medium. Twenty-four hours later 4 ml of complete medium is added to each dish to give a total volume of 8 ml, and the cells are treated with the test compounds.

A variety of chemicals have been used for *in vitro* oncogenesis. It should be noted that the reactivities of each compound and its stability in heat, light, serum, and certain diluents should be known, because each of these factors will determine proper handling and will affect the outcome of the tests. In addition, it is important to check the purity of the compounds by appropriate means.

Depending on its solubility properties, the test compound is dissolved in acetone, dimethyl sulfoxide, or PBS at the desired concentration. A 20-μl volume of this stock solution is added to the dishes, followed by a gentle swirling to assure even distribution of the compound. The stock solution is prepared such that 20 μl when added to the 8 ml of medium in the dish gives the desired final concentration of test compound. Control dishes are treated with a 20-μl volume of the appropriate solvent. The duration of treatment is either 7 days or 1 day, in which case the medium containing the test compound is removed and replaced with fresh medium.

Depending on colony size, as determined by examination of growing cultures on an inverted microscope, 6 to 10 days after treatment with the test compound the experiment is terminated. The medium is removed, the dishes washed with PBS, and then absolute methanol is added for 20 minutes to fix the colonies. After removal of the methanol the colonies are stained with a solution of Giemsa in 0.01 *M* phosphate buffer (1:10) for at least 40 minutes, rinsed with water, and dried. Determination of cytotoxicity and examination for transformed colonies are done with the same dish. Cytotoxicity is determined by counting the total colonies per dish and expressing this number as a percentage of the total number of cells plated. This percentage is called the plating efficiency. Good control plating efficiencies range from 10 to 20%.

Scoring of stained colonies is done with a dissecting microscope. As has been shown,[1,2,10-12] and as illustrated in Fig. 1, there are distinct morphological differ-

[11] J. A. DiPaolo, R. L. Nelson, and P. J. Donovan, *Cancer Res.* **31,** 1118 (1971).

[12] E. Huberman, T. Kuroki, H. Marquardt, J. K. Selkirk, C. Heidelberger, P. L. Grover, and P. Sims, *Cancer Res.* **32,** 1391 (1972).

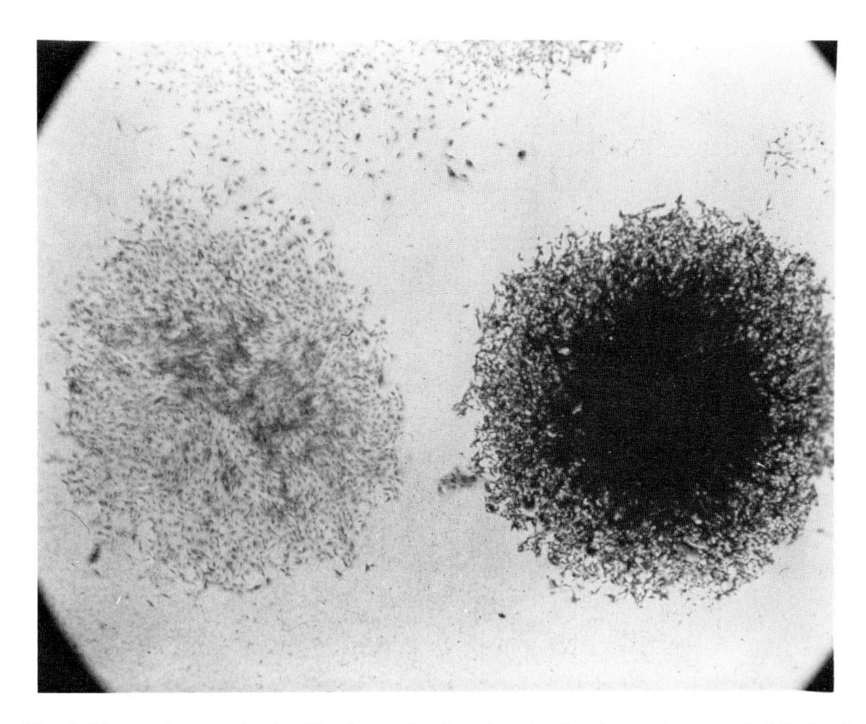

Fig. 1. Photomicrographs (×5) of two fixed and stained colonies in a methylcholanthrene-treated dish containing hamster embryo cells in the cloning assay described in the test. The colony on the left is normal; the cells are in a strict monolayer and their cytoplasm is evident. The colony on the right is transformed; the cells are smaller, more fibroblastic in appearance, and are heavily piled up.

ences between a normal colony as seen on the left and a transformed colony on the right. The normal colony is composed of slower growing and less "fibroblastic" appearing cells. The cells in the transformed colony exhibit extensive criss-crossing and piling up. Although normal colonies may also be thick, criss-crossing is a characteristic of and specific only for transformed cells. Since it can be difficult to determine cell morphology and interactions in piled up central portions of a colony, the less dense edges more accurately indicate whether the colony is normal or transformed (Fig. 2).

The malignancy of transformed colonies can be determined by isolating colonies from a treated dish and then growing up large numbers of cells derived from these colonies for injection into hamsters. Generally 10^5 to 10^7 cells are injected into irradiated weanling hamsters either intradermally or intracranically. Morphologically altered "transformed" clones produce tumors, while normal colonies from treated dishes do not give rise to tumors.[10]

Mouse Cell Lines

The selection of suitable lines of mouse cells for *in vitro* chemical oncogenesis is of great importance. Lines should have a low saturation density (be contact

Fig. 2. Photomicrograph ($\times 30$) of the edges of the same two colonies shown in Fig. 1. The criss-cross piling up of the transformed colony on the right is clearly shown.

sensitive), have a clear background (free of focal areas of heavier density), and should not give rise to tumors when inoculated into isologous mice. Recently isolated clones tend to have more homogeneous backgrounds than mass populations of cell lines, and, therefore, are most suitable for transformation experiments. Because of the well-known tendency of mouse embryo cell lines to undergo spontaneous malignant transformation,[13] suitable cell lines are not readily available and so we have found it necessary to establish and use new lines of cells, mostly derived from adult mouse ventral prostate.

Culture of Mouse Cells. All stock cultures of mouse cell lines are maintained at 37°C in 25-cm² closed Falcon tissue culture flasks in the absence of antibiotics. The pH in the flask is adjusted by gassing with 5% CO_2 in air. The culture medium for all mouse cells consists of Eagle's Basal Medium with phenol red plus 10% heat-inactivated fetal calf serum. The medium is changed as needed, usually twice weekly. Cells are routinely subcultured when confluent by dispersion with 0.1% trypsin at 37°C for 10 minutes. All cell lines are tested every fifth passage for mycoplasma contamination by the procedure of Allen *et al.*[14] Cell lines are also routinely tested every fifth passage for tumor production by subcutaneous inoculation of six weanling isologous male mice (6–8 weeks old) with

[13] K. K. Sanford, *Nat. Cancer Inst. Monogr.* **26**, 387 (1968).
[14] V. Allen, S. Sveltmann, and C. Lawson, *Health Lab. Sci.* **4**, 90 (1967).

10^6 cells each in 0.25 ml of PBS. Test animals are observed for tumor growth for a minimum of 1 year.

In the past, lines of prostate fibroblasts were established by the dispersion of organ cultures of adult C3H mouse ventral prostates.[3] Recently, however, the method of Gey and Gey[15] has been successfully used to establish lines of prostate fibroblasts from individual mice.[6] According to this method, one or more lobes of either the ventral or anterior prostate of a 6- to 8-week-old C3H male mouse are removed and washed in PBS. The organ is then placed in a glass Petri dish containing complete medium, where each lobe is cut by scalpels into about eight small pieces. These pieces of the explant are then placed directly onto the surface of a 25-cm^2 plastic tissue culture flask, which has been preconditioned by wetting with complete medium. The flask is then gassed with 5% CO_2 and incubated without additional medium for 30 minutes at 37°C. During this period the explants usually attach firmly to the plastic surface, and 5 ml of medium can then be gently added without dislodging the explants from the plastic surface. If the explants do not adhere to the surface, the incubation period without medium should be extended or repeated.

As early as 24 hours after the cultures are set up, cells migrate from the explant onto the plastic surface. Usually in a week or two at most, the area around each of the explants is heavily populated with cells, and the first subculture can be done following removal of the primary explant. The new cell culture is then given regular weekly medium changes and subcultured when confluent. Following the initial rapid growth usually observed in primary and secondary cultures, there is a period where growth does not occur. This has been observed by others.[16–18] Following this decline, the growth rate then steadily increases and eventually stabilizes. At this time cloning is done by the ring isolation technique[19] and the suitability of the individual clones for *in vitro* oncogenesis is determined by studying their properties as described earlier.

The establishment of C3H mouse embryo lines is done according to the procedure of Todaro and Green[17] and Aaronson and Todaro[18] with some slight modifications. Whole mouse embryos are minced with fine surgical blades and disaggregated with 0.25% trypsin. The trypsin is removed by centrifugation and the cells resuspended in medium. Primary cultures are established by plating 3×10^6 cells per 75-cm^2 plastic flask. After the primary cultures attain confluence, secondary cultures are made in the same manner by seeding 25-cm^2 plastic flasks with 3×10^5 cells each. The cultures are then put on a regular schedule of twice weekly subculture with each seeding at the same cell density. Between the fifth and tenth transfer, there is usually a period where no increase in cell number occurs. At this time, one of the weekly subcultures is replaced by a simple medium change. This adjustment is made because a number of cultures do not survive frequent transfers during this period. This adaption of

[15] G. Gey and M. K. Gey, *Amer. J. Cancer* **27**, 45 (1936).
[16] L. Hayflick and P. S. Moorhead, *Exp. Cell Res.* **25**, 585 (1961).
[17] G. J. Todaro and H. Green, *J. Cell Biol.* **17**, 299 (1963).
[18] S. A. Aaronson and G. J. Todaro, *J. Cell. Physiol.* **72**, 141 (1968).
[19] T. T. Puck and P. I. Marcus, *Proc. Nat. Acad. Sci. U. S.* **41**, 432 (1955).

the original procedure of Todaro and Green[17] does not contradict their basic principle of the frequent transfer schedule designed to minimize cell-to-cell contact, because the omission of regular subculturing is done only during the period when an increase in cell number does not occur. By the twentieth passage, the growth rate of the cells usually stabilizes, and cloning is done to select lines for *in vitro* oncogenesis. By using a modification of this procedure, a line of C3H mouse embryo cells has been established, which is suitable for quantitative studies and has been designated T½.[6,7]

Plating of Mouse Cells for Transformation and Cytotoxicity Tests. For transformation tests, 10^3 cells are plated per 60-mm plastic Petri dish in 5 ml of medium. For cytotoxicity tests, 200 cells are plated per dish in the same manner.

Twenty-four hours after seeding, the test chemical is added directly to the 5 ml of medium in the Petri dishes using a 20-μl pipette, as described above. The duration of the treatment is usually 24 hours, at which time a medium change is made.

Cytotoxicity Tests for Mouse Cells. The cytotoxicity exerted by a chemical is related to its concentration and to the duration of the exposure. Dose-response curves for cytotoxicity should be done for each cell line and for each chemical. There are two ways to determine cytotoxicity. The first and most commonly used method is to count colonies to determine whether the compound being tested has caused a reduction in plating efficiency. Another method to measure cytotoxicity is to count the cells in order to determine whether there is a reduction in the total cell number in treated cultures compared to the controls. This second method is particularly useful when there has been little or no reduction of the plating efficiency by the compound.

Careful microscopic observation is necessary to determine the optimal time to stain the cells for colony counts. Usually between the 7th and 10th days, discrete and well-defined colonies can be seen in the dishes. At this time these dishes are washed with saline, and fixed and stained as described above. Colonies are counted with a binocular dissection microscope or by eye. For cell counts, the dishes are trypsinized and the resulting cell suspensions are counted with a Coulter counter. Toxicity, as expressed by a slowing of cell growth, can then be determined by comparing the average number of cells per dish in the treated groups with the controls. It is also possible to calculate the average number of cells per colony. To make this comparison, the cell counts should be done at the same time as the colony counts.

Transformation Tests in the Mouse Systems. Confluence is attained in the dishes seeded with 1000 cells on approximately the tenth day after seeding. Optimal results in terms of detection of transformed foci are obtained when these dishes are kept for approximately 5 weeks postconfluence and given a weekly medium change. The best results are obtained when the level of calf serum in the medium is maintained at 10%. Addition of 100 units/ml of penicillin, 50 μg/ml of streptomycin, and 1 μg/ml of Fungizone after confluence

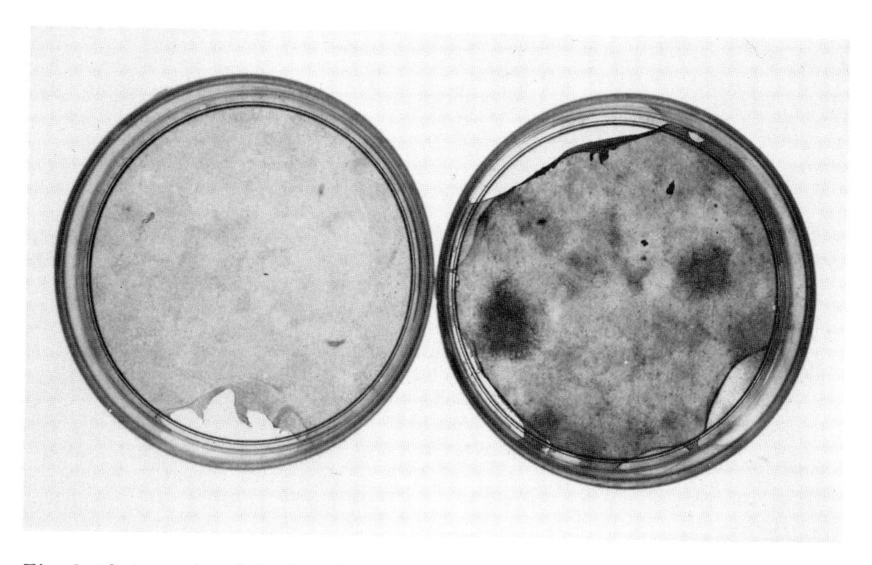

Fig. 3. Photographs of fixed and stained dishes of confluent prostate fibroblasts. The dish on the left is a control, and the monolayer of lightly stained cells is only faintly seen. The dish on the right was treated with methylcholanthrene (MCA), and distinct transformed colonies are seen. (From T. T. Chen and C. Heidelberger.[5])

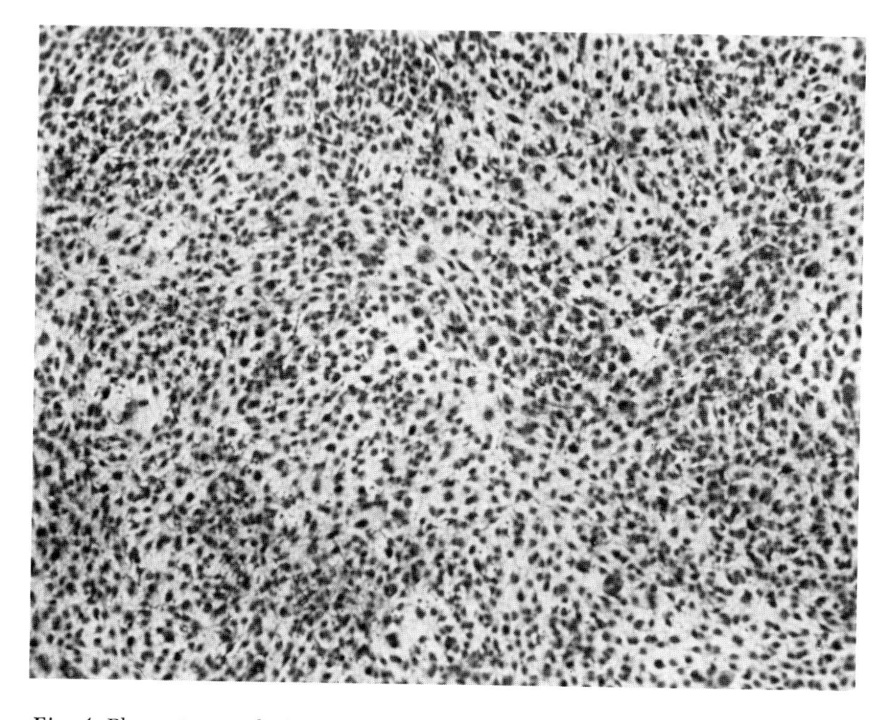

Fig. 4. Photomicrograph (×45) of a monolayer area of the MCA-treated dish in Fig. 3. The appearance of untreated or solvent control dishes is identical. (From T. T. Chen and C. Heidelberger.[5])

is attained does not affect the transformation assay and effectively prevents contamination of Petri dishes during the long duration of these experiments. No antibiotics are used in the stock cell cultures because contamination is not a problem in closed tissue culture flasks.

At 6 weeks, the dishes in the transformation group are fixed and stained with Giemsa as described. Transformation causes piling up and an alteration in regular cell morphology resulting eventually in dense foci of cells arranged in criss-cross irregular patterns. The initial scoring of many of the large transformed foci can be done by eye, since the dense areas of transformed cells take up the stain more deeply than the normal background, as shown in Fig. 3. However, the entire dish is carefully scanned microscopically for additional less prominent foci. Each focus is then checked microscopically for altered cell morphology (Figs. 4 and 5). The necessity for suitable cell lines is most evident during scoring, because if the background is not homogeneous and the controls are not negative scoring becomes subjective and difficult.

Although staining and final scoring of transformed foci is not done until approximately 6 weeks after plating, it is possible microscopically with experience and skill to identify transformed foci in the living cultures as early as 10–14 days, using an inverted microscope. However, many transformed foci do not appear until later, and some are small and not well developed in the first few

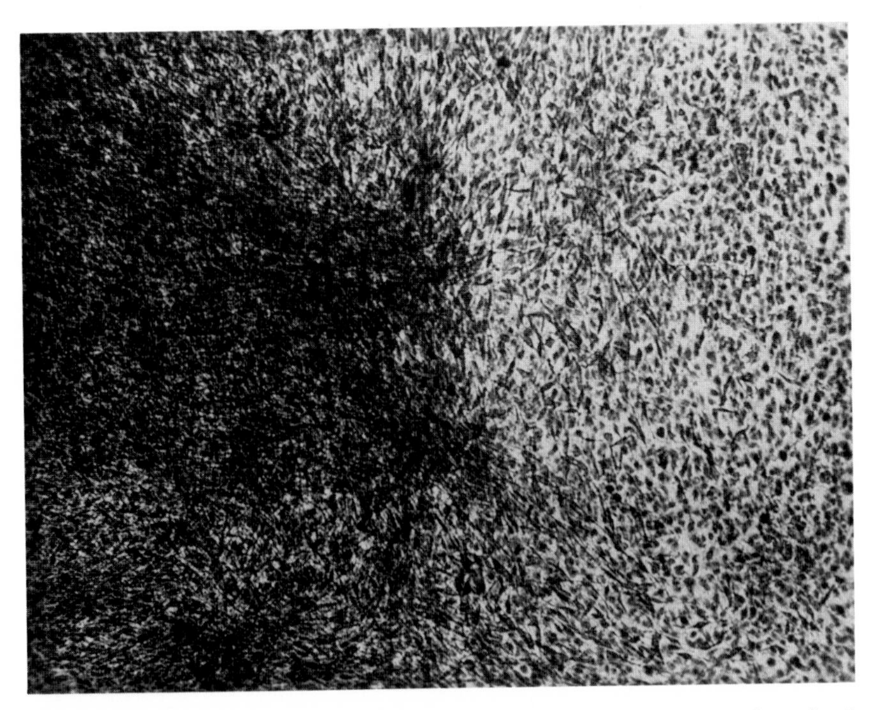

Fig. 5. Photomicrograph (×45) of the edge of one of the piled up transformed colonies in the MCA-treated dish in Fig. 3. The deeply stained array of cells and their criss-cross orientation in the transformed colony is evident and contrasts with the strict monolayer orientation of the cells outside the transformed colony. (From T. T. Chen and Heidelberger.[5])

weeks of the experiment. This can cause uncertainty in scoring. Therefore, we consider that the best results in terms of accurate and quantitative scoring are obtained when the full 6 weeks is allowed.

Tests of Transformed Cells for Tumor Production. It is always necessary to correlate *in vitro* morphological transformation with *in vivo* malignancy. Before staining, some transformed foci are picked by the ring isolation technique[19] and grown in culture. Areas of control cells are isolated in the same manner. These are inoculated into mice as described for the routine testing of the stock cultures. If it is desired to shorten the latent period before tumor production, animals to be inoculated can be preirradiated. This is usually accomplished by X-irradiation with 450 rads. Intracerebral inoculation requires fewer cells than subcutaneous injection and also has a latent period for tumor production.[5]

CHAPTER 8

Malignant Transformation of Cells in Culture

F. Rapp

Transformation of cells from a variety of animal species has been effected *in vitro* by numerous viruses. The most widely employed cells for this purpose have been embryonic cells derived from rodent and avian species, but other cell types, including human and monkey cells, have also been found useful. It is critical that the cell to be employed is nonpermissive for virus replication or that the virus is defective in its ability to cause cytopathic effects in those instances wherein virus multiplication would be accompanied by cell destruction. Cytopathology generally represents a greater problem for tumor viruses containing deoxyribonucleic acid than those cancer viruses containing ribonucleic acids as a source of their genetic information.

Preparation of Cell Culture

Cell cultures used for transformation studies can be derived from various organs taken from the intact animal, from cells undergoing secondary passages, or from established cell lines. The most sensitive systems are often primary cultures obtained by dissociating cells from organs freshly removed from chicken embryos or from mouse or hamster fetuses. The embryos are generally treated with 0.25% trypsin for approximately 20 minutes and then centrifuged at low speed to sediment the cells. The trypsin supernatant is removed and the cells are suspended in appropriate medium (often Eagle's Basal Medium) supplemented with growth factor most appropriate for the cells being utilized. Often, these include 10% fetal bovine serum, sometimes 10% tryptose phosphate broth, and sodium bicarbonate to yield a pH of approximately 7.4. In addition, many investigators add 100 units of penicillin and 100 μg of streptomycin per milliliter of fluid to inhibit bacterial contamination.

When secondary cell cultures or cell lines are employed, these cells are also dissociated with either trypsin or a chelating agent (such as Versene) and then handled as above. A sample of the resuspended cells is removed for a cell count in a hemocytometer and the cells are diluted with additional growth medium to adjust them to the concentration desired. Typically, the cells are placed into an 8-ounce (about 240-ml) glass prescription bottle at a cell density of about 5×10^6 cells per bottle; however, Petri dish cultures and bottles of other sizes can also be employed. The cells can then be exposed to the virus either as a cell suspension (prior to placing them into the culture vessel) or after they have attached to either the glass or plastic vessel into which they were plated.

Exposure of Cell Cultures to Transforming Virus

It is critical that the cells exposed to the virus can replicate after exposure. For this reason, the use of confluent cell cultures is undesirable and if used, such cultures should be dissociated by enzymatic digestion or Versene (as above) a few hours after virus adsorption. They should then be replated into fresh vessels with fresh medium at a density enabling cell multiplication.[1]

The amount of virus to be added will be dependent on the virus–cell system. In general, DNA-containing viruses should be added at a high multiplicity of virus particles per cell. For example, it would be desirable to add at least 50–100 plaque-forming units of simian papovavirus 40 per cell for which transformation attempts are being made. However, this is true only for nonpermissive systems (such as SV40 in mouse or hamster cells). In permissive systems for these viruses, it is more desirable to keep the multiplicity low (less than one plaque-forming unit per cell) and to prevent the spread of infection by placing into the medium some serum containing specific neutralizing antibody against the virus. With viruses containing RNA, low multiplicities will suffice to transform the cells.[2,3] If most of the cells in the culture are to be transformed, higher multiplicities of infection will be required. After the adsorption of the virus (generally carried out at 37°C), growth medium for the cells is added and the cells are incubated at 37°C; if grown in Petri dishes, incubation should take place in an incubator maintained in an atmosphere of 3–5% CO_2. This is generally

[1] R. Duff and F. Rapp, *J. Virol.* **8**, 469 (1971).
[2] F. Rapp and R. Duff, *Perspect. Virol.* **7**, 37 (1971).
[3] H. M. Temin and H. Rubin, *Virology* **6**, 669 (1958).

the most suitable system for transformation and for quantitation of the effects observed.

The medium should be changed at least twice weekly although more frequent changes may be necessary if cell growth is so rapid as to cause acidic conditions in the medium; these are generally monitored by adding a small amount of phenol red to the medium at the time of preparation. Thus, the appearance of the yellow color signifies a drop in pH and the fluid should be changed when this occurs. It is also a requirement that overcrowding be prevented as the cells will have to undergo at least one cycle of replication to fix the transformed state. In addition, they will need space in which to grow so that the transformation can be observed. This means that at least once or more often (usually twice a week), cells will have to be dissociated and replated in smaller quantities in fresh culture vessels in fresh medium. These passages should be continued indefinitely until the cultures have been well established and growth characteristics can be evaluated.

Criteria for Transformation

The transformation of cells from higher organisms in culture can be monitored in a number of ways. The most commonly used criterion is either a change in morphology or a change in the social behavior of the transformed cells (Fig. 1). The two often accompany each other and can generally be observed microscopically within 2 to 3 weeks after transformation has occurred (Fig. 2).

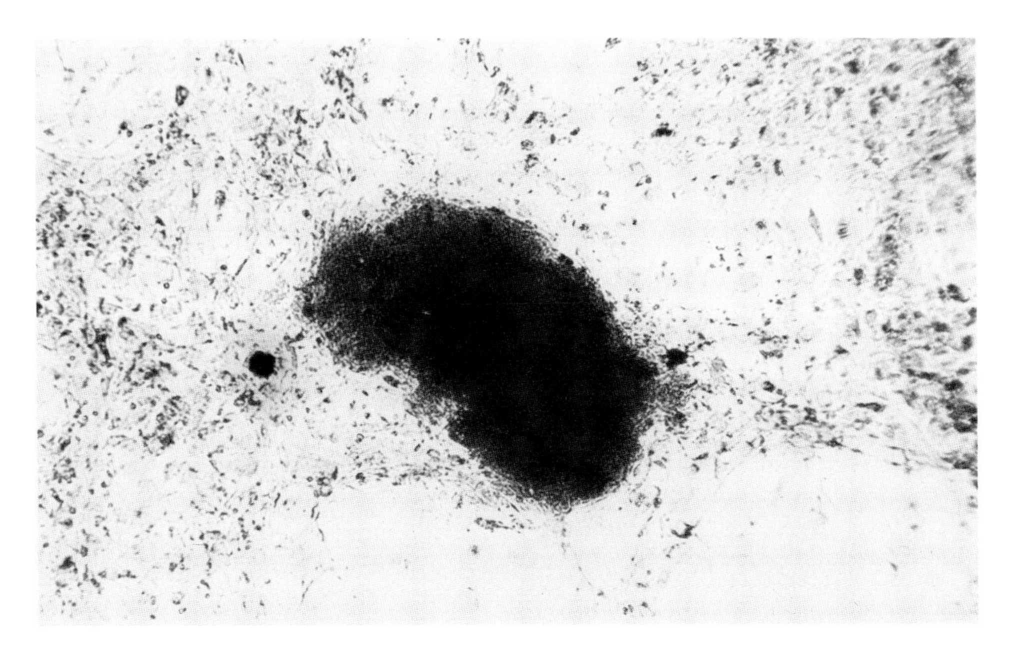

Fig. 1. Focus of transformed cells following exposure of hamster cells to PARA (defective SV40)-adenovirus type 7 (a DNA-containing tumor virus). ×35.

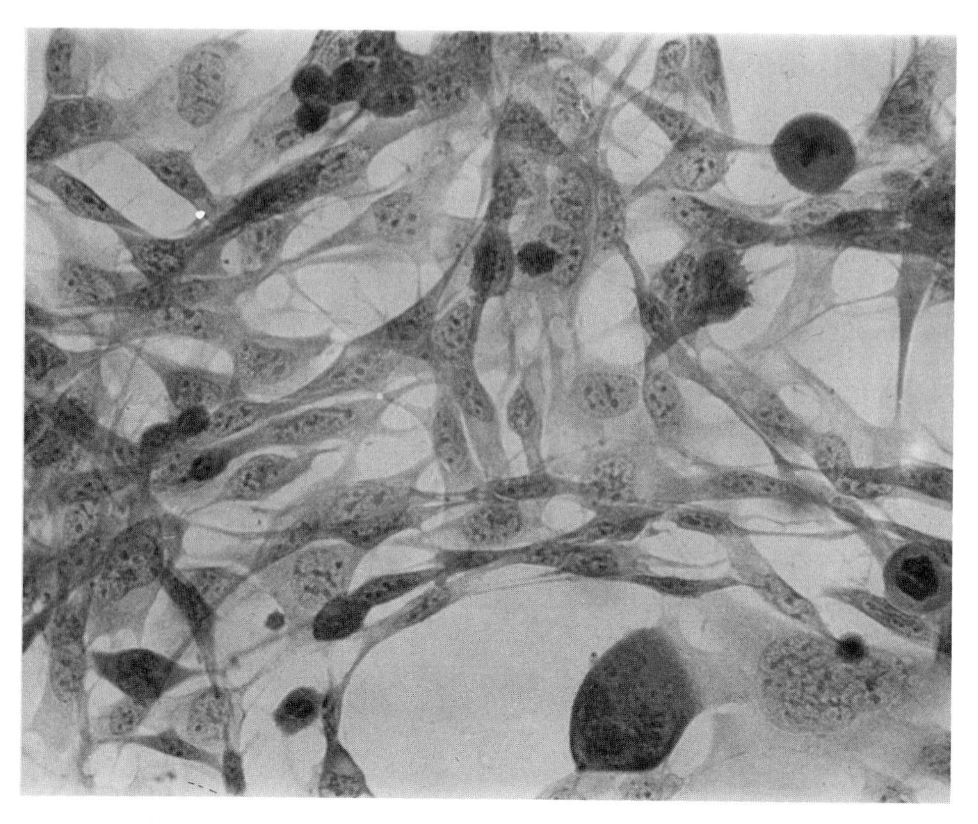

Fig. 2. Transformed cells showing loss of contact inhibition. Note that the cells are beginning to exhibit the criss-cross growth pattern observed in many cultures following transformation by tumor viruses. ×450.

The changes involve either the conversion of fibroblastlike to epitheliumlike cells or the converse. This is often accompanied by a palisading of the cells and involves a three-dimensional growth, a phenomenon not often observed in normal cell cultures. Because of this, the transformed cells will achieve a higher saturation density (number of cells to which the culture will grow in a given area) than will normal cells. Such changes, if the cells are dispersed, give rise to colonies of transformed cells which may resemble fungal colonies when viewed macroscopically (see Fig. 1). These colonies can easily be counted and a quantitative measure of the transforming ability of the virus in that system can thus be obtained.[4]

However, other markers for transformation have also been employed. These include changes in the surface characteristics of the cells (expressed by ability of plant lectins to agglutinate the cells), synthesis of new enzymes, synthesis of new antigens at the surface of the cells, synthesis of new antigens in the nucleus (Fig. 3), and synthesis of either virus-specific messenger RNA or virus-specific DNA (when DNA viruses are involved). The last criteria are especially

[4] J. W. Hartley and W. P. Rowe, *Proc. Nat. Acad. Sci. U. S.* **55**, 780 (1966).

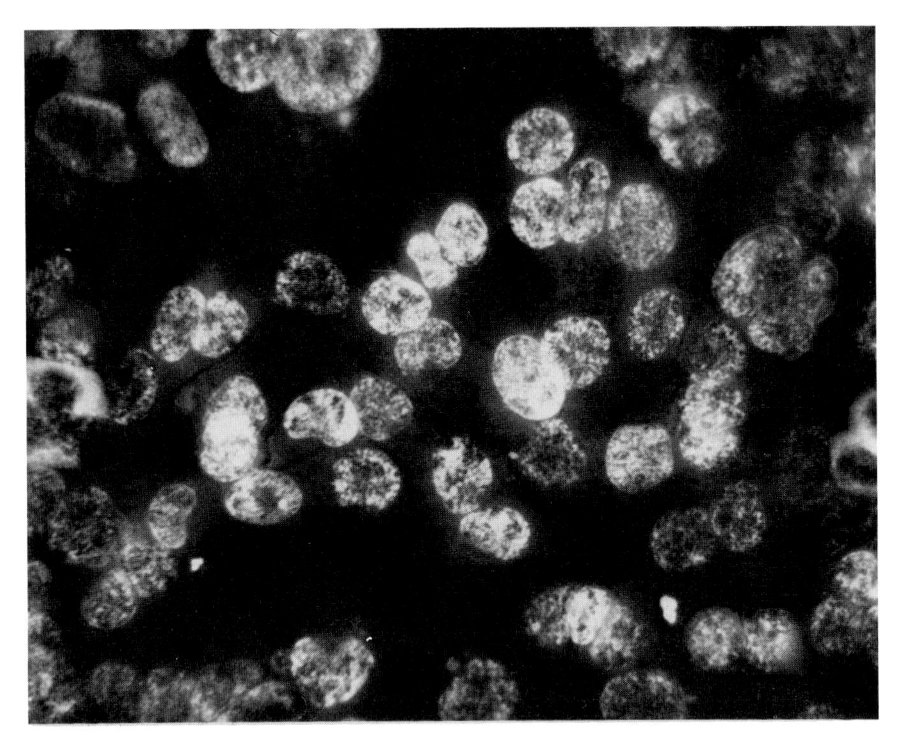

Fig. 3. Immunofluorescence micrograph of tumor (T) antigen (white areas) in nuclei of hamster cells transformed by Simian virus 40. Note that all cells in the culture contain this virus-specific antigen, characteristic of cultures transformed by papovaviruses and by adenoviruses.

useful for attempts to relate transforming events directly to the virus but require fairly sophisticated biochemical and biophysical equipment and techniques. The amounts of virus-specific nucleic acids present in the cells are often minimal and it is generally easier to detect the proteins directed by these nucleic acids.

The properties mentioned above can all be used to follow transformation of cells by viruses. However, to determine the malignancy of the cells it is necessary to transplant the cells to an appropriate animal host. It is for this reason that studies in avian and rodent species have been uniquely important. Generally, less than 100,000 normal cells will not transplant into the same species in the absence of treatments of the animal desired to prevent an immune response. Therefore, cultures of transformed cells are dissociated as above and the cells are then inoculated subcutaneously into an animal of the same species from which the cells were originally derived. Under special circumstances, it is also possible to inoculate such animals intramuscularly, intracerebrally, intraperitoneally, or into some other site. Often, increments of cells ranging from 10 to 100,000 in tenfold steps are inoculated in efforts to quantitate the number of cells required to produce tumors. Most often, young animals are used as they appear to be somewhat more sensitive for this purpose. The animals are inoculated with the cells and then are observed at weekly intervals for development of tumors at the site of implantation. Animals inoculated subcutaneously can be easily followed because the tumors can be palpated and their development followed. It is important to stress that the only criterion that determines whether the cells are malignant is ability to produce a tumor in the appropriate host.

Transformation implies that the cells have acquired new heritable properties which are passed to daughter cells during cell multiplication. Often, not all the cells in the culture will express any given property and it is therefore important that cells be cloned (plated out in numbers so that each cell can give rise to a colony); such colonies can then be monitored for the property under investigation. Most often, this property will be present in cultures derived from each cell and, if so, it is generally assumed that the new properties are being heritably passed during mitosis. This does not rule out the possibility that infectious agents are present in the culture, although in many transformed cells, viruses can neither be detected nor rescued using a variety of techniques. Transformation of animal cells by viruses is generally thought to be accompanied by integration of virus-specific nucleic acids with that of host cell chromosomes but this has not yet been demonstrated for most systems that have been developed. However, such integration would have to be stable and if so, would be transmitted during cell division so that every cell in the culture would have the potential to express the properties being examined (Fig. 3). However, not every cell may be able to express those properties at any given time. It is also important to note that many of the changed properties in the cell may be due to secondary consequences of the integration of virus-specific nucleic acids.

Caution Concerning Maintenance of Transformed Cells

Many of the cultures established following transformation of cells by viruses undergo "crisis" at sometime following the passage of the cells. This varies from cell to cell and from culture to culture but when it occurs, the cells will take on a granular appearance and appear to be deteriorating. During this time, the medium should be changed regularly (two times per week) and no attempt should be made to transfer the cells by dissociation with either trypsin or Versene. This period of crisis may last for weeks or even months but once the cells start to multiply again, they will do so continuously and the cell lines established will continue to multiply for indefinite periods of time as long as the cultures are supported with medium changes and overcrowding is prevented.

CHAPTER 9

Some in Vitro Models of Normal and Malignant Cell Interaction

G. Barski and J. Belehradek, Jr.

Time-Lapse Microcinematography Studies of Locomotory Behavior of Malignant and Normal Cells Associated in Mixed Cultures

Abercrombie and Heaysman[1] first formulated in a pertinent way the concept that "social" behavior of malignant and normal cells may be fundamentally different; this difference can be observed and analyzed *in vitro.* According to their findings,[1,2] the locomotion of a malignant cell growing *in vitro* on a glass substrate is not stopped by contact with sister malignant cells or with normal cells, whereas the locomotion of a normal cell is inhibited by contact with other cells.

Following these observations, the question arose as to whether the "non-inhibited" behavior of transformed, malignant cells is an intrinsic property of these cells, which will be expressed similarly in contact with any malignant or normal cell *in vitro.* Therefore, we designed experiments to study these phenomena with mouse and human cells, using two different techniques for association of malignant and normal cells *in vitro:* (a) simultaneous explantations of malignant and normal cells juxtaposed in plasma clot cultures; (b) dropping of malignant cells on preestablished monolayer cultures of normal cells.

Studies on Malignant and Normal Cell Outgrowths Juxtaposed in Mixed Plasma Clot Cultures. Mixed cultures associating malignant mouse cells permanent line N1 (deriving from the clone NCTC 2472)[3] and fragments of newborn mouse heart were prepared in chick plasma clot, on glass coverslips, in the following manner. Freshly excised hearts maintained in PBS solution, were aseptically cut into fragments of about 0.5 to 1 mm diameter and deposited in a drop of chicken plasma spread on a standard 12 × 32 mm coverslip. Cell clumps were scraped from the N1 monolayer with an incurved Pasteur Pipette and placed in the plasma on the same coverslip, at a distance of 3 to 5 mm from the heart fragments. One microdrop of chick embryo extract was then

[1] M. Abercrombie and J. E. M. Heaysman, *Exp. Cell Res.* **6**, 293 (1954).
[2] M. Abercrombie, J. E. M. Heaysman, and H. M. Karthauser, *Exp. Cell Res.* **13**, 276 (1957).
[3] K. K. Sanford, G. D. Likely, and W. R. Earle, *J. Nat. Cancer Inst.* **15**, 215 (1954).

added to the plasma and the excess of plasma and extract mixture was immediately withdrawn from the coverslip, so that, after clotting, only a thin film of plasma clot coated the glass and embedded the fragments and cell clumps. The coverslips bearing the cultures were placed in flattened culture tubes, and 2 ml of 199 Synthetic Medium supplemented with 15% heat-inactivated horse serum and antibiotics (penicillin 100 units/ml and streptomycin 100 μg/ml) were introduced in each tube. The tubes were then stoppered and maintained at 37°C.

After 4 to 8 days, cell outgrowths from the heart fragments and the N1 clumps came into contact with each other. The contact and interpenetration areas were observed and submitted to time-lapse microcinematography. For this purpose, the coverslips with the cultures were removed aseptically from the culture tubes, assembled in a specially adapted perfusion chamber,[4] and then placed on a thermostatic plate[5] maintained at 37°C. The microcinematographical device used was the Kurt Michel type, time-lapse equipment (Carl Zeiss, Oberkochen, West Germany), equipped with phase-contrast and an incandescent lamp light source, filtered through a green filter (Carl Zeiss VG 9) and an anticaloric filter (Carl Zeiss KG 1). Gevaert Sciencia 45 C 62 negative film was chosen for its fine grain and good contrast.

Detailed examination of the mixed cultures before filming allowed us to distinguish and localize: (a) areas of malignant cell outgrowth, appearing as a monolayer of refringent, mutually overlapping spindle-shaped cells, dense in the center, irregular and loose at the periphery; (b) areas of typical fibroblast outgrowth from the heart fragments, i.e., monolayers of large spindle-shaped, well spread and fairly transparent cells, with limited mutual overlapping; and (c) zones of presumed endothelial cell outgrowth from the heart explants, i.e., cells appearing smaller than the fibroblasts, rather refringent and composing compact pavementlike monolayers having well delimited borders.

Numerous fields were chosen for time-lapse recording, showing at different stages the contact areas between the malignant cell outgrowth and both types of normal cell outgrowth. Usually, records were made during 10 to 15 hours with a 10× objective and a time interval of 60 seconds between frames, and also during periods of 4 to 6 hours with the 40× objective, and 15-second intervals for recording of interactions at an individual cell level. The essential observations in this system were as follows.[6] Infiltration of normal cell sheets by malignant cells occurred easily in areas of loose arrangements of normal fibroblasts presenting gaps and interstices, where individual malignant cells (termed by us "sentinel cells") detached from the periphery of the malignant cell colony. Conversely, the progression of malignant cells was restricted or stopped, even for several days, when opposed by more compact sheets of normal cells forming pavementlike endothelial structures. Malignant cells, when arrested in their progression, shifted occasionally along the borderline of normal cells, accordingly changing their own orientation.

[4] G. Barski and R. Robineaux, *Ann. Inst. Pasteur (Paris)* **90**, 514 (1956).
[5] R. Robineaux and G. Barski, *Ann. Inst. Pasteur (Paris)* **91**, 106 (1956).
[6] G. Barski and J. Belehradek, Jr., *Exp. Cell Res.* **37**, 464 (1965).

Studies with Malignant Cells Dropped onto Preestablished Monolayers of Normal Cells. In order to study the cell locomotory phenomena at the very beginning of the mutual contact between the malignant and normal cells, we developed another technique of confrontation of these cells *in vitro*. Normal cells were harvested by trypsinization from primary cultures of human kidney cortex or, alternatively, from primary cultures of human amnion cells, and seeded on glass coverslips in flattened tubes in a medium composed of a 1:1 mixture of MEM and NCTC 109 Synthetic Media supplemented with 20% heat-inactivated calf serum and with the usual antibiotics. One to three days later, when monolayers of these cells reached an adequate density, a trypsinized suspension of the malignant human KB line[7] containing single cells and small clumps was added for attachment on and penetration into the preformed normal cell monolayer. After incubation at 37°C for 15 to 60 minutes, the cultures were assembled in perfusion chambers for observation and recording with the aid of time-lapse phase-contrast microcinematography as described above. The attached KB cells could be easily recognized since they appeared more refringent and less flattened than the normal kidney or amnion cells and also because of the distinct aspect of their nuclei. Pertinent observations could be made on the behavior of the added cancer cells.[8] During the first hour after "dropping," the KB cells attached and spread more easily and more rapidly on glass than on the sheets of normal cells, whether kidney or amnion. Later, about 24 hours after "dropping," the KB cells were filling up gaps and interstices in the sheets of kidney cells, overlapping them only occasionally. They infiltrated the kidney cell monolayer, especially in older mixed cultures, as a dispersed network of laces and ramifications. Yet, in the frame by frame analysis, it appeared that the human kidney cell monolayer was not a static, permanent structure. These cells were moving and their mutual junctions were continually formed and dissolved. Temporary gaps in the kidney cell sheets appeared frequently and the kidney cells' own movements were obviously contributing to the process of interpenetration with KB cell structures.

The situation was quite different in the mixed KB and human amnion cultures. When KB cells, monodispersed or in small clumps, were dropped onto amnion cells, they reached the glass surface rapidly through the few gaps present in the monolayer, forming well-delimited, rounded islands which were separated from the amnion cells by a narrow free area. These islands grew progressively in size as the number of KB cells increased following mitotic divisions. However, the KB colonies seldom overlapped the amnion cells and their expansion occurred by shifting and compression of the surrounding amnion cells. Meanwhile, no appreciable changes in cell number or relative cell position were observed in the amnion cell area, and the shrinkage of "living space" occupied by these cells occurred essentially by their contraction.

In conclusion, the system of continuous observation with the aid of time-lapse

[7] H. Eagle, *Proc. Soc. Exp. Biol. Med.* **89**, 362 (1955).

[8] G. Barski and J. Belehradek, Jr., *In* "The Proliferation and Spread of Neoplastic Cells" (Anderson Tumor Institute), p. 511. Williams & Wilkins Company, Baltimore, Maryland, 1968.

microcinematography revealed its great utility in the detailed analysis of the variable kinetic aspects of normal and malignant cell interaction.

Quantitative Model of Proliferation of Malignant and Normal Cells Associated in Mixed Cultures

Not only locomotory behavior *in vitro*, but also growth and multiplication can be reciprocally influenced by the mutual contacts between normal and malignant cells in mixed cultures. This aspect of cell interaction was reviewed thoroughly by Stoker.[9]

Both stimulatory and inhibitory factors may act, simultaneously or otherwise, and the final result of their interplay will depend on parameters such as relative cell number and density, local cell population topography, age of culture, and composition of medium.

To study the respective growth activities in mixed cultures of malignant and normal cells, we chose malignant cells of the human KB strain associated *in vitro* with normal cells from primary cultures of human kidney cortex. As already mentioned, these two cell types are easily identifiable by their morphology. The associated cultures, as well as control pure cultures, were established in flattened tubes on 32×12 mm glass coverslips. They were seeded with trypsinized monodispersed cell suspensions, containing 5×10^4 KB cells and 5×10^4 or 2×10^5 kidney cells per tube. Kidney cells (2×10^5 per tube) irradiated with 5000 r were used as an experimental variant. The medium was MEM + NCTC 109 in equal parts, supplemented with 20% decomplemented calf serum and with the usual antibiotics. The cultures were kept at 37°C and fed every 48 hours. After 8 days of growth, three to five cultures from each group were dissociated with 0.1 *M* citric acid and crystal violet solutions according to Sanford's technique,[10] and cell nuclei were counted in a hemocytometer. We performed counts on four to eight samples from each culture and calculated the mean value of the total number of cells per tube for each experimental group. Two or more parallel tubes from each group of mixed cultures were used for differential counts of KB and kidney cells. The cultures were fixed and stained by the May-Grünwald-Giemsa method and the respective proportions of the two types of cells evaluated at a 500× magnification and with the aid of camera lucida drawings. Twenty fields along two lines in the center of the preparations were examined; the number of cells counted in each preparation was of the order of 3 to 7×10^3. The results[8] of this experiment can be summarized as follows. The presence of living normal kidney cells impeded to some extent–by 20 to 25%–the growth of the KB cells in mixed cultures, as compared with the proliferation of pure control KB cultures. When the kidney cells were previously irradiated, they not only did not inhibit the proliferation of the KB cells, but, on the contrary, they enhanced it by approximately 40%. As far as the

[9] M. Stoker, *Curr. Topics Develop. Biol.* (1967).

[10] K. K. Sanford, W. R. Earle, V. J. Evans, H. K. Waltz, and J. E. Shannon, *J. Nat. Cancer Inst.* **11**, 773 (1951).

growth activity of the kidney cells was concerned, we found that the presence of the KB cells reduced it by 66 to 71%.

INFLUENCE OF MUTUAL CONTACT OF MALIGNANT AND NORMAL CELLS *in Vitro* ON THEIR DNA AND RNA SYNTHESIS

The findings concerning the interactions between malignant and normal cells in mixed cultures *in vitro* in terms of cell locomotion and proliferation led to the question dealing with the effect of mutual contact between malignant and normal cells on the DNA and RNA synthesis of the partners. We studied this problem with the aid of autoradiography after incorporation of tritiated thymidine and uridine in mixed cultures of KB malignant cells and normal human kidney cells. As emphasized above, the striking difference in their respective appearance made possible the easy recognition of sheets composed of the KB or the kidney cells, as well as the distinction between cells in direct contact with the "adverse" cells or with their own sister cells.

Mixed cultures were established on coverslips in flattened tubes with trypsinized cell suspensions containing 5×10^4 KB cells and 5×10^4 or 2×10^5 cells from primary cultures of normal human kidney cells, in 2 ml of medium comprising equal parts of the MEM and NCTC 109 Synthetic Media, supplemented with 20% heat-inactivated calf serum and antibiotics.

For the thymidine incorporation studies, cultures were incubated, after 5 or 9 days of growth, with tritiated thymidine ([^{3}H]methyl thymidine, specific activity 4.9 Ci/m*M*) added to the medium yielding a final concentration of 2 μCi/ml, during 25 hours. To study the uridine incorporation, tritiated uridine ([5-^{3}H]uridine, specific activity 5 Ci/m*M*) was added to the culture medium at the concentration of 1 μCi/ml, for a pulse of 30 minutes after 6 or 10 days of growth. At the end of the incubation period, the cultures were washed with warm medium containing 100 μg of unlabeled corresponding nucleoside in order to eliminate the nonincorporated labeled compound. The cultures were: (1) washed with warm PBS; (2) fixed with a 1:3 mixture of acetic acid and ethanol at 2°C for 20 minutes; (3) treated with 70% ethanol at 2°C for 15 minutes; (4) rinsed with cold distilled water; (5) treated with 1% $HClO_4$ water solution at 2°C for 30 minutes, in order to eliminate the labeled acid-soluble compounds; (6) washed thoroughly in running water at room temperature for 30 minutes; and (7) dried overnight in a 37°C incubator. The coverslips bearing the cultures (the cells on the upper side) were then mounted by means of Eukitt balsam on histological glass slides and, after the balsam had hardened, the preparations, handled in the dark room, were dipped into melted radiosensitive emulsion (Ilford K2, Ilford Ltd, Essex, G.B.) for coating. They were allowed to dry and kept in a refrigerator for exposure. After 10 to 18 days of exposure, the preparations were developed and stained with Giemsa stain.

For quantitative evaluation of the effect of mutual contact on the biosynthetic activity of each of the associated cell partners, we chose fields where colonies of malignant and normal cells were touching one another. Then, starting from the borderline, we established categories of topographical position of the cells. As a first category were considered rows of cells in direct contact with the opposite colony border. The second category was composed of rows of cells separated from the borderline by one row of similar cells, and, the third category, cells situated far from the opposite colony. The RNA synthesis was evaluated by counting the silver grains on the nuclear areas of at least 100 cells in each topographical category for each [^{3}H]uridine-treated preparation. The DNA synthesis was evaluated by counting the cells with a silver grain labeled nucleus among at least 1000 cells in each topographical category, on each [^{3}H]thymidine-treated preparation.

The findings[11,12] of these analyses showed that, concerning both the DNA and RNA syntheses in the KB cells, the cells of all topographical categories had similar activities. The behavior of the normal kidney cells was quite different; the RNA synthesis in the cells of the first category, i.e., cells in direct contact with KB cells, was markedly inhibited as compared with the second and third categories of normal cells. Similarly, concerning the DNA synthesis, the normal cells of the first category were significantly less frequently labeled than those of the second and third categories.

Moreover, to assess whether or not the local cell density was responsible for the differences in biosynthetic activities of different cell topographical categories, we cut out pictures representing 300 to 400 cells of each category from photographic prints on paper and weighed them. The relative density index, i.e., the ratio of the weight of the print cuts divided by the number of cells was similar for the first and second categories. Thus, the reduction of both DNA and RNA synthesis in normal kidney cells was clearly related to their direct contact with the malignant KB cells.

[11] F. Fakan and G. Barski, *C. R. Acad. Sci.* (*Paris*) **266**, 2519 (1968).
[12] G. Barski, F. Fakan, and J. Belehradek, Jr., *Bull. Cancer* (*Paris*) **55**, 39 (1968).

CHAPTER 10

Growth and Isolation of Nutritional Variants

M. K. Patterson, Jr. and E. Conway

Early studies on the nutritional requirements of cells in culture revealed that whereas certain cells have an apparent requirement for a nutrient, segments of the population show no requirement for the nutrient. The result of such observations is that nutritional variants have been isolated and have provided cell populations with inherent nutritional markers. This was the case in the requirement for L-asparagine by the Jensen sarcoma.[1] The procedures to be described have been used in attempts to elucidate the "developed" resistances of tumor cells to enzyme therapy using L-asparaginase. Populations, which do not require L-asparagine and are resistant to L-asparaginase, have been isolated from the Jensen sarcoma,[2] Walker 256 carcinoma,[3] L-5178Y mouse leukemia,[4] and the mouse 6C3HED lymphosarcoma,[5] which require the amino acid and are sensitive to the enzyme.

Cell Culture Procedures

Jensen sarcoma cells are cultured as described in this volume[6] using 400,000 cell inoculum in T-15 flasks. After the cells have been established 48 hours in McCoy's 5a Complete Medium, supplemented with 5% Sephadexed bovine serum,[7] the medium is changed to McCoy's 5a Medium prepared minus L-asparagine or the complete medium containing 0.1 IU L-asparaginase (E.C. 3.5.1.1) per milliliter medium. Subsequent medium changes are made at 24-hour intervals. When no pH change is observed with a 24-hour period (approximately 3 to 5 days on experimental medium), it is necessary to "acidify" the experimental medium by bubbling CO_2 into it until a color change (orange to yellow) of the indicator (phenol red) is noted. As the cell population decreases the cells remaining become more nonrefractile and, therefore, more difficult to observe microscopically. The "acidified" media changes are continued until cell proliferation is observed

[1] T. A. McCoy, M. D. Maxwell, E. Irvine, and A. C. Sartorelli, *Proc. Soc. Exp. Biol. Med.* **100**, 862 (1959).
[2] M. K. Patterson, Jr., M. D. Maxwell, and E. Conway, *Cancer Res.* **29**, 296 (1969).
[3] M. K. Patterson, Jr. and G. Orr, *Biochem. Biophys. Res. Commun.* **26**, 228 (1967).
[4] W. P. Summers and R. E. Handschumacher, *Biochem. Pharmacol.* **20**, 2213 (1971).
[5] J. D. Broome, *J. Exp. Med.* **118**, 121 (1963).
[6] M. K. Patterson, Jr., this volume, Section III.
[7] M. K. Patterson, Jr., this volume, Section XIV.

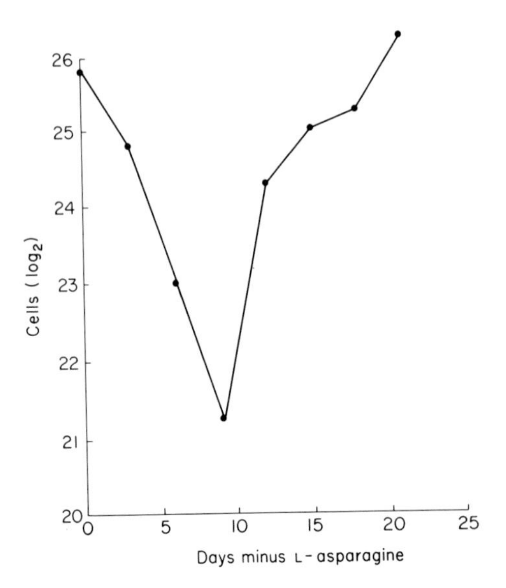

Fig. 1. Growth curve of Jensen sarcoma cultured in media minus L-asparagine.

(approximately 12 days on experimental medium). Cells can be harvested by placing the flask in a refrigerator for 1 hour followed by vigorous shaking of the flask.[6] A typical growth curve obtained for the Jensen sarcoma under the above conditions is shown in Fig. 1. With cells in suspension, e.g., L5178Y cells, 15-ml centrifuge tubes are used and medium changes made by centrifuging (200 *g*) the cells and resuspending them in fresh experimental medium.

Similar procedures have been used for individual colony isolation of Jensen sarcoma[2] and Walker 256 carcinoma,[8] where Petri dishes (50 mm) were used as culture flasks.[9] The medium change intervals were the same as above but the dishes were incubated in a 5% CO_2-95% air incubator.

Applications

These procedures have been used to determine that cells resistant to L-asparaginase are derived primarily as the result of cell selection[2] and that the selected cell population arises as the result of spontaneous mutations.[2,8] Moreover, the variant cells have high levels of asparagine synthetase (E.C. 6.3.1.1).[10] Recently five variants that do not appear to require arginine, leucine, lysine, phenylalanine, or tryptophan have been isolated[11] using similar procedures.

While these procedures are applicable to isolating variant cells that proliferate in deficient media, variant cells that fail to proliferate in such a media ("nutritionally deficient mutants") have been isolated by procedures introduced by

[8] J. Morrow, *J. Cell. Physiol.* **77**, 423 (1971).
[9] See also Section V this volume.
[10] M. K. Patterson, Jr., *Recent Results Cancer Res.* **33**, 22 (1970).
[11] M. K. Patterson, Jr., M. D. Maxwell, and E. Conway, *In Vitro* **7**, 152 (1971).

Puck and Kao.[12] 5-Bromodeoxyuridine (BUdR) was added to the medium and allowed to be incorporated by proliferating cells. Subsequent exposure of the population to fluorescent light results in cell death to those cells incorporating the BUdR, leaving only the nonproliferating population which is rescued by addition of the specified nutrient.

[12] T. T. Puck and F. Kao, *Proc. Nat. Acad. Sci. U. S.* **58**, 1227 (1967).

SECTION XIV

Quality Control Measures

Editors' Comments

Nothing is more important in the practice of cell and tissue culture than the subject matter of this section. In that sense, then, the best has been saved for last. These 30 protocols and discussions deal with cleanliness, containers, nutrients, preservations, contaminations, identifications (serological, karyological, and biochemical), and transport of and shipping instructions for cells in culture, all matters of vital interest and importance to high quality work *in vitro*.

CHAPTER 1

Glassware Preparation, Sterilizations, and Use of Laminar Flow Systems

Lewis L. Coriell

Glassware Preparation

Disposable polystyrene plastic and soda glass containers are in common use for cell cultures. They both have the advantage of low initial cost and can therefore be discarded after one use. This also reduces the number of personnel and facilities required for washing and reprocessing of tissue culture vessels, but it requires more storage space. Plastic flasks have the additional advantage that they are wrapped in cellophane and sterilized with ethylene oxide or ultraviolet light by the manufacturer and can therefore be used directly from the carton without further preparation or sterilization. Soda glass vessels should be washed before sterilization, but for some purposes it is satisfactory to rinse in distilled water before sterilization. The latter procedure has been extensively used, for example, in mass titration of neutralizing antibody in the metabolic inhibition test for poliovirus antibodies using a rapidly metabolizing cell such as primary monkey kidney or HeLa and a well buffered growth medium containing serum. Neither plastic nor soda glass is satisfactory for cultivation of cells in chemically defined media, because the media cannot neutralize minute traces of chemicals or toxic substances. Borosilicate glass is preferred for growth of many delicate cells, for research on nutrition of cells, and for cells grown in chemically defined media. It can be cleaned with strong solvents, is resistant to heat, breakage, scratching, and can be reprocessed repeatedly. Large containers can be handled without breakage.

Washing Borosilicate Glassware

Three washing procedures are in use in this institution: (1) machine washing with a strong caustic detergent, used for all glassware that will fit in the washer; (2) hand washing with a milder cleaning mixture for large pieces, funnels, cylinders, and odd pieces that do not fit into the washer; and (3) acid cleaning by the method of Earle[1] and Evans[2] when cells are to be grown in chemically defined medium.

Machine Washing. Contaminated glassware is immersed immediately after use in water containing ½ ounce of RX concentrate per gallon (Rochester Germicide Co., Rochester, New York). In the central kitchen, contaminated glassware is autoclaved, the contents emptied into a flush sink, and then loaded into a Better Built Turbomatic Washer Model 3000 (Better Built Machinery Corp., New York), where it is run through two cycles. The first cycle uses a cleaner and the second full cycle, without cleaner, removes traces of detergent and caustic. One pound of Simbal [caustic soda and mono(trichloro)-tetra-(monopotassium dichloro)penta-*s*-triazine trione] is added to the first cycle (Wyandotte Chemical Corp., J. B. Ford Division, Wyandotte, Michigan). The machine is programmed for a prewash rinse of 3 minutes, a wash period of 10 minutes, 4 rinses in tap water, and one rinse in demineralized water. This strong cleaner removes all organic matter and obviates the need for hand scrubbing to remove adherent cell cultures or protein coagulated by the autoclave. It has the disadvantage that it may sometimes leave a trace of caustic or detergent on some pieces of glassware. To prevent this the whole cycle is repeated without cleaner which provides a total of twelve rinses after the detergent and caustic wash period. The clean glassware is dried in a Better Built dryer.

Hand Washing. Combine three parts Calgolac and 1 part Microsolv and mix the dry powders thoroughly. (Microsolv detergent is obtained from Microbiological Associates, Inc., 4813 Bethesda Ave., Bethesda, Maryland; Calgolac is obtained from Calgon Corp., P.O. Box 1346, Pittsburgh, Pennsylvania.) Measure 2 ounces of the mixed cleaner for each 5 gallons of hot water, 180°C, and soak dirty glassware for 30 minutes. Scrub with a brush or sponge, rinse fourteen times in tap water, and four times in distilled water. These large items are dried on wooden racks.

Acid Cleaning. When cells are to be grown in chemically defined media, for cloning of diploid cells, and for quantitative reproduction of small numbers of isolated cells, it is recommended to use borosilicate glass and to clean it with acid as follows. Wearing gloves and a rubber apron, submerge the clean glassware at room temperature for 48 hours in a mixture of 80% concentrated H_2SO_4, 5% concentrated nitric acid, and 15% distilled water in a container of heavy porcelain, glass, polyethylene, or stainless steel. Remove individual pieces with tongs, wash twice in tap water, and submerge in Calgolac, 1 tablespoon per gallon of distilled water. Heat to 90°–95°C (not boiling) for 30 minutes, cool to 50°C, remove and rinse four times in tap water and eight times in distilled water, and dry in oven.[2]

[1] W. R. Earle, *J. Nat. Cancer Inst.* **4**, 131 (1943).
[2] V. Evans, personal communication, 1970.

Sterilization

Before sterilization glassware must be covered so it will remain sterile until used. Aluminum foil is the most frequent cover with or without a cotton plug. Cotton plugs tend to shed fibers and may leave a deposit on glassware. Vegetable parchment paper is preferred to brown wrapping paper. All packages are sealed with 3M autoclave indicator tape No. 1222 (Minnesota Mining Co., Medical Products Div., St. Paul, Minnesota), and liquids receive a temptube indicator, Propper No. SCB (Propper Manuf. Co., Inc., Long Island City, New York). All dry glassware is sterilized in a dry heat oven for 4 hours at 350°F. Liquids, rubber, and plastics are sterilized in the autoclave for 10 to 30 minutes at 15 lbs. pressure and 250°F. Each sterilizer has a recording thermometer, a written record is made of each run, and the date of sterilization is marked on each package.

Laminar Flow Systems

The most recent advance in prevention of airborne contamination while manipulating cultures is the use of filtered laminar air ventilation. The transfer area is ventilated by continuous pistonlike displacement of air that has been passed through a high efficiency particulate air (HEPA) filter. These filters are 99.97% efficient in removal of particles 0.3 μm in diameter and for practical purposes they sterilize the ambient air of the transfer room.[3-5] Aerosols generated by technical manipulations or from the skin and clothing of the operator do not fall into open culture vessels because they are flushed out of the area in seconds. The filtered air enters through the whole ceiling (or one wall) and exits at the floor (or opposite wall) and any airborne particles ride the airstream directly to the nearest exit. It is strongly recommended that all tissue culture manipulations be carried out within a vertical laminar flow enclosure to reduce airborne contamination of cell cultures. If the cell culture contains or may contain a potentially pathogenic virus, cell manipulations should be carried out in a biohazard hood such as the one developed by The Baker Company, Sanford, Maine.[6] This type laminar flow hood is designed to prevent airborne infection of cell cultures while the culture vessels are open during transfer, and in addition, to prevent airborne infection of the technician through inhalation of aerosols generated during manipulation of infected cell cultures. All laminar flow equipment must be checked for leaks, balanced, and certified after installation. A magnehelic gauge shows the air flow rate and indicates when the throwaway prefilters should be changed. The HEPA filters in our laboratory have lasted 3 to 5 years.

[3] J. B. Harstad, H. Decker, L. M. Buchanan, and M. E. Filler, *Amer. J. Public Health* **57**, 2186 (1963).

[4] L. Coriell, G. McGarrity, and J. Horneff, *Amer. J. Public Health* **57**, 1824 (1967).

[5] L. Coriell and G. McGarrity, *Appl. Microbiol.* **20**, 274 (1970).

[6] L. Coriell and G. McGarrity, *Appl. Microbiol.* **16**, 1895 (1968).

CHAPTER 2

Purification and Standardization of Water for Tissue Culture

R. W. Pumper

Fundamental to, and the main constituent of, all tissue culture media is the diluent, water. It therefore stands to reason that reasonably great changes in its ionic or material content can make consistency in all phases of tissue culture work difficult to achieve. There is no great panacea to which the cell culture worker can turn to produce high purity water. The end product will depend to a great extent on the feed water not only according to geographical location, but also from day to day in a given location.

Most hospital or research laboratories are equipped with a central water purification system. When this is the case, it is important that the end product user be familiar with the system and its delivery. The contamination of high purity water from piping during delivery is more often the rule than the exception. In most cases, it is advisable to retreat the water at the end point. If or when a new central water purifying facility is being constructed, it would be wise to investigate reverse osmosis, one of the newest commercial processes for removing water from solutions.

General Feed Water Impurities (Tap Water); Dissolved Ionic Compounds

Cations: Magnesium, calcium, sodium, potassium, ammonium, iron, manganese, etc.

Anions: Sulfate, phosphate, nitrate, chloride, hydroxide, bicarbonate, carbonate, and certain organic materials.

Suspended Matter: Organic material, silt, dirt, silica, pipe corrosion products, oils, metallic oxides.

Microorganisms: Bacteria, molds, algae, fungi.

Gases: Chlorine, oxygen, ammonia, hydrogen sulfide, carbon dioxide.

Preparation of High Purity Water

The most common method for purifying water in small laboratories is demineralization or distillation or a combination of both. It is reasonable to expect that since all mammalian cell culture media contains calcium, magnesium, potassium, sodium, and bicarbonate as well as other normal water contaminants, the strict removal of these substances is not necessary. In addition, the contamination of cell culture media by serum, which is normally added to mammalian cell media, is excessive. In the case of serum-free media, this problem is averted.

Demineralization. The process of demineralizing water by ion exchange involves the conversion of salts to their corresponding acids by hydrogen cation exchange and consequent removal of these acids by anion exchange. This involves the use of strong, weak, or intermediate base anion or cation exchangers. For small laboratories the usual deionizer unit consists of a mixture of an equivalent quantity of strong base anion exchanger and strong acid cation exchanger so that complete deionization is achieved in a single operation. Properly operated units of this type are highly efficient in removing ionized substances from water. Water having a resistance (ohms/cm at 25°C) of between 1 and 10 million ohms is possible with these units. It must be remembered that conductivity is only a measurement of ionized particle removal and gives no indication of the suspended nonionics and organic materials which may be present and are not retained by this method. In addition, the ion-exchange resins and breakdown products are themselves present in the effluent. A thorough review of ion-exchange demineralization is contained in a recent text by Applebaum.[1]

Distillation. Evaporation of water from a solution and recovering the desired constituent by condensation is an efficient method of purifying water. The properly operating still will remove all materials, microorganisms, pyrogens, and other contaminants which may be important in cell culture. The end product of distillation may only have a conductivity of 5×10^5 or 10^6 ohms/cm, but will in reality be a much purer water than that obtained from a mixed bed deionizer which has a resistance of 10×10^6 ohms. A word of caution regarding small laboratory glass stills is in order. Proper distillation and still design is a highly complicated matter. It is strongly advised that the investigator consult the technical personnel of those companies selling stills before purchasing one. For information concerning companies selling small stills, reference should be made to the "Guide to Scientific Instruments," published by the American Association for the Advancement of Science.

Practical Functional Systems

For the average tissue culture laboratory an efficient system for obtaining high purity water would be a mixed bed ion exchanger followed by a glass still. This, properly operated, will result in a high purity water which is free of pyrogens, gasses, organics, and has a conductivity of 1 to 2 million ohms. In a recent survey of a large number of tissue culture laboratories there was no case found where such a system, properly operating, produced water which

[1] S. B. Applebaum, "Demineralization by Ion Exchange." Academic Press, New York, 1968.

was inhibitory to cell growth. The problem areas to watch for are exceeding the flow rate or capacity of the exchanger, foaming, steam droplet carryover, and improper venting and cooling of the condenser in distillation.

Storage of High Purity Water

Plastic carboys are the most common water storage vessels in tissue culture laboratories. These have been the source of a wide range of problems. There are a wide variety of plastics in use but they all have in common the disadvantage of being porous. This makes them extremely difficult to clean both prior to and after use. In addition, some plastics elute contaminants, which are toxic to cells in culture, into stored water. The porosity in some cases also allows atmospheric gasses to permeate the plastic thereby lowering conductivity. If it is necessary to use plastic containers it is advised that they be "aged" in several changes of high purity water for a period of several weeks. An exception to these plastic problems is Teflon, but this material is quite expensive. The most practical storage vessel is a borosilicate glass carboy, tightly sealed. Since most tissue culture is done in glass, the small amount of silica dissolved by stored water from the carboy is not a problem. It must be kept in mind that atmospheric air overlying the stored water will contribute CO_2 and air pollutants to pure water. There are methods of obviating this problem by means of an inert gas overlay or filter vents but they are not practical for most small laboratories. One must also be aware that it is possible to have bacteria grow in stored high purity water.[2] For all of the above reasons it is not advisable to store large quantities of water at room temperature for long periods of time. A practical method for overcoming this problem in small laboratories is to store water in 1-liter containers and use the entire amount at one time. A source of ideal bottles of this type is the "water for irrigation" bottles from hospitals. These bottles are considered disposable and may be obtained free after hospital use.

Water Purity Determinations

Conductivity is the easiest and most frequently used method of determining water purity. An indicator light, or series of lights which go off at certain levels of resistivity, or a monitoring meter may be used. Conductivity is, except for the evaluation of ionized particle removal, a poor method of determining water purity. Particulate matter (undissolved substances) may be present in relatively large amounts even though resistance is high; however, as a method of continuous monitoring of deionizer function it is highly recommended. It must be remembered that most references to resistivity are based on a water temperature of 25°C, and that the resistance decreases as the temperature increases.

[2] M. S. Favero, L. A. Carson, W. W. Bond, and N. J. Peterson, *Science* **173**, 836 (1971); see also P. J. Thomas, *Science* **175**, 8 (1972).

It is totally impractical for small laboratories to consider pyrogen testing of water. A properly operating still can be counted on to produce pyrogen-free water and when pyrogens become a problem it is usually through water storage. In the case of other determinations of organics or gasses, again routine testing is usually not done. If water is suspect as a contribution of toxicity, or if ion exchanger–still function is to be tested, any of several standard methods texts available in most libraries may be consulted.

CHAPTER 3

A Surveillance Procedure Applied to Sera

Charles W. Boone

A 1-year quality control study was conducted between 1968 and 1969 on commercial fetal bovine serum (FBS) produced for tissue culture purposes in the United States, and also on some special FBS which was produced under conditions of maximal sterility, freedom from whole cells, and rapid processing in the cold.[1]

A special facility was set up to produce FBS according to a set of specifications designed to achieve high quality. One hundred eleven lots of special FBS, 20 liters or more each, were prepared by this facility. In addition, a total of 125-lot samples of FBS were obtained from eight different commercial suppliers by purchasing 100-ml bottles through conventional channels. The volume of the lots from which each of these samples came was 20 liters or more. The lot samples of "special" and "commercial" sera were sent to a Coding Laboratory, which processed them into coded "unknowns" that were completely unrecognizable as to original source. These were then shipped to a Testing Laboratory which performed the panel of quality control tests described below.

Production of FBS with Improved Cell Growth-Supporting Capacity Compared to Commercial FBS

Production Method. The procedure used to produce the special FBS was as follows: (1) blood from the fresh fetus was drained by cardiac puncture into

[1] C. W. Boone, N. Mantel, T. D. Caruso, Jr., E. Kazam, and R. E. Stevenson, *In Vitro* **7**, 174 (1972).

a 500-ml blood donor bottle; (2) the clotted blood was centrifuged in the donor bottle at 0°C; (3) the serum was aspirated into a 500-ml Fenwall plastic blood donor bag and centrifuged at 5000 *g* to sediment residual red and white cells; (4) the cell-free serum was transferred to sterile bottles and frozen after a sample was removed for sterility testing; (5) after the sterility check, the unit bleedings were thawed, pooled to form lots of 20 liters, and coarse-filter clarified to remove fibrin particles; (6) a sample was taken for final sterility testing, and the lot was bottled.

The main features of the special method were as follows. (1) Fetuses were processed within 30 minutes after the mother was killed. (2) No sterile filtration step was performed to prove that perfectly sterile techniques had been used. The success rate in producing 20-liter lots was 85%. (3) Removal of residual red and white cells before freezing was done to prevent contamination of the serum with the products of hemolysis and particularly leukocytolysis. The prevention of leukocytolysis was considered important because proteases and other hydrolases released into the serum from leukocyte lysosomes could act on serum components either to lower their nutritional value or to produce toxic products such as free fatty acids. In view of the high rate of contamination of FBS with mycoplasma,[2] bovine herpes viruses, and myxoviruses (see below), this probably also removes many of these cell-associated microorganisms. (4) A maximal time of 5 hours was permitted from arrival of the fetus to freezing of the cell-free serum. (5) All steps were carried out at 0°C whenever possible.

Method of Testing for Cell Growth-Supporting Capacity. A large pool of third passage human fetal lung cells, explanted and grown in Eagle's Minimal Essential Medium (MEM) with 10% FBS, was stored frozen in vials containing 5×10^6 cells, 4 ml of medium, and 7.5% dimethyl sulfoxide (DMSO). These cells were used for all assays. The sample of serum to be tested was added to Eagle's MEM to make a concentration of 10% v/v. This was called "test medium." Freshly thawed human fetal lung cells were diluted in 20 ml of test medium, centrifuged, and the supernatant discarded. The cells were resuspended in 20 ml of test medium, counted with a Coulter electronic cell counter, and centrifuged again. They were again suspended in test medium, planted in glass prescription bottles in half-confluency, and allowed to grow to confluency using two changes of test medium. The cells were then trypsin-detached, counted, washed once by centrifugation in 20 ml of test medium, and planted in Falcon 60-mm Petri dishes in triplicate at two different inoculation densities: 10,000 and 20,000 cells per cm^2. At 30 and 72 hours after planting, the cells were trypsin-detached and counted. A standard serum sample of known cell growth-supporting capacity was tested in parallel with each test serum.

Comparison of the Cell Growth-Supporting Capacity of the Specially Produced FBS and Commercial FBS. Table I compares the cell growth-supporting

[2] M. F. Barile and J. Kern, personal communication (1971), National Institutes of Health, Bethesda, Maryland.

TABLE I
Cell Growth-Supporting Capacity of Specially Produced FBS Compared to Commercial FBS

Group	No. lot samples tested (>20 liters per lot)	Initial cell plant (cells/cm²)	Mean cell growth 30 hours (cells/cm²)	Mean cell growth 72 hours (cells/cm²)
Specially produced FBS	81	10,000	14,400	30,700
Commercially produced FBS	117	10,000	12,000	24,800
Specially produced FBS	81	20,000	22,900	52,200
Commercially produced FBS	117	20,000	18,500	42,900

capacity of the specially produced and commercial sera. The specially produced sera supported cell growth better at the $p < 0.01$ level of significance.

Chemical Tests of the Specially Produced and Commercial Sera (Table II)

Protein Panel. The total protein and gamma globulin content of the special and commercial sera did not differ significantly. However, the variability of the commercial sera was greater.

Cell Lysis Panel. The hemoglobin and lactic dehydrogenase content of special sera was significantly lower than the commercial sera. This was expected because residual RBS and WBC had been removed from the special sera before freezing.

TABLE II
Chemical Tests of Specially Produced FBS Compared to Commercial FBS

Chemical test[a]	Specially produced Mean	Specially produced S.D.	Commercially produced Mean	Commercially produced S.D.
Total protein, g/100 ml	3.91	±0.26	3.81	±0.040
Gamma globulin, mg/100 ml	49.0	±32.0	60.0	±68.0
Hemoglobin, mg/100 ml	17.3	±6.5	30.6	±15.0
Lactic dehydrogenase, units[b]	1067	±257	1478	±537
Total lipids, mg/100 ml	294	±27	308	±53
Neutral fats, mg/100 ml	190	±23	204	±42
Phospholipids, mgP/100 ml	34.0	±3.6	34.8	±8.4
Cholesterol, mg/100 ml	45.0	±5.0	47.0	±11.0
Free fatty acids,[c] mEq/liter	0.176	±.06	0.242	±0.10

[a] The total protein, hemoglobin, lactic dehydrogenase, and gamma globulin determinations were done on more than 90 lots at greater than 20 liters per lot. The remaining assays were done on more than 33 lots at greater than 20 liters per lot.
[b] P. G. Caband and F. Wroblewski, *Amer. J. Clin. Pathol.* **30**, 234 (1958).
[c] The differences shown are significant at the level of $p < .001$, Student t test.

Lipid Panel. Total lipids, neutral fats, phospholipids, and cholesterol did not differ significantly between the special and commercial sera. Free fatty acid content was higher, however, in the commercial sera at the $p < 0.001$ level. Free fatty acid content was the only chemical test which correlated with poor cell growth-supporting capacity, as discussed below.

Correlations between Chemical Tests and Cell Growth-Supporting Capacity

From both special group and commercial group sera there were selected subgroups of 10-lot samples having the highest (subgroup a) and the lowest (subgroup b) values for the mean of the cell counts 72 hours after planting 10,000 and 20,000 cells per cm^2. The differences in growth-supporting capacity between subgroups a and b was significant at the $p < 0.01$ level in all cases. Groups a and b were then compared in regard to the chemical measurements discussed above. No significant differences were observed between the two subgroups for all of the chemical tests with one exception: free fatty acid content. This test correlated with poor cell growth-supporting capacity at the $p < 0.05$ level.

Contamination with Viruses, Mycobacteria, Bacteria, and Fungi

Viruses. Lots from only one commercial supplier were tested. About 10% of over 100 lots, 50 liters per lot, were contaminated with bovine viruses.[3] The most common viruses found were infectious bovine rhinotracheitis virus, bovine diarrhea virus, and parainfluenza type 3 virus.

Bacteria and Fungi. Approximately 10% of 125 lots of commercial sera from eight different suppliers were contaminated with either bacteria and/or fungi. The test media used were thioglycolate, brain heart infusion, trypticase soy agar, blood agar, and Sabourand agar. Two-thirds of the positive tests were picked up on blood agar plates alone.

Mycoplasma. No mycoplasma organisms were found in 111 lot samples of special group sera, by the culture method of Hayflick.[4] This finding is consistent with the statement of Barile and Kern[2] that all published attempts to isolate mycoplasma species from FBS have been reported as unsuccessful. However, recently Barile and Kern[2] found that, when they took 25-ml samples (instead of the conventional 0.5-ml samples taken in our study), bovine mycoplasma species were indeed present in 17 of 139 lots of FBS from commercial sources.

[3] C. W. Molander, A. J. Kniazeff, C. W. Boone, A. Paley, and D. T. Imagawa, *In Vitro* **7**, 168 (1972).
[4] L. Hayflick, *Tex. Rep. Biol. Med.* **23**, 285 (1969).

"Minimal" and "Stringent" Specification for Chemical Tests of FBS

The panel of tests for chemical composition is summarized in Table III. Two sets of values are shown, those which apply to 95% of the commercial group samples, and those which apply to 95% of the special group samples. These two sets could well be the basis for "minimal" or for "stringent" specifications required by the purchaser of FBS.

Summary of Results of the Quality Control Survey of Commercial and Specially Produced FBS

1. FBS with improved cell growth-supporting capacity compared to commercial sera was produced by using fresh fetuses and aseptic collection techniques, by removing residual red and white cells before freezing, and by processing rapidly in the cold.
2. Free fatty acid concentration was the only chemical test that correlated (inversely) with cell growth-supporting capacity.
3. The free fatty acid concentration of commercial FBS was significantly ($p < 0.001$) higher than that of the specially produced sera. This finding may be offered as one possible reason for the relatively better cell growth-supporting capacity of the specially produced FBS.
4. Approximately 10% of commercial samples were contaminated with bacteria, fungi, or bovine viruses. No contamination with mycoplasma was found. Barile and Kern[2] reported significant contamination of FBS samples with mycoplasma when 25-ml aliquots were taken for testing instead of the conventional 0.5-ml aliquots used in this study.

TABLE III
Ninety-Five Percent Ranges of Chemical Tests on Fetal Bovine Serum[a]

Chemical test	Composition of 95% of lots	No. of lots tested	Mean
Commercial sera			
Hemoglobin, mg/100 ml	Less than 68	117	31.12
LDH, units	Less than 2600	69	1,542
Total protein %	Greater than 3.10	116	3.86
γ-Globulin, mg/100 ml	Less than 280	117	50.6
Free fatty acids, mEq/liter	Less than 0.440	46	0.229
Specially produced sera			
Hemoglobin, mg/100 ml	Less than 28	108	17.31
LDH, units	Less than 1650	96	1,067
Specific gravity	Greater than 1.019	103	1.020
Total protein, %	Greater than 3.51	108	3.91
γ-Globulin, mg/100 ml	Less than 100	90	49.0
Free fatty acids, mEq/liter	Less than 0.22	33	0.176

[a] From C. W. Boone *et al.*[1] Reproduced with permission of the Tissue Culture Association, Inc.

5. Two-thirds of the positive tests for bacteria contamination were picked up on blood agar plates alone.

6. "Minimal" and "stringent" specifications for chemical testing of FBS were established.

CHAPTER 4

Preparation of Sera for Nutritional Studies

M. K. Patterson, Jr. and M. D. Maxwell

Most unadapted cells grown in culture require serum as a component of the medium to maintain their proliferative capacity. While whole serum is permissible for routine purposes, studies involving nutritional parameters require that the constituent under study be removed. The most commonly used procedure for removal of these constituents is dialysis of the whole serum. More recently passage of the serum over Sephadex columns has been employed. The procedures to be described have been used in this laboratory primarily for amino acid nutritive studies but, as the analysis of the treated serum indicates, should be applicable to other nutritional studies.

Sephadex Treatment

Fifty grams of dry Sephadex G-50 (coarse)[1] is allowed to swell overnight at room temperature in 1 liter of saline (0.85% NaCl) prepared in glass-distilled water. The column (2.5 × 100 cm)[1] is packed by pouring all of the Sephadex slurry onto the column at one time. A funnel is attached to the top of the column to accommodate the excess slurry. The packed column is washed with 5 liters of saline with the operating hydrostatic pressure adjusted so that an effluent flow rate of approximately 4–5 ml per minute is obtained. Up to 200 ml of whole serum can be embedded on such a column, which is then eluted with saline. The elution profile of the serum proteins and free amino acids is shown in Fig. 1. The protein elution pattern is established either by the Lowry procedure[2] or

[1] Pharmacia Fine Chemicals, Inc., 800 Centennial Avenue, Piscataway, New Jersey 08854.

[2] O. H. Lowry, N. J. Rosebrough, A. L. Farr, and R. J. Randall, *J. Biol. Chem.* **193**, 265 (1951).

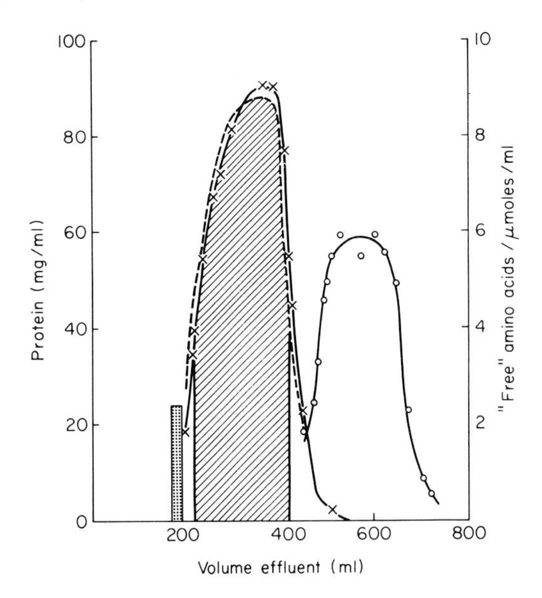

Fig. 1. Protein and amino acid elution pattern of whole serum embedded on Sephadex G-50. Solid bar indicates elution of dextran blue 2000. Protein determined by Lowry method, ×—×; protein determined spectrophotometrically, - - - -; amino acids determined by ninhydrin, ○—○. Hatched area represents fraction used for culture medium supplement.

spectrophotometrically[3]; the amino acid pattern is established using a ninhydrin procedure.[4] Once the pattern is established the early and late protein fractions are discarded and the middle fractions pooled. After some experience, the decision of the pool limits can be made by visual evaluation of the serum color intensity. A protein assay of this pooled serum fraction relative to the whole serum protein gives a dilution factor. The treated serum is then added to medium on the basis of this factor.

At the end of each "run" the column is washed with 5 liters of saline. Periodically the Sephadex is sterilized by rinsing thoroughly with distilled water and the slurry autoclaved for 15 minutes at 121°C and 18 pounds/square inch. The Sephadex is then ready for repacking the column as described above.

Dialysis Treatment

Serum is dialyzed against 100 volumes of saline or Earle's solution at 4°C with a change of solution every 24 hours for 48 hours. Sufficient cellulose casing[5] to contain the volume of serum is soaked in several changes of glass-distilled water to remove impurities and to make the tubing more pliable. A double knot

[3] E. Layne, *In* "Methods in Enzymology" (S. P. Colowick and N. Kaplan, eds.), Vol. 3, p. 447. Academic Press, New York, 1957.

[4] S. Moore and W. H. Stein, *J. Biol. Chem.* **211**, 907 (1954).

[5] Visking Company, Division of Union Carbide Corporation, 6733 West 65 Street, Chicago, Illinois 60638.

is tied in one end, the serum poured in the tubing, and the end sealed with another double knot. For a recent review on dialysis techniques, see McPhie.[6]

Comparative Studies

Serum Components. Table I shows the levels of free amino acids in serum before and after treatment. Levels were determined, after deproteinization with picric acid, with a Technicon Corporation automatic amino acid analyzer. The levels of free amino acids in whole serum suggest that they contribute significantly to the exogenous supply of amino acids, even at a 5% serum level. Essentially complete removal by either treatment was achieved.

Analysis of other constituents of the sera[7] showed three classes of constituents (Table II). Protein levels, other than those resulting from treatment dilution, were

TABLE I
Amino Acid Analysis of Freshly Prepared and Stored Serum Before and After Treatment

	Serum treatment[a]					
	Whole		Sephadexed[b]		Dialyzed[b]	
Free amino acid	(I)[c]	(II)[d]	(I)[c]	(II)[d]	(I)[c]	(II)[d]
Alanine	0.672	0.592	0	Tr	Tr	Tr
Arginine	0.246	0.330	Tr	0.001	0	0
Aspartic acid	0.146	0.123	Tr	0.002	Tr	Tr
Glycine	0.408	0.505	Tr	Tr	Tr	Tr
Glutamic acid	0.309	0.568	0	0.002	Tr	0
Histidine	0.216	0.105	Tr	0.002	Tr	0
Isoleucine	0.121	0.121	Tr	0	0	0
Leucine	0.347	0.298	0	Tr	0	0
Lysine	0.129	0.141	0	Tr	Tr	0
Methionine	0.015	0.012	0	0	0.005	0
Phenylalanine	0.126	0.061	0	Tr	0.001	0
Proline	0.387	0.229	Tr	0	0	0
Serine	0.295	0.397	Tr	Tr	Tr	Tr
Threonine	0.261	0.200	Tr	Tr	Tr	0
Tyrosine	0.498	0.592	0	Tr	0	0
Valine	0.336	0.401	0	Tr	0	0

[a] Values (average of two separate analyses) are expressed as μmoles/ml of sera. Trace (Tr) amounts indicate that values are less than 0.001 μmoles/ml.

[b] Amino acid analysis was made on 20 ml treated serum, deproteinized, lyophilized, and the entire sample applied to column.

[c] Values before storage.

[d] Values after storage at 4°C for 2 weeks.

[6] P. McPhie, *In* "Methods in Enzymology" (S. P. Colowick and N. Kaplan, eds.), Vol. 22, p. 23. Academic Press, New York, 1971.

[7] Analysis was performed on a Hycel Mark X Clinical Analyzer by Mr. Mel Schene, Physician's Laboratory, Ardmore, Oklahoma 73401.

TABLE II
Serum Constituents Before and After Sephadex Treatment and Dialysis

Constituent	Serum treatment[a,b]		
	Whole	Sephadexed	Dialyzed
Total protein (g%)	7.3	6.2 (15.0)	6.5 (10.9)
Globulin (g%)	3.8	3.6 (5.2)	3.8 (0)
Alkaline phosphatase (units)	15.9	13.2 (16.9)	14.0 (11.9)
SGOT (units)	122	101 (17.2)	104 (14.8)
Cholesterol (mg%)	135	113 (16.3)	122 (9.6)
Uric acid (mg%)	1.4	0.8 (42.8)	0.7 (50.0)
Total bilirubin (mg%)	0.7	0.3 (57.1)	0.3 (57.1)
PBI (μg%)	>15	4.2 (>75)	8.1 (>47)
Urea nitrogen (mg%)	15.1	1.0 (93.4)	1.1 (92.7)
Glucose (mg%)	132	6 (95.4)	7 (94.6)
Potassium (mEq/liter)	5.8	0 (100)	0 (100)
Calcium (mg%)	7.2	2.8 (61.1)	0.2 (97.2)

[a] Values represent average of three analyses.
[b] Figures in parenthesis represent percentage of whole serum values removed by treatment.

maintained. Uric acid, total bilirubin, and plasma-bound iodine, those constituents bound to protein, were removed by approximately 50% by treatment. Essentially complete removal of low molecular weight compounds, e.g., urea nitrogen and glucose, and the cations was accomplished by treatment of the sera.

Since it is advantageous to prepare large quantities of sera at one time, its storage properties are of importance. Piez *et al.*[8] reported that Sephadex human serum underwent less proteolysis during storage for 11 days at 5°C or incuba-

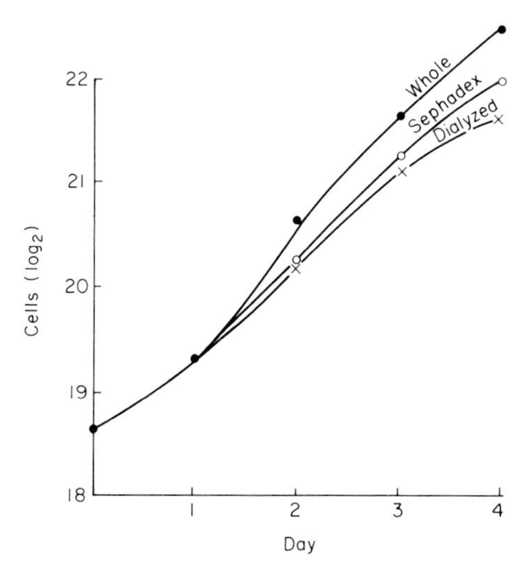

Fig. 2. Growth curve of established Jensen sarcoma cells in McCoy's 5a Medium supplemented with 5% whole (—●—), Sephadexed (—○—), or dialyzed (—×—) bovine serum.

[8] K. A. Piez, V. I. Oyama, L. Levintow, and H. Eagle, *Nature* (*London*) **188,** 59 (1960).

tion for 5 or 28 days at 37°C than did dialyzed serum. Only limited proteolysis has been observed, however, in whole fetal calf serum stored at 37°C up to 21 days.[9] In the current studies on bovine serum, no significant increases were noted in the treated sera after storage at 4°C for 14 days (Table I).

Cell Growth. Figure 2 shows the comparative growth curves of the established Jensen sarcoma cells using McCoy's 5a Medium supplemented with 5% whole or treated sera. The protein concentration of the treated sera was adjusted to the equivalent of that found in whole serum. While the differences in the final cell counts were reproducible, they did not appear to result from removal of a growth factor since with increased serum concentrations, e.g., 10, 15, or 20%, the results were the same.

[9] W. L. Ryan and C. Cardin, *Proc. Soc. Exp. Biol. Med.* **123**, 27 (1966).

CHAPTER 5

Sample Preparations of Media

A. Vertebrate Cells

Helen J. Morton and Joseph F. Morgan

Introduction

Before preparing medium for the first time the worker should read other parts of Section XIV thoroughly, particularly those on water, osmolarity, glassware, sterilization, and pH. It is essential to develop a system of dishwashing suitable specifically for tissue culture, and to decide whether glass or plastic containers will be used to store the finished medium. Since some ingredients such as iron[1] and insulin[2] may be preferentially adsorbed on glass or plastic, they cannot be used interchangeably. Moreover, plastic containers should be chosen with care since some are toxic.[3] None of these systems should be changed without thorough comparative tests since even small changes may affect the final quality of the medium. The time and temperature of storage

[1] M. D. Smith and T. Kernaghan, *J. Cell Pathol.* **22**, 690 (1969).
[2] C. Waymouth and D. E. Reed, *Tex. Rep. Biol. Med.* **23**, (Suppl. 1), 413 (1965).
[3] R. L. DeHaan, *Nature* (*London*) **231**, 85 (1971).

of stock solutions and finished media should also be standardized and adhered to carefully.

Many minor variations can be tolerated by cells growing in a mixture of synthetic medium and serum or other biological or proteinaceous supplements. However, when cells are grown entirely in synthetic media very minor variants can be critical. An extreme example occurred in the authors' laboratory. Primary chick embryo heart fragments failed to adhere to glass test tubes obtained from the usual supplier. On investigation it was discovered that the supplier had introduced new automated equipment, and the lower CO_2 content of the air used (compared to a human glass blower) seemed to have altered the glass surface. For a detailed discussion of glass surfaces consult Rappaport *et al.*[4,4a]

This section cannot cover variations in formulas. Unfortunately some media exist in variants under the same name. Careful comparison of formulas and reference to the appropriate supplier is recommended. Attempts are now being made to standardize all formulas.[5]

Quality Controls

It cannot be overemphasized that all procedures, sources of reagents, etc., should be standardized, empirically if need be, and adhered to rigorously. The highest quality of ingredients should be used initially but if a better quality becomes available at a later date it must be tested before converting to its use, since trace contaminants are sometimes beneficial. Water, being the largest single component of all media, is the most important. Salts with detailed analysis of trace contaminants and A.C.S. standard amino acids should be used if possible. L-and DL-amino acids should not be interchanged, even with appropriate weight changes, since at least one D-amino acid can be utilized by some cells.[6] To maintain high quality in chemicals it is also advisable to date reagent bottles on their receipt in the laboratory and label them for media use only. The practice of rinsing the spatula in double-distilled water between each weighing is also recommended. Cross-contamination of reagent bottles may become quite significant, and it is better to antagonize one's co-workers by refusing to lend "media only" reagents than to acquire a trace of arsenic in the medium.

All organic solvents used for media should be redistilled immediately before use. Standard laboratory alcohol is unsuitable since absolute alcohol usually contains benzene and 95% alcohol contains a variety of impurities. Emulsifiers, such as the Tween series, may break down to free fatty acid after prolonged storage. These fatty acids are not always harmful but all such sources of variation should be tested and controlled. In some cases Methocel can be used as an alternative carrier for lipid-soluble compounds such as steroids.[7]

[4] C. Rappaport, J. P. Poole, and H. P. Rappaport, *Exp. Cell Res.* **20**, 465 (1960).
[4a] C. Rappaport and C. B. Bishop, *Exp. Cell Res.* **20**, 580 (1960).
[5] H. J. Morton, *In Vitro* **6**, 89 (1970).
[6] J. F. Morgan and H. J. Morton, *J. Biol. Chem.* **221**, 529 (1956).
[7] J. W. Byron, *Nature (London)* **228**, 1204 (1970).

The particular salt forms and water of crystallization should also be chosen with care. For example, $MgSO_4 \cdot 7\ H_2O$ is strongly efflorescent and can lose sufficient water to produce a magnesium toxicity when the normal weight is used to prepare balanced salt solution. Similarly, $MgCl_2 \cdot 6\ H_2O$ is to be preferred over other forms of magnesium chloride which are deliquescent and may thus cause a deficiency of magnesium. Such salts can, of course, be dispensed into small, tightly sealed, containers which are discarded after one or two exposures to the atmosphere. These problems are particularly acute in very humid or very dry areas.

Stock Solutions and Storage

Routine medium preparation is facilitated by the preparation of concentrated stock solutions. Components of similar solubility can often be grouped together for convenience. All stock solutions should be stored in concentrated solution and at the lowest convenient temperature to minimize decomposition; many can be dispensed in small aliquots and stored frozen for up to 1 year. However, once thawed, they should be used within a few days and never refrozen. Exceptions to this rule are compounds of low solubility such as tyrosine, cystine, and any mixture containing high concentrations of both calcium and phosphate. Regardless of storage temperature, no solution containing good bacterial growth factors (such as amino acids or carbohydrates) should ever be stored nonsterile. Stability should also be considered when grouping ingredients in stock solutions. For example, ascorbic acid is best stored in solution with cysteine, glutathione, or other reducing agents.

The amounts and types of acid and alkali used to dissolve ingredients, and the time and temperature of any heat required, should also be standardized. It is preferable to use NaOH and HCl wherever possible since NaCl is the largest component of balanced salt solutions and these agents, therefore, produce the smallest percentage change in the respective ions. All acids, alkalis, salt forms, and trace metals should be considered carefully, but particular attention should be given to NH_4^+,[8] K^+,[9,10] Ca^{2+},[11] and Fe^{2+} and Zn^{2+}.[12]

Storage of complete media is necessary for sterility testing and for practical use. There are no quantitative rules on temperature and duration, therefore it is best to select an arbitrary period, as short as possible, and adhere to it. Medium M150[13] (a minor variant of Medium 199[14]), has proved entirely satisfactory for at least 2 weeks at room temperature in the authors's laboratories. Room tempera-

[8] E. M. Jensen and O. C. Liu, *Proc. Soc. Exp. Biol. Med.* **107**, 834 (1961).

[9] C. Waymouth, *In* "Growth, Nutrition and Metabolism of Cells in Culture" (V. Cristofalo and G. Rothblat, eds.), p. 18. Academic Press, New York, 1972.

[10] R. L. DeHaan, *Develop. Biol.* **16**, 217 (1967).

[11] H. J. Morton, *Proc. Soc. Exp. Biol. Med.* **128**, 112 (1968).

[12] K. Higuchi, *J. Cell. Physiol.* **75**, 65 (1970).

[13] J. F. Morgan, M. E. Campbell, and H. J. Morton, *J. Nat. Cancer Inst.* **16**, 557 (1955).

[14] J. F. Morgan, H. J. Morton, and R. C. Parker, *Proc. Soc. Exp. Biol. Med.* **73**, 1 (1950).

ture was selected in this case because, under these conditions, the medium was, itself, better than any conventional sterility test.

Any ingredient whose stability is uncertain should, of course, not be prepared as a stock solution but be added to the rest of the medium immediately before sterilization. It has been accepted convention to treat glutamine as unstable but it is not certain that this is necessary.[15]

Physical Properties

Phenol red is incorporated in most media, at 15 to 20 mg/liter, as a pH indicator. Higher concentrations may be toxic. Its use is recommended whenever possible to provide early warning of many practical problems, such as, errors in media preparation, contamination, and faulty glassware. Similarly, it is advisable to check the osmolarity of the final medium since wide variations can occur and may be critical.[16] The redox potential should also be considered.[17,18]

Miscellaneous

Some filters contain soluble reagents (e.g., Triton X100 in Millipore filters or salts in Seitz filters). Some autoclaves—especially those operating on city tap water, or those which do not drain completely after each run—deposit undesirable material on filters. These problems can be minimized by discarding the first 5–10% of the volume of medium to be filtered. This technique is mandatory with asbestos filter pads.[19] Paradoxically, more rapid filtration is often obtained if a minimum amount of suction is used. However, regardless of the rate of filtration all media containing bicarbonate buffers will change their pH during filtration and should be readjusted with sterile CO_2 gas.

The sterilization of tissue culture media by autoclaving is now under intensive investigation.[20,21] The beginner is advised to avoid autoclaving complete media until the techniques are fully worked out. If individual stock solutions are autoclaved, they must never contain both carbohydrates and amino acids since these interact to produce brown pigments (carbonyl amino or Maillard reaction[22]).

Antibiotics, if used, should be added only to the complete medium and never incorporated in stock or stored solutions. Pure, bulk, antibiotics, rather than hospital preparations, should be used to avoid preservatives and fillers. Only

[15] G. Stoner and D. J. Merchant, *In Vitro* **7**, 330 (1972).
[16] C. Waymouth, *In Vitro* **6**, 109 (1970); cf. Chapter 6, this section.
[17] G. M. Healy, D. C. Fisher, and R. C. Parker, *Can. J. Biochem. Physiol.* **32**, 327 (1954).
[18] W. F. Daniels, L. H. Garcia, and J. F. Rosensteel, *Biotechnol. Bioeng.* **12**, 409 (1970).
[19] W. House, *Nature (London)* **201**, 1242 (1964).
[20] S. Nagle and B. L. Brown, *J. Cell. Physiol.* **77**, 259 (1971).
[21] I. Yamane, Y. Matsuya, and K. Jimbo, *Proc. Soc. Exp. Biol. Med.* **127**, 335 (1968).
[22] N. A. M. Eskin, H. M. Henderson, and R. J. Townsend, "Biochemistry of Foods," pp. 84–92. Academic Press, New York, 1971.

special formula white rubber stoppers, or silicone stoppers, should be used to seal media bottles or tubes, since the common black laboratory stoppers are highly toxic.[23] Bottles or tubes for holding media should be sterilized, unplugged, in closed, stainless steel containers since cotton plugs release toxic, volatile lipids when heated. Plastic and paper containers used in hospital autoclaves also produce toxic agents. These are not removed by normal tissue culture dishwashing and their effects are seen only after repeated use.

Examples

The preparation of Medium M150[13] provides some practical examples of these general rules.

Solution 1. L-Arginine·HCl, 140 mg; L-histidine·HCl, 40 mg; L-lysine·HCl, 140 mg; DL-tryptophan, 40 mg; DL-phenylalanine, 100 mg; DL-methionine, 60 mg; DL-serine, 100 mg; DL-threonine, 120 mg; DL-leucine, 240 mg; DL-isoleucine, 80 mg; DL-valine, 100 mg; DL-glutamic acid monohydrate, 300 mg; DL-aspartic acid, 120 mg; DL-α-alanine, 100 mg; L-proline, 80 mg; L-hydroxyproline, 20 mg; glycine, 100 mg; sodium acetate trihydrate, 166 mg; phenol red (water soluble), 40 mg. These ingredients are dissolved by thorough mixing in 1000 ml of double-strength Hanks' balanced salt solution[24] containing neither glucose nor sodium bicarbonate.

Solution 2. L-Tyrosine, 200 mg; L-cystine, 100 mg. These materials are dissolved in 100 ml of 0.075 *N* HCl at 98°–100°C.

Solution 3. B Vitamin Group. Niacin, 25 mg; niacinamide, 25 mg; pyridoxine·HCl, 25 mg; pyridoxal·HCl, 25 mg; thiamine·HCl, 10 mg; riboflavin, 10 mg; calcium pantothenate, 10 mg; *i*-inositol, 50 mg; *p*-aminobenzoic acid, 50 mg; choline chloride, 500 mg; ion-exchange water to 200 ml.

Solution 4. Fat Soluble Group. Prepare unsterile solutions in absolute or 95% ethyl alcohol of cholesterol (10 mg/ml) and menadione [vitamin K (10 mg/ml)]. Weigh out 10 mg of calciferol (vitamin D_2) and dissolve in 2 ml of the alcoholic cholesterol solution. Then add 0.1 ml of alcoholic vitamin K solution. Use this combined alcoholic solution to dissolve 10 mg of vitamin A (as the alcohol or acetate). Then add 10 ml of 5% aqueous Tween 80 solution and dilute the turbid mixture to 100 ml with ion-exchange water. This solution remains opalescent but becomes clear when incorporated in the final medium.

Solution 5. Vitamin C Mixture. L-Cysteine·HCl, 100 mg; glutathione (GSH), 50 mg; ascorbic acid, 50 mg; ion-exchange water to 100 ml.

[23] R. C. Parker, J. F. Morgan, and H. J. Morton, *Proc. Soc. Exp. Biol. Med.* **76**, 444 (1951).
[24] J. H. Hanks and R. E. Wallace, *Proc. Soc. Exp. Biol. Med.* **71**, 196 (1949).

Solution 6. Disodium α-tocopherol phosphoric acid, 10 mg; ion-exchange water to 100 ml.

Solution 7. Folic acid, 10 mg; ion-exchange water to 100 ml with the inclusion of just sufficient $N/100$ NaOH to effect solution (approx. 0.2 ml).

Solution 8. D-Biotin, 10 mg; 1.0 ml of 1.0 N HCl; ion-exchange water to 100 ml.

Solution 9. Adenine sulfate, 100 mg; dissolve with heat in 100 ml of ion-exchange water containing 0.5–0.7 ml of concentrated NH_4OH.

Solution 10. Purine and Pyrimidine Mixture. Guanine·HCl, 10 mg; monosodium xanthine, 11.4 mg; hypoxanthine, 10 mg; thymine, 10 mg; uracil, 10 mg; dissolve, at about 70°C, in 200 ml of ion-exchange water containing 0.5 ml of concentrated NH_4OH.

Solution 11. Adenosine triphosphate, disodium, 27 mg; ion-exchange water to 50 ml. Prepare fresh each week.

Solution 12. Iron Solution. Dissolve 36.0 mg $Fe(NO_3)_3 \cdot 9\ H_2O$ in 100 ml of ion-exchange water containing one drop of concentrated HNO_3.

Solution 13. Muscle adenylic acid, 10 mg; ion-exchange water to 100 ml.

Solution 14. D-Ribose, 100 mg; D-2-deoxyribose, 100 mg; ion-exchange water to 100 ml.

TABLE I

Method of Preparation of Synthetic Medium M150[a]

Solution no.	Composition	Dilution of stock solution	Milliliters per 1000 ml of final medium
1	Amino acids in 2× Hanks' BSS	Undil.	500
2	Tyrosine and cystine	Undil.	20
3	Vitamin B group	1/50	10
4	Fat-soluble group	1/10	10
5	Vitamin C mixture	1/100	10
6	Vitamin E	1/100	10
7	Folic acid	1/100	10
8	Biotin	1/100	10
9	Adenine sulfate	Undil.	10
10	Purines and pyrimidines	Undil.	6
11	Adenosine triphosphate	Undil.	2
12	Iron solution	Undil.	2
13	Adenylic acid	Undil.	2
14	Ribose and deoxyribose	1/4	2

[a] To the solutions above add double-glass distilled ion-exchange water to 1000 ml; sodium bicarbonate, 1400 mg; glucose, 1000 mg; glutamine, 100 mg.

Solutions 1 and 14 are stored sterile at 4°C. Solutions 2, 9, and 10 are stored sterile at room temperature. Other solutions are stored nonsterile at 4°C. These stock solutions are diluted and combined in the proportions shown in Table I. Solid ingredients are added and the complete medium sterilized by filtration through a 0.2-μm filter.

Note added in proof: Investigators should be made aware that membrane filters that do not contain surface active agents are available.

CHAPTER 5

Sample Preparations of Media

B. Invertebrate Cells[1]

James L. Vaughn

The media for invertebrate cell culture are based partly on the composition of the hemolymph of the animal from which tissues are taken for culture and partly on information gained from the experience with other animal cells in culture. The hemolymphs of invertebrates differ from the bloods of vertebrates mainly in the composition of inorganic ions, amino acids, and organic acids. The first two of these probably are the most important when choosing a medium for invertebrate cell culture.

It is possible to group the invertebrates on the basis of the inorganic ion content of their bloods and these groupings can be a guide in selecting media for cell culture. For example, the inorganic ion concentrations in the bloods of marine invertebrates are similar to those found in sea water. The bloods of freshwater invertebrates have a ratio of one ion to another similar to that of sea water, but the concentration of each ion is lower. Among the insects, the largest group of terrestrial invertebrates, the more specialized Lepidoptera, and Hymenoptera, have a $Na^+:K^+$ ion ratio of less than 1. More primitive insects have $Na^+:K^+$ ion ratios similar to other terrestrial animals, that is, greater than 1.

Perhaps the greatest difference between the bloods of vertebrates and those of invertebrates is in the levels of free amino acids found in the hemolymphs of insects. The total free amino acids in most insect hemolymph varies from 300 to 2100 mg/ml and can account for up to 40% of the osmotic pressure of some insect hemolymphs. Therefore, most insect culture media contain high levels of free amino acids. Aspartic acid, glutamine, glutamic acid, leucine, isoleucine, cystine,

[1] Mention of a proprietary name in this publication does not constitute an endorsement by the U. S. Department of Agriculture.

and tyrosine have been shown to be important for cells from Lepidoptera and Orthoptera, but the minimal and optimal levels of these amino acids have not yet been determined.

Insects are one of the few animals whose blood contains free organic acids. Citrate, α-ketoglutarate, malate, fumarate, succinate, oxaloacetate, and pyruvate are present in amounts ranging from a few tenths of a millimole to 25–35 mmole. In the insect they are important chelating agents and, as such, have an important role in the cation balance of the hemolymph. Many media for the culture of cells from invertebrates contain one or more of these organic acids. However, their role in cell culture is unknown. In fact, there is some evidence that they are of little or no importance.

In choosing the media listed in this section, an attempt was made to illustrate not only how these differences in blood composition are reflected in media composition, but also to illustrate the wide range of formulations which are satisfactory for one or more groups of invertebrates. Some of these media were formulated from the known composition of a particular hemolymph; others were prepared by combining components from media used for the growth of other cells from both invertebrates and vertebrates. It is hoped that these examples will provide some guidelines in choosing a suitable medium for a variety of invertebrate cells.

Media Preparation

Invertebrate cell culture media are normally prepared by dissolving similar ingredients together and then combining these solutions to prepare the final medium. The customary solutions are the inorganic salts, the organic acids, the amino acids, the sugars, and the vitamins. Among the amino acids, cystine and tyrosine are dissolved separately in a small amount of 0.1 *N* HCl and then added in turn to the amino acid solution. The organic acids are neutralized with 5% KOH or other base after they have been dissolved. It is possible to prepare 8X stock solutions of the sugars, the amino acids, and the organic acids and store them frozen for several months to simplify the preparation of individual lots of media.

After all the solutions are prepared, they are combined, usually in the following order: inorganic salts, sugars, amino acids, and vitamins. The volume then is adjusted to 90% of the final volume and the $CaCl_2$ is added slowly with continuous stirring. This step is critical in the preparation of the Grace media because the level of calcium is very near the limit of solubility for the sulfate and phosphate salts and care must be taken to avoid irreversible precipitation of these calcium salts.

After the calcium has been added, the pH is adjusted with KOH. The pH of the blood of most marine invertebrates is between 7.2 and 7.8. Insect hemolymphs have a broad range in pH, both interspecifically and intraspecifically. For insects in the family Coleoptera, the range is 5.9–7.3, for Diptera from 6.3–7.7, for Lepidoptera from 6.2–7.6, and for Orthoptera from 6.0–7.6. The pH of the hemolymph of a single species may vary by as much as 0.7 of a pH unit during different stages of development. Generally, the pH of tissue culture media tend to be in the lower part of a range, but little is known about the effects of pH on the growth of cells from invertebrate animals.

Following adjustment of the pH, the medium is brought up to final volume. At this point the osmotic pressure of the medium should be checked and adjusted if necessary. Most insect hemolymphs have a freezing point depression in the range of −0.5 to −0.9. The hemolymphs of most marine invertebrates are nearly isotonic with sea water (freezing point depression −1.8). These figures can serve as a guide in making adjustments on culture media but it is worthwhile to test the effect of a range of osmotic pressures on each cell culture to determine the optimum level for a particular cell type. Instruments are available in which the osmotic pressure can be determined quickly and with only 2 or 3 ml of medium. Adjustments can be made by the addition of KCl, sugar, or distilled water.

Supplementation

For the successful culture of invertebrate cells the minimal media must be supplemented with sera of some type. For cultures of insect cells, hemolymph is often used. The hemolymph need not be from the same stage, larvae, pupae, or adult, as was the tissue. It also need not even be from the same species. However, all untreated hemolymph, when exposed to air, forms toxic quinones as a result of reactions involving the enyzme polyphenol oxidase. The most common treatment to prevent this is to heat the hemolymph to 60°C for 5 minutes and then to freeze it. After 24 hours the hemolymph is thawed and the precipitated protein, including the polyphenol oxidase, is removed by centrifugation. The heat-treated hemolymph can be sterilized by filtration and stored frozen for long periods. It is added to the minimal medium at levels ranging from 2 to 20%.

Because insect hemolymph is difficult to obtain, many vertebrate sera, which can be easily purchased, have been tested as substitutes. The most suitable such serum is fetal calf serum which should be heated to 60°C for 30 minutes before using. It is used alone or in combination with bovine serum albumin as a substitute for insect hemolymph. Concentrations ranging from 5 to 20% have been used successfully for many types of invertebrate cells. If bovine serum is also used, concentrations of 1% are most common.

Sample Media

Grace's Antheraea Medium.[2] This medium was developed for the culture of cells from Lepidoptera and was used by Grace in developing the first invertebrate cell line. It is now widely used for many Lepidoptera cell lines. Frequently used supplements are: 5% insect hemolymph; 10% fetal calf serum, 10% whole egg

[2] T. D. C. Grace, *Nature* (*London*) **195,** 788 (1962).

ultrafiltrate, 1% bovine serum albumin or 7.5% fetal bovine serum, 7.5% egg ultrafiltrate, 0.47 mg/100 ml bovine serum albumin, 0.28 mg/100 ml yeastolate, and 0.28 mg/100 ml lactalbumin hydrolysate. The formulation is tabulated below:

Component[a]	Amount (mg/100 ml)	Component[a]	Amount (mg/100 ml)
A. Salts		C. Sugars	
$NaH_2PO_4 \cdot 2H_2O$	114.0	Sucrose	2668.0
$NaHCO_3$	35.0	Fructose	40.0
KCl	224.0	Glucose	70.0
$CaCl_2$	100.0	D. Organic acids	
$MgCl_2 \cdot 6H_2O$	228.0	Malic acid	67.0
$MgSO_4 \cdot 7H_2O$	278.0	α-Ketoglutaric acid	37.0
B. Amino acids		Succinic acid	6.0
L-Arginine·HCl	70.0	Fumaric acid	5.5
L-Aspartic acid	35.0	E. Vitamins	
L-Asparagine	35.0	Thiamine·HCl	0.002
β-Alanine	20.0	Riboflavin	0.002
L-Alanine	22.5	Calcium pantothenate	0.002
L-Cystine·HCl	2.5	Pyridoxine·HCl	0.002
L-Glutamic acid	60.0	*p*-Aminobenzoic acid	0.002
L-Glutamine	60.0	Folic acid	0.002
L-Glycine	65.0	Niacin	0.002
L-Histidine	250.0	*Iso*-Inositol	0.002
L-Isoleucine	5.0	Biotin	0.001
L-Leucine	7.5	Choline chloride	0.020
L-Lysine·HCl	62.5	F. Antibiotics	
L-Methionine	5.0	Penicillin G (sodium salt)	3.0
L-Proline	35.0	Streptomycin sulfate	10.0
L-Phenylalanine	15.0		
DL-Serine	110.0		
L-Tyrosine	5.0		
L-Tryptophan	10.0		
L-Threonine	17.5		
L-Valine	10.0		

[a] A, dissolved in 20 ml H_2O; B, dissolved in 30 ml H_2O; C, dissolved in 20 ml H_2O; D, dissolved in 20 ml H_2O. The calcium chloride is dissolved separately in a small amount of H_2O and added after A, B, C, and D are thoroughly mixed; the vitamins then are added. The pH is adjusted to 6.5 with potassium hydroxide, and the volume made to 100 ml and antibiotics added if desired.

Mitsuhashi and Maramorosch Medium.[3] The use of lactalbumin hydrolysate and yeastolate in place of the purified amino acids and the vitamins used in the Grace medium simplifies the preparation of this medium. The medium was developed by the authors for use in culturing cells from several leafhoppers (Homoptera). However, it was also used by Singh[4] to develop cell lines from two species of *Aedes* mosquitoes and has since been widely used for the culture of several mosquito cells. It is normally supplemented with 20% fetal bovine serum. The formulation is as follows:

[3] J. Mitsuhashi and K. Maramorosch, *Contrib. Boyce Thompson Inst.* **22**, 435 (1964).
[4] K. R. P. Singh, *Curr. Sci.* **36**, 506 (1967).

Component[a]	Amount (mg/100 ml)	Component[a]	Amount (mg/100 ml)
A. $NaH_2PO_4 \cdot H_2O$	20.0	Glucose	400.0
$MgCl_2 \cdot 6\ H_2O$	10.0	Yeastolate	500.0
KCl	20.0	Lactalbumin hydrolyzate	650.0
$CaCl_2 \cdot 2\ H_2O$	20.0	Penicillin	5000 units
NaCl	700.0	Streptomycin	5.0
B. $NaHCO_3$	12.0		

[a] Stock solutions A and B of inorganic salts are stored separately in the refrigerator and the medium is prepared fresh shortly before use. The pH was adjusted to 6.5 by Mitsuhashi and Maramorosch for leaf-hopper cells and to 7.0 by Singh for mosquito cells.

Echalier and Ohanessian Medium for Drosophila Cells.[5] This medium is one example of the design of cell culture medium based on the composition of the hemolymph of the insect from which the cells were taken. The carbohydrate level, the osmotic pressure, the pH, and the ionic concentrations (both cation and anion) are very near those values determined for the hemolymph of the insect. Levels of other medium components were based on previous insect cell culture media. Stock solutions of the medium are prepared as follows:

Solution 2. Glutamic acid, 7.35 g; glycine 3.74 g. Dissolve in a small amount of H_2O and neutralize with 10 *N* KOH. Bring volume to 100 ml with distilled H_2O.

Solution 2. Glutamic acid, 7.35 g; glycine 3.74 g. Dissolve in a small amount of H_2O and neutralize with 10 *N* NaOH. Bring volume to 100 ml with distilled H_2O.

Following the preparation, 54 ml of solution 1 and 94 ml of solution 2 are combined with solutions containing the following components.

Component[a]	Amount (g/liter)	Component[a]	Amount (g/liter)
Salts		Vitamins	
$MgCl_2 \cdot 6\ H_2O$	1.0	Yeastolate	1.5
$MgSO_4 \cdot 7\ H_2O$	3.7	Vitamins (as in Grace's Antheraea medium)	
$NaH_2PO_4 \cdot 2\ H_2O$	0.47		
$CaCl_2$ (dissolve separately)	0.89	Sugar	
Organic acids		Glucose	2.0
Malic acid	0.670	Lactalbumin hydrolysate	15.0
Succinic acid	0.060		
Sodium acetate·3 H_2O	0.025		

[a] Bring final volume to 1000 ml with distilled H_2O. Adjust pH to 6.7 with KOH. Medium should have a freezing point depression of $-0.66°C$. Supplement with 10 to 20% fetal calf serum.

Schneider's Medium for Culture of Mosquito Cells.[6] This medium is a modified Grace medium used for the culture of cells from the mosquito, *Anopheles*

[5] G. Echalier and A. Ohanessian, *In Vitro* **6**, 162 (1970).

[6] I. Schneider, *J. Cell Biol.* **42**, 603 (1969).

Stephensi. The sugar and inorganic salt concentrations have been changed. Sodium phosphate, fructose, vitamins, and organic acids were omitted. However, a broad spectrum of possible growth factors was added in the abbreviated formulation of NCTC-135 used (lacks amino acids, inorganic salts, and sugars). Since this formulation is available commercially, the preparation of the medium is simplified. The final medium contains the following components.

Component[a]	Amount (mg/100 ml)
Salts	
$MgCl_2 \cdot 6\ H_2O$	114.0
$MgSO_4 \cdot 7\ H_2O$	40.0
KCl	110.0
NaCl	300.0
$CaCl_2$	40.0
Sugars	
Glucose	100.0
Sucrose	1600.0
Trehalose	50.0
Cholesterol[b]	0.2

[a] This medium was supplemented with 1% 10× abbreviated NCTC-135 Medium and with 15% heat-treated fetal bovine serum.

[b] Cholesterol was dissolved in Tween 80 and 95% ethanol.

Flandre and Vago Medium for the Culture of Cells of Snails.[7] The authors have used this medium to establish primary cultures of heart, foot muscle, and mantle tissue from *Helix* species in chicken plasma clots. The formulation is given below:

Component[a]	Amount (mg/100 ml)
Salts	
NaCl	720
KCl	40
$CaCl_2$	30
NaH_2PO_4	20
$MgSO_4$	20
$NaHCO_3$	55
Sugar	
Glucose	50
Amino acids	
Lactalbumin hydrolysate	100

[a] The pH of the medium is 7.6 and it is supplemented with 10% snail serum and 40% chick embryo extract. The plasma clot was prepared by mixing 20 parts of supplemented medium with 1 part of chicken plasma.

[7] O. Flandre and C. Vago, *Ann. Epiphyt.* **14,** 161 (1963).

CHAPTER 5

Sample Preparations of Media
C. Plant Cultures

T. Murashige

A great diversity exists in the composition of plant culture media. This diversity is to a large extent a necessary and desirable feature of working with the multitude of species, organs, and research objectives. Indeed, little can be gained through experiments with cell and organ cultures if all plants and their parts possessed common nutritional requirements. As a routine first step in research, plant culturists should systematically develop or refine formulations to best suit their specific needs. It is unrealistic to expect widespread success by simply applying an existing formula. Existing media should serve only as reference. In developing or refining the formula, it will be helpful to consider the inorganic salt portion separately from the organic addenda. Fortunately, the salt requirement has been very consistent among a range of plants and applications. In most instances it has been sufficient to simply compare the effectiveness of a few presently available major formulations. The formula in Table I has demonstrated superiority over others in a variety of situations and therefore has been adopted as standard in this laboratory.

In contrast to the inorganic salts, extensive variability has been the case of the organic constituents. This is as expected, inasmuch as many of them appear to serve a growth-regulating role. The only ingredients common to many media are sucrose and thiamine·HCl; even then, only the sucrose concentration has remained constant (2–3%), and the thiamine·HCl content has varied between 0.1 and 30 mg/liter. Additional vitamins, especially *myo*-inositol, nicotinic acid, and pyridoxine, have been used in many instances. Auxin and/or cytokinin may be critical for some applications; in such instances, tests should include kind of auxin or cytokinin, as well as concentration of each. Certain amino acids or their amides and nucleic acid derivatives have also been beneficial. Some investigators find it necessary to use diverse extracts of undefined compositions, e.g., endosperm fluids, extracts of yeast or malt, protein hydrolysates, and fruit juices.

In the development of nutrient formulations, possible antagonisms and synergisms should be recognized and chemicals should be tested accordingly. Indeed, interactions should be expected between auxin and cytokinin and among certain amino acids.

In working with diverse plant species it may be helpful to compare the relative effectiveness between agar-solidified and liquid media. Experience in this laboratory has shown that the medium form can be a critical factor in the culture

TABLE I
Inorganic Salt Formulation for Plant Cultures

Salt	mg/liter medium
NH_4NO_3	1650
KNO_3	1900
$MgSO_4 \cdot 7\ H_2O$	370
$CaCl_2 \cdot 2\ H_2O$	440
KH_2PO_4	170
$FeSO_4 \cdot 7\ H_2O$	27.84
Na_2EDTA[a]	37.24
$MnSO_4 \cdot H_2O$	16.9
$ZnSO_4 \cdot 7\ H_2O$	8.6
H_3BO_3	6.2
KI	0.83
$CoCl_2 \cdot 6\ H_2O$	0.025
$CuSO_4 \cdot 5\ H_2O$	0.025
$Na_2MoO_4 \cdot 2\ H_2O$	0.25

[a] Disodium (ethylenedinitrilo)tetraacetate.

of certain species. For example, bromeliads develop only in liquid culture, whereas citrus does better on an agar medium.

Preparation of Stock Solutions

Media preparations can be expedited greatly by having on hand many of the ingredients in the form of stock solutions. Constituents required in large quantities, e.g., sucrose and agar, can be weighed in at the time of medium preparation. Desired aliquots of stocks can be mixed and diluted to arrive at the final medium. Glass-redistilled or demineralized distilled water should be used to prepare any component of culture media. Chemicals of the highest purity should be used as is possible. Weighings for stock solutions should be done accurately. Unless stability has been established, stock solutions should be refrigerated. Iron-containing solutions should be stored in amber-colored bottles. Some organic compounds, e.g., auxins and cytokinins, require an organic cosolvent to achieve initial solution. Dimethyl sulfoxide has been very useful for this because of its low toxicity. The organic substance is dissolved first in a minimum volume of cosolvent (gentle warming may be necessary), and the solution is diluted quickly with a suitable quantity of water. Most of the inorganic salts can be combined into a few stock solutions. Coprecipitatable ions should be avoided in the combinations. Table II contains stock solutions of the salts used in this laboratory.

Sterilization of Media

Most ingredients can be sterilized by autoclaving 15 minutes at 15 pounds/square inch. Substances subject to modification by heat can be added to the autoclaved portion of the medium following filtration (membrane, fritted glass, or

TABLE II
Stock Solutions[a] of Inorganic Salts

Component	Amount (g/liter)
Nitrate stock	
NH_4NO_3	165.0
KNO_3	190.0
Sulfate stock	
$MgSO_4 \cdot 7\ H_2O$	37.0
$MnSO_4 \cdot H_2O$	1.69
$ZnSO_4 \cdot 7\ H_2O$	0.86
$CuSO_4 \cdot 5\ H_2O$	0.0025
Halide stock	
$CaCl_2 \cdot 2\ H_2O$	44.0
KI	0.083
$CoCl_2 \cdot 6\ H_2O$	0.0025
PBMo stock	
KH_2PO_4	17.0
H_3BO_3	0.62
$Na_2MoO_4 \cdot 2\ H_2O$	0.025
NaFeEDTA stock[b]	
$FeSO_4 \cdot 7\ H_2O$	2.784
Na_2EDTA[c]	3.724

[a] Stocks contain 100× final medium concentration of each salt.
[b] Store iron-containing solution in amber-colored bottle to retard deterioration.
[c] Disodium (ethylenedinitrilo)tetraacetate.

comparable ultrafilters) of their stock solutions. The filtration can be carried out with either funnel-type apparatuses or syringes. The autoclaved portion should be cooled to about 35°C before adding the filtrate. Premature gelling of agar-containing media can be prevented by holding the media in a constant 35°C water bath while the filter-sterilized solutions are being added.

pH Adjustment

Plant cultures tolerate a range of pH's, although a value between 5 and 6 is most commonly provided. Usually, the pH adjustment is made just prior to final dilution during medium preparation, using weak solutions of NaOH and HCl. The influence of pH on solidification of agar media should be recognized. A pH below 5.5 is likely to result in poor gelation, whereas one above 6.0 will give an excessively hard medium. During the course of culture, drifts from the initially set pH are expected; these drifts are minimal in highly buffered media and are generally ignored.

Culture Vessels and Closures

Vessels of diverse sizes and shapes, chosen to meet individual needs, have been employed in plant culture. For liquid suspension cultures flasks with baffles or tabulations are sometimes used; culture tubes are also applicable when rotating apparatuses are employed. Agar media can be contained in Erlenmyer or

TABLE III
Nutrient Medium for the Development of Rooted Plants from Shoot Apical Meristems of Herbaceous Angiosperms

Ingredients	mg/liter medium
Inorganic salts (See Table I)	
Organic constituents	
Sucrose	30,000.0
myo-Inositol	100.0
Thiamine, HCl	0.4
Indole-3-acetic acid	1.0
Addenda	
Bacto-agar	8,000.0

Delong flasks, or they can be prepared as slants in culture tubes. Indeed, there is no restriction regarding the type of culture vessel one chooses to use. In this laboratory 125-ml Delong flasks and 25 × 150 mm Pyrex tubes have been standard. When agar medium is employed it is distributed as 25 ml per tube and 50 ml per flask. Liquid media are distributed as 5 ml per tube and 25 ml per flask. It should be remembered that the amount of medium and vessel size are proportionate with the explant size.

The nonabsorbent cotton used with plant cultures for many years is being replaced by other types of vessel closure. Polypropylene and stainless steel caps are more durable, can be cleaned readily for reuse, and provide more uniform gas exchange. These caps are available for culture tubes and Delong flasks. Autoclavable sponge plugs are also superior to cotton plugs and are sometimes convenient, particularly with Erlenmeyer flasks.

SAMPLE MEDIA

Listed in Table III through VII are some reference media used in this laboratory. The medium in Table III is employed for the development of rooted plants from excised shoot apical meristems of herbaceous angiosperms. That in

TABLE IV
Nutrient Medium for Rapid Asexual Multiplication of Herbaceous Angiosperms through Stem Tip Cultures

Ingredients	mg/liter medium
Inorganic salts	
Table I salts supplemented with 170 mg $NaH_2PO_4 \cdot H_2O$	
Organic constituents	
Sucrose	30,000.0
myo-Inositol	100.0
Thiamine·HCl	0.4
Indole-3-acetic acid	2.0
Kinetin	2.0
Adenine sulfate dihydrate	80.0
Addenda	
Bacto-agar (if gelled medium is desired)	8,000.0

TABLE V

Nutrient Medium Composition for Formation of Plantlets from Shoot Apex Explants of Asparagus

Ingredients	mg/liter medium
Inorganic salts	
Table I salts plus 170 mg $NaH_2PO_4 \cdot H_2O$	
Organic constituents	
NAA	0.3
Kinetin	0.1
Thiamine·HCl	1.0
Pyridoxine·HCl	5.0
Nicotinic acid	5.0
myo-Inositol	100.0
Adenine sulfate dihydrate	40.0
Sucrose	25,000.0
Addenda	
Bacto-agar	6,000.0
Bacto malt extract	500.0

TABLE VI

Nutrient Formula for Citrus Embryo and Nucellus Cultures

Ingredients	mg/liter medium
Inorganic salts (See Table I)	
Organic constituents	
Thiamine·HCl	10.0
Nicotinic acid	5.0
Pyridoxine·HCl	10.0
myo-Inositol	100.0
Glycine	2.0
Sucrose	50,000.0
Addenda	
Bacto-agar	10,000.0
Bacto malt extract	500.0

TABLE VII

A Medium for the Initiation of Haploid Embryos as in Tobacco Anther Cultures

Ingredients	mg/liter medium
Inorganic salts (See Table I)	
Organic constituents	
Sucrose	30,000.0
myo-Inositol	100.0
Thiamine·HCl	0.4
Addenda	
Bacto agar	8,000.0
Bacto malt extract	500.0

Table IV has served in the multiplication of shoots, or rapid asexual propagation, of many herbaceous angiosperms by stem tip cultures. Table V contains a medium found satisfactory for the initiation of plantlets and rapid asexual propagation of asparagus. In Table VI is a formulation employed successfully in the culture of citrus embryos and in the initiation of adventive embryos in citrus nucellus cultures. That shown in Table VII is a medium we employ for the initiation of haploid embryos in anther cultures of tobacco. In all above, the pH of agar containing media should be set at 5.7; that of liquid media may be in the neighborhood of 5.0.

CHAPTER 6

Determination and Survey of Osmolality in Culture Media

Charity Waymouth

Living cells maintain, by a combination of active and passive transport, an internal composition very different from, but highly dependent upon, their milieu. In particular, cells respond by changes in size and shape to changes in the osmotic activity of the environment.[1] The systems of active transport, especially for ions, generally respond by feedback mechanisms to changes in the external medium, thus compensating, at least in part, for imbalances. Somatic cells are resistant to a somewhat wider range of osmolality than erythrocytes, but they are nevertheless sensitive to rather small osmotic changes. Ion transport and changes in ionic concentrations outside and inside cells affect the transport of other nutrients, e.g., amino acids and sugars, in and out of cells, and these changes in turn affect, directly and indirectly, basic cellular synthetic systems. Proper ion balance is therefore an essential condition for the optimal nutrition and metabolism of cells in culture. Osmolality depends upon the total osmotic activity contributed by ions and nonionized molecules to the solution. Since the major part of this activity in a tissue culture medium is contributed by ions, its control is important, not only for maintaining tonicity, but also for regulating cell metabolism. Changes in osmolality affect, for example, not only Na^+, K^+-ATPase (one

[1] A. Ames, III, J. B. Isom, and F. B. Nesbett, *J. Physiol.* **177**, 246 (1965).

of the principal enzyme systems for active cation transport),[2] and other important enzymes such as alkaline phosphatase[3] but also RNA synthesis and polymerization in cells,[4] and in viruses[5] for which the cells may be a vehicle.

Different cell types respond optimally to different osmolalities.[6] For example, lymphocytes survive best at low (about 230 mOsm),[7,8] and granulocytes at higher osmolalities (about 330 mOsm).[9] Mouse and rabbit eggs develop optimally *in vitro* at around 270 mOsm, 250–280 mOsm being satisfactory, while above 280 mOsm development is retarded.[10-13] *Drosophila* embryos require 332 mOsm.[14] Little is known about differences in optimal osmolalities between cells from different species, though differences in blood osmolalities between species[6] might suggest that this matter should be more carefully examined.

Osmolality is an important parameter in the quality control of tissue culture media. Chemically defined media should, if carefully prepared, be reproducible from batch to batch. Errors in composition can often be traced from variations in osmolality. Fifty consecutive batches of each of two chemically defined media routinely used in the author's laboratory had osmolalities of 299.8 (SD = 9.4) mOsm/kg H_2O (MAV50/1) and 284.5 (SD = 8.9) mOsm/kg H_2O (MAP954/1), respectively. The spread of osmolalities (grouped in blocks for each 10 mOsm spread) is shown in Fig. 1. Presumably similar media compounded by different commercial suppliers may differ in osmolality,[6] perhaps because of differences in interpretation of original formulas, e.g., with respect to differences in water of crystallization of a salt. In Table I are listed quality control osmolality values on a number of commercially available tissue culture media.

Definitions

The difference between osmolality (measurable in absolute terms) and tonicity (a relative term), and a discussion of the osmole, osmolarity, osmolality, osmosity, and the osmotic coefficient are presented elsewhere.[6] For the present purposes, it is sufficient to state that one *osmole* is the mass of 6.023×10^{23} osmotically active particles in an aqueous solution; one *milliosmole per kilogram water* (mOsm/kg H_2O) produces a freezing point depression of 0.001858°C; and 1 mOsm at 38°C is equivalent to an osmotic pressure of 19 mm Hg. *Osmolality*, like molality, relates to *weight* of solvent (mOsm/kg H_2O); *osmolarity*, like

[2] J. C. Alexander and J. B. Lee, *Amer. J. Physiol.* **219**, 1742 (1970).
[3] H. M. Nitowsky, F. Herz, and S. Geller, *Biochem. Biophys. Res. Commun.* **12**, 293 (1963).
[4] J. Stolkowski and A. Reinberg, *C. R. Acad. Sci. (Paris)* **248**, 2400 (1959).
[5] M. R. F. Waite and E. R. Pfefferkorn, *J. Virol.* **2**, 759 (1968).
[6] C. Waymouth, *In Vitro* **6**, 109 (1970).
[7] D. R. Lucas, *Exp. Cell Res.* **40**, 112 (1965).
[8] O. A. Trowell, *Exp. Cell Res.* **29**, 220 (1963).
[9] J. L. Tullis, *Amer. J. Physiol.* **148**, 708 (1947).
[10] R. L. Brinster, *J. Exp. Zool.* **158**, 49 (1965).
[11] R. L. Brinster, *J. Reprod. Fertil.* **10**, 227 (1965).
[12] D. L. Naglee, R. R. Maurer, and R. H. Foote, *Exp. Cell Res.* **58**, 331 (1969).
[13] W. K. Whitten, *Advan. Biosci.* **6**, 129 (1971).
[14] G. Eschalier and A. Ohanessian, *In Vitro* **6**, 162 (1970).

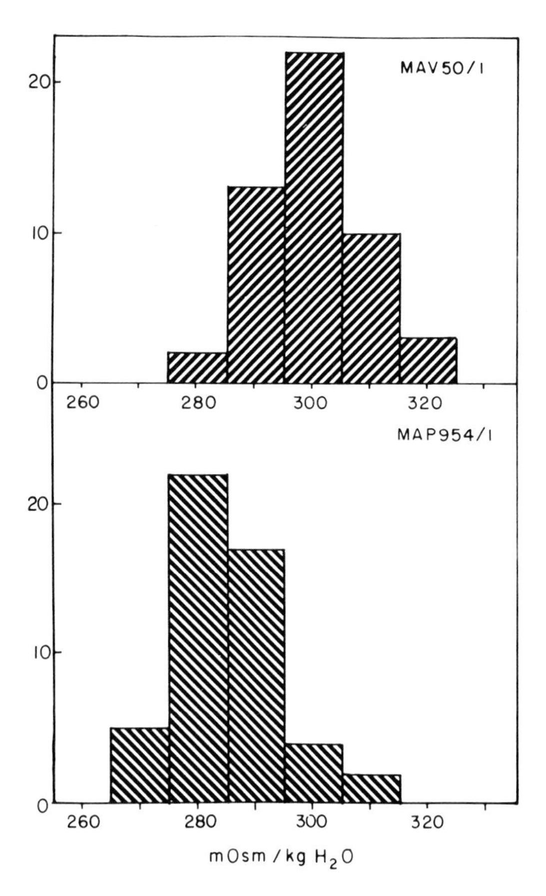

Fig. 1. Osmolalities of 50 consecutive batches of each of two chemically defined media, MAV50/1 and MAP954/1, used in the Jackson Laboratory.

molarity, relates to *volume* (mOsm/liter solution). In dilute aqueous solutions, such as tissue culture media, the differences are small.

Methods of Determining Osmolality

Since the osmotic activity of a solution depends upon the number of active ions plus nonionized molecules contained in it, the ideal or theoretical osmolality may be calculated (assuming complete ionization of salts). The ideal osmolality, in milliosmoles, of a solution containing several solutes with millimolalities $m_1, m_2, m_3 \ldots m_n$, which dissociate, respectively, into $x_1, x_2, x_3 \ldots x_n$ osmotically active particles, is $\Sigma mx = m_1x_1 + m_2x_2 + m_3x_3 \ldots m_nx_n$. For example, one may calculate the ideal osmolality of Earle's solution as in the tabulation on page 708. The real (from freezing point depression) osmolality of Earle's solution is 283 mOsm.[6]

Osmotic pressure may be measured directly with a membrane osmometer. Vapor pressure depression and freezing point depression are proportional to the

TABLE I

Osmolalities of Some Commercially Available Tissue Culture Media

	No. of samples	Osmolality mOSm/kg	High value	Low value	Standard deviation	Supplier[a]
Dulbecco's phosphate-buffered saline, PBS	17	294	311	280		D
Dulbecco's phosphate-buffered saline, w/o Ca and Mg	25	284	316	277		D
Earle's balanced salt solution, EBSS		281			10	A
	20	290	308	271		D
		289			9	C
Earle's balanced salt solution, w/o Ca and Mg	24	287	305	264		D
Gey's balanced salt solution		292			8	C
for slides	7	286	305	265		D
Hanks' balanced salt solution, HBSS		286			6	A
		285			9	C
Hanks' balanced salt solution, w/o Ca and Mg	26	283	310	250		D
CMRL-1066		310			9	C
	25	299	318	288		D
CMRL-1066, w/o glutamine	26	300	321	280		D
CMRL-H597[b]	14	283			5.31	B
CMRL-HB597[c]	32	300			4.66	B
CMRL-1415 ATM[d]	14	262			9.27	B
CMRL-1969[e]	40	284			6.19	B
Dulbecco's modified Eagle medium		338			10	C
	39	342	358	320		D
Eagle's basal medium, Earle's salts,		276			9	A
EBME		295			9	C
with glutamine	30	298	318	282		D
w/o glutamine	27	298	312	284		D
Eagle's basal medium, Hanks' salts,		287			8	A
HBME		292			9	C
with glutamine	25	290	310	270		D
w/o glutamine	21	290	314	270		D
Eagle's basal medium for cells in suspension		303			9	C
with glutamine, with Earle's salts	25	318	339	307		D
Eagle's minimal essential medium,		289			7	A
Earle's salts, EMEM		300			9	C
	50	297	314	278		D
w/o leucine	20	295	312			D
w/o leucine and glutamine	20	301	316			D
Eagle's minimal essential medium,		297			9	A
Hanks' salts, HMEM		298			9	C
	34	292	317	274		D
Eagle's minimal essential medium for		285			13	A
cells in suspension with Earle's salts	34	315	330	297		D
Fischer's medium for leukemic cells		315			9	C
	30	314	332	292		D
Grace's insect culture medium	25	364	380	340		D
Ham's F10 medium		285			5	A
		294			9	C

TABLE I (*Continued*)

	No. of samples	Osmolality mOsm/kg	High value	Low value	Standard deviation	Supplier[a]
Ham's F10 medium (*cont'd.*)	34	292	314	278		D
Ham's F12 medium		296			9	C
	44	309	332	285		D
HeLa maintenance medium		338			10	C
Lactalbumin hydrolysate, Earle's salts		305			8	A
	27	323	340	302		D
Lactalbumin hydrolysate, Hanks' salts		304			6	A
	29	329	335	297		D
Leibovitz' medium L-15		315			9	A
		297			9	C
	25	322	337	310		D
McCoy's medium 5A		304			9	C
		302			5	A
	56	300	320	275		D
Melnick's monkey kidney medium A		312			9	C
	11	322	337	306		D
Melnick's monkey kidney medium B		310			9	C
	14	324	337	308		D
Macpherson and Stoker's BHK-21 medium	30	316	334	292		D
NCTC-109		302			9	C
NCTC-135		302			9	C
	28	299	322	280		D
Parker's medium 199		318			8	A
		323			10	C
With Earle's salts	25	304	324	292		D
With Hanks' salts	25	295	316	275		D
Puck's medium N-15	16	304	376	288		D
Puck's medium N-16		297			9	C
	23	298	323	278		D
RPMI 1603 medium	20	283	303	260		D
RPMI 1629		306			9	C
	20	302	325	284		D
RPMI 1630	20	277	318	238		D
RPMI 1634	21	272	309	265		D
RPMI 1640		300			9	C
	108	287	319	268		D
Scherer's maintenance medium		332			10	C
	9	336	345	321		D
Swim's 67-G medium	50	290	308	275		D
Trowell's T8 medium		307			9	C
Waymouth's MB752/1 medium		319			10	C
	39	316	335	294		D
Waymouth's MD705/1 medium	20	310	326	288		D
Waymouth's MAB87/3 medium	10	298			8.5	D

[a] Contributing suppliers: Connaught Medical Research Laboratories; Flow Laboratories Inc.; Grand Island Biological Company; Microbiological Associates Inc. Letters are used to denote the different suppliers listed here but do not correspond to order in which they are listed.

[b] Modified Medium 199, Hanks' salts. pH 3.5 + 0.1, requires bicarbonate for pH adjustment.

[c] HB597 = H597 after addition of 0.87 g/liter $NaHCO_3$.

[d] Bicarbonate-free medium, for use in unsealed containers. Requires NaOH for pH adjustment.

[e] Based on unadjusted (pH) fluid.

	Millimoles m	x	mx
NaCl	116.5	2	233.0
KCl	5.35	2	10.7
$CaCl_2$	1.80	3	5.4
$MgSO_4$	0.83	2	1.7
$NaHCO_3$	26.10	2	52.2
NaH_2PO_4	1.01	2	2.0
Glucose	5.56	1	5.6
			Σmx = 310.6

number of osmotically active particles in solution. Both vapor pressure and freezing point methods have been used. Vapor pressure, e.g., by the Hill or Hill-Baldes method,[15-17] can be used for biological fluids. Osmometry of solutions of substances of high molecular weight, e.g., Dextrans or methylcellulose, sometimes used as adjuvants to tissue culture media, and which may prevent freezing or give anomalous freezing point readings, can be done with Sephadex beads,[18,19] provided that the solute does not penetrate the beads.

Aside from these special cases, however, the most practical routine procedures for osmotic measurements on tissue culture media are those depending upon depression of freezing point of the solution. Modern osmometers, based upon the freezing point principle, are designed to read directly in milliosmoles. Such instruments are available from several manufacturers.[20]

Precautions

Dissolved gases (O_2 and CO_2) affect the osmolality of plasma[21] and other fluids. For very precise measurements on tissue culture media in which the pH is equilibrated by gassing with, e.g., 5 or 10% CO_2 in air, or other gas mixtures, the osmolality should be read quickly before gas loss has occurred by exposure of the sample in an open tube. For routine checks, the difference due to gas loss may be neglected.

Plasma or serum prepared from preserved blood may have raised osmolalities, due either to breakdown of high molecular weight components (e.g., proteins and polysaccharides) into larger numbers of smaller osmotically active molecules, and to leakage of ions from the cells before separation of the plasma.[22]

[15] A. V. Hill, *Proc. Roy. Soc.* **A127**, 9 (1930).
[16] E. J. Baldes and A. F. Johnson, *Biodynamica* **47**, 1 (1939).
[17] P. R. Steinmetz and D. B. Ludlum, *J. Lab. Clin. Med.* **63**, 687 (1964).
[18] B. N. Preston, M. Davis, and A. G. Ogston, *Biochem. J.* **96**, 449 (1968).
[19] A. G. Ogston and J. D. Wells, *Biochem. J.* **119**, 67 (1970).
[20] Guide to Scientific Instruments 1971-1972, *Science* **174A** (1971).
[21] G. Meschia and D. H. Barron, *Quart. J. Exp. Physiol.* **41**, 180 (1956).
[22] D. Popescu, H. Cristea, and S. Negreanu, *Chirurgia* **7**, 457 (1963).

FREEZING POINT OSMOMETRY

A measured sample is placed in a small tube, so designed that a thermistor probe on the osmometer will be in the center of the sample. The sample is cooled in a freezing bath held at −6° or −7°C. Near the freezing point, the sample is stirred sharply with a vibrating stirrer, to induce rapid freezing. During the plateau phase of freezing, i.e., while the latent heat of fusion is being absorbed by the sample, the temperature remains stable for a time long enough for a reading to be taken. A galvanometer connected to the thermistor is balanced, and a direct reading of the osmolality, corresponding to the freezing temperature, is obtained. Standard solutions, e.g., of 100 and 500 mOsm, are used to calibrate the instrument in the range required for tissue culture medium measurements.

Samples of 2.0 ml are commonly used. Each sample takes about 1 minute to freeze. Reproducibility is about ±2 mOsm/kg H_2O. Osmometers equipped for smaller samples and with special small-sample tubes (0.2 to 0.3 ml) are convenient for routine use, since freezing takes place more rapidly. Reproducibility with the small samples is about ±5 mOsm/kg H_2O.

CHAPTER 7

Control of Culture pH with Synthetic Buffers

Charles Shipman, Jr.

Sodium bicarbonate-carbon dioxide buffer systems have two important disadvantages, namely: (a) a carbon dioxide enriched atmosphere is essential if adequate pH stability is to be achieved, and (b) the 6.1 pK_a of $NaHCO_3$ results in suboptimal buffering throughout the physiological pH range.

Recently Good and his associates[1] developed a series of hydrogen ion buffers covering the pK_a range 6.15–8.60. Most of these buffers are amino acids. Although the buffers were tested originally in studies on phosphorylation-coupled oxidation of succinate by mitochondrial preparations, a number of these synthetic buffers

[1] N. E. Good, G. D. Winget, W. Winter, T. N. Connolly, S. Izawa, and R. M. M. Singh, *Biochemistry* 5, 467 (1966).

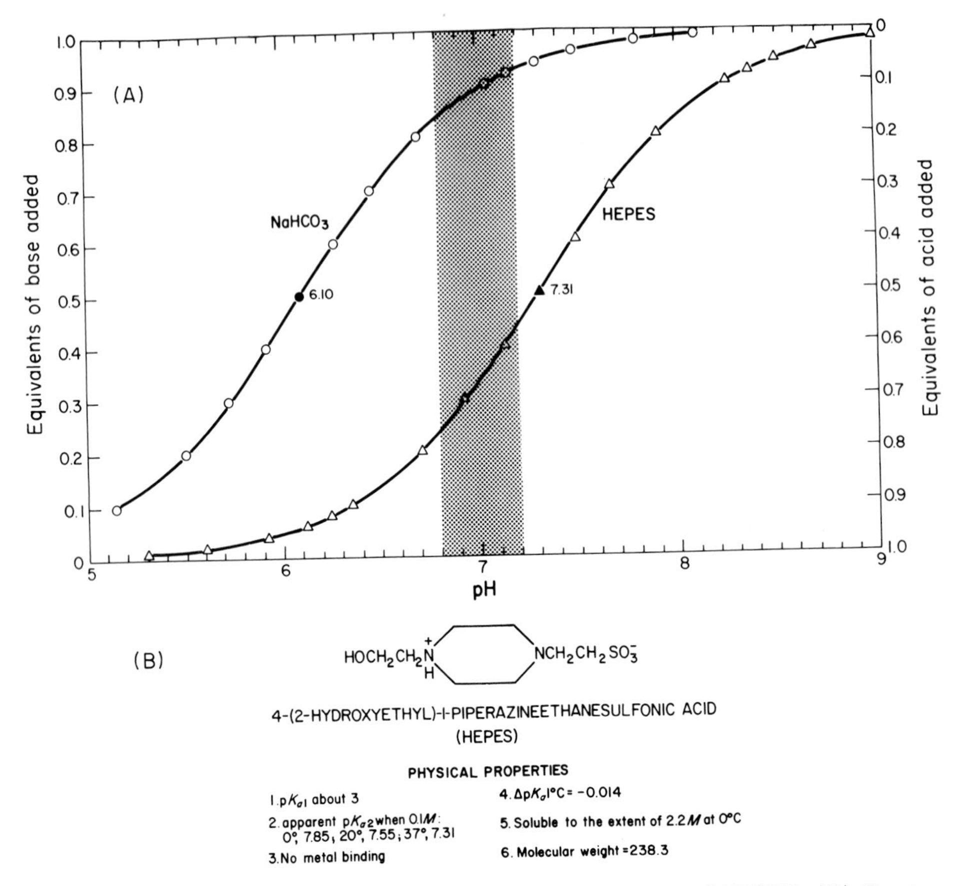

Fig. 1. (A) Theoretical titration curves at 37°C for $NaHCO_3$ and HEPES. (B) Structure, physical, and chemical characteristics of HEPES.

have now been evaluated for their suitability in controlling pH in tissue and organ culture systems.[2-7]

The synthetic buffer 4-(2-hydroxyethyl)-1-piperazineethanesulfonic acid (HEPES) has gained widespread use at this time and HEPES-buffered media are currently available from a number of commercial supply houses. Figure 1 illustrates theoretical titration curves of HEPES and $NaHCO_3$ and lists important physicochemical characteristics of HEPES. It can be quickly noted from Fig. 1 that whereas near optimal buffering occurs with HEPES in the physiological pH range (i.e., 6.8–7.2) the buffering available from $NaHCO_3$ is minimal in this

[2] A. Richter, *Appl. Microbiol.* **15,** 1507 (1967).
[3] J. D. Williamson and P. Cox, *J. Gen. Virol.* **2,** 309 (1968).
[4] C. Shipman, Jr., *Proc. Soc. Exp. Biol. Med.* **130,** 305 (1969).
[5] A. Fisk and S. Pathak, *Nature (London)* **224,** 1030 (1969).
[6] R. S. Gardner, *J. Cell Biol.* **42,** 320 (1969).
[7] H. Eagle, *Science* **174,** 500 (1971).

range. At pH 7.2 the addition of 0.09 equivalents of acid will depress the pH to 6.8 in a $NaHCO_3$-buffered system. In contrast, 0.21 equivalents of acid are required to depress the pH to 6.8 in a HEPES-buffered system.

HEPES has recently been shown to support the long-term division potential of WI-38 cells.[8] WI-38 cells were carried beyond the 60th passage in medium containing 0.01 *M* HEPES as compared with 32 passages in $NaHCO_3$-buffered medium. However, since air was used for the gas phase of both systems, these results are not as meaningful as those which would have been obtained had the $NaHCO_3$-buffered system utilized a CO_2-enriched atmosphere for increased pH stability. Nevertheless, the finding that WI-38 cells can be successfully passaged for 60+ generations is a very important observation and indicates that HEPES can be substituted for sodium bicarbonate for the long-term cultivation of human embryonic diploid cell cultures.

Preparation of HEPES-Buffered Media

Ordinarily, concentrations of 10 m*M* (2.38 g/liter) to 25 m*M* (5.95 g/liter) are utilized. The choice is primarily dependent upon the buffering capacity needed to maintain the pH in a physiological range and the tonicity of the medium after the addition of the buffer.

HEPES can be autoclaved in distilled water for periods of time up to 45 minutes without a serious degradation of the molecule. Thus, some laboratories find it convenient to prepare 1 *M* stock solutions of HEPES. The pH of these solutions is adjusted with NaOH.

In our laboratory we find it more convenient to add powdered HEPES to the medium, adjust the pH with NaOH, and then sterilize the medium by means of membrane filtration. When preparing HEPES-buffered media, the $-0.014 \Delta pK/°C$ must be considered in determining both the buffering capacity and the operating pH. Thus a HEPES-buffered medium adjusted to pH 7.3 at 22°C would have a pH of approximately 7.6 at 0°C and 7.1 at 37°C. This phenomenon is, of course, not unique with HEPES and is exhibited by most buffer systems. By comparison, the $\Delta pK/°C$ of $NaHCO_3$ and Tris is -0.009 and -0.031, respectively.

The decision as to whether or not to add $NaHCO_3$ to the medium depends upon the cell line being utilized and the culture conditions. Although the bicarbonate ion is an essential metabolite for many cell lines,[9,10] we have never found it necessary in our laboratory to supplement the medium with even small amounts of $NaHCO_3$. Presumably sufficient bicarbonate ion is present in the serum or is generated by the CO_2 in the atmosphere to meet the metabolic requirements of the cells. The addition of $NaHCO_3$ also defeats one of the primary advantages of a synthetic buffer system, namely, it is unnecessary to enrich the atmosphere of the cultures with CO_2.

[8] H. R. Massie, H. V. Samis, and M. B. Baird, *In Vitro* **7**, 191 (1972).
[9] H. E. Swim and R. F. Parker, *J. Biophys. Biochem. Cytol.* **4**, 525 (1958).
[10] R. P. Geyer and R. S. Chang, *Arch. Biochem. Biophys.* **73**, 500 (1958).

Buffer Combinations for Mammalian Cell Culture

The effective pK_a of a buffer system can be broadened by combining buffers of different pK_a's. Eagle has recently described a series of buffer systems using $NaHCO_3$, HEPES, and other organic buffers.[7] These systems may offer advantages to investigators who wish to operate at varying pH's in the range of 6.8 to 8.2 without changing the composition of the medium.

Interference in the Lowry Method for Protein Determination

Gregory and Sajdera[11] have reported that in the Lowry reaction, HEPES will give color in the absence of protein. This color has the same absorption spectrum as the color produced by protein. Five micromoles of HEPES is equivalent to 300 μg of bovine serum albumin.

[11] J. D. Gregory and S. W. Sajdera, *Science* **169**, 97 (1970).

CHAPTER 8

Freezing, Storage, and Recovery of Cell Stocks

John E. Shannon and Marvin L. Macy

The preservation of animal cells in liquid nitrogen is now a routine practice.[1-3] Perhaps the two most salient events leading to the development of the current highly successful techniques for the long-term preservation of animal cells in the frozen state were: (1) the accidental discovery by Polge *et al.*[4] in 1949 that glycerol protected animal cells against damage associated with freezing, and (2) the technological development of improved methods for extraction of gases from the air. The discovery of the cryoprotective properties of glycerol in the freezing of fowl spermatozoa[4] was later extended to red blood

[1] C. S. Stulberg, W. D. Peterson, and L. Berman, *Nat. Cancer Inst. Monogr.* **7,** 17 (1962).
[2] V. J. Evans, H. Montes de Oca, J. C. Bryant, E. L. Schilling, and J. E. Shannon, *J. Nat. Cancer Inst.* **29,** 749 (1962).
[3] L. L. Coriell, A. E. Greene, and R. D. Silver, *Cryobiology* **1,** 72 (1964).
[4] C. Polge, A. U. Smith, and A. S. Parkes, *Nature (London)* **164,** 666 (1949).

cells, tissue fragments, and eventually to animal cell cultures.[5-8] The development of improved methods for extraction of nitrogen gas from the air was a practical matter that made liquid nitrogen available in large quantities and at reasonable cost. Indeed, this has been a valuable spin-off from space-age technology.

Although some animal cells can be stored for 5 years in a dry ice chest[5,6] the percentage of viable cells that can be recovered is very low. We have conducted carefully controlled experiments at different storage temperature levels using a common pool of NCTC 929 (L) cells (ATCC-CCL 1)[9] that were frozen at the same time and in the same manner. A high percentage of viable cells (80–90% according to trypan blue dye exclusion) were recovered after storage for 6 months both at dry ice (−79°C) and liquid nitrogen (−196°C) temperatures. At the same time interval no viable cells were recovered from ampoules stored at −65°C. In assays at the end of 1 and 2 years the percentage of viable cells recovered from dry ice temperatures dropped to only 2 to 3%, while those stored in liquid nitrogen remained at the 80 to 90% level. The viability of a wide variety of animal cells after storage in liquid nitrogen is uniformly high.[10] We recently recovered CCL 1 cells stored over 10 years in liquid nitrogen and the viability index was within 5% of the initial high level.

The procedures we currently employ for the freezing, storage, and recovery of reference animal cells are described below.

The Cryoprotective Agents and Their Preparation for Use

Although glycerol and dimethyl sulfoxide (DMSO) appear to be almost equally effective in preserving many cell lines, glycerol is best for certain lines and DMSO for others. Of approximately 150 animal cell lines in the Animal Cell Culture Collection,[10] 80% are preserved with glycerol at concentrations (v/v) of 5 or 10%. The balance of the lines are preserved with DMSO at concentrations of 5, 7.5, or 10%. The number of lines in the Collection preserved with DMSO is fewer, largely because many lines were preserved in glycerol before DMSO was widely accepted as a suitable cryoprotective agent. DMSO, in particular, is preferred by many investigators for the preservation of primary tissue culture cells. In our experience, glycerol or DMSO at the relatively low level of 5% (v/v) is quite adequate for the preservation of most animal cell lines. Use of the cryoprotective agent at this low level has the advantage that the removal of the agent is facilitated upon dilution with fresh culture medium at the time of thawing.

[5] W. F. Scherer and A. C. Hoogasian, *Proc. Soc. Exp. Biol. Med.* **87**, 480 (1954).
[6] W. E. Scherer, *Exp. Cell Res.* **19**, 175 (1960).
[7] H. E. Swim, R. F. Haff, and R. F. Parker, *Cancer Res.* **18**, 711 (1958).
[8] T. S. Hauschka, J. T. Mitchell, and D. J. Niederpruem, *Cancer Res.* **19**, 643 (1959).
[9] J. E. Shannon, H. E. Den Beste, M. L. Macy, and J. L. Jackson, *Excerpta Med.* **18**, 37 (1964).
[10] J. E. Shannon and M. L. Macy, Registry of Animal Cell Lines, 2nd ed., American Type Culture Collection (1972).

We recommend that the cryoprotective agent employed be of reagent grade. Also, because of the accumulation of oxidative breakdown products we avoid using an opened bottle for too long a period of time. To minimize this problem we sterilize the glycerol and DMSO in units of small volume that are used only once. The volume of glycerol required for a particular freeze is dispensed into a glass vessel of appropriate size that is stoppered temporarily with a cotton-gauge stopper and is autoclaved for 15 minutes at slow exhaust. The DMSO is filtered, using 03 porosity Selas filter candles, collected in 10- to 15-ml quantities in test tubes, and then stored, frozen, at +5°C (DMSO freezes at +18°C).

Labeling and Sterilization of the Ampoules

We use heavy-walled borosilicate (bull semen) ampoules[11] of 1.2-ml capacity. No washing of the ampoules is necessary. We label the ampoules by means of a Markem Labeling Machine[12] (Model 135A) using a specially formulated ink (Markem No. 7224K) that is resistant to ultralow temperatures and brief exposure to ordinary laboratory solvents such as alcohol. The labeling is accomplished by the indirect transfer of ink from interchangeable printers type. With this machine one can conveniently and rapidly mark small (1-ml), large (10-ml) ampoules, or even bottles up to about 4 inches in diameter. After marking, the ampoules are put in an aluminum ampoule rack and are placed in a hot-air oven for 60 minutes at 120°C to anneal the ink. Upon cooling, specially fabricated glass ampoule caps are placed over the neck of each ampoule in the rack, and then the ampoules are sterilized by dry-heat sterilization (270°C for 2 hours).

If it is necessary to label the ampoules by hand this may be done by using an ordinary stick pen and most laboratory inks, provided the ink is subsequently annealed. Although the lab pen marking may not be as legible as that by machine, it is quite satisfactory for identification purposes and, in our experience, is far superior to any type of gum paper or tape label.

Liquid Nitrogen Refrigerators and Storage Procedures

Liquid nitrogen refrigerators in a wide variety of sizes are available from commercial sources such as the Linde Division of the Union Carbide Corporation, New York; Minnesota Valley Engineering Inc., New Prague, Minnesota; Cryenco Engineering Company, Denver, Colorado, and others. In terms of 1-ml ampoules the storage capacities of the refrigerators range from approximately 300 ampoules to 30,000 ampoules or more. Some of the smaller containers with a capacity of 30 liters, or so, of liquid nitrogen have been ruggedly designed for field work in the bull semen industry. These units may also be used for the bulk shipment of 300 to 600 ampoules as well as for storage purposes.

[11] Wheaton Cryule No. 12483 (unscored) and No. 12523 (prescored), Celstir Unit, No. 356676-79 Wheaton Scientific Co., Millville, New Jersey 08332.

[12] Markem Corporation, 150 Congress St., Keene, New Hampshire 03431.

The refrigerators should be kept in well-ventilated areas because of the constant release of the nitrogen gas. The liquid nitrogen evaporation rate may vary from approximately 10–15 liters per day for a refrigerator of 300- to 600-liter capacity to less than 1 liter per day for one of 30-liter capacity. Liquid nitrogen costs generally run from 25 cents to 50 cents per liter depending upon the volume used and the geographical location. Maintenance of the proper level of liquid nitrogen in the larger refrigerators may be monitored electronically by heat-sensitive thermistors. In the smaller refrigerators a dip stick showing a frost line may be used, or the refrigerators may be periodically weighed on large-capacity scales. While liquid nitrogen refrigerators are extremely reliable, the vacuum insulation in very old refrigerators is sometimes abruptly lost. For this reason, especially with very valuable material, it is wise to make some provision for a back-up system.

A carefully organized but simple retrieval system is a must. Safety precautions, as described later, should be exercised at all times upon retrieval of the ampoules. The location of specific canes[13] may be mapped out by the position in a certain area of the refrigerator and the location of specific ampoules may be designated by their position on a specific cane. The tops of the canes may be marked with ink or they may be color coded.

Cultivation of Stock Cultures

Stock cultures are maintained as monolayers in T-60 flasks, plastic T-75 flasks, or other vessels, such as roller bottles. The culture medium is changed routinely two or three times a week, and the cells are permitted to reach confluency before subcultures are prepared. At this time, samples of the cell suspensions are taken for viability assay (trypan blue dye exclusion), cell density, and bacteriological tests. The routine assay of cell yield per flask enables the technician to determine the number of culture vessels necessary to prepare a freeze of 100 to 200 ampoules with 2–6 million viable cells per ampoule. In all instances, active log phase cultures (which have had the culture medium renewed 24 hours previously) are selected for freezing.

Harvesting the Cells

In most instances, fresh trypsin or a trypsin–Versene solution is used in dislodging the cells from surfaces of the culture vessels. The culture medium is removed, and 5 to 10 ml of the trypsinizing solution is added to each flask. The cultures are then allowed to stand at room temperature for 5 to 10 minutes, or until the cell sheet begins to slough off. At this time, an equal amount of fresh culture medium is added to each flask to inhibit further tryptic action, and the crude suspension is gently aspirated to break up the large clumps of cells. The resulting suspensions of single cells and small aggregates of cells are

[13] Nasco (No. A545) Fort Atkinson, Wisconsin 53538.

then collected and dispensed into centrifuge tubes, and centrifuged at approximately 200–300 g for 20 to 25 minutes at room temperature.

Suspension of the Cells in the Freeze Medium

While the cells are being centrifuged, the freeze medium is prepared. This involves the simple admixture by pipette of 5 ml of the sterile cryoprotective agent with 95 ml of fresh culture medium. After the cryoprotective agent is added to the culture medium, the solution is agitated vigorously, and, if necessary, the pH is adjusted to approximately 7.2 with a humidified mixture of 5–10% CO_2 in air.

Dispensing and Ampoulization

After centrifugation, the cells are resuspended in the freshly prepared freeze medium by gentle aspiration with a pipette and are then introduced into a dispensing apparatus for distribution into ampoules. The dispensing apparatus consists of a Wheaton Celstir unit,[11] modified with a sidearm near the base for attaching a Cornwall pipetting unit. This apparatus contains a Teflon-coated magnetic stirring bar for gentle agitation of the cell suspension, and is also equipped with sidearms near the top for continuous gassing with 5–10% CO_2 in air mixture to maintain the proper pH. One-milliliter aliquots of the cell suspension are dispensed into the ampoules by means of the Cornwall syringe fitted near the base of the dispensing apparatus. During the entire dispensing procedure, samples are taken at random for bacteriological tests, viability assays, and cell counts. The ampoules are immediately sealed, using a Kahlenberg-Globe semiautomatic ampoule sealer.[14] This unit is designed to seal ampoules using the pull-seal method. The pull-seal method is preferred over the tip-seal method because it minimizes the occurrence of pinhole leaks in the tip of the ampoule. After sealing, the ampoules are placed on aluminum canes and submerged in a 0.05% methylene blue solution for approximately 30–45 minutes at +5°C. This allows time for the cryoprotective agent to equilibrate with the cells, and also time for the dye to seep into improperly sealed ampoules. The ampoules, now racked on canes, are removed from the dye solution, rinsed in tap water, and all improperly sealed ampoules are discarded.

The proper sealing of ampoules and testing for leaks is a must if the ampoules are to be stored completely immersed in the liquid phase of a liquid nitrogen refrigerator. Such safety procedures obviate the potential hazard of improperly sealed ampoules exploding when they are rapidly removed from the liquid nitrogen and are placed at room temperature. This hazard can also be circumvented by storing the ampoules in the vapor phase of a liquid nitrogen refrigerator.

[14] K-6 ampoule sealer Bench Model 161 Kahlenberg-Globe Equipment Co., P. O. Box 3636, Sarasota, Florida 33577.

FREEZING THE CELLS

We routinely use a Linde[15] BF3-2 freezer (which operates on a differential thermocouple principle) because the freezing rate can be easily varied from ½° per minute to approximately 20°C per minute. The ampoules are placed into the chamber of the freezing unit and cooled at a programmed rate of a 1°–3°C drop in temperature per minute from about +10° to −30°C. At this point, a more rapid drop in temperature is programmed (20°–30°C per minute) until the temperature of the ampoules is below −150°C. The ampoules are then moved from the freezing unit chamber, and immediately transferred to storage in liquid nitrogen at −196°C or liquid nitrogen vapor storage at a temperature range of about −170° to −180°C.

THAWING THE CELLS

The cells are thawed by removing the ampoule rapidly from the liquid nitrogen refrigerator and plunging it immediately into a water bath at 37°–40°C (special precautions such as a protective face mask and heavy asbestos gloves should be used when removing ampoules from liquid nitrogen). The thawing of the cells should be accomplished as rapidly as possible (with moderately vigorous agitation the ice usually melts within 40–60 seconds). Immediately after thawing, the ampoules are removed from the water bath and immersed in 70% ethanol at room temperature. All of the operations from this point on should be carried out under strictly aseptic conditions in a sterile room, cubicle, or hood.

Standard ampoules must be scored on the neck with a small file that has been immersed in ethanol. A definitive sharp nick about ⅛ inch in length on one side is all that is necessary. If the ampoule is prescored simply break the neck of the ampoule between several folds of a sterile towel and transfer the contents of the ampoule into a single flask containing at least ten volumes of culture medium. A sterile Pasteur-type pipette with a small bore or a large-gauge needle attached to a syringe is suitable for this purpose. Since it is important to avoid excessive alkalinity of the culture medium during recovery of the cells, the pH should be adjusted to approximately 7.2 *prior* to the addition of the ampoule contents. The tenfold dilution of the ampoule contents lowers the concentration of the cryoprotective agent to a level that does not necessitate its complete removal. In order to expedite the removal of the freezing additive the culture medium should be changed the day after the cells are thawed. If it is desirable that the freezing additive be removed immediately, or that a more concentrated cell suspension be obtained, centrifuge the diluted suspension at approximately 125 *g* for 10 minutes, discard the fluid and resuspend the cells in the desired volume of culture medium. In recovering the cells we find it best (although not absolutely necessary) to introduce the ampoule contents into a

[15] Union Carbide Corporation, Linde Division, P. O. Box 766, New York 10019.

single flask. With cell lines that have a high minimum inoculum size this helps to ensure the establishment of a vigorously growing culture.

Additional Considerations

Mechanical refrigerators that are capable of maintaining a temperature of −90°C are quite adequate for the storage of cell stocks for several years and undoubtedly for longer periods of time. The efficiency of recovery of the cells, however, is not as great as is the case with cells stored at liquid nitrogen temperatures.

For laboratories where dry ice is available but liquid nitrogen is not, stocks containing a high percentage of viable cells can be perpetuated by preparing a new batch of frozen cells every 6 months. This procedure reduces the selection of cells that would otherwise occur because of deterioration of viability during storage in the dry ice for longer intervals.

Programmed freezing units provide a controlled rate of freezing, and a permanent record of the freezing curve may be conveniently obtained. For critical experiments and repository production procedures controlled freezing rates are quite necessary. In other instances, however, simpler procedures are just as suitable. For instance, many cells can be successfully frozen by simply placing the ampoules in a dry ice chest or mechanical refrigerator at −65°C for several hours, or even overnight. This provides an adequate rate of freezing then, after the cells are frozen, they may be placed in a liquid nitrogen refrigerator for long-term storage.

CHAPTER 9

Detection, Elimination, and Prevention of Bacteria and Fungi in Tissue Cultures

Lewis L. Coriell

Nutrient media used to support the growth of mammalian cell cultures also provide excellent nutrition for bacteria and fungi. Frequently, such contaminations are readily recognized by gross inspection. However, some contaminants produce only minor changes in the culture medium and are compatible with continued growth of the cell culture. Some that grow quite well in tissue cultures fail to grow if subcultured only on the more frequently used media, i.e., nutrient agar, blood agar, or thioglycollate broth. An acceptable procedure for detection of contamination of cell cultures must therefore be adequate to detect all the known contaminants. The following procedures are designed to meet this objective. They can be applied to cell cultures and to culture media, serum, trypsin, or other additives.

Routine Sterility Cultures

It saves time and expense to use multiple culture media and incubation temperatures at the outset. Sterility tests should be made the first time a new cell culture is opened and the inoculum should include both cells and culture medium. If the sterility tests are negative but the cell culture contains antibiotics the following procedure is recommended. When the cell culture is subcultured, initiate a subline with the same medium without antibiotics. After several passages without antibiotics repeat the sterility tests.

If a cell culture is contaminated with bacteria or fungi, there are usually many organisms present by the time the cell culture is ready for subculture. It is therefore reasonable to test a small sample, including cells and culture fluid. We recommend an inoculum of 0.5 ml into duplicate tubes of the following broth media: brain-heart infusion, tryptose phosphate, trypticase soy, yeast and mold, and Sabouraud dextrose. One tube of each media is incubated at 37°C and the other at 30°C. A blood agar plate is inoculated with 0.1 ml and incubated at 37°C for 14 days. If culture tubes become cloudy or develop a precipate, examine a wet mount, gram stain, subculture, and identify.

Microscopic Examination

At the time the cell culture is opened for refeeding or subculture, remove the spent media and centrifuge (1400 *g*) for 20 minutes, and prepare smears and cultures of the sediment. Smears are fixed in methyl alcohol, stained with Gram's stain freshly filtered through Whatman No. 2 filter paper, and examined under an oil immersion lens. The staining procedure will reveal the presence of microbial contaminants that fail to grow under the culture conditions provided, or, that are suppressed by antibiotics in the cell culture media, and the results are available the same day without disturbing the cell culture. It will detect dead organisms if such are present in any of the media components, in the gram stains, or on the slides. The advantages of culturing the sediment are two. It concentrates the contaminant and greatly dilutes the antibiotic if such is present in the cell culture media.

Elimination

We recommend that infected cell cultures be autoclaved at once and start over again with a frozen stock of clean cells stored previously in liquid nitrogen.

Microbial contamination of cell culture media can be eliminated by filtration or heat sterilization, but these measures cannot be applied to infected cell cultures because they would also eliminate the cells. Many investigators[1,2] use antibiotics to prevent infection or to free cell cultures of contamination, but in our experience it is a futile exercise.

The recommended procedure is to test the sensitivity of the contaminant to a battery of antibiotics to identify those most lethal for the contaminant, then determine the minimal toxic dose of the most effective ones for the cell culture under study. The antibiotic most lethal for the contaminant and least toxic for the cells is then used as follows: Grow the cells for two to three passages in the maximum tolerated dose of the antibiotic then remove all antibiotics for three or more passages and repeat the sterility tests.

This procedure can succeed but the success rate is not high, and considerable time and expense is involved. The antibiotics can have an adverse effect on the cell culture itself, and can encourage the growth of L forms and mycoplasma. Each time the contaminated culture is opened there is danger of spreading the contaminant to other clean cell cultures carried in the same laboratory.

Prevention

Since elimination of bacterial and fungal contamination is so unsatisfactory and detection is quite laborious, it is recommended that emphasis be placed on prevention. Prevention is practical and offers other fringe benefits. Recommended procedures are as follows:

1. Remove antibiotics from tissue culture media. This encourages the practice of meticulous aseptic techniques, and permits contaminants to grow promptly. The ensuing turbidity is an automatic indicator to the technician that something has gone wrong, and that the cell culture is no longer useful for most types of study. One exception is that antibiotics (preferably bactericidal ones) are recommended for primary cell cultures from potentially contaminated tissues such as biopsies of skin, cervix, or tumors. Antibiotics should be removed at the time of the first subculture. If the antibiotics have not killed the contaminants in 1-week exposure they are unlikely to do so thereafter, and if the culture is in fact sterile this should be confirmed by removing the antibiotics. If it is still contaminated it should be discarded.

2. Store a frozen seed stock of important cell cultures in liquid nitrogen.[3,4]

3. Pretest all culture medium components including serum as described above and observe for 14 days before use.[5]

[1] W. F. Sherer, L. L. Coriell, T. C. Hsu, D. W. King, S. H. Madin, H. T. Meryman, H. R. Morgan, H. M. Rose, K. K. Sanford, and J. E. Shannon, *Science* **146**, 241 (1964).

[2] L. L. Coriell, *Nat. Cancer Inst. Monogr.* **7**, 33 (1962).

[3] A. Greene, R. K. Silver, M. Krug, and L. Coriell, *Proc. Soc. Exp. Biol. Med.* **116**, 462 (1964).

[4] A. Greene, B. Athreya, H. B. Lehr, and L. Coriell, *Proc. Soc. Exp. Biol. Med.* **124**, 1302 (1967); cf. Chapter 8, this section.

[5] G. McGarrity and L. Coriell, *In Vitro* **6**, 257 (1971).

4. Work in an enclosure that is flushed with air passed through high efficiency filters (HEPA).[5]

5. Exclude street clothing and shoes from the tissue culture transfer rooms, cover the hair, wear masks, or do not talk while transferring cultures.

6. Wash down the laboratory floor with disinfectant solution daily, and wash the work bench between each different cell culture.[5]

7. Install bacterial filters on the discard air from aspirating pumps used to remove spent media from cell cultures (DFA Filter Assembly from Pall Corp., Glen Cove, New York 11542).

8. Maintain adequate sterility surveillance of all apparatus.[5]

9. Sterility tests of medium components.

(a) Serum and purchased liquid media. Each bottle is cultured in five broth media, a blood agar plate, and an undiluted sample of the medium. Cultures of serum are incubated at 30° and 37°C for 14 days. Cultures of media are incubated at 37°C for at least 7 days. The serum and media bottles are stored at 4°C until the cultures are completed. When preparing media for use on any given day, this pretested serum and media are combined aseptically to prepare 10% more than the quantity of complete media needed. The residual medium remaining after all cultures are fed is placed in the incubator at 37°C and observed for contamination.

(b) Media prepared in the laboratory is sterilized by Millipore filtration into a 10-liter media storage and dispensing jar (Bellco Glass Co., Vineland, New Jersey). From the 10-liter jar it is aseptically dispensed into smaller bottles for storage at 4°C and the following sterility tests are made. The first and the last 100 ml dispensed from the 10-liter bottle are put into milk dilution (MD) bottles. The two MD bottles are incubated at 37°C for at least 7 days. If both MD bottles remain sterile it is assumed that the other bottles are sterile.

(c) Trypsin solution made in the laboratory is sterilized by Millipore filtration as above and dispensed in 100-ml quantities in milk dilution bottles for storage at 4°C. The first, last, and three additional bottles picked at random are incubated for 7 days at 37°C as a sterility test. If all five bottles remain sterile it is assumed the other bottles are sterile. This pretested trypsin is used without further sterility tests but any residual trypsin left in a MD bottle at the end of a days work is not used again, but is incubated at 37°C as an additional control of aseptic procedures in the transfer room.

CHAPTER 10

Screening Tissue Cultures for Mycoplasma Infections[1]

Leonard Hayflick

Experimental results utilizing tissue cultures are usually evaluated on the premise that such cultures are free from unwanted microbial contaminants. Indeed, the most important technical advancement which allowed the powerful technique of cell cultivation to be used in even the most primitively equipped laboratory was the use of antibiotics in cell cultures. Ordinary microbial contamination—long a major problem for cell culturists—has now been reduced to a minor role. In recent years, however, two categories of microorganisms have been detected as contaminants despite the use of penicillin and streptomycin, the antibiotics classically used in tissue culture. One of these groups of possible contaminants is viruses found in animal serum and latent in primary cell cultures and against which antibiotics are ineffective. In practice the presence of latent viruses has usually been limited to studies employing primary cell cultures. In the absence of overt cytopathology passaged diploid cell strains and heteroploid cell lines are, generally, free of contaminating viruses. At least, this kind of contaminant has been found much less frequently in serially passaged cultures than has contamination by microörganisms of the class Mollicutes. For a detailed account of the biology of the mycoplasmas see "The Mycoplasmatales and the L-Phase of Bacteria."[2] Mycoplasmas have the following properties:

1. They are the smallest free-living organisms with a size similar to that of the myxoviruses. The smallest reproductive units have a size range of 125 to 150 nm.
2. They can reproduce in a cell-free medium where, on agar, they exhibit a characteristic colonial morphology with the center of the colony embedded beneath the surface.
3. They lack a rigid cell wall and, consequently, are highly pleomorphic. The membrane surrounding them is triple layered.
4. Most species require sterol for growth.
5. All species exhibit an absolute resistance to penicillin whereas many strains are inhibited by low concentrations of tetracyclines.
6. The growth of mycoplasmas can be inhibited by specific antibody.
7. In contradistinction to the L-phase of bacteria, mycoplasmas have no history of reversion to or from a bacterial parental form.

[1] Supported, in part, by research contract NIH 69-2053 from the National Cancer Institute, National Institutes of Health, Bethesda, Maryland.

[2] L. Hayflick (ed.). "The Mycoplasmatales and the L-Phase of Bacteria," 731 pp. Appleton-Century-Crofts, New York, 1969.

Detection of Mycoplasmas

The presence of mycoplasmas in the great majority of contaminated cell cultures cannot be detected by any macro-, or microscopic change in appearance of the culture medium or of the cells themselves. Thus it is essential that detection of mycoplasma contamination be based on something other than visualization of the culture. In fact, proof of the presence of mycoplasmas in any material can only be made by demonstrating their characteristic growth on agar.

The medium to be described is now perhaps the most widely used. Its popularity resulted from its use as the medium on which I isolated and identified a mycoplasma which was subsequently found to be biologically identical with the Eaton agent.[3,4] This mycoplasma, now called *Mycoplasma pneumoniae*[4,5] is the cause of cold agglutinin-positive primary atypical pneumonia in man.[3] Its identification is the first unequivocal demonstration of the pathogenicity of a mycoplasma for man.

This medium is a modification of formulas described by Edward[6] and by Morton *et al.*[7] and has been found to support the growth of most mycoplasma species. Its use has resulted in the detection of other new fastidious mycoplasmas incapable of growth on many other media formulas. One of these is *M. orale*, first isolated in 1961 as a tissue culture contaminant, and known to be a part of the human oral flora.[4]

Mycoplasma Broth (*1 Liter*). Difco beef heart for infusion, 50 g; Difco Bacto-peptone, 10 g; NaCl, 5 g; water, 900 ml.

The beef heart for infusion preparation is soaked for 1 hour in distilled water brought to 50°C. This solution is brought to a boil and filtered through two sheets of Whatman No. 12 filter paper (50-cm diameter, folded). The Bacto-peptone and NaCl are added to this infusion and the pH raised to 7.8 by addition of 1.6 ml of 10 *N* NaOH. The solution is then brought to a boil and filtered through two more sheets of filter paper. It is made up to 1 liter by adding distilled water and 0.2 ml 10 *N* HCl added before autoclaving. The final pH should be 7.6–7.8 and the broth should be clear with no precipitate.

Before use, the broth must be supplemented with 20% unheated horse serum and 10% yeast extract prepared as described below. It is often convenient to divide the unsupplemented broth into small volumes for storage at 5°C, where it is stable for a few months.

[3] R. M. Chanock, L. Hayflick, and M. F. Barile, *Proc. Nat. Acad. Sci. U. S.* **48**, 41–49 (1962).

[4] L. Hayflick and R. M. Chanock, *Bacteriol. Rev.* **29**, 185–221 (1965).

[5] R. M. Chanock, L. Dienes, M. D. Eaton, D. G. ff. Edward, E. A. Freundt, L. Hayflick, J. F. Hers, K. E. Jensen, C. Liu, B. P. Marmion, H. E. Morton, M. A. Mufson, P. F. Smith, N. L. Somerson, and D. Taylor-Robinson, *Science* **140**, 662 (1963).

[6] D. G. ff. Edward, *J. Gen. Microbiol.* **1**, 238–243 (1947).

[7] H. E. Morton, P. F. Smith, and P. R. Leberman, *J. Syph. Gonor. Ven. Dis.* **35**, 361–369 (1951).

Mycoplasma Agar. Difco Bacto-mycoplasma agar, dehydrated, is prepared according to the directions accompanying it. Ordinarily, the agar is made up in quantities of 70 ml. Prior to use, the agar is melted, allowed to cool sufficiently so that the bottle can be handled, and supplemented with 20% unheated horse serum and 10% yeast extract. About twelve 50-mm Petri dishes can be poured from 100 ml of supplemented agar.

Unsupplemented agar can be stored indefinitely at 5°C. Although supplemented agar plates can be held for a few weeks at 5°C, the superior growth of mycoplasmas on freshly prepared agar plates has often been observed.

Yeast Extract. Add 250 g of active dry bakers' yeast (Fleischmann's, a division of Standard Brands, Inc.) to 1 liter of distilled water and heat until boiling. Filter the mixture through two sheets of Whatman No. 12 filter paper, using a number of filtration setups since the process is slow. Add sufficient NaOH to raise the pH to 8.0. The yield from 1 liter is about 400 ml. Dispense in 10-ml aliquots, autoclave, and store at −20°C. The stability of this material at −20°C is about 2 months.

Horse Serum. Agamma horse serum ordinarily used for the cultivation of tissue cultures is used. The serum can be stored for months at 5°C.

Technique for Detecting Mycoplasmas in Cell Cultures. A few drops of suspected tissue culture supernatant fluid (at least 3 days on the culture and containing cells) are added to 2 ml of supplemented broth. The tube is incubated for 6 days at 37°C. Ordinarily, it is not possible to see growth in broth containing mycoplasmas. A second tissue culture sample is streaked directly on agar with a bacteriological loop after depositing a few drops on the agar with a 1-ml pipette. At least 4 plates should be prepared; two incubated aerobically at 37°C and two incubated in an atmosphere of 5% CO_2–95% N_2 at 37°C for 6 to 14 days.

After about 6 days' incubation, samples of the broth specimen are streaked onto agar plates as indicated above and incubated at 37°C aerobically and anaerobically. Mycoplasmas grow well only in a humid atmosphere and provisions must be made for this requirement when incubating agar plates.

Identification of Mycoplasma Colonies. The plates are inverted (without removing the cover) on the microscope stage so that one looks for the colonies of mycoplasmas (which are only observable microscopically) by focusing through the agar. It is possible to orient the plane of focus to the surface of the agar by focusing on the lines that were created when the loop was moved over the surface. Use a 10X objective and a 10, 12.5, or 15X ocular.

Most mycoplasma colonies appear as round colonies with a dense center and a less dense periphery, giving the appearance of a fried egg. Mycoplasma colonies have been isolated from tissue cultures, however, that do not conform strictly to this appearance on primary isolation. They may appear to lack a distict periphery and to be totally embedded in the agar. These colonies are usually very small and look "granular" or "feathery." Mycoplasma colonies vary from 10 to 500 μm in diameter and characteristically the center only, or all of the colony, is embedded in the agar (see footnote 12). Individual organisms cannot be resolved in the light microscope. After locating the colonies, their position is marked off on the underside of the Petri dish with a glass marking pencil.

Identification of mycoplasma colonies depends, in addition to morphological features, on the following criteria:

1. Inability to remove the embedded portion of the colony from the agar surface by stroking the colony with a bacteriological loop. This demonstrates the fact that part or all of the colony is embedded. With the exception of some actinomycetes, bacterial colonies will rub off.

2. The nonreversion to bacteria which subsequent passages of the colonies will reveal. Reversion to a bacterial form would be typical of the L-phase of bacteria.

3. A requirement for sterol.

4. Reaction with the Dienes stain.[8,9]

The Dienes stain is prepared by dissolving 2.5 g of methylene blue, 1.25 g of azure II, 10.0 g of maltose, and 0.25 g of sodium carbonate in 100 ml of distilled water. With a cotton swab moistened in the stain, an area of the agar is stroked just adjacent to the suspected colony. The stain will diffuse to the colony which is then examined under the microscope as described above. The mycoplasma colonies stand out distinctly with dense blue staining centers and light blue peripheries. Bacterial colonies are also stained but these are decolorized in 30 minutes. The mycoplasma colonies *never* decolorize the stain. A source of error could be a colony of stained microorganisms composed of dead bacteria.

Subculturing Mycoplasmas

Subcultivation of mycoplasma colonies can be made by cutting out 1-cm^2 agar blocks with a scalpel sterilized by flaming in alcohol. The area is previously marked off with a glass marking pencil to locate the colonies. The scalpel is slipped under the block of cut agar and the block inverted onto a fresh plate. The block is then pushed over the surface of the new plate with the aid of the scalpel and the block finally pushed to one side of the plate. It is also possible to initiate subcultivations by placing a strip of agar-containing colonies directly into broth. Broth to broth transfers are done in the usual manner.

Artifacts Simulating Mycoplasmas

Common errors made when attempting to identify mycoplasma colonies on agar usually are attributable to:

1. Air bubbles. Small bubbles trapped on the agar surface when the plates are poured can mimic mycoplasma colonies since they form craters in the agar

[8] L. Dienes, M. W. Ropes, W. E. Smith, S. Madoff, and W. Bauer, *N. Engl. J. Med.* **238**, 509–515, 563–567 (1948).

[9] L. Dienes and H. W. Weinberger, *Bacteriol. Rev.* **15**, 245–288 (1951).

and often reflect the light as do mycoplasma colonies. Bubble formation can be prevented by stroking the agar surface with a Bunsen burner flame when plates are prepared and before the agar has solidified.

2. Water condensation. This is created by syneresis of the agar gel and deposition of microdroplets on the agar surface during incubation of plates under humid conditions. This condensation artifact cannot reasonably be controlled; however, application of criteria detailed above should easily discriminate between microscopic water droplets and mycoplasma colonies.

3. Tissue culture cells. Depending on the cell type being tested, their resemblance to mycoplasma colonies can be very great. A cell spread on the agar surface can easily be confused with a mycoplasma colony in which the nucleus appears to represent the embedded portion of the colony. Clusters of small cell aggregates can also be misleading. Serial subcultivation of the suspicious area can usually circumvent this pitfall.

4. Pseudocolonies. These artifacts probably represent the most serious source of error. They form on agar plates having a high serum content similar to the aforementioned formula. The detailed description of these artifacts by Brown *et al.*[10] and by Hayflick (see footnote 12) should be consulted by anyone first attempting studies with the mycoplasmas. The characteristics of these pseudocolonies are such that their mimicry of mycoplasma colonies is truly extraordinary. The pseudocolonies are composed of calcium and magnesium soaps which form crystalline structures on the agar surface. They form slowly at 37°C and by crystalline "growth" reach maximum size in about 6 to 10 days. They range in size from 50 to 150 μm in diameter. The dense central area of the pseudocolonies, where the crystalline outgrowth originates, is easily confused with the embedded central portion of mycoplasma colonies. A further source of error is the fact that pseudocolonies can be indefinitely subcultivated on agar exactly as are mycoplasma colonies. Presumably the pseudocolonies break up on a fresh agar plate during the subculturing procedure and create nidi for new crystal growth only on those areas that were streaked. This fact could easily mislead the investigator to assume that a replicating microorganism had been subcultivated.

When Is a Mycoplasma a Mycoplasma?

If a suspicious organism has not been isolated can we still call it a mycoplasma or is the optimum medium for its growth unknown? If we agree that mycoplasmas must be grown on agar in order to be identified as mycoplasmas then this issue becomes a central one. Yet it really poses a dilemma. For if you are convinced a mycoplasma exists, then you must grow it on agar. If it cannot be grown on agar then the advocates of the existence of a mycoplasma might take refuge in the position that we do not as yet have an optimum medium for the growth of all mycoplasmas, in particular, the one he is looking for. That argument may be valid but it is no less valid for other microorganisms. Consequently it can be

[10] T. M. Brown, H. F. Swift, and R. F. Watson, *J. Bacteriol.* **40**, 857–866 (1940).

argued that optimum media for the growth of all bacteria, fungi, viruses, and rickettsia are also not known. The point is that condition X could be caused by a mycoplasma that cannot be grown in conventional culture, or by a bacterium that cannot be grown in conventional culture, or by a virus that cannot be grown in conventional culture or by a rickettsia that cannot be grown in conventional culture. But why single out a mycoplasma when any organism that could not be grown would qualify? If you agree that mycoplasmas might exist that cannot be grown on agar then you must be prepared to agree that some viruses or rickettsia might exist that can. We find ourselves on the horns of this dilemma because an optimum medium is *ipso facto* defined as optimum when we can grow the organism we want on it! If we cannot, it is not! Some advocates of this principle of nihilism will further postulate that there exists a class of mycoplasmas that simply cannot, under any circumstances, be cultivated, but that they are mycoplasmas nonetheless. Again the same thing can be said of any other microorganism. Such a negativistic approach will get us nowhere, for it can be said with absolute certainty that all mycoplasms can be cultured on agar. I will take the position, with the full support of most mycoplasmologists, that until it is cultured on agar it cannot be called a mycoplasma.

Ecology of Mycoplasmas Found as Cell Culture Contaminants

From the time of the first reported mycoplasma contamination of cell cultures by Robinson *et al.*[11] in 1956, and the subsequent finding of these microorganisms as widespread cell culture contaminants, the source of this contamination has been a mystery. Nevertheless, several clues as to source may be provided by knowledge of the ecology of the mycoplasma species contaminating cell cultures. From the time of the first report in 1956 until approximately 9 years later in 1965, the predominant species found were virtually all human species and in particular *M. hominis* and *M. orale*.[12] Since both of these species are normal inhabitants of the human oral cavity, it was concluded that faulty aseptic technique accounted for the finding of human mycoplasma species in a variety of cell cultures. In the next 5 years, from 1966 to 1971, the number of human mycoplasma species found in the approximately 7000 cell cultures examined by us diminished considerably as they were largely eclipsed by animal species, predominantly *M. hyorhinis*.[12,13] Since *M. hyorhinis* is a swine mycoplasma, we theorized that the widespread finding of *M. hyorhinis* in cell cultures might be attributed to contamination of trypsin which is universally used by cell culturists. This conjecture still remains unproved despite our negative finding of mycoplasmas in several lots of crude trypsin tested for mycoplasma content prior to filtration.

In the past year another change has taken place in the ecology of mycoplasma species contaminating cell cultures. The predominant species found by us in

[11] L. B. Robinson, R. H. Wichelhausen, and B. Roisman, *Science* **124**, 1147 (1956).

[12] L. Hayflick, *Tex. Rep. Biol. Med.* **23** (Suppl. 1), 285–303 (1965).

[13] R. H. Purcell, N. L. Somerson, H. Fox, D. Wong, H. C. Turner, and R. M. Chanock, *J. Nat. Cancer Inst.* **37**, 251–253 (1966).

over 200 contaminated cell cultures from dozens of laboratories is *A. laidlawii* and occasionally *M. arginini*. *Mycoplasma hyorhinis* and the human species have all but disappeared. Simultaneous with this observation has come the important contribution of Barile and Kern,[14] 1971, in which they have shown that a substantial proportion of commercial calf serum was contaminated with *M. arginini* and *A. laidlawii*. They have shown further that the reason these contaminants have gone unrecognized until this time is that their presence in serum is at such a low concentration that sample sizes of 25 ml or greater must be used in order to detect growth. Of further interest in this regard is our finding that a single lot of commercial horse serum used by us as the supplement in our mycoplasma growth medium[12] was found to be contaminated with *A. laidlawii*.[15] It is, therefore, of great importance that the horse serum supplement usually used by mycoplasmologists be tested itself for contamination prior to use or false positives will be obtained.

Prevention of Contamination

In view of the almost total absence of mycoplasma contamination in cell cultures carried in laboratories not using antibiotics, this approach is probably the most effective. It is likely that the absence of antibiotics from culture media does not per se contribute to the absence of mycoplasmas but that the greater skill necessary for maintaining asepsis in antibiotic-free cultures prevents mycoplasma contamination from the worker himself or from already contaminated cultures received from other laboratories.

An alternative to cell cultivation in antibiotic-free media is (1) cultivation in the presence of antibiotics known to be effective against the mycoplasmas and (2) use of sera pretested for the absence of mycoplasmas. Few resistant strains have been encountered when using chlortetracycline.[16,17] Chlortetracycline (Aureomycin·HCl, crystalline intravenous, Lederle product No. 4691-96, 500-mg vial) is used at a concentration of 50 μg/ml in cell culture media. This product is reconstituted in 50 ml of warm (37°C) sterile distilled water and agitated to ensure a clear amber solution. Five-milliliter aliquots, each sufficient for 1 liter of medium, are dispensed and stored at -20°C.

[14] M. F. Barile and J. Kern, *Proc. Soc. Exp. Biol. Med.* **138**, 432 (1971); cf. next chapter.
[15] L. Hayflick and N. Pleibel, unpublished observations, 1971.
[16] L. Hayflick, unpublished results.
[17] L. Hayflick, *Exp. Cell Res.* **37**, 614–636 (1965).

CHAPTER 11

Mycoplasma Contamination of Cell Cultures: Incidence, Source, Prevention, and Problems of Elimination

Michael F. Barile

A test for mycoplasmas is required for viral vaccines produced in cell cultures. Consequently, our laboratory has maintained a continuing study to examine the incidence, source, prevalence, and spread of contamination and also studies to evaluate procedures for the primary isolation of, the prevention of, and the elimination of mycoplasmas from contaminated cell cultures.

Isolation of Mycoplasmas

Several procedures are used, including: (a) a standard culture procedure for isolating mycoplasmas from contaminated cell cultures; (b) a more sensitive large specimen–broth culture procedure for isolating small numbers of mycoplasmas from minimally infected or contaminated specimens; and (c) a semisolid broth culture procedure for screening cell cultures for microbial (mycoplasma, bacteria, fungi) contamination.

Standard Culture Procedure.[1] The procedure used successfully for isolating mycoplasmas from contaminated cell cultures is as follows: The Edward-Hayflick formula consisting of 70 parts basal medium (3% brain-heart infusion or trypticase soy), 20 parts of horse serum (Baltimore Biological Laboratories, Tissue Culture Select), and 10 parts of a 25% autoclaved extract of fresh yeast (Standard Brand Yeast No. 2040) is used. This basic medium is supplemented with energy sources (0.5% dextrose and/or 0.5% l-arginine hydrochloride), 0.002% thymic nucleic acid (DNA) (Mann), or diphosphopyridine nucleotides (DPN) (Mann), vitamins (as used in Eagle's tissue culture medium), antibacterial agents (1:2000 parts thallium acetate and 1000 units/ml of penicillin G), and a dye indicator (0.002% phenol red) buffered at pH 7.2 to 7.4. The following modifications may be made: (a) add 0.05% Noble agar for semisolid broth medium; or (b) add 1.0% Noble agar for solid medium; or (c) add 0.5% urea, delete dextrose and arginine, and adjust pH to 6.0 for growth of T-strain mycoplasmas.

[1] M. F. Barile, *In* "Contamination in Tissue Culture" (J. Fogh, ed.), in press. Academic Press, New York.

For culture, 1 ml and 0.1 ml of exhausted cell culture medium ("pour-off") fluids or of cell suspensions are inoculated into 10 ml of standard broth and onto 10 ml of agar medium, respectively. The broth cultures are incubated aerobically at 36° ± 1°C, and then subcultured to agar media when the culture becomes turbid, or at weekly intervals for 4 weeks. Duplicate agar media are inoculated; one is incubated aerobically and the other in a 5% carbon dioxide in nitrogen (5% CO_2 in N_2) atmosphere.[1] The agar cultures are examined (at 50–100X magnification) for mycoplasma colonies for at least 3 weeks before being considered negative. For subculture, blocks (1-cm^2) of agar culture are inoculated into broth or onto agar media by "friction transfer;" that is, the agar culture is placed colony face down and pushed across the surface of the agar medium. Agar plates are placed in jars and/or incubated in humidified incubators to prevent desiccation. Since mycoplasmas lose viability on prolonged incubation, early (2- to 7-day-old) broth and agar cultures are used for subculture.

Large Specimen–Broth Culture Procedure.[1] Twenty-five milliliters of liquid specimen (e.g., commercial bovine sera or tissue culture suspension) are inoculated into 100 ml of standard broth medium. To increase sensitivity, 50 or 100 ml of specimen are inoculated into 400 ml or more of broth medium. The broth culture is incubated at 36° ± 1°C and then subcultured to agar media, in duplicate, when the broth becomes turbid or at weekly intervals for 4 weeks. The agar media are incubated aerobically and in a 5% CO_2 in N_2 atmosphere. Each primary isolation broth culture is stored frozen at −70°C and maintained for mycoplasma identification studies.

Semisolid Broth Culture Procedure.[1] We have developed a semisolid broth (SSB) culture procedure for screening cell cultures for microbial (mycoplasma, bacteria, and fungi) contamination. To prepare SSB medium, a small amount (0.05%) of Noble agar is added to the broth medium described above, except that the antibacterial agents (penicillin and thallium acetate) are deleted from the medium. The SSB medium is dispensed in screw-capped tubes and stored at refrigerated temperatures until used. For culture, 1 ml of tissue culture specimen is inoculated into 10 ml of medium. To increase sensitivity, larger quantities (25 ml or more) of specimen are inoculated into 100 ml or more of medium. The cultures are incubated at 36 ± 1°C and observed periodically by visual examination for the presence of microbial growth during a 4-week period. Uninoculated broth tubes are included in each test and serve as negative controls. The additional agar produces an oxygen gradient, providing an oxygen tension suitable for the growth of most contaminants. It also permits mycoplasmas to produce dense layers of luxuriant turbid growth with microcolony formation. The arginine-using mycoplasmas produce an alkaline shift and the fermentors an acid shift in pH, each with corresponding changes in the color of the culture. The development of turbidity and color changes in the culture aids in detecting the presence of growth by visual examination. Thus, the SSB procedure has several advantages; it requires only a few basic materials (media and tubes), provides an oxygen gradient for the growth of most mycoplasmas, bacteria, and fungi which contaminate cell cultures, and growth is detected readily by visual examination. In addition, the SSB medium provides for a very sensitive culture procedure since large quantities of specimen can be examined. This medium also provides a good stabilizing menstruum for storage of mycoplasma cultures, i.e., either at refrigerated (2°–5°C) or freezing (−70°C) temperatures. In fact, some mycoplasma cultures can maintain their viability in SSB at 37°C for many weeks. On the other hand, SSB has some disadvantages: (a) it cannot be used for identification of mycoplasmas. Mycoplasmas are identified on the basis of colony morphology and agar cultures must be used; and (b) it will not support the growth of all species of mycoplasmas. However, no one medium can support the growth of all mycoplasmas. Thus, the SSB culture procedure can grow most microbial contaminants and it provides an improved broth culture procedure for detection and surveillance of microbial contamination.

Incidence of Mycoplasma Contamination

Since mycoplasmas are frequently present in oral and genital tissues of man and animals, and in neoplastic tissues, cell cultures prepared from these tissues would theoretically have a greater contamination risk. Nonetheless, primary cell cultures are rarely contaminated (1–3%), and the original tissues are apparently not a major source of contamination. On the other hand, continuous cell cultures are frequently contaminated (45–92%). Most contamination occurs during cell propagation, and comes from outside sources.[1]

Contamination occurs in cell cultures derived from all animal species tested, including human, nonhuman primates (rhesus, vervet and grivet monkey, chimpanzee), swine, dog, rabbit, hamster, mouse, rat, chicken, quail, duck, snake, fish, and insect. Cell cultures derived from plant tissues and from all types of animal cells and tissue cultures are subject to mycoplasma contamination, including primary, continuous, diploid and heteroploid cell cultures, tissue cultures, fibroblast and epithelial cell cultures, lymphocyte and macrophage cultures, cells grown in monolayer and suspension, and cells derived from normal, infected, or neoplastic tissues. The mycoplasma titers in contaminated cell cultures range from 10^5 to 10^8 colony-forming units/ml (CFU/ml) of medium fluids, regardless of cell type and origin. Occasionally a cell culture line is found which may be resistant to a given mycoplasma.[1]

The lowest incidence of contamination has been found in cells grown in small volumes and observed carefully for morphological changes consistent with contamination. The highest incidence of contamination has been found in cells used to propagate microorganisms or in cells supplied by the cell culture producer.[2] These cells are produced on a large scale, grown in large volumes, and in large numbers of containers, conditions which provide a high contamination risk. Contamination occurs more frequently in cell cultures grown in media containing antibiotics and/or serum. Antibiotics provide a false sense of security, and may

[2] M. F. Barile, W. F. Malizia, and D. B. Riggs, *J. Bacteriol.* 84, 130 (1962).

be used in lieu of rigid sterile procedures. Serum may provide a better medium for the growth of mycoplasmas, or serum may be a source of contamination.

Sources of Contamination

The 1063 contaminants isolated in our study[1] were identified as either human, bovine, swine, murine, avian, or canine species of mycoplasmas. Thus, the sources of contamination are of human, bovine, swine, murine, avian, and canine origin.[1] However, most of the contaminations (99%) were caused by either human, bovine, or swine mycoplasmas.[1]

Human Sources. The human oral and genital species of mycoplasmas represent one of the largest group of contaminants and, therefore, the investigator and his environs are a major source of contamination. Of the 1063 mycoplasma contaminants identified,[1] 442 were human oral or genital strains of mycoplasma: 370 (35%) were identified as *M. orale,* type 1, 67 (6%) were *M. hominis,* and 5 were *M. fermentans.* The most common human mycoplasma contaminant of cell cultures was *M. orale,* type 1, which caused more than one-third (35%) of the contaminations. The major vehicle of *M. orale,* type 1 contamination is the mouth pipette. Saliva is introduced into the cell culture and the antibiotics present destroy the bacteria and allow the mycoplasma to fluorish. In medium without antibiotics, the cell culture becomes grossly contaminated with bacteria, is destroyed and discarded. Contamination by the human genital mycoplasmas is probably due to the use of faulty or inadequate sterile procedures.

Bovine Sources. Of 400 (38%) bovine mycoplasmas isolated,[1] 239 (22%) were identified as *M. arginini,* 75 (7%) were *A. laidlawii,* and 86 (8%) were unclassified strains which represent several new species. Contaminated commercial bovine serum is the major source of bovine mycoplasma contamination.[3,4]

Swine Sources. The third major group of contaminants are the swine species of mycoplasmas: 204 (19%) of the 1063 contaminants were identified as *M. hyorhinis.* Although the source of swine contamination is unknown, trypsin has been incriminated because it is prepared from swine tissues and is "universally" used in the preparation of cell cultures. However, all attempts to isolate mycoplasmas from commercial trypsin have failed. Commercial bovine sera may be a source of swine mycoplasma contamination. Since swine and cattle are frequently processed through the same slaughter houses, swine myocplasmas may contaminate bovine sera during collection and manufacture and, in fact, one strain of *M. hyorhinis* was isolated from a lot of commercial bovine serum.[1]

Original Tissue Source. The original tissues used to prepare the primary cell culture can be a source of contamination. Three canine and one avian myco-

[3] M. F. Barile and J. Kern, *Proc. Soc. Exp. Biol. Med.* **138,** 432 (1971).

[4] M. F. Barile and R. A. DelGiudice, *In* "Pathogenic Mycoplasmas," Ciba Foundation Symposium, pp. 165–185. Elsevier, Amsterdam, 1972.

plasmas were isolated from three *primary* canine kidney and one *primary* chick embryo cell culture lots, respectively. In these cases, the original tissues were the probable source of contamination. However, the original tissues are overall a minor source (1%) of contamination.

Prevalence of Mycoplasma Contamination

Human, bovine, and swine species of mycoplasmas caused 1046 (99%) of the contaminations observed.[1] The prevalence of mycoplasma contamination determined for the periods 1960–1964, 1965–1968, and 1969–1972 indicated that there have been three waves or "epidemics" of cell culture contamination; (1) the first wave of contamination occurred in the early 1960's and was caused by the human mycoplasmas. The major vehicle of contamination was the mouth pipette; (2) the second wave occurred during the mid-1960's and was caused by the swine mycoplasmas (primarily *M. hyorhinis*). Of 204 *M. hyorhinis* isolated, 192 (94%) were isolated from 1965 to 1968. By 1970 the incidence of swine contamination dropped drastically to 1% of the contaminants; and (3) the third wave of contamination began during the late 1960's and was caused by the bovine mycoplasmas. Presently, more than 50% of all mycoplasma contaminants are of bovine origin.

Spread of Contamination

Contamination can be spread from cell to cell either by contaminated virus pools, antisera, or by other reagents commonly used in cell culture studies. Contamination can be spread also by aerosols, either of contaminated cell culture materials, or of contaminated secretions (that is, discharges from either oral, nasal, genital, or intestinal tissues) of man and animals.[1]

Prevention of Mycoplasma Contamination

The measures used are designed to control the sources and the spread of contamination. The major sources of contamination are of human, bovine, and swine origin. The major vehicles of spread are aerosols and contaminated reagents. The measures used successfully for reducing the risk of mycoplasma contamination include the following: (a) use primary cell cultures whenever feasible; (b) eliminate the use of mouth pipetting which is the major vehicle of human oral mycoplasma contamination; (c) use rigid sterile procedures, aseptic technique, and good personal habits to prevent contamination by human genital mycoplasmas; (d) treat commercial bovine sera, prior to use, to remove or inactivate potential contaminants; for example, either heat serum at 56°C for 30 minutes, or filter serum twice through a 220-nm filter, or add serum to tissue culture medium and then filter the complete medium twice through a 220-nm filter. Whenever feasible, use a serum substitute instead; (e) treat trypsin, prior to use, by filtering prepared trypsin through either a 100-nm filter once or a 220-nm filter twice. Whenever feasible, use a trypsin substitute instead; (f) use absolute Cambridge-type filters to monitor the affluent and effluent air; (g) use laminar flow hoods; (h) decontaminate the working area daily and the room weekly; (i) discard contaminated cell cultures immediately and replace with mycoplasma-free cultures; (j) monitor and treat contaminated virus pools, antisera, and other reagents to inactivate or remove the mycoplasma contaminants. The treatment of choice, when feasible, is either ether or chloroform inactivation; for example, mix 1 ml of ether with 1 ml of the contaminated virus pool, incubate the mixture at 22° to 25°C overnight, remove ether by evaporation at 37°C, and test the virus pool for mycoplasma by culture procedures. Another procedure used successfully is the combined use of an antibiotic [which was pretested using growth (disk) inhibition and found effective] at high concentrations and filtration, using either a 100-nm filter once (when feasible) or a 220-nm filter twice; for example, add 10,000 μg/ml of kanamycin to the pretested contaminated virus pool, incubate at 22° to 25°C overnight, filter as above, and test the virus pool for mycoplasmas.

Elimination of Mycoplasma Contamination

The published procedures[1] used include: (a) antibiotics (tetracycline, chlorotetracycline, oxytetracycline, kanamycin, tylosin, and novobiocin); (b) prolonged heat at 41°C for 18 hours; (c) specific high-titered neutralizing antiserum; (d) aurothiomalate; (e) Tricine; (f) Triton X; (g) photodynamic inactivation; and (h) the combined use of an antiserum and an antibiotic. Although many reports[1] conclude that mycoplasmas are readily eliminated from contaminated cells, this widespread optimism is unfounded. To the best of my knowledge and experience, there is no simple, rapid, and universally effective procedure for eliminating mycoplasma from contaminated cells. Moreover, mycoplasma contaminants develop resistance to antibiotics. Treatment is prolonged, toxic to cells, and may induce selection with serious consequences since surviving cells may not retain all of the original properties of the culture. Thus, attempts to eliminate mycoplasma from contaminated cells must be considered only as the last resort. Instead, the investigator must use preventive procedures to reduce the risk of contamination, must examine cell cultures routinely for mycoplasma contamination, and must discard contaminated cells immediately and replace them with mycoplasma-free cell cultures. When a cell culture cannot be replaced and must be treated, the treatment of choice, in our experience, is the combined use of high-titered neutralizing antiserum and an antibiotic. The antibiotic and antiserum must be pretested using the disk growth inhibition procedure and shown to be effective before use. We recommend that either tetracycline or kanamycin be considered. The concentration of antibiotic is predetermined as that concentration which falls between the effective dose for the mycoplasma and the toxic dose for the cell culture. We have used either 10 μg/ml of pretested tetracycline or 100 μg/ml of pretested kanamycin. The antibiotic and a 5% concentration of high-titered neutralizing antiserum are added to medium, and the contaminated cell culture is grown in medium containing the antibiotic and antiserum for at least

4 weeks with weekly subcultures and/or changes of medium. The "pour-off" medium fluids or cell suspension are examined periodically for the presence of mycoplasma. In most cases, the treatment causes a rapid drop in the mycoplasma titer, but low orders of mycoplasma contamination can persist for extended periods of time. Treatment must be continued even though the results of culture may appear to be negative. To establish an effective "cure," treatment is stopped, the cell culture is grown in medium without antibiotic and antiserum for 4 weeks, and the cell culture shown to be negative for mycoplasma using the large specimen–broth culture procedure; that is, a 25 ml of cell culture suspension is inoculated into 100 ml of broth medium (see the "large specimen–broth culture procedure"). If the mycoplasma contamination persists, the procedure is repeated using an increased concentration of the pretested antibiotic and antiserum. The persisting mycoplasma contaminant must be identified since mixed contamination of cell cultures with two mycoplasma species is not uncommon.[1]

CHAPTER 12

Elimination of Pathogens from Shoot Tip Cultures

Ralph Baker and Harold Kinnaman

Shoot tip culture to eliminate pathogens from foundation plants, especially carnations, has been practiced at Colorado State University for over a decade.[1] Propagative material derived from shoot tips is now used almost exclusively in the Colorado carnation industry. Losses due to plant pathogens have been reduced from 1 to 65%, before the program was initiated, to approximately 0.07% in 1970–1971. While the methods are simple and have been successfully accomplished by part-time student employees, the overall operation requires dedication and training in plant pathology, plant culture, and greenhouse management, as well as a knowledge of tissue culture techniques.

Eliminating Pathogens from Plants

Many cultivars of vegetatively propagated plants harbor one or more pathogens carried either externally[2] or internally.[3] The sterile procedures used in the

[1] R. Baker and D. J. Phillips, *Phytopathology* **52**, 1242 (1962).
[2] L. J. Petersen, R. Baker, and R. E. Skiver, *Plant Dis. Rep.* **43**, 1204 (1959).
[3] P. E. Nelson, J. Tammen, and R. Baker, *Phytopathology* **50**, 356 (1960).

shoot tip culture method ensures elimination of all of these pathogens except viruses; in the unlikely event that a bacterial or fungal pathogen is present, the contamination in the nutrient solution supporting the developing shoot tip is readily detected.

Some viruses are distributed erratically in plants. Opportune removal of the small piece of virus-free tissue from the apex of an infected plant thus might eliminate such a virus. If this does not occur, however, other procedures are necessary.

While chemotherapy may be effective in eliminating some viruses from plants, heat therapy has been the most successful and widely used method.[4] Research indicates that virus concentration is lowest in the apical meristem. If the host is grown at high temperature, the pathogen may not be present in this area. In practice, actively growing plants are placed in a thermotherapy chamber and exposed to temperatures of 35° to 40°C. Time of exposure varies from a few hours to months depending upon the characteristics of the virus. We have consistently eliminated all viruses from carnation shoot tips with continuous heat treatment of 38°C for 2 months. Preconditioning of the plants is necessary. Over a period of 2 weeks, the temperature is gradually raised to 38°C. Relative humidity must be at least 85–95%. Plants also are grown in inert media (gravel) supplemented with peat and automatically watered with nutrient solution every 6 hours.

Shoot Tip Culture

Carnation shoot tip culture has been described in detail by Phillips.[5] All leaves from a cutting are removed by hand except for the last 0.5 cm at the tip. All remaining leaves except for the last pair are removed with a flamed sterile scalpel (Bard-Parker No. 1.1 blade). The shoot tip is excised by cutting immediately underneath the remaining leaf pair with the tip of the scalpel blade. This tip (approximately 0.25 mm) is then placed in a previously autoclaved culture tube.

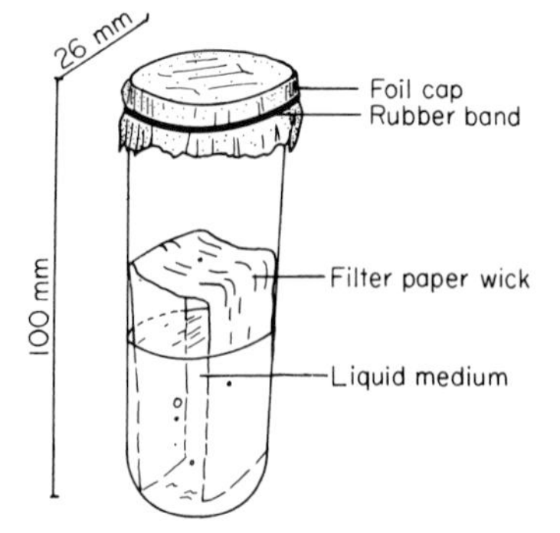

Fig. 1. Tube used for shoot tip culture.

[4] M. Hollings, *Annu. Rev. Phytopathol.* **3**, 367 (1965).
[5] D. J. Phillips, *Colorado St. Univ. Exp. Sta. Tech. Bull.* **102**, 22 (1968).

The culture tube is shown in Fig. 1. It contains liquid nutrient medium, a filter paper wick for support of the shoot tip, and is capped with foil securely fastened by a rubber band. The foil covering not only prevents desiccation but is most effective in reducing contamination.

The composition of the nutrient medium is dictated by the requirements of the species of plant used. We have found a modified Morel's medium very satisfactory for growth of carnation shoot tips (tabulated below):

Component	Amount	Component	Amount
$Ca(NO_3)_2 \cdot 4\ H_2O$	0.5 g	Thiamine	1 mg
KNO_3	0.125 g	Adenine sulfate	8 mg
$MgSO_4 \cdot 7\ H_2O$	0.125 g	Indole	1 mg
KH_2PO_4	0.125 g	Berthelot solution[a]	0.5 ml
$Fe_2(SO_4)_3 \cdot X\ H_2O$	0.025 g	Glucose	40 g
Naphthaleneacetic acid	1 mg	Water (distilled)	1000 ml

[a] Berthelot's solution is prepared by adding the following chemicals to a liter of water: $MnSO_4 \cdot 4\ H_2O$, 2 g; KI, 0.5 g; $NiCl_2 \cdot 6\ H_2O$, 0.05 g; $CoCl_2 \cdot 6\ H_2O$, 0.05 g; $ZnSO_4 \cdot 7\ H_2O$, 0.1 g; $CuSO_4 \cdot 5\ H_2O$, 0.05; H_3BO_3, 0.05 g; concentrated H_2SO_4, 1 ml.

After the tips have rooted and achieved maximum weight (usually approximately 25 mg dry weight) the plant may continue to grow slowly but often is chlorotic and difficult to transplant. Maximum weight is correlated with a shift in pH from 4.8, of the original nutrient solution, to about 6.5. Thus a dye solution may be incorporated in the medium which changes color at the appropriate pH and indicates when the shoot tips should be transplanted. This pH indicator is prepared by adding 400 mg of bromthymol blue to 1000 ml of weak NaOH (0.7 g NaOH/liter). This is added to the nutrient solution at 10 ml/liter.

Desiccation of the tip and excess callus formation may occur when agar is used as a support. For this reason we use a folded filter paper strip inserted into the tube forming a bridge supporting the shoot tip above the liquid medium.

It is possible to premix large quantities of the dry ingredients of the nutrient medium, place them in storage, and then reconstitute small amounts of the mixture as required. All chemicals necessary for 10 or 15 liters of complete medium, except $Ca(NO_3)_2$, are weighed and placed in a desiccator for 1 week over Drierite or other suitable desiccant. When dry, $Ca(NO_3)_2$ (anhydrous) is added in an amount equivalent to the concentration of $Ca(NO_3)_2 \cdot 4\ H_2O$ used in Morel's medium. The mixture is homogenized for 30 to 45 minutes into a powder in a porcelain ball mill containing one-third volume of ½ inch "burundum balls." The milled medium may be placed in moisture-proof plastic envelopes and stored at 5°C until needed.

Transplanting and Storage

When the shoot tips have achieved maximum growth in the culture tubes and are well rooted (Fig. 2), they may be transplanted. If desired, however, shoot tips may be stored for as long as a year at 2°C before transplanting as long as the tubes are securely capped so that there is no water loss.

Fig. 2. Tip grown for 6 to 8 weeks. Some leaves and roots are out of focus.

The young plants may be transferred into any well-drained medium. We use a potting mixture containing equal amounts of peat and horticultural additives such as Perlite (to improve drainage) and Terragreen (to neutralize the acidity of peat). Greenhouse temperature should be kept at about 21°C and certainly not below 10°C for carnations. Plants should be watered with a conventional nutrient solution. Supplemental light and high relative humidity furnished by an intermittent mist system may also be beneficial to young plants.

Virus Indexing

Not all plants obtained from shoot tips are virus free. Hollings[4] has reviewed means for detecting viruses if they occur in such stock. The most sensitive methods involve transfer of possible viruses to suitable indicator hosts. We use *Chenopodium amaranthicolor* or *Saponaria vaccaria*[6] for the detection of mottle which is the most difficult virus in carnation to eliminate with a combination of heat treatment and shoot tip culture.

[6] F. A. Hakkaart and M. Van Olphen, *Neth. J. Plant Pathol.* **77,** 127 (1971).

EFFICIENCY

Use of these procedures for carnations has resulted in nucleus blocks of plants free from all known pathogens. Shoot tips have produced vigorous plants throughout the year. Indeed, tips derived from most cultivars have a survival rate of 90 to 100%; of these 90–95% have been successfully transplanted. Since most propagative material is known to produce occasional "sports" and the effect of shoot tip culture and heat treatment on mutation frequency is not known, it is desirable to couple the procedures described above with an adequate selection program to screen out undesirable genetic aberrations.[7]

[7] W. D. Holley and R. Baker, "Carnation Production," p. 142. Wm. C. Brown Co., Dubuque, Iowa, 1963.

CHAPTER 13

Contamination of Tissue Culture and Its Cure: Some Examples

W. House

Faulty aseptic technique and overconfidence in antibiotics are the main causes of contamination in tissue cultures. Contamination may be confined to one laboratory or be connected with the individual technician, or may be more widespread. The following examples of contamination, their investigation, and the steps taken to right the situation were obtained from the records of two large research institutes during the past decade.

CONTAMINATION CONFINED TO ONE LABORATORY

Description. A mixed flora of bacteria and fungi were present in the cultures in one laboratory, but not in others in the building.

Investigation. An abnormally high bacterial count was found in swabs taken from floor, benches, incubators, and water baths. Agar plates were exposed at various sites in the laboratory for various times during the day. The bacterial counts were low in the morning, rose during the day, and fell after working hours. Contamination was heaviest in one particular work place, suggesting that the person working in this place was in some way responsible, and that contamination spread from this focus during the day.

Tests of the taps, sinks, and water pumps proved that the vacuum system (flask, tubing, and water pump) was heavily contaminated with a mixed flora of bacteria and fungi.

Remedy. The vacuum system was used for removing tissue culture medium from dishes of cell monolayers. This medium was held in the vacuum flask for periods of up to 3 days so that an excellent food supply for microorganisms was available. Continuous variation in pressure within the system would foster the creation of bacterial and fungal aerosols. It had been the practice to discard the contents of the flask when it was full, and occasionally to cover the bottom of the flask with a mild disinfectant. This system was obviously inefficient.

The contamination ceased when the vacuum system was renewed daily with washed and sterilized components.

Contamination from Animal Tissue

Description. On two successive occasions, the production batch of secondary cultures of mouse embryo cells was contaminated with yeast.

Investigation. Contamination became apparent 5–7 days after subculturing from primary cells. There was no evidence of similar contamination in the continuous cell lines carried by the cell production laboratory, or in the routine control checks of tissue culture medium, serum, and trypsin. The possibility that the contamination arose from the work area, the operator, equipment, or the animals themselves was then investigated.

A batch of pregnant mice was divided between the regular operator and another technician. One set of equipment and instruments was sterilized by autoclave, the other set by hot-air oven. Both technicians used an identical preparation method, and the tissue culture medium, serum, and trypsin, etc., were from the same batches.

Samples were collected by each worker at each stage of the technique, e.g., washings from the embryos, first tissue mince, tissue mince washing, and so on through trypsinization to the suspension of cells in medium prior to dispensing. Nutrient broth was added to each of the collected samples and these were incubated at 37°C.

The incubated samples collected by both workers were contaminated with yeast after 3 days incubation, contamination occurring throughout the process beginning with the first washings from the embryos.

The cell cultures were checked daily from day 2 onward by removing an aliquot of medium from each bottle of the primary cells, and centrifuging and streaking the sediment on blood agar plates. The plates were incubated at 37°C for 24 hours, and at room temperature for a further 24 hours. Of 48 bottles checked, 14 were shown to be contaminated with yeast by the fifth day.

We informed our mouse supplier of the results, and he sent us a batch of pregnant mice from a reserve breeding colony. This batch was processed as a single entity, samples were taken at each stage of the operation, and also from each bottle of primary cells as before. There was no contamination by yeast or any other organism.

Remedy. It was arranged with the mouse supplier to meet our future needs from the reserve breeding colony, and, if possible, to discard the yeast contaminated colony.

The testing procedures described in this investigation were retained as a routine precautionary measure. Samples taken at each stage of the procedure were held for 3 weeks before discarding, and primary cell cultures were tested 24–48 hours before subculturing into secondary cell cultures. Bottles of primary cultures which showed evidence of contamination were discarded, and if more than 10% of the cultures showed contamination the testing was repeated to establish whether we should use any of this batch of cells. If the contamination rate was greater than 35%, none of that batch was used.

Widespread Contamination

Description. Contamination of primary and secondary cell cultures as well as continuous cell lines occurred in seven out of ten laboratories.

Investigation. The contaminant was a gram-negative bacillus. It became obvious on examination that, although most of the laboratories were involved, most of the contaminated cell cultures had been recently supplied by a central laboratory. The environment, technique, and personnel of the central laboratory were, therefore, investigated.

Sterile batches of serum, trypsin, and Versene, which were reserved for the cell production laboratory, were tested again and still found to be sterile. Floor, benches, air inlets, water baths, and incubator were examined. The water baths were heavily contaminated with a gram-negative bacillus, identical (by simple fermentation tests) with the tissue culture contaminant.

In the absence of a member of the cell production staff, the daily, routine, cleaning of the water baths, and refilling with sterile water had been neglected.

Remedy. It was assumed that the contaminant entered the bottles of serum while they were thawing in the water bath. In some cases, the level of the water may have been high enough to seep below the caps by capillary action. Alternatively, the contaminated water might seep under the cap when the bottle was inverted on removal from the water bath in order to mix the serum.

The daily routine of replacing sterile water in the bath was resumed, and the outside of the bottles were dried on their removal from the bath—contamination ceased. A recommendation was that the medium should in future be thawed in an incubator instead of water bath.

Contamination through Filtration Failure

Description. In our routine procedure when medium or serum is sterilized by filtration, every tenth bottle is tested for contamination. On this occasion eight out of ten of such test bottles were contaminated—the medium contained gram-negative bacilli and the serum contained gram-positive cocci and gram-positive and -negative bacilli.

Investigation. The contamination clearly arose from failure of sterilization when the bottles and filters were autoclaved, or from faults in the filters themselves. The autoclaved bottles proved to be sterile when sterile broth was incubated in them for 72 hours at 37°C or at room temperature. The autoclaved filters, however, did not prevent the passage through them of *E. coli* suspended in nutrient broth; *E. coli* grew in the filtrate when it was incubated for 24 hours at 37°C.

Close inspection of the filters showed no obvious flaws, so their porosity was checked. This indicated a porosity of greater than 0.7 μm instead of the required 0.2 μm.

Remedy. It was never established how the wrong filters came into use, whether through faulty labeling by the manufacturers or switching of the labeled lids by one of our own staff. The manufacturers justifiably said that contamination would not have occurred if we had used the "bubble test," which they recommend as standard procedure.

The routine was altered by using two 0.2 μm filters in series in separate filter holders; parts of the second filter were tested for contamination by incubating in nutrient broth and on agar plates at different temperatures suitable for the growth of likely contaminants.

This was a more convenient method than incorporation of the "bubble test," although it increased the cost. However, it had the advantage of testing all, instead of only a part, of the batch, and the effects of antibiotics on the test system were eliminated.

Contamination by Inefficient Sterilization

Description. Many cell monolayer cultures were contaminated within 24 hours of seeding.

Investigation. Gram-positive cocci, gram-negative bacilli, and fungi were all present, but the predominant organism was a gram-positive, spore-bearing bacillus. This last organism appeared in Pyrex Petri dishes, but not in plastic ones.

Fifty Pyrex dishes were tested by adding 5 ml of nutrient broth to each and incubating at 37°C for 72 hours—the cultures remained sterile. This confirmed the evidence from the indicator tubes placed routinely in the hot-air oven each sterilizing cycle. However, on testing incubators, gram-positive bacilli were found in small numbers in nine out of twenty-three incubators.

This suggested that the bacilli present in the incubator entered the Pyrex dishes, but not the plastic ones—possibly because the lids of the latter made a better fit. However when the incubators were disinfected with formaldehyde and washed with 80% methanol, contamination continued.

The system of sterilizing the dishes was then examined. Routinely the dishes were placed in tins each holding fifty dishes, and the tins were put into a hot-air oven. It was found that an overenthusiastic worker in the sterilizing unit had removed the wire shelves from the oven in order to pack in more tins. The packed tins formed a solid mass in the oven, and the center of the mass was not sterilized at the temperature and duration of time used; indicator tubes placed at the center of the mass confirmed this.

Remedy. An indicator tube was routinely placed in a dish in the center of each tin and rules for the loading of many tins into the oven were laid down.

Other Contaminants

These examples have all described contamination with bacteria and/or fungi, which are the commonest known forms of contamination, probably because they are the easiest to observe.

I have previously described[1] instances of viral contamination in primary mouse embryo cell cultures arising from mouse poxvirus (ectromelia) and polyoma virus. In each case it was clearly established that the source of contamination was the mice.

Mycoplasma contamination is also common in cell cultures. Although it is not difficult to identify such contamination, it is extremely difficult to pin down the source. Barile and Kern[2] recently demonstrated the presence of mycoplasma in commercially supplied serum.

Another form of contamination is the introduction of cells of other lines. This is a hazard which is often overlooked, possibly because the techniques involved in study are not routinely available.

Conclusion

None of the contaminations described should have occurred—all were caused by faulty techniques, or a disregard of standard procedure. However, they all

[1] W. House, *Lab. Pract.* **17**, 587 (1968).

[2] M. F. Barile and J. Kern, *Proc. Soc. Exp. Biol. Med.* **138**, 432 (1971); cf. Chapter 11, this Section.

did occur and, although personnel may change, are likely to recur whether they are the result of carelessness, lack of supervision, or misplaced initiative.

The lesson to be learned from the above examples is that on media, sera, and cell cultures continuous routine checks are essential if a tissue culture laboratory is to function efficiently.

CHAPTER 14

Serological Identifications of Cells in Culture

A. Animal Cells by Fluorescent-Labeled Antibody[1]

C. S. Stulberg and W. F. Simpson

During the past decade it has become well recognized that there is a constant need to identify cell cultures. This is especially so where such cultures are used as reference materials.[2] Since the recognition that cell cultures could readily become contaminated by cultures of other species, several investigators have clearly shown that cell species could be identified by serological as well as by other procedures.[2-12] Recently these procedures have been extensively reviewed by Stulberg.[13] Cytogenetic methods, cytotoxicity, hemagglutination, mixed agglutination, biochemical-genetic procedures, as well as other immunological methods

[1] Supported by USPHS Grant CA-02947 from the National Cancer Institute.

[2] C. S. Stulberg, L. L. Coriell, A. J. Kniazeff, and J. E. Shannon, *In Vitro* **5**, 1 (1970).

[3] K. H. Rothfels, A. A. Axelrad, L. Siminovitch, E. A. McCulloch, and R. C. Parker, *Proc. 3rd Can. Cancer Res. Conf.*, p. 189 (1959).

[4] V. Defendi, R. E. Billingham, W. K. Silvers, and P. Moorhead, *J. Nat. Cancer Inst.* **25**, 359 (1960).

[5] K. G. Brand and J. T. Syverton, *J. Nat. Cancer Inst.* **24**, 1007 (1960).

[6] R. R. A. Coombs, B. W. Gurner, A. J. Beale, E. Christofinis, and Z. Page, *Exp. Cell Res.* **24**, 604 (1961).

[7] C. S. Stulberg, W. F. Simpson, and L. Berman, *Proc. Soc. Exp. Biol. Med.* **108**, 343 (1961).

[8] W. F. Simpson and C. S. Stulberg, *Nature (London)* **189**, 616 (1963).

[9] D. Franks, B. W. Gurner, R. R. A. Coombs, and R. Stevenson, *Exp. Cell Res.* **28**, 608 (1963).

[10] A. E. Greene, L. L. Coriell, and J. Charney, *J. Nat. Cancer Inst.* **32**, 779 (1964).

[11] S. M. Gartler, *Nat. Cancer Inst. Monogr.* **26**, 1167 (1967).

[12] J. Fogh, N. Holmgren, and P. P. Ludovici, *In Vitro* **7**, 26 (1971).

[13] C. S. Stulberg, *In* "Contamination in Tissue Culture" (J. Fogh, ed.). Academic Press, New York, in press.

will be considered by other authors in this volume. We will be concerned only with species identification of animal cells by fluorescein-labeled antibody.

Cells cultivated *in vitro* are readily identified with regard to their species of origin by the immunofluorescence method. Once labeled antisera have been prepared against cells of different species, the method is simple to perform. An added advantage of this procedure is the detection of very small numbers of cells of a different species in a much larger cell population.

Preparation of Antisera

Antisera against verified cell strains[14] can be made by inoculating guinea pigs or rabbits. Monolayers or suspension cultures can be prepared according to the method used in a particular laboratory. With monolayers, the medium is then discarded and they are washed in four changes of Hanks' balanced salt solution (BSS). The washed monolayers are then scraped from the flask or removed with a dispersing agent (e.g., Versene-trypsin) and the cells resuspended in Hanks' BSS.

The suspended cells are washed three times in Hanks' BSS, resuspended, and, depending on the injection schedule to be used, may be injected directly into animals, stored at $-20°C$, or stored viably in ampoules at liquid nitrogen temperatures[15] for subsequent inoculation. If used directly or as thawed viable cells, they are washed again and diluted to the proper density for injections as subsequently described. The washing procedure removes all serum that might stimulate antibody.

If guinea pigs are used for immunization, usually six animals are injected for preparation of a given antiserum because the potency of the antisera produced may vary among animals. Each animal yields approximately 5–8 ml of serum. The numbers of cells used for immunization may differ somewhat depending on their animal source, presumably because of antigenic dosage per cell. For example, with human heteroploid epitheliumlike lines and monkey lines, we have used about 27×10^6 cells for intraperitoneal injections. On the other hand, with Chinese hamster lines, half this number of cells produces equally satisfactory antisera. Similar amounts of cells are used for subcutaneous injections.

Guinea pigs are injected both intraperitoneally and subcutaneously on day one. For intraperitoneal inoculations, the appropriate number of cells are suspended in 1.0 ml of Hanks' BSS. For subcutaneous inoculations, 0.5 ml of cell suspension and 0.5 ml of Freund's complete adjuvant are mixed and 0.25 ml of the mixture is injected into each of four sites. Three weeks later, another intraperitoneal injection of the same number of cells (as day one) is given. The following week the animals are bled by cardiac puncture.

For rabbit immunization, usually two animals are used. The appropriate

[14] American Type Culture Collection: Registry of Animal Cell Lines. 2nd Ed., American Type Culture Collection, Rockville, Maryland (1972).

[15] C. S. Stulberg, W. D. Peterson, Jr., and L. Berman, *Nat. Cancer Inst. Monogr.* **7**, 17 (1962).

number (see below) of washed cells is suspended in Hanks' BSS and injected twice weekly into the marginal ear vein. The injections consist of a graded series and start with about 3×10^5 cells. Then the number of cells given is approximately doubled with each injection until an inoculum of 6×10^6 cells has been reached. After this, three or four weekly injections of 6×10^6 cells are injected as boosters until test bleedings from the ear vein give a high enough titer (serum dilutions of 1:16–1:32) by the indirect fluorescent test as subsequently described. Three or four times as many cells appear to be necessary when cultures of mouse, Syrian hamster, and canine cells are used and the schedule adjusted accordingly. Concentrations as high as 24×10^6 cells in 3 ml of Hanks' BSS have been given with no ill effects.

All antisera can be stored at −20°C.

Preparation of Fluorescent-Labeled Antibody

The fluorescein-labeled antibody is prepared by a relatively simple procedure.[8] The sera are treated with 18% sodium sulfate (one volume serum/two volumes 27% sodium sulfate added dropwise). Then this is held at 37°C for 16 hours and resulting globulin precipitates are centrifuged at 3000 rpm for 30 minutes at room temperature. These precipitates are then dissolved in a minimum of distilled water (1–3 ml), and the total globulin is dialyzed for 1 hour against two changes of distilled water and then for 4 to 5 hours against several changes of physiological saline. Following dialysis, the protein content of the resulting globulin solution is determined by the Biuret reaction. Such solutions are then adjusted to pH 9 with 0.5 *M* carbonate-bicarbonate buffer by adding one part buffer to nine parts globulin solution. Dry fluorescein isothiocyanate (FITC) and cellulose powder[16] are mixed. This mixture contains one part FITC (0.025 mg FITC/mg protein) and nine parts cellulose powder. The proper concentration of globulin solution is then added directly to the dry mixture and shaken carefully (to minimize denaturation) and then agitated gently for 5 minutes.[17] The undissolved cellulose powder is removed by centrifugation and the unbound dye from the conjugated serum by Sephadex gel.[18]

Sephadex columns are prepared in the following way. Sephadex is washed twice with physiological saline. The resulting gel is poured into a 0.5-inch bore glass tube (a wide-mouth 25-ml pipette with the top cut off works well) that has a constricted end loosely plugged with glass wool. The excess saline is allowed to drain out while the Sephadex is settling. After settling, a small circular piece of filter paper is placed on the top surface of the Sephadex and the conjugated antiserum is added carefully by pipette over the filter paper. When the conjugated globulin drains down to the top of the Sephadex more saline is added so that all the colored coupled globulin freely passes through. As soon as the colored drops begin to drip from the constricted end of the tube, they are

[16] Celite, Johns-Mansville.

[17] H. Rinderknecht, *Nature* (*London*) **193**, 167 (1962).

[18] G-25, fine. Pharmacia Uppsala, Sweden.

collected. This is continued until the column is free of color and only a band of uncoupled dye remains near the top of the tube.

Fluorescent Staining of Cells

Species identification of cultures is accomplished as follows. Monolayers of cultures to be tested are trypsinized or frozen viable cells can be used. The cell suspensions are washed three times in buffered saline (pH 7.5), resuspended to yield 3–4 $\times$ 10^6 cells per milliliter, and 0.1 ml of cell suspension is mixed with 0.1 ml of fluorescein-labeled antibody in Kahn tubes, which are agitated for 30 minutes (mechanically on a rotator and shaken manually at intervals). The cells are again washed three times in buffered saline to remove unbound antibody. A drop of the final suspension of *living* cells is placed on a slide and sealed under a coverslip with fingernail lacquer. Care should be taken so that the cells are not crushed when the coverslip is applied. The preparation is then observed by fluorescence microscopy. The presence of "apple green" peripheral fluorescence on the surface membrane of individual cells treated with specific labeled antibody indicates a positive reaction (Fig. 1).

The dilution of labeled antibody that is used is determined by staining titers. With most of our labeled antisera, dilutions of 1:4 to 1:8 give specific fluorescent antigen–antibody reactions of brilliant intensity. With some conjugates, even higher dilutions give satisfactory reactions. The specificities and sensitivities of different fluorescein-labeled antibody preparations have been reported.[8,13,19]

In our experience, the staining reactions are specific except in three instances. First, human and monkey antibody may cross-react with monkey and human cells, respectively, but the potencies of such antisera are different enough in guinea pig antisera so that appropriate dilutions place them on a specific basis. Second, antiserum prepared in rabbits against monkey cells will give strong cross-reactions with human cells and a less strong reaction with bovine cells. Third, another exception to the rule of an "all or none" reaction is found when labeled antisera to some rabbit cells are prepared in guinea pigs. These rabbit antisera may cross-react with other cell species. Otherwise, we have found that all cellular antisera that we have tried are specific whenever labeled antisera are used.

Indirect Fluorescence

The indirect fluorescent antibody test can also be used. The principle of indirect fluorescent tests is well known. However, for the identification of cell species in culture, we use the indirect method for several purposes. Previously, it was mentioned that as rabbits were being immunized, small amounts of blood could be drawn at intervals and the sera tested for their antibody content by

[19] C. S. Stulberg and W. D. Peterson, Jr., *Quart. Rev. Biol.* **41**, 124 (1966).

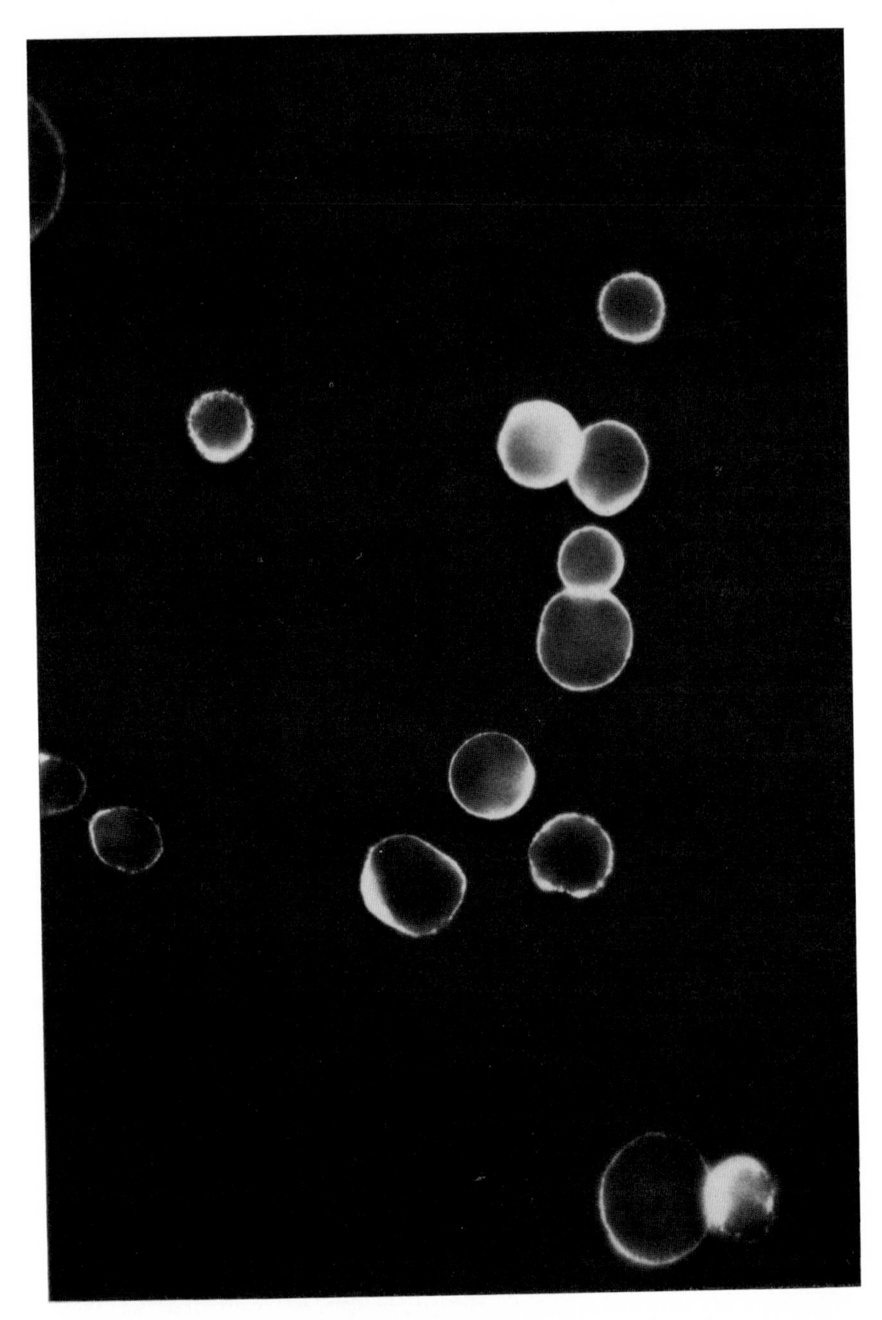

Fig. 1. MDBK cells of bovine origin (American Type Culture Collection, Rockville, Maryland) stained with bovine fluorescein-labeled antibody illuminated with near-ultraviolet light. Note peripheral fluorescence of these cells.

the indirect test. With guinea pigs, after immunization, an antiserum can be tested by indirect fluorescence to determine its antibody potency so that only those with the desired antibody content may be labeled. In addition, if a specific fluorescent antibody to a given species is unavailable, the indirect test can provide a method to approach identification by fluorescence.

In either case the procedure is carried out as follows. Usually 0.1 ml of sera is used undiluted followed by twofold dilutions to 1:128. Such serum dilutions are reacted for 30 minutes with cells suspended in 0.1 ml buffer, pH 7.5. The cells are then washed three times with buffer, and resuspended in 0.1 ml buffer together with 0.1 ml of fluorescein-coupled antiglobulin made in another animal against the globulin of the species used to produce the cellular antibodies. The procedure then follows that of the direct fluorescent antibody test. However, the indirect test, except when necessary, is time-consuming and has the added disadvantage of losing cells because of the extra washings involved.

Fluorescent Staining of Cell Mixtures

Individual cells of a different species can be detected by the visualization of such cells when as few as one in 10,000 are present.[8,13] Thus, contamination of cultures by cells of diverse origins can be detected early or conversely the investigator can be assured that he is working with a line that is not contaminated with another.

Using antisera of differing specificities, the numbers of peripherally stained and unstained cells in dark fields observed alternatively with tungsten and near ultraviolet light and the percentage of stained (contaminating) cells is determined. Experiments[8] have indicated how cell mixtures were first studied and illustrations[13] show the appearance of such mixtures.

CHAPTER 14

Serological Identifications of Cells in Culture

B. Poikilotherm Cell Identity by Cytotoxic Antibody Test

Arthur E. Greene and Jesse Charney

For a number of years our laboratory has been engaged in tissue culture programs requiring the characterization and preservation of many valuable cell lines.[1] One of the major problems encountered has been the contamination of cell lines with cells of a different species. This required the development of methods for the identification of the species of origin of tissue culture cell lines. The use of poikilothermic species as necessary tools for research in viral diseases, molecular biology, and oncology of cold-blooded vertebrates increased the need for constant surveillance and characterization of these cultures in the laboratory in order to assure such freedom from contamination. The cytotoxic antibody dye exclusion test, because of the rapidity with which the test may be carried out and its freedom from requirements for expensive equipment or experienced personnel, is a useful procedure for the species identification of individual poikilothermic,[2] mammalian,[3] and avian cell cultures.

Cultivation of Cell Lines

Most amphibian, reptilian, and fish cell lines can be grown in culture media suitable for mammalian cells, such as Minimum Essential Medium Eagle[4] in Earle's or Hanks' balanced salt solution plus 10% fetal bovine serum. A few cell lines including the bullfrog (ATCC-CCL 41)[5] and the toad, *Xenopus laevis* (ATCC-CCL 102)[6] do not thrive in the above media and require that the culture media be adjusted with distilled water to the osmolarity for amphibian tissue. Saltwater fish cultures are grown in media also adjusted to their normal marine osmolarity.[7]

[1] C. S. Stulberg, L. L. Coriell, A. J. Kniazeff, and J. E. Shannon, *In Vitro* **5**, 1 (1970).
[2] A. E. Greene, H. Goldner, and L. L. Coriell, *Growth* **30**, 305 (1966).
[3] A. E. Greene, L. L. Coriell, and J. Charney, *J. Nat. Cancer Inst.* **32**, 779 (1964).
[4] H. Eagle, *Science* **130**, 432 (1959).
[5] K. Wolf and M. C. Quimby, *Science* **144**, 1065 (1962).
[6] K. A. Rafferty, Jr., *In* "Biology of Amphibian Tumors" (M. Mizell, ed.), p. 52. Springer-Verlag, New York, 1969.
[7] L. W. Clem, M. M. Sigel, and R. R. Friis, *Ann. N. Y. Acad. Sci.* **126**, 343 (1965); cf. Section II, this volume.

Preparation of Antisera

Antisera is prepared by immunizing rabbits with muscle tissue from fish, amphibians, and reptiles. The tissue is minced, washed, and suspended, following which it is homogenized in a Serval Omni-mixer at top speed for 5 minutes. The homogenate is distributed into vials and stored at $-20°C$. When it is not possible to obtain the desired species of poikilothermic animal tissue for antiserum preparation, cell cultures must be used as the source of antigen; if cell cultures are used, they must be authenticated as to species. Poikilothermic cells are grown in glass bottles or plastic flasks at their optimal temperature until the cell sheet is confluent. The cells are removed from the surface of the vessel with 0.25% trypsin, counted in a hemocytometer, washed three times in Hanks' balanced salt solution, and suspended in culture medium without serum. A minimum of 10^6 cells per milliliter are ampoulized in sterile vials and stored at $-20°C$.

Rabbits are inoculated with a series of at least six intramuscular injections of 1 ml of cell suspension or homogenate at three-day intervals. In certain situations where it was difficult to obtain sufficient amounts of antigenic material for inoculation or if the material was found to be a poor antigen, we have found it necessary to homogenize the antigen with Freund's complete adjuvant and inoculate 0.5 ml intramuscularly into each thigh of the rabbit. A second inoculation with adjuvant is given 1 month later, and test bleedings to determine the presence of antibody are taken 1 month after the second dose. If satisfactory cytotoxic antibody levels are present, the animal is bled. If not, additional injections are given. All test sera are inactivated at 56°C for 30 minutes and stored at $-20°C$.

Absorption of Antisera

Absorption of antisera is sometimes required to ensure specificity. This is accomplished by absorbing antisera three times at 37°C with an equal volume of packed cell suspensions of the heterologous tissue with which the antiserum cross-reacts nonspecifically. Cells grown in tissue culture or trypsinized or washed cell preparations from animal tissue are employed.

The Cytotoxicity Test

Viable cells, 1×10^5 to 2×10^5 in 1 ml of Eagle's Medium, are placed in Wassermann tubes. The various poikilothermic and homeothermic animal antisera, 0.1 ml, to be tested are added to the tubes and the cell–antiserum mixtures incubated at 37°C for 1 hour. Guinea pig serum, 0.1 ml, as a source of complement, is added to the tubes which are again incubated at 37°C for 30 minutes. Fresh guinea pig serum can be ampoulized and stored in liquid nitrogen at

TABLE I

Poikilothermic Cell Culture Species Identification by the Cytotoxic Antibody Test[a]

Cell lines tested	Viable cells (%) after exposure to antisera									
	Grunt[b] fin	Car[c] (goldfish)	Fathead[b,c] minnow	Rainbow[b,c] trout	Human	Mouse	Syrian hamster	Chinese hamster	Bovine	Control[d]
Grunt fin	0	90	89	91	89	90	89	83	88	91
Goldfish	89	5	92	82	90	94	89	95	93	94

[a] From A. E. Greene, L. L. Coriell, and J. Charney.[8]

[b] Antisera to fathead minnow, rainbow trout, and grunt fin were absorbed with car (goldfish) tissue, car (*Carassius auratus*)–goldfish (ATCC-CCL 71), grunt fin (ATCC-CCL 58).

[c] Antisera to fathead minnow, car (goldfish) and rainbow trout were absorbed with grunt fin tissue.

[d] Control consisted of preimmunization normal rabbit serum and complement.

−196°C for a minimum of 2 years[8] or lyophilized and reconstituted for the test with sterile distilled water.[3] Approximately 0.1 ml of 0.25% trypan blue in phosphate-buffer saline (pH 7.2) is added to the tubes and samples of each cell–serum specimen are placed in a hemocytometer and the stained (killed) and unstained (viable) cells counted. The death of 50% or more of the cells, as shown by trypan blue uptake, is considered evidence of antiserum cytotoxicity.[3] A cell control containing normal rabbit serum with and without complement must be included in the test. Table I demonstrates the use of the cytotoxic antibody test to distinguish goldfish cells (ATCC-CCL 71) from the blue-striped grunt fin cells (ATCC-CCL 58).[7]

[8] A. E. Greene, L. L. Coriell, and J. Charney, *In* "Biology of Amphibian Tumors" (M. Mizell, ed.), p. 112. Springer-Verlag, New York, 1969.

CHAPTER 14

Serological Identifications of Cells in Culture

C. Invertebrate Cell Cultures

Arthur E. Greene and Jesse Charney

The successful establishment of a number of insect cell lines by Grace[1] and later by Singh[2] made it possible to apply tissue culture techniques to the study of insect biology. Although these cells are usually incubated below 30°C, which excludes cross-contamination with most mammalian cells, contamination with different insects cells carried in the same laboratory has frequently occurred. In our laboratory one of the most effective methods for correctly identifying the species of mislabeled insect cultures has been the agar gel microimmunodiffusion technique.[3]

Cell Culture Antigens

Insect cell cultures grow as suspension or as monolayer cultures at 30°C. Grace's *Antheraea eucalypti* (moth) cells grow as suspension cultures; some of

[1] T. D. C. Grace, *Nature (London)* **195**, 788 (1962).
[2] K. R. P. Singh, *Curr. Sci.* **36**, 506 (1967).
[3] A. E. Greene and J. Charney, *In* "Current Topics in Microbiology and Immunology" (E. Weiss, ed.), Vol. 55, p. 51, Springer-Verlag, New York, 1971.

the cells attach to the surface but are easily removed by shaking. The cells vary greatly in size, are round or spindle-shaped and may contain bizarre giant forms. Although *Antheraea eucalypti* cells require insect hemolymph as a growth factor, they have been adapted to hemolymph-free medium.[4] After growth, suspension cells are harvested by centrifugation at 570 *g* for 10 minutes to remove the culture medium and serum.

Singh's mosquito cells, *Aedes albopictus* and *Aedes aegypti,* are examples of insect monolayer cultures. The cells grow in Mitsuhashi and Maramorsch insect medium[5] with 20% uninactivated fetal bovine serum, attach to glass or plastic surfaces, and form complete monolayers of small epithelial cells as described by Singh. When growth is confluent, 0.1% trypsin in Rinaldini's salt solution[2] is added to the flasks to remove the cells adhering to the surface and centrifuged.

Antigen extracts for the immunodiffusion test are prepared by washing the cells three times in saline followed by three freeze–thaw cycles of the cell pellet.[3] The preparations can be stored at $-20°C$ indefinitely without significant loss of potency.

Preparation of Antisera

Antisera may be prepared by immunizing rabbits with cell cultures or with tissues from the insect; the latter is preferred as it eliminates the chance of making antisera to mislabeled cell cultures. Antisera must be prepared to cell cultures when it is impossible to obtain insects for antigen preparation due to exclusion by Federal Law, or extreme difficulty in obtaining sufficient insect tissue for incubation.

Preparation of Antigens for Immunization from Suspension Cultures

The cells are cultured in 75-cm^2 plastic flasks and harvested after approximately 14 days. A small aliquot of cells is first withdrawn for counting in a hemocytometer and the remaining cells are centrifuged at 570 *g* for 10 minutes to remove the culture medium. The cells are then washed five times in Grace's insect medium without serum or hemolymph. One-milliliter aliquots containing a minimum of 2×10^7 cells in Grace's medium are placed in sterile Wassermann tubes and stored at $-20°C$.

Preparation of Antigens for Immunization from Monolayer Cultures

Monolayer cultures are grown in milk dilution bottles. When cell growth is observed to be confluent, the culture medium is removed and 0.1% trypsin in

[4] C. E. Yunker, J. L. Vaughn, and J. Cory, *Science* **155**, 1565 (1967).

[5] J. Mitsuhashi and K. Maramorosch, *Contrib. Boyce Thompson Inst.* **22**, 435 (1964); cf. Chapter 5B, this section.

Rinaldini's salt solution is added to the cells and incubated at 37°C. The dissociated cells are centrifuged at 570 *g* to remove trypsin and then washed five times in Grace's insect medium without serum or hemolymph. The cells containing a minimum of 2×10^7 viable cells are placed in sterile Wassermann tubes and frozen at −20°C.

Preparation of Larval Antigens

Fourth instar larvae are homogenized in a sterile tissue grinder with four volumes of Grace's insect medium and 1-ml aliquots of the antigen homogenate are dispensed into sterile Wassermann tubes. The antigens are stored at −20°C.

Immunization of Rabbits

Insect antigen preparations are homogenized with an equal volume of Freund's complete adjuvant and injected intramuscularly in rabbits.[6] The antigen-adjuvant mixture, 0.5 ml, is inoculated intramuscularly into each thigh of the rabbit. A second inoculation with adjuvant is given 1 month later. A test bleeding to determine the presence of antibody is taken a month after the second dose and if satisfactory antibody levels are present the animal is bled. If not, additional injections are given as required.

When antibody is present at undesirably low levels, it may be necessary to concentrate the antiglobulins from the antisera by ammonium sulfate precipitation. The serum is precipitated with an equal volume of saturated ammonium sulfate. The globulin precipitate is sedimented and washed with ten to twenty volumes of 0.5 saturated ammonium sulfate solution. The sedimented pellet is dissolved in 0.02 *M* phosphate buffer and dialyzed twice against fifty volumes of the same buffer. The dialyzed globulin, in one-fifth to one-tenth the original serum volume is centrifuged for 15 minutes at 25,000 *g* and the clarified supernatant is stored at −20°C.[7]

Absorption of Antisera

All antisera are absorbed with newborn bovine serum to remove nonspecific reactions to media constituents by adding one-fifth volume of newborn bovine serum to the test antiserum and incubating the mixture with periodic shaking at room temperature for 1 hour. Nonspecific reactions present in the antiserum due to insect hemolymph may be prevented by absorbing the serum with *Antheraea pernyi* or other related hemolymph in the same way.

[6] A. E. Greene, J. Charney, W. W. Nichols, and L. L. Coriell, *In Vitro* **7**, 313 (1972).
[7] J. Charney and L. L. Coriell, *J. Nat. Cancer Inst.* **33**, 285 (1964).

MICROIMMUNODIFFUSION MATERIALS[7]

Glass slides, frosted on one end for labeling, are stacked in metal racks and submerged in a large metal pan containing a glass cleaning detergent (equal parts of Microsolv[8] and Calgolax[9]) in distilled water. The pan is covered with aluminum foil and autoclaved for 15 minutes at 15 lbs. pressure. The fluid is poured off and the slides rinsed in at least five changes of distilled water, keeping the slides wet at all times. If the slides cannot be coated with agar on the same day as they are washed, they may be stored overnight in the washing pan submerged in distilled water.

The slides are coated with 0.2% Ionagar No. 2[10] which is prepared by dissolving 2.0 g of Ionagar in 1000 ml of boiling distilled water. The agar is heated until clear. The slides are dipped into the hot agar solution for 2 minutes, drained on a slant toward the frosted side, and then placed in a 60°C hot-air oven until dry.

The 0.8% Ionagar No. 2 setting agar is made by adding 8.0 g of Ionagar No. 2 to 950 ml of 0.02 *M* phosphate buffer and the agar boiled until it is clear. Fifty milliliters of a 1:1000 dilution of Merthiolate is added to prevent microbial contamination and the agar is dispensed in 10-ml aliquots and stored at refrigerator temperature.

Stock phosphate buffer 0.4 *M*, at pH 7.2 is prepared with 27.2 g KH_2PO_4 and 34.8 g of K_2HPO_4 made up to 1000 ml with distilled water. Phosphate buffer 0.02 *M*, at pH 7.2 is prepared by adding 100 ml 0.4 *M* phosphate buffer to 1900 ml of distilled water.

Seven-holed templates (size $1 \times 1 \times \frac{1}{8}$ inch)[7] are used in the test (Fig. 1).

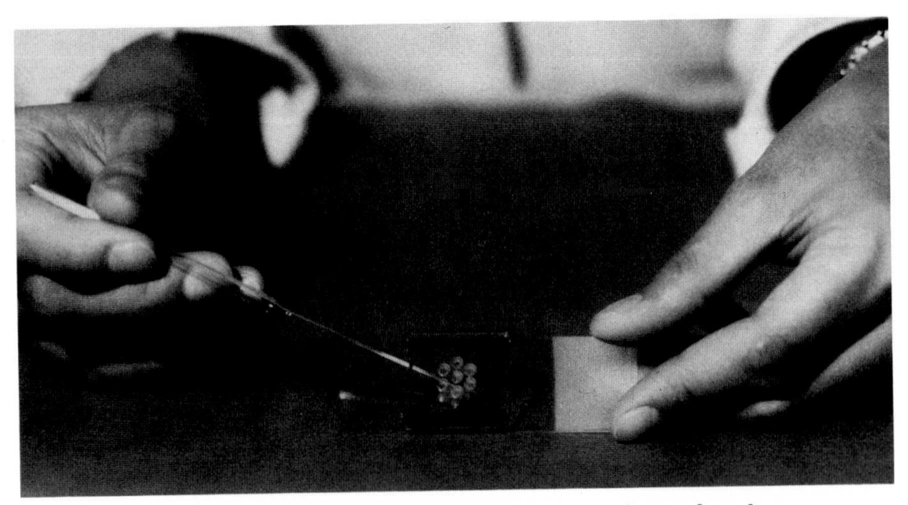

Fig. 1. Agar gel microimmunodiffusion slide with template and applicator.

[8] Available from Microbiological Associates, 4813 Bethesda Avenue, Bethesda 14, Maryland.

[9] Available from Calgon Corporation, P. O. Box 1346, Pittsburgh, Pennsylvania.

[10] Available from Colab Labs Inc., Box 66, Chicago Heights, Illinois 60411.

The holes are 1/16 inch in size and 0.4 cm apart. A single thickness of electrical tape cut to a width of 3/8 inch is attached to the upper and lower horizontal edges of the template and remains permanently attached to it during cleaning and reuse. A single thickness of electrical tape is used which results in an agar film approximately 0.2 mm thick.

The template is placed tape downward, on the agar-coated slide. Melted setting agar (0.8%) is added with a 4-mm capillary pipette to the open sides of the template so that the agar fills the space between slide and template. Just enough agar is applied to fill the space without flooding the area under the wells.

The Test Procedure[7]

The prepared slides are placed into large Petri dishes on filter paper moistened with a 1% solution of Merthiolate in distilled water to prevent microbial growth and drying of the agar.

Rabbit antiserum to insect cell culture or larval tissue is placed in the central well and the insect cell antigens in the peripheral wells. The reactants are added with capillary pipettes drawn from 4-mm glass tubing to a fine-bore tip. Care must be taken to avoid bubbles in the well. The filled slides in the closed Petri dishes are developed at 4°C for 3 days. The templates are then removed from the slides under running distilled water and the slides leached first for 2 hours in 0.2 *M* phosphate buffer, then for 2 hours in distilled water, and rinsed

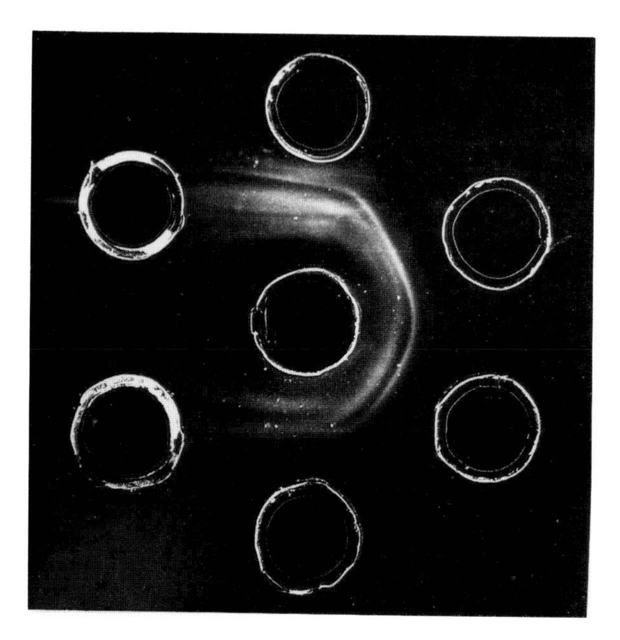

Fig. 2. Central well: rabbit anti-*Antheraea eucalypti* serum; bottom well, *Antheraea eucalypti* cells; lower left well, *Aedes albopictus* cells; upper left well, *Aedes aegypti* cells; upper well, Suitor's clone of *Aedes aegypti* cells; upper right well, *Aedes vexans* cells; lower right well, *Culiseta inornata* cells.

with distilled water until bright indirect illumination shows no soiling particles. Excess agar is removed from the slides before they are dried at 60°C.

The slides are stained for 2 minutes in a solution of 0.25% amido black 10B made up in 5:5:1 methanol: distilled water: glacial acetic acid. They are then destained in three washes in distilled water containing 1% glacial acetic acid and 1% glycerol, drained, and redried at 60°C.

The stained immunodiffusion lines may be read directly against a diffusely lit background. If the lines are faint they can be better visualized by bright oblique illumination against a dark background.

Fusion of the precipitin lines given by an antigen preparation in one well with that of an antigen in a neighboring well provides a test of identity, i.e., proof that both preparations contain a common antigen.

Interpretation

Figure 2 shows the immunodiffusion patterns observed when rabbit anti-*Antheraea eucalypti* (moth) serum (central well) is reacted against the test antigen (peripheral wells). Identity lines appear between *Antheraea eucalypti* (moth) in the bottom well and three mislabeled mosquito cell lines; a clone of *Aedes aegypti, Aedes vexans,* and *Culiseta inornata,* in the top, upper right, and lower right wells. Identity lines are not observed with monolayer cell cultures antigens of *Aedes albopictus* (lower left well) and *Aedes aegypti* (upper left well). The serological identification of the mislabeled mosquito cell lines as moth cells was confirmed by chromosome and isozyme analysis.

CHAPTER 14

Serological Identifications of Cells in Culture

D. Detection of *HL-A* Antigens on Cultured Normal Human Diploid Fibroblasts[1]

Chaim Brautbar

Human histocompatibility (*HL-A*) antigens have been described as membrane-associated transplantation antigens.[2] These antigens are present on the

[1] Supported in part by Research Grant HD 04004 from the National Institute of Child Health and Human Development, National Institutes of Health, Bethesda, Maryland.

surface membrane of all nucleated cells that have been studied thus far.[3-5] The *HL-A* antigens are associated with a variety of biological phenomena, and have the property of eliciting an immune response by the host following transplantation of allogenic organs and tissues.[2] At present, the consensus is that the *HL-A* system consists of two segregant series, namely the first and the second, to each of which belong a substantial number of mutually exclusive alleles.[6] Each individual has the potential of having four of these specificities: two of the first segregant series and two of the second. These specificities are inherited in a Mendelian pattern.[6] Several laboratories have reported that *HL-A* antigens remain stable on human diploid fibroblasts after prolonged cultivation *in vitro*.[7-9] These observations were recently confirmed and extended to Phase III cells.[10] Considering the potential value of the *HL-A* antigen system as a marker for cell identification, the following protocol describes two methods for the detection of histocompatibility antigens on suspended human diploid fibroblasts.

Direct Cytotoxicity

The following procedure is a modification of the fluorochromatic cytotoxicity assay of human lymphocytes, described by Bodmer *et al.*[11] and adapted for the detection of *HL-A* antigens on suspended human fibroblasts. The cytotoxicity reaction depends upon alteration in permeability induced by the action of complement on cells which have been exposed to anti-*HL-A* antibodies. It has been shown by electron microscope studies that the major alterations induced in lymphocytes by the action of alloantibodies and complement include disappearance of ribosomes, fragmentation of mitochondria, and extensive dilation and bleb formation of the nuclear membrane.[12] Similar alterations probably occur in fibroblasts.

Materials and Equipment

Basal Medium Eagle (BME)[13] supplemented with 10% calf serum and 50 μg/ml aureomycin[14]

---

[2] B. D. Kahan and R. A. Reisfeld, *Bacteriol. Rev.* **35**, 59 (1971).

[3] C. J. M. Melief, M. Van der Hart, C. P. Engelfriet, V. P. Eysvoogel, and J. J. Van Loghem, *Vox Sanguinis* **15**, 187 (1968).

[4] R. L. Walford, *Ser. Haematol.* **2**, 2 (1969).

[5] F. Milgrom and F. Kano, *In* "Histocompatibility Testing" (A. Videbaek, ed.), p. 179. Munksgaard, Copenhagen, 1965.

[6] J. Dausset, J. Colombani, L. Legrand, and M. Fellows, *In* "Histocompatibility Testing" (P. I. Terasaki, ed.), p. 53. Munksgaard, Copenhagen, 1970.

[7] V. C. Miggiano, M. Nabholz, and W. F. Bodmer, *In* "Histocompatibility Testing" (P. I. Terasaki, ed.), p. 623. Munksgaard, Copenhagen, 1970.

[8] D. Lapeyre, J. Hors, J. Dausset, J. Colombani, and R. Netter, *Symp. Human Diploid Cells, Zagreb, September 23–24,* 1970.

[9] D. Lapeyre, J. Hors, J. Colombani, J. Dausset, and R. Netter, *12th Congr. Intern. Standartisation Biol., Annecy, France, September 20–24,* 1971.

[10] C. Brautbar, R. Payne, and L. Hayflick, *Exp. Cell Res.* **75**, 31 (1972).

Fluorescein diacetate, 0.5% stock solution: 2 ml of reagent grade acetone is added to 10 mg fluorescein diacetate (FDA).[15] Use a tightly stoppered bottle to prevent evaporation, avoid frequent exposure to light, and store at −20°C.

Tissue typing plates[16]

Fluorescence microscope

Establishment of Human Diploid Cell Cultures from Skin Biopsies.[17] Obtain a full thickness skin biopsy, 2 mm in diameter. Cut the skin into small fragments, and place the tissue fragments between a coverslip and the bottom of a 60-mm diameter Flacon plastic Petri dish. Add 4 ml of BME and incubate at 37°C in a CO_2 atmosphere. Change the medium weekly. When a confluent monolayer is formed, in approximately 25–30 days, subculture the cells into one 4-ounce prescription bottle after dispersal with 0.25% trypsin. When cells reach confluency, subcultivate the fibroblasts at a 1:2 split ratio.

HL-A Alloantisera. The sources of antisera reacting with human histocompatibility antigens are multiparous women, transfused and sensitized patients, and human volunteers immunized with skin grafts, leukocytes, or platelets. These alloantisera are available from the serum bank of the National Institutes of Health and from tissue typing laboratories at major medical centers.

Alloantisera should be examined for their reactivity with *HL-A* antigens on human diploid fibroblasts to obtain those most potent for fibroblast typing. This is done by employing the fluorochromatic cytotoxicity assay as outlined in the following paragraphs. At least two alloantisera should be used for the detection of each *HL-A* specificity.

Cell Preparation for Cytotoxicity Assay. Harvest the fibroblasts by trypsinization. Aspirate vigorously, by pipette, several times in BME in order to obtain a single cell suspension. Treat the fibroblasts ($1–10 \times 10^6$/ml) with 0.1 ml BME containing 2 μg FDA per 1 ml of cell suspension. The FDA should be diluted, just prior to treatment of the cells, in BME supplemented with 20% heat-inactivated calf serum (BME 20%). Incubate the mixture at room temperature for 30 minutes in the dark and agitate gently at intervals of 5 to 10 minutes. After incubation dilute the fibroblast suspension in four volumes of BME 20%. Centrifuge the cell suspension for 10 to 15 minutes at 1500 rpm in the cold and resuspend the pellet in BME 20% at 4°C to prevent attachment of

---

[11] W. Bodmer, M. Tripp, and J. Bodmer, *In* "Histocompatibility Testing" (E. S. Curtoni, P. L. Mattiuz, and R. M. Tosi, eds.), p. 341. Munksgaard, Copenhagen, 1967.

[12] R. L. Walford, H. Latta, and G. M. Troup, *Ann. N. Y. Acad. Sci.* **129**, 490 (1966).

[13] H. Eagle, *Science* **122**, 501 (1955).

[14] L. Hayflick, *Exp. Cell Res.* **37**, 614 (1965).

[15] Available from Nutritional Biochemicals Corporation, Cleveland, Ohio.

[16] Available from G. D. Searle and Co., Ltd., High Wycombe, Bucks, England.

[17] This procedure is also applicable to cell cultures derived from various human adult and embryonic organs.

the cells to the tube. Adjust the density of the cell suspension to 2000 cells per mm^3 using a hemocytometer.

Cytotoxicity Assay. Add 2 μl of typing serum to each well of a typing plate filled with 3 ml of heavy duty mineral oil. Add 1 μl of 2000 FDA-treated fresh fibroblasts to each well containing the typing alloantiserum. Incubate the plates in the dark, at room temperature for 30 minutes, after which time add 6 μl of absorbed rabbit complement[18] to each well and continue the incubation at room temperature in the dark for 4 more hours. To each typing plate add the following controls: 6 μl BME 20% + 1 μl of 2000 FDA-treated cells, 2 μl BME 20% + 1 μl of 2000 FDA-treated cells + 6 μl absorbed rabbit complement, and 2 μl heat-inactivated human AB serum + 1 μl of 2000 FDA-treated cells + 6 μl absorbed rabbit complement. Results are read after 4 hours of incubation by examining the plates under a fluorescence microscope for the disappearance of fluorescing fibroblasts, and are graded on an arbitrary scale from zero to 4+ in inclusive intervals of 25%. At 3+ or 4+ results should be considered positive, that is, when more than 50% of the fluorescing cells disappear.

Preparation of the Complement. The choice of complement (C′) is essential to provide a reliable cytotoxic assay. Rabbit serum is known to be the most effective C′ source for the cytotoxic reaction *in vitro*.[19] However, this serum is toxic to the fibroblasts and causes nonspecific killing of target cells. Normal rabbit sera were found to contain a natural cytotoxic C′-dependent antibody that reacts with antigen(s) present on cultured human lymphoid cells.[20,21] Therefore, the normal rabbit serum should be routinely absorbed as follows with packed human fibroblasts. To avoid loss of complement activity, absorptions are performed in the presence of EDTA, as described by Boyse *et al.*[22]

Add one volume of 0.1 *M* EDTA to nine volumes of normal rabbit serum. Absorb the EDTA-treated rabbit serum with packed fibroblasts that have been washed with Earle's balanced salt solution lacking calcium and magnesium, in the ratio of 7:1 (EDTA-treated serum/cells:v/v). The absorption step is performed in an ice bath for 40 minutes, with mixing at 10-minute intervals. Free divalent cations are restored by the addition of one volume of 0.1 *M* $CaCl_2$. Store the absorbed C′ at −70°C in 300-μl aliquots, and thaw only once before use.

Quantitative Absorption

The following procedure is a modification of the microabsorption technique for *HL-A* typing of human lymphocytes, recently described by Pellegrino *et al.*[23]

[18] Normal rabbit serum is available from Pel-Freez Biologicals, Inc., Rogers, Arkansas.

[19] R. L. Walford, R. Gallagher, and J. R. Sjaarda, *Science* **144**, 868 (1964).

[20] J. C. McDonald, L. Jacobbi, and R. W. Williams, *Transplantation* **10**, 499 (1970).

[21] S. Ferrone, N. R. Cooper, M. A. Pellegrino, and R. A. Reisfeld, *J. Immunol.* **107**, 939 (1971).

[22] E. A. Boyse, L. Hubbard, E. Stockert, and M. E. Lamm, *Transplantation* **10**, 446 (1970).

It has been adapted for the quantitation of histocompatibility antigens on suspended human diploid fibroblasts.[24]

This technique relies upon the absorption of *HL-A* alloantisera with fibroblasts containing the corresponding *HL-A* antigen. After the absorption step, the serum is assayed in the direct cytotoxicity test with standardized lymphocyte target cells.

The absorption technique is more sensitive than the direct cytotoxicity assay, as evidenced by the CYNAP (cytotoxicity negative absorption positive) phenomenon. This reaction has been reported by many laboratories, and refers to the phenomenon in which cells do not react with a particular *HL-A* alloantiserum and complement in the cytotoxicity assay, but are capable of absorbing this antibody.[25,26] The role of rabbit serum in the cytotoxic reaction is as a source of complement components and natural heteroantibody which facilitates the binding of C′ by *HL-A* antibody.[21] Removal of this natural antibody by absorption with normal human fibroblasts may decrease the efficiency of the rabbit serum as a C′ source for the cytotoxic reaction of specific *HL-A* alloantisera with certain cultured human fibroblasts. Therefore, a negative result in the direct test, especially when a large panel of typing sera is not available, should be confirmed by the absorption method.

An additional advantage of the quantitative absorption of *HL-A* alloantisera is that it affords a comparison of the densities of different *HL-A* determinants located on the same cell and of the same *HL-A* specificity present on fibroblasts derived from different individuals. A further application of this technique might be for the quantitation of *HL-A* antigens on normal and transformed human diploid fibroblasts.

Materials and Equipment

Standardized target lymphocytes containing the appropriate *HL-A* specificities
Hanks' balanced salt solution (HBSS)
Beckman plastic microtubes
Beckman microfuge

Preparation of HL-A Alloantisera. Titrate the antisera against the target cells which contain the corresponding *HL-A* antigen. The dilution which corresponds to the 95% lytic end point is defined as a 0 cytotoxic unit. For the absorption procedure use the serum at 1 cytotoxic unit, which corresponds to a two-fold

---

[23] M. A. Pellegrino, S. Ferrone, and A. Pellegrino, *Proc. Soc. Exp. Biol. Med.* **139**, 484 (1972).

[24] C. Brautbar, M. A. Pellegrino, S. Ferrone, R. A. Reisfeld, R. Payne, and L. Hayflick, *Exp. Cell Res.* in press (1973).

[25] R. L. Walford, *In* "Histocompatibility Testing," Publ. 1229, Nat. Acad. Sci., p. 41. Washington, D. C., 1965.

[26] S. Ferrone, R. M. Tosi, and D. Centis, *In* "Histocompatibility Testing" (E. S. Curtoni, P. L. Mattiuz, and R. M. Tosi, eds.), p. 357. Munksgaard, Copenhagen, 1967.

greater concentration than the 0 cytotoxic unit. To each of eight Beckman plastic microtubes add 5-μl alloantiserum at a dilution corresponding to 1 cytotoxic unit.

Cell Preparation for Absorption. Harvest the fibroblasts by trypsinization. Aspirate vigorously, by pipette, several times in BME to obtain a single cell suspension. Centrifuge the cell suspension for 5 minutes at 400 *g*. Resuspend the pellet by vibrating the tube on a Vortex mixer while slowly adding HBSS. Wash the cells in HBSS three times, and resuspend in HBSS. Using a hemocytometer, adjust the density of the cell suspension to the desired concentration (usually 50,000 cells/μl). Prepare eight twofold dilutions of the fibroblasts for absorption.

Absorption Assay. Add 5 μl cell suspension from each dilution to the corresponding Beckman microtubes which contain 5 μl alloantiserum. Prepare a control tube by adding 5 μl HBSS to 5 μl antiserum. After the addition of the fibroblasts, mix for a few seconds with a Vortex mixer. Incubate the suspensions at room temperature for 60 minutes, and mix with a Vortex mixer at intervals of 10 minutes. After incubation, centrifuge the tubes a full speed in a Beckman microfuge for 5 minutes. Collect the supernatant of each tube separately and transfer to another Beckman plastic tube. Assay the serum by direct cytotoxicity against the selected target cells. The supernatant may be stored at $-20°C$ for several days. The percentage of absorption = 100 − [(percentage cells killed by absorbed serum divided by the percentage cells killed by nonabsorbed serum) × 100]. To compare results from different quantitative absorptions, use the (AD_{50}) parameter, i.e., the number of cells required to inhibit by 50% the cytotoxicity of a selected alloantiserum.

CHAPTER 15

Karyology of Cells in Culture

A. Preparation and Analysis of Karyotypes and Idiograms

T. C. Hsu

The chromosome constitution of cells *in vitro* has been one of the most useful criteria for monitoring cell populations. In a number of cases, contamination of one cell culture with another was detected by karyological analysis. When the chromosomal features of the two cell cultures are very different, even casual microscopic inspection will suffice.

It is well known that cell cultures *in vitro* frequently change their chromosomal constitutions. Karyotypical analysis is useful for monitoring the "normalcy" of the cell populations under consideration.

There are many procedures for karyological preparations, each slightly different from others in some minor details. For details one may find the book edited by Yunis[1] useful. The procedure described below is probably the easiest to follow, since not much "art" is involved.

Procedure

1. Use cell cultures in the active (exponential) growth phase, preferably feeding them the day before scheduled harvest.
2. Add Colcemid or Velban (final concentration of 0.06 μg/ml) to the culture and incubate for 2 hours.
3. Decant medium, rinse the monolayer culture with physiological saline, and dislodge the cells with a Pronase solution (Pronase 0.01% dissolved in physiological saline). Trypsin solution can be used in lieu of the pronase solution. For suspension cultures, this step is omitted.
4. As soon as the monolayer cells are dislodged, they should be centrifuged to pack into a pellet. The speed of centrifugation is immaterial as long as the cells can be pelletted within a few minutes. Decant the Pronase or trypsin solution, and introduce a hypotonic solution (one part of growth medium and two parts of distilled water).
5. Suspend the cells in the hypotonic solution (5–7 ml) with a Pasteur pipette, leave for 5 to 10 minutes, and centrifuge.
6. Decant the hypotonic solution and introduce a fixative (three parts methyl

[1] J. J. Yunis, "Human Chromosome Methodology." Academic Press, New York, 1965.

alcohol and one part glacial acetic acid, freshly prepared, 5 ml to each sample) to the centrifuge tube without disturbing the pellet for at least 20 minutes.

7. Suspend cells with a Pasteur pipette; recentrifuge.

8. Decant the old fixative, add fresh fixative, resuspend the cells, and recentrifuge the cells. Repeat this process three times. Finally, reduce the fixative to 1 ml or less, and suspend the cells.

9. Dip alcohol-cleaned slides in a mixture (ice cold) containing 40% methanol and 60% distilled water. Pull out the slide and drain off the excess amount of fluid but do not shake it off.

10. Hold the slide at a 45° angle toward the floor. Place a drop of the final cell suspension on the upper portion of the slide and let the cell suspension spread downward. If the cell suspension is too thin, place more drops on each slide. Dry the slides in air.

11. Stain with acetic orcein (2% orcein in 45% acetic acid, 3–5 minutes) or Giemsa (10% commercial Giemsa stock solution in phosphate buffer, pH 7, 10–15 minutes). Rinse, air-dry, and mount in any conventional mounting medium.

Karyotyping

In all cytological preparations of animal cells, regardless of the method used, there is a certain proportion of ruptured cells. When one tries to determine the diploid number (or the number of a particular specimen), metaphases losing or gaining one or more chromosomes as artifacts are not formidable obstacles since counts of 20 to 30 cells would invariably reveal the predominant number. Furthermore, there would be no particular chromosome or chromosomes involved in the loss or gain. The modal number is the real number. However, in tissues or individuals with chromosome mosaicism, two or more modal numbers may be recorded. Although the two chromosome numbers are not always represented by equal number of metaphases, the second number is usually significantly higher in the distribution curve than the remaining artifacts. Further, karyotype analysis should reveal that the extra chromosome or the missing chromosome invariably involves the same element. Table I gives some examples of data collection with several cases taken from preparations made from human lymphocyte cultures.

In heteroploid cell populations such as long-term cell lines (HeLa, L, etc.)

TABLE I
Chromosome Counts of Four Samples of Human Lymphocyte Cultures

Subject	Chromosome number							Number of cells
	43	44	45	46	47	48	49	
Normal volunteer	1	—	2	28	1	—	—	32
Down's syndrome	—	1	—	2	25	1	—	29
Normal/Down's syndrome mosaic	—	1	1	32	12	—	1	47
Turner's syndrome	—	1	34	1	—	—	—	36

where cells with many different chromosome numbers may exist, the elimination of artifacts becomes a difficult task. Karyotype analysis will not be of great help either, because cells with the same number of chromosomes do not always have the same chromosomes.

Contemporary cytologists use cut-out photomicrographs of metaphase chromosomes for constructing karyotypes. This practice is preferred over camera lucida drawings because photographs greatly reduce the subjectivity in the part of the investigators, especially inexperienced investigators.

In conventional preparations, the metaphase chromosomes exhibit rather limited features for differentiation and recognition: length, position of the centromere (thereby arm ratio), and the presence of secondary constrictions which may occur in special chromosomes. Therefore, in most mammalian species, chromosome pairs with similar morphology are common. When constructing karyotypes, it is necessary to group the chromosomes of similar length and shape if such grouping is feasible. The best example is the karyotype of man, where seven groups are recognized. Within each group, identification of particular chromosome pairs is uncertain with conventional staining procedures. In the karyotypes of some species, e.g., the chromosomes of the laboratory mice, even grouping is not possible because all chromosomes possess the same morphology and there is no distinct break in the length gradation. Only in a few exceptional species, such as the Indian muntjac and several species of Australian marsupials, every pair of chromosomes can be identified without ambiguity.

Figures 1–3 show two extreme and an intermediate case. In Fig. 1 (a karyotype of an Indian muntjac) there is no problem pairing the elements. In Fig. 2 (a karyotype of the laboratory mouse) it is not possible to identify individual pairs. In Fig. 3 (a karyotype of a striped hyena) grouping according to morphological features is feasible, but identifying pairs within each group is subjective.

Idiogram Construction

An idiogram is an idealized karyotype or a statistically representative karyotype. Therefore no single metaphase, whether it be the actual photograph or length measurements, can be considered as an idiogram. An idiogram must be arrived at by the normalization of many karyotypes. It is a time-consuming process if an idiogram is constructed manually. With the aid of computers, the task can be simplified a great deal.[2-4] However, it is pointless to construct idiograms just for the sake of constructing them. It is worth doing only if the chromosomes of a species are widely used in cytological investigations.

As can be seen, the utility of the conventional karyotype and idiogram analysis has many limitations. It is virtually impossible to determine paracentric

[2] M. Mendelsohn, D. Hungerford, B. Mayall, B. Perry, T. Conway, and J. Prewitt, *Ann. N. Y. Acad. Sci.* **157**, 376 (1969).

[3] C. J. Hilditch and D. Rutovitch, *Ann. N. Y. Acad. Sci.* **157**, 333–364 (1969).

[4] J. W. Butler, M. K. Butler, and A. Stroad, *In* "Data Acquisition and Processing in Biology and Medicine" (K. Enslein, ed.), pp. 261–275, Pergamon Press, New York, 1964.

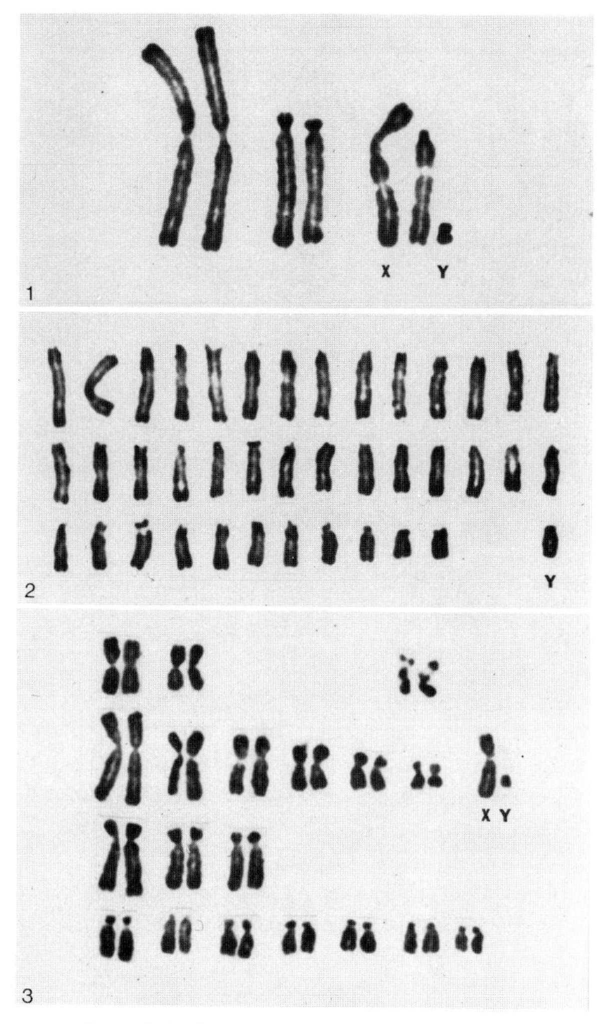

Fig. 1. Karyotype of a male Indian muntjac (*Muntiacus muntjak*), $2n = 7$ (♀ $2n = 6$). A Robertsonian fusion has apparently occurred between the *X* chromosome and the acrocentric autosome, but the homologous autosome was not translocated onto the Y chromosome, thus giving an impression that there are two Y chromosomes. Every chromosome pair of this karyotype is morphologically distinct.

Fig. 2. Karyotype of a male C3H mouse (*Mus musculus*), $2n = 40$. All chromosomes are acrocentric and their lengths form a continuous gradation. Pairing is not possible. The only recognizable elements are the two smallest autosomes and the Y chromosome, which form a group, but identification of the Y is usually arbitrary.

Fig. 3. Karyotype of a male striped hyena (*Hyaena hyaena*), $2n = 40$. The chromosomes can be classified into several major groups, but identification of each pair within a group is equivocal.

inversions or reciprocal translocations of nearly equal portions without examining meiotic behavior. The difficulty has been partially removed by the recent inventions of many new procedures which would reveal characteristic chromosome bands (see next few procedures).

CHAPTER 15

Karyology of Cells in Culture

B. Harvesting Human Leukocyte Cultures for Chromosome Cytology

P. S. Moorhead

A procedure for culturing human blood leukocytes is given in this volume, Section II. The methodology for determining the chromosomal constitution from a sample of dividing leukocytes obtained from blood is relatively simple but the profusion of published procedures which differ in certain small details confuses or puzzles the investigator beginning such work. For a more extended consideration the reader may wish to examine the more widely used published procedures together with certain other key references.[1-4] Although many substitutions can be made within a proved procedure, only with experience can one safely judge which "detail" may be modified. There are several critical points to be kept in mind in dealing with hypotonically swollen cells and with the very plastic nature of the fixed cell in suspension. Quite subtle factors can drastically affect the morphology of the chromosomes themselves. This is best illustrated by noting the long list of special treatments which are capable of inducing a differential pattern of banding along the metaphase chromosomes. Among these are excessive temperature, high pH of staining, and partial trypsin digestion.

The importance of healthy growth cannot be overemphasized. No single factor contributes more to the success of any metaphase chromosome technique. A multitude of sins can be covered when the mitotic rate is high. In scanning a preparation one can visually reject in a few seconds dozens of poorly spread or poorly fixed metaphase cells and examine those few which happen to prove suitable. Even with optimal technique, when growth has been poor the results will be correspondingly poor, or so sparse in metaphase cells of suitable quality that much time would be wasted on such material. The necessity of selecting the better metaphase cells for chromosome cytology studies makes valid comparisons of minor changes in technique difficult. Thus, much of the seeming "witchcraft" does have a basis in experience.

[1] W. J. Mellman, *In* "Human Chromosome Methodology" (J. J. Yunis, ed.), p. 22. Academic Press, New York, 1965.

[2] P. S. Moorhead and P. C. Nowell, *In* "Methods in Medical Research" (H. N. Eisen, ed.), p. 310. Year Book, Chicago, Illinois, 1964.

[3] D. T. Arakaki and R. S. Sparkes, *Cytogenetics* **2**, 57 (1963).

[4] D. A. Hungerford, *Stain Technol.* **40**, 333 (1965).

HARVESTING AND FIXATION

Materials Needed

Colcemid stock (12.5 μg/ml)
Centrifuge tubes, conical, graduated, 12 or 15 ml, nonsterile
Pasteur pipette, with rubber bulbs
Centrifuge, International Clinical Centrifuge, table-top type
Sidearm vacuum flask, with two-hole stopper, tubing for inlet/outlet and sidearm
Hanks' balanced salt solution, pH 7.0–7.2
Centrifuge tube rack
Distilled water
Glacial acetic acid, pint, reagent grade
Methyl alcohol, absolute, reagent grade
Graduated cylinder, or flask, for mixing fixative

Procedure. At the end of the 3-day incubation period (cf. Section II, this volume), colchicine or Colcemid[5] (or other spindle-inhibiting drug such as vincaleukoblastine) is added to the culture during the final hours of incubation. These drugs disorganize the metaphase spindle yielding an accumulation of metaphase cells. However, in longer treatments, such as 10 to 18 hours, many cells are injured, some escape from the effect, and the majority of the metaphase cells accumulated possess overcontracted stubby chromosomes. Therefore, it is recommended that treatment with a spindle-inhibitor should not exceed 6 hours. The most important aspect of Colcemid or colchicine pretreatment is the dispersal of the chromosomes apart from one another, as a result of the destruction of the spindle.

One drop of the stock solution of Colcemid (12.5 μg/ml) is added to each 5-ml culture, which yields approximately 0.05 μg/ml in final concentration. If colchicine is used, the final concentration should be 0.4 μg/ml. A useful colchicine stock is 1.0 mg dissolved in 10 ml distilled water and one drop (approximately 0.02 ml) is added per 5-ml culture.

After 2 to 6 hours of exposure to Colcemid the culture is harvested by swirling and vigorous pipetting of any clumped or attached cells on the floor of the culture vessel. Centrifuge the cells from the medium in a graduated conical centrifuge tube at 600 to 900 rpm for 5 minutes (International Clinical Centrifuge,

[5] Due to a worldwide shortage of colchicine from which Colcemid is derived, it has been impossible lately to obtain Colcemid. Colcemid is about ten times stronger in its action on cells in cultivation than is colchicine and is less toxic in animal experiments. Colcemid has been supplied through the courtesy of the CIBA Pharmaceutical Co. in the past and is now commercially available from Calbiochem. (Dr. E. A. Shneour, Director of Research, P. O. Box 12087, San Diego, California 92112.) Although CIBA has recommended putting Colcemid into solution using ethanol, propylene glycol, and phosphates followed by autoclaving, for *in vitro* work we have found that Colcemid is easily dissolved in distilled water, it can be frozen in small aliquots, thawed for use, and discarded after brief storage at 5°C. It might be a better practice to filter sterilize the solution before preparing the aliquots.

table-top type).[6] Using a vacuum flask arrangement as a trap, and sidearm tubing which accepts a Pasteur pipette, aspirate nearly all of the fluid from above the pellet of cells. Add 2–3 ml of Hanks' balanced salt solution (HBSS) and with Pasteur pipette suspend cells thoroughly to wash cells free of the medium. Do not create bubbles in pipetting as cells become trapped in the fluid films and upon touching the upper walls of the centrifuge tube cells adhere and rupture and are thus lost from the sample. When cells are well suspended, add excess of HBSS, approximately 8–10 ml. Centrifuge as before and discard nearly all of supernatant. Add HBSS to the 0.5 ml mark on the tube and resuspend cells carefully with Pasteur pipette. Add 2.0 ml of distilled water to this, pipette briefly, and let stand in this hypotonic solution at room temperature for 8 to 10 minutes. The timing of this hypotonic pretreatment is critical. Factors concerned with fixation and with air drying can affect spreading so that increasing the time in hypotonic solution may only cause collapse of cell membranes and no improvement in the apparent underspreading. Resuspend cells lightly in the hypotonic solution at end of 8- to 10-minute period and centrifuge as before. Aspirate nearly all of fluid above the pellet of cells leaving only a meniscus wetting the upper surface of the cell pellet.

Fix cells by addition of 0.2 to 0.5 ml of freshly made[7] methyl–acetic fixative so as not to disrupt the cell pellet. It is important to adjust the amount of fixative used in relation to the size of the pellet of cells. The volume of fixative should not exceed three to four times the volume of the cell pellet, especially in the initial fixation. If too much fixative is used, the cells will tend to clump irretrievably. Let cells remain in this first fixative for 5 to 20 minutes, depending upon size of the pellet. A barely noticeable pellet should stand for only 5 or 10 minutes, whereas larger pellets approaching 0.1 ml require 20 minutes standing before disruption. Suspend cells using Pasteur pipette with care and control so that cells are not needlessly lost by adherence to upper walls of the pipette or centrifuge tube. If the cell pellet is of moderate or large size, add 1.0 to 2.0 ml more fixative following suspension in original volume. If pellet is very small do not add any more fixative. Centrifuge as before and aspirate excess fixative from above the cells. Add 0.2–0.5 ml of fresh fixative to cell pellet depending upon size of pellet. Resuspend cells with careful pipetting so as not to create bubbles.[8] Obtain a fine suspension by pipetting using only the terminal 2–3 inches of the pipette. If clumps are stubborn, soften the Pasteur pipette in flame of burner, draw into a finer tip, and break off so as to reduce size of orifice. When cells are

[6] Avoid slant-head centrifuges unless all glassware is well siliconized, since leukocytes tend to adhere to the wall of the centrifuge tube.

[7] Freshly made fixative means that it was mixed not more than 30 minutes or so before use. Addition of one volume of glacial acetic acid (reagent grade) *to* three volumes of absolute methyl alcohol (reagent grade) results in thorough mixing without pipetting. Do not substitute ethyl for methyl alcohol. Although results with ethyl–acetic are sometimes satisfactory, in general, methyl alcohol–acetic fixative is superior. As the fumes of the fixative are quite caustic, avoid breathing these; good ventilation is recommended.

[8] Fine control of volume delivered can be achieved by grasping pipette entirely in palm of hand with the bulb extending toward the thumb and bulb is squeezed between thumb and forefinger.

very well suspended add excess of fresh fixative, 5–10 ml. Centrifuge as before and remove all of supernatant fluid. Again add 0.2–0.5 ml of fresh fixative and by gently pipetting obtain a fine cell suspension. A bluish cast or haziness indicates that individual cells are well suspended and that further pipetting or changes of fixative are not necessary.

Slide Preparation

Materials Needed

Microslides, precleaned, 3 × 1 inch, thickness 0.96–1.06 mm (as Clay Adams, Rite-On Microslides)
Phase contrast microscope
Warming tray
Bunsen burner
Distilled water in beaker
Dry ice
Filter paper, or other absorbent lint-free paper

Procedure. Dip precleaned slide into beaker of distilled water containing lump of dry ice. If slide is not completely wettable across its entire surface, discard it. Grease from fingers or elsewhere interferes with film of water over slide and can definitely affect spreading results. Accepting this necessity of discarding some slides, commercially available precleaned slides are found to be quite superior to lab cleaned slides. Hold slide in horizontal position and drop one or two large drops of fixative containing suspended leukocytes onto the wet surface. *Immediately* turn slide on edge and press its long edge against filter paper on table surface while blowing directly onto slide surface to hasten evaporation. Blow repeatedly for 7 to 8 seconds from a distance of about 8 to 10 inches from slide. It is also satisfactory to give one short explosive blast to the wet slide surface from about 2 inches distance. Take a deep breath before dropping material onto slide as delays of even a few seconds can affect the quality of results. Immediately place slide on a warming tray to complete the drying process. Temperature of warming tray is not critical for ordinary staining with Giemsa but is quite important to certain methods for producing banding patterns (cf. this section). No blowing may be necessary if 5 parts distilled water plus 1 part HBSS is used for hypotonic step.

Ignition Drying. As an alternative to blowing slides dry, reproducibly good spreading can be obtained by igniting the fixative on the surface of the wet slide. *Immediately* on addition of the fixative-suspended cells touch the edge of the slide to burner flame and let fixative burn off. Complete drying on warming tray. If resulting flame is weak or fixative fails to ignite, this is probably due to improper strength or purity of the methyl alcohol or acetic acid. Dipping the slide in 20–30% ethyl alcohol instead of water permits use of the ignition technique in such a case. Do not use ignition drying if orcein staining is to be used (unless only faint orcein is required, as for oil immersion phase contrast microscopy). Do not use ignition drying if banding or quinicrine fluorescent staining is to be applied.

Examine the dried test slide for cells in metaphase using a phase contrast microscope (or after 5 minutes in Giemsa stain). If mitoses are scarce on the slide, the cells may be further concentrated in a smaller volume of fresh fixative. If any dried salt crystals are observed, change the fixative again. If mitoses are too close together, dilute by adding fresh fixative. If insufficient spreading is evident or if chromosomes appear indistinct, resuspend the cells in fresh changes of fixative once, twice, or three times more. The fixed button of cells can be kept refrigerated at 5°C for months but it is usually best to make slides within a day or so of initial fixation. On removal from 5°C storage several successive changes with fresh fixative are necessary before making any slides.

Staining with Ordinary Giemsa

Materials Needed

Giemsa blood stain
Hydrochloric acid, 5 *N*
Wheaton staining dishes
Coplin staining jars
Forceps, 5–7 inch
Pasteur pipettes with bulbs
Distilled water
Xylol
Cover glasses, thinness No. 1 (as Gold Seal)
Mountant fluid (as Permount or DPX)
Ammonium hydroxide, 0.15 *N*

Procedure. Handle all slides so that surface with cell material always faces the operator. Place slides in trays or in dishes in order corresponding to source boxes or lists of identification numbers so that penciled labels can be restored if they become obscured during staining manipulations. For orcein or ordinary Giemsa staining, acid hydrolysis is recommended. Acid hydrolysis with HCl "clears" the background by reducing uptake of stain in the cytoplasm but interferes with certain staining procedures and may be eliminated if desired. Place slides in 5 *N* HCl at room temperature for 5 minutes. Remove slides from HCl and place in Wheaton staining dishes, five or six per dish. Flush slides in Wheaton dishes with running tap water for at least 10 minutes. Rinse slides in dishes with three or four changes of distilled water.

Before introducing slides into Giemsa stain as prepared here,[9] the pH should be about 8.0 and after processing of five to six slides, the pH falls to about 7.0. Remove slides with forceps to Coplin jars containing Giemsa stain, five to six

[9] Giemsa stain. Fill Coplin jar to neck with distilled water (80 ml). Add 1 full Pasteur pipette of Giemsa blood stain to jar (1½ ml). Add one drop of 0.15 *N* NH_4OH from Pasteur pipette.

slides per jar. Stain for 5 to 6 minutes. Discard stain after each batch of slides. Note that very intense staining is not desirable since the interpretation of overlapping chromatids depends upon comparisons of density. On the other hand, if stain is bluish or too faint, microscope work is fatiguing. Faint staining is usually the result of too low a pH. Before removing stained slides, pour an excess of distilled water into each staining jar so as to float away any flocculant surface film. Remove each slide with forceps and immerse it in a beaker of distilled water. Holding the slide firmly, scrape the long edge of the slide against sides of the beaker to produce a vibration which dislodges any flocculant precipitate present. Rinse each slide in a beaker of fresh distilled water and then stand slides on edge to permit draining. Slides should be protected from dust. When thoroughly dry, dip each slide in xylol and drain off excess. Mount cover glass of No. 1 thinness using one drop of commercial mountant. A proper sized drop will eliminate the need for any cleaning of the slide after mounting. Complete the mounting process by placing slides on the surface of the warming tray. Immediate oil immersion examination is feasible if objective is kept away from tacky mountant at edges of coverglass.

CHAPTER 15

Karyology of Cells in Culture

C. Constitutive Heterochromatin (C-Bands)

Frances E. Arrighi

Microscopists observed, in the last century, heavily stained (condensed) chromatin areas in interphase nuclei. Such condensed chromatin (chromocenter) was studied in considerable detail by Heitz[1-3] who suggested the term heterochromatin.

Extensive cytological studies showed that heterochromatin represents chromosome segments which are permanently condensed. Thus in interphase when regular chromosomes and chromosome segments (euchromatin) decondense and become diffuse, heterochromatin areas are particularly conspicuous. However, it is not possible to determine the precise locations of heterochromatin at this stage

[1] E. Heitz, *Jahrb. Wiss. Bot.* **69**, 762 (1928).
[2] E. Heitz, *Bedeutung. Z. Zellforsch.* **20**, 237 (1934).
[3] E. Heitz, *Biol. Zentrabl.* **54**, 588 (1934).

because individual chromosome morphology cannot be delineated at interphase. Conversely, at metaphase and anaphase where chromosome morphology is at its best, differentiation between euchromatin and heterochromatin again is not feasible because both are fully condensed. The only stage in which both differential condensation and chromosome morphology can be observed is the short stage prometaphase. In *Drosophila*, for example, where the diploid number is low and the chromosomes are relatively small, competent cytologists can localize heterochromatin at prometaphase. In vertebrates, where most species possess high diploid numbers, the existence of chromocenters could be observed in interphase, but their location or locations were not critically determined in most karyotypes until recently.

When Pardue and Gall[4] and Gall and Pardue[5] performed experiments on *in situ* DNA/RNA hybridization, they found that in the chromosomes of the laboratory mouse the centromeric regions were more deeply stained than the chromosome arms. Applying the *in situ* hybridization procedure to human cells, Arrighi and Hsu[6] and Chen and Ruddle[7] also found that the series of treatments revealed deeply stained areas in the centromeric regions of human chromosomes and that the amount and location of the heterochromatin were characteristic for each chromosome pair. These authors considered that such deeply stained areas are equivalent to heterochromatin. Yunis *et al.*,[8] using a slightly different procedure, reached a similar conclusion.

In order to confirm that the classic term heterochromatin can indeed be applied to vertebrate chromosomes, Hsu[9] applied the same procedure to cytological preparations of neuroblast metaphase chromosomes of *Drosophila melanogaster*. *Drosophila* chromosomes responded to the treatment in the same manner as the mammalian chromosomes and the areas of heterochromatin were localized as described by Heitz.[3] Hsu and Arrighi[10] found that constitutive heterochromatin of mammals, though primarily located in the centromeric areas, may be located at the terminal segments, interstitial segments, or entire chromosome arms, depending upon the species and the specific chromosomes under consideration.

Procedure

1. Harvest cultures with Colcemid and hypotonic solution pretreatments. For squash preparations, fix the cells in 50% acetic acid; for flame- or air-dried preparations, fix the cells in fresh Carnoy's fixative (3 methanol:1 acetic acid) and wash the cells with fresh fixative three times. We prefer the squash preparations,

[4] M. L. Pardue and J. G. Gall, *Science* **168**, 1356 (1970).

[5] J. G. Gall and M. L. Pardue, *In* "Methods in Enzymology," Nucleic Acids (L. Grossman and K. Moldave, eds.). Vol. 21, p. 470. Academic Press, New York, 1971.

[6] F. E. Arrighi and T. C. Hsu, *Cytogenetics* **10**, 81 (1971).

[7] T. R. Chen and F. H. Ruddle, *Chromosoma* **34**, 51 (1971).

[8] J. J. Yunis, L. Roldan, W. G. Yasmineh, and J. C. Lee, *Nature* (*London*) **231**, 532 (1971).

[9] T. C. Hsu, *J. Hered.* **62**, 285 (1971).

[10] T. C. Hsu and F. E. Arrighi, *Chromosoma* **34**, 243 (1971).

particularly for materials whose distribution of heterochromatin is not known. For human chromosomes flame-dried preparations give good results (Craig-Holmes and Shaw[11]).

2. Prepare "subbed" slides by dipping alcohol or acid-cleaned slides (preferably with frosted edge for pencil labeling) in a solution of 0.1% gelatin and 0.01% chromium potassium sulfate. The slides are allowed to dry. The subbed slides can be prepared in large numbers and be stored. The purpose of using "subbed" slides is to prevent cellular loss during the NaOH treatment, which is necessary for this procedure. For flame-dried preparations, subbing is not required.

3. Squash fixed cells without stain and remove the coverslips using the dry ice method. For flame- or air-dried preparations, steps 3 and 4 are omitted.

4. Rinse slides with 95% alcohol twice and air dry.

5. Treat the slides with 0.2 *N* HCl at room temperature for 30 minutes and rinse several times with distilled water. This step may be omitted.

6. Treat the cells with pancreatic RNase (100 μg/ml in 2xSSC) in a moist chamber for 60 minutes. Commercially available RNase may contain DNase. Therefore the stock solution should be heated in a boiling water bath for 5 to 10 minutes. For SSC, prepare a 10× stock solution (Marmur[12]) of the following and dilute appropriately: 0.15 *M* trisodium citrate; 1.5 *M* NaCl. The working SSC solution should be at pH 7.0.

7. Rinse in several changes of 2xSSC, 70% ethanol, and 95% ethanol.

8. Treat the slides with 0.07 *N* NaOH. This step is critical. The original instructions treat the slides for 2 minutes but it was found by us and others (Gagné *et al.*[13]) that the concentration of NaOH as well as the duration of treatment can be greatly reduced for some species. The best way, for mammalian chromosomes at least, is to retain the original concentration and test the preparations for several durations, e.g., 30, 60, 90, and 120 seconds. One of them should give the desired results.

9. Rinse in several changes of 70% ethanol and several changes of 95% ethanol and air dry. It is important to remove NaOH as rapidly as possible to prevent further denaturation.

10. Incubate the slides, in moist chambers, in 2xSSC or 6xSSC at 65° overnight.

11. Rinse several times in 2xSSC, 70%, and 95% ethanol.

12. Stain for 15 to 30 minutes in a Giemsa solution (10% stock Giemsa staining solution in phosphate buffer, pH 6.8–7.2). Rinse with distilled water, air dry, and mount. The phosphate buffer is 0.01 *M*.

13. Additional notes. M. L. Pardue (personal communication) found that the 2x or 6xSSC may become very alkaline during the renaturation period (65°C overnight) when certain types of slides are used. This alkalinity may interfere with the C-band staining. Such conditions can be overcome by using slides of high quality.

Wet chambers can be constructed in a number of ways. We use plastic Petri

[11] A. P. Craig-Holmes and M. W. Shaw, *Science* **174**, 702 (1971).
[12] J. Marmur, *J. Mol. Biol.* 3, 208 (1961).
[13] R. Gagné, R. Tanguay, and C. Laberge, *Nature N. Biol.* **232**, 29 (1971).

dishes and add 15–18 ml of 2x or 6xSSC in the bottom half. Plastic dishes are superior to glass Petri dishes because with plastic the moisture that condenses on the lid does not drop off onto the slides. The slides are placed on a rack or rubber grommets. The solution (2xSSC, 6xSSC, or RNase) is placed on the slides and a cleaned coverglass placed over the solution. The total volume used depends on the area covered by cells. For flame- or air-dried preparations a 22 × 50 mm coverglass is useful with 0.3 to 0.35 ml of solution. If a 22 × 40 coverglass is used 0.2 ml is sufficient.

Placing the slides in Coplin jars instead of moist chambers is undesirable. Treating with RNase in this manner is financially impracticable, and slides immersed in 2xSSC or 6xSSC at 65°C overnight are stained as well as the cells.

As mentioned, constitutive heterochromatin is not confined to the centromeric areas. Terminal heterochromatin, interstitial heterochromatin, and in some species, total heterochromatic arms, have been observed. Figure 1 presents a karyotype of the Syrian (golden) hamster (*Mesocricetus auratus*) showing the distribution of heterochromatin. In addition to centromeric heterochromatin, many pairs of autosomes as well as the X chromosomes possess totally heterochromatic arms. The Y chromosomes of this species is almost entirely heterochromatic.

In some groups of animals, e.g., the birds (Stefos and Arrighi[14]) and *Drosophila* (Hsu[9]), the original procedure proved to be too strong. The chromosome thus treated appeared swollen with little differentiation. A modified proce-

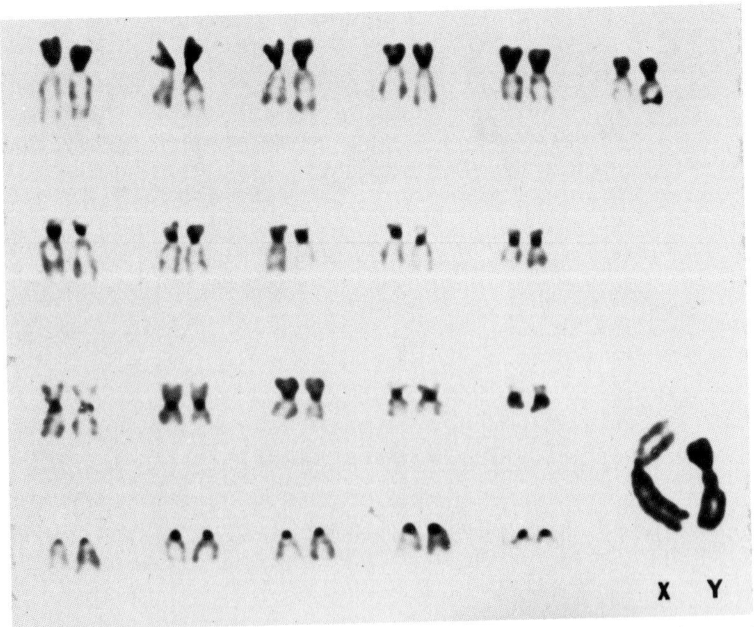

Fig. 1. Karyotype of male Syrian hamster (*Mesocricetus auratus*) metaphase cell stained for C-bands. In addition to the centromeric heterochromatin, note the entire heterochromatic arms of many submetacentric chromosomes. The entire Y chromosome and the long arm of the X chromosome are also heterochromatic.

[14] K. Stefos and F. E. Arrighi, *Exp. Cell Res.* 68, 228 (1971).

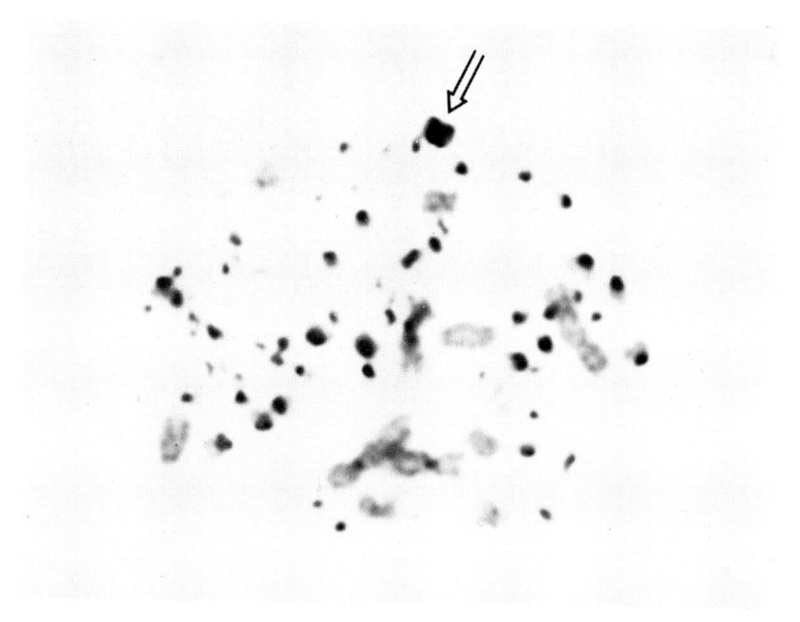

Fig. 2. Metaphase cell of a male domestic pigeon, *Columba livia*, stained for C-bands showing the heterochromatic W chromosome (arrow). The centromeric areas of the microchromosomes are heterochromatic. In addition, some of the macrochromosomes of this species possess small amounts of centromeric heterochromatin. (Photograph through the courtesy of Miss K. Stefos.)

dure, omitting the HCl and RNase treatment and preparing NaOH solution in SSC (adjusting pH to 12) works well. The W chromosome of the birds stains deeply as a totally heterochromatic element (Fig. 2).

The more recent G-banding techniques (terms, G-band, C-band, Q-band and R-band, suggested by the IVth International Conference on Standardization in Human Cytogenetics) reveal cross-bands in the chromosomes, thus enabling cytologists to identify most, if not all, chromosomes of a karyotype. However, G-bands do not completely replace C-bands because they do not always coincide. For example, the centromeric heterochromatin of the human chromosome 9, a very conspicuous segment, is unstained in G-band preparations. Variability in the size of the C-bands is extremely common, and their detection is definitely simpler with the C-band technique than with the G-band techniques.

CHAPTER 15

Karyology of Cells in Culture

D. Fluorescent Banding of Chromosomes (Q-bands)

C. C. Lin and Irene A. Uchida

The quinacrine fluorescent technique was the first technique developed to demonstrate specific banding patterns (Q-bands) of chromosomes. The procedures for quinacrine dihydrochloride staining have been published[1] but in less detail. An alternative technique is available using quinacrine mustard as the fluorochrome.[2]

Quinacrine Dihydrochloride Fluorescent Technique

Chromosome Preparation. The essential prerequisite for more detailed Q-bands is the preparation of chromosomes of high quality. We follow the conventional techniques for preparing air-dried slides[3] with the following points of emphasis. Colcemid, with a final concentration of 1 μg/ml, is added 3 hours prior to harvest. The hypotonic solution and fixative used are 0.075 *M* KCl and 3:1 methanol:acetic acid, respectively. The cells in fixative are kept in the refrigerator overnight before the slides are made. The slides are stored at room temperature and satisfactory Q-bands can still be produced, in some instances, several years later. However, high quality fluorescent karyotypes are most easily obtained from newly prepared slides.

Staining. The staining solution consists of 0.5% (w/v) quinacrine dihydrochloride (Atebrin, G. T. Gurr Ltd., England) in glass-distilled water. It can be kept for a few weeks in a refrigerator and reused. For staining freshly prepared slides, the solution (about 40 to 50 ml) is adjusted to pH 4.5 with 0.1 *N* HCl. The distilled water used throughout the procedure, approximately 200 ml, is also adjusted to pH 4.5. Precise control of the pH is crucial. A lower pH usually produces a brighter fluorescence but the differentiation of the bands is sometimes poor. A higher pH will improve the differentiation but reduce the brightness, and the chromosomes tend to "melt" easily after a short time of exposure to UV light. For older preparations, experimentation is needed to determine the optimal pH (usually lower than 4.5).

[1] C. C. Lin, I. A. Uchida, and E. Byrnes, *Can. J. Genet. Cytol.* **13**, 361 (1971).
[2] T. Caspersson, L. Zech, C. Johnsson, and E. J. Modest, *Chromosoma* **30**, 215 (1970).
[3] P. S. Moorhead, P. C. Nowell, W. J. Mellman, D. M. Battips, and D. A. Hungerford, *Exp. Cell Res.* **20**, 613 (1960); cf. Chapter 15.B, this section.

Procedure. Five Coplan jars are used. One jar contains the staining solution while the others contain distilled water.

1. Dip the slides into distilled water for a few seconds.
2. Transfer into the staining solution for 15 minutes.
3. Wash the slides by passing them through the three remaining jars of distilled water for a total of 10 minutes.
4. Allow the slides to air dry by propping them up against the Coplan jars.
5. Mount with a drop of distilled water (the coverslip used is No. 1 or 1½).
6. Blot to remove excess water. The water layer should be kept as thin as possible to prevent the scattering of light which produces a hazy image.
7. Seal the edges of the coverslip with paraffin wax or clear nail polish.
8. To avoid deterioration of the chromosomes the slides should be examined immediately after preparation. If this is not possible, they may be kept in the refrigerator, but they should be examined on the day they are made.

Microscopy. Both a Zeiss standard fluorescence microscope and a photomicroscope equipped with a BHO 200 W/4 mercury burner and an ultradark-field condenser for transmitted illumination are used in our laboratory. Excitation filter No. 1 (BG 12) and barrier filter No. 47 are inserted. The red-suppression filter BG 38 may be removed to increase the light intensity. Only two objectives are required: both are planapochromat oil and equipped with iris, with a 40× objective mainly for scanning and 100× for photography. The photomicroscope must have a cable release to move the beam-splitting prism out of the light path for photographic use.

Photography. The choice of cell to be karyotyped is important. The chromosomes should be fairly long with the chromatids held closely together (Fig. 1). Photographs are taken with Kodak high contrast copy film (ASA 64). For the standard fluorescent microscope with camera attachment, the exposure time is about 2 to 2½ minutes with an 8× eyepiece and 100× oil objective. For the photomicroscope the exposure time is about 1½ minutes.[4] Prints are made on Kodabromide paper F2 or F3. The karyotype shown in Fig. 2 was prepared by mounting the chromosomes on an exposed sheet of Kodabromide paper.

When the analysis has been completed, the coverslip is removed by freezing. The slide can then be reexamined with conventional stains.

Quinacrine Mustard Fluorescent Technique

The staining techniques described are based on Caspersson *et al.*[2] Quinacrine mustard is supplied in limited amounts by Sterling Winthrop Research Institute, Rensselaer, New York.

The staining solution consists of 0.5 mg of quinacrine mustard per 10 ml of

[4] Recently, an HBO 200 exciter unit for DC operation has become available from Carl Zeiss, New York Office. The main advantage of this DC-unit is a much more stable light source and substantial increase in intensity. The appropriate exposure time is about ¾ minute.

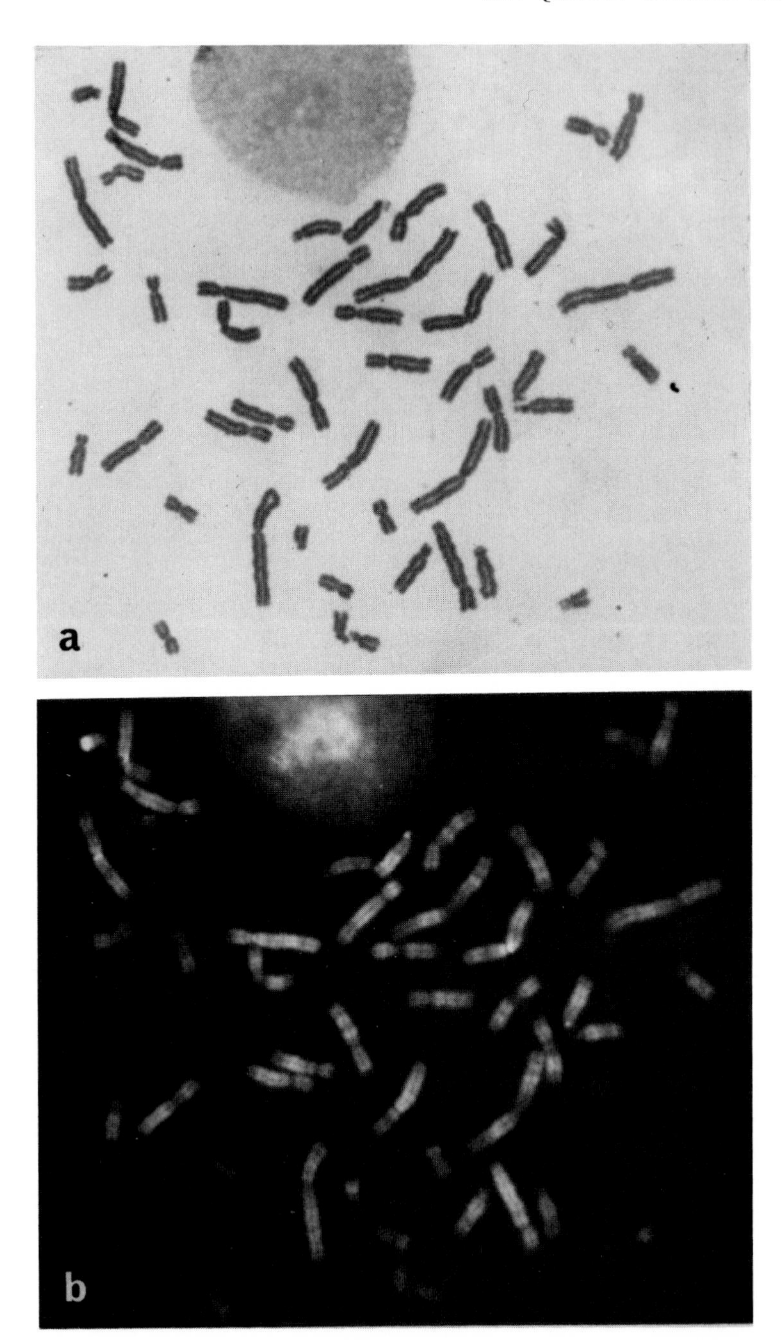

Fig. 1. (a) Metaphase plate of leukocyte stained with orcein after fluorescent studies completed. Note fairly long chromosomes with chromatids close together. (b) The same metaphase plate stained with quinacrine dihydrochloride.

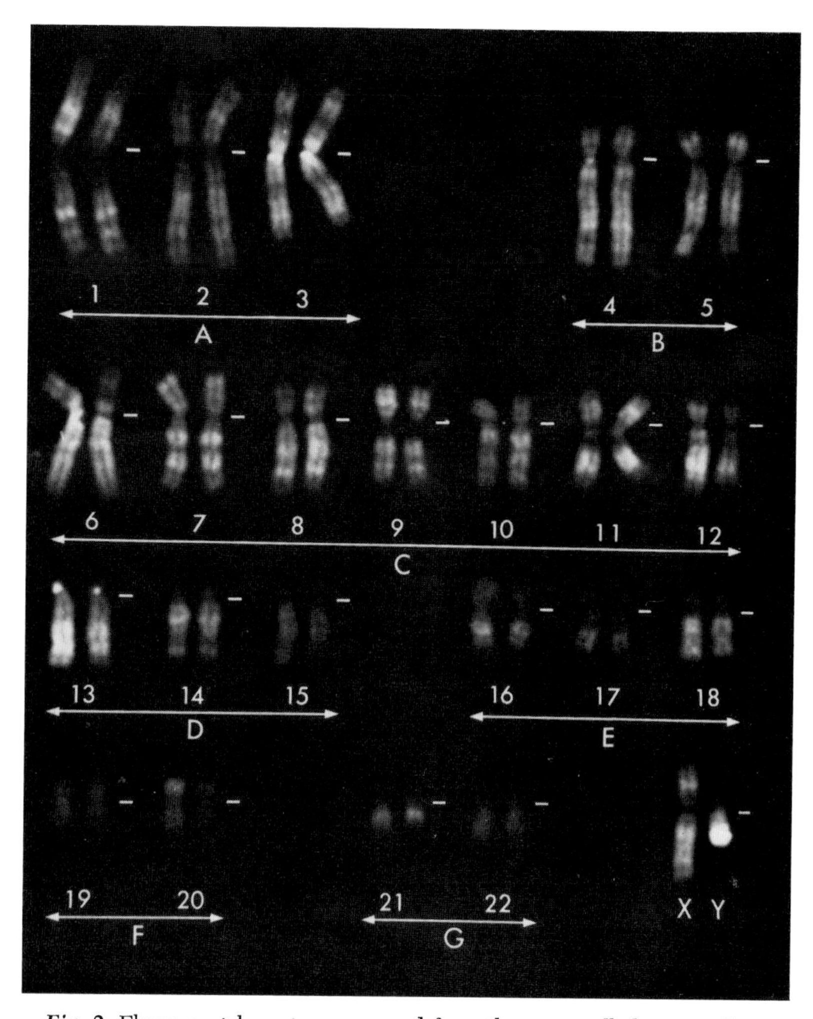

Fig. 2. Fluorescent karyotype prepared from the same cell shown in Fig. 1.

MacIlvaine's disodium phosphate/citric acid buffer. This buffer is prepared by diluting 86.3 ml of 0.1 *M* citric acid ($C_6H_8O_7$) and 453.7 ml of 0.2 disodium phosphate (Na_2HPO_4) to 1 liter with distilled water. The pH is adjusted to 7.0. The air-dried slides are placed in a horizontal position, flooded with staining solution, and left for 20 minutes. The slides are washed by dipping through three changes of buffer solution or distilled water (pH 7.0) for a total of 3 minutes and then mounted with buffer or distilled water.

CHAPTER 15

Karyology of Cells in Culture

E. Trypsin Technique to Reveal G-Bands

H. C. Wang and S. Fedoroff

The recently developed trypsin technique to reveal the so-called "Giemsa stained banding patterns–G-bands" on chromosomes is based on the removal of nucleoproteins from fixed chromosomes by proteolytic enzymes.[1-3] Even before staining, the banding patterns are observable with phase-contrast microscopy (Fig. 1c), and may represent regions in which chromatin is accumulated due to either increased folding or coiling of DNA. This is supported by observation of the chromosomes with Nomarski optics in which the bands appear as elevations on the chromosome surfaces (Fig. 1d and e). For purposes of light microscopy the bands can be intensified by staining the chromosomes with Giemsa stain or one of its modifications (Fig. 1b).

So far our experience with this method is limited to human and mouse cells. We found that the ease with which good banding patterns could be produced and the optimal conditions necessary varied with the type of cells. For example, good banding was easy to produce on human lymphocytes, whereas it was quite difficult when working with mouse bone marrow cells. Moreover, two sublines of clone 929-L cells grown under identical conditions differed in their sensitivity to trypsin treatment.

The advantage of using banding patterns in addition to well established gross morphological criteria for chromosome identification is that the bands provide additional information about individual chromosomes. For example, the HeLa cells in our laboratory generally have four No. 2 chromosomes instead of the two present in human diploid cells. Although the No. 2 chromosomes in HeLa cells have similar gross morphology, they have different banding patterns. This suggests that during the development of cell lines profound intrachromosomal changes, as revealed by the banding patterns, may accompany changes in chromosome number. The changes in banding patterns may be due to translocations, inversions, or other known chromosomal aberrations, or, due to structural or functional changes in specific regions of the chromosomes. In any case, the banding patterns of the chromosomes are very useful as markers for identification of specific chromosomes.

[1] M. Seabright, *Lancet* **ii**, 971 (1971).

[2] B. A. Chiarelli, M. Sarti-Chiarelli, and D. A. Shafer, *Proc. 10th Annu. Somatic Cell Genet. Conf.* (Abst. 1972).

[3] H. C. Wang and S. Fedoroff, *Nature N. Biol.* **235**, 52 (1972).

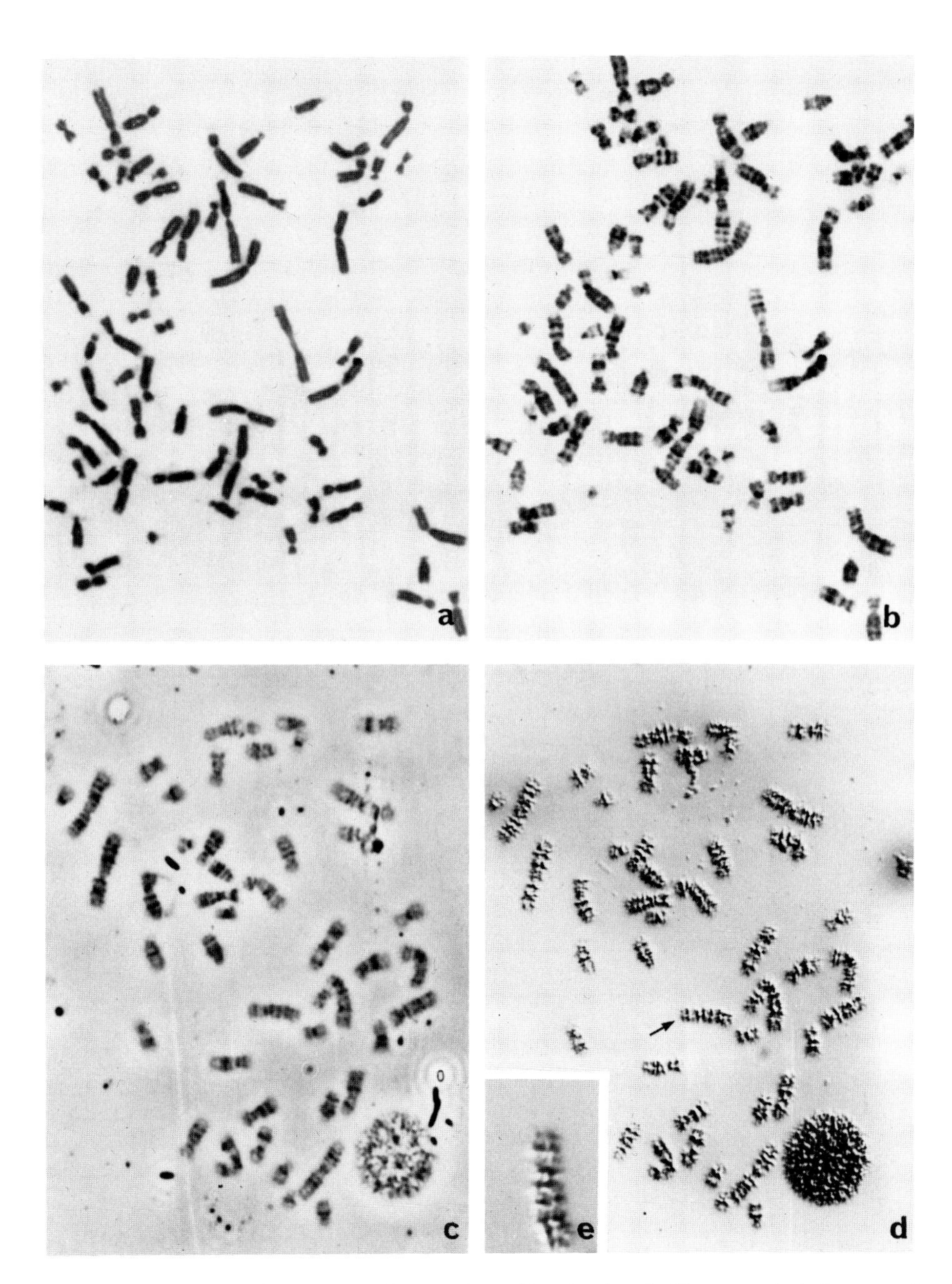

Fig. 1. Metaphase chromosomes from HeLa cell stained with Giemsa stain before trypsin treatment (a) and with Wright's-Giemsa stain after trypsin treatment (b). Metaphase chromosomes from normal human lymphocyte after trypsin treatment photographed under phase contrast optics (c) and Nomarski optics (d), and one chromosome under higher magnification (e).

Procedures

Chromosome Preparation. The cells used for chromosome studies are pretreated with a mitostatic agent such as Colcemid (0.06 μg/ml of media, CIBA) for 1 to 2 hours followed by treatment with hypotonic solution for 5 to 15 minutes at 37°C. The hypotonic solution can be either 0.075 *M* KCl or 0.9% sodium citrate, or mixtures of both in differing proportions depending upon the type of cell used.

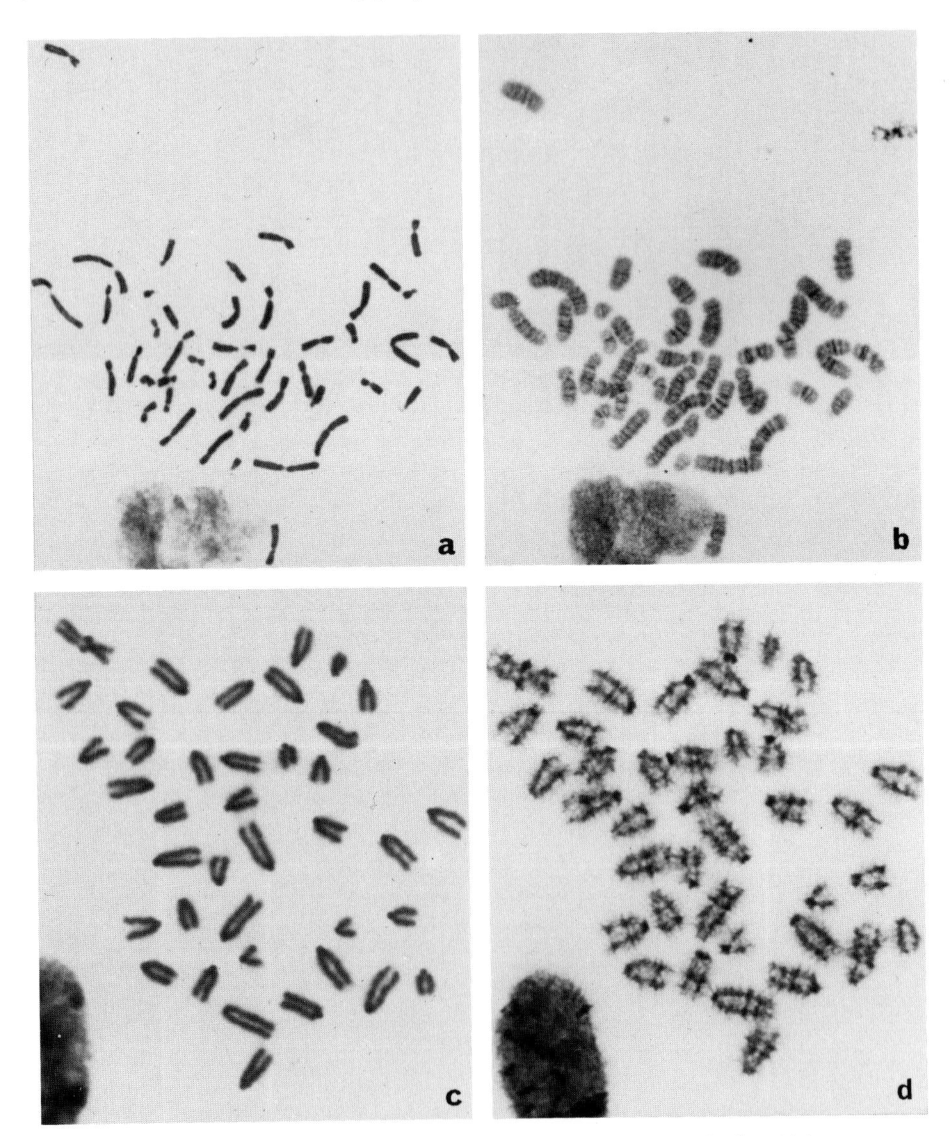

Fig. 2. Metaphase chromosomes from WI-38 cell before (a) and after (b) trypsin treatment. Note that after trypsin treatment the chromosomes have swollen and therefore some overlap is seen. Metaphase chromosomes from normal mouse embryo fibroblast before (c) and after (d) trypsin treatment. The chromosomes are overtreated with Colcemid, and therefore the chromatids did not fuse after trypsin treatment.

After these pretreatments, the cells are fixed in fresh fixative made of three parts methanol to one part glacial acetic acid. Chromosome preparations are then flame-dried.

In such preparations the chromosomes should not be too contracted, and the two chromatids should not be too widely separated, otherwise the bands will be less distinct because two neighboring bands may appear as one, due to their close proximity. In order to avoid this, the concentration of Colcemid and the duration of its application to the cells must be carefully adjusted.

The chromosomes should be well separated from each other, as the subsequent trypsin treatment will cause them to swell. If they lie too close to each other, neighboring chromosomes may overlap due to this swelling (Fig. 2a and b).

Trypsin Treatment. The chromosome preparations are treated with 0.025–0.1% trypsin solution at 22°–30°C for 10 to 15 minutes. The trypsin solution is made from powdered stock (Difco 1:250) by dissolving the powder in Puck's BSS (Ca^{+2} and Mg^{+2} free) or in a mixture of equal amounts of Puck's BSS and Versene (0.02% dibasic EDTA in 0.85% saline). The pH of these solutions is adjusted to between 7.0 and 8.0 using 1.4% sodium bicarbonate.

After trypsin treatment, the slides are rinsed in two changes each of 70 and 100% ethanol and left to air dry.

The concentration of trypsin, the temperature during treatment, and the duration of the treatment have to be carefully adjusted for each cell type.[4] Figures 3 and 4a and b show the effects of variation of these factors. We found it most practical to keep the temperature and concentration of trypsin constant, and to vary the time of treatment as required, monitoring the end point by observing the appearance of the bands under the phase microscope.

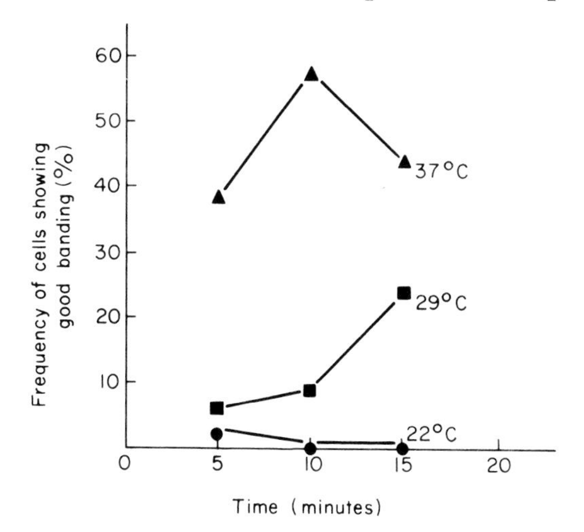

Fig. 3. Effect of duration of treatment with 0.25% trypsin (Difco 1:250) on the quality of G-bands at different temperatures. (From H. C. Wang *et al.*[4])

[4] H. C. Wang, S. Fedoroff, and C. Dickinson, *Cytobios* **6** (No. 21) in press.

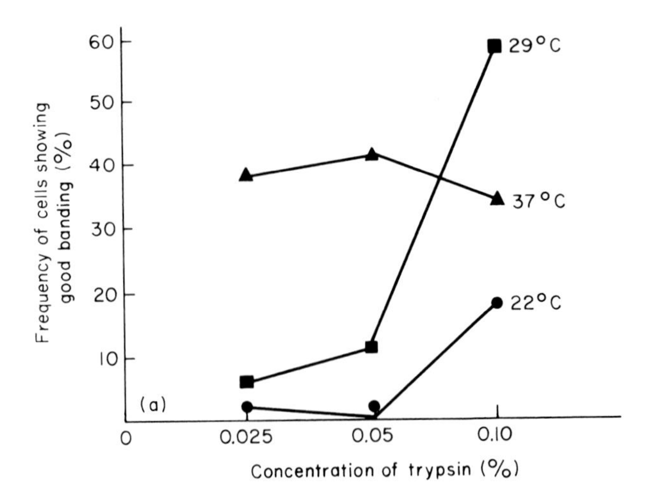

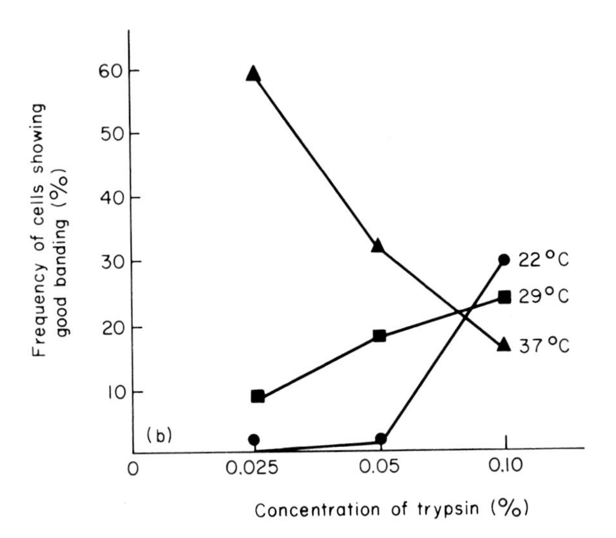

Fig. 4. Effect of concentration of trypsin (Difco 1:250) on the quality of G-bands after 5 minutes (a) and after 10 minutes (b) treatment at various temperatures. (From H. C. Wang *et al.*[4])

The chromosomes swell when treated with trypsin and the two chromatids fuse together and appear as a unit. The light-staining bands appear most distinctly when chromatids are thus fused. We consider the end point of the trypsin treatment to be the time when fusion occurs. If the treatment is not carried to this point, and the chromatids remain apart, then the banding pattern is not as distinct as when chromatids are fused (Fig. 2c and d). If the treatment is carried past this point, the morphology of the chromosomes becomes distorted and only the coarse and darkly stained bands are visible. If Colcemid treatment was excessive, the chromatids are too far apart and the optimal trypsin treatment may not bring them together.

The chromosomes should all lie on the same plane and be as free of cytoplasm as possible, otherwise the banding quality within the same cell may vary from one chromosome to another.

Staining. Preparations showing good banding by the phase-contrast microscope may be stained for 2 minutes in Wright's stain followed by 1 minute in Giemsa stain. The slides are given a quick rinse in 10% methanol and then in distilled water, air-dried, and mounted in either immersion oil or a mounting medium such as Permount. The staining solutions are prepared as given below.

1. *Wright's stock solution.* 0.8 g Wright's stain (Fisher Scientific Company), 480 ml absolute methanol. (Ripen for 2 days.)
2. *Wright's working solution.* 1 part Wright's stock solution, 1 part Sörensen buffer.
3. *Giemsa.* 1 part Giemsa stock solution (Fisher Scientific Company), 20 parts Sörensen buffer.
4. *Sörensen buffer.* KH_2PO_4, 6.63 g; $Na_2HPO_4 \cdot 7\ H_2O$, 2.56 g; distilled water to 1000 ml.

The advantage of staining chromosomes first with Wright's followed by Giemsa stain is that the pH of the stains is not critical, the working solutions can be used for several days, and the differentiation of banding patterns is superior to that of preparations stained with Giemsa or Wright's stain only.

Photography. Stained chromosomes can be photographed before and after trypsin treatment (Fig. 1a and b). Sometimes this is desirable because the swelling caused by trypsin treatment may obscure the centromeric region of the chromosomes making identification of the chromosomes based on established gross morphology difficult, especially if the karyotype of the cells is not well known. To photograph stained chromosomes before trypsin treatment it is best to stain chromosomes with Giemsa stain only, because it is easier to remove the Giemsa stain before trypsin treatment than to remove the combined Wright's and Giemsa stains. We use ASA 64 high contrast copy film for photographing chromosomes.

CHAPTER 15

Karyology of Cells in Culture

F. Characteristics of Insect Cells

Imogene Schneider

The procedure for enumerating the chromosomes of cultured insect cells is very similar to that employed for vertebrate cells. Because of the lower temperature at which the former are grown, it is customary to incubate the cells in Colcemid for a much longer interval than is recommended for vertebrate cells. Distilled water rather than a sodium citrate solution is used to expand the cells and Giemsa stain usually gives better results than acetic orcein.

PROCEDURE

Once the cells are in mid-log phase of growth, Colcemid is added to the culture medium to give a final concentration of 0.06 μg/ml. After being incubated for 18 hours at 22° to 27°C, the cells are flushed from the bottom of the flask with a Pasteur pipette and transferred to a 15-ml conical centrifuge tube. An equal volume of distilled water is added and the cells gently agitated. The tube is placed in a 32°C water bath for a total time of 20 minutes with a second equal volume of distilled water added after the first 10 minutes and a third equal volume 5 minutes thereafter. A few drops of fixative (glacial acetic acid and methanol, 1:3) are added, the cells mixed very gently, and then spun for 5 minutes at 150 g. The supernatant is removed with a small-bore pipette and 0.1 ml of fixative added to the pellet. This is allowed to stand without mixing for 20 minutes at room temperature. An additional 2.9 ml of fixative is then added and the cells mixed by gently blowing air into the fluid. The cells are centrifuged at 150 g for 10 minutes and all but 0.1–1.0 ml of fixative removed from the pellet. The cells are resuspended in the remaining fixative and dropped on precleaned slides which have been dipped in distilled water. The slides should be held in a nearly vertical position and not more than two to three drops used per slide. The slides are placed in a slanted position and allowed to dry for at least an hour. The cells are then stained for 30 minutes with a freshly made 5% Giemsa solution, buffered at pH 7.2.

KARYOTYPES OF CULTURED INSECT CELLS

An accurate count of the chromosomal complement in dipteran cells is readily made; in hemipteran and orthopteran cells it is somewhat more difficult and in lepidopteran cells it is virtually impossible.

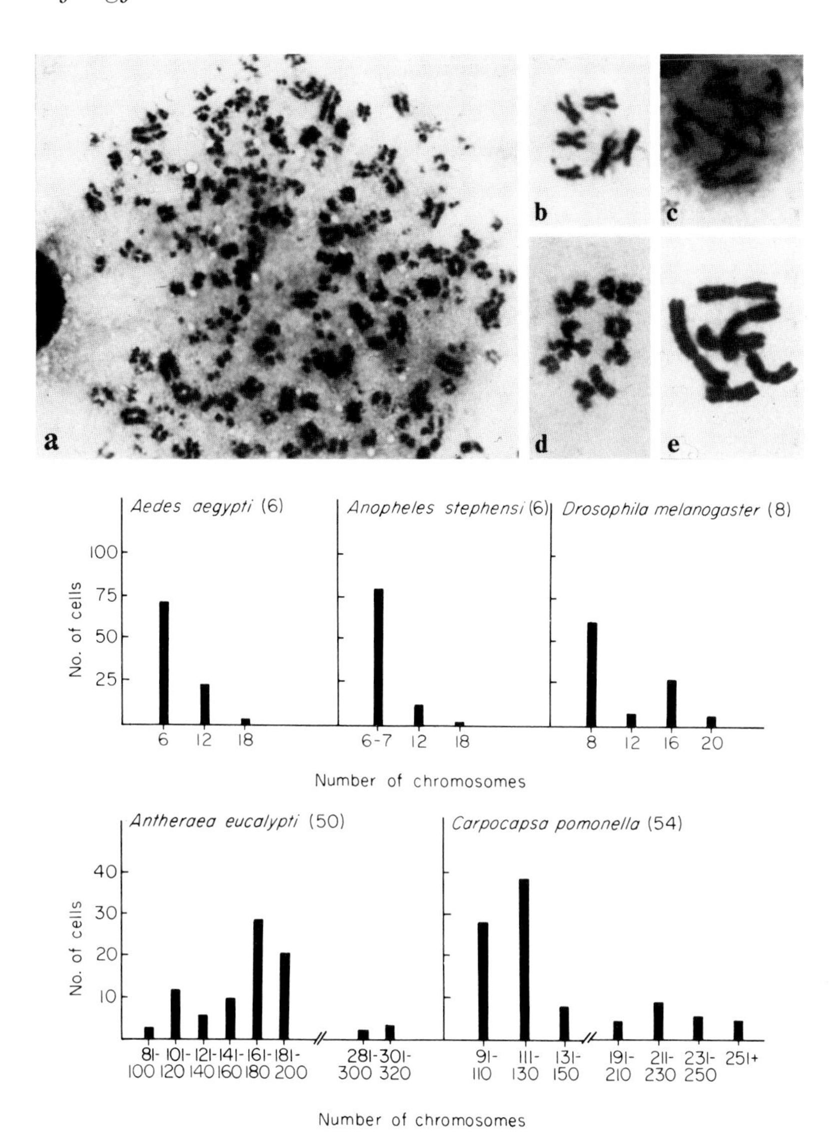

Fig. 1. (Upper portion). Metaphase plates from lepidopteran and dipteran cell lines: (a) *Antheraea eucalypti,* (b) and (c) diploid and aneuploid cells, respectively, of *Drosophila melanogaster,* (d) *Anopheles stephensi,* and (e) *Aedes aegypti.* ×2100. (Lower portion). Chromosome counts of cells from three dipteran and two lepidopteran species. Note the extensive range of chromosome numbers in the latter two as compared to minimal variations in the former. The figure in parenthesis after each cell line indicates the diploid number for that species.

An inspection of Fig. 1a indicates the difficulty of working with cells from the order Lepidoptera. The cell is typical of those found in Grace's line[1] of *Antheraea eucalypti* (a saturniid moth). The haploid number of this species is 25 but in somatic tissues the ploidy may range from $2n$ (50) to $128n$ (3200).[2]

[1] T. D. C. Grace, *Nature* (*London*) **195,** 788 (1962).

[2] J. A. Thomson and T. D. C. Grace, *Aust. J. Biol. Sci.* **16,** 869 (1963).

There is no possibility of identifying the individual microchromosomes because of their minute size and little possibility of identifying the macrochromosomes since the latter have diffuse rather than localized centromeres. Hence, the chromosomes can not be distinguished as acrocentrics, metacentrics, and the like. In the lower half of Fig. 1 are histograms of the chromosome numbers found in various insect cell lines. The two lepidopteran lines, *A. eucalypti* and *Carpocapsa pomonella* (the codling moth) show a wide range, with the former having a greater percentage of highly mixoploid cells than the latter. The counts probably are accurate, at best, to within ±5%.

In contrast, the chromosomes of dipteran cell lines (Fig. 1b–e) are readily counted due to their low diploid number ($2n = 6$ for mosquitoes; $2n = 8$ for *Drosophila* species) and the lack of high ploidy. Chromosome counts of cells from three dipteran lines are also given in Fig. 1. Diploid cells comprise no less than 60% of the population in any one line and rarely are cells containing more than $4n$ chromosomes encountered.

Between these extremes are the cell lines from hemipteran and orthopteran species. Hirumi[3] reported the presence of predominantly 4–8n cells in a line of *Agallia constricta* (a leafhopper), a species having a normal diploid number of 22 for the female and 21 for the male. Accurate counts are apparently possible even though the chromosomes in the preparations tend to have a somewhat fuzzy appearance and can be classified only to the extent of being "large" or "small." Detailed studies on karyotypes of orthopteran cell lines have not been carried out. However, Landureau[4] reported that cells from the cockroach, *Blabera fusca*, ($2n = 74$ for the female and 73 for the male) invariably become polyploid after being cultured a few weeks and that these are the only cells which continue to multiply *in vitro*.

[3] H. Hirumi, personal communication.

[4] J. C. Landureau, *Exp. Cell Res.* **50**, 323 (1968).

CHAPTER 15

Karyology of Cells in Culture

G. Characteristics of Plant Cells

Carl R. Partanen

The assessment of the karyological state of plant cells in culture is, of course, a primary objective for those who are interested principally in nuclear mechanisms and who are using the *in vitro* method of cell propagation. However, when plant cells *in vitro* are used for any other purpose, e.g., to study the responses of cells or tissues to alterations of media, and added factors, this is still an important assessment to be made. The almost general observation is that the chromosome number in plant tissue or cell cultures is not stable at the diploid level but is subject to variation, usually through the process of endoreduplication.[1] On the other hand, the relative paucity of information on how other cell processes may be related to level of ploidy suggests that the karyological state of the system being studied is one of the parameters that needs to be defined in any careful study.

However, a proper assessment of karyology is not made casually; it does not suffice to have looked at the chromosomal status once, or even at sporadic intervals, since it is frequently variable over time. Instead, for this parameter to be defined precisely, it needs to be examined regularly, during each passage, at some defined time. This becomes mandatory if the system displays any instability; for example, some systems may manifest cyclic fluctuations in level of ploidy.[2] With sporadic observations, such fluctuations may pass unnoticed, and the tissue may be erroneously characterized.

Other important considerations in making a proper assessment are cytological technique,[3] including pretreatment of cells, and the question of sampling. Either of these, casually treated, can lead to possible erroneous conclusions.

Handling of Cells or Tissues

The procedures outlined here will be concerned only with the preparation of squash preparations. For most general purposes, with adequate precautions, this method is quite satisfactory as well as expedient. If large numbers of samples are to be analyzed it is the only practicable approach, the alternative of embed-

[1] C. R. Partanen, *Intern. Rev. Cytol.* **15**, 215 (1963).

[2] C. F. Demoise and C. R. Partanen, *Amer. J. Bot.* **56**, 147 (1969).

[3] C. R. Partanen, *Proc. Intern. Conf. Plant. Tissue Culture.* McCutchan, Berkeley, California, 1965.

ding and sectioning being tedious, time-consuming, and not suited for making quick assessments.

The procedures call for passing the tissue or cells through several treatments. Coherent tissues can simply be transferred from one treatment to the next with forceps. Extremely friable callus, or cell suspensions, can be conveniently handled in centrifuge tubes. The cells are collected by light centrifugation, e.g., in a simple clinical model, decanting off the supernatant or removing it with a Pasteur pipette, and resuspending the cell pellet in the next treatment.

Pretreatment of Cells

The advisability of pretreating cells will depend upon the objectives of the study. If the sole objective is to count chromosomes, and there is no concern for other aspects of nuclear behavior, then a prefixation treatment of the cells with colchicine is not only permissible, but desirable. The resultant hypercontraction and dispersion of arrested metaphase chromosomes greatly facilitates counting. Since materials vary somewhat in their responsiveness to colchicine, and if difficulties are encountered in obtaining a response, it may be necessary to determine a suitable concentration and time empirically. However, a reasonable starting point is an 0.02% aqueous solution of colchicine for 1 to 2 hours. Too high a concentration can inhibit mitoses completely, and too long an exposure will, of course, lead to chromosome doubling.

Callus may be transferred into a small volume of colchicine solution. To treat an aliquot of a cell suspension, an appropriately more concentrated solution may be added to the volume of medium containing the cells to make the desired concentration.

However, if there is any concern for other aspects of nuclear behavior, such as anaphase behavior, or possible endoreduplication, then it is mandatory that at least some part of each lot to be sampled be examined without prior treatment with colchicine, or other substances with similar effects, since these tend to obscure the observations.

Sampling of the Cell Population

Sampling is a question of both place and time. In any sort of coherent system it can be assumed that divisions are not occurring uniformly throughout the tissue mass. Instead, they are characteristically localized, usually peripherally, although the exact location and distribution must be determined. Ideally this would be done by embedding and serial sectioning. However, at least two acceptable shortcuts are expedient, if not quite as precise. One is to make careful localized small samplings from the callus after it has been carried through the entire staining procedure. The other is to examine the Feulgen-stained intact callus; areas of high mitotic activity are usually more intensely stained when examined under low magnification in the dissecting microscope. In cell suspensions, and in friable callus which dissociates in treatment, one is in effect sampling the total cell population and there is no opportunity to be selective for areas of high mitotic activity.

The other, and sometimes frustrating, problem is when to sample. In defining a new system there is no alternative to establishing the approximate time of first divisions, if these be of particular interest, by frequent sampling. Systems vary considerably in this respect, and if daily samplings do not yield results, then closer intervals, even hourly from the time of inoculation, may be necessary.

Fixation of Cells

For most karyological purposes acetic alcohol is usually quite effective. It is best freshly made at the time of fixation, or shortly before, using three volumes of absolute (or 95%) ethanol to one volume of glacial acetic acid. Cells are usable within minutes, but may be stored in the fixative for days if necessary. Callus is simply placed into a vial of fixative, or fixative may be added to the centrifuged and decanted cell pellet of a cell suspension sample.

Hydrolysis and Maceration

For the Feulgen reaction, hydrolysis in 1 *N* HCl at 60°C is necessary. However, if the tissue is at all coherent, this treatment has the added advantage of partially breaking down the intercellular cohesion, thereby facilitating cell maceration in the final squash preparation. Again, the period of hydrolysis is somewhat variable among different plant materials, but an approximate hydrolysis curve can be run simply by hydrolyzing samples at 2-minute intervals, up to about 12 minutes, and visually estimating the intensity of the subsequent Feulgen staining reaction. For quantitative cytophotometric estimation of DNA in individual nuclei the procedures of the Feulgen reaction must be quite precisely defined, but when it is used simply as a staining reaction for karyology the conditions need not be as rigidly defined. For example, if on the first trial a 6-minute hydrolysis yields satisfactory staining, there is no need to define a hydrolysis curve more precisely.

Staining

For most purposes the Feulgen technique is preferable to all others. The reagent is made as follows: to 1 g of basic fuchsin add 100 ml of boiling distilled water. Stir well and cool to approximately 50°C. Filter, and add 15 ml of 1 *N* HCl and 1.5 g of $K_2S_2O_5$. Store for 24 hours in a tightly stoppered bottle in the dark. Add 0.5 g Norit, shake well, and filter through rapid paper. Store in a tightly stoppered bottle in the refrigerator.

The hydrolyzed cells are placed into a small volume of Feulgen reagent for about 1 hour. They are then passed through three 10-minute changes of SO_2 water (5 ml of 1 *N* HCl, 5 ml of 10% aqueous solution of $K_2S_2O_5$, in 90 ml of distilled water). After this, the cells are best placed into a small volume of water, in which they may be stored for days if necessary, until they are used for slide preparations. A frequently permissible shortcut for ordinary karyology is the elimination of the SO_2 water treatment, going directly from Feulgen reagent to water.

Slide Preparation

If only temporary preparations are desired, then ordinary clean slides may be used. If it is desirable to make the better preparations permanent, the slides should be albumenized on one side (very thinly coated by rubbing a minute drop of egg albumen with the finger, and air-dried or passed through an alcohol flame). Alternatively they may be dipped into a "subbing" solution (0.5% aqueous solution of gelatin and 0.1% chrome alum) and air-dried, dust-free, for at least 2 days. Either of these treatments makes the cells more adherent to the slide, and minimizes loss of cells in subsequent treatment.

To make the slide preparation, remove a small piece of tissue and place it into a small drop of 45% acetic acid on the slide. With the broad side of a spear point needle, mash the tissue until it is no longer visible to the naked eye. Apply a coverslip over the drop and press firmly on the table between the pages of a bibulous pad or between paper towels. Examine the preparation in the microscope; a good one consists of a single layer of separated, well-flattened cells. If any coherent masses of tissue remain, they may be dispersed by gently tapping the coverslip over the tissue with the flat end of a pencil, or something similar, and pressing the preparation again. Air under the coverslip can be eliminated by introducing a drop of 45% acetic acid to the edge of the coverslip and subsequent gentle blotting. It is often helpful to heat the slide gently over an alcohol flame, either before first flattening it, or subsequently; this helps to affix the cells to the slide and to clear the cytoplasm. However, it must be done with caution.

If the slide is being made from a dissociated callus or a cell suspension, a small drop of water containing the cells is placed on the slide. Excess water is removed with the edge of a paper towel, 45% acetic acid added, and the above procedure is followed.

If the preparation is a good one, it can be kept for several days by sealing the edges of the coverslip. Any of a number of sealing substances can be used; a satisfactory one is a mixture of paraffin and beeswax. It is applied with a hot wire, e.g., a bent nail with a small cork as a handle. This seal dissolves in the process of making permanent preparations.

If, as is the case in some materials, the Feulgen reaction does not give sufficiently intense staining of chromosomes, the cells may be mounted in any of a number of acetic stains, e.g., aceto-carmine, aceto-orcein, or aceto-lacmoid,[4] instead of the 45% acetic acid. Or, alternatively, they may be examined in phase-contrast microscopy.

[4] C. D. Darlington and L. F. La Cour, "The Handling of Chromosomes," 3rd ed. George Allen and Unwin Ltd., London, 1960.

To make a preparation permanent, the dry ice method[5] is quick and simple. Place the slide on a piece of dry ice for a few minutes, until it is frozen completely. Pop off the coverslip with a scalpel and pass both slide and coverslip through 95% alcohol, absolute alcohol, a 1:1 mixture of absolute alcohol and xylol, and, finally, xylol. Allow a couple of minutes in each. Remove slide and coverslip from the xylol, drain, add a small drop of permanent mounting medium over the cells, and recombine slide and coverslip. If the solvent for the mounting medium is other than xylol, e.g., toluene, this may be substituted for xylol in the series. If an albumenized or subbed slide was used, and most of the cells have adhered to the slide, the original coverslip may be discarded and a clean one used. The slides should initially be stored flat until the mounting medium has hardened.

Characterization of Karyology

From good preparations, a number of determinations can be made. These include chromosome counts, karyotyping, determination of mitotic index, and analysis of nuclear cycle deviations and mitotic abnormalities. This basic technique can also be adapted to studies of nuclear cycle timing, which is discussed in another section.[6]

Chromosome Counting

Unless working with extremely well flattened preparations, with relatively low chromosome numbers, and few confusing overlaps of chromosomes, this is best done directly from the slides rather than from photographs. In the more usual case, a camera lucida can be an indispensible adjunct. The chromosomes are sketched on the drawing pad, e.g., with a felt tip pen, and the count is made from the sketch. In unravelling a complex overlap of chromosomes, a set of contrasting colored pens will help in making the distinctions between separate chromosomes in the final count. The greatest of caution must be exercised to assure that only intact cells are being included. This is particularly important if counts appear to deviate from euploid levels.

Karyotyping

This is best done from those cells which are extremely flat and have no overlaps. A photograph is made and the print enlarged to a large size. The chromosomes are cut out, and in a diploid system, arranged systematically in morphological pairs. Aneuploids may be analyzed for the presence or absence of specific chromosomes. Structural alterations of chromosomes can also be detected. From these, idiograms can be made, in which chromosomes are drawn straightened

[5] A. D. Conger and L. M. Fairchild, *Stain Technol.* **28**, 289 (1953).

[6] J. Van't Hof, this volume, Section VIII.

out, and arranged in an orderly array of pairs with respect to position of centromere, i.e., lengths of respective arms, and presence or absence of satellites. Where a previously published karyotype of the species exists, comparisons can be made, and the established designations of specific chromosomes should be followed.

Mitotic Index

This is a frequently misused determination; it refers to the ratio of mitotic cells (M) to total cells, interphase + mitotic (I + M), expressed on a percentage basis $[M/(I + M) \times 100]$. There are at least two complicating factors which, if not considered, can render the determination questionable, if not meaningless. Both have to do with sampling, the time and the place. Time: if there is any degree of synchrony in the system, then obviously the time in the cycle at which the sample is fixed can have a profound effect upon the ratio. This can only be determined by close-interval sampling. Place: if, as is the case in most coherent tissues, mitoses are localized, then the ratio will vary as the sampling of tissue varies to include more or less of dividing and nondividing portions, respectively. Unless these problems can be handled with certainty, the designation "mitotic index" should be treated with ample reservation. In cell suspensions under agitation this is not a source of error if the aliquot can be considered to be representative. Presumably the concern would be with the whole cell population, as indeed it can be in all cases if the frame of reference is the whole population, the sampled cells are randomized, and the qualifying statement is made. One other sampling problem concerns which cells to count. Two approaches to this are possible; selection of numbered grids from a table of random numbers, or simply running several transects across the slide and including all of the cells thus encountered into the count. The most difficult distinction to be made is between early prophases and interphases, but initial study of the material and the establishment of appropriate criteria can help to remove at least some of the subjectivity from this distinction.

Nuclear Cycle Deviations

If there appear to be any deviations in chromosome numbers from the basic level of ploidy, then further analysis of the slides may reveal diagnostic features. If polyploidy is manifest, the question of mechanism arises. As long as replication, karyokinesis, cytokinesis, replication, etc., follow in sequential order, no deviations at the level of ploidy should occur. Where polyploidy exists, the question is whether it arose through endoreduplication, additional replication cycle(s), without the intervention of mitosis, or whether it came about through failure of cytokinesis, resulting in binucleate cells.[3] Other features to note are evidences of differential behavior between nuclei in single cells, and fusion of nuclei. The diagnostic characteristic of endoreduplication in the division immediately following the additional replication cycle is the presence of diplochromosomes, i.e., prophase chromosomes with four chromatids, or metaphases in which identical pairs of chromosomes are adjacent to one another and which have the increased chromosome number.

CHAPTER 16

Biochemical Identification of Cells in Culture

A. Human Cell Lines by Enzyme Polymorphism

S. M. Gartler and R. A. Farber

There have been a number of demonstrable instances of cross-culture contamination.[1] This occurs by manipulative errors, and, though rare, the introduction of a single foreign cell with a slight growth rate advantage can be sufficient to swamp the host population. The morphological similarity of mammalian cells in culture prevents the early identification of cross-culture contamination in most cases. Errors leading to contamination probably cannot be completely avoided; therefore, the ability to identify a culture with specificity is of the utmost importance.

Biological identification is basically a problem in reading the cell's genotype through its phenotype. Only a small fraction of the cell's genotype is expressed phenotypically; measurements include cytological, immunological, and electrophoretic enzyme assays. The electrophoretic enzyme assays are simple and direct and may be used for both interspecific and intraspecific contamination problems. Since most, if not all, enzymes have species-specific amino acid sequences, it is possible with one or two enzymes to identify definitively the species derivation of a cell line. Intraspecific identification problems are more difficult in that, in order for the enzyme marker to be useful, it must be polymorphic, and there are only a few human enzymes known at the present time which are both polymorphic and detectable in fibroblasts in culture.

The expression of electrophoretic markers is quite stable; rare somatic mutations should be expected, but no instance of a spontaneous mutation affecting an electrophoretic variant in normal cultures has been recorded. In established lines with chromosome instability, chromosome loss in a heterozygous culture could result in a change of electrophoretic pattern (i.e., loss of a band). There is no

[1] S. M. Gartler, *Nat. Cancer Inst. Monogr.* **26**, 167 (1967).

evidence, however, that new or additional electrophoretic bands can appear in such cultures.

Starch Gel Electrophoresis

A number of methods are available for the electrophoretic separation of different enzymes, including cellulose acetate strips,[2] Cellogel,[3] and starch gel. Each technique has its special advantage; we have worked almost exclusively with starch gel and find it a reliable and widely applicable system for separating all enzymes and their variants that are pertinent to cell culture identification problems.

The technique we will describe for horizontal starch gel electrophoresis is a modification of the original method of Smithies.[4] Commercial systems are available, but the apparatus we use can be constructed of Lucite in any plastic shop or university instrument facility. The gel tray consists of a bottom, with a lip around the edge, and a separate frame. This is illustrated with dimensions in Fig. 1. The frame is placed on the bottom and secured temporarily at each end with rubber bands until the gel has solidified. With a single frame, the total depth of the well will be 0.8 cm. More than one frame can be used if the gel is to be sliced into three or more sections for staining of several enzymes.

The two buffer trays are made of 0.3-cm Lucite and are divided longitudinally into two compartments, as diagrammed in Fig. 2. Electrodes consist of 30 cm of platinum wire wound around a glass rod.

Gels are prepared by thoroughly mixing 36 g of hydrolyzed starch (Connaught Medical Research Laboratories, Toronto, Canada) with 300 ml of the gel buffer in a 1-liter filter flask. The starch is cooked by constantly swirling the flask at approximately a 30° angle over the flame of a burner until the suspension becomes viscous and translucent (around 1½ to 2 minutes). The starch is then degassed with a vacuum for 30 to 60 seconds, until most of the small bubbles have disappeared.

The starch is poured into the gel tray and allowed to solidify at room temperature. It is then covered with Saran wrap and placed in the cold for at least 40 minutes or left at room temperature for no less than 3 or 4 hours.

For applying samples, slots are made in the gel with a form which consists of a row of plastic prongs, varying in number from 8 to 16, and separated by at least 0.2 cm. The base of the form is sandwiched between two pieces of Lucite to form a handle. A diagram of this device is shown in Fig. 3. A convenient origin position for most enzymes is 4.5 cm from one end of the gel. From 10 to 20 μl of sample is applied to a piece of Whatman No. 3 filter paper (slightly smaller than the precut slot) and inserted into the gel.

One liter of bridge buffer (see below) is distributed equally among the four compartments of the buffer trays and placed in the cold. The electrophoresis is

[2] R. S. Sparkes, M. C. Baluda, and D. E. Townsend, *J. Lab. Clin. Med.* **73**, 531 (1969).

[3] P. Meera Khan and M. C Rattazzi, *Biochem. Genet.* **2**, 231 (1968).

[4] O. Smithies, *Biochem. J.* **61**, 629 (1955).

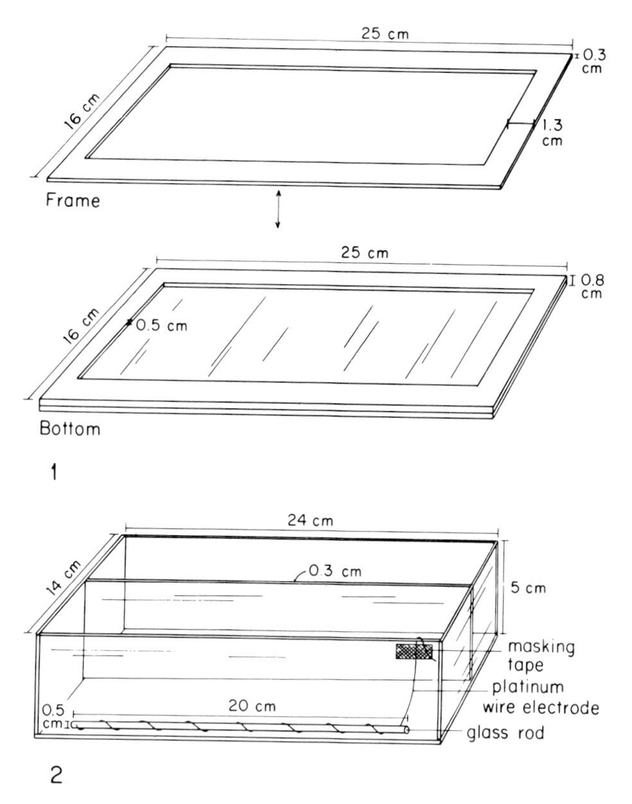

Fig. 1. Gel tray.
Fig. 2. Buffer tray.

carried out at 4°C. Each end of the gel tray is supported on the edge of a buffer tray (the edge away from the electrode). Wicks (three layers of Whatman No. 1 chromatography paper, 15.5 × 10 cm), soaked in buffer, are placed with one edge over the gel; the remainder of each is suspended into the buffer compartment below. Another set of wicks is draped across the divider in each tray so that it is in contact with the buffer in both compartments. The gel is covered with Saran wrap, and the electrodes are connected to a power supply; one satisfactory model is an EC454 (E-C Apparatus Corp., 755 St. Marks St., Philadelphia, Pennsylvania 19104). The complete setup is shown in Fig. 4.

The voltage should be checked 30–60 minutes after the current is initially turned on, since a short period is required for the system to reach equilibrium. Measurements are made directly in the gel.

When the gel is ready for staining, the frame is removed. The gel is then sliced horizontally at the appropriate position with a cheese slicer or piano wire strung tautly on a coping saw frame (the wire should be at least as long as the width of the gel). It should be sliced as uniformly and smoothly as possible. The top layer of the gel is removed, and one of the resulting cut surfaces is stained. It is possible to stain both halves in cases where the same buffer system can be used for two different enzymes.

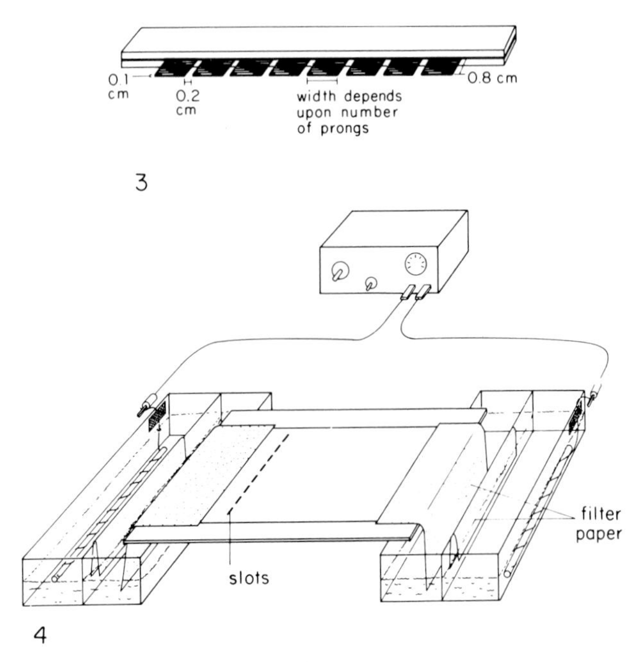

Fig. 3. Slot maker.
Fig. 4. Electrophoresis apparatus assembled.

The staining mixture is combined with a solution of 1% molten Noble agar at 50°C and poured over the surface of the gel. The gel is then incubated in the dark at 37°C until the stain appears. The incubation time varies from 15 minutes to 2 hours, depending on the activity of the enzymes. Gels can be fixed for permanent storage by soaking in 50% methanol overnight.

Enzyme Markers

The enzymes which will be discussed are ones whose electrophoretic patterns can be detected in a few million cells or less. We have included only those which are polymorphic in the major population groups.

To prepare cells for electrophoresis, confluent monolayers are harvested with trypsin. The action of trypsin is stopped by adding an equal amount of medium containing 10–20% serum. The cells are centrifuged at 200 g for 10 minutes, washed with physiological saline, and recentrifuged. The pellet is suspended in 20 μl of the appropriate buffer and is frozen and thawed three times. This homogenate is used directly for electrophoresis.

The enzymes considered here all migrate toward the anode.

Glucose-6-Phosphate Dehydrogenase (G6PD). There are over eighty-five known mutants at the X-linked G6PD locus. Many of these are separable electrophoretically, whereas some differ only quantitatively. Most of these variants occur infrequently and are not important for this discussion. The most common allele is *Gd-B*. One variant, B−, which has about 15% of the normal enzyme activity in fibroblasts, but the same electrophoretic mobility as *B*, is found with a fairly high frequency in certain Mediterranean populations. Approximately 30% of American Negro males have a faster migrating variant, *A* (two alleles, *Gd-A* with normal activity, and *Gd-A−* with slightly reduced activity, comigrate electrophoretically).[5]

Even though this enzyme is a dimer, *AB* heterozygotes express only two bands (*A* and *B*). The formation of a hybrid band requires that both alleles be expressed in each cell, and, because of X chromosome inactivation, only a single X-linked allele is expressed in a somatic cell. The relative quantities of *A* and *B* enzyme in heterozygotes will vary between individuals and possibly with time in culture, depending on the proportion of cells with a given inactive X. Males and homozygous females show only a single band. These patterns are diagrammed in Fig. 5. Cells from the HeLa line are *Gd-A*.

The bridge buffer for G6PD is 0.1 *M* phosphate, pH 7.0; this buffer is diluted to 0.015 *M* for the gel. NADP (2.5 mg) is added before degassing. An equal amount of NADP is added to the cathode buffer. Cell extracts are prepared in 0.1 *M* Tris, pH 8.0, with 0.001 *M* mercaptoethanol and 1×10^{-5} *M* NADP. The enzyme can be detected in as few as 1×10^{4} cells. (Cells can be stored frozen as a pellet for several days prior to use.) The gel is run at 2.5 volts/cm for 18 hours.

For staining, the gel should be sliced between 1 and 6 cm beyond the origin; 100 mg of agar are melted in 5 ml 0.2 *M* Tris, pH 8.0, plus 3.5 ml H_2O. The staining mixture includes: 0.4 ml NADP (5 mg/ml), 0.1 ml $MgCl_2$ (0.1 *M*), 0.5 ml glucose-6-phosphate (20 mg/ml), 0.1 ml phenazine methyl sulfate (PMS) (4 mg/ml), 0.4 ml MTT tetrazolium (5 mg/ml).[6]

6-Phosphogluconate Dehydrogenase (6PGD). An autosomal locus controls 6PGD. The common allele is *PGD-A*. Approximately 4–5% of European Caucasians are heterozygous for *PGD-A* and *PGD-C*; these individuals show three bands on starch gels: *A*, *C*, and a hybrid band. Homozygotes for *PGD-C* occur

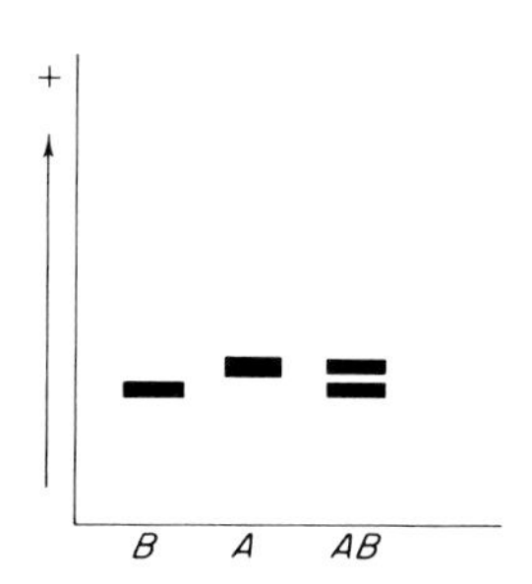

Fig. 5. G6PD phenotypes. Relative intensities of the *A* and *B* bands will vary between *AB* individuals because of mosaicism resulting from X inactivation.

[5] E. R. Giblett, "Genetic Markers in Human Blood." F. A. Davis, Philadelphia, Pennsylvania, 1969.

[6] R. A. Fildes and C. W. Parr, *Biochem. J.* **87**, 45P (1963).

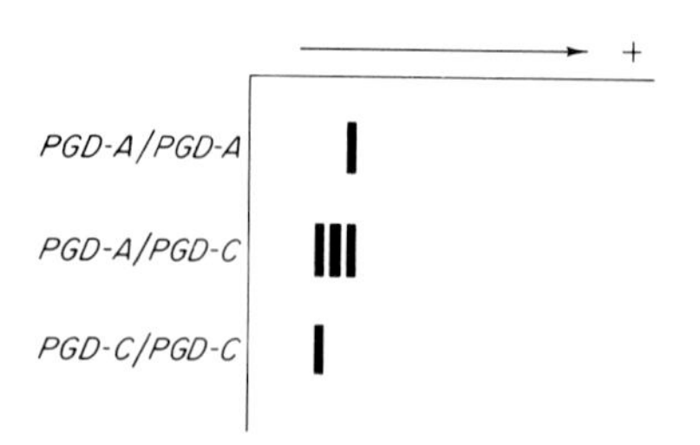

Fig. 6. 6PGD types.

at a frequency of less than 1%.[7] The frequency of heterozygotes is somewhat higher in certain populations in Africa and Asia.[5] The electrophoretic patterns are shown in Fig. 6.

This enzyme can be assayed on the same gel with G6PD, but it is necessary to use at least four times as many cells for 6PGD. (If NADP is included in the gel, one or more additional bands may appear cathodally.[5]) The enzyme migrates a few centimeters. The reaction mixture is identical to that for G6PD, except that 6-phosphogluconate is substituted for glucose-6-phosphate.

Phosphoglucomutase (*PGM*). PGM is so far the most useful enzyme for tracing cell contamination, since its variants occur at very high frequencies. There are three independent *PGM* loci, two of which (PGM_1 and PGM_3) are polymorphic. Variants for PGM_2 are rare, and this isozyme is only weakly expressed in cultured cells.

At the PGM_1 locus, PGM_1^1 is present in approximately 60% of Caucausian and Negro individuals. PGM_1^{2-1} heterozygotes represent about 35% of the population and PGM_1^2 homozygotes around 5%. At the PGM_3 locus, the PGM_3^1 allele is most common among Caucausians and the PGM_3^2 allele among Negroes. Heterozygote frequencies are 40–45% in both populations.[8] Figure 7 shows the patterns for PGM_1 and PGM_3 phenotypes. HeLa cells are PGM_1^1 and PGM_3^1.

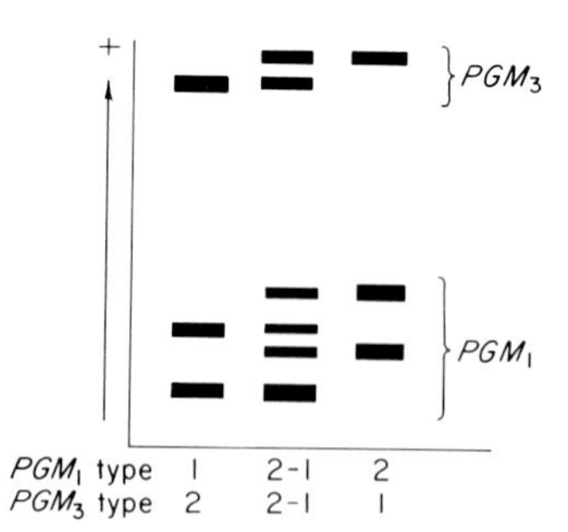

Fig. 7. PGM phenotypes. The PGM_1 and PGM_3 loci are independent. They are illustrated in the above combinations for convenience only. An additional band may appear anodally for PGM_3, depending upon the conditions of the electrophoresis (how long the gels are run and how many cells are used).

[7] C. W. Parr, *Nature* (*London*) **210**, 487 (1966).

[8] H. Harris, "The Principles of Human Biochemical Genetics." North-Holland, Amsterdam, 1970.

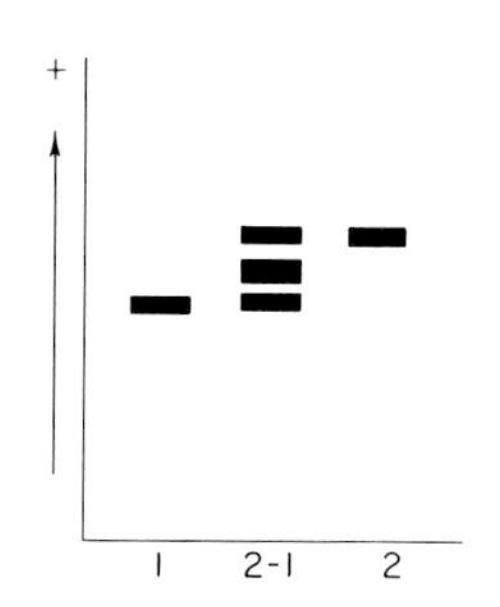

Fig. 8. Peptidase A phenotypes.

The bridge buffer for PGM consists of 0.1 *M* Tris, 0.1 *M* maleic acid, 0.01 *M* EDTA, and 0.01 *M* $MgCl_2$, brought to pH 7.5 with NaOH. This buffer is diluted 1 in 10 for the gel. Cell extracts are prepared in saline. PGM_1 can be detected in as few as 10^5 cells. PGM_3 activity is much lower, however, and requires a minimum of 10^6 cells. The cell pellet can be stored frozen for several days. The gel is run at 4.5 to 5 v/cm for 18 hours.

For visualization of both enzymes, the gel should be sliced between the origin and 14 cm. Agar, 200 mg, is melted in 10.0 ml Tris (0.06 *M*, pH 8.0) plus 5.0 ml H_2O. Stain components are as follows: 0.4 ml NADP (5 mg/ml), 0.2 ml $MgCl_2$ (0.1 *M*), 4.0 ml glucose-1-phosphate (0.05 M) (Mann[9]), 0.3 ml PMS (4 mg/ml), 0.4 ml MTT tetrazolium (5 mg/ml), 0.8 units G6PD.[10]

Peptidase (*Pep*). Several different peptidases, as defined by their electrophoretic mobilities and substrate specificities, are known in humans. *Pep A* and *Pep D* are polymorphic, but we will consider only *Pep A*, since good substrate for *Pep D* is difficult to obtain. The most common phenotype, *Pep A 1*, is present in approximately 80–85% of Negroes; 11–16% are heterozygous *Pep A 2–1*. *Pep A* is not polymorphic in Caucausians, *Pep A 1* being the only frequent allele.[11] These patterns are diagrammed in Fig. 8. (Some activity for *Pep B* and *Pep C* may appear on the gel with *Pep A*, but *A* is the most slowly migrating of these enzymes.)

The buffer system for peptidase is 0.1 *M* phosphate, pH 7.0; the gel buffer is 0.01 *M*. Approximately 1–2 × 10^6 cells per sample are required for *Pep A*. Extracts are prepared in saline. The gel is run at 5 v/cm for 16 hours.

For staining, the gel is sliced between 5 cm and 14 cm. Agar, 150 mg, is melted in 15 ml H_2O. The stain components are as follows (those listed in milligram amounts should be weighed out just prior to use): 12.0 mg peptide (L-valyl-L-leucine for *Pep A*), 3.0 mg *Bothrops atrox* or *Crotalus adamanteus* venom, 6.0 mg horseradish peroxidase (Boehringer), 3.0 mg *o*-dianisidine-diHCl (Eastman), 15.0 ml buffer (0.2 *M* phosphate HCl, 0.002 *M* $MnCl_2$ pH 7.5).[11]

One, or at most two, of these enzymes may be used to detect interspecies

[9] Glucose-1,6-diphosphate, which is necessary for the reaction, occurs as a contaminant in Mann G-1-P.

[10] N. Spencer, D. A. Hopkinson, and H. Harris, *Nature* (*London*) **204,** 742 (1964).

[11] W. H. P. Lewis and H. Harris, *Nature* (*London*) **215,** 351 (1967).

contamination. Intraspecies contamination problems require polymorphic systems; with the increasing tempo of work in human biochemical and population genetics, it is almost certain that the power of the above method for the specific identification of human cell cultures will increase considerably.

CHAPTER 16

Biochemical Identification of Cells in Culture

B. Enzymatic "Fingerprinting"

John E. Shannon and Marvin L. Macy

In 1969 we initiated studies[1] on all of the animal cell lines in the Animal Cell Culture Collection[2] to determine whether electrophoretic analysis of isoenzymes might be a reliable and practical means of species identification. This was suggested in 1962 by Vesell[3] *et al.*, and in 1967 Gartler[4] pointed out that electrophoretic variants of polymorphic enzymes might also serve as genetic markers useful in the intraspecific contamination of cell cultures. The latter aspect of the subject is discussed in the preceding chapter.

We chose the starch-gel method for the study of glucose-6-phosphate dehydrogenase (G6PD) and lactic dehydrogenase (LDH) electrophoretic patterns. Mrs. Frida Montes de Oca, a former member of our staff, worked out the necessary procedures which are described below.

Preparations of Extracts

The extracts are prepared by either (1) treatment with *n*-octyl alcohol, or (2) by freezing the cells in liquid nitrogen and thawing them at room temperature. Cell lines to be assayed are selected from active log phase monolayer cultures or from freshly thawed ampoules. Each sample should contain approximately 1×10^7 viable cells.

[1] F. Montes de Oca, M. L. Macy, and J. E. Shannon, *Proc. Soc. Exp. Biol. Med.* **132,** 462 (1969).

[2] J. E. Shannon and M. L. Macy, Registry of Animal Cell Lines, 2nd ed. American Type Culture Collection (1972).

[3] E. S. Vesell, J. Phillip, and A. G. Bearn, *J. Exp. Med.* **116,** 797 (1962).

[4] S. M. Gartler, *Nat. Cancer Inst. Monogr.* **26,** 167 (1967).

If monolayer cultures are used the cells are harvested and centrifuged at 125 *g* for 10 minutes at 4°C. If ampoules are used the cells are thawed, diluted with 10 ml of culture medium, and centrifuged in the same manner. The cells are then washed three times with 0.9% NaCl solution containing 6.6×10^{-4} *M* ethylenediaminotetraacetate. For treatment with *n*-octyl alcohol the supernatant from the third wash is decanted and approximately 0.3 ml of the salt solution and 0.2 ml of *n*-octyl alcohol is added to each sample. A Pasteur pipette is used to gently resuspend the cells in this mixture and the centrifuge tubes are then stoppered and placed in a +5°C refrigerator overnight. Extracts prepared by the freeze-thaw procedure are resuspended in the salt solution following the third washing at a concentration of approximately 1×10^{7} cells/ml. The cells are then frozen in liquid nitrogen and thawed at room temperature three times.

The crude extract is then centrifuged at 20,000 *g* for 30 minutes at 4°C, assayed without purification, and when not used immediately, stored at −95°C. In our experience the mobility of the isoenzymes is the same regardless of the method employed for preparation of the extracts.

Electrophoresis

We routinely use the vertical starch-gel technique described by Kirkman[5] using a vertical plexiglass electrophoresis apparatus.[6] Extracts are assayed for G6PD and or LDH activity by a modification of the methods of Glock and McLean,[7] and Vesell and Bearn.[8]

Development of Zymograms

After electrophoresis is completed (overnight), the gel is sliced in half horizontally, inverted with the cut surface facing up, and covered with a thinner gel consisting of 3.5 g of starch in 30 ml of 0.25 *M* Tris-HCl buffer, pH 7.5, to which the incubation mixture is added. For G6PD this mixture contains 2 mg of phenazine methosulfate and 5 ml each of nitroblue tetrazolium (2 mg/ml), 0.025 *M* glucose-6-phosphate, 0.005 *M* nicotinamide adenine dinucleotide phosphate, and 0.1 *M* $MgCl_2$. For LDH, 1.0 M sodium lactate and nicotinamide adenine dinucleotide (10 mg/ml) is substituted for the glucose-6-phosphate and NADP and the $MgCl_2$ is omitted. The gels are incubated at 37°C for a period of time depending on the activity of the extracts (usually from 30 minutes to 1 hour). Immediately thereafter the gels are washed with cold tap water and a photographic record made. If desired, the gels can be treated by immersion in a 5:5:1:1 mixture of methanol, water, acetic acid, and glycerol, and stored at room temperature for long intervals of time.

[5] H. N. Kirkman and E. Hendrickson, *Amer. J. Human Genet.* **15**, 241 (1963).
[6] Buchler Instruments, Inc., 1327 16th St., Fort Lee, New Jersey 07065.
[7] G. E. Glock and P. McLean, *Biochem. J.* **55**, 400 (1953).
[8] E. S. Vessel and A. G. Bearn, *J. Gen. Physiol.* **45**, 553 (1962).

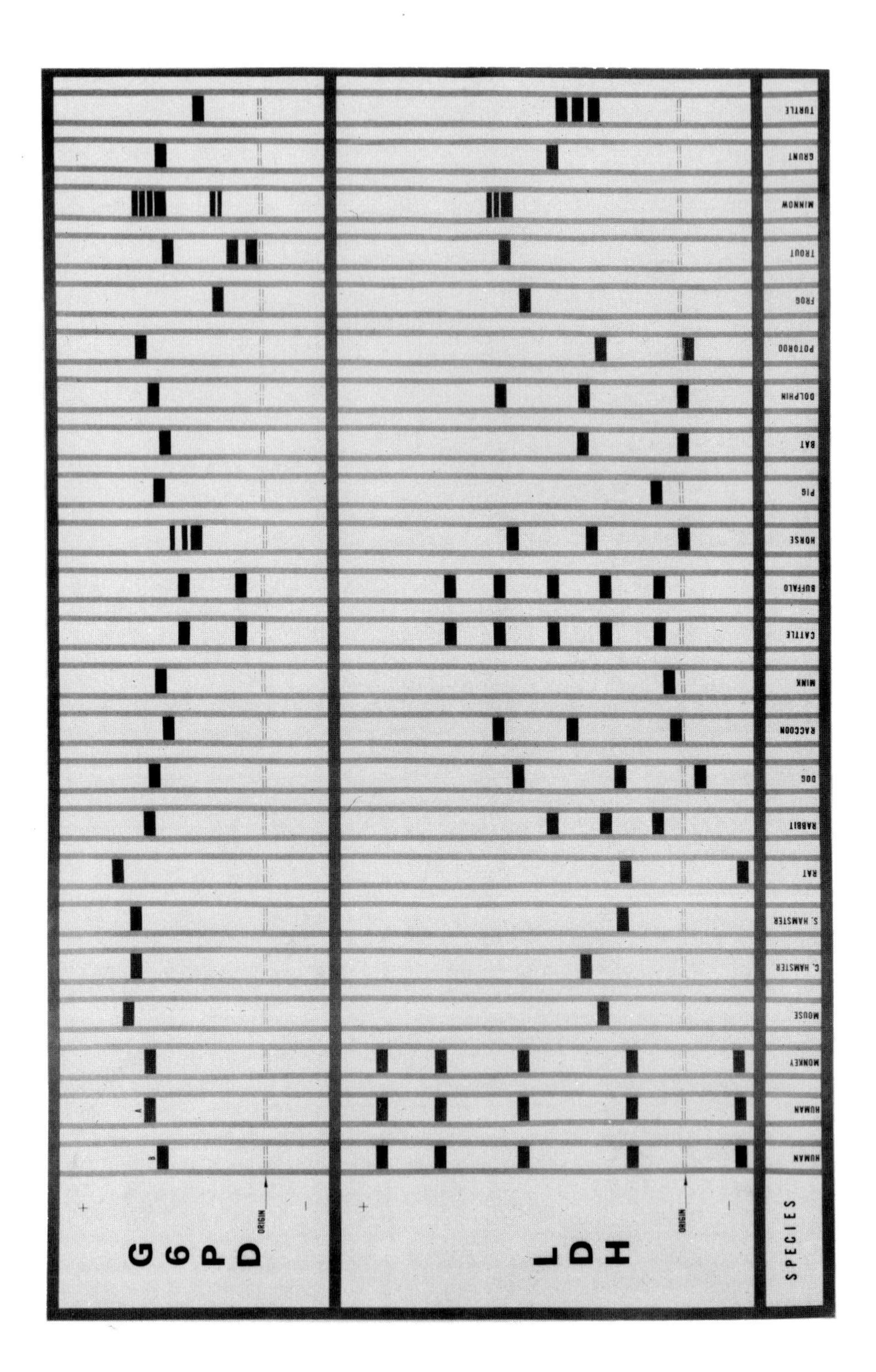
G6PD
ORIGIN
LDH
ORIGIN
SPECIES
HUMAN
HUMAN
MONKEY
MOUSE
C. HAMSTER
S. HAMSTER
RAT
RABBIT
DOG
RACCOON
MINK
CATTLE
BUFFALO
HORSE
PIG
BAT
DOLPHIN
POTOROO
FROG
TROUT
MINNOW
GRUNT
TURTLE

Results and Discussion

Using the above procedures we found that the determinations of the electrophoretic patterns of only two polymorphic enzymes was sufficient to provide an effective method for the identification of cells cultured *in vitro*, at least as far as the generic, and possibly the species level. Cell cultures representing 20 out of 22 taxonomic groups can be easily distinguished from each other by comparison of their G6PD and LDH electrophoretic patterns (Fig. 1). Of the two exceptions, similarity in the patterns of the two isoenzymes exists at the family level in one case and in the other, at the generic level. The study of additional polymorphic enzymes would undoubtedly clearly differentiate cells from these taxonomic groups.

Isoenzyme analysis provides a means of complementing or supplanting immunological techniques for a species identification. As we have shown, once the electrophoretic patterns for several isoenzymes have been determined for a large number of species and a "fingerprint" identification chart constructed the species or genera of an unknown can be readily pinpointed. If the isoenzyme tests show that the species of the cells is *other* than that which was originally supposed, then it is not necessary to subject the cells to immunological tests. This conserves both time and valuable antisera. Another advantage is that a permanent photographic record of zymograms may be made for future comparisons.

In addition, isoenzyme analysis has been useful in determining the species of presumably "transformed" cells that have been submitted for identification. In all cases thus far, determination of the G6PD and LDH patterns has shown that such cultures contained cells predominantly (if not all) from another species. Cultures containing cells from two different species clearly exhibit the isoenzyme patterns characteristic of both species but we have not determined the lower limits of sensitivity of the isoenzyme tests. For detection of low-level contaminations with cells of another species more refined techniques would be necessary.

Fig. 1. Diagrammatic comparison of the LDH and G6PD zymograms of 86 animal cell lines representing 22 taxonomic groups. For the human cells the B band of G6PD represents 10 cell lines and the A band represents 24 cell lines. The zymograms of the monkey were obtained from 6 cell lines; the mouse from 16; the Chinese hamster from 5; the Syrian hamster and the rat from 3; and the potoroo from 2. All of the zymograms for the remaining species were obtained from a single cell line. Not that all cells may be readily differentiated from each other except for those originating from the human (A) and monkey, and the buffalo and cattle. (From F. Montes de Oca, M. L. Macy, and J. E. Shannon.[1]) Reproduced with permission of *Proc. Soc. Exp. Biol. Med.*

CHAPTER 17

Introduction to Cytoenzymological Methods and Isozymes[1]

P. J. Melnick

The vast literature on cell cultures describes studies with many types of biological techniques, but cytoenzymology and isozymes have been used infrequently because of uniquely difficult technical problems, which will be described subsequently. The author was confronted with these problems in a study of the enzyme patterns of leukemic cells using cytoenzymology[2]; they were successfully resolved for each enzyme examined (more than thirty). Later, a similar study of five established human cell lines was carried out; in addition, isozyme patterns of these cell lines were determined. The results were published in a monograph[3] which contains complete description of the techniques, as well as a complete review of the literature of cytoenzymology and isozymes of cultured cells to 1970. In the present short chapter only a summary of these studies can be presented, and the reader is referred to the above cited references for complete details of the methods and findings.

Cytoenzymology

Cytohistochemistry has grown rapidly since 1939 when Gomori[4] and Takamatsu[5] devised the technique for demonstrating phosphatases in tissue sections; before 1939 its growth was slow and sporadic. The techniques of cytoenzymology consist of chemical reactions produced in cells and tissue sections that yield insoluble end products easily seen with the microscope, and which furnishes morphological expressions of many enzymes and of many chemical components of cells. They are distinguished from stains, which yield a wealth of morphological information but little or nothing about their chemical reactions with cell elements. In contrast, the reactions of cytoenzymology have special meaning

[1] This research was supported in part by the Veterans Administration Research Service, and by U. S. Public Health Service Research Grants No. CA-05202 and No. 07468.

[2] P. J. Melnick, *In* "Pathology of Leukemia" (G. D. Amromin, ed.), pp. 125–160. Hoeber Medical Division, Harper and Row, New York, 1968.

[3] P. J. Melnick, *In* "Progress in Histochemistry and Cytochemistry" (W. Graumann, Z. Lojda, A. G. E. Pearse, and T. H. Schiebler, eds.), Volume 2, No. 1. Gustav Fischer, Stuttgart, 1971.

[4] G. Gomori, *Proc. Soc. Exp. Biol. Med.* **42**, 23 (1939).

[5] H. Takamatsu, *Trans. Soc. Pathol. Jap.* **29**, 429 (1939).

because their chemical principles are understood; they are of increasing interest for all biologists who work with the microscope. When the techniques are well designed and properly performed, cytoenzymology (indeed all of cytohistochemistry) yields precise localization of enzymes and of other chemical components, with the cell structure remaining intact. Other biological sciences have methods to separate cell components for study, with less interest in structure as a whole. Cytohistochemistry contributes precise localization with nondestructive methods. Reaction products are estimated semiquantitatively, meaningful because of precise localization, and supplementing biochemistry which contributes precise quantitation. Barka and Anderson[6] stated that it ". . . provides the only means of examining chemical processes in relation to cell structure." Siekevitz[7] speculated ". . . what we do to the structural aspects of the system will discourage our biochemical examination of it, and vice versa."

Basic Problems

Free-living cells have intact plasma membranes that are highly impermeable to many enzyme substrates in cytohistochemical techniques (with the exception of alkaline phosphatase and a few other enzymes), because of their high electrical resistance[8] and their highly complex structure and organization of the cell periphery.[7,9-14] The composition, structural, physical, chemical, and other properties of the plasma membrane also influence permeability.[15-17] Of the several groups of mechanisms involving membrane transport in living cells, five require no metabolic energy[18] and thus are involved in cytoenzymology of cells spread on glass slides and treated as described below.

In contrast, cells in tissue sections are cut across, allowing direct access of substrates, capture reagents, and cofactors, etc., to their interiors. Therefore, the patterns of histochemical enzyme methods developed during the past 30 years,

[6] T. Barka and P. J. Anderson, "Histochemistry. Theory, Practice and Bibliography." Hoeber Medical Division, Harper and Row, New York, 1963.

[7] P. Siekevitz, *New Engl. J. Med.* **283**, 1035 (1970).

[8] W. R. Loewenstein, *Ann. N. Y. Acad. Sci.* **137**, 403 (1966).

[9] F. Bronner and A. Kleinzeller, (eds.), "Current Topics in Membrane Transport," Academic Press, New York, 1971.

[10] H. D. Brown (ed.), "Chemistry of the Cell Interface, Parts A and B." Academic Press, New York, 1971.

[11] D. Chapman (ed.), "Biological Membranes." Academic Press, New York, 1968.

[12] A. J. Dalton and F. Haguenau (eds.), "The Membranes." Academic Press, New York, 1972.

[13] L. J. Rothfield (ed.), "Structure and Function of Biological Membranes." Academic Press, New York, 1971.

[14] L. Weiss, "The Cell Periphery." North-Holland, Amsterdam, 1967.

[15] B. J. Clarris and J. R. Fraser, *Exp. Cell Res.* **49**, 181 (1968).

[16] A. S. G. Curtis, "The Cell Surface: Its Molecular Role in Morphogenesis." Academic Press, New York, 1967.

[17] J. Paul, "Cell Biology." Heinemann Educational Books, Ltd, London, 1967.

[18] C. R. Park, *In* "Membrane Transport and Metabolism" (A. Kleinzeller and A. Kotyk, eds.), p. 19. Academic Press, New York, 1961.

TABLE I

Hydrolases[a]

Hydrolase	Detroit 6				Detroit 6, clone 12				HeLa				HeLa clone 229				Chang liver			
	Stock	Unfrozen	Slowly frozen	Rapidly frozen	Stock	Unfrozen	Slowly frozen	Rapidly frozen	Stock	Unfrozen	Slowly frozen	Rapidly frozen	Stock	Unfrozen	Slowly frozen	Rapidly frozen	Stock	Unfrozen	Slowly frozen	Rapidly frozen
1. Esterase																				
Naphthyl AS-D chloroacetate	4+	4+	4+	4+	4+	4+	4+	4+	4+	4+	4+	4+	4+	4+	4+	4+	4+	4+	4+	4+
Naphthyl acetate	2+	2+	2+	3+	2+	2+	2+	3+	2+	2+	2+	3+	2+	2+	2+	3+	2+	2+	2+	2+
6-Bromo-2-napthyl acetate	2+	2+	2+	3+	2+	2+	2+	3+	2+	2+	2+	3+	2+	2+	2+	3+	2+	2+	2+	2+
Barka and Anderson	4+	4+	4+	4+	4+	4+	4+	4+	4+	4+	4+	4+	4+	4+	4+	4+	4+	4+	4+	4+
5-Bromoindoxyl acetate	—	—	—	—	—	—	—	—	—	—	—	—	—	—	—	—	—	—	—	—
2. Lipases																				
Tweens: 20, 40, 60, and 80	±	±	±	±	±	±	±	±	±	±	±	±	±	±	±	±	±	±	±	±
Naphthyl esters	—	—	—	—	—	—	—	—	—	—	—	—	—	—	—	—	—	—	—	—
3. Cholinesterase																				
6-Bromo-2-naph.-carbo-naph. choline iod	4+	4+	4+	4+	4+	4+	4+	4+	4+	4+	4+	4+	4+	4+	4+	4+	4+	4+	4+	4+
Gomori	±	±	±	±	±	±	±	±	±	±	±	±	±	±	±	±	±	±	±	±
Thiocholine substrates	±	±	±	±	±	±	±	±	±	±	±	±	±	±	±	±	±	±	±	±

4. Acid phosphatase																				
Naphthyl AS-MX phosphate	4+	4+	4+	4+	4+	4+	4+	4+	4+	4+	4+	4+	4+	4+	4+	4+	4+	4+	4+	4+
Alpha-naphthyl phosphate	4+	4+	4+	4+	4+	4+	4+	4+	4+	4+	4+	4+	4+	4+	4+	4+	4+	4+	4+	4+
Gomori	4+	4+	4+	4+	4+	4+	4+	4+	4+	4+	4+	4+	4+	4+	4+	4+	4+	4+	4+	4+
5. Alkaline phosphatase																				
Naphthyl AS-MX phosphate	0–4+	0–4+	0–4+	0–4+	0–4+	0–4+	0–4+	0–4+	0–4+	0–4+	0–4+	0–4+	0–4+	0–4+	0–4+	0–4+	0–4+	0–4+	0–4+	0–4+
Gomori	0–4+	0–4+	0–4+	0–4+	0–4+	0–4+	0–4+	0–4+	0–4+	0–4+	0–4+	0–4+	0–4+	0–4+	0–4+	0–4+	0–4+	0–4+	0–4+	0–4+
6. ATPase																				
Wachstein and Meisel	0–4+	0–4+	0–4+	0–4+	0–4+	0–4+	0–4+	0–4+	0–4+	0–4+	0–4+	0–4+	0–4+	0–4+	0–4+	0–4+	0–4+	0–4+	0–4+	0–4+
Padykula and Herman	0–2+	0–2+	0–2+	0–2+	0–2+	0–2+	0–2+	0–2+	0–2+	0–2+	0–2+	0–2+	0–2+	0–2+	0–2+	0–2+	0–2+	0–2+	0–2+	0–2+
7. Glucose-6-phosphatase	0–3+	0–3+	0–3+	0–3+	0–3+	0–3+	0–3+	0–3+	0–2+	0–2+	0–2+	0–2+	0–2+	0–2+	0–2+	0–2+	0–2+	0–2+	0–2+	0–2+
8. 5′-Nucleotidase	0–4+	0–4+	0–4+	0–4+	0–4+	0–4+	0–4+	0–4+	0–2+	0–2+	0–2+	0–2+	0–2+	0–2+	0–2+	0–2+	0–2+	0–2+	0–2+	0–2+
9. Aminopeptidase																				
Leucine	4+	4+	4+	4+	4+	4+	4+	4+	4+	4+	4+	4+	4+	4+	4+	4+	4+	4+	4+	4+
Alanine	4+	4+	4+	4+	4+	4+	4+	4+	4+	4+	4+	4+	4+	4+	4+	4+	4+	4+	4+	4+
10. β-Glucuronidase																				
Pearse	0–1+	0–1+	0–1+	0–2+	0–1+	0–1+	0–1+	0–2+	0–1+	0–1+	0–1+	0–2+	0–1+	0–1+	0–1+	0–2+	0–1+	0–1+	0–1+	0–2+
Seligman	±	±	±	±	±	±	±	±	+	+	+	0–1+	0–1+	0–1+	0–1+	0–2+	0–1+	0–1+	0–1+	0–2+
11. Aryl sulfatase	1–4+	1–4+	1–4+	1–4+	1–4+	1–4+	1–4+	1–4+	0–3+	0–3+	0–3+	1–4+	0–3+	0–3+	0–3+	1–4+	1–4+	1–4+	1–4+	1–4+

[a] Results of the comparative evaluation of enzyme reaction products for eleven hydrolases in five cell lines, including evaluation of four modalities for each (1) frozen seed stock, (2) living cell suspensions, (3) slowly frozen, and (4) rapidly frozen living cell suspensions, with 7½% PVP as cryoprotective agent. (Reproduced by courtesy of Gustav Fischer Verlag.)

TABLE II
Oxidative Enzymes[a]

Oxidative enzyme	Detroit 6				Detroit 6, clone 12				HeLa				HeLa clone 229				Chang liver			
	Stock	Un-frozen	Slowly frozen	Rapidly frozen	Stock	Un-frozen	Slowly frozen	Rapidly frozen	Stock	Un-frozen	Slowly frozen	Rapidly frozen	Stock	Un-frozen	Slowly frozen	Rapidly frozen	Stock	Un-frozen	Slowly frozen	Rapidly frozen
1. Succinic	4+	3+	3+	4+	4+	2+	2+	4+	4+	1+	1+	4+	4+	2+	1+	4+	4+	2+	1+	4+
2. Malic	4+	1+	2+	4+	4+	2+	2+	4+	4+	2+	2+	4+	4+	2+	2+	4+	4+	3+	3+	4+
3. Oxalacetic	4+	4+	4+	4+	4+	4+	4+	4+	4+	4+	4+	4+	4+	4+	4+	4+	4+	4+	4+	4+
4. α-Ketoglutaric	4+	4+	4+	4+	4+	4+	4+	4+	4+	4+	4+	4+	4+	4+	4+	4+	4+	4+	4+	4+
5. Isocitric	2+	2+	4+	4+	1+	1+	4+	4+	1+	1+	4+	4+	2+	2+	2+	4+	2+	2+	2+	4+
6. Glutamic	3+	3+	4+	4+	2+	2+	4+	4+	3+	3+	4+	4+	3+	3+	3+	4+	2+	2+	4+	4+
7. β-Hydroxybutyric	1+	1+	1+	2+	1+	1+	1+	2+	±	±	1+	2+	1+	1+	1+	2+	1+	1+	1+	2+
8. α-Glycero-phosphate	4+	4+	4+	4+	4+	4+	4+	4+	4+	4+	4+	4+	4+	4+	4+	4+	4+	4+	4+	4+
9. Pyruvic	4+	4+	4+	4+	4+	4+	4+	4+	4+	4+	4+	4+	4+	4+	4+	4+	4+	4+	4+	4+
10. Lactic	4+	4+	4+	4+	4+	4+	4+	4+	4+	4+	4+	4+	4+	4+	4+	4+	4+	4+	4+	4+
11. Ethanol	2+	2+	2+	4+	2+	2+	2+	4+	3+	3+	3+	4+	3+	3+	3+	4+	2+	2+	2+	4+
12. Glucose-6-Phosphate	1+	1+	1+	2+	±	±	1+	2+	±	±	±	2+	±	±	±	2+	±	±	±	2+
13. 6-Phospho-gluconate	±	±	±	2+	±	±	±	2+	±	±	±	2+	3+	3+	3+	3+	3+	3+	3+	4+
14. Isocitric (TPN)	3+	2+	2+	2+	3+	2+	2+	2+	3+	2+	2+	2+	3+	2+	2+	2+	2+	1+	1+	2+
15. DPNH Diaphorase	±	±	±	2+	±	±	±	2+	±	±	±	2+	±	±	±	2+	1+	1+	1+	2+
16. TPNH Diaphorase	±	±	±	2+	±	±	±	2+	±	±	±	2+	±	±	±	2+	1+	1+	1+	2+
17. Cytochrome c oxidase	4+	4+	4+	4+	4+	4+	4+	4+	4+	4+	4+	4+	4+	4+	4+	4+	4+	4+	4+	4+

[a] Similar to Table I with the exception that seventeen oxidative enzymes are evaluated. (Reproduced by courtesy of Gustav Fischer Verlag.)

that yield such brilliant results in tissue sections, are not generally effective for free-living cells such as those of the blood, marrow, and cultured cell suspension.

Prolonged Incubation Time

It was found that prolonging the incubation time, beyond the relatively short periods suitable for enzyme histochemistry of tissue sections, enabled enough increments of substrate to permeate the intact plasma membranes of free-living cells so that the total became sufficient to yield ample amounts of enzyme reaction products. Very few reports of prolonged incubation time for cytoenzymology of free-living cells could be found; Rodova[19] described fragmentary results after 30 minutes but better results after 2 to 4 hours incubation for alkaline phosphatase in cultured cells. Balogh[20] incubated 22 hours to demonstrate aminopeptidase in blood platelets. In a number of papers prolonged incubation is mentioned, but in very few is it clearly specified.

It became evident to this author that prolonging the incubation time was important, because the small increments of permeation are cumulative. Sixteen to twenty-four hours incubation were optimal for most enzymes; 6 to 8 hours were often adequate for some of the phosphatases. However, a difficult problem was presented by the long incubation time, because it allowed leaching of soluble enzymes and soluble components of the reaction products out of the cells into the incubating solutions, as well as diffusion and displacement of loosely structured enzymes and reaction products. These were defects that required correction, because the valuable feature of cytohistochemical techniques is precise localization.

Protective Adjuvants

There has been preoccupation in histochemistry with the strong fixatives used in histology to achieve immobilization, perhaps because of the influence of more than a century of development of histological stains. Such strong fixatives cross-link, denature, or coagulate enzyme proteins, tending to obliterate or mask active sites of enzymes. Tentative trials have been made with milder methods: hypertonic media[21,22]; agar overlay[23]; various gums and gelatin[24,25]; and dextran.[26]

There is a group of substances used as cryoprotectives that includes such materials as polyvinylpyrrolidone (PVP), polyvinyl alcohol (PVA), dimethyl sulf-

[19] H. Rodova, *J. Anat.* **82**, 175 (1948).

[20] K. Balogh, Jr., *Nature (London)* **199**, 1196 (1963).

[21] D. G. Scarpelli, R. Hess, and A. G. E. Pearse, *J. Biophys. Biochem. Cytol.* **4**, 747 (1958).

[22] J. L. Conklin, M. M. Dewey, and R. H. Kahn, *Amer. J. Anat.* **110**, 19 (1962).

[23] N. Kaufman and R. Hill, *J. Histochem. Cytochem.* **7**, 144 (1959).

[24] W. H. Fishman, K. T. Ladue, and P. R. F. Borges, *J. Histochem. Cytochem.* **9**, 424 (1961).

[25] A. B. Novikoff, J. Drucker, W. Y. Shin, and S. Goldfischer, *J. Histochem. Cytochem.* **9**, 434 (1961).

[26] B. F. Woolfrey, *Lab. Invest.* **13**, 581 (1969).

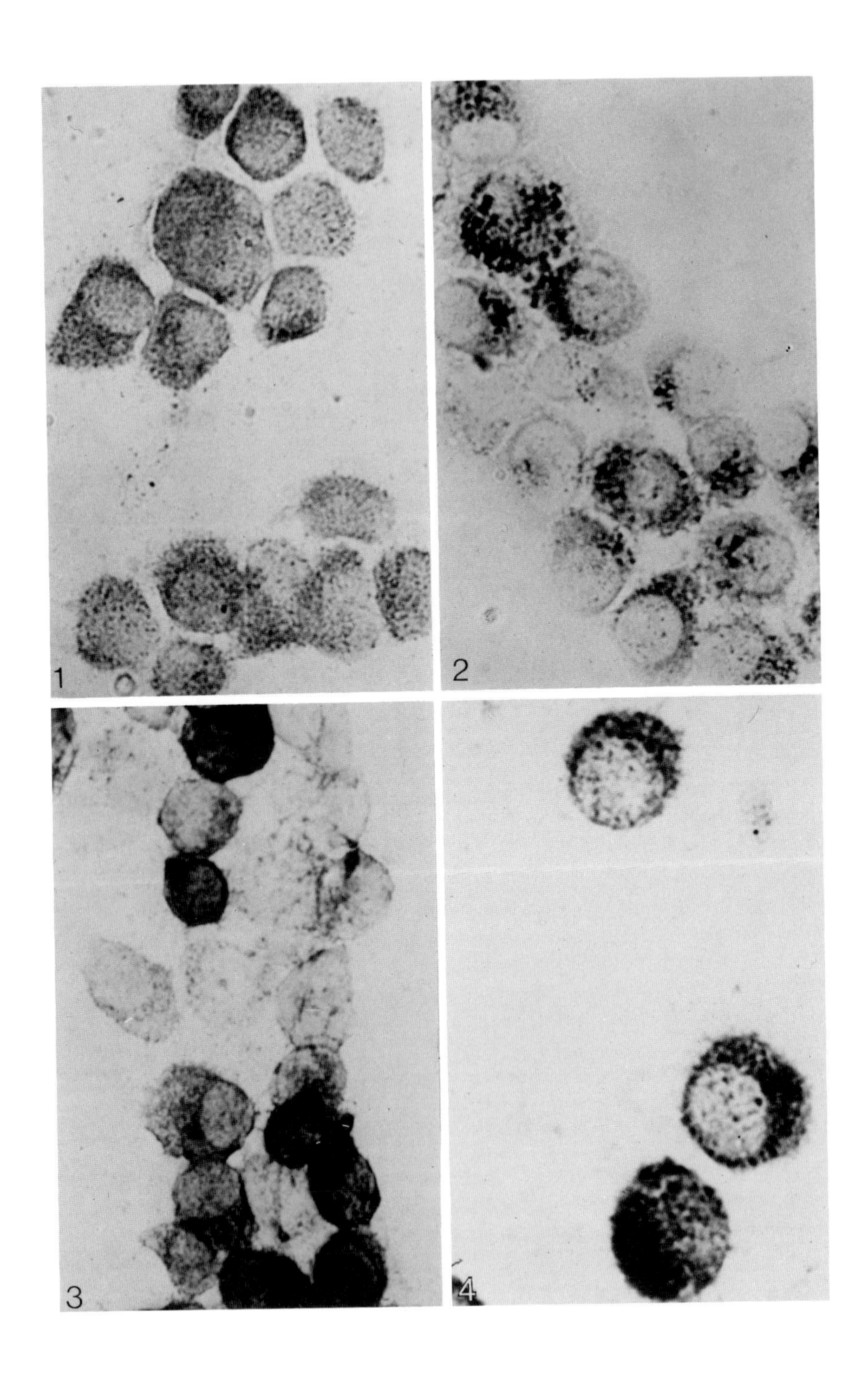
1
2
3
4

oxide (DMSO), glycerol, etc. They do not denature, crosslink, or coagulate proteins, as do the stronger fixatives. They bind by hydrogen bonds to the structured layer of water integral with and lining ultrastructural membranes, leaving physiological functions essentially intact. Hydrogen bonding polymers have been used with considerable success in preventing freezing injury, chiefly of cell suspensions, such as blood for transfusion, spermatozoa, and cell cultures.[27-29] A few reports have appeared that indicate improved results for enzyme cytohistochemistry with these and other cryoprotectives.[21-24,26,30-36]

Systematic studies have been made by the author during the past 10 years to study the effects of various freezing rates and of cryoprotective adjuvants to obtain the full histochemical expression of enzyme activity in cryostat frozen sections of fresh tissue, and to retain enzymes and their reaction products within cells with precise localization.[2,3,37-41] Twenty-two different substances were tested at fast and slow freezing rates.

Ample evidence was obtained that the cryoprotectives facilitate permeation of substrates, and retain enzymes and their reaction products within the cells. Further, the reaction products tend to be discretely granular, appearing to be

[27] T. Malinin, "The Processing and Storage of Viable Human Tissues." Public Health Service, Department of Health, Education and Welfare, Washington, D. C., 1966.

[28] H. T. Meryman, "Textbook of Cryobiology." Academic Press, New York, 1966.

[29] H. T. Meryman, *Cryobiology* **8**, 173 (1971).

[30] A. B. Novikoff, *J. Biophys. Biochem. Cytol.* **2** (Suppl. 4) 65 (1956).

[31] R. Hess, D. G. Scarpelli, and A. G. E. Pearse, *J. Biophys. Biochem. Cytol.* **4**, 753 (1958).

[32] M. S. Burstone and T. J. Fleming, *J. Histochem. Cytochem.* **7**, 203 (1959).

[33] P. Fortelius, E. Levonen, and E. Saxen, *Acta Pathol. Microbiol. Scand.* **52**, 23 (1961).

[34] F. P. Altman and J. Chayen, *Nature (London)* **207**, 1205 (1965).

[35] R. Klen, V. Srb, and T. Husak, *Soc. Biol.* **162**, 1876 (1968).

[36] J. Chayen, L. Bitensky, R. G. Butcher, and L. Poulter, "A Guide to Practical Histochemistry." Lippincott, Philadelphia, Pennsylvania, 1969.

[37] P. J. Melnick, *Cryobiology* **1**, 140 (1964).

[38] P. J. Melnick, *Ann. N. Y. Acad. Sci.* **125**, 689 (1965).

[39] P. J. Melnick, *Fed. Proc.* **24**, S-259–S-268 (1965).

[40] P. J. Melnick, *In* "Cryosurgery" (R. W. Rand, A. P. Rinfret, and H. von Leden, eds.), pp. 52–77. Thomas, Springfield, Illinois, 1968.

[41] P. J. Melnick, *Proc. 13th Intern. Congr. Refrig.* **10**, 1 (1971).

Fig. 1. Esterase with naphthyl AS-D chloroacetate. Example of finely granular reaction product, suggesting localization in ultrastructural elements. This photomicrograph and those that follow are of Detroit 6 cells, and representative of all five cell lines examined. (Reprinted courtesy of Gustav Fischer Verlag.)

Fig. 2. Acid phosphatase with napthyl As-MX phosphate. Example of granular reaction product, suggesting localization in lysosomes; the preponderant juxtanuclear location suggests a relation to the Golgi. (Reprinted courtesy of Gustav Fischer Verlag.)

Fig. 3. Alkaline phosphatase with naphthyl AS-MX phosphate. Example of great variability of enzyme reactivity among individual cells. Variability may represent differently active clones, variations of enzyme induction, or time of synthesis in the mitotic cycle.

Fig. 4. Isocitric dehydrogenase (DPN). The discrete granular reaction product suggests mitochondrial localization; a similar result was observed in all the mitochrondrial or mitochondrial-related oxidative enzymes. A similar result was observed in all the nonmitochondrial oxidative enzymes, suggesting localization in organized locales of the cells.

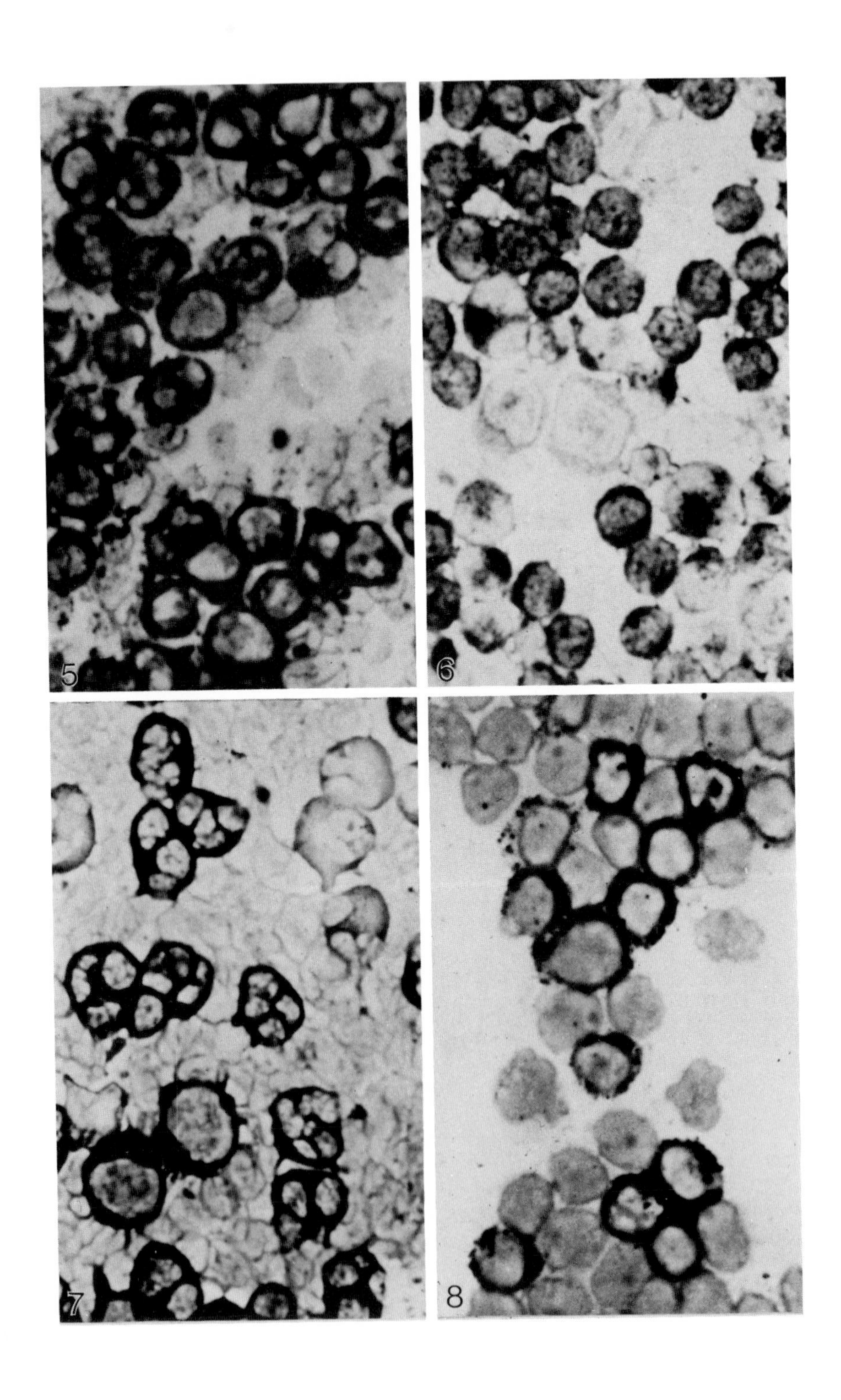
5
6
7
8

related to cell organelles or cell organization. The results have been very useful because, although some hydrolases appear to withstand brief cold formalin or other aldehyde fixation (even these are partially inhibited), the dehydrogenases and oxidative enzymes, in general, are very labile and generally will not withstand any form of fixation.

Acetone Treatment

It was also found that brief immersion in cold (4°C) absolute acetone added precision to the histological detail, and also helped to retain cells on the glass slides. Not as much is known about acetone as about many other substances used to obtain precision of cell structure,[42] but it is of special interest that brief exposure to cold acetone tends to leave the reactive groups of enzymes to a large extent intact and unmasked. Cold acetone appears to have come into increasing use for enzyme cytohistochemistry.[36,43–54] References to similar reports by Ackerman, Balogh, Lillie, and others are contained in the previously mentioned studies.[2,3] Eletsky[55] pointed out that cold acetone (at −10°C) dissolves lipids within 30 to 60 seconds. In cell membranes the choline or ethanolamine component of the phospholipid layer is sterically turned so that it leaves a space into

[42] A. G. E. Pearse, "Histochemistry, Theoretical and Applied," 3rd ed. Little, Brown and Company, Boston, Massachusetts, 1968.
[43] R. O. Stafford and W. B. Atkinson, *Science* **107**, 279 (1948).
[44] A. M. Seligman and A. M. Rutenburg, *Science* **113**, 317 (1951).
[45] G. Gomori, "Microscopic Histochemistry. Principles and Practice." The University of Chicago Press, Chicago, Illinois, 1952.
[46] W. L. Doyle and R. Liebelt, *Anal. Rec.* **118**, 384 (1954).
[47] M. S. Burstone, *J. Histochem. Cytochem.* **6**, 322 (1958).
[48] M. S. Burstone, *J. Histochem. Cytochem.* **9**, 146 (1961).
[49] R. G. J. Willighagen and H. T. Planteydt, *Nature (London)* **183**, 263 (1951).
[50] D. Quaglino and F. G. J. Hayhoe, *Nature (London)* **187**, 85 (1960).
[51] A. B. Novikoff, W. Y. Shin, and J. Drucker, *J. Histochem. Cytochem.* **8**, 37 (1960).
[52] L. S. Kaplow and M. S. Burstone, *Nature (London)* **200**, 690 (1963).
[53] J. Wieckowski and Z. Darzynkiewicz, *Experientia* **21**, 387 (1965).
[54] E. H. Fowler, L. Kasza, and A. Koestner, *Cancer Res.* **26**, 2409 (1966).
[55] Y. K. Eletsky, *Ark. Patol.* **27**, 68 (1965).

Fig. 5. Lactic dehydrogenase. Case of granulocytic leukemia (specimen No. 44). Intense activity is seen in all the cells; such intense activity was generally seen in acute and subacute leukemias and in chronic forms with high leukocyte counts, before antileukemic therapy.

Fig. 6. Succinic dehydrogenase. Case of granulocytic leukemia (specimen No. 21). Myeloblasts are negative or feebly positive; absence or feeble activity of the more undifferentiated cells was observed in all the leukemias for many enzymes, seen better in marrow preparations.

Fig. 7. Succinic dehydrogenase. Case of granulocytic leukemia after busulfan therapy (specimen No. 114). Markedly diminished enzyme activity is seen in a number of the leukemic cells, a phenomenon observed rather regularly during antileukemic chemotherapy.

Fig. 8. Cholinesterase (Gomori method). Case of granulocytic leukemia (specimen No. 28). Complete absence of enzyme reaction product is seen in cells of the same stage of differentiation as others that contain it. Since this patient had received no therapy at the time of this examination, it is indicative of a primary mosaicism. (Figs. 5–8 reprinted courtesy of Hoeber Medical Division of Harper & Row.)

which cholesterol molecules are inserted.[17] It is conceivable that these cholesterol molecules are rapidly dissolved out by cold acetone, so that permeation by substrate is even further facilitated in the cytoenzymological procedures to be described below.

Materials, Methods, and Results

Cultured Cell Lines. Five cultured cell lines were studied, furnished by courtesy of Dr. John E. Shannon, American Type Culture Collection, Rockville, Maryland. They consisted of both frozen seed stock, and suspensions of live cultures in liquid media, of the following: (1) HeLa; (2) HeLa clone 229; (3) Detroit 6; (4) Detroit 6, clone 12; and (5) Chang liver cells. Aliquots of each cell line were studied as follows: (1) frozen seed stock; (2) live cultures in liquid media; (3) slowly frozen cultures; and (4) rapidly frozen cultures; as well as other aspects of the material not pertinent to this report. All material was gently centrifuged at 250 *g* for 20 minutes in a refrigerated centrifuge. PVP (mol. wt. 40,000, 7½% by weight) was added to the live specimens in liquid media before they were slowly or rapidly frozen; the frozen seed stock had already been cryoprotected.

Smears were made of the sediment on glass slides, and dried quickly for 4 or 5 minutes. They were then immersed for exactly 60 seconds in cold (4°C) absolute acetone, drained for a minute or two, immersed for exactly 60 seconds in 5% PVP, then allowed to drain 4 or 5 minutes. They could then be safely stored in a deep freeze at −85°C until such time that they could be conveniently processed with the incubating solutions for enzymes. Space does not permit including the technical details beyond this point, but they are presented in detail elsewhere.[3] It should be mentioned, however, that the timing described above was worked out carefully; less exposure to acetone and PVP was ineffective and more was inhibitory. Also, the specimens were at room temperature during the processing for much less than 30 minutes to prevent possible denaturing by thermal agitation.

Tables I and II (see pp. 810–812) contain in detail the results observed for eleven hydrolases and seventeen oxidative enzymes. Space does not permit discussion of the results in detail; however, it appears that the enzyme patterns of the five cell lines are very similar, and may represent adjustment to *in vitro* conditions. It will be noticed that for some of the hydrolases several different substrates were evaluated.

Figures 1 to 4 contain representative illustrations of the results. The discrete granular reaction products and the sharp cell borders give a strong impression of precise localization in cell organelles or in organized areas of the cell structure. The technical excellence encourages the possibility that meaningful applications of these techniques could be used on cultured cells altered by metabolic, viral, immunological, and other modalities.[56]

[56] P. Fortelius, *Acta Pathol. Microbiol. Scand. Suppl.* **164**, 1 (1963).

Blood and Marrow Cells. Cytoenzymology of blood and marrow cells was studied in 21 normal persons, 76 patients with leukocytosis, 20 with granulocytic leukemia, 12 with lymphocytic leukemia, and 5 with monocytic leukemia.[2] The methods described above were first developed and applied on this material, and are the same as applied to the cultured cell lines. Figures 5 to 8 illustrate some representative findings from cases of granulocytic leukemia. They illustrate enhancement of enzyme activity in leukocytosis; absence or decreased activity in undifferentiated cells; diminution of enzyme activity in the course of antileukemic chemotherapy; occasional mosaicisms, both spontaneous and induced by chemotherapy; and occasional enzyme deletions. If enzyme defects can be consistently related with cytogenetic chromosome abnormalities, perhaps basic information can thereby be obtained about chromosome functions.

Isozymes

Isozymes are the multiple molecular forms of enzymes that catalyze conversion of the same substrate. In free-living cells and in tissue sections, cytohistochemical enzyme techniques produce reaction products that are the sum total of the different isozymes. Thus, their identification is difficult, requiring differential specific inhibitors, immunological techniques, and examinations for substrate preference, thermal stability, aerobic, or anerobic requirements, etc.

Fortunately, the various isozymes have different electrophoretic mobilities that enable them to be individually isolated in gel media by electrophoresis, and to be readily demonstrated by application of histochemical enzyme incubating solutions to the gels. A kind of chemical dissection and individual study of each is thereby possible. The historical development and full details of the techniques involved in isozyme demonstration in starch gel electrophoresis are contained in the author's monograph,[3] reporting an isozyme study for eighteen enzymes of sonicated aliquots of the five cell lines, with starch gel electrophoresis. Acrylamide and other types of supporting media, and other electrophoretic methods, are also useful. However, zone electrophoresis in starch gels[57-60] examined with histochemical enzyme techniques[61] is widely used because it is a relatively easy (although meticulous) method, generally yielding good separation of the isozyme bands. Also, being widely used it is more likely to relate to studies by other workers.

A varied and interesting perspective was furnished by the isozyme preparations of the eighteen enzymes examined in the five cell lines. Applied to cell cultures, isozymes are very useful for cell line characterization, for determining contaminating cell lines, and for studying the effects of metabolic, immunological,

[57] O. Smithies, *Nature (London)* **175**, 307 (1955).
[58] O. Smithies, *Biochem. J.* **61**, 629 (1955).
[59] O. Smithies, *Advan. Protein Chem.* **14**, 65 (1959).
[60] O. Smithies, *Biochem. J.* **71**, 585 (1959).
[61] R. L. Hunter and C. L. Markert, *Science* **125**, 1294 (1957).

and other factors such as thermal stability, substrate preference, specific inhibitors, and aerobic or anaerobic properties, as mentioned above. They also open up other perspectives such as conformational variants, and insights into the regulatory machinery of the cell.

Although there is a general relatedness, there is a high degree of specificity of the isozyme patterns of each cell line for each enzyme. Figures 9 and 10 are representative examples of the usefulness of isozymes. For example, those for glucose-6-phosphate dehydrogenase demonstrate the faster A-band for HeLa and HeLa clone 229, but the slightly slower B-band in Detroit 6, Detroit 6 clone 12

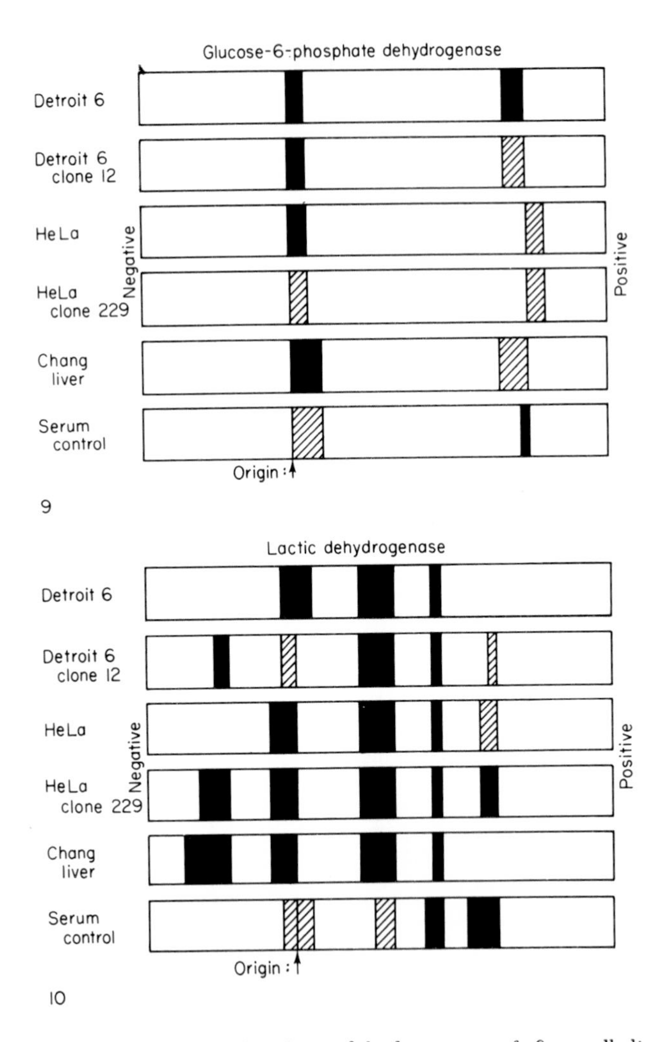

Fig. 9. Isozymes of glucose-6-phosphate dehydrogenase of five cell lines. HeLa and HeLa clone 229 contain the somewhat faster anionic A-band; and the others contain the slower B-band, indicating that they are not contaminated by HeLa.

Fig. 10. Isozymes of lactic dehydrogenase of five cell lines. The unique pattern of each cell line indicates the value of isozymes as one of the methods for characterizing cultured cell lines. (Figures 9 and 10 reprinted courtesy of Gustav Fischer Verlag.)

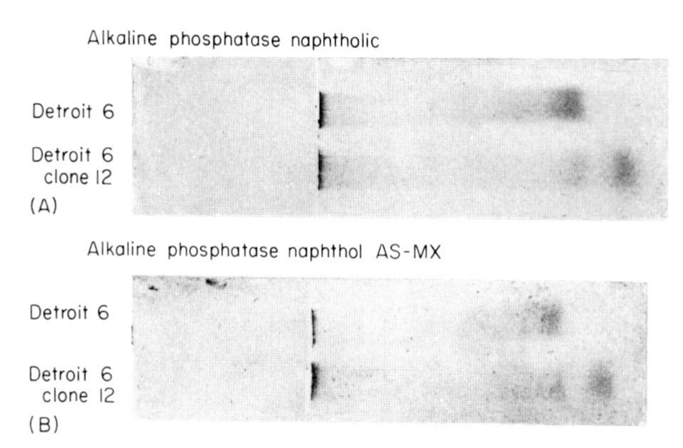

Fig. 11. Isozymes of alkaline phosphatase determined by two different substrates, naphthol phosphate (A) and Naphthol AS-MX phosphate (B), in Detroit 6 and Detroit 6 clone 12. Both have a moderately fast anionic band, but Detroit 6 clone 12 has an additional faster anionic band. Checking with more than one substrate helps rule out technical error. These unique patterns are another example of isozymes in cell line characterization.

and Chang liver, indicating that the latter three are not contaminated by HeLa. Figure 11 (A and B) illustrate the need for probing and controlling technical factors, to obtain valid results. Figure 12 is one of the many types of problems in pathology and biology that can be explored with isozymes.

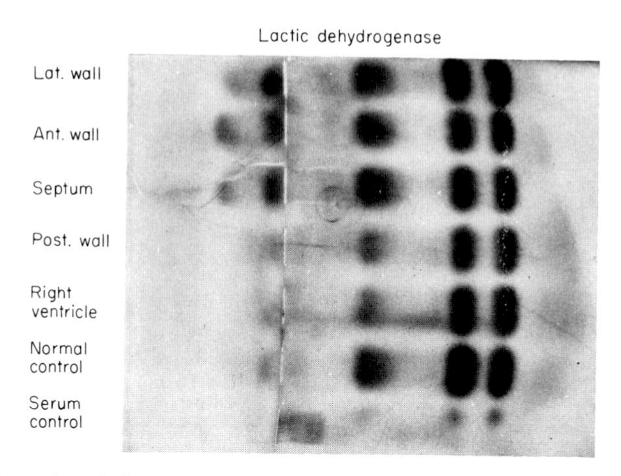

Fig. 12. Effect of pathological states on isozyme patterns. Lactic dehydrogenase isozymes in five representative portions of myocardium from an autopsied case of early acute myocardial infarction, included to suggest the potential use of isozymes in many basic biological problems. Band 5, not present in the uninvolved myocardium and normal control, has appeared in the infarcted area and in the relatively ischemic areas on either side of it; in these same areas band 4 is very intense. This and other enzymes of the glycolytic pathway and pentose shunt were found greatly activated in a study of early infarcts, indicating mobilization of anaerobic bioenergetic enzymes. From a basic standpoint, such findings may yield insights into problems of enzyme synthesis, latent or lost genomes, induction, repression and derepression, and epigenetic or conformational variants of enzymes, which could be studied in cell cultures. (Reprinted by courtesy of Gustav Fischer Verlag and Charles C Thomas, Publisher, 1972).

CHAPTER 18

Transport and Shipping of Cultures, Including International Regulations[1]

Leonard Hayflick

Monolayer cell cultures are best shipped by one of two methods: (1) Sufficient medium is allowed to remain in the vessel so as to keep the monolayer wet during shipment. (2) The culture vessel is completely filled with medium. Both methods require that stoppers be securely taped to the vessel in order to prevent medium loss when lower air pressures are encountered during air transport.

SHIPPING REQUIREMENTS OF STATE AND FEDERAL AGENCIES

Postal Service. (1) Domestic shipments: The general provisions of the United States Postal Service (18 U.S.C., Sec. 1716) state that harmful matter—that which is likely to destroy, deface or otherwise damage mail or other property, or that which may kill or injure another—is nonmailable. Cell cultures may be sent through the mail if they conform to the special preparation and handling described below. (2) International shipments: Cultures may be sent in the postal service with proper packaging and permits. Postal Manual 221.325 states:

a. Mailing Restrictions
   If a country prohibits perishable biological materials, this is shown in the Directory of International Mail under *Prohibitions.* (Information is available from local postmasters.)
b. Qualifications of Mailers
   (i) Only officially recognized laboratories may send or receive letter packages containing perishable biological materials. Laboratories of the following categories are so designated:
      Laboratories of local, state, and federal government agencies.
      Laboratories of federally licensed manufacturers of biologic substances derived from bacteria and viruses.
      Laboratories affiliated with or operated by hospitals, universities, research facilities, and other teaching institutions. Private laboratories licensed, certified, recognized, or approved by a public authority.
   (ii) A laboratory desiring to mail letter packages containing materials of this kind shall make written application on its letterhead stationery to the Classification and Special Services Division, Bureau of Operations, Post Office Department, Washing-

[1] Supported, in part, by research grant HD 04004 from the National Institute of Child Health and Human Development, National Institutes of Health, Bethesda, Maryland.

ton, D. C. 20260, explaining its qualifications and those of the prospective addressee to send and receive such materials, and stating how many packages are to be mailed. On approval, the mailer will receive a sufficient number of the violet labels for the contemplated shipments . . .

Bureau of Customs–Imports. It is imperative that living cultures be properly packaged so there will be no harm to any officials who may inspect the package. Manifests should be accurate to speed inspection at the customs office. Generally, living cultures are duty free under the Tariff Schedules of the United States. However, cultures, whether dutiable or duty free, may not be released by Customs until entry is made and a permit for their release granted.

United States Department of Commerce—Office of Export Control. A "general license" is a published general authorization which allows export from the United States under certain conditions without applying for a validated license. The authority to export in such an instance is given in the Comprehensive Export Schedule, published by the Department of Commerce, which specifies the conditions under which each general license may be used.

A "validated license" must be applied for and obtained before making any exportation, regardless of value, to foreign destinations, other than Canada, of either technical data or commodities, unless the particular exportation is authorized under the terms of one of the general licenses printed in export regulations.

According to the Current Export Bulletins of 1968 concerning the Commodity Control List, living cell cultures fit into the following description:

CCL #29100 (Processing Code 248)—Organs, therapeutic glands or other organs and their extracts, bulk (includes cell cultures). Countries which require a validated license are: Southern Rhodesia, Communist China, North Korea, Communist-controlled area of Vietnam, Cuba and East Germany.

Exports to all other countries can be shipped under the general license with the notation "G-DEST, Export License Not Required" on the address label. Application for a validated license may be made on the Application for Export License Form FC-419 and the Application Processing Card Form FC-420 obtainable from the United States Department of Commerce, Washington, D. C. 20230 or any field office thereof. The validated license and shipper's export declaration must be surrendered to the Postmaster or forwarding agent upon shipment. The number of this license must appear on the address label of the package.

A destination control statement is required for all exports (except Canada).

In regard to cell cultures that require shipment under cryogenic conditions (liquefied gas), note should be made that the cryogenic container (liquefied gas type) requires a validated license for shipment to eastern European countries.

Destination Control Statement. In addition to the licenses mentioned above, a destination control statement is required for exports. This regulation is designed to prevent the shipment of United States origin goods to unauthorized destinations by transshipment or diversion. Under the destination control statement provisions, most commercial shipments leaving the United States must show one of three "anti-diversion" notices, as appropriate, on one of the following documents: (a) export declaration, (b) bill of lading or air way bill, and (c) commercial invoice. In addition, a shipper's export declaration may be required.

Shipper's Export Declaration. Effective October 1, 1969, no shipper's export declaration is required for shipments where the value of the commodities classified under a single Schedule B number does not exceed $250.00, provided the shipment does not require an Export License. The Form 7525-V shipper's export declaration may also include the required destination control statement.

The Schedule B commodity number required in column 13 of the shipper's export declaration is a seven digit number which may be found in the Bureau of Census publication (1965) "Schedule B, Statistical Classification of Domestic and Foreign Commodities Exported from the United States and amendments thereto." The Schedule B number is the Commodity Control List number with further identifying digits for statistical purposes.

For living cell cultures the Schedule B Commodity number is 2910080.

Because of changes in the Department of Commerce regulations from time to time, it is important to check with them when in doubt: United States Department of Commerce, Office of Export Control, Bureau of International Programs, Washington, D. C. 20226.

It must be remembered that several natural materials used in animal and plant cell cultures may be vectors of disease agents forbidden to be imported to this country (e.g., bovine sera from foot and mouth disease enzootic areas). Thus importation of cell cultures from such areas is expressly forbidden. For information write to the following:

United States Public Health Service
Foreign Quarantine Program
Center for Disease Control
Atlanta, Georgia 30333

United States Treasury Department
Bureau of Customs
Washington, D. C. 20226

United States Department of Agriculture
USDA—Agricultural Research Service
Plant Quarantine Division
Federal Center Building
Hyattsville, Maryland 20782

USDA—Agricultural Research Service
Animal Health Division
Federal Center Building
Hyattsville, Maryland 20782

Requirements for Packaging

United States Postal Manual. Inasmuch as the shipper bears the greatest responsibility for safe delivery of biological specimens, it is imperative that every precaution be observed in order to prevent accidental contamination of the mail and personnel involved. The United States Postal Manual indicates the following minimum requirements for proper packaging of perishable biological materials in the international postal service (221.325c Packaging Dated 8-30-68):

(1) Perishable biological material not of a pathogenic nature must be packed in a nonporous container surrounded by sufficient absorbent material to take up all the liquid and must be placed in an outer protective container where it should fit tightly to avoid any shifting.
(2) Perishable biological material of a pathogenic nature must be packed in a tightly closed bottle or tube of heavy glass wrapped in thick absorbent material rolled several times around the bottle or tube and tied at the ends, sufficient in quantity to absorb all the liquid; the wrapped container must be placed in a strong well-closed metal box constructed to prevent any contamination outside of it. This metal box must be wrapped in cushioning material and placed in an outer protective box where it should fit tightly to avoid shifting. The outer container must consist of a hollow block of strong wood, metal, or other equally strong material with a tight lid so fitted that it cannot open during transportation.

Public Health Service. The Public Health Service regulation No. 72.25 for transport of etiological agents specifies somewhat the same requirements as noted above for proper packaging. It states:

(b)[2] A person shall not knowingly transport or cause to be transported in interstate traffic any etiologic agent unless the agent is packaged in a minimum of two sealed containers, and each such double container is enclosed in a third container, hereinafter referred to as an individual shipping container, in accordance with the following requirements:
(1) (i) The materials shall be placed in a watertight and airtight container which shall then be enclosed in a second durable watertight and airtight container. In the case of liquid (including frozen materials), the intervening space between the containers shall be provided with sufficient absorbent material so placed as to absorb the entire contents in case of breakage or leakage. Each such double container shall then be enclosed in an individual shipping container constructed of corrugated cardboard, fiberglass, wood, or other materials of equivalent strength.
(ii) If dry ice is used as a refrigerant, the individual shipping container shall be vented, and if an outside shipping container is used, it shall also be vented.
(2) (i) The maximum amount of etiologic agent which may be shipped in an individual shipping container shall not exceed one U. S. gallon[3] provided that two or more individual shipping containers may be overpacked in a single outside shipping container.
(ii) All containers and closures are so designed and constructed of such materials that they are capable of withstanding without rupture or leakage of contents, all shocks, pressure changes, or other conditions ordinarily incident to transportation handling.
(3) The shipping documents and the manifest accompanying the shipment include statements that the shipment contains infectious material and identifies the etiologic agent involved. The shipment itself shall be appropriately labeled.
(4) The requirements of this paragraph are in addition to and not in lieu of any other packaging or labeling requirements for the interstate shipment of etiologic agents established by the Interstate Commerce Commission and Civil Aeronautics Board.

[2] Sec. 72.25(b) Amended August 29, 1961, 26 FR 8073.
[3] Recently changed (January, 1973): not to exceed 50 ml.

Methods of Packing

Unfrozen, Living Cell Cultures. Cultures are individually wrapped in an absorbent tissue and then in Air Cap (two sheets of saran-coated polyethylene film enclosing bubbles of air) (Sealed Air Corporation, Fairlawn, New Jersey). This is placed into an all metal, soldered, screw-cover can with sufficient cotton to prevent movement. These metal cans are placed into a fiberboard container with a wad of cotton to prevent rattle. The fiberboard carton is a one-piece body tube with a crimped metal bottom and a metal screw cap. These mailing tubes may be purchased in assorted sizes from most laboratory supply companies. A packing slip is enclosed, the screw cap is taped with an imprinted label stating "IMPORTANT, MUST BE OPENED IN THE LABORATORY," and the address label is affixed to the carton. If permits are required they are placed near the label.

Frozen, Preserved Cell Cultures for Shipment in Dry Ice. Civil Aeronautics Board No. 82, Page 127, Packaging Note 803 states: "Carbon dioxide, solid (dry ice) must be packed so as to permit release of carbon dioxide gas and *must be plainly marked* on each package: "ORA. Group A—Dry Ice."

Postal Manual 125.323 states: "Dry ice (carbon dioxide solid) is acceptable when wrapped securely in heavy paper. Dry ice must not be packed in glass, metal, or other air-tight containers. Sufficient insulation is necessary if a fiberboard box is used in order to prevent condensation and wetting of the shipping carton."

Cell-culture ampoules are taken from liquid-nitrogen storage and placed in a paper tube (about 0.6 inch in diameter) with a small amount of dry ice. Both ends are stapled and placed into a metal can and then into the dry-ice shipper with the required amount of dry ice sufficient to last 72 hours for domestic shipments and seven days for international shipments.

A packing slip is enclosed. The lid is taped closed lightly with masking tape. Containers must be marked "Packed in Dry Ice."

Frozen, Preserved Cell Cultures for Shipment in Liquid Nitrogen. Ampoules are attached to metal canes and placed in canisters designed for this purpose. The canisters are placed in a vacuum dewar containing a 30-day supply of liquid nitrogen which in turn is placed in a heavy plywood box or other suitable protection specifically made to prevent damage to the vacuum container. The liquid is properly vented so that there is no pressure build-up. The box is clearly marked, "DO NOT TILT, LIQUID NITROGEN, NONPRESSURIZED," "THIS END UP."

Shipping of Cell Cultures

Shipment of Unfrozen Cultures. (1) Domestic: Cultures are shipped first class or air mail; special delivery or via air forwarder. (2) International: All shipments are sent by air mail or via an air forwarder. The regulations for shipment are stated in the United States Postal Manual in International Postal Service

Section 221.325 as follows: "Perishable Biological Materials. Perishable biological materials, including those of pathogenic nature, when sent in the postal union mail are accepted only as LETTER PACKAGES. The packages must be packed as prescribed in 221.325c (see above, U. S. Postal Manual) and must bear distinctive violet labels by which they can be readily recognized and receive careful handling and prompt delivery." In addition a green customs label procured at a local post office must be placed on the outside of the package along with the violet label.

Shipment of Frozen Cultures. (1) Dry-Ice Refrigerant: Dry-ice shipments are sent by an air forwarder where the following regulations apply:

Civil Aeronautics Board
C.A.B. No. 82, Page 23, Rule 9c:
Not more than 440 lbs. of dry ice will be carried in any cargo pit or bin on any aircraft except by special arrangements with the transporting carrier.
C.A.B. No. 82, Page 23, Rule 9d:
Dry ice will not be carried in the same belly or tail pit with any live animal.

(2) Liquid-Nitrogen Refrigerant: Shipments are sent by Railway Express although air shipments may be made with special arrangements.

Author Index

Numbers in parentheses are reference numbers and indicate that an author's work is referred to although his name is not cited in the text.

C

D

E

F

G

I

J

K

L

M

N

O

P

Q

R

S

Y

Z

Subject Index

N

Q

R

T

U

V

W

X

Y

Z

C 4
D 5
E 6
F 7
G 8
H 9
I 0
J 1
2
3